STUDENT'S SOLUTIONS MANUAL

JUDITH A. PENNA

Indiana University Purdue University Indianapolis

INTERMEDIATE ALGEBRA

TENTH EDITION

Marvin L. Bittinger

Indiana University Purdue University Indianapolis

PEARSON

Addison
Wesley

Boston San Francisco New York
London Toronto Sydney Tokyo Singapore Madrid
Mexico City Munich Paris Cape Town Hong Kong Montreal

Reproduced by Pearson Addison-Wesley from electronic files supplied by the author.

Copyright © 2007 Pearson Education, Inc.
Publishing as Pearson Addison-Wesley, 75 Arlington Street, Boston, MA 02116.

ISBN   0-321-30579-5

1 2 3 4 5 6 BB 09 08 07 06

Contents

Chapter R

Review of Basic Algebra

Exercise Set R.1

In Exercises 1-5, consider the following numbers: $-6, 0, 1,$ $-\frac{1}{2}, -4, \frac{7}{9}, 12, -\frac{6}{5}, 3.45, 5\frac{1}{2}, \sqrt{3}, \sqrt{25}, -\frac{12}{3},$ $0.131331333133331\ldots$.

1. The natural numbers are numbers used for counting. The natural numbers in the list above are 1, 12, and $\sqrt{25}$ ($\sqrt{25} = 5$).

3. The rational numbers can be named as quotients of integers with nonzero divisors. The rational numbers in the list above are $-6, 0, 1, -\frac{1}{2}, -4, \frac{7}{9}, 12, -\frac{6}{5}, 3.45$ ($3.45 = \frac{345}{100}$), $5\frac{1}{2}$ ($5\frac{1}{2} = \frac{11}{2}$), $\sqrt{25}$ ($\sqrt{25} = 5$), and $-\frac{12}{3}$.

5. The real numbers consist of the rational numbers and the irrational numbers. All of the numbers in the list are real numbers.

In Exercises 7-11, consider the following numbers: $-\sqrt{5}, -3.43,$ $-11, 12, 0, \frac{11}{34}, -\frac{7}{13}, \pi, -3.565665666566665\ldots$

7. The whole numbers consist of the natural numbers and 0. The whole numbers in the list above are 12 and 0.

9. The integers consist of the whole numbers and their opposites. The integers in the list above are -11, 12, and 0.

11. The irrational numbers are the numbers that are not rational. The irrational numbers in the list above are $-\sqrt{5}$, π, and $-3.565665666566665\ldots$.

13. We list the members of the set.
$\{m,a,t,h\}$

15. We list the members of the set.
$\{1, 2, 3, 4, 5, 6, 7, 8, 9, 10, 11, 12\}$

17. We list the members of the set.
$\{2, 4, 6, 8, \ldots\}$

19. We specify conditions by which we know whether a number is in the set.
$\{x | x \text{ is a whole number less than or equal to 5}\}$

21. We specify conditions by which we know whether a number is in the set.
$\left\{ \frac{a}{b} \middle| a \text{ and } b \text{ are integers and } b \neq 0 \right\}$

23. We specify conditions by which we know whether a number is in the set.
$\{x | x \text{ is a real number and } x > -3\}$, or $\{x | x > -3\}$

25. Since 13 is to the right of 0 on the number line, we have $13 > 0$.

27. Since -8 is to the left of 2 on the number line, we have $-8 < 2$.

29. Since -8 is to the left of 8, we have $-8 < 8$.

31. Since -8 is to the left of -3, we have $-8 < -3$.

33. Since -2 is to the right of -12, we have $-2 > -12$.

35. Since -9.9 is to the left of -2.2, we have $-9.9 < -2.2$.

37. Since $37\frac{1}{5}$ is to the right of $-1\frac{67}{100}$, we have $37\frac{1}{5} > -1\frac{67}{100}$.

39. We convert to decimal notation: $\frac{6}{13} = 0.461538\ldots$ and $\frac{13}{25} = 0.52$. Thus $\frac{6}{13} < \frac{13}{25}$.

41. $-8 > x$

The inequality $x < -8$ has the same meaning.

43. $-12.7 \leq y$

The inequality $y \geq -12.7$ has the same meaning.

45. $6 \leq -6$ False since neither $6 < -6$ nor $6 = -6$ is true.

47. $5 \geq -8.4$ True since $5 > -8.4$ is true.

49. $x < -2$

We shade all numbers less than -2. We indicate that -2 is not a solution by using a parenthesis at -2.

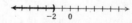

51. $x \leq -2$

We shade all the numbers to the left of -2 and use a bracket at -2 to indicate that it is also a solution.

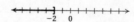

53. $x > -3.3$

We shade all the numbers to the right of -3.3 and use a parenthesis at -3.3 to indicate that it is not a solution.

55. $x \geq 2$

We shade all the numbers to the right of 2 and use a bracket at 2 to indicate that it is also a solution.

57. The distance of -6 from 0 is 6, so $|-6| = 6$.

59. The distance of 28 from 0 is 28, so $|28| = 28$.

61. The distance of -35 from 0 is 35, so $|-35| = 35$.

63. The distance of $-\dfrac{2}{3}$ from 0 is $\dfrac{2}{3}$, so $\left|-\dfrac{2}{3}\right| = \dfrac{2}{3}$.

65. The distance of $\dfrac{0}{7}$, or 0, from 0 is 0, so $\left|\dfrac{0}{7}\right| = 0$.

67. Discussion and Writing Exercise

69. $|-3| = 3$, so $|-3| \leq 5$.

71. $|-7| = 7$, so $|4| \leq |-7|$.

73. For comparison, we first write each number in decimal notation.

$$\frac{1}{11} = 0.090909\ldots$$

$$1.1\% = 0.011$$

$$\frac{2}{7} = 0.285714285714\ldots$$

$$0.3\% = 0.003$$

$$0.11 = 0.11$$

$$\frac{1}{8}\% = 0.00125 \qquad \left(\frac{1}{8} = 0.125\right)$$

$$0.009 = 0.009$$

$$\frac{99}{1000} = 0.099$$

$$0.286 = 0.286$$

$$\frac{1}{8} = 0.125$$

$$1\% = 0.01$$

$$\frac{9}{100} = 0.09$$

Then these rational numbers listed from least to greatest are $\dfrac{1}{8}\%$, 0.3%, 0.009, 1%, 1.1%, $\dfrac{9}{100}$, $\dfrac{1}{11}$, $\dfrac{99}{1000}$, 0.11, $\dfrac{1}{8}$, $\dfrac{2}{7}$, and 0.286.

Exercise Set R.2

1. $-10 + (-18)$

The sum of two negative numbers is negative. We add their absolute values, $10 + 18 = 28$, and make the answer negative.

$$-10 + (-18) = -28$$

3. $7 + (-2)$

We find the difference of their absolute values, $7 - 2 = 5$. Since the positive number has the larger absolute value, the answer is positive.

$$7 + (-2) = 5$$

5. $-8 + (-8)$

The sum of two negative numbers is negative. We add the absolute values, $8 + 8 = 16$, and make the answer negative.

$$-8 + (-8) = -16$$

7. $7 + (-11)$

We find the difference of their absolute values, $11 - 7 = 4$. Since the negative number has the larger absolute value, the answer is negative.

$$7 + (-11) = -4$$

9. $-16 + 6$

We find the difference of their absolute values, $16 - 6 = 10$. Since the negative number has the larger absolute value, the answer is negative.

$$-16 + 6 = -10$$

11. $-26 + 0$

One number is 0. The sum is -26.

$$-26 + 0 = -26$$

13. $-8.4 + 9.6$

We find the difference of their absolute values, $9.6 - 8.4 = 1.2$. Since the positive number has the larger absolute value, the answer is positive.

$$-8.4 + 9.6 = 1.2$$

15. $-2.62 + (-6.24)$

The sum of two negative numbers is negative. We add the absolute values, $2.62 + 6.24 = 8.86$, and make the answer negative.

$$-2.62 + (-6.24) = -8.86$$

17. $-\dfrac{5}{9} + \dfrac{2}{9}$

We find the difference of their absolute values, $\dfrac{5}{9} - \dfrac{2}{9} = \dfrac{3}{9} = \dfrac{1}{3}$. Since the negative number has the larger absolute value, the answer is negative.

$$-\frac{5}{9} + \frac{2}{9} = -\frac{1}{3}$$

19. $-\dfrac{11}{12} + \left(-\dfrac{5}{12}\right)$

The sum of two negative numbers is negative. We add their absolute values, $\dfrac{11}{12} + \dfrac{5}{12} = \dfrac{16}{12}$, or $\dfrac{4}{3}$, and make the answer negative.

$$-\frac{11}{12} + \left(-\frac{5}{12}\right) = -\frac{4}{3}$$

21. $\dfrac{2}{5} + \left(-\dfrac{3}{10}\right)$

We find the difference of their absolute values.

$\dfrac{2}{5} - \dfrac{3}{10} = \dfrac{2}{5} \cdot \dfrac{2}{2} - \dfrac{3}{10} = \dfrac{4}{10} - \dfrac{3}{10} = \dfrac{1}{10}$

Since the positive number has the larger absolute value, the answer is positive.

$\dfrac{2}{5} + \left(-\dfrac{3}{10}\right) = \dfrac{1}{10}$

23. $-\dfrac{2}{5} + \dfrac{3}{4}$

We find the difference of their absolute values.

$\dfrac{3}{4} - \dfrac{2}{5} = \dfrac{3}{4} \cdot \dfrac{5}{5} - \dfrac{2}{5} \cdot \dfrac{4}{4} = \dfrac{15}{20} - \dfrac{8}{20} = \dfrac{7}{20}$

Since the positive number has the larger absolute value, the answer is positive.

$-\dfrac{2}{5} + \dfrac{3}{4} = \dfrac{7}{20}$

25. When $a = -4$, then $-a = -(-4) = 4$.

(The opposite, or additive inverse, of -4 is 4.)

27. When $a = 3.7$, then $-a = -3.7$.

(The opposite, or additive inverse, of 3.7 is -3.7.)

29. The opposite, or additive inverse, of 10 is -10, because $10 + (-10) = 0$.

31. The opposite, or additive inverse, of 0 is 0, because $0 + 0 = 0$.

33. $3 - 7 = 3 + (-7) = -4$

35. $-5 - 9 = -5 + (-9) = -14$

37. $23 - 23 = 23 + (-23) = 0$

39. $-23 - 23 = -23 + (-23) = -46$

41. $-6 - (-11) = -6 + 11 = 5$

43. $10 - (-5) = 10 + 5 = 15$

45. $15.8 - 27.4 = 15.8 + (-27.4) = -11.6$

47. $-18.01 - 11.24 = -18.01 + (-11.24) = -29.25$

49. $-\dfrac{21}{4} - \left(-\dfrac{7}{4}\right) = -\dfrac{21}{4} + \dfrac{7}{4} = -\dfrac{14}{4} = -\dfrac{7}{2}$

51. $-\dfrac{1}{3} - \left(-\dfrac{1}{12}\right) = -\dfrac{1}{3} + \dfrac{1}{12} = -\dfrac{4}{12} + \dfrac{1}{12} = -\dfrac{3}{12} = -\dfrac{1}{4}$

53. $-\dfrac{3}{4} - \dfrac{5}{6} = -\dfrac{3}{4} + \left(-\dfrac{5}{6}\right) = -\dfrac{9}{12} + \left(-\dfrac{10}{12}\right) = -\dfrac{19}{12}$

55. $\dfrac{1}{3} - \dfrac{4}{5} = \dfrac{1}{3} + \left(-\dfrac{4}{5}\right) = \dfrac{5}{15} + \left(-\dfrac{12}{15}\right) = -\dfrac{7}{15}$

57. $3(-7)$

The product of a positive number and a negative number is negative. We multiply their absolute values, $3 \cdot 7 = 21$, and make the answer negative.

$3(-7) = -21$

59. $-2 \cdot 4$

The product of a negative number and a positive number is negative. We multiply their absolute values, $2 \cdot 4 = 8$, and make the answer negative.

$-2 \cdot 4 = -8$

61. $-8(-3)$

The product of two negative numbers is positive. We multiply their absolute values, $8 \cdot 3 = 24$, and make the answer positive.

$-8(-3) = 24$

63. $-7 \cdot 16$

The product of a negative number and a positive number is negative. We multiply their absolute values, $7 \cdot 16 = 112$, and make the answer negative.

$-7 \cdot 16 = -112$

65. $-6(-5.7)$

The product of two negative numbers is positive. We multiply their absolute values, $6(5.7) = 34.2$, and make the answer positive.

$-6(-5.7) = 34.2$

67. $-\dfrac{3}{5} \cdot \dfrac{4}{7}$

The product of a negative number and a positive number is negative. We multiply their absolute values, $\dfrac{3}{5} \cdot \dfrac{4}{7} = \dfrac{12}{35}$, and make the answer negative.

$-\dfrac{3}{5} \cdot \dfrac{4}{7} = -\dfrac{12}{35}$

69. $-3\left(-\dfrac{2}{3}\right)$

The product of two negative numbers is positive. We multiply their absolute values, $3 \cdot \dfrac{2}{3} = \dfrac{6}{3} = 2$, and make the answer positive.

$-3\left(-\dfrac{2}{3}\right) = 2$

71. $-3(-4)(5)$

$= 12(5)$ The product of two negative numbers is positive.

$= 60$ The product of two positive numbers is positive.

73. $-4.2(-6.3)$

The product of two negative numbers is positive. We multiply the absolute values, $4.2(6.3) = 26.46$, and make the answer positive.

$-4.2(-6.3) = 26.46$

75. $-\dfrac{9}{11} \cdot \left(-\dfrac{11}{9}\right)$

The product of two negative numbers is positive. We multiply their absolute values, $\dfrac{9}{11} \cdot \dfrac{11}{9} = \dfrac{99}{99} = 1$, and make the answer positive.

$-\dfrac{9}{11} \cdot \left(-\dfrac{11}{9}\right) = 1$

77. $-\dfrac{2}{3}\cdot\left(-\dfrac{2}{3}\right)\cdot\left(-\dfrac{2}{3}\right)$

$=\dfrac{4}{9}\cdot\left(-\dfrac{2}{3}\right)$ The product of two negative numbers is positive.

$=-\dfrac{8}{27}$ The product of a positive number and a negative number is negative.

79. When a negative number is divided by a positive number, the answer is negative.

$\dfrac{-8}{4}=-2$

81. When a positive number is divided by a negative number, the answer is negative.

$\dfrac{56}{-8}=-7$

83. When a negative number is divided by a negative number, the answer is positive.

$-77\div(-11)=\dfrac{-77}{-11}=7$

85. When a negative number is divided by a negative number, the answer is positive.

$\dfrac{-5.4}{-18}=\dfrac{5.4}{18}=0.3$

87. $\dfrac{5}{0}$ Not defined: Division by 0.

89. $\dfrac{0}{32}=0$ because $0\cdot32=0$.

91. $\dfrac{9}{y-y}$ Not defined: $y-y=0$ for any y.

93. The reciprocal of $\dfrac{3}{4}$ is $\dfrac{4}{3}$, because $\dfrac{3}{4}\cdot\dfrac{4}{3}=1$.

95. The reciprocal of $-\dfrac{7}{8}$ is $-\dfrac{8}{7}$, because $-\dfrac{7}{8}\cdot\left(-\dfrac{8}{7}\right)=1$.

97. The reciprocal of 25 is $\dfrac{1}{25}$, because $25\cdot\dfrac{1}{25}=1$.

99. The reciprocal of 0.2 is $\dfrac{1}{0.2}$.
This can also be expressed as follows:
$\dfrac{1}{0.2}=\dfrac{1}{0.2}\cdot\dfrac{10}{10}=\dfrac{10}{2}=5$.

101. The reciprocal of $-\dfrac{a}{b}$ is $-\dfrac{b}{a}$, because $-\dfrac{a}{b}\cdot\left(-\dfrac{b}{a}\right)=1$.

103. $\dfrac{2}{7}\div\left(-\dfrac{11}{3}\right)=\dfrac{2}{7}\cdot\left(-\dfrac{3}{11}\right)=-\dfrac{6}{77}$

105. $-\dfrac{10}{3}\div-\dfrac{2}{15}=-\dfrac{10}{3}\cdot\left(-\dfrac{15}{2}\right)=\dfrac{150}{6}$, or 25

107. $18.6\div(-3.1)=\dfrac{18.6}{-3.1}=-\dfrac{18.6}{3.1}=-6$

109. $(-75.5)\div(-15.1)=\dfrac{-75.5}{-15.1}=\dfrac{75.5}{15.1}=5$

111. $-48\div0.4=\dfrac{-48}{0.4}=-\dfrac{48}{0.4}=-120$

113. $\dfrac{3}{4}\div\left(-\dfrac{2}{3}\right)=\dfrac{3}{4}\cdot\left(-\dfrac{3}{2}\right)=-\dfrac{9}{8}$

115. $-\dfrac{5}{4}\div\left(-\dfrac{3}{4}\right)=-\dfrac{5}{4}\cdot\left(-\dfrac{4}{3}\right)=\dfrac{20}{12}=\dfrac{5}{3}$

117. $-\dfrac{2}{3}\div\left(-\dfrac{4}{9}\right)=-\dfrac{2}{3}\cdot\left(-\dfrac{9}{4}\right)=\dfrac{18}{12}=\dfrac{3}{2}$

119. $-\dfrac{3}{8}\div\left(-\dfrac{8}{3}\right)=-\dfrac{3}{8}\cdot\left(-\dfrac{3}{8}\right)=\dfrac{9}{64}$

121. $-6.6\div3.3=\dfrac{-6.6}{3.3}=-2$

123. $\dfrac{-12}{-13}=\dfrac{12}{13}$, or $0.\overline{923076}$

125. $\dfrac{48.6}{-30}=\dfrac{16.2}{-10}=-1.62$

127. $\dfrac{-9}{17-17}$ Not defined: $17-17=0$

129. $\dfrac{2}{3}:$ The opposite of $\dfrac{2}{3}$ is $-\dfrac{2}{3}$, because $\dfrac{2}{3}+\left(-\dfrac{2}{3}\right)=0$.

The reciprocal of $\dfrac{2}{3}$ is $\dfrac{3}{2}$, because $\dfrac{2}{3}\cdot\dfrac{3}{2}=1$.

$-\dfrac{5}{4}:$ The opposite of $-\dfrac{5}{4}$ is $\dfrac{5}{4}$, because $-\dfrac{5}{4}+\dfrac{5}{4}=0$.

The reciprocal of $-\dfrac{5}{4}$ is $-\dfrac{4}{5}$, because $-\dfrac{5}{4}\cdot\left(-\dfrac{4}{5}\right)=1$.

$0:$ The opposite of 0 is 0, because $0+0=0$.

The reciprocal of 0 does not exist. (Only nonzero numbers have reciprocals.)

$1:$ The opposite of 1 is -1, because $1+(-1)=0$.

The reciprocal of 1 is 1, because $1\cdot1=1$.

$-4.5:$ The opposite of -4.5 is 4.5, because $-4.5+4.5=0$.

The reciprocal of -4.5 is $-\dfrac{1}{4.5}$, because $-4.5\cdot\left(-\dfrac{1}{4.5}\right)=1$. (Note that $-\dfrac{1}{4.5}=-\dfrac{1}{4.5}\cdot\dfrac{10}{10}=-\dfrac{10}{45}$, or $-\dfrac{2}{9}$.)

$x, x\neq0:$ The opposite of x is $-x$, because $x+(-x)=0$.

The reciprocal of x, $x\neq0$, is $\dfrac{1}{x}$, because $x\cdot\dfrac{1}{x}=1$.

131. Discussion and Writing Exercise

133. The set of whole numbers is $\{0, 1, 2, 3, \ldots\}$. The whole numbers in the given list are 26 and 0.

135. The set of integers is $\{\ldots, -3, -2, -1, 0, 1, 2, 3, \ldots\}$. The integers in the given list are -13, 26 and 0.

137. The set of rational numbers is
$$\left\{ \frac{p}{q} \,\middle|\, p \text{ is an integer, } q \text{ is an integer, and } q \neq 0 \right\}.$$
Decimal notation for rational numbers either terminates or has a repeating block of digits. The integers in the given list are -12.47, -13, 26, 0, $-\frac{23}{32}$, and $\frac{7}{11}$.

139. Since -7 is to the left of 8 on the number line, we have $-7 < 8$.

141. Since -45.6 is to the left of -23.8 on the number line, we have $-45.6 < -23.8$.

143.
$$\frac{1}{r_1} + \frac{1}{r_2}$$
$$= \frac{1}{12} + \frac{1}{6} \quad \text{Substituting 12 for } r_1 \text{ and 6 for } r_2$$
$$= \frac{1}{12} + \frac{2}{12}$$
$$= \frac{3}{12}, \text{ or } \frac{1}{4}$$
The conductance is $\frac{1}{4}$.

145. We want to find a number c such that $-625 = -0.02 \cdot c$. From the definition of division we know that $c =$
$$\frac{-625}{-0.02} = 31,250$$

Exercise Set R.3

1. $\underbrace{4 \cdot 4 \cdot 4 \cdot 4 \cdot 4}_{5 \text{ factors}} = 4^5$

3. $\underbrace{5 \cdot 5 \cdot 5 \cdot 5 \cdot 5 \cdot 5}_{6 \text{ factors}} = 5^6$

5. $\underbrace{m \cdot m \cdot m}_{3 \text{ factors}} = m^3$

7. $\underbrace{\frac{7}{12} \cdot \frac{7}{12} \cdot \frac{7}{12} \cdot \frac{7}{12}}_{4 \text{ factors}} = \left(\frac{7}{12}\right)^4$

9. $\underbrace{(123.7)(123.7)}_{2 \text{ factors}} = (123.7)^2$

11. $2^7 = \underbrace{2 \cdot 2 \cdot 2 \cdot 2 \cdot 2 \cdot 2 \cdot 2}_{7 \text{ factors}} = 128$

13. $(-2)^5 = \underbrace{(-2) \cdot (-2) \cdot (-2) \cdot (-2) \cdot (-2)}_{5 \text{ factors}} = -32$

15. $\left(\frac{1}{3}\right)^4 = \underbrace{\frac{1}{3} \cdot \frac{1}{3} \cdot \frac{1}{3} \cdot \frac{1}{3}}_{4 \text{ factors}} = \frac{1}{81}$

17. $(-4)^3 = \underbrace{(-4) \cdot (-4) \cdot (-4)}_{3 \text{ factors}} = -64$

19. $(-5.6)^2 = \underbrace{(-5.6)(-5.6)}_{2 \text{ factors}} = 31.36$

21. $5^1 = 5 \qquad$ (For any number a, $a^1 = a$.)

23. $34^0 = 1 \qquad$ (For any nonzero number a, $a^0 = 1$.)

25. $(\sqrt{6})^0 = 1 \qquad$ (For any nonzero number a, $a^0 = 1$.)

27. $\left(\frac{7}{8}\right)^1 = \frac{7}{8} \qquad$ (For any number a, $a^1 = a$.)

29. $\left(\frac{1}{4}\right)^{-2} = \frac{1}{\left(\frac{1}{4}\right)^2} = \frac{1}{\frac{1}{16}} = 1 \cdot \frac{16}{1} = 16$

31. $\left(\frac{2}{3}\right)^{-3} = \frac{1}{\left(\frac{2}{3}\right)^3} = \frac{1}{\frac{8}{27}} = 1 \cdot \frac{27}{8} = \frac{27}{8}$

33. $y^{-5} = \frac{1}{y^5}$

35. $\frac{1}{a^{-2}} = a^2$

37. $(-11)^{-1} = \frac{1}{(-11)^1} = \frac{1}{-11} = -\frac{1}{11}$

39. $\frac{1}{3^4} = 3^{-4}$

41. $\frac{1}{b^3} = b^{-3}$

43. $\frac{1}{(-16)^2} = (-16)^{-2}$

45. $[12 - 4(5 - 1)] = [12 - 4(4)]$
$$= [12 - 16]$$
$$= -4$$

47. $9[8 - 7(5 - 2)] = 9[8 - 7 \cdot 3]$
$$= 9[8 - 21]$$
$$= 9[-13]$$
$$= -117$$

49. $[5(8 - 6) + 12] - [24 - (8 - 4)] = [5 \cdot 2 + 12] - [24 - 4]$
$$= [10 + 12] - [24 - 4]$$
$$= 22 - 20$$
$$= 2$$

51. $[64 \div (-4)] \div (-2) = -16 \div (-2)$
$= 8$

53. $19(-22) + 60 = -418 + 60$
$= -358$

55. $(5 + 7)^2 = 12^2 = 144$
$5^2 + 7^2 = 25 + 49 = 74$

57. $2^3 + 2^4 - 20 \cdot 30 = 8 + 16 - 600$
$= 24 - 600$
$= -576$

59. $5^3 + 36 \cdot 72 - (18 + 25 \cdot 4)$
$= 5^3 + 36 \cdot 72 - (18 + 100)$
$= 5^3 + 36 \cdot 72 - 118$
$= 125 + 36 \cdot 72 - 118$
$= 125 + 2592 - 118$
$= 2717 - 118$
$= 2599$

61. $(13 \cdot 2 - 8 \cdot 4)^2 = (26 - 32)^2$
$= (-6)^2$
$= 36$

63. $4000 \cdot (1 + 0.12)^3 = 4000(1.12)^3$
$= 4000(1.404928)$
$= 5619.712$

65. $(20 \cdot 4 + 13 \cdot 8)^2 - (39 \cdot 15)^3$
$= (80 + 104)^2 - (585)^3$
$= 184^2 - 585^3$
$= 33,856 - 200,201,625$
$= -200,167,769$

67. $18 - 2 \cdot 3 - 9 = 18 - 6 - 9$
$= 12 - 9$
$= 3$

69. $(18 - 2 \cdot 3) - 9 = (18 - 6) - 9$
$= 12 - 9$
$= 3$

71. $[24 \div (-3)] \div \left(-\frac{1}{2}\right)$
$= -8 \div \left(-\frac{1}{2}\right)$
$= -8 \cdot (-2)$
$= 16$

73. $15 \cdot (-24) + 50 = -360 + 50$
$= -310$

75. $4 \div (8 - 10)^2 + 1 = 4 \div (-2)^2 + 1$
$= 4 \div 4 + 1$
$= 1 + 1$
$= 2$

77. $6^3 + 25 \cdot 71 - (16 + 25 \cdot 4)$
$= 6^3 + 25 \cdot 71 - (16 + 100)$
$= 6^3 + 25 \cdot 71 - 116$
$= 216 + 25 \cdot 71 - 116$
$= 216 + 1775 - 116$
$= 1991 - 116$
$= 1875$

79. $5000 \cdot (1 + 0.16)^3 = 5000 \cdot (1.16)^3$
$= 5000(1.560896)$
$= 7804.48$

81. $4 \cdot 5 - 2 \cdot 6 + 4 = 20 - 12 + 4$
$= 8 + 4$
$= 12$

83. $4 \cdot (6 + 8)/(4 + 3) = 4 \cdot 14/7$
$= 56/7$
$= 8$

85. $[2 \cdot (5 - 3)]^2 = [2 \cdot 2]^2$
$= 4^2$
$= 16$

87. $8(-7) + 6(-5) = -56 - 30$
$= -86$

89. $19 - 5(-3) + 3 = 19 + 15 + 3$
$= 34 + 3$
$= 37$

91. $9 \div (-3) + 16 \div 8 = -3 + 2$
$= -1$

93. $7 + 10 - (-10 \div 2) = 7 + 10 - (-5)$
$= 7 + 10 + 5$
$= 17 + 5$
$= 22$

95. $5^2 - 8^2 = 25 - 64$
$= -39$

97. $20 + 4^3 \div (-8) = 20 + 64 \div (-8)$
$= 20 + (-8)$
$= 12$

99. $-7(3^4) + 18 = -7 \cdot 81 + 18$
$= -567 + 18$
$= -549$

101. $9[(8 - 11) - 13] = 9[-3 - 13]$
$$= 9[-16]$$
$$= -144$$

103. $256 \div (-32) \div (-4) = -8 \div (-4)$
$$= 2$$

105. $\dfrac{5^2 - |4^3 - 8|}{9^2 - 2^2 - 1^5} = \dfrac{5^2 - |64 - 8|}{81 - 4 - 1}$
$$= \dfrac{5^2 - |56|}{77 - 1}$$
$$= \dfrac{5^2 - 56}{77 - 1}$$
$$= \dfrac{25 - 56}{76}$$
$$= \dfrac{-31}{76}$$
$$= -\dfrac{31}{76}$$

107. $\dfrac{30(8 - 3) - 4(10 - 3)}{10|2 - 6| - 2(5 + 2)} = \dfrac{30 \cdot 5 - 4 \cdot 7}{10|-4| - 2 \cdot 7}$
$$= \dfrac{150 - 28}{10 \cdot 4 - 2 \cdot 7}$$
$$= \dfrac{122}{40 - 14}$$
$$= \dfrac{122}{26}$$
$$= \dfrac{61}{13}$$

109. Discussion and Writing Exercise

111. The distance of $-\dfrac{9}{7}$ from 0 is $\dfrac{9}{7}$, so $\left| -\dfrac{9}{7} \right| = \dfrac{9}{7}$.

113. The distance of 0 from 0 is 0, so $|0| = 0$.

115. $23 - 56 = 23 + (-56) = -33$

117. $-23 - (-56) = -23 + 56 = 33$

119. $(-10)(2.3) = -23$

(The product of a negative number and a positive number is negative.)

121. $10(-2.3) = -23$

(The product of a positive number and a negative number is negative.)

123. $(-2)^0 - (-2)^3 - (-2)^{-1} + (-2)^4 - (-2)^{-2}$
$$= 1 - (-8) - \left(-\dfrac{1}{2} \right) + 16 - \dfrac{1}{(-2)^2}$$
$$= 1 - (-8) - \left(-\dfrac{1}{2} \right) + 16 - \dfrac{1}{4}$$
$$= 1 + 8 + \dfrac{1}{2} + 16 - \dfrac{1}{4}$$
$$= 25\dfrac{1}{4}$$

125. $9 \cdot 5 + 2 - (8 \cdot 3 + 1) = 22$

127. $(0.2)^{(-0.2)^{-1}} = (0.2)^{1/-0.2} = (0.2)^{-5} =$
$$\dfrac{1}{(0.2)^5} = \dfrac{1}{0.00032} = 3125$$

129. $(2 + 3)^{-1} = 5^{-1} = \dfrac{1}{5}$; $\quad 2^{-1} + 3^{-1} = \dfrac{1}{2} + \dfrac{1}{3} = \dfrac{3}{6} + \dfrac{2}{6} = \dfrac{5}{6}$;
thus $(2 + 3)^{-1} \neq 2^{-1} + 3^{-1}$.

Exercise Set R.4

1. 8 more than b

We have $b + 8$, or $8 + b$.

3. 13.4 less than c

We have $c - 13.4$.

5. 5 increased by q

We have $5 + q$, or $q + 5$.

7. b more than a

We have $a + b$, or $b + a$.

9. x divided by y

We have $x \div y$, or $\dfrac{x}{y}$.

11. x plus w

We have $x + w$, or $w + x$.

13. m subtracted from n

We have $n - m$.

15. The sum of p and q

We have $p + q$, or $q + p$.

17. Three times q

We have $3q$.

19. -18 multiplied by m

We have $-18m$.

21. The product of 17% and your salary

Let s represent your salary. We have $17\%s$, or $0.17s$.

23. We have $75t$.

25. We have $\$40 - x$.

27. Substitute -4 for z and carry out the multiplication.
$$23z = 23(-4) = -92$$

29. Substitute -24 for a and -8 for b and carry out the division.
$$\dfrac{a}{b} = \dfrac{-24}{-8} = 3$$

31. Substitute 36 for m and 4 for n and carry out the calculations.

$$\frac{m-n}{8} = \frac{36-4}{8} = \frac{32}{8} = 4$$

33. Substitute 9 for z and 2 for y and carry out the calculations.

$$\frac{5z}{y} = \frac{5 \cdot 9}{2} = \frac{45}{2}, \text{ or } 22\frac{1}{2}, \text{ or } 22.5$$

35. Substitute 4 for b and 6 for c and carry out the calculations.

$$
\begin{aligned}
2c \div 3b &= 2 \cdot 6 \div 3 \cdot 4 \\
&= 12 \div 3 \cdot 4 \\
&= 4 \cdot 4 \\
&= 16
\end{aligned}
$$

37. Substitute 3 for r and 27 for s and carry out the calculations.

$$
\begin{aligned}
25 - r^2 + s \div r^2 &= 25 - 3^2 + 27 \div 3^2 \\
&= 25 - 9 + 27 \div 9 \\
&= 25 - 9 + 3 \\
&= 16 + 3 \\
&= 19
\end{aligned}
$$

39. Substitute 15 for m and 3 for n and carry out the calculations.

$$
\begin{aligned}
m + n(5 + n^2) &= 15 + 3(5 + 3^2) \\
&= 15 + 3(5 + 9) \\
&= 15 + 3 \cdot 14 \\
&= 15 + 42 \\
&= 57
\end{aligned}
$$

41. Substitute \$7345 for P, 6% or 0.06 for r, and 1 for t and carry out the calculations.

$$
\begin{aligned}
I &= Prt \\
I &= \$7345(0.06)(1) \\
I &= \$440.70
\end{aligned}
$$

43. Substitute 3.14 for π and 6 for r in each formula and carry out the calculations.

$$
\begin{aligned}
A &= \pi r^2 \\
&= 3.14(6)^2 \\
&= 3.14(36) \\
&= 113.04 \text{ cm}^2 \\
C &= 2\pi r \\
&= 2(3.14)(6) \\
&= 37.68 \text{ cm}
\end{aligned}
$$

45. Discussion and Writing Exercise

47. $3^5 = 3 \cdot 3 \cdot 3 \cdot 3 \cdot 3 = 243$

49. $(-10)^2 = (-10)(-10) = 100$

51. $(-5.3)^2 = (-5.3)(-5.3) = 28.09$

53. $(4.5)^0 = 1$ (For any nonzero number a, $a^0 = 1$.)

55. $(3x)^1 = 3x$ (For any number a, $a^1 = a$.)

57. Distance = speed × time

$$d = r \cdot t$$

59. Substitute 2 for x and 4 for y and carry out the calculations.

$$\frac{y+x}{2} + \frac{3y}{x} = \frac{4+2}{2} + \frac{3 \cdot 4}{2} = \frac{6}{2} + \frac{12}{2} = 3 + 6 = 9$$

Exercise Set R.5

1. Substitute and find the value of each expression. For example, for $x = -2$, $2x + 3x = 2(-2) + 3(-2) = -4 - 6 = -10$.

	$2x + 3x$	$5x$	$2x - 3x$
$x = -2$	-10	-10	2
$x = 5$	25	25	-5
$x = 0$	0	0	0

The values of $2x + 3x$ and $5x$ are the same for the given values of x and, indeed, for any allowable replacement for x. Thus, they are equivalent.

The values of $2x + 3x$ and $2x - 3x$ do not agree for $x = -2$ and for $x = 5$. Since one disagreement is sufficient to show that two expressions are not equivalent, we know that these expressions are not equivalent. Similarly, $5x$ and $2x - 3x$ are not equivalent.

3. Substitute and find the value of each expression. For example, for $x = -1$, $4x + 8x = 4(-1) + 8(-1) = -4 - 8 = -12$.

	$4x + 8x$	$4(x + 3x)$	$4(x + 2x)$
$x = -1$	-12	-16	-12
$x = 3.2$	38.4	51.2	38.4
$x = 0$	0	0	0

The values of $4x + 8x$ and $4(x + 2x)$ are the same for the given values of x and, indeed, for any allowable replacement for x. Thus, they are equivalent.

The values of $4x + 8x$ and $4(x + 3x)$ do not agree for $x = -1$ and for $x = 3.2$. Since one disagreement is sufficient to show that two expressions are not equivalent, we know that these expressions are not equivalent. Similarly, $4(x + 3x)$ and $4(x + 2x)$ are not equivalent.

5. Since $8x = 8 \cdot x$, we multiply by 1 using x/x as a name for 1.

$$\frac{7}{8} = \frac{7}{8} \cdot 1 = \frac{7}{8} \cdot \frac{x}{x} = \frac{7x}{8x}$$

7. Since $8a = 4 \cdot 2a$, we multiply by 1 using $\dfrac{2a}{2a}$ as a name for 1.

$$\frac{3}{4} = \frac{3}{4} \cdot 1 = \frac{3}{4} \cdot \frac{2a}{2a} = \frac{6a}{8a}$$

9. $\dfrac{25x}{15x} = \dfrac{5 \cdot 5x}{3 \cdot 5x}$ We look for the largest common factor of the numerator and denominator and factor each.

$\qquad = \dfrac{5}{3} \cdot \dfrac{5x}{5x}$ Factoring the expression

$\qquad = \dfrac{5}{3} \cdot 1 \qquad \left(\dfrac{5x}{5x} = 1 \right)$

$\qquad = \dfrac{5}{3}$

11. $\quad -\dfrac{100a}{25a}$

$\quad = -\dfrac{25a \cdot 4}{25a \cdot 1}$ Factoring numerator and denominator

$\quad = -\dfrac{25a}{25a} \cdot \dfrac{4}{1}$ Factoring the expression

$\quad = -1 \cdot \dfrac{4}{1} \qquad \left(\dfrac{25a}{25a} = 1 \right)$

$\quad = -\dfrac{4}{1}$

$\quad = -4$

13. $w + 3 = 3 + w$ Commutative law of addition

15. $rt = tr$ Commutative law of multiplication

17. $4 + cd = cd + 4$ Commutative law of addition

or

$4 + cd = cd + 4 = dc + 4$ Commutative laws of addition and multiplication

or

$4 + cd = 4 + dc$ Commutative law of multiplication

19. $yz + x = x + yz$ Commutative law of addition

or

$yz + x = x + yz = x + zy$ Commutative laws of addition and multiplication

or

$yz + x = zy + x$ Commutative law of multiplication

21. $m + (n + 2) = (m + n) + 2$ Associative law of addition

23. $(7 \cdot x) \cdot y = 7 \cdot (x \cdot y)$ Associative law of multiplication

25. $(a + b) + 8 = a + (b + 8)$ Associative law

$\qquad = a + (8 + b)$ Commutative law

$(a + b) + 8 = a + (b + 8)$ Associative law

$\qquad = a + (8 + b)$ Commutative law

$\qquad = (a + 8) + b$ Associative law

$(a + b) + 8 = (b + a) + 8$ Commutative law

$\qquad = b + (a + 8)$ Associative law

Other answers are possible.

27. $7 \cdot (a \cdot b) = 7 \cdot (b \cdot a)$ Commutative law

$\qquad = (7 \cdot b) \cdot a$ Associative law

$7 \cdot (a \cdot b) = (7 \cdot a) \cdot b$ Associative law

$\qquad = b \cdot (7 \cdot a)$ Commutative law

$\qquad = b \cdot (a \cdot 7)$ Commutative law

$7 \cdot (a \cdot b) = 7 \cdot (b \cdot a)$ Commutative law

$\qquad = (b \cdot a) \cdot 7$ Commutative law

Other answers are possible.

29. $4(a + 1) = 4 \cdot a + 4 \cdot 1$

$\qquad = 4a + 4$

31. $8(x - y) = 8 \cdot x - 8 \cdot y$

$\qquad = 8x - 8y$

33. $-5(2a + 3b) = -5 \cdot 2a + (-5) \cdot 3b$

$\qquad = -10a - 15b$

35. $2a(b - c + d) = 2a \cdot b - 2a \cdot c + 2a \cdot d$

$\qquad = 2ab - 2ac + 2ad$

37. $2\pi r(h + 1) = 2\pi r \cdot h + 2\pi r \cdot 1$

$\qquad = 2\pi rh + 2\pi r$

39. $\dfrac{1}{2}h(a + b) = \dfrac{1}{2}h \cdot a + \dfrac{1}{2}h \cdot b$

$\qquad = \dfrac{1}{2}ha + \dfrac{1}{2}hb$

41. $4a - 5b + 6 = 4a + (-5b) + 6$

The terms are $4a$, $-5b$, and 6.

43. $2x - 3y - 2z = 2x + (-3y) + (-2z)$

The terms are $2x$, $-3y$, and $-2z$.

45. $24x + 24y = 24 \cdot x + 24 \cdot y$

$\qquad = 24(x + y)$

47. $7p - 7 = 7 \cdot p - 7 \cdot 1$

$\qquad = 7(p - 1)$

49. $7x - 21 = 7 \cdot x - 7 \cdot 3$

$\qquad = 7(x - 3)$

51. $xy + x = x \cdot y + x \cdot 1$

$\qquad = x(y + 1)$

53. $2x - 2y + 2z = 2 \cdot x - 2 \cdot y + 2 \cdot z$

$\qquad = 2(x - y + z)$

55. $3x + 6y - 3 = 3 \cdot x + 3 \cdot 2y - 3 \cdot 1$

$\qquad = 3(x + 2y - 1)$

57. $ab + ac - ad = a \cdot b + a \cdot c - a \cdot d$

$\qquad = a(b + c - d)$

59. $\dfrac{1}{4}\pi rr + \dfrac{1}{4}\pi rs = \dfrac{1}{4}\pi r \cdot r + \dfrac{1}{4}\pi r \cdot s$

$\qquad\qquad\qquad = \dfrac{1}{4}\pi r(r + s)$

61. Discussion and Writing Exercise

63. Let x and y represent the numbers. Then we have

$\qquad (x + y)^2$.

65. $-34.2 + -67.8 = -34.2 + (-67.8) = -102$

67. $-\dfrac{1}{4}\left(-\dfrac{1}{2}\right) = \dfrac{1 \cdot 1}{4 \cdot 2} = \dfrac{1}{8}$

69. Let $x = 1$ and $y = -1$.

$\qquad x^2 + y^2 = 1^2 + (-1)^2 = 1 + 1 = 2$

$\qquad (x + y)^2 = (1 + (-1))^2 = 0^2 = 0$

Since the expressions have different values, they are not equivalent.

71. For $x = -2$, $x^2 \cdot x^3 = (-2)^2 \cdot (-2)^3 = 4(-8) = -32$ and $x^5 = (-2)^5 = -32$.

For $x = 0$, $x^2 \cdot x^3 = 0^2 \cdot 0^3 = 0 \cdot 0 = 0$ and $x^5 = 0^5 = 0$.

For $x = 1$, $x^2 \cdot x^3 = 1^2 \cdot 1^3 = 1 \cdot 1 = 1$ and $x^5 = 1^5 = 1$.

Indeed, the expressions have the same value for all values of x, so they are equivalent.

Exercise Set R.6

1. $7x + 5x = (7 + 5)x$

$\qquad\qquad = 12x$

3. $8b - 11b = (8 - 11)b$

$\qquad\qquad = -3b$

5. $14y + y = 14y + 1y$

$\qquad\qquad = (14 + 1)y$

$\qquad\qquad = 15y$

7. $12a - a = 12a - 1a$

$\qquad\qquad = (12 - 1)a$

$\qquad\qquad = 11a$

9. $t - 9t = 1t - 9t$

$\qquad\qquad = (1 - 9)t$

$\qquad\qquad = -8t$

11. $5x - 3x + 8x = (5 - 3 + 8)x$

$\qquad\qquad\qquad = 10x$

13. $3x - 5y + 8x = (3 + 8)x - 5y$

$\qquad\qquad\qquad = 11x - 5y$

15. $3c + 8d - 7c + 4d = (3 - 7)c + (8 + 4)d$

$\qquad\qquad\qquad\qquad = -4c + 12d$

17. $4x - 7 + 18x + 25 = (4 + 18)x + (-7 + 25)$

$\qquad\qquad\qquad\qquad = 22x + 18$

19. $\quad 1.3x + 1.4y - 0.11x - 0.47y$

$= (1.3 - 0.11)x + (1.4 - 0.47)y$

$= 1.19x + 0.93y$

21. $\quad \dfrac{2}{3}a + \dfrac{5}{6}b - 27 - \dfrac{4}{5}a - \dfrac{7}{6}b$

$= \left(\dfrac{2}{3} - \dfrac{4}{5}\right)a + \left(\dfrac{5}{6} - \dfrac{7}{6}\right)b - 27$

$= \left(\dfrac{10}{15} - \dfrac{12}{15}\right)a + \left(-\dfrac{2}{6}\right)b - 27$

$= -\dfrac{2}{15}a - \dfrac{1}{3}b - 27$

23. $P = 2l + 2w$

$\quad P = 2 \cdot l + 2 \cdot w$

$\quad P = 2(l + w)$

25. $-(-2c) = -1(-2c)$

$\qquad\qquad = [-1(-2)]c$

$\qquad\qquad = 2c$

27. $-(b + 4) = -1(b + 4)$

$\qquad\qquad = -1 \cdot b + (-1)4$

$\qquad\qquad = -b - 4$

29. $-(b - 3) = -1(b - 3)$

$\qquad\qquad = -1 \cdot b - (-1) \cdot 3$

$\qquad\qquad = -b + [-(-1)3]$

$\qquad\qquad = -b + 3$, or $3 - b$

31. $-(t - y) = -1(t - y)$

$\qquad\qquad = -1 \cdot t - (-1) \cdot y$

$\qquad\qquad = -t + [-(-1)y]$

$\qquad\qquad = -t + y$, or $y - t$

33. $-(x + y + z) = -1(x + y + z)$

$\qquad\qquad\qquad = -1 \cdot x + (-1) \cdot y + (-1) \cdot z$

$\qquad\qquad\qquad = -x - y - z$

35. $\quad -(8x - 6y + 13)$

$= -8x + 6y - 13 \quad$ Changing the sign of every term inside parentheses

37. $\quad -(-2c + 5d - 3e + 4f)$

$= 2c - 5d + 3e - 4f \qquad$ Changing the sign of every term inside parentheses

39. $\quad -\left(-1.2x + 56.7y - 34z - \dfrac{1}{4}\right)$

$= 1.2x - 56.7y + 34z + \dfrac{1}{4} \qquad$ Changing the sign of every term inside parentheses

41. $a + (2a + 5) = a + 2a + 5$
$ = 3a + 5$

43. $4m - (3m - 1) = 4m - 3m + 1$
$ = m + 1$

45. $5d - 9 - (7 - 4d) = 5d - 9 - 7 + 4d$
$ = 9d - 16$

47. $ -2(x + 3) - 5(x - 4)$
$= -2(x + 3) + [-5(x - 4)]$
$= -2x - 6 + [-5x + 20]$
$= -2x - 6 - 5x + 20$
$= -7x + 14$

49. $ 5x - 7(2x - 3) - 4$
$= 5x + [-7(2x - 3)] - 4$
$= 5x + [-14x + 21] - 4$
$= 5x - 14x + 21 - 4$
$= -9x + 17$

51. $ 8x - (-3y + 7) + (9x - 11)$
$= 8x + 3y - 7 + 9x - 11$
$= 17x + 3y - 18$

53. $ \frac{1}{4}(24x - 8) - \frac{1}{2}(-8x + 6) - 14$
$= \frac{1}{4}(24x - 8) + \left[-\frac{1}{2}(-8x + 6)\right] - 14$
$= 6x - 2 + [4x - 3] - 14$
$= 6x - 2 + 4x - 3 - 14$
$= 10x - 19$

55. $7a - [9 - 3(5a - 2)] = 7a - [9 - 15a + 6]$
$ = 7a - [15 - 15a]$
$ = 7a - 15 + 15a$
$ = 22a - 15$

57. $5\{-2 + 3[4 - 2(3 + 5)]\} = 5\{-2 + 3[4 - 2(8)]\}$
$\phantom{5\{-2 + 3[4 - 2(3 + 5)]\}} = 5\{-2 + 3[4 - 16]\}$
$\phantom{5\{-2 + 3[4 - 2(3 + 5)]\}} = 5\{-2 + 3[-12]\}$
$\phantom{5\{-2 + 3[4 - 2(3 + 5)]\}} = 5\{-2 - 36\}$
$\phantom{5\{-2 + 3[4 - 2(3 + 5)]\}} = 5\{-38\}$
$\phantom{5\{-2 + 3[4 - 2(3 + 5)]\}} = -190$

59. $ [10(x + 3) - 4] + [2(x - 1) + 6]$
$= [10x + 30 - 4] + [2x - 2 + 6]$
$= 10x + 26 + 2x + 4$
$= 12x + 30$

61. $ [7(x + 5) - 19] - [4(x - 6) + 10]$
$= [7x + 35 - 19] - [4x - 24 + 10]$
$= [7x + 16] - [4x - 14]$
$= 7x + 16 - 4x + 14$
$= 3x + 30$

63. $ 3\{[7(x - 2) + 4] - [2(2x - 5) + 6]\}$
$= 3\{[7x - 14 + 4] - [4x - 10 + 6]\}$
$= 3\{[7x - 10] - [4x - 4]\}$
$= 3\{7x - 10 - 4x + 4\}$
$= 3\{3x - 6\}$
$= 9x - 18$

65. $ 4\{[5(x - 3) + 2^2] - 3[2(x + 5) - 9^2]\}$
$= 4\{[5(x - 3) + 4] - 3[2(x + 5) - 81]\}$
$= 4\{[5x - 15 + 4] - 3[2x + 10 - 81]\}$
$= 4\{[5x - 11] - 3[2x - 71]\}$
$= 4\{5x - 11 - 6x + 213\}$
$= 4\{-x + 202\}$
$= -4x + 808$

67. $ 2y + \{8[3(2y - 5) - (8y + 9)] + 6\}$
$= 2y + \{8[6y - 15 - 8y - 9] + 6\}$
$= 2y + \{8[-2y - 24] + 6\}$
$= 2y + \{-16y - 192 + 6\}$
$= 2y + \{-16y - 186\}$
$= 2y - 16y - 186$
$= -14y - 186$

69. Discussion and Writing Exercise

71. $17 + (-54)$

We find the difference of the absolute values: $54 - 17 = 37$. Since the negative number has the larger absolute value, the answer is negative.

$17 + (-54) = -37$

73. $-13.78 + (-9.32)$

The sum of two negative numbers is negative. We add their absolute values, $13.78 + 9.32 = 23.1$, and make the answer negative.

$-13.78 + (-9.32) = -23.1$

75. When a negative number is divided by a positive number, the answer is negative.

$-256 \div 16 = -16$

77. When a positive number is divided by a negative number, the answer is negative.

$256 \div (-16) = -16$

79. $8(a - b) = 8 \cdot a - 8 \cdot b = 8a - 8b$

81. $6x(a-b+2c) = 6x \cdot a - 6x \cdot b + 6x \cdot 2c =$
$6ax - 6bx + 12cx$

83. $24a - 24 = 24 \cdot a - 24 \cdot 1 = 24(a-1)$

85. $ab - ac + a = a \cdot b - a \cdot c + a \cdot 1 = a(b-c+1)$

87. $(3-8)^2 + 9 = 34$

89. $5 \cdot 2^3 \div (3-4)^4 = 40$

91.
$$\begin{aligned}&[11(a-3)+12a] - \{6[4(3b-7)-(9b+10)]+11\}\\ &= [11a-33+12a] - \{6[12b-28-9b-10]+11\}\\ &= [23a-33] - \{6[3b-38]+11\}\\ &= 23a-33-\{18b-228+11\}\\ &= 23a-33-\{18b-217\}\\ &= 23a-33-18b+217\\ &= 23a-18b+184\end{aligned}$$

93.
$$\begin{aligned}&z - \{2z+[3z-(4z+5x)-6z]+7z\}-8z\\ &= z-\{2z+[3z-4z-5x-6z]+7z\}-8z\\ &= z-\{2z-7z-5x+7z\}-8z\\ &= z-\{2z-5x\}-8z\\ &= z-2z+5x-8z\\ &= -9z+5x\end{aligned}$$

95.
$$\begin{aligned}&x-\{x+1-[x+2-(x-3-\{x+4-\\ &\qquad [x-5+(x-6)]\})]\}\\ &= x-\{x+1-[x+2-(x-3-\{x+4-[2x-11]\})]\}\\ &= x-\{x+1-[x+2-(x-3-\{x+4-2x+11\})]\}\\ &= x-\{x+1-[x+2-(x-3-\{-x+15\})]\}\\ &= x-\{x+1-[x+2-(x-3+x-15)]\}\\ &= x-\{x+1-[x+2-(2x-18)]\}\\ &= x-\{x+1-[x+2-2x+18]\}\\ &= x-\{x+1-[-x+20]\}\\ &= x-\{x+1+x-20\}\\ &= x-\{2x-19\}\\ &= x-2x+19\\ &= -x+19\end{aligned}$$

Exercise Set R.7

1. $3^6 \cdot 3^3 = 3^{6+3} = 3^9$

3. $6^{-6} \cdot 6^2 = 6^{-6+2} = 6^{-4} = \dfrac{1}{6^4}$

5. $8^{-2} \cdot 8^{-4} = 8^{-2+(-4)} = 8^{-6} = \dfrac{1}{8^6}$

7. $b^2 \cdot b^{-5} = b^{2+(-5)} = b^{-3} = \dfrac{1}{b^3}$

9. $a^{-3} \cdot a^4 \cdot a^2 = a^{-3+4+2} = a^3$

11.
$$\begin{aligned}(2x)^3(3x)^2 &= 8x^3 \cdot 9x^2\\ &= 8 \cdot 9 \cdot x^3 \cdot x^2\\ &= 72x^{3+2}\\ &= 72x^5\end{aligned}$$

13.
$$\begin{aligned}(14m^2n^3)(-2m^3n^2) &= 14 \cdot (-2) \cdot m^2 \cdot m^3 \cdot n^3 \cdot n^2\\ &= -28m^{2+3}n^{3+2}\\ &= -28m^5n^5\end{aligned}$$

15.
$$\begin{aligned}(-2x^{-3})(7x^{-8}) &= -2 \cdot 7 \cdot x^{-3} \cdot x^{-8}\\ &= -14x^{-3+(-8)}\\ &= -14x^{-11} = -\frac{14}{x^{11}}\end{aligned}$$

17.
$$\begin{aligned}(15x^{4t})(7x^{-6t}) &= 15 \cdot 7 \cdot x^{4t} \cdot x^{-6t}\\ &= 105x^{4t+(-6t)}\\ &= 105x^{-2t}\\ &= \frac{105}{x^{2t}}\end{aligned}$$

19.
$$\begin{aligned}(2y^{3m})(-4y^{-9m}) &= 2 \cdot (-4) \cdot y^{3m} \cdot y^{-9m}\\ &= -8y^{3m+(-9m)}\\ &= -8y^{-6m}\\ &= -\frac{8}{y^{6m}}\end{aligned}$$

21. $\dfrac{8^9}{8^2} = 8^{9-2} = 8^7$

23. $\dfrac{6^3}{6^{-2}} = 6^{3-(-2)} = 6^{3+2} = 6^5$

25. $\dfrac{10^{-3}}{10^6} = 10^{-3-6} = 10^{-3+(-6)} = 10^{-9} = \dfrac{1}{10^9}$

27. $\dfrac{9^{-4}}{9^{-6}} = 9^{-4-(-6)} = 9^{-4+6} = 9^2$

29. $\dfrac{x^{-4n}}{x^{6n}} = x^{-4n-6n} = x^{-10n} = \dfrac{1}{x^{10n}}$

31. $\dfrac{w^{-11q}}{w^{-6q}} = w^{-11q-(-6q)} = w^{-11q+6q} = w^{-5q} =$
$\dfrac{1}{w^{5q}}$

33. $\dfrac{a^3}{a^{-2}} = a^{3-(-2)} = a^{3+2} = a^5$

35. $\dfrac{2yx^7z^5}{-9x^2z} = \dfrac{27}{-9}x^{7-2}z^{5-1} = -3x^5z^4$

37.
$$\begin{aligned}\frac{-24x^6y^7}{18x^{-3}y^9} &= \frac{-24}{18}x^{6-(-3)}y^{7-9}\\ &= -\frac{24}{18}x^{6+3}y^{-2}\\ &= -\frac{4}{3}x^9y^{-2} = -\frac{4x^9}{3y^2}\end{aligned}$$

39. $\dfrac{-18x^{-2}y^3}{-12x^{-5}y^5} = \dfrac{-18}{-12}x^{-2-(-5)}y^{3-5}$

$\qquad\qquad = \dfrac{18}{12}x^{-2+5}y^{-2}$

$\qquad\qquad = \dfrac{3}{2}x^3y^{-2} = \dfrac{3x^3}{2y^2}$

41. $(4^3)^2 = 4^{3\cdot2} = 4^6$

43. $(8^4)^{-3} = 8^{4(-3)} = 8^{-12} = \dfrac{1}{8^{12}}$

45. $(6^{-4})^{-3} = 6^{-4(-3)} = 6^{12}$

47. $(5a^2b^2)^3 = 5^3(a^2)^3(b^2)^3$

$\qquad\qquad = 125a^{2\cdot3}b^{2\cdot3}$

$\qquad\qquad = 125a^6b^6$

49. $(-3x^3y^{-6})^{-2} = (-3)^{-2}(x^3)^{-2}(y^{-6})^{-2}$

$\qquad\qquad = \dfrac{1}{(-3)^2}x^{3(-2)}y^{-6(-2)}$

$\qquad\qquad = \dfrac{1}{9}x^{-6}y^{12}$

$\qquad\qquad = \dfrac{y^{12}}{9x^6}$

51. $(-6a^{-2}b^3c)^{-2} = (-6)^{-2}(a^{-2})^{-2}(b^3)^{-2}c^{-2}$

$\qquad\qquad = \dfrac{1}{(-6)^2}a^{-2(-2)}b^{3(-2)}c^{-2}$

$\qquad\qquad = \dfrac{1}{36}a^4b^{-6}c^{-2} = \dfrac{a^4}{36b^6c^2}$

53. $\left(\dfrac{4^{-3}}{3^4}\right)^3 = \dfrac{(4^{-3})^3}{(3^4)^3} = \dfrac{4^{-3\cdot3}}{3^{4\cdot3}} = \dfrac{4^{-9}}{3^{12}} = \dfrac{1}{4^9\cdot3^{12}}$

55. $\left(\dfrac{2x^3y^{-2}}{3y^{-3}}\right)^3 = \dfrac{(2x^3y^{-2})^3}{(3y^{-3})^3}$

$\qquad\qquad = \dfrac{2^3(x^3)^3(y^{-2})^3}{3^3(y^{-3})^3}$

$\qquad\qquad = \dfrac{8x^9y^{-6}}{27y^{-9}}$

$\qquad\qquad = \dfrac{8x^9y^{6-(-9)}}{27}$

$\qquad\qquad = \dfrac{8x^9y^3}{27}$

57. $\left(\dfrac{125a^2b^{-3}}{5a^4b^{-2}}\right)^{-5} = \left(\dfrac{5a^4b^{-2}}{125a^2b^{-3}}\right)^5$

$\qquad\qquad = \left(\dfrac{a^{4-2}b^{-2-(-3)}}{25}\right)^5$

$\qquad\qquad = \left(\dfrac{a^2b}{25}\right)^5$

$\qquad\qquad = \dfrac{(a^2)^5(b)^5}{(25)^5} = \dfrac{a^{2\cdot5}b^{1\cdot5}}{25^{1\cdot5}}$

$\qquad\qquad = \dfrac{a^{10}b^5}{25^5}, \text{ or } \dfrac{a^{10}b^5}{5^{10}}$

$\qquad\qquad\qquad [25^5 = (5^2)^5 = 5^{10}]$

59. $\left(\dfrac{-6^5y^4z^{-5}}{2^{-2}y^{-2}z^3}\right)^6 = (-1\cdot6^5\cdot2^2y^{4-(-2)}z^{-5-3})^6$

$\qquad\qquad = (-1\cdot6^5\cdot2^2y^6z^{-8})^6$

$\qquad\qquad = (-1)^6(6^5)^6(2^2)^6(y^6)^6(z^{-8})^6$

$\qquad\qquad = 1\cdot6^{5\cdot6}2^{2\cdot6}y^{6\cdot6}z^{-8\cdot6}$

$\qquad\qquad = 6^{30}2^{12}y^{36}z^{-48}$

$\qquad\qquad = \dfrac{6^{30}2^{12}y^{36}}{z^{48}}$

61. $[(-2x^{-4}y^{-2})^{-3}]^{-2} = [(-2)^{-3}(x^{-4})^{-3}(y^{-2})^{-3}]^{-2}$

$\qquad\qquad = [(-2)^{-3}x^{12}y^6]^{-2}$

$\qquad\qquad = [(-2)^{-3}]^{-2}(x^{12})^{-2}(y^6)^{-2}$

$\qquad\qquad = (-2)^6x^{-24}y^{-12}$

$\qquad\qquad = \dfrac{64}{x^{24}y^{12}}$

63. $\left(\dfrac{3a^{-2}b}{5a^{-7}b^5}\right)^{-7} = \left(\dfrac{5a^{-7}b^5}{3a^{-2}b}\right)^7$

$\qquad\qquad = \left(\dfrac{5a^{-7-(-2)}b^{5-1}}{3}\right)^7$

$\qquad\qquad = \left(\dfrac{5a^{-5}b^4}{3}\right)^7$

$\qquad\qquad = \dfrac{5^7(a^{-5})^7(b^4)^7}{3^7}$

$\qquad\qquad = \dfrac{5^7a^{-35}b^{28}}{3^7}$

$\qquad\qquad = \dfrac{5^7b^{28}}{3^7a^{35}}$

65. $\dfrac{10^{2a+1}}{10^{a+1}} = 10^{2a+1-(a+1)} = 10^{2a+1-a-1} = 10^a$

67. $\dfrac{9a^{x-2}}{3a^{2x+2}} = \dfrac{9}{3}\cdot\dfrac{a^{x-2}}{a^{2x+2}} = 3a^{x-2-(2x+2)} =$

$3a^{x-2-2x-2} = 3a^{-x-4}$

69. $\dfrac{45x^{2a+4}y^{b+1}}{-9x^{a+3}y^{2+b}} = \dfrac{45}{-9}\cdot\dfrac{x^{2a+4}y^{b+1}}{x^{a+3}y^{2+b}} =$

$-5x^{2a+4-(a+3)}y^{b+1-(2+b)} = -5x^{2a+4-a-3}y^{b+1-2-b} =$

$-5x^{a+1}y^{-1}, \text{ or } \dfrac{-5x^{a+1}}{y}$

71. $(8^x)^{4y} = 8^{x\cdot4y} = 8^{4xy}$

73. $(12^{3-a})^{2b} = 12^{(3-a)(2b)} = 12^{6b-2ab}$

75. $(5x^{a-1}y^{b+1})^{2c} = 5^{2c}x^{(a-1)(2c)}y^{(b+1)(2c)} =$

$5^{2c}x^{2ac-2c}y^{2bc+2c}, \text{ or } (5^2)^cx^{2ac-2c}y^{2bc+2c} =$

$25^cx^{2ac-2c}y^{2bc+2c}$

77. $\dfrac{4x^{2a+3}y^{2b-1}}{2x^{a+1}y^{b+1}} = \dfrac{4}{2}\cdot\dfrac{x^{2a+3}y^{2b-1}}{x^{a+1}y^{b+1}} =$

$2x^{2a+3-(a+1)}y^{2b-1-(b+1)} = 2x^{2a+3-a-1}y^{2b-1-b-1} =$

$2x^{a+2}y^{b-2}$

79. $4\underset{\underline{\qquad\qquad}}{.7,000,000,000.}$

10 places

Large number, so the exponent is positive.

$47,000,000,000 = 4.7 \times 10^{10}$

81. $0.00000001\underset{\underline{\qquad\qquad}}{.6}$

8 places

Small number, so the exponent is negative.

$0.000000016 = 1.6 \times 10^{-8}$

83. $7\underset{\underline{\qquad\qquad}}{.585,000,000.}$

9 places

Large number, so the exponent is positive.

$7,585,000,000 = 7.585 \times 10^{9}$

85. 200 billionths $= 200 \times 0.000000001 = 0.0000002$

$0.0000002\underset{\underline{\qquad\qquad}}{.}$

7 places

Small number, so the exponent is negative.

$0.0000002 = 2 \times 10^{-7}$

87. $6\underset{\underline{\qquad\qquad}}{.73000000.}$

8 places

Positive exponent so the number is large.

$6.73 \times 10^{8} = 673,000,000$

89. $0.00006\underset{\underline{\qquad}}{.6}$ cm

5 places

Negative exponent, so the number is small.

6.6×10^{-5} cm $= 0.000066$ cm

91. $1\underset{\underline{\qquad\qquad}}{.1000000.}$

7 places

Positive exponent, so the number is large.

$1.1 \times 10^{7} = 11,000,000$

93. $(2.3 \times 10^{6})(4.2 \times 10^{-11})$

$= (2.3 \times 4.2)(10^{6} \times 10^{-11})$

$= 9.66 \times 10^{-5}$

95. $(2.34 \times 10^{-8})(5.7 \times 10^{-4})$

$= (2.34 \times 5.7)(10^{-8} \times 10^{-4})$

$= 13.338 \times 10^{-12}$

$= (1.3338 \times 10^{1}) \times 10^{-12}$

$= 1.3338 \times (10^{1} \times 10^{-12})$

$= 1.3338 \times 10^{-11}$

97. $\dfrac{8.5 \times 10^{8}}{3.4 \times 10^{5}} = \dfrac{8.5}{3.4} \times \dfrac{10^{8}}{10^{5}}$

$\phantom{\dfrac{8.5 \times 10^{8}}{3.4 \times 10^{5}}} = 2.5 \times 10^{3}$

99. $\dfrac{4.0 \times 10^{-6}}{8.0 \times 10^{-3}} = \dfrac{4.0}{8.0} \times \dfrac{10^{-6}}{10^{-3}}$

$\phantom{\dfrac{4.0 \times 10^{-6}}{8.0 \times 10^{-3}}} = 0.5 \times 10^{-3}$

$\phantom{\dfrac{4.0 \times 10^{-6}}{8.0 \times 10^{-3}}} = (5 \times 10^{-1}) \times 10^{-3}$

$\phantom{\dfrac{4.0 \times 10^{-6}}{8.0 \times 10^{-3}}} = 5 \times (10^{-1} \times 10^{-3})$

$\phantom{\dfrac{4.0 \times 10^{-6}}{8.0 \times 10^{-3}}} = 5 \times 10^{-4}$

101. $3290 = 3.29 \times 10^{3}$; $12,426 = 1.2426 \times 10^{4}$

Then we have

$(3.29 \times 10^{3})(1.2426 \times 10^{4})$

$= (3.29 \times 1.2426)(10^{3} \times 10^{4})$

$\approx 4.09 \times 10^{7}$

The movie earned about 4.09×10^{7} in its first weekend.

103.   2000 yr

$= 2000 \text{ yr} \times \dfrac{365 \text{ days}}{1 \text{ yr}} \times \dfrac{24 \text{ hr}}{1 \text{ day}} \times \dfrac{60 \text{ min}}{1 \text{ hr}} \times \dfrac{60 \text{ sec}}{1 \text{ min}}$

$= 63,072,000,000 \text{ sec}$

$= 6.3072 \times 10^{10} \text{ sec}$

105. We divide.

$\dfrac{2.4 \times 10^{13}}{5.88 \times 10^{12}} = \dfrac{2.4}{5.88} \times \dfrac{10^{13}}{10^{12}}$

$\phantom{\dfrac{2.4 \times 10^{13}}{5.88 \times 10^{12}}} \approx 0.408 \times 10$

$\phantom{\dfrac{2.4 \times 10^{13}}{5.88 \times 10^{12}}} \approx (4.08 \times 10^{-1}) \times 10$

$\phantom{\dfrac{2.4 \times 10^{13}}{5.88 \times 10^{12}}} \approx 4.08 \text{ light-years}$

107. $\dfrac{10,000}{300,000} = \dfrac{10^{4}}{3 \times 10^{5}}$

$\phantom{\dfrac{10,000}{300,000}} \approx 0.333 \times 10^{-1}$

$\phantom{\dfrac{10,000}{300,000}} \approx (3.33 \times 10^{-1}) \times 10^{-1}$

$\phantom{\dfrac{10,000}{300,000}} \approx 3.33 \times 10^{-2}$

The part of the total number of words in the English language that an average person knows is about 3.33×10^{-2}.

109. First we divide to find the number of $5 bills in $4,540,000.

$\dfrac{\$4,540,000}{\$5} = \dfrac{4.54 \times 10^{6}}{5} = 0.908 \times 10^{6} =$

$(9.08 \times 10^{-1}) \times 10^{6} = 9.08 \times 10^{5}$

Then we divide 1 ton, or 2000 lb, by the number of bills to find the weight of a single bill.

$\dfrac{2000}{9.08 \times 10^{5}} = \dfrac{2 \times 10^{3}}{9.08 \times 10^{5}} = \dfrac{2}{9.08} \times \dfrac{10^{3}}{10^{5}} \approx$

$0.220 \times 10^{-2} \approx (2.2 \times 10^{-1}) \times 10^{-2} \approx 2.2 \times 10^{-3}$

A five-dollar bill weighs about 2.2×10^{-3} lb.

111. Discussion and Writing Exercise

113. $9x - (-4y + 8) + (10x - 12)$

$= 9x + 4y - 8 + 10x - 12$

$= 19x + 4y - 20$

115. $4^2 + 30 \cdot 10 - 7^3 + 16 = 16 + 30 \cdot 10 - 343 + 16$
$$= 16 + 300 - 343 + 16$$
$$= 316 - 343 + 16$$
$$= -27 + 16$$
$$= -11$$

117. $20 - 5 \cdot 4 - 8 = 20 - 20 - 8$
$$= 0 - 8$$
$$= -8$$

119. $\dfrac{(2^{-2})^{-4} \times (2^3)^{-2}}{(2^{-2})^2 \cdot (2^5)^{-3}} = \dfrac{2^8 \times 2^{-6}}{2^{-4} \cdot 2^{-15}}$
$$= \dfrac{2^{8+(-6)}}{2^{-4+(-15)}}$$
$$= \dfrac{2^2}{2^{-19}}$$
$$= 2^{2-(-19)}$$
$$= 2^{21}$$

121. $\left[\left(\dfrac{a^{-2}}{b^7}\right)^{-3} \cdot \left(\dfrac{a^4}{b^{-3}}\right)^2\right]^{-1} = \left[\dfrac{(a^{-2})^{-3}}{(b^7)^{-3}} \cdot \dfrac{(a^4)^2}{(b^{-3})^2}\right]^{-1}$
$$= \left[\dfrac{a^6}{b^{-21}} \cdot \dfrac{a^8}{b^{-6}}\right]^{-1}$$
$$= \left[\dfrac{a^{6+8}}{b^{-21+(-6)}}\right]^{-1}$$
$$= \left[\dfrac{a^{14}}{b^{-27}}\right]^{-1}$$
$$= \dfrac{a^{-14}}{b^{27}}$$
$$= \dfrac{1}{a^{14}b^{27}}$$

123. $\left[\dfrac{(2x^a y^b)^3}{(-2x^a y^b)^2}\right]^2 = \left[\dfrac{(2x^a y^b)^3}{(2x^a y^b)^2}\right]^2 \quad [(-2x^a y^b)^2 = (2x^a y^b)^2]$
$$= [(2x^a y^b)^{3-2}]^2$$
$$= (2x^a y^b)^2$$
$$= 2^2(x^a)^2(y^b)^2$$
$$= 4x^{2a}y^{2b}$$

Chapter R Review Exercises

1. The rational numbers can be named as quotients of integers with nonzero divisors. Of the given numbers, the rational numbers are 2, $-\dfrac{2}{3}$, $0.45\overline{45}$, and -23.788.

2. We specify the conditions by which we know whether a number is in the set.

$\{x | x$ is a real number less than or equal to 46$\}$

3. Since -3.9 is to the left of 2.9, we have $-3.9 < 2.9$.

4. $19 > x$

The inequality $x < 19$ has the same meaning.

5. $-13 \geq 5$ is false since neither $-13 > 5$ nor $-13 = 5$ is true.

6. $7.01 \leq 7.01$ is true since $7.01 = 7.01$ is true.

7. $x > -4$

We shade all numbers to the right of -4 and use a parenthesis at -4 to indicate it is not a solution.

8. $x \leq 1$

We shade all the numbers to the left of 1 and use a bracket at 1 to indicate it is also a solution.

9. The distance of -7.23 from 0 is 7.23, so $|-7.23| = 7.23$.

10. The distance of $9 - 9$, or 0, from 0 is 0, so $|9 - 9| = 0$.

11. $6 + (-8)$

We find the difference of the absolute values, $8 - 6 = 2$. Since the negative number has the larger absolute value, the answer is negative.

$6 + (-8) = -2$

12. $-3.8 + (-4.1)$

The sum of two negative numbers is negative. We add the absolute values, $3.8 + 4.1 = 7.9$, and make the answer negative.

$-3.8 + (-4.1) = -7.9$

13. $\dfrac{3}{4} + \left(-\dfrac{13}{7}\right)$

We find the difference of the absolute values, $\dfrac{13}{7} - \dfrac{3}{4} = \dfrac{52}{28} - \dfrac{21}{28} = \dfrac{31}{28}$. Since the negative number has the larger absolute value, the answer is negative.

$\dfrac{3}{4} + \left(-\dfrac{13}{7}\right) = -\dfrac{31}{28}$

14. $-8 - (-3) = -8 + 3 = -5$

15. $-17.3 - 9.4 = -17.3 + (-9.4) = -26.7$

16. $\dfrac{3}{2} - \left(-\dfrac{13}{4}\right) = \dfrac{3}{2} + \dfrac{13}{4} = \dfrac{6}{4} + \dfrac{13}{4} = \dfrac{19}{4}$

17. $(-3.8)(-2.7)$

The product of two negative numbers is positive. We multiply the absolute values, $3.8(2.7) = 10.26$, and make the answer positive.

$(-3.8)(-2.7) = 10.26$

18. $-\dfrac{2}{3}\left(\dfrac{9}{14}\right)$

The product of a negative number and a positive number is negative. We multiply the absolute values and make the answer negative.

$$\dfrac{2}{3}\cdot\dfrac{9}{14}=\dfrac{2\cdot 9}{3\cdot 14}=\dfrac{2\cdot 3\cdot 3}{3\cdot 2\cdot 7}=\dfrac{2\cdot 3}{2\cdot 3}\cdot\dfrac{3}{7}$$

Thus, $-\dfrac{2}{3}\left(\dfrac{9}{14}\right)=-\dfrac{3}{7}$

19. $-6(-7)(4)=42(4)=168$

20. When a negative number is divided by a positive number, the answer is negative.

$$-12\div 3=\dfrac{-12}{3}=-4$$

21. When a negative number is divided by a negative number, the answer is positive.

$$\dfrac{-84}{-4}=21$$

22. When a positive number is divided by a negative number, the answer is negative.

$$\dfrac{49}{-7}=-7$$

23. $\dfrac{5}{6}\div\left(-\dfrac{10}{7}\right)=\dfrac{5}{6}\cdot\left(-\dfrac{7}{10}\right)=-\dfrac{5\cdot 7}{6\cdot 10}=-\dfrac{5\cdot 7}{6\cdot 2\cdot 5}=$

$-\dfrac{7}{6\cdot 2}\cdot\dfrac{5}{5}=-\dfrac{7}{6\cdot 2}=-\dfrac{7}{12}$

24. $-\dfrac{5}{2}\div\left(-\dfrac{15}{16}\right)=-\dfrac{5}{2}\cdot\left(-\dfrac{16}{15}\right)=-\dfrac{5\cdot 16}{2\cdot 15}=\dfrac{5\cdot 2\cdot 8}{2\cdot 3\cdot 5}=$

$\dfrac{2\cdot 5}{2\cdot 5}\cdot\dfrac{8}{3}=\dfrac{8}{3}$

25. $\dfrac{25}{0}$ Not defined: Division by 0.

26. $-108\div 4.5=\dfrac{-108}{4.5}=-24$

27. When $a=-7$, then $-a=-(-7)=7$.

28. When $a=2.3$, then $-a=-2.3$.

29. When $a=0$, then $-a=-0=0$.

30. $\underbrace{a\cdot a\cdot a\cdot a\cdot a}_{5\text{ factors}}=a^5$

31. $\underbrace{\left(-\dfrac{7}{8}\right)\left(-\dfrac{7}{8}\right)\left(-\dfrac{7}{8}\right)}_{3\text{ factors}}=\left(-\dfrac{7}{8}\right)^3$

32. $a^{-4}=\dfrac{1}{a^4}$

33. $\dfrac{1}{x^6}=x^{-6}$

34. $2^3-3^4+(13\cdot 5+67)=8-81+(13\cdot 5+67)$

$=8-81+(65+67)$

$=8-81+132$

$=-73+132$

$=59$

35. $64\div(-4)+(-5)(20)=-16-100=-116$

36. Let x represent the number. Then we have $5x$.

37. Let y represent the number. Then we have $28\%y$, or $0.28y$.

38. We have $t-9$.

39. Let a and b represent the numbers. Then we have $\dfrac{a}{b}-8$.

40. Substitute -2 for x and carry out the calculations.

$$5x-7=5(-2)-7=-10-7=-17$$

41. Substitute 4 for x and 20 for y and carry out the calculations.

$$\dfrac{x-y}{2}=\dfrac{4-20}{2}=\dfrac{-16}{2}=-8$$

42. We usually consider length to be longer than width so we substitute 12 for l and 7 for w and carry out the calculation. (The result is the same if we substitute in the opposite order.)

$$A=lw=12\cdot 7=84\text{ ft}^2$$

43. Substitute to find the value of each expression.

	x^2-5	$(x+5)^2$	$(x-5)^2$	x^2+5
$x=-1$	-4	16	36	6
$x=10$	95	225	25	105
$x=0$	-5	25	25	5

There is no pair of expressions that is the same for the given values of x, so there are no equivalent expressions.

44.

	$2x-14$	$2x-7$	$2(x-7)$	$2x+14$
$x=-1$	-16	-9	-16	12
$x=10$	6	13	6	34
$x=0$	-14	-7	-14	14

The values of $2x-14$ and $2(x-7)$ are the same for the given values of x and, indeed, for any allowable replacement for x. Thus, they are equivalent.

There is no other pair of expressions that is the same for the given values of x. Thus, there are no other equivalent expressions.

45. Since $9x=3\cdot 3x$, we multiply by 1 using $\dfrac{3x}{3x}$ as a name for 1.

$$\dfrac{7}{3}=\dfrac{7}{3}\cdot\dfrac{3x}{3x}=\dfrac{21x}{9x}$$

46. $\dfrac{-84x}{7x} = \dfrac{7x(-12)}{7x \cdot 1} = \dfrac{7x}{7x} \cdot \dfrac{-12}{1} = \dfrac{-12}{1} = -12$

47. $11 + a = a + 11$ Commutative law of addition

48. $8y = y \cdot 8$ Commutative law of multiplication

49. $(9 + a) + b = 9 + (a + b)$ Associate law of addition

50. $8(xy) = (8x)y$ Associate law of multiplication

51. $-3(2x - y) = -3 \cdot 2x - 3(-y) = -6x + 3y$

52. $4ab(2c + 1) = 4ab \cdot 2c + 4ab \cdot 1 = 8abc + 4ab$

53. $5x + 10y - 5z = 5 \cdot x + 5 \cdot 2y - 5 \cdot z = 5(x + 2y - z)$

54. $ptr + pts = pt \cdot r + pt \cdot s = pt(r + s)$

55. $2x + 6y - 5x - y = (2 - 5)x + (6 - 1)y = -3x + 5y$

56. $7c - 6 + 9c + 2 - 4c = (7 + 9 - 4)c + (-6 + 2) = 12c - 4$

57. $-(-9c + 4d - 3)$
$= 9c - 4d + 3$ Changing the sign of every
term inside parentheses

58. $4(x - 3) - 3(x - 5) = 4x - 12 - 3x + 15 = x + 3$

59. $12x - 3(2x - 5) = 12x - 6x + 15 = 6x + 15$

60. $7x - [4 - 5(3x - 2)]$
$= 7x - [4 - 15x + 10]$
$= 7x - [14 - 15x]$
$= 7x - 14 + 15x$
$= 22x - 14$

61. $4m - 3[3(4m - 2) - (5m + 2) + 12]$
$= 4m - 3[12m - 6 - 5m - 2 + 12]$
$= 4m - 3[7m + 4]$
$= 4m - 21m - 12$
$= -17m - 12$

62. $(2x^4y^{-3})(-5x^3y^{-2}) = 2(-5) \cdot x^4 \cdot x^3 \cdot y^{-3} \cdot y^{-2}$
$= -10x^7y^{-5}$
$= -\dfrac{10x^7}{y^5}$

63. $\dfrac{-15x^2y^{-5}}{10x^6y^{-8}} = \dfrac{-15}{10}x^{2-6}y^{-5-(-8)}$
$= -\dfrac{3}{2}x^{-4}y^{-5+8}$
$= -\dfrac{3}{2}x^{-4}y^3$
$= -\dfrac{3y^3}{2x^4}$

64. $(-3a^{-4}bc^3)^{-2} = (-3)^{-2}(a^{-4})^{-2}b^{-2}(c^3)^{-2}$
$= \dfrac{1}{(-3)^2}a^{-4(-2)}b^{-2}c^{3(-2)}$
$= \dfrac{1}{9}a^8b^{-2}c^{-6}$
$= \dfrac{a^8}{9b^2c^6}$

65. $\left[\dfrac{-2x^4y^{-4}}{3x^{-2}y^6}\right]^{-4} = \left[\dfrac{3x^{-2}y^6}{-2x^4y^{-4}}\right]^4$
$= \left[\dfrac{3x^{-2-4}y^{6-(-4)}}{-2}\right]^4$
$= \left[\dfrac{3x^{-6}y^{10}}{-2}\right]^4$
$= \dfrac{3^4(x^{-6})^4(y^{10})^4}{(-2)^4}$
$= \dfrac{81x^{-6 \cdot 4}y^{10 \cdot 4}}{16}$
$= \dfrac{81x^{-24}y^{40}}{16}$
$= \dfrac{81y^{40}}{16x^{24}}$

66. $\dfrac{2.2 \times 10^7}{3.2 \times 10^{-3}} = \dfrac{2.2}{3.2} \times \dfrac{10^7}{10^{-3}}$
$= 0.6875 \times 10^{10}$
$= (6.875 \times 10^{-1}) \times 10^{10}$
$= 6.875 \times (10^{-1} \times 10^{10})$
$= 6.875 \times 10^9$

67. $(3.2 \times 10^4)(4.1 \times 10^{-6}) = (3.2 \times 4.1)(10^4 \times 10^{-6})$
$= 13.12 \times 10^{-2}$
$= (1.312 \times 10) \times 10^{-2}$
$= 1.312 \times (10 \times 10^{-2})$
$= 1.312 \times 10^{-1}$

68. First we convert 0.00015 and 79 to scientific notation.
$0.00015 = 1.5 \times 10^{-4}$
$79 = 7.9 \times 10$
Now we find the volume.
$V = lwh$
$= (7.9 \times 10)(1.2)(1.5 \times 10^{-4})$
$= (7.9 \times 1.2 \times 1.5)(10 \times 10^{-4})$
$= 14.22 \times 10^{-3}$
$= (1.422 \times 10) \times 10^{-3}$
$= 1.422 \times (10 \times 10^{-3})$
$= 1.422 \times 10^{-2} \text{ m}^3$

69. First we convert 5.0 mils to dollars.

$5.0 \times \$0.001 = \$0.005$

Next we convert 0.005 and 13.4 million to scientific notation.

$0.005 = 5 \times 10^{-3}$

$13.4 \text{ million} = 13,400,000 = 1.34 \times 10^7$

Finally we multiply to find the revenue.

$(1.34 \times 10^7)(5 \times 10^{-3})$

$= (1.34 \times 5)(10^7 \times 10^{-3})$

$= 6.7 \times 10^4$

The revenue will be $\$6.7 \times 10^4$.

70. *Discussion and Writing Exercise.* The commutative laws deal with order and tell us that we can change the order when adding or when multiplying without changing the result. The associative laws deal with grouping and tell us that numbers can be grouped in any manner when adding or multiplying without affecting the result. The distributive laws deal with multiplication together with addition or subtraction. They tell us that adding or subtracting two numbers b and c and then multiplying the sum or difference by a number a is equivalent to first multiplying b and c by a and then adding or subtracting those products. Each of the commutative and associate laws involves only one operation. Each of the distributive laws involves two operations.

71. *Discussion and Writing Exercise.* The expressions $2x + 3x$ and $5x$ are equivalent. (We can show this using the distributive law.) The expressions $(x+1)^2$ and $x^2 + 1$ are not equivalent, because when $x = 3$, for example, we have $(x+1)^2 = (3+1)^2 = 16$ but $x^2 + 1 = 3^2 + 1 = 10$.

72. $(x^y \cdot x^{3y})^3 = (x^{y+3y})^3 = (x^{4y})^3 = x^{12y}$

73. $a^{-1}b = (2^x)^{-1}(2^{x+5}) = (2^{-x})(2^{x+5}) = 2^{-x+x+5} = 2^5 = 32$

74. $3x - 3y = 3(x - y)$, so (a) and (i) are equivalent.

$(x^{-2})^5 = x^{-10}$, so (d) and (f) are equivalent.

$x(y + z) = xy + xz$, so (h) and (j) are equivalent.

There are no other equivalent expressions.

Chapter R Test

1. The irrational numbers are the numbers that cannot be named as quotients of integers with nonzero divisors. That is, the irrational numbers are the numbers that are not rational. They are $\sqrt{7}$ and π.

2. We specify the conditions by which we know whether a number is in the set.

$\{x | x \text{ is a real number greater than } 20\}$

3. Since -4.5 is to the right of -8.7, we have $-4.5 > -8.7$.

4. $a \leq 5$

The inequality $5 \geq a$ has the same meaning.

5. $-6 \geq -6$ is true since $-6 = -6$ is true.

6. $-8 \leq -6$ is true since $-8 < -6$ is true.

7. $x > -2$

We shade all numbers to the right of -2 and use a parenthesis at -2 to indicate that it is not a solution.

8. The distance of 0 from 0 is 0, so $|0| = 0$.

9. The distance of $-\frac{7}{8}$ from 0 is $\frac{7}{8}$, so $\left| -\frac{7}{8} \right| = \frac{7}{8}$.

10. $7 + (-9)$

We find the difference of the absolute values, $9 - 7 = 2$. Since the negative number has the larger absolute value, the answer is negative.

$7 + (-9) = -2$

11. $-5.3 + (-7.8)$

The sum of two negative numbers is negative. We add the absolute values, $5.3 + 7.8 = 13.1$, and make the answer negative.

$-5.3 + (-7.8) = -13.1$

12. $-\frac{5}{2} + \left(-\frac{7}{2} \right)$

The sum of two negative numbers is negative. We add the absolute values, $\frac{5}{2} + \frac{7}{2} = \frac{12}{2} = 6$ and make the answer negative.

$-\frac{5}{2} + \left(-\frac{7}{2} \right) = -6$

13. $-6 - (-5) = -6 + 5 = -1$

14. $-18.2 + (-11.5) = -29.7$

15. $\frac{19}{4} - \left(-\frac{3}{2} \right) = \frac{19}{4} + \frac{3}{2} = \frac{19}{4} + \frac{6}{4} = \frac{25}{4}$

16. $(-4.1)(8.2)$

The product of a negative number and a positive number is negative. We multiply the absolute values, $(4.1)(8.2) = 33.62$, and make the answer negative.

$(-4.1)(8.2) = -33.62$

17. $-\frac{4}{5} \left(-\frac{15}{16} \right)$

The product of two negative numbers is positive. We multiply the absolute values and make the answer positive.

$\frac{4}{5} \left(\frac{15}{16} \right) = \frac{4 \cdot 15}{5 \cdot 16} = \frac{4 \cdot 3 \cdot 5}{5 \cdot 4 \cdot 4} = \frac{4 \cdot 5}{4 \cdot 5} \cdot \frac{3}{4} = \frac{3}{4}$

Thus, $-\frac{4}{5} \left(-\frac{15}{16} \right) = \frac{3}{4}$.

18. $-6(-4)(-11)2 = 24(-11)2 = (-264)2 = -528$

19. When a negative number is divided by a negative number, the answer is positive.

$$-75 \div (-5) = \frac{-75}{-5} = 15$$

20. When a negative number is divided by a positive number, the answer is negative.

$$\frac{-10}{2} = -5$$

21. $-\dfrac{5}{2} \div \left(-\dfrac{15}{16}\right) = -\dfrac{5}{2} \cdot \left(-\dfrac{16}{15}\right) = \dfrac{80}{30} = \dfrac{8}{3}$

22. $-459.2 \div 5.6 = \dfrac{-459.2}{5.6} = -82$

23. $\dfrac{-3}{0}$ Not defined: Division by 0.

24. When $a = -13$, then $-a = -(-13) = 13$.

25. When $a = 0$, then $-a = -0 = 0$.

26. $\underbrace{q \cdot q \cdot q \cdot q}_{\text{4 factors}} = q^4$

27. $\dfrac{1}{a^9} = a^{-9}$

28.
$$1 - (2-5)^2 + 5 \div 10 \cdot 4^2$$
$$= 1 - (-3)^2 + 5 \div 10 \cdot 4^2$$
$$= 1 - 9 + 5 \div 10 \cdot 16$$
$$= 1 - 9 + 0.5 \cdot 16$$
$$= 1 - 9 + 8$$
$$= -8 + 8$$
$$= 0$$

29.
$$\frac{7(5 - 2 \cdot 3) - 3^2}{4^2 - 3^2} = \frac{7(5-6) - 3^2}{16 - 9}$$
$$= \frac{7(-1) - 3^2}{7}$$
$$= \frac{7(-1) - 9}{7}$$
$$= \frac{-7 - 9}{7}$$
$$= \frac{-16}{7}, \text{ or } -\frac{16}{7}$$

30. $t + 9$, or $9 + t$

31. Let x and y represent the numbers. Then we have $\dfrac{x}{y} - 12$.

32. Substitute 2 for x and -4 for y and carry out the calculations.

$$3x - 3y = 3 \cdot 2 - 3(-4) = 6 + 12 = 18$$

33. Substitute 3 for b and 2.5 for h and carry out the calculations.
$$A = \frac{1}{2}bh$$
$$= \frac{1}{2}(3)(2.5)$$
$$= (1.5)(2.5)$$
$$= 3.75 \text{ cm}^2$$

34.

	$x(x-3)$	$x^2 - 3x$
$x = -1$	4	4
$x = 10$	70	70
$x = 0$	0	0

The values of $x(x-3)$ and $x^2 - 3x$ are the same for the given values of x and, indeed, for any allowable replacement for x. Thus, they are equivalent.

35.

	$3x + 5x^2$	$8x^2$
$x = -1$	2	8
$x = 10$	530	800
$x = 0$	0	0

Although the expressions have the same value for $x = 0$, they are not the same for all of the given values of x, so they are not equivalent.

36. Since $36 = 4 \cdot 9x$, we multiply by 1 using $\dfrac{9x}{9x}$ as a name for 1.
$$\frac{3}{4} = \frac{3}{4} \cdot \frac{9x}{9x} = \frac{27x}{36x}$$

37. $\dfrac{-54x}{-36x} = \dfrac{-18x \cdot 3}{-18x \cdot 2} = \dfrac{-18x}{-18x} \cdot \dfrac{3}{2} = \dfrac{3}{2}$

38. $pq = qp$ Commutative law of multiplication

39. $t + 4 = 4 + t$ Commutative law of addition

40. $3 + (t + w) = (3 + t) + w$ Associative law of addition

41. $(4a)b = 4(ab)$ Associative law of multiplication

42. $-2(3a - 4b) = -2 \cdot 3a - 2 \cdot (-4b) = -6a + 8b$

43. $3\pi r(s + 1) = 3\pi r \cdot s + 3\pi r \cdot 1 = 3\pi rs + 3\pi r$

44. $ab - ac + 2ad = a \cdot b - a \cdot c + a \cdot 2d = a(b - c + 2d)$

45. $2ah + h = h \cdot 2a + h \cdot 1 = h(2a + 1)$

46. $6y - 8x + 4y + 3x = (6 + 4)y + (-8 + 3)x = 10y - 5x$

47. $4a - 7 + 17a + 21 = (4 + 17)a + (-7 + 21) = 21a + 14$

48. $-(-9x + 7y - 22) = 9x - 7y + 22$ Changing the sign of every term inside parentheses

49. $-3(x+2) - 4(x-5) = -3x - 6 - 4x + 20 = -7x + 14$

50. $\quad 4x - [6 - 3(2x - 5)]$
$= 4x - [6 - 6x + 15]$
$= 4x - [21 - 6x]$
$= 4x - 21 + 6x$
$= 10x - 21$

51. $\quad 3a - 2[5(2a - 5) - (8a + 4) + 10]$
$= 3a - 2[10a - 25 - 8a - 4 + 10]$
$= 3a - 2[2a - 19]$
$= 3a - 4a + 38$
$= -a + 38$

52. $\dfrac{-12x^3y^{-4}}{8x^7y^{-6}} = \dfrac{-12}{8}x^{3-7}y^{-4-(-6)}$
$= -\dfrac{3}{2}x^{-4}y^{-4+6}$
$= -\dfrac{3}{2}x^{-4}y^2$
$= -\dfrac{3y^2}{2x^4}$

53. $(3a^4b^{-2})(-2a^5b^{-3}) = 3(-2)\cdot a^4 \cdot a^5 \cdot b^{-2} \cdot b^{-3}$
$= -6a^9b^{-5}$
$= -\dfrac{6a^9}{b^5}$

54. $(5a^{4n})(-10a^{5n}) = 5(-10)\cdot a^{4n}\cdot a^{5n}$
$= -50a^{4n+5n}$
$= -50a^{9n}$

55. $\dfrac{-60x^{3t}}{12x^{7t}} = \dfrac{-60}{12}x^{3t-7t} = -5x^{-4t},$ or $-\dfrac{5}{x^{4t}}$

56. $(-3a^{-3}b^2c)^{-4} = (-3)^{-4}(a^{-3})^{-4}(b^2)^{-4}c^{-4}$
$= \dfrac{1}{(-3)^4}a^{-3(-4)}b^{2(-4)}c^{-4}$
$= \dfrac{1}{81}a^{12}b^{-8}c^{-4}$
$= \dfrac{a^{12}}{81b^8c^4}$

57. $\left[\dfrac{-5a^{-2}b^8}{10a^{10}b^{-4}}\right]^{-4} = \left[\dfrac{10a^{10}b^{-4}}{-5a^{-2}b^8}\right]^4$
$= \left[\dfrac{10}{-5}a^{10-(-2)}b^{-4-8}\right]^4$
$= [-2a^{12}b^{-12}]^4$
$= (-2)^4(a^{12})^4(b^{-12})^4$
$= 16a^{48}b^{-48}$
$= \dfrac{16a^{48}}{b^{48}}$

58. $0.00004.37$
$\quad\underset{\text{5 places}}{\lfloor\qquad\uparrow}$

Small number, so the exponent is negative.
$0.0000437 = 4.37 \times 10^{-5}$

59. $(8.7 \times 10^{-9})(4.3 \times 10^{15}) = (8.7 \times 4.3)(10^{-9} \times 10^{15})$
$= 37.41 \times 10^6$
$= (3.741 \times 10) \times 10^6$
$= 3.741 \times 10 \times 10^6$
$= 3.741 \times 10^7$

60. $\dfrac{1.2 \times 10^{-12}}{6.4 \times 10^{-7}} = \dfrac{1.2}{6.4} \times \dfrac{10^{-12}}{10^{-7}}$
$= 0.1875 \times 10^{-5}$
$= (1.875 \times 10^{-1}) \times 10^{-5}$
$= 1.875 \times (10^{-1} \times 10^{-5})$
$= 1.875 \times 10^{-6}$

61. First we convert 0.002 to scientific notation.
$0.002 = 2 \times 10^{-3}$
Now we multiply to find the mass of Pluto.
$(2 \times 10^{-3})(5.98 \times 10^{24}) = (2 \times 5.98)(10^{-3} \times 10^{24})$
$= 11.96 \times 10^{21}$
$= (1.196 \times 10) \times 10^{21}$
$= 1.196 \times (10 \times 10^{21})$
$= 1.196 \times 10^{22} \text{ kg}$

62. $600 = 6 \times 10^2$
1 million $= 1,000,000 = 1 \times 10^6$
We multiply to find the answer.
$6 \times 10^2 \text{ megabytes} \times \dfrac{1 \times 10^6 \text{ bytes}}{1 \text{ megabyte}} \times \dfrac{8 \text{ pits}}{1 \text{ byte}}$
$= 6 \times 10^2 \times 1 \times 10^6 \times 8 \times \dfrac{\text{megabytes}}{\text{megabytes}} \times \dfrac{\text{bytes}}{\text{bytes}} \times \text{pits}$
$= 48 \times 10^8 \text{ pits}$
$= (4.8 \times 10) \times 10^8 \text{ pits}$
$= 4.8 \times 10^9 \text{ pits (pieces of information)}$

63. $(x^{-3})^{-4} = x^{12}$, so (b) and (e) are equivalent.
$5 + 5x = 5x + 5 = 5(x + 1)$, so (d), (f), and (h) are equivalent.
$5(xy) = (5x)y$, so (i) and (j) are equivalent.
There are no other equivalent expressions.

Chapter 1

Solving Linear Equations and Inequalities

Exercise Set 1.1

1. $\underline{x + 23 = 40}$ Writing the equation

   $17 + 23 \ ? \ 40$ Substituting 17 for x

      $40 \ \bigm|$ TRUE

Since the left-hand and the right-hand sides are the same, 17 is a solution of the equation.

3. $\underline{2x - 3 = -18}$ Writing the equation

  $2(-8) - 3 \ ? \ -18$ Substituting -8 for x

   $-16 - 3 \ \bigm|$

     $-19 \ \bigm|$ FALSE

Since the left-hand and the right-hand sides are not the same, -8 is not a solution of the equation.

5. $\dfrac{-x}{9} = -2$ Writing the equation

  $\dfrac{-45}{9} \ ? \ -2$ Substituting

   $-5 \ \bigm|$ FALSE

Since the left-hand and the right-hand sides are not the same, 45 is not a solution of the equation.

7. $\underline{2 - 3x = 21}$

  $2 - 3 \cdot 10 \ ? \ 21$ Substituting

   $2 - 30 \ \bigm|$

    $-28 \ \bigm|$ FALSE

Since the left-hand and the right-hand sides are not the same, 10 is not a solution of the equation.

9. $\underline{5x + 7 = 102}$

  $5 \cdot 19 + 7 \ ? \ 102$

   $95 + 7 \ \bigm|$

    $102 \ \bigm|$ TRUE

Since the left-hand and the right-hand sides are the same, 19 is a solution of the equation.

11. $\underline{7(y - 1) = 84}$

  $7(-11 - 1) \ ? \ 84$

   $7(-12) \ \bigm|$

    $-84 \ \bigm|$ FALSE

Since the left-hand and the right-hand sides are not the same, -11 is not a solution of the equation.

13. $y + 6 = 13$

  $y + 6 - 6 = 13 - 6$ Subtracting 6 on both sides

    $y + 0 = 7$ Simplifying

      $y = 7$ Using the identity property of 0

Check: $\underline{y + 6 = 13}$

       $7 + 6 \ ? \ 13$

         $13 \ \bigm|$ TRUE

The solution is 7.

15. $-20 = x - 12$

  $-20 + 12 = x - 12 + 12$ Adding 12 on both sides

     $-8 = x + 0$ Simplifying

     $-8 = x$ Using the identity property of 0

Check: $\underline{-20 = x - 12}$

     $-20 \ ? \ -8 - 12$

        $\bigm| \ -20$ TRUE

The solution is -8.

17. $-8 + x = 19$

  $8 - 8 + x = 8 + 19$ Adding 8

    $0 + x = 27$

      $x = 27$ Using the identity property of 0

Check: $\underline{-8 + x = 19}$

     $-8 + 27 \ ? \ 19$

        $19 \ \bigm|$ TRUE

The solution is 27.

19. $-12 + z = -51$

  $12 + (-12) + z = 12 + (-51)$ Adding 12

    $0 + z = -39$

     $z = -39$

The number -39 checks, so it is the solution.

21. $p - 2.96 = 83.9$

  $p - 2.96 + 2.96 = 83.9 + 2.96$

    $p + 0 = 86.86$

     $p = 86.86$

The number 86.86 checks, so it is the solution.

23.
$$-\frac{3}{8} + x = -\frac{5}{24}$$

$$\frac{3}{8} + \left(-\frac{3}{8}\right) + x = \frac{3}{8} + \left(-\frac{5}{24}\right)$$

$$0 + x = \frac{3}{8} \cdot \frac{3}{3} + \left(-\frac{5}{24}\right)$$

$$x = \frac{9}{24} + \left(-\frac{5}{24}\right)$$

$$x = \frac{4}{24}$$

$$x = \frac{1}{6}$$

The number $\frac{1}{6}$ checks, so it is the solution.

25.
$$3x = 18$$

$$\frac{3x}{3} = \frac{18}{3} \quad \text{Dividing by 3 on both sides}$$

$$1 \cdot x = \frac{18}{3} \quad \text{Simplifying}$$

$$x = 6$$

Check:
$$\begin{array}{c} 3x = 18 \\ \hline 3 \cdot 6 \ ? \ 18 \\ 18 \ \bigg| \qquad \text{TRUE} \end{array}$$

The solution is 6.

27.
$$-11y = 44$$

$$\frac{-11y}{-11} = \frac{44}{-11}$$

$$1 \cdot y = \frac{44}{-11}$$

$$y = -4$$

Check:
$$\begin{array}{c} -11y = 44 \\ \hline -11(-4) \ ? \ 44 \\ 44 \ \bigg| \qquad \text{TRUE} \end{array}$$

The solution is -4.

29.
$$-\frac{x}{7} = 21$$

$$-\frac{1}{7}x = 21$$

$$-7\left(-\frac{1}{7}\right)x = -7 \cdot 21 \quad \text{Multiplying by } -7$$
$$\text{on both sides}$$

$$1 \cdot x = -147$$

$$x = -147$$

Check:
$$\begin{array}{c} -\dfrac{x}{7} = 21 \\ \hline -\dfrac{-147}{7} \ ? \ 21 \\ -(-21) \ \bigg| \\ 21 \ \bigg| \qquad \text{TRUE} \end{array}$$

The solution is -147.

31.
$$-96 = -3z$$

$$\frac{-96}{-3} = \frac{-3z}{-3} \quad \text{Dividing by } -3$$

$$\frac{-96}{-3} = 1 \cdot z$$

$$32 = z$$

Check:
$$\begin{array}{c} -96 = -3z \\ \hline -96 \ ? \ -3 \cdot 32 \\ \bigg| \ -96 \qquad \text{TRUE} \end{array}$$

The solution is 32.

33.
$$4.8y = -28.8$$

$$\frac{4.8y}{4.8} = \frac{-28.8}{4.8}$$

$$1 \cdot y = -\frac{28.8}{4.8}$$

$$y = -6$$

The number -6 checks, so it is the solution.

35.
$$\frac{3}{2}t = -\frac{1}{4}$$

$$\frac{2}{3} \cdot \frac{3}{2}t = \frac{2}{3} \cdot \left(-\frac{1}{4}\right)$$

$$1 \cdot t = -\frac{2}{12}$$

$$t = -\frac{1}{6}$$

The number $-\frac{1}{6}$ checks, so it is the solution.

37.
$$6x - 15 = 45$$

$$6x - 15 + 15 = 45 + 15 \quad \text{Adding 15}$$

$$6x = 60$$

$$\frac{6x}{6} = \frac{60}{6} \quad \text{Dividing by 6}$$

$$x = 10$$

Check:
$$\begin{array}{c} 6x - 15 = 45 \\ \hline 6 \cdot 10 - 15 \ ? \ 45 \\ 60 - 15 \ \bigg| \\ 45 \ \bigg| \qquad \text{TRUE} \end{array}$$

The solution is 10.

39.
$$5x - 10 = 45$$

$$5x - 10 + 10 = 45 + 10$$

$$5x = 55$$

$$\frac{5x}{5} = \frac{55}{5}$$

$$x = 11$$

Check:
$$\begin{array}{c} 5x - 10 = 45 \\ \hline 5 \cdot 11 - 10 \ ? \ 45 \\ 55 - 10 \ \bigg| \\ 45 \ \bigg| \qquad \text{TRUE} \end{array}$$

The solution is 11.

41.
$$9t + 4 = -104$$
$$9t + 4 - 4 = -104 - 4$$
$$9t = -108$$
$$\frac{9t}{9} = \frac{-108}{9}$$
$$t = -12$$

Check:
$$\begin{array}{c|c} 9t + 4 = -104 \\ \hline 9(-12) + 4 \ ? \ -104 \\ -108 + 4 \\ -104 & \text{TRUE} \end{array}$$

The solution is -12.

43.
$$-\frac{7}{3}x + \frac{2}{3} = -18, \text{ LCM is 3}$$
$$3\left(-\frac{7}{3}x + \frac{2}{3}\right) = 3(-18) \quad \begin{array}{l}\text{Multiplying by 3 to}\\ \text{clear fractions}\end{array}$$
$$-7x + 2 = -54$$
$$-7x = -56 \quad \text{Subtracting 2}$$
$$x = \frac{-56}{-7} \quad \text{Dividing by } -7$$
$$x = 8$$

The number 8 checks. It is the solution.

45.
$$\frac{6}{5}x + \frac{4}{10}x = \frac{32}{10}, \text{ LCM is 10}$$
$$10\left(\frac{6}{5}x + \frac{4}{10}x\right) = 10 \cdot \frac{32}{10} \quad \begin{array}{l}\text{Multiplying by 10 to}\\ \text{clear fractions}\end{array}$$
$$12x + 4x = 32$$
$$16x = 32 \quad \text{Collecting like terms}$$
$$x = \frac{32}{16} \quad \text{Dividing by 16}$$
$$x = 2$$

The number 2 checks. It is the solution.

47.
$$0.9y - 0.7y = 4.2$$
$$10(0.9y - 0.7y) = 10(4.2) \quad \begin{array}{l}\text{Multiplying by 10 to clear}\\ \text{fractions}\end{array}$$
$$9y - 7y = 42$$
$$2y = 42 \quad \text{Collecting like terms}$$
$$y = \frac{42}{2}$$
$$y = 21$$

The number 21 checks, so it is the solution.

49.
$$8x + 48 = 3x - 12$$
$$5x + 48 = -12 \quad \text{Subtracting } 3x$$
$$5x = -60 \quad \text{Subtracting 48}$$
$$x = \frac{-60}{5} \quad \text{Dividing by 5}$$
$$x = -12$$

The number -12 checks, so it is the solution.

51.
$$7y - 1 = 27 + 7y$$
$$-1 = 27 \quad \text{Subtracting } 7y$$

The equation $-1 = 27$ is false. No matter what number we try for x we get a false sentence. Thus, the equation has no solution.

53.
$$3x - 4 = 5 + 12x$$
$$-4 = 5 + 9x \quad \text{Subtracting } 3x$$
$$-9 = 9x \quad \text{Subtracting 5}$$
$$-1 = x \quad \text{Dividing by 9}$$

The number -1 checks, so it is the solution.

55.
$$5 - 4a = a - 13$$
$$5 = 5a - 13 \quad \text{Adding } 4a$$
$$18 = 5a \quad \text{Adding 13}$$
$$\frac{18}{5} = a \quad \text{Dividing by 5}$$

The number $\frac{18}{5}$ checks. It is the solution.

57.
$$3m - 7 = -7 - 4m - m$$
$$3m - 7 = -7 - 5m \quad \text{Collecting like terms}$$
$$3m = -5m \quad \text{Adding 7}$$
$$8m = 0 \quad \text{Adding } 5m$$
$$m = \frac{0}{8} \quad \text{Dividing by 8}$$
$$m = 0$$

The number 0 checks, so it is the solution.

59.
$$5x + 3 = 11 - 4x + x$$
$$5x + 3 = 11 - 3x \quad \text{Collecting like terms}$$
$$8x + 3 = 11 \quad \text{Adding } 3x$$
$$8x = 8 \quad \text{Subtracting 3}$$
$$x = \frac{8}{8} \quad \text{Dividing by 8}$$
$$x = 1$$

The number 1 checks, so it is the solution.

61.
$$-7 + 9x = 9x - 7$$
$$-7 = -7 \quad \text{Subtracting } 9x$$

The equation $-7 = -7$ is true. Replacing x by any real number gives a true sentence. Thus, all real numbers are solutions.

63.
$$6y - 8 = 9 + 6y$$
$$-8 = 9 \quad \text{Subtracting } 6y$$

The equation $-8 = 9$ is false. No matter what number we try for x we get a false sentence. Thus, the equation has no solution.

65.
$$2(x + 7) = 4x$$
$$2x + 14 = 4x \quad \text{Multiplying to remove parentheses}$$
$$14 = 2x \quad \text{Subtracting } 2x$$
$$7 = x \quad \text{Dividing by 2}$$

Check: $\begin{array}{r} 2(x+7) = 4x \\ \hline 2(7+7) \text{ ? } 4\cdot 7 \\ 2\cdot 14 \quad | \quad 28 \\ 28 \quad | \end{array}$ TRUE

The solution is 7.

67. $80 = 10(3t + 2)$

$80 = 30t + 20$

$60 = 30t$

$2 = t$

Check: $\begin{array}{r} 80 = 10(3t+2) \\ \hline 80 \text{ ? } 10(3\cdot 2 + 2) \\ 10(6+2) \\ 10\cdot 8 \\ 80 \end{array}$ TRUE

The solution is 2.

69. $180(n - 2) = 900$

$180n - 360 = 900$

$180n = 1260$

$n = 7$

Check: $\begin{array}{r} 180(n-2) = 900 \\ \hline 180(7-2) \text{ ? } 900 \\ 180\cdot 5 \\ 900 \end{array}$ TRUE

The solution is 7.

71. $5y - (2y - 10) = 25$

$5y - 2y + 10 = 25$

$3y + 10 = 25$

$3y = 15$

$y = 5$

Check: $\begin{array}{r} 5y - (2y - 10) = 25 \\ \hline 5\cdot 5 - (2\cdot 5 - 10) \text{ ? } 25 \\ 25 - (10-10) \\ 25 - 0 \\ 25 \end{array}$ TRUE

The solution is 5.

73. $7(3x + 6) = 11 - (x + 2)$

$21x + 42 = 11 - x - 2$

$21x + 42 = 9 - x$

$22x + 42 = 9$

$22x = -33$

$x = \dfrac{-33}{22}$

$x = -\dfrac{3}{2}$

The number $-\dfrac{3}{2}$ checks, so it is the solution.

75. $2[9 - 3(-2x - 4)] = 12x + 42$

$2[9 + 6x + 12] = 12x + 42$

$2[6x + 21] = 12x + 42$

$12x + 42 = 12x + 42$

$42 = 42$

We get a true equation. Replacing x with any real number gives a true sentence. Thus, all real numbers are solutions.

77. $\dfrac{1}{8}(16y + 8) - 17 = -\dfrac{1}{4}(8y - 16)$

$2y + 1 - 17 = -2y + 4$

$2y - 16 = -2y + 4$

$4y - 16 = 4$

$4y = 20$

$y = 5$

The number 5 checks, so it is the solution.

79. $3[5 - 3(4 - t)] - 2 = 5[3(5t - 4) + 8] - 26$

$3[5 - 12 + 3t] - 2 = 5[15t - 12 + 8] - 26$

$3[-7 + 3t] - 2 = 5[15t - 4] - 26$

$-21 + 9t - 2 = 75t - 20 - 26$

$9t - 23 = 75t - 46$

$-23 = 66t - 46$

$23 = 66t$

$\dfrac{23}{66} = t$

The number $\dfrac{23}{66}$ checks, so it is the solution.

81. $\dfrac{2}{3}\left(\dfrac{7}{8} + 4x\right) - \dfrac{5}{8} = \dfrac{3}{8}$

$$\dfrac{2}{3}\left(\dfrac{7}{8} + 4x\right) = 1$$

$$\dfrac{7}{12} + \dfrac{8}{3}x = 1$$

$$12\left(\dfrac{7}{12} + \dfrac{8}{3}x\right) = 12 \cdot 1$$

$$7 + 32x = 12$$

$$32x = 5$$

$$x = \dfrac{5}{32}$$

The number $\dfrac{5}{32}$ checks, so it is the solution.

83. $5(4x - 3) - 2(6 - 8x) + 10(-2x + 7) = -4(9 - 12x)$

$$20x - 15 - 12 + 16x - 20x + 70 = -36 + 48x$$

$$16x + 43 = -36 + 48x$$

$$43 = -36 + 32x$$

$$79 = 32x$$

$$\dfrac{79}{32} = x$$

The number $\dfrac{79}{32}$ checks, so it is the solution.

85. Discussion and Writing Exercise

87. $a^{-9} \cdot a^{23} = a^{-9+23} = a^{14}$

89. $(6x^5 y^{-4})(-3x^{-3}y^{-7})$

$$= 6 \cdot (-3) \cdot x^5 \cdot x^{-3} \cdot y^{-4} \cdot y^{-7}$$

$$= -18x^{5+(-3)}y^{-4+(-7)}$$

$$= -18x^2 y^{-11}$$

$$= -\dfrac{18x^2}{y^{11}}$$

91. $2(6 - 10x) = 2 \cdot 6 - 2 \cdot 10x = 12 - 20x$

93. $-4(3x - 2y + z) = -4 \cdot 3x - 4(-2y) - 4 \cdot z$

$$= -12x + 8y - 4z$$

95. $2x - 6y = 2 \cdot x - 2 \cdot 3y$

$$= 2(x - 3y)$$

97. $4x - 10y + 2 = 2 \cdot 2x - 2 \cdot 5y + 2 \cdot 1$

$$= 2(2x - 5y + 1)$$

99. $\{1, 2, 3, 4, 5, 6, 7, 8, 9\}$;

$\{x \,|\, x$ is a positive integer less than 10$\}$

101. $4.23x - 17.898 = -1.65x - 42.454$

$$5.88x - 17.898 = -42.454$$

$$5.88x = -24.556$$

$$x \approx -4.176$$

The solution is approximately -4.176.

103. $\dfrac{3x}{2} + \dfrac{5x}{3} - \dfrac{13x}{6} - \dfrac{2}{3} = \dfrac{5}{6}$, LCM is 6

$$6\left(\dfrac{3x}{2} + \dfrac{5x}{3} - \dfrac{13x}{6} - \dfrac{2}{3}\right) = 6 \cdot \dfrac{5}{6}$$

Multiplying by 6 to clear fractions

$$3 \cdot 3x + 2 \cdot 5x - 1 \cdot 13x - 2 \cdot 2 = 1 \cdot 5$$

$$9x + 10x - 13x - 4 = 5$$

$$6x - 4 = 5 \quad \text{Collecting like terms}$$

$$6x = 9 \quad \text{Adding 4}$$

$$x = \dfrac{9}{6} \quad \text{Dividing by 6}$$

$$x = \dfrac{3}{2}$$

The number $\dfrac{3}{2}$ checks, so it is the solution.

105. $x - \{3x - [2x - (5x - (7x - 1))]\} = x + 7$

$$x - \{3x - [2x - (5x - 7x + 1)]\} = x + 7$$

$$x - \{3x - [2x - (-2x + 1)]\} = x + 7$$

$$x - \{3x - [2x + 2x - 1]\} = x + 7$$

$$x - \{3x - [4x - 1]\} = x + 7$$

$$x - \{3x - 4x + 1\} = x + 7$$

$$x - \{-x + 1\} = x + 7$$

$$x + x - 1 = x + 7$$

$$2x - 1 = x + 7$$

$$x - 1 = 7$$

$$x = 8$$

The number 8 checks, so it is the solution.

Exercise Set 1.2

1. $d = rt$

$$\dfrac{d}{t} = r \quad \text{Dividing by } t$$

3. $A = bh$

$$\dfrac{A}{b} = h \quad \text{Dividing by } b$$

5. $P = 2l + 2w$

$$P - 2l = 2w \quad \text{Subtracting } 2l$$

$$\dfrac{P - 2l}{2} = w \quad \text{Dividing by 2}$$

or $\dfrac{P}{2} - l = w$

7. $A = \dfrac{1}{2}bh$

$$2A = bh \quad \text{Multiplying by 2}$$

$$\dfrac{2A}{h} = b \quad \text{Dividing by } h$$

9. $A = \dfrac{a + b}{2}$

$$2A = a + b \quad \text{Multiplying by 2}$$

$$2A - b = a \quad \text{Subtracting } b$$

11. $F = ma$

$\dfrac{F}{a} = m$ Dividing by a

13. $I = Prt$

$\dfrac{I}{Pr} = t$ Dividing by Pr

15. $E = mc^2$

$\dfrac{E}{m} = c^2$ Dividing by m

17. $Q = \dfrac{p - q}{2}$

$2Q = p - q$ Multiplying by 2

$2Q + q = p$ Adding q

19. $Ax + By = c$

$By = c - Ax$ Subtracting Ax

$y = \dfrac{c - Ax}{B}$ Dividing by B

21. $I = 1.08\dfrac{T}{N}$

$IN = 1.08T$ Multiplying by N

$N = \dfrac{1.08T}{I}$ Dividing by I

23. $C = \dfrac{3}{4}(m + 5)$

$\dfrac{4}{3} \cdot C = \dfrac{4}{3} \cdot \dfrac{3}{4}(m + 5)$ Multiplying by $\dfrac{4}{3}$

$\dfrac{4}{3}C = m + 5$

$\dfrac{4}{3}C - 5 = m$, or   Subtracting 5

$\dfrac{4C - 15}{3} = m$

25. $n = \dfrac{1}{3}(a + b - c)$

$3n = a + b - c$ Multiplying by 3

$3n - a + c = b$ Subtracting a and adding c

27. $d = R - Rst$

$d = R(1 - st)$ Factoring out R

$\dfrac{d}{1 - st} = R$ Dividing by $1 - st$

29. $T = B + Bqt$

$T = B(1 + qt)$ Factoring

$\dfrac{T}{1 + qt} = B$ Dividing by $1 + qt$

31. a) We substitute 87 for w, 185 for h, and 60 for a and calculate K.

$K = 19.18w + 7h - 9.52a + 92.4$

$K = 19.18(87) + 7(185) - 9.52(60) + 92.4$

$K = 2484.86$

Marv needs about 2485 calories per day.

b) Solve for a:

$$K = 19.18w + 7h - 9.52a + 92.4$$

$$K + 9.52a = 19.18w + 7h + 92.4$$

$$9.52a = 19.18w + 7h + 92.4 - K$$

$$a = \dfrac{19.18w + 7h + 92.4 - K}{9.52}$$

Solve for h:

$$K = 19.18w + 7h - 9.52a + 92.4$$

$$K - 19.18w + 9.52a - 92.4 = 7h$$

$$\dfrac{K - 19.18w + 9.52a - 92.4}{7} = h$$

Solve for w:

$$K = 19.18w + 7h - 9.52a + 92.4$$

$$K - 7h + 9.52a - 92.4 = 19.18w$$

$$\dfrac{K - 7h + 9.52a - 92.4}{19.18} = w$$

33. a) We substitute 8.5 for d and 24.1 for a and calculate P.

$P = 9.337da - 299$

$P = 9.337(8.5)(24.1) - 299$

$P = 1912.68445 - 299$

$P = 1613.68445$

The projected birth weight is about 1614 g.

b) $P = 9.337da - 299$

$P + 299 = 9.337da$

$\dfrac{P + 299}{9.337d} = a$

35. a) We substitute 3 for a and 250 for d and calculate c.

$c = \dfrac{ad}{a + 12}$

$c = \dfrac{3 \cdot 250}{3 + 12}$

$c = \dfrac{750}{15}$

$c = 50$

The child's dosage is 50 mg.

b) $c = \dfrac{ad}{a + 12}$

$c(a + 12) = ad$

$\dfrac{c(a + 12)}{a} = d$

This result can also be expressed as follows:

$$d = \dfrac{c(a + 12)}{a} = \dfrac{ac + 12c}{a} = c + \dfrac{12c}{a}$$

37. Discussion and Writing Exercise

39. $\dfrac{80}{-16} = -\dfrac{80}{16} = -5$

41. $-\dfrac{1}{2} \div \dfrac{1}{4} = -\dfrac{1}{2} \cdot \dfrac{4}{1} = -\dfrac{4}{2} = -2$

43. $-\dfrac{2}{3} \div \left(-\dfrac{5}{6}\right) = -\dfrac{2}{3} \cdot \left(-\dfrac{6}{5}\right) = \dfrac{2 \cdot 6}{3 \cdot 5} = \dfrac{2 \cdot 2 \cdot 3}{3 \cdot 5} =$

$\dfrac{2 \cdot 2 \cdot \cancel{3}}{\cancel{3} \cdot 5} = \dfrac{4}{5}$

45. $\dfrac{-90}{15} = -\dfrac{90}{15} = -6$

47. $\qquad A = \pi r s + \pi r^2$

$\qquad A - \pi r^2 = \pi r s$

$\qquad \dfrac{A - \pi r^2}{\pi r} = s$

49. Solve for V_1:

$\qquad \dfrac{P_1 V_1}{T_1} = \dfrac{P_2 V_2}{T_2}$

$\qquad V_1 = \dfrac{T_1 P_2 V_2}{P_1 T_2} \quad$ Multiplying by $\dfrac{T_1}{P_1}$

Solve for P_2:

$\qquad \dfrac{P_1 V_1}{T_1} = \dfrac{P_2 V_2}{T_2}$

$\qquad \dfrac{P_1 V_1 T_2}{T_1 V_2} = P_2 \quad$ Multiplying by $\dfrac{T_2}{V_2}$

51. We substitute \$75 for P, \$3 for I, and 5%, or 0.05, for r in the formula $t = \dfrac{I}{Pr}$.

$\qquad t = \dfrac{I}{Pr}$

$\qquad t = \dfrac{\$3}{\$75(0.05)} \quad$ Substituting

$\qquad t = 0.8$

It will take 0.8 yr.

53. $H = W\left(\dfrac{v}{234}\right)^3$

a) $H = 2700\left(\dfrac{83}{234}\right)^3$

$\quad H \approx 120.5$ horsepower

b) $H = 3100\left(\dfrac{73}{234}\right)^3$

$\quad H \approx 94.1$ horsepower

Exercise Set 1.3

1. Familiarize. Let d = Benoit's distance from the starting point, in miles. Then $3d$ = her distance from the end of the course. The total length of the course is 26.2 mi.

Translate.

Distance from start	plus	Distance from end	is	26.2 mi
d	$+$	$3d$	$=$	26.2

Solve. We solve the equation.

$\qquad d + 3d = 26.2$

$\qquad\quad 4d = 26.2$

$\qquad\quad\ d = 6.55$

If $d = 6.55$, then $3d = 19.65$.

Check. The distance from the end of the course, 19.65 mi, is three times 6.55 mi, the distance from the starting point. Also, $6.55 + 19.65 = 26.2$ mi, the total length of the race. The answer checks.

State. Benoit had run 6.55 mi.

3. Familiarize. Let l = the insurance loss from Hurricane Charley, in billions of dollars.

Translate.

Loss from Hurricane Charley	plus	\$13.5 billion more	is	Loss from Hurricane Andrew
l	$+$	13.5	$=$	20.3

Solve. We solve the equation.

$\qquad l + 13.5 = 20.3$

$\qquad\quad\ \ l = 6.8 \quad$ Subtracting 13.5

Check. Since $6.8 + 13.5 = 20.3$, we see that \$13.5 billion more than \$6.8 billion is \$20.3 billion and the answer checks.

State. The insurance loss from Hurricane Charley was \$6.8 billion.

5. Familiarize. Let x = the first angle. Then $5x$ = the second angle and $x - 2$ = the third angle. The sum of the measures of the angles is $180°$.

Translate.

First angle	$+$	Second angle	$+$	Third angle	is	$180°$.
x	$+$	$5x$	$+$	$(x-2)$	$=$	180

Solve. We solve the equation.

$\qquad x + 5x + (x - 2) = 180$

$\qquad\qquad\quad 7x - 2 = 180$

$\qquad\qquad\qquad 7x = 182$

$\qquad\qquad\qquad\ x = 26$

If $x = 26$, then $5x = 5 \cdot 26 = 130$ and $x - 2 = 26 - 2 = 24$.

Check. The second angle, $130°$, is five times the first and the third angle, $24°$, is $2°$ less than the first. Also, $26° + 130° + 24° = 180°$. The numbers check.

State. The first angle is $26°$, the second is $130°$, and the third is $24°$.

7. Familiarize. Let p = the price of the computer before the sales tax is added. Then the sales tax is 6% of p, or $0.06p$.

Translate.

Cost of computer	$+$	Sales tax	is	Total price
p	$+$	$0.06p$	$=$	1187.20

Solve. We solve the equation.

$$p + 0.06p = 1187.20$$
$$1p + 0.06p = 1187.20$$
$$1.06p = 1187.20$$
$$p = 1120$$

Check. 6% of $1120 is $67.20 and $1120 + $67.20 = $1187.20. The answer checks.

State. The price of the computer is $1120.

9. **Familiarize.** Let w = the width of the court. Then $w + 44$ = the length. Recall that the perimeter of a rectangle with length l and width w is $2l + 2w$.

Translate.

$$\underbrace{\text{Perimeter}} \quad \text{is} \quad \underbrace{288 \text{ ft.}}$$
$$2(w + 44) + 2w = 288$$

Solve. We solve the equation.

$$2(w + 44) + 2w = 288$$
$$2w + 88 + 2w = 288$$
$$4w + 88 = 288$$
$$4w = 200$$
$$w = 50$$

If $w = 50$, then $w + 44 = 50 + 44 = 94$.

Check. The length, 94 ft, is 44 ft longer than the width. Also $2 \cdot 94 + 2 \cdot 50 = 188 + 100 = 288$. The numbers check.

State. The length of the court is 94 ft and the width is 50 ft.

11. **Familiarize.** Recall that the perimeter of a square is 4 times the length of a side. Let s = the length of a side of the smaller square. Then $2s$ = the length of a side of the larger square. The sum of the two perimeters is 100 cm.

Translate.

$$\underbrace{\text{Perimeter of smaller square}} \quad \text{plus} \quad \underbrace{\text{perimeter of larger square}} \quad \text{is} \quad \underbrace{100 \text{ cm}}.$$
$$4s + 4 \cdot 2s = 100$$

Solve. We solve the equation.

$$4s + 4 \cdot 2s = 100$$
$$4s + 8s = 100$$
$$12s = 100$$
$$s = \frac{100}{12} = \frac{25}{3}$$

If $s = \frac{25}{3}$, then $4s = 4 \cdot \frac{25}{3} = \frac{100}{3}$, or $33\frac{1}{3}$;

$2s = 2 \cdot \frac{25}{3} = \frac{50}{3}$ and $4 \cdot \frac{50}{3} = \frac{200}{3}$, or $66\frac{2}{3}$.

Check. The length of a side of the larger square, $\frac{50}{3}$, is twice the length of a side of the smaller square, $\frac{25}{3}$. Also $33\frac{1}{3} + 66\frac{2}{3} = 100$. The numbers check.

State. The wire should be cut so that one piece is $33\frac{1}{3}$ cm long. Then the other piece will be $66\frac{2}{3}$ cm long.

13. **Familiarize.** Let p = the selling price of the house. Then $p - 100,000$ is the amount that exceeds $100,000. Also 7% of $100,000 is $7000.

Translate.

$$\underbrace{\begin{matrix}7\% \text{ of} \\ \$100,000\end{matrix}} + 5\% \text{ of} \underbrace{\begin{matrix}\text{the amount that} \\ \text{exceeds } \$100,000\end{matrix}} \text{ is } \$15,250.$$
$$7000 + 0.05 \cdot (p - 100,000) = 15,250$$

Solve. We solve the equation.

$$7000 + 0.05 \cdot (p - 100,000) = 15,250$$
$$7000 + 0.05p - 5000 = 15,250$$
$$0.05p + 2000 = 15,250$$
$$0.05p = 13,250$$
$$p = 265,000$$

Check. If the selling price is $265,000, then the amount that exceeds $100,000 is $265,000 - $100,000, or $165,000. Then the commission would be $7000 + 0.05($165,000), or $15,250. The answer checks.

State. The selling price of the house was $265,000.

15. **Familiarize.** Let x = the first odd integer. Then $x + 2$ and $x + 4$ are the next two odd integers.

Translate.

$$\underbrace{\begin{matrix}\text{First} \\ \text{integer}\end{matrix}} + 2 \text{ times} \underbrace{\begin{matrix}\text{Second} \\ \text{integer}\end{matrix}} +$$
$$x + 2 \cdot (x + 2) +$$

$$3 \text{ times} \underbrace{\begin{matrix}\text{Third} \\ \text{integer}\end{matrix}} \text{ is } 70.$$
$$3 \cdot (x + 4) = 70$$

Solve. We solve the equation.

$$x + 2 \cdot (x + 2) + 3 \cdot (x + 4) = 70$$
$$x + 2x + 4 + 3x + 12 = 70$$
$$6x + 16 = 70$$
$$6x = 54$$
$$x = 9$$

If $x = 9$, then $x + 2 = 9 + 2$, or 11, and $x + 4 = 9 + 4$, or 13.

Check. The numbers 9, 11, and 13 are consecutive odd integers. Also $9 + 2 \cdot 11 + 3 \cdot 13 = 9 + 22 + 39 = 70$. The numbers check.

State. The numbers are 9, 11, and 13.

17. **Familiarize.** Let x = the first number. Then $x + 1$ = the second number. The sum of the numbers is 697.

Translate.

$$\underbrace{\text{First number}} + \underbrace{\text{Second number}} \text{ is } 697.$$
$$x + (x + 1) = 697$$

Solve. We solve the equation.

$$x + (x+1) = 697$$
$$2x + 1 = 697$$
$$2x = 696$$
$$x = 348$$

If $x = 348$, then $x + 1 = 348 + 1 = 349$.

Check. The numbers 348 and 349 are consecutive integers. Also, $348 + 349 = 697$. The numbers check.

State. The numbers are 348 and 349.

19. *Familiarize*. Let $c =$ the number of square feet of carpet the customer had cleaned. Then the square footage that exceeds 200 sq ft is $c - 200$. The cost for cleaning the stairs is $1.40(13)$, or $18.20.

Translate.

$$
\underbrace{75}_{\downarrow\;75} \; \underbrace{+}_{\downarrow\;+} \; \underbrace{0.25}_{\downarrow\;0.25} \; \underbrace{\text{times}}_{\downarrow\;\cdot} \; \underbrace{\substack{\text{footage} \\ \text{that exceeds} \\ 200 \text{ sq ft}}}_{\downarrow\;(c-200)} \; \underbrace{+}_{\downarrow\;+} \; \underbrace{\substack{\text{cost of} \\ \text{cleaning} \\ \text{stairs}}}_{\downarrow\;18.20} \; \underbrace{= \$253.95}_{\downarrow\;=\;253.95}
$$

Solve. We solve the equation.

$$75 + 0.25 \cdot (c - 200) + 18.20 = 253.95$$
$$75 + 0.25c - 50 + 18.20 = 253.95$$
$$0.25c + 43.20 = 253.95$$
$$0.25c = 210.75$$
$$c = 843$$

Check. If the total square footage is 843 sq ft, then the area in excess of 200 sq ft is $843 - 200$, or 643 sq ft. The cost of cleaning this excess area is $160.75. Then the total cost is $75 + \$160.75 + \$18.20$, or $253.95. The answer checks.

State. The customer had 843 sq ft of carpet cleaned.

21. *Familiarize*. Let $s =$ the original salary. Then the amount of the raise is 8% of s, or $0.08s$.

Translate.

$$
\underbrace{\text{Original salary}}_{\downarrow\;s} \; \underbrace{\text{plus}}_{\downarrow\;+} \; \underbrace{\text{Raise}}_{\downarrow\;0.08s} \; \underbrace{\text{is}}_{\downarrow\;=} \; \underbrace{\$42,066.}_{\downarrow\;42,066}
$$

Solve. We solve the equation.

$$s + 0.08s = 42,066$$
$$1s + 0.08s = 42,066$$
$$1.08s = 42,066$$
$$s = 38,950$$

Check. 8% of $38,950 is $3116, and $38,950 + \$3116$ is $42,066. The answer checks.

State. The salary before the raise was $38,950.

23. *Familiarize*. Let $a =$ the number of millions of pounds of Arabica coffee beans purchased by Starbucks in 2004. Then 650% of this amount is $650\%a$, or $6.5a$.

Translate.

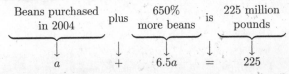

$$
\underbrace{\substack{\text{Beans purchased} \\ \text{in 2004}}}_{\downarrow\;a} \; \underbrace{\text{plus}}_{\downarrow\;+} \; \underbrace{\substack{650\% \\ \text{more beans}}}_{\downarrow\;6.5a} \; \underbrace{\text{is}}_{\downarrow\;=} \; \underbrace{\substack{225 \text{ million} \\ \text{pounds}}}_{\downarrow\;225}
$$

Solve. We solve the equation.

$$a + 6.5a = 225$$
$$7.5a = 225$$
$$a = 30$$

Check. 650% of 30 is 195 and $30 + 195 = 225$, so the answer checks.

State. Starbucks purchased 30 million pounds of Arabica coffee beans in 2004.

25. a) In 2003, $x = 2003 - 2000 = 3$. We substitute 3 for x and compute y.

$$y = 19.5x + 17,380.2$$
$$y = 19.5(3) + 17,380.2$$
$$y = 58.5 + 17,380.2$$
$$y = 17,438.7$$

There were about 17,439 alcohol-related traffic fatalities in 2003.

In 2006, $x = 2006 - 2000 = 6$. We substitute 6 for x and compute y.

$$y = 19.5x + 17,380.2$$
$$y = 19.5(6) + 17,380.2$$
$$y = 117 + 17,380.2$$
$$y = 17,497.2$$

There were about 17,497 alcohol-related traffic fatalities in 2006.

b) We substitute 17,458 for y and solve for x.

$$y = 19.5x + 17,380.2$$
$$17,458 = 19.5x + 17,380.2$$
$$77.8 = 19.5x$$
$$4 \approx x$$

About 17,458 alcohol-related traffic deaths occurred about 4 yr after 2000, or in 2004.

27. *Familiarize*. Let $t =$ the number of minutes it will take the plane to reach the cruising altitude. The plane needs to travel a distance of $29,000 - 8000$, or 21,000 ft.

Translate. We use the motion formula.

$$d = rt$$
$$21,000 = 3500t$$

Solve. We solve the equation.

$$21,000 = 3500t$$
$$\frac{21,000}{3500} = t$$
$$6 = t$$

Check. The distance the plane travels at a speed of 3500 ft/min for 6 min is $3500 \cdot 6$, or 21,000 ft. The answer checks.

State. It will take 6 min for the plane to reach the cruising altitude.

29. First we will find how long it takes for Jen to travel 15 mi downstream.

Familiarize. We will use the formula $d = rt$. Let $t =$ the time, in hours, it will take Jen to travel 15 mi downstream. The speed of the boat traveling downstream is $10 + 2$, or 12 mph.

Translate.
$$d = rt$$
$$15 = 12t$$

Solve. We solve the equation.
$$15 = 12t$$
$$1.25 = t$$

Check. If the boat travels at a speed of 12 mph for 1.25 hr, it travels $12(1.25)$, or 15 mi. The answer checks.

State. It will take Jen 1.25 hr to travel 15 mi downstream.

Now we find how long it will take Jen to travel 15 mi upstream.

Familiarize. We will use the formula $d = rt$. Let $t =$ the time, in hours, it will take Jen to travel 15 mi upstream. The speed of the boat traveling upstream is $10 - 2$, or 8 mph.

Translate.
$$d = rt$$
$$15 = 8t$$

Solve. We solve the equation.
$$15 = 8t$$
$$1.875 = t$$

Check. If the boat travels at a speed of 8 mph for 1.875 hr, it travels $8(1.875)$, or 15 mi. The answer checks.

State. It will take Jen 1.875 hr to travel 15 mi upstream.

31. *Familiarize*. We will use the formula $d = rt$. Let $t =$ the time, in hours, it takes the *Delta Queen* to cruise 2 mi upstream. The speed of the boat traveling upstream is $7 - 3$, or 4 mph.

Translate. We substitute in the formula.
$$d = rt$$
$$2 = 4t$$

Solve. We solve the equation.
$$2 = 4t$$
$$\frac{1}{2} = t$$

Check. If the boat travels at a speed of 4 mph for $\frac{1}{2}$ hr, it travels $4 \cdot \frac{1}{2}$, or 2 mi. The answer checks.

State. It will take the boat $\frac{1}{2}$ hr to cruise 2 mi upstream.

33. Discussion and Writing Exercise

35. $\quad -44 \cdot 55 - 22 = -2420 - 22$
$$= -2442$$

37. $\quad (5 - 12)^2 = (-7)^2$
$$= 49$$

39. $\quad 5^2 - 2 \cdot 5 \cdot 12 + 12^2 = 25 - 2 \cdot 5 \cdot 12 + 144$
$$= 25 - 10 \cdot 12 + 144$$
$$= 25 - 120 + 144$$
$$= -95 + 144$$
$$= 49$$

41. $\quad \dfrac{12|8 - 10| + 9 \cdot 6}{5^4 + 4^5} = \dfrac{12|-2| + 9 \cdot 6}{625 + 1024}$
$$= \dfrac{12 \cdot 2 + 9 \cdot 6}{1649}$$
$$= \dfrac{24 + 54}{1649}$$
$$= \dfrac{78}{1649}$$

43. $\quad [-64 \div (-4)] \div (-16) = 16 \div -16$
$$= -1$$

45. $\quad 2^{13} \div 2^5 \div 2^3 = 2^8 \div 2^3$
$$= 2^5, \text{ or } 32$$

47. *Familiarize*. Let $p =$ the price of the house in 2002. In 2003 real estate prices increased 6%, so the house was worth $p + 0.06p$, or $1.06p$. In 2004 prices increased 2%, so the house was then worth $1.06p + 0.02(1.06p)$, or $1.02(1.06p)$. In 2005 prices dropped 1%, so the value of the house became $1.02(1.06p) - 0.01(1.02)(1.06p)$, or $0.99(1.02)(1.06p)$.

Translate.

$\underbrace{\text{The price of the house in 2005}}$ was \$117,743.

$$0.99(1.02)(1.06p) \quad = \quad 117,743$$

Solve. We carry out some algebraic manipulation.
$$0.99(1.02)(1.06p) = 117,743$$
$$p = \dfrac{117,743}{0.99(1.02)(1.06)}$$
$$p \approx 110,000$$

Check. If the price of the house in 2002 was \$110,000, then in 2003 it was worth $1.06(\$110,000)$, or \$116,600. In 2004 it was worth $1.02(\$116,600)$, or \$118,932, and in 2005 it was worth $0.99(\$118,932)$, or \$117,743. Our answer checks.

State. The house was worth \$110,000 in 2002.

49. *Familiarize*. Let n represent the number of romance novels. Then 65%n, or $0.65n$, represents the number of science fiction novels; 46%$(0.65n)$, or $(0.46)(0.65n)$, or $0.299n$, represents the number of horror novels; and 17%$(0.299n)$, or $0.17(0.299n)$, or $0.05083n$, represents the number of mystery novels.

Translate.

$$\underbrace{\text{Romance}\atop\text{novels}} + \underbrace{\text{Science}\atop\text{fiction}\atop\text{novels}} + \underbrace{\text{Horror}\atop\text{novels}} + \underbrace{\text{Mystery}\atop\text{novels}} \text{ is } 400.$$

$$n \quad + \quad 0.65n \quad + \quad 0.299n + 0.05083n = \quad 400$$

Solve.

$$n + 0.65n + 0.299n + 0.05083n = 400$$
$$1.99983n = 400$$
$$n \approx 200$$

Check. If the number of romance novels is 200, then the other novels are:

Science fiction: $0.65(200) = 130$

Horror: $0.299(200) \approx 60$

Mystery: $0.05083(200) \approx 10$

The total number of novels is $200 + 130 + 60 + 10$, or 400. The numbers check.

State. The professor has 130 science fiction novels.

51. *Familiarize.* Let p = the total percent change, represented in decimal notation. We represent the original population as 1 (100% of the population). After a 20% increase the population is $1 + 0.2 \cdot 1$, or 1.2. When this new population is increased by 30%, we have $1.2 + 0.3(1.2)$, or $1.3(1.2)$. When this population is decreased by 20% we have $1.3(1.2) - 0.2[1.3(1.2)]$, or $0.8(1.3)(1.2)$.

Translate.

Percent of original population after changes	minus	100% of original population	is	total percent change
$0.8(1.3)(1.2)$	$-$	1	$=$	p

Solve.

$$0.8(1.3)(1.2) - 1 = p$$
$$1.248 - 1 = p \qquad \text{Multiplying}$$
$$0.248 = p$$
$$0.25 \approx p \qquad \text{Rounding}$$

Since p is a positive number, it represents an increase in population.

Check. We repeat the computations. The result checks.

State. The total percent change is a 25% increase.

53. *Familiarize.* Let s = the number of seconds after which the watches will show the same time again. The difference in time between the two watches is

$$2.5\,\frac{\text{sec}}{\text{hr}} = 2.5\,\frac{\text{sec}}{\text{hr}} \times \frac{1\ \text{hr}}{60\ \text{min}} \times \frac{1\ \text{min}}{60\ \text{sec}} = \frac{2.5\ \text{sec}}{3600\ \text{sec}}.$$

The watches will show the same time again when the difference in time between them is

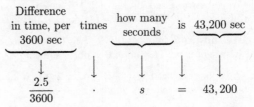

$$12\ \text{hr} = 12\ \text{hr} \times \frac{60\ \text{min}}{1\ \text{hr}} \times \frac{60\ \text{sec}}{1\ \text{min}} = 43{,}200\ \text{sec}.$$

Translate.

Difference in time, per 3600 sec	times	how many seconds	is	43,200 sec
$\dfrac{2.5}{3600}$	$\cdot$	s	$=$	$43{,}200$

Solve.

$$\frac{2.5}{3600}s = 43{,}200$$
$$\frac{3600}{2.5} \cdot \frac{2.5}{3600}s = \frac{3600}{2.5} \cdot 43{,}200$$
$$s = 62{,}208{,}000$$

Check. At a rate of $\dfrac{2.5\ \text{sec}}{3600\ \text{sec}}$, in 62,208,000 sec the difference in time will be $\dfrac{2.5}{3600} \cdot 62{,}208{,}000$, or $43{,}200$ sec, or 12 hr. The result checks.

State. The watches will show the same time again after 62,208,000 sec.

55. Answers may vary. One possibility is: A piece of material 75 cm long is to be cut into two pieces, one of them $\dfrac{2}{3}$ as long as the other. How should the material be cut?

57. *Familiarize.* Using the properties of parallel lines intersected by a transversal, we know that $m\angle 8 = m\angle 4$. Also $m\angle 4 + m\angle 2 = 180°$ and $m\angle 4 = m\angle 1$.

Translate. First we find x and use it to find $m\angle 4$.

$$m\angle 8 = m\angle 4$$
$$5x + 25 = 8x + 4$$

Solve.

$$5x + 25 = 8x + 4$$
$$21 = 3x \qquad \text{Subtracting } 5x \text{ and } 4$$
$$7 = x$$

If $x = 7$, we have $m\angle 4 = 8 \cdot 7 + 4 = 60°$.

Then $m\angle 4 + m\angle 2 = 180°$, so

$$60° + m\angle 2 = 180°$$
$$m\angle 2 = 120°. \qquad \text{Subtracting } 60°$$

Also, $m\angle 4 = m\angle 1 = 60°$.

Check. We go over the computations. The results check.

State. $m\angle 2 = 120°$ and $m\angle 1 = 60°$.

Exercise Set 1.4

1. $x - 2 \geq 6$

-4 : We substitute and get $-4 - 2 \geq 6$, or $-6 \geq 6$, a false sentence. Therefore, -4 is not a solution.

0 : We substitute and get $0 - 2 \geq 6$, or $-2 \geq 6$, a false sentence. Therefore, 0 is not a solution.

4 : We substitute and get $4 - 2 \geq 6$, or $2 \geq 6$, a false sentence. Therefore, 4 is not a solution.

8 : We substitute and get $8 - 2 \geq 6$, or $6 \geq 6$, a true sentence. Therefore, 8 is a solution.

3. $t - 8 > 2t - 3$

0 : We substitute and get $0 - 8 > 2 \cdot 0 - 3$, or $-8 > -3$, a false sentence. Therefore, 0 is not a solution.

-8 : We substitute and get $-8 - 8 > 2(-8) - 3$, or $-16 > -19$, a true sentence. Therefore, -8 is a solution.

-9 : We substitute and get $-9 - 8 > 2(-9) - 3$, or $-17 > -21$, a true sentence. Therefore, -9 is a solution.

-3 : We substitute and get $-3 - 8 > 2(-3) - 3$, or $-11 > -9$, a false sentence. Therefore, -3 is not a solution.

$-\dfrac{7}{8}$: We substitute and get $-\dfrac{7}{8} - 8 > 2\left(-\dfrac{7}{8}\right) - 3$, or $-\dfrac{71}{8} > -\dfrac{38}{8}$, a false sentence. Therefore, $-\dfrac{7}{8}$ is not a solution.

5. Interval notation for $\{x | x < 5\}$ is $(-\infty, 5)$.

7. Interval notation for $\{x | -3 \leq x \leq 3\}$ is $[-3, 3]$.

9. $\{x | -4 > x > -8\} = \{x | -8 < x < -4\} = (-8, -4)$

11. Interval notation for the given graph is $(-2, 5)$.

13. Interval notation for the given graph is $(-\sqrt{2}, \infty)$.

15. $\quad x + 2 > 1$

$x + 2 - 2 > 1 - 2 \qquad$ Subtracting 2

$\qquad x > -1$

The solution set is $\{x | x > -1\}$, or $(-1, \infty)$.

17. $\quad y + 3 < 9$

$y + 3 - 3 < 9 - 3 \qquad$ Subtracting 3

$\qquad y < 6$

The solution set is $\{y | y < 6\}$, or $(-\infty, 6)$.

19. $\quad a - 9 \leq -31$

$a - 9 + 9 \leq -31 + 9 \qquad$ Adding 9

$\qquad a \leq -22$

The solution set is $\{a | a \leq -22\}$, or $(-\infty, -22]$.

21. $\quad t + 13 \geq 9$

$t + 13 - 13 \geq 9 - 13 \qquad$ Subtracting 13

$\qquad t \geq -4$

The solution set is $\{t | t \geq -4\}$, or $[-4, \infty)$.

23. $\quad y - 8 > -14$

$y - 8 + 8 > -14 + 8 \qquad$ Adding 8

$\qquad y > -6$

The solution set is $\{y | y > -6\}$, or $(-6, \infty)$.

25. $\quad x - 11 \leq -2$

$x - 11 + 11 \leq -2 + 11 \qquad$ Adding 11

$\qquad x \leq 9$

The solution set is $\{x | x \leq 9\}$, or $(-\infty, 9]$.

27. $\quad 8x \geq 24$

$\dfrac{8x}{8} \geq \dfrac{24}{8} \qquad$ Dividing by 8

$\qquad x \geq 3$

The solution set is $\{x | x \geq 3\}$, or $[3, \infty)$.

29. $\quad 0.3x < -18$

$\dfrac{0.3x}{0.3} < \dfrac{-18}{0.3} \qquad$ Dividing by 0.3

$\qquad x < -60$

The solution set is $\{x | x < -60\}$, or $(-\infty, -60)$.

31. $\quad \dfrac{2}{3}x > 2$

$\dfrac{3}{2} \cdot \dfrac{2}{3}x > \dfrac{3}{2} \cdot 2 \qquad$ Multiplying by $\dfrac{3}{2}$

$\qquad x > 3$

The solution set is $\{x | x > 3\}$, or $(3, \infty)$.

33. $-9x \geq -8.1$

$\dfrac{-9x}{-9} \leq \dfrac{-8.1}{-9}$ Dividing by -9 and reversing the inequality symbol

$x \leq 0.9$

The solution set is $\{x | x \leq 0.9\}$, or $(-\infty, 0.9]$.

35. $-\dfrac{3}{4}x \geq -\dfrac{5}{8}$

$-\dfrac{4}{3}\left(-\dfrac{3}{4}x\right) \leq -\dfrac{4}{3}\left(-\dfrac{5}{8}\right)$ Multiplying by $-\dfrac{4}{3}$ and reversing the inequality symbol

$x \leq \dfrac{20}{24}$

$x \leq \dfrac{5}{6}$

The solution set is $\left\{x \middle| x \leq \dfrac{5}{6}\right\}$, or $\left(-\infty, \dfrac{5}{6}\right]$.

37. $2x + 7 < 19$

$2x + 7 - 7 < 19 - 7$ Subtracting 7

$2x < 12$

$\dfrac{2x}{2} < \dfrac{12}{2}$ Dividing by 2

$x < 6$

The solution set is $\{x | x < 6\}$, or $(-\infty, 6)$.

39. $5y + 2y \leq -21$

$7y \leq -21$ Collecting like terms

$\dfrac{7y}{7} \leq \dfrac{-21}{7}$ Dividing by 7

$y \leq -3$

The solution set is $\{y | y \leq -3\}$, or $(-\infty, -3]$.

41. $2y - 7 < 5y - 9$

$-5y + 2y - 7 < -5y + 5y - 9$ Adding $-5y$

$-3y - 7 < -9$

$-3y - 7 + 7 < -9 + 7$ Adding 7

$-3y < -2$

$\dfrac{-3y}{-3} > \dfrac{-2}{-3}$ Dividing by -3 and reversing the inequality symbol

$y > \dfrac{2}{3}$

The solution set is $\left\{y \middle| y > \dfrac{2}{3}\right\}$, or $\left(\dfrac{2}{3}, \infty\right)$.

43. $0.4x + 5 \leq 1.2x - 4$

$-1.2x + 0.4x + 5 \leq -1.2x + 1.2x - 4$ Adding $-1.2x$

$-0.8x + 5 \leq -4$

$-0.8x + 5 - 5 \leq -4 - 5$ Subtracting 5

$-0.8x \leq -9$

$\dfrac{-0.8x}{-0.8} \geq \dfrac{-9}{-0.8}$ Dividing by -0.8 and reversing the inequality symbol

$x \geq 11.25$

The solution set is $\{x | x \geq 11.25\}$, or $[11.25, \infty)$.

45. $5x - \dfrac{1}{12} \leq \dfrac{5}{12} + 4x$

$12\left(5x - \dfrac{1}{12}\right) \leq 12\left(\dfrac{5}{12} + 4x\right)$ Clearing fractions

$60x - 1 \leq 5 + 48x$

$60x - 1 - 48x \leq 5 + 48x - 48x$ Subtracting $48x$

$12x - 1 \leq 5$

$12x - 1 + 1 \leq 5 + 1$ Adding 1

$12x \leq 6$

$\dfrac{12x}{12} \leq \dfrac{6}{12}$ Dividing by 12

$x \leq \dfrac{1}{2}$

The solution set is $\left\{x \middle| x \leq \dfrac{1}{2}\right\}$, or $\left(-\infty, \dfrac{1}{2}\right]$.

47. $4(4y - 3) \geq 9(2y + 7)$

$16y - 12 \geq 18y + 63$ Removing parentheses

$16y - 12 - 18y \geq 18y + 63 - 18y$ Subtracting $18y$

$-2y - 12 \geq 63$

$-2y - 12 + 12 \geq 63 + 12$ Adding 12

$-2y \geq 75$

$\dfrac{-2y}{-2} \leq \dfrac{75}{-2}$ Dividing by -2 and reversing the inequality symbol

$y \leq -\dfrac{75}{2}$

The solution set is $\left\{y \middle| y \leq -\dfrac{75}{2}\right\}$, or $\left(-\infty, -\dfrac{75}{2}\right]$.

49. $3(2 - 5x) + 2x < 2(4 + 2x)$

$6 - 15x + 2x < 8 + 4x$

$6 - 13x < 8 + 4x$ Collecting like terms

$6 - 17x < 8$ Subtracting $4x$

$-17x < 2$ Subtracting 6

$x > -\dfrac{2}{17}$ Dividing by -17 and reversing the inequality symbol

The solution set is $\left\{x \middle| x > -\dfrac{2}{17}\right\}$, or $\left(-\dfrac{2}{17}, \infty\right)$.

51. $5[3m - (m + 4)] > -2(m - 4)$

$5(3m - m - 4) > -2(m - 4)$

$5(2m - 4) > -2(m - 4)$

$10m - 20 > -2m + 8$

$12m - 20 > 8$ Adding $2m$

$12m > 28$ Adding 20

$m > \dfrac{28}{12}$

$m > \dfrac{7}{3}$

The solution set is $\left\{ m \middle| m > \dfrac{7}{3} \right\}$, or $\left(\dfrac{7}{3}, \infty \right)$.

53. $3(r - 6) + 2 > 4(r + 2) - 21$

$3r - 18 + 2 > 4r + 8 - 21$

$3r - 16 > 4r - 13$ Collecting like terms

$-r - 16 > -13$ Subtracting $4r$

$-r > 3$ Adding 16

$r < -3$ Multiplying by -1 and reversing the inequality symbol

The solution set is $\{r | r < -3\}$, or $(-\infty, -3)$.

55. $19 - (2x + 3) \leq 2(x + 3) + x$

$19 - 2x - 3 \leq 2x + 6 + x$

$16 - 2x \leq 3x + 6$ Collecting like terms

$16 - 5x \leq 6$ Subtracting $3x$

$-5x \leq -10$ Subtracting 16

$x \geq 2$ Dividing by -5 and reversing the inequality symbol

The solution set is $\{x | x \geq 2\}$, or $[2, \infty)$.

57. $\dfrac{1}{4}(8y + 4) - 17 < -\dfrac{1}{2}(4y - 8)$

$2y + 1 - 17 < -2y + 4$

$2y - 16 < -2y + 4$ Collecting like terms

$4y - 16 < 4$ Adding $2y$

$4y < 20$ Adding 16

$y < 5$

The solution set is $\{y | y < 5\}$, or $(-\infty, 5)$.

59. $2[4 - 2(3 - x)] - 1 \geq 4[2(4x - 3) + 7] - 25$

$2[4 - 6 + 2x] - 1 \geq 4[8x - 6 + 7] - 25$

$2[-2 + 2x] - 1 \geq 4[8x + 1] - 25$

$-4 + 4x - 1 \geq 32x + 4 - 25$

$4x - 5 \geq 32x - 21$

$-28x - 5 \geq -21$

$-28x \geq -16$

$x \leq \dfrac{-16}{-28}$ Dividing by -28 and reversing the inequality symbol

$x \leq \dfrac{4}{7}$

The solution set is $\left\{ x \middle| x \leq \dfrac{4}{7} \right\}$, or $\left(-\infty, \dfrac{4}{7} \right]$.

61. $\dfrac{4}{5}(7x - 6) < 40$

$5 \cdot \dfrac{4}{5}(7x - 6) < 5 \cdot 40$ Clearing the fraction

$4(7x - 6) < 200$

$28x - 24 < 200$

$28x < 224$

$x < 8$

The solution set is $\{x | x < 8\}$, or $(-\infty, 8)$.

63. $\dfrac{3}{4}(3 + 2x) + 1 \geq 13$

$4\left[\dfrac{3}{4}(3 + 2x) + 1 \right] \geq 4 \cdot 13$ Clearing the fraction

$3(3 + 2x) + 4 \geq 52$

$9 + 6x + 4 \geq 52$

$6x + 13 \geq 52$

$6x \geq 39$

$x \geq \dfrac{39}{6}$, or $\dfrac{13}{2}$

The solution set is $\left\{ x \middle| x \geq \dfrac{13}{2} \right\}$, or $\left[\dfrac{13}{2}, \infty \right)$.

65. $\dfrac{3}{4}\left(3x - \dfrac{1}{2} \right) - \dfrac{2}{3} < \dfrac{1}{3}$

$\dfrac{9x}{4} - \dfrac{3}{8} - \dfrac{2}{3} < \dfrac{1}{3}$

$24\left(\dfrac{9x}{4} - \dfrac{3}{8} - \dfrac{2}{3} \right) < 24 \cdot \dfrac{1}{3}$ Clearing fractions

$54x - 9 - 16 < 8$

$54x - 25 < 8$

$54x < 33$

$x < \dfrac{33}{54}$, or $\dfrac{11}{18}$

The solution set is $\left\{ x \middle| x < \dfrac{11}{18} \right\}$, or $\left(-\infty, \dfrac{11}{18} \right)$.

67. $0.7(3x + 6) \geq 1.1 - (x + 2)$

$10[0.7(3x + 6)] \geq 10[1.1 - (x + 2)]$ Clearing decimals

$7(3x + 6) \geq 11 - 10(x + 2)$

$21x + 42 \geq 11 - 10x - 20$

$21x + 42 \geq -9 - 10x$

$31x + 42 \geq -9$

$31x \geq -51$

$x \geq -\dfrac{51}{31}$

The solution set is $\left\{ x \middle| x \geq -\dfrac{51}{31} \right\}$, or $\left[-\dfrac{51}{31}, \infty \right)$.

69. $a + (a - 3) \leq (a + 2) - (a + 1)$

$a + a - 3 \leq a + 2 - a - 1$

$2a - 3 \leq 1$

$2a \leq 4$

$a \leq 2$

The solution set is $\{a | a \leq 2\}$, or $(-\infty, 2]$.

71. **Familiarize**. We will use the formula $I = \dfrac{704.5W}{H^2}$. Recall that $H = 73$ in.

Translate.

$$I < 25, \text{ or } \dfrac{704.5W}{H^2} < 25$$

We replace H with 73.

$$\dfrac{704.5W}{73^2} < 25$$

Solve. We solve the inequality.

$$\dfrac{704.5W}{73^2} < 25$$

$$\dfrac{704.5W}{5329} < 25$$

$$704.5W < 133,225$$

$$W < 189.1 \qquad \text{Rounding}$$

Check. As a partial check we can substitute a value of W less than 189.1 and a value greater than 189.1 in the formula.

For $W = 189$: $I = \dfrac{704.5(189)}{73^2} \approx 24.97$

For $W = 190$: $I = \dfrac{704.5(190)}{73^2} \approx 25.12$

Since a value of W less than 189.1 gives a body mass index less than 25 and a value of W greater than 189.1 gives an index greater than 25, we have a partial check.

State. Weights of approximately 189.1 lb or less will keep Marv's body mass index below 25. In terms of an inequality we write $\{W | W < \text{(approximately) } 189.1 \text{ lb}\}$.

73. **Familiarize**. List the information in a table. Let $x = $ the score on the fourth test.

Test	Score
Test 1	89
Test 2	92
Test 3	95
Test 4	x
Total	360 or more

Translate. We can easily get an inequality from the table.

$$89 + 92 + 95 + x \geq 360$$

Solve.

$276 + x \geq 360$ Collecting like terms

$x \geq 84$ Adding -276

Check. If you get 84 on the fourth test, your total score will be $89 + 92 + 95 + 84$, or 360. Any higher score will also give you an A.

State. A score of 84 or better will give you an A. In terms of an inequality we write $\{x | x \geq 84\}$.

75. **Familiarize**. Let $v = $ the blue book value of the car. Since the car was not replaced, we know that \$9200 does not exceed 80% of the blue book value.

Translate. We write an inequality stating that \$9200 does not exceed 80% of the blue book value.

$$9200 \leq 0.8v$$

Solve.

$9200 \leq 0.8v$

$11,500 \leq v$ Multiplying by $\dfrac{1}{0.8}$

Check. We can do a partial check by substituting a value for v greater than 11,500. When $v = 11,525$, then 80% of v is $0.8(11,525)$, or \$9220. This is greater than \$9200; that is, \$9200 does not exceed this amount. We cannot check all possible values for v, so we stop here.

State. The blue book value of the car is \$11,500 or more. In terms of an inequality we write $\{v | v \geq \$11,500\}$.

77. **Familiarize**. We make a table of information.

Plan A: Monthly Income	Plan B: Monthly Income
\$400 salary	\$610 salary
8% of sales	5% of sales
Total: 400 + 8% of sales	Total: 610 + 5% of sales

Translate. We write an inequality stating that the income from Plan A is greater than the income from Plan B. We let $S = $ gross sales.

$$400 + 8\% S > 610 + 5\% S$$

Solve.

$400 + 0.08S > 610 + 0.05S$

$400 + 0.03S > 610$

$0.03S > 210$

$S > 7000$

Check. We calculate for $S = \$7000$ and for some amount greater than \$7000 and some amount less than \$7000.

Plan A: Plan B:

$400 + 8\%(7000)$ $610 + 5\%(7000)$

$400 + 0.08(7000)$ $610 + 0.05(7000)$

$400 + 560$ $610 + 350$

$\$960$ $\$960$

When $S = \$7000$, the income from Plan A is equal to the income from Plan B.

Plan A: Plan B:

$400 + 8\%(8000)$ $610 + 5\%(8000)$

$400 + 0.08(8000)$ $610 + 0.05(8000)$

$400 + 640$ $610 + 400$

$\$1040$ $\$1010$

When $S = \$8000$, the income from Plan A is greater than the income from Plan B.

Plan A: Plan B:

$400 + 8\%(6000)$ $610 + 5\%(6000)$

$400 + 0.08(6000)$ $610 + 0.05(6000)$

$400 + 480$ $610 + 300$

$\$880$ $\$910$

When $S = \$6000$, the income from Plan A is less than the income from Plan B.

State. Plan A is better than Plan B when gross sales are greater than \$7000. In terms of an inequality we write $\{S | S > \$7000\}$.

79. Familiarize. Let $c =$ the number of checks per month. Then the Anywhere plan will cost $\$0.20c$ per month and the Acu-checking plan will cost $\$2 + \$0.12c$ per month.

Translate. We write an inequality stating that the Acu-checking plan costs less than the Anywhere plan.

$$2 + 0.12c < 0.20c$$

Solve.

$$2 + 0.12c < 0.20c$$
$$2 < 0.08c$$
$$25 < c$$

Check. We can do a partial check by substituting a value for c less than 25 and a value for c greater than 25. When $c = 24$, the Acu-checking plan costs $\$2 + \$0.12(24)$, or \$4.88, and the Anywhere plan costs $\$0.20(24)$, or \$4.80, so the Anywhere plan is less expensive. When $c = 26$, the Acu-checking plan costs $\$2 + \$0.12(26)$, or \$5.12, and the Anywhere plan costs $\$0.20(26)$, or \$5.20, so Acu-checking is less expensive. We cannot check all possible values for c, so we stop here.

State. The Acu-checking plan costs less for more than 25 checks per month. In terms of an inequality we write $\{c | c > 25\}$.

81. Familiarize. Let $p =$ the number of guests at the wedding party. Then the number of guests in excess of 25 is $p - 25$.

The cost under plan A is $30p$, and the cost under plan B is $1300 + 20(p - 25)$.

Translate. We write an inequality stating that plan B costs less than plan A.

$$1300 + 20(p - 25) < 30p$$

Solve. We solve the inequality.

$$1300 + 20(p - 25) < 30p$$
$$1300 + 20p - 500 < 30p$$
$$800 + 20p < 30p$$
$$800 < 10p$$
$$80 < p$$

Check. We calculate for $p = 80$ and for some number less than 80 and some number greater than 80.

Plan A: Plan B:

$30 \cdot 80$ $1300 + 20(80 - 25)$

$\$2400$ $\$2400$

When 80 people attend, plan B costs the same as plan A.

Plan A: Plan B:

$30 \cdot 79$ $1300 + 20(79 - 25)$

$\$2370$ $\$2380$

When fewer than 80 people attend, plan B costs more than plan A.

Plan A: Plan B:

$30 \cdot 81$ $1300 + 20(81 - 25)$

$\$2430$ $\$2420$

When more than 80 people attend, plan B costs less than plan A.

State. For parties of more than 80 people, plan B will cost less. In terms of an inequality we write $\{p | p > 80\}$.

83. Familiarize. We want to find the values of s for which $I > 36$.

Translate. $2(s + 10) > 36$

Solve.

$$2s + 20 > 36$$
$$2s > 16$$
$$s > 8$$

Check. For $s = 8$, $I = 2(8 + 10) = 2 \cdot 18 = 36$. Then any U.S. size larger than 8 will give a size larger than 36 in Italy.

State. For U.S. dress sizes larger than 8, dress sizes in Italy will be larger than 36. In terms of an inequality we write $\{s | s > 8\}$.

85. a) Substitute 0 for t and carry out the calculation.

$$N = 0.733(0) + 8.398$$
$$N = 0 + 8.398$$
$$N = 8.398$$

Each person drank 8.398 gal of bottled water in 1990.

Substitute 5 for t and carry out the calculation.

$$N = 0.733(5) + 8.398$$
$$N = 3.665 + 8.398$$
$$N = 12.063$$

Each person drank 12.063 gal of bottled water in 1995.

In 2000, $t = 2000 - 1990 = 10$. Substitute 10 for t and carry out the calculation.

$$N = 0.733(10) + 8.398$$
$$N = 7.33 + 8.398$$
$$N = 15.728$$

Each person will drink 15.728 gal of bottled water in 2000.

b) **Familiarize**. The amount of bottled water that each person drinks t years after 1990 is given by $0.733t + 8.398$.

Translate.

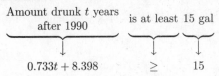

Amount drunk t years after 1990	is at least	15 gal
↓	↓	↓
$0.733t + 8.398$	$\geq$	15

Solve. We solve the inequality.

$$0.733t + 8.398 \geq 15$$
$$0.733t \geq 6.602 \qquad \text{Subtracting } 8.398$$
$$t \geq 9 \qquad \text{Rounding}$$

Check. We calculate for 9, for some number less than 9, and for some number greater than 9.

For $t = 9$: $0.733(9) + 8.398 \approx 15$.

For $t = 8$: $0.733(8) + 8.398 \approx 14.3$.

For $t = 10$: $0.733(10) + 8.398 \approx 15.7$.

For a value of t greater than or equal to 9, the number of gallons of bottled water each person drinks is at 15. We cannot check all the possible values of t, so we stop here.

State. Each person will drink at least 15 gal of bottled water 9 or more years after 1990, or for all years after 1999.

87. Discussion and Writing Exercise

89. $3a - 6(2a - 5b) = 3a - 12a + 30b$
$$= -9a + 30b$$

91. $\quad 4(a - 2b) - 6(2a - 5b)$
$$= 4a - 8b - 12a + 30b$$
$$= -8a + 22b$$

93. $30x - 70y - 40 = 10 \cdot 3x - 10 \cdot 7y - 10 \cdot 4$
$$= 10(3x - 7y - 4)$$

95. $-8x + 24y - 4 = -4 \cdot 2x - 4(-6y) - 4 \cdot 1$
$$= -4(2x - 6y + 1)$$

97. $-2.3 - 8.9 = -2.3 + (-8.9) = -11.2$

99. $-2.3 + (-8.9) = -11.2$

101. a) **Familiarize**. We will use

$$S = 460 + 94p \quad \text{and} \quad D = 2000 - 60p.$$

Translate. Supply is to exceed demand, so we have

$$S > D, \text{ or}$$
$$460 + 94p > 2000 - 60p.$$

Solve. We solve the inequality.

$$460 + 94p > 2000 - 60p$$
$$460 + 154p > 2000 \qquad \text{Adding } 60p$$
$$154p > 1540 \qquad \text{Subtracting } 460$$
$$p > 10 \qquad \text{Dividing by } 154$$

Check. We calculate for $p = 10$, for some value of p less than 10, and for some value of p greater than 10.

For $p = 10$: $\quad S = 460 + 94 \cdot 10 = 1400$
$\qquad\qquad\quad D = 2000 - 60 \cdot 10 = 1400$

For $p = 9$: $\quad S = 460 + 94 \cdot 9 = 1306$
$\qquad\qquad\quad D = 2000 - 60 \cdot 9 = 1460$

For $p = 11$: $\quad S = 460 + 94 \cdot 11 = 1494$
$\qquad\qquad\quad D = 2000 - 60 \cdot 11 = 1340$

For a value of p greater than 10, supply exceeds demand. We cannot check all possible values of p, so we stop here.

State. Supply exceeds demand for values of p greater than 10. In terms of an inequality we write $\{p | p > 10\}$.

b) We have seen in part (a) that $D = S$ for $p = 10$, $S < D$ for a value of p less than 10, and $S > D$ for a value of p greater than 10. Since we cannot check all possible values of p, we stop here. Supply is less than demand for values of p less than 10. In terms of an inequality we write $\{p | p < 10\}$.

103. True

105. $\quad x + 5 \leq 5 + x$
$$5 \leq 5 \qquad \text{Subtracting } x$$

We get a true inequality, so all real numbers are solutions.

107. $x^2 + 1 > 0$

$x^2 \geq 0$ for all real numbers, so $x^2 + 1 \geq 1 > 0$ for all real numbers.

Exercise Set 1.5

1. $\{9, 10, 11\} \cap \{9, 11, 13\}$

The numbers 9 and 11 are common to the two sets, so the intersection is $\{9, 11\}$.

3. $\{a, b, c, d\} \cap \{b, f, g\}$

Only the letter b is common to the two sets. The intersection is $\{b\}$.

5. $\{9, 10, 11\} \cup \{9, 11, 13\}$

The numbers in either or both sets are 9, 10, 11, and 13, so the union is $\{9, 10, 11, 13\}$.

7. $\{a, b, c, d\} \cup \{b, f, g\}$

The letters in either or both sets are a, b, c, d, f, and g, so the union is $\{a, b, c, d, f, g\}$.

9. $\{2, 5, 7, 9\} \cap \{1, 3, 4\}$

There are no numbers common to the two sets. The intersection is the empty set, $\emptyset$.

11. $\{3, 5, 7\} \cup \emptyset$

The numbers in either or both sets are 3, 5, and 7, so the union is $\{3, 5, 7\}$.

13. $-4 < a$ *and* $a \leq 1$ can be written $-4 < a \leq 1$. In interval notation we have $(-4, 1]$.

The graph is the intersection of the graphs of $a > -4$ and $a \leq 1$.

15. We can write $1 < x < 6$ in interval notation as $(1, 6)$.

The graph is the intersection of the graphs of $x > 1$ and $x < 6$.

17. $-10 \leq 3x + 2$ *and* $3x + 2 < 17$

$\quad -12 \leq 3x \quad$ *and* $\quad 3x < 15$

$\quad -4 \leq x \quad$ *and* $\quad x < 5$

The solution set is the intersection of the solution sets of the individual inequalities. The numbers common to both sets are those that are greater than or equal to -4 *and* less than 5. Thus the solution set is $\{x| -4 \leq x < 5\}$, or $[-4, 5)$.

19. $3x + 7 \geq 4 \quad$ *and* $\quad 2x - 5 \geq -1$

$\quad 3x \geq -3 \quad$ *and* $\quad 2x \geq 4$

$\quad x \geq -1 \quad$ *and* $\quad x \geq 2$

The solution set is $\{x|x \geq -1\} \cap \{x|x \geq 2\} = \{x|x \geq 2\}$, or $[2, \infty)$.

21. $4 - 3x \geq 10 \quad$ *and* $\quad 5x - 2 > 13$

$\quad -3x \geq 6 \quad$ *and* $\quad 5x > 15$

$\quad x \leq -2 \quad$ *and* $\quad x > 3$

The solution set is $\{x|x \leq -2\} \cap \{x|x > 3\} = \emptyset$.

23. $\quad -4 < x + 4 < 10$

$\quad -4 - 4 < x + 4 - 4 < 10 - 4 \qquad$ Subtracting 4

$\quad -8 < x < 6$

The solution set is $\{x| -8 < x < 6\}$, or $(-8, 6)$.

25. $\quad 6 > -x \geq -2$

$\quad -6 < x \leq 2 \qquad$ Multiplying by -1

The solution set is $\{x| -6 < x \leq 2\}$, or $(-6, 2]$.

27. $\quad 1 < 3y + 4 \leq 19$

$\quad 1 - 4 < 3y + 4 - 4 \leq 19 - 4 \quad$ Subtracting 4

$\quad -3 < 3y \leq 15$

$\quad \dfrac{-3}{3} < \dfrac{3y}{3} \leq \dfrac{15}{3} \quad$ Dividing by 3

$\quad -1 < y \leq 5$

The solution set is $\{y| -1 < y \leq 5\}$, or $(-1, 5]$.

29. $\quad -10 \leq 3x - 5 \leq -1$

$\quad -10 + 5 \leq 3x - 5 + 5 \leq -1 + 5 \quad$ Adding 5

$\quad -5 \leq 3x \leq 4$

$\quad \dfrac{-5}{3} \leq \dfrac{3x}{3} \leq \dfrac{4}{3} \quad$ Dividing by 3

$\quad -\dfrac{5}{3} \leq x \leq \dfrac{4}{3}$

The solution set is $\left\{x| -\dfrac{5}{3} \leq x \leq \dfrac{4}{3}\right\}$, or $\left[-\dfrac{5}{3}, \dfrac{4}{3}\right]$.

31. $\quad 2 < x + 3 \leq 9$

$\quad 2 - 3 < x + 3 - 3 \leq 9 - 3 \quad$ Subtracting 3

$\quad -1 < x \leq 6$

The solution set is $\{x| -1 < x \leq 6\}$, or $(-1, 6]$.

33. $\quad -6 \leq 2x - 3 < 6$

$\quad -6 + 3 \leq 2x - 3 + 3 < 6 + 3$

$\quad -3 \leq 2x < 9$

$\quad \dfrac{-3}{2} \leq \dfrac{2x}{2} < \dfrac{9}{2}$

$\quad -\dfrac{3}{2} \leq x < \dfrac{9}{2}$

The solution set is $\left\{x| -\dfrac{3}{2} \leq x < \dfrac{9}{2}\right\}$, or $\left[-\dfrac{3}{2}, \dfrac{9}{2}\right)$.

35. $\quad -\dfrac{1}{2} < \dfrac{1}{4}x - 3 \leq \dfrac{1}{2}$

$\quad -\dfrac{1}{2} + 3 < \dfrac{1}{4}x - 3 + 3 \leq \dfrac{1}{2} + 3$

$\quad \dfrac{5}{2} < \dfrac{1}{4}x \leq \dfrac{7}{2}$

$\quad 4 \cdot \dfrac{5}{2} < 4 \cdot \dfrac{1}{4}x \leq 4 \cdot \dfrac{7}{2}$

$\quad 10 < x \leq 14$

The solution set is $\{x|10 < x \leq 14\}$, or $(10, 14]$.

37.
$$-3 < \frac{2x-5}{4} < 8$$
$$4(-3) < 4\left(\frac{2x-5}{4}\right) < 4 \cdot 8$$
$$-12 < 2x-5 < 32$$
$$-12+5 < 2x-5+5 < 32+5$$
$$-7 < 2x < 37$$
$$\frac{-7}{2} < \frac{2x}{2} < \frac{37}{2}$$
$$-\frac{7}{2} < x < \frac{37}{2}$$

The solution set is $\left\{x\middle| -\frac{7}{2} < x < \frac{37}{2}\right\}$, or $\left(-\frac{7}{2}, \frac{37}{2}\right)$.

39. $x < -2$ *or* $x > 1$ can be written in interval notation as $(-\infty, -2) \cup (1, \infty)$.

The graph is the union of the graphs of $x < -2$ and $x > 1$.

41. $x \le -3$ *or* $x > 1$ can be written in interval notation as $(-\infty, -3] \cup (1, \infty)$.

The graph is the union of the graphs of $x \le -3$ and $x > 1$.

43.
$$x+3 < -2 \quad or \quad x+3 > 2$$
$$x+3-3 < -2-3 \quad or \quad x+3-3 > 2-3$$
$$x < -5 \quad or \quad x > -1$$

The solution set is $\{x | x < -5 \text{ or } x > -1\}$, or $(-\infty, -5) \cup (-1, \infty)$.

45.
$$2x-8 \le -3 \quad or \quad x-1 \ge 3$$
$$2x-8+8 \le -3+8 \quad or \quad x-1+1 \ge 3+1$$
$$2x \le 5 \quad or \quad x \ge 4$$
$$\frac{2x}{2} \le \frac{5}{2} \quad or \quad x \ge 4$$
$$x \le \frac{5}{2} \quad or \quad x \ge 4$$

The solution set is $\left\{x\middle| x \le \frac{5}{2} \text{ or } x \ge 4\right\}$, or $\left(-\infty, \frac{5}{2}\right] \cup [4, \infty)$.

47. $7x+4 \ge -17$ *or* $6x+5 \ge -7$
$$7x \ge -21 \quad or \quad 6x \ge -12$$
$$x \ge -3 \quad or \quad x \ge -2$$
The solution set is $\{x | x \ge -3\}$, or $[-3, \infty)$.

49.
$$7 > -4x+5 \quad or \quad 10 \le -4x+5$$
$$7-5 > -4x+5-5 \quad or \quad 10-5 \le -4x+5-5$$
$$2 > -4x \quad or \quad 5 \le -4x$$
$$\frac{2}{-4} < \frac{-4x}{-4} \quad or \quad \frac{5}{-4} \ge \frac{-4x}{-4}$$
$$-\frac{1}{2} < x \quad or \quad -\frac{5}{4} \ge x$$

The solution set is $\left\{x\middle| x \le -\frac{5}{4} \text{ or } x > -\frac{1}{2}\right\}$, or
$$\left(-\infty, -\frac{5}{4}\right] \cup \left(-\frac{1}{2}, \infty\right).$$

51. $3x-7 > -10$ *or* $5x+2 \le 22$
$$3x > -3 \quad or \quad 5x \le 20$$
$$x > -1 \quad or \quad x \le 4$$

All real numbers are solutions. In interval notation, the solution set is $(-\infty, \infty)$.

53.
$$-2x-2 < -6 \quad or \quad -2x-2 > 6$$
$$-2x-2+2 < -6+2 \quad or \quad -2x-2+2 > 6+2$$
$$-2x < -4 \quad or \quad -2x > 8$$
$$\frac{-2x}{-2} > \frac{-4}{-2} \quad or \quad \frac{-2x}{-2} < \frac{8}{-2}$$
$$x > 2 \quad or \quad x < -4$$

The solution set is $\{x | x < -4 \text{ or } x > 2\}$, or $(-\infty, -4) \cup (2, \infty)$.

55.
$$\frac{2}{3}x-14 < -\frac{5}{6} \quad or \quad \frac{2}{3}x-14 > \frac{5}{6}$$
$$6\left(\frac{2}{3}x-14\right) < 6\left(-\frac{5}{6}\right) \quad or \quad 6\left(\frac{2}{3}x-14\right) > 6 \cdot \frac{5}{6}$$
$$4x-84 < -5 \quad or \quad 4x-84 > 5$$
$$4x-84+84 < -5+84 \quad or \quad 4x-84+84 > 5+84$$
$$4x < 79 \quad or \quad 4x > 89$$
$$\frac{4x}{4} < \frac{79}{4} \quad or \quad \frac{4x}{4} > \frac{89}{4}$$
$$x < \frac{79}{4} \quad or \quad x > \frac{89}{4}$$

The solution set is $\left\{x\middle| x < \frac{79}{4} \text{ or } x > \frac{89}{4}\right\}$, or
$$\left(-\infty, \frac{79}{4}\right) \cup \left(\frac{89}{4}, \infty\right).$$

57.
$$\frac{2x-5}{6} \le -3 \quad or \quad \frac{2x-5}{6} \ge 4$$

$$6\left(\frac{2x-5}{6}\right) \le 6(-3) \quad or \quad 6\left(\frac{2x-5}{6}\right) \ge 6\cdot 4$$

$$2x-5 \le -18 \quad or \quad 2x-5 \ge 24$$

$$2x-5+5 \le -18+5 \quad or \quad 2x-5+5 \ge 24+5$$

$$2x \le -13 \quad or \quad 2x \ge 29$$

$$\frac{2x}{2} \le \frac{-13}{2} \quad or \quad \frac{2x}{2} \ge \frac{29}{2}$$

$$x \le -\frac{13}{2} \quad or \quad x \ge \frac{29}{2}$$

The solution set is $\left\{ x \middle| x \le -\dfrac{13}{2} \ or \ x \ge \dfrac{29}{2} \right\}$, or

$\left(-\infty, -\dfrac{13}{2}\right] \cup \left[\dfrac{29}{2}, \infty\right)$.

59. **Familiarize.** We will use the formula $P = 1 + \dfrac{d}{33}$.

Translate. We want to find those values of P for which

$$1 \le P \le 7$$

or

$$1 \le 1 + \frac{d}{33} \le 7.$$

Solve. We solve the inequality.

$$1 \le 1 + \frac{d}{33} \le 7$$

$$0 \le \frac{d}{33} \le 6$$

$$0 \le d \le 198$$

Check. We could do a partial check by substituting some values for d in the formula. The result checks.

State. The pressure is at least 1 atm and at most 7 atm for depths d in the set $\{d | 0 \text{ ft} \le d \le 198 \text{ ft}\}$.

61. **Familiarize.** Let $b =$ the number of beats per minute. Note that 10 sec $= 10$ sec $\times \dfrac{1 \text{ min}}{60 \text{ sec}} = \dfrac{10}{60} \times \dfrac{\text{sec}}{\text{sec}} \times 1 \text{ min} = \dfrac{1}{6}$ min. Then in 10 sec, or $\dfrac{1}{6}$ min, the woman should have between $\dfrac{1}{6} \cdot 138$ and $\dfrac{1}{6} \cdot 162$ beats.

Translate. We want to find the value of b for which

$$\frac{1}{6} \cdot 138 < b < \frac{1}{6} \cdot 162$$

Solve. We solve the inequality.

$$\frac{1}{6} \cdot 138 < b < \frac{1}{6} \cdot 162$$

$$23 < b < 27$$

Check. If the number of beats in 10 sec, or $\dfrac{1}{6}$ min, is between 23 and 27, then the number of beats per minute is between $6 \cdot 23$ and $6 \cdot 27$, or between 138 and 162. The answer checks.

State. The number of beats should be between 23 and 27.

63. **Familiarize.** We will use the formula $I = \dfrac{704.5W}{H^2}$, where $H = 73$. That is, $I = \dfrac{704.5W}{73^2}$.

Translate. We want to find those values of W for which

$$18.5 < I < 25$$

or

$$18.5 < \frac{704.5W}{73^2} < 25.$$

Solve. We solve the inequality.

$$18.5 < \frac{704.5W}{73^2} < 25$$

$$18.5 < \frac{704.5W}{5329} < 25$$

$$98,586.5 < 704.5W < 133,225$$

$$139.9 < W < 189.1 \qquad \text{Rounding}$$

Check. We could do a partial check by substituting some values for W in the formula. The result checks.

State. Weights between approximately 139.9 lb and 189.1 lb will allow Marv to keep his body mass index between 18.5 and 25. These weights are the values of W in the set $\{W | 139.9 \text{ lb} < W < 189.1 \text{ lb}\}$.

65. **Familiarize.** We will use the formula $c = \dfrac{ad}{a+12}$, where $a = 8$. That is, $c = \dfrac{8d}{8+12}$, or $c = \dfrac{8d}{20} = \dfrac{2d}{5}$.

Translate. We want to find the values of d for which

$$100 < C < 200$$

or

$$100 < \frac{2d}{5} < 200.$$

Solve. We solve the inequality.

$$100 < \frac{2d}{5} < 200$$

$$500 < 2d < 1000$$

$$250 < d < 500$$

Check. We could do a partial check by substituting some values for d in the formula. The result checks.

State. The equivalent adult dosage is between 250 mg and 500 mg. The dosages are the values of d in the set $\{d | 250 \text{ mg} < d < 500 \text{ mg}\}$.

67. Discussion and Writing Exercise

69. $|-3.2| = 3.2$

(The distance of -3.2 from 0 is 3.2.)

71. $|-5+7| = |2| = 2$

73. $(-2x^{-4}y^6)^5 = (-2)^5(x^{-4})^5(y^6)^5$
$$= -32x^{-20}y^{30}$$
$$= -\frac{32y^{30}}{x^{20}}$$

75. $\dfrac{-4a^5b^{-7}}{5a^{-12}b^8} = -\dfrac{4a^{5-(-12)}b^{-7-8}}{5}$
$$= -\frac{4a^{17}b^{-15}}{5}$$
$$= -\frac{4a^{17}}{5b^{15}}$$

77. $\left(\dfrac{56a^5b^{-6}}{28a^7b^{-8}}\right)^{-3} = (2a^{-2}b^2)^{-3}$

$\qquad\qquad\qquad = 2^{-3}a^{-2(-3)}b^{2(-3)}$

$\qquad\qquad\qquad = \dfrac{1}{2^3}a^6b^{-6}$

$\qquad\qquad\qquad = \dfrac{a^6}{8b^6}$

79. $x - 10 < 5x + 6 \le x + 10$

$\qquad -10 < 4x + 6 \le 10 \qquad$ Subtracting x

$\qquad -16 < 4x \le 4$

$\qquad -4 < x \le 1$

The solution set is $\{x| -4 < x \le 1\}$, or $(-4, 1]$.

81. $-\dfrac{2}{15} \le \dfrac{2}{3}x - \dfrac{2}{5} \le \dfrac{2}{15}$

$\qquad -\dfrac{2}{15} \le \dfrac{2}{3}x - \dfrac{6}{15} \le \dfrac{2}{15}$

$\qquad \dfrac{4}{15} \le \dfrac{2}{3}x \le \dfrac{8}{15}$

$\qquad \dfrac{3}{2} \cdot \dfrac{4}{15} \le \dfrac{3}{2} \cdot \dfrac{2}{3}x \le \dfrac{3}{2} \cdot \dfrac{8}{15}$

$\qquad \dfrac{2}{5} \le x \le \dfrac{4}{5}$

The solution set is $\left\{x \Big| \dfrac{2}{5} \le x \le \dfrac{4}{5}\right\}$, or $\left[\dfrac{2}{5}, \dfrac{4}{5}\right]$.

83. $3x < 4 - 5x < 5 + 3x$

$\qquad 0 < 4 - 8x < 5 \qquad$ Subtracting $3x$

$\qquad -4 < -8x < 1$

$\qquad \dfrac{1}{2} > x > -\dfrac{1}{8}$

The solution set is $\left\{x \Big| -\dfrac{1}{8} < x < \dfrac{1}{2}\right\}$, or $\left(-\dfrac{1}{8}, \dfrac{1}{2}\right)$.

85. $x + 4 < 2x - 6 \le x + 12$

$\qquad 4 < x - 6 \le 12 \qquad$ Subtracting x

$\qquad 10 < x \le 18$

The solution set is $\{x|10 < x \le 18\}$, or $(10, 18]$.

87. If $-b < -a$, then $-1(-b) > -1(-a)$, or $b > a$, or $a < b$. The statement is true.

89. Let $a = 5$, $c = 12$, and $b = 2$. Then $a < c$ and $b < c$, but $a \not< b$. The given statement is false.

91. The numbers in either the set of all rational numbers or the set of all irrational numbers are all real numbers, so the union is all real numbers.

There are no numbers common to the set of all rational numbers and the set of all irrational numbers, so the intersection is $\emptyset$.

Exercise Set 1.6

1. $|9x| = |9| \cdot |x| = 9|x|$

3. $|2x^2| = |2| \cdot |x^2|$

$\qquad\quad = 2|x^2|$

$\qquad\quad = 2x^2 \qquad$ Since x^2 is never negative

5. $|-2x^2| = |-2| \cdot |x^2|$

$\qquad\quad = 2|x^2|$

$\qquad\quad = 2x^2 \qquad$ Since x^2 is never negative

7. $|-6y| = |-6| \cdot |y| = 6|y|$

9. $\left|\dfrac{-2}{x}\right| = \dfrac{|-2|}{|x|} = \dfrac{2}{|x|}$

11. $\left|\dfrac{x^2}{-y}\right| = \dfrac{|x^2|}{|-y|}$

$\qquad\quad = \dfrac{x^2}{|-y|}$

$\qquad\quad = \dfrac{x^2}{|y|} \qquad$ The absolute value of the opposite of a number is the same as the absolute value of the number.

13. $\left|\dfrac{-8x^2}{2x}\right| = |-4x| = |-4| \cdot |x| = 4|x|$

15. $\left|\dfrac{4y^3}{-12y}\right| = \left|\dfrac{y^2}{-3}\right| = \dfrac{|y^2|}{|-3|} = \dfrac{y^2}{3}$

17. $|-8 - (-46)| = |38| = 38$, or

$\qquad |-46 - (-8)| = |-38| = 38$

19. $|36 - 17| = |19| = 19$, or

$\qquad |17 - 36| = |-19| = 19$

21. $|-3.9 - 2.4| = |-6.3| = 6.3$, or

$\qquad |2.4 - (-3.9)| = |6.3| = 6.3$

23. $|-5 - 0| = |-5| = 5$, or

$\qquad |0 - (-5)| = |5| = 5$

25. $|x| = 3$

$\qquad x = -3 \ or \ x = 3 \quad$ Absolute-value principle

The solution set is $\{-3, 3\}$.

27. $|x| = -3$

The absolute value of a number is always nonnegative. Therefore, the solution set is $\emptyset$.

29. $|q| = 0$

The only number whose absolute value is 0 is 0. The solution set is $\{0\}$.

31. $|x - 3| = 12$

$\qquad x - 3 = -12 \ or \ x - 3 = 12 \quad$ Absolute-value principle

$\qquad\qquad x = -9 \ or \qquad x = 15$

The solution set is $\{-9, 15\}$.

33. $|2x - 3| = 4$

$2x - 3 = -4 \quad or \quad 2x - 3 = 4 \qquad$ Absolute-value
$\qquad\qquad\qquad\qquad\qquad\qquad\qquad\qquad$ principle

$2x = -1 \quad or \qquad 2x = 7$

$x = -\dfrac{1}{2} \quad or \qquad x = \dfrac{7}{2}$

The solution set is $\left\{ -\dfrac{1}{2}, \dfrac{7}{2} \right\}$.

35. $|4x - 9| = 14$

$4x - 9 = -14 \quad or \quad 4x - 9 = 14$

$4x = -5 \quad or \qquad 4x = 23$

$x = -\dfrac{5}{4} \quad or \qquad x = \dfrac{23}{4}$

The solution set is $\left\{ -\dfrac{5}{4}, \dfrac{23}{4} \right\}$.

37. $|x| + 7 = 18$

$|x| + 7 - 7 = 18 - 7 \qquad$ Subtracting 7

$|x| = 11$

$x = -11 \quad or \quad x = 11 \quad$ Absolute-value principle

The solution set is $\{-11, 11\}$.

39. $574 = 283 + |t|$

$291 = |t| \qquad$ Subtracting 283

$t = -291 \quad or \quad t = 291 \qquad$ Absolute-value principle

The solution set is $\{-291, 291\}$.

41. $|5x| = 40$

$5x = -40 \quad or \quad 5x = 40$

$x = -8 \quad or \quad x = 8$

The solution set is $\{-8, 8\}$.

43. $|3x| - 4 = 17$

$|3x| = 21 \qquad$ Adding 4

$3x = -21 \quad or \quad 3x = 21$

$x = -7 \quad or \quad x = 7$

The solution set is $\{-7, 7\}$.

45. $7|w| - 3 = 11$

$7|w| = 14 \qquad$ Adding 3

$|w| = 2 \qquad$ Dividing by 7

$w = -2 \quad or \quad w = 2 \qquad$ Absolute-value principle

The solution set is $\{-2, 2\}$.

47. $\left| \dfrac{2x - 1}{3} \right| = 5$

$\dfrac{2x - 1}{3} = -5 \quad or \quad \dfrac{2x - 1}{3} = 5$

$2x - 1 = -15 \quad or \quad 2x - 1 = 15$

$2x = -14 \quad or \qquad 2x = 16$

$x = -7 \quad or \qquad x = 8$

The solution set is $\{-7, 8\}$.

49. $|m + 5| + 9 = 16$

$|m + 5| = 7 \qquad$ Subtracting 9

$m + 5 = -7 \quad or \quad m + 5 = 7$

$m = -12 \quad or \qquad m = 2$

The solution set is $\{-12, 2\}$.

51. $10 - |2x - 1| = 4$

$-|2x - 1| = -6 \qquad$ Subtracting 10

$|2x - 1| = 6 \qquad$ Multiplying by -1

$2x - 1 = -6 \quad or \quad 2x - 1 = 6$

$2x = -5 \quad or \qquad 2x = 7$

$x = -\dfrac{5}{2} \quad or \qquad x = \dfrac{7}{2}$

The solution set is $\left\{ -\dfrac{5}{2}, \dfrac{7}{2} \right\}$.

53. $|3x - 4| = -2$

The absolute value of a number is always nonnegative. The solution set is $\emptyset$.

55. $\left| \dfrac{5}{9} + 3x \right| = \dfrac{1}{6}$

$\dfrac{5}{9} + 3x = -\dfrac{1}{6} \quad or \quad \dfrac{5}{9} + 3x = \dfrac{1}{6}$

$3x = -\dfrac{13}{18} \quad or \qquad 3x = -\dfrac{7}{18}$

$x = -\dfrac{13}{54} \quad or \qquad x = -\dfrac{7}{54}$

The solution set is $\left\{ -\dfrac{13}{54}, -\dfrac{7}{54} \right\}$.

57. $|3x + 4| = |x - 7|$

$3x + 4 = x - 7 \quad or \quad 3x + 4 = -(x - 7)$

$2x + 4 = -7 \quad or \quad 3x + 4 = -x + 7$

$2x = -11 \quad or \quad 4x + 4 = 7$

$x = -\dfrac{11}{2} \quad or \qquad 4x = 3$

$x = -\dfrac{11}{2} \quad or \qquad x = \dfrac{3}{4}$

The solution set is $\left\{ -\dfrac{11}{2}, \dfrac{3}{4} \right\}$.

59. $|x + 3| = |x - 6|$

$x + 3 = x - 6 \quad or \quad x + 3 = -(x - 6)$

$3 = -6 \qquad or \quad x + 3 = -x + 6$

$3 = -6 \qquad or \qquad 2x = 3$

$3 = -6 \qquad or \qquad x = \dfrac{3}{2}$

The first equation has no solution. The second equation has $\dfrac{3}{2}$ as a solution. There is only one solution of the original equation. The solution set is $\left\{ \dfrac{3}{2} \right\}$.

61. $|2a + 4| = |3a - 1|$

$2a + 4 = 3a - 1 \quad or \quad 2a + 4 = -(3a - 1)$

$-a + 4 = -1 \quad or \quad 2a + 4 = -3a + 1$

$-a = -5 \quad or \quad 5a + 4 = 1$

$a = 5 \quad or \quad 5a = -3$

$a = 5 \quad or \quad a = -\dfrac{3}{5}$

The solution set is $\left\{5, -\dfrac{3}{5}\right\}$.

63. $|y - 3| = |3 - y|$

$y - 3 = 3 - y \quad or \quad y - 3 = -(3 - y)$

$2y - 3 = 3 \quad or \quad y - 3 = -3 + y$

$2y = 6 \quad or \quad -3 = -3$

$y = 3 \qquad\qquad \text{True for all real values of } y$

All real numbers are solutions.

65. $|5 - p| = |p + 8|$

$5 - p = p + 8 \quad or \quad 5 - p = -(p + 8)$

$5 - 2p = 8 \quad or \quad 5 - p = -p - 8$

$-2p = 3 \quad or \quad 5 = -8$

$p = -\dfrac{3}{2} \qquad\qquad \text{False}$

The solution set is $\left\{-\dfrac{3}{2}\right\}$.

67. $\left|\dfrac{2x - 3}{6}\right| = \left|\dfrac{4 - 5x}{8}\right|$

$\dfrac{2x - 3}{6} = \dfrac{4 - 5x}{8} \quad or \quad \dfrac{2x - 3}{6} = -\left(\dfrac{4 - 5x}{8}\right)$

$24\left(\dfrac{2x - 3}{6}\right) = 24\left(\dfrac{4 - 5x}{8}\right) \quad or \quad \dfrac{2x - 3}{6} = \dfrac{-4 + 5x}{8}$

$8x - 12 = 12 - 15x \quad or \quad 24\left(\dfrac{2x - 3}{6}\right) = 24\left(\dfrac{-4 + 5x}{8}\right)$

$23x - 12 = 12 \quad or \quad 8x - 12 = -12 + 15x$

$23x = 24 \quad or \quad -7x - 12 = -12$

$x = \dfrac{24}{23} \quad or \quad -7x = 0$

$\qquad\qquad\qquad\qquad x = 0$

The solution set is $\left\{\dfrac{24}{23}, 0\right\}$.

69. $\left|\dfrac{1}{2}x - 5\right| = \left|\dfrac{1}{4}x + 3\right|$

$\dfrac{1}{2}x - 5 = \dfrac{1}{4}x + 3 \quad or \quad \dfrac{1}{2}x - 5 = -\left(\dfrac{1}{4}x + 3\right)$

$\dfrac{1}{4}x - 5 = 3 \quad or \quad \dfrac{1}{2}x - 5 = -\dfrac{1}{4}x - 3$

$\dfrac{1}{4}x = 8 \quad or \quad \dfrac{3}{4}x - 5 = -3$

$x = 32 \quad or \quad \dfrac{3}{4}x = 2$

$x = 32 \quad or \quad x = \dfrac{8}{3}$

The solution set is $\left\{32, \dfrac{8}{3}\right\}$.

71. $|x| < 3$

$-3 < x < 3$

The solution set is $\{x| -3 < x < 3\}$, or $(-3, 3)$.

73. $|x| \geq 2$

$x \leq -2 \text{ or } x \geq 2$

The solution set is $\{x| x \leq -2 \text{ or } x \geq 2\}$, or $(-\infty, -2] \cup [2, \infty)$.

75. $|x - 1| < 1$

$-1 < x - 1 < 1$

$0 < x < 2$

The solution set is $\{x| 0 < x < 2\}$, or $(0, 2)$.

77. $5|x + 4| \leq 10$

$|x + 4| \leq 2 \qquad\qquad \text{Dividing by 5}$

$-2 \leq x + 4 \leq 2$

$-6 \leq x \leq -2 \qquad \text{Subtracting 4}$

The solution set is $\{x| -6 \leq x \leq -2\}$, or $[-6, -2]$.

79. $|2x - 3| \leq 4$

$-4 \leq 2x - 3 \leq 4$

$-1 \leq 2x \leq 7 \qquad\qquad \text{Adding 3}$

$-\dfrac{1}{2} \leq x \leq \dfrac{7}{2} \qquad\qquad \text{Dividing by 2}$

The solution set is $\left\{x \middle| -\dfrac{1}{2} \leq x \leq \dfrac{7}{2}\right\}$, or $\left[-\dfrac{1}{2}, \dfrac{7}{2}\right]$.

81. $|2y - 7| > 10$

$2y - 7 < -10 \quad or \quad 2y - 7 > 10$

$2y < -3 \quad or \quad 2y > 17 \qquad \text{Adding 7}$

$y < -\dfrac{3}{2} \quad or \quad y > \dfrac{17}{2} \qquad \text{Dividing by 2}$

The solution set is $\left\{y \middle| y < -\dfrac{3}{2} \text{ or } y > \dfrac{17}{2}\right\}$, or

$\left(-\infty, -\dfrac{3}{2}\right) \cup \left(\dfrac{17}{2}, \infty\right)$.

83. $|4x - 9| \geq 14$

$\qquad 4x - 9 \leq -14 \quad or \quad 4x - 9 \geq 14$

$\qquad\quad 4x \leq -5 \quad or \qquad 4x \geq 23$

$\qquad\qquad x \leq -\dfrac{5}{4} \quad or \qquad x \geq \dfrac{23}{4}$

The solution set is $\left\{ x \middle| x \leq -\dfrac{5}{4} \ or \ x \geq \dfrac{23}{4} \right\}$, or

$\left(-\infty, -\dfrac{5}{4} \right] \cup \left[\dfrac{23}{4}, \infty \right)$.

85. $|y - 3| < 12$

$\qquad -12 < y - 3 < 12$

$\qquad\quad -9 < y < 15 \qquad\qquad$ Adding 3

The solution set is $\{ y | -9 < y < 15 \}$, or $(-9, 15)$.

87. $|2x + 3| \leq 4$

$\qquad -4 \leq 2x + 3 \leq 4$

$\qquad -7 \leq 2x \leq 1 \qquad\qquad$ Subtracting 3

$\qquad -\dfrac{7}{2} \leq x \leq \dfrac{1}{2} \qquad$ Dividing by 2

The solution set is $\left\{ x \middle| -\dfrac{7}{2} \leq x \leq \dfrac{1}{2} \right\}$, or $\left[-\dfrac{7}{2}, \dfrac{1}{2} \right]$.

89. $|4 - 3y| > 8$

$\qquad 4 - 3y < -8 \quad or \quad 4 - 3y > 8$

$\qquad -3y < -12 \quad or \quad -3y > 4 \qquad$ Subtracting 4

$\qquad\quad y > 4 \qquad or \qquad y < -\dfrac{4}{3} \quad$ Dividing by -3

The solution set is $\left\{ y \middle| y < -\dfrac{4}{3} \ or \ y > 4 \right\}$, or

$\left(-\infty, -\dfrac{4}{3} \right) \cup (4, \infty)$.

91. $|9 - 4x| \geq 14$

$\qquad 9 - 4x \leq -14 \quad or \quad 9 - 4x \geq 14$

$\qquad\quad -4x \leq -23 \quad or \quad -4x \geq 5 \qquad$ Subtracting 9

$\qquad\qquad x \geq \dfrac{23}{4} \quad or \qquad x \leq -\dfrac{5}{4} \quad$ Dividing by -4

The solution set is $\left\{ x \middle| x \leq -\dfrac{5}{4} \ or \ x \geq \dfrac{23}{4} \right\}$ or

$\left(-\infty, -\dfrac{5}{4} \right] \cup \left[\dfrac{23}{4}, \infty \right)$.

93. $|3 - 4x| < 21$

$\qquad -21 < 3 - 4x < 21$

$\qquad -24 < -4x < 18 \qquad$ Subtracting 3

$\qquad\quad 6 > x > -\dfrac{9}{2} \qquad$ Dividing by -4 and simplifying

The solution set is $\left\{ x \middle| 6 > x > -\dfrac{9}{2} \right\}$, or

$\left\{ x \middle| -\dfrac{9}{2} < x < 6 \right\}$, or $\left(-\dfrac{9}{2}, 6 \right)$.

95. $\left| \dfrac{1}{2} + 3x \right| \geq 12$

$\qquad \dfrac{1}{2} + 3x \leq -12 \quad or \quad \dfrac{1}{2} + 3x \geq 12$

$\qquad\quad 3x \leq -\dfrac{25}{2} \quad or \qquad 3x \geq \dfrac{23}{2} \quad$ Subtracting $\dfrac{1}{2}$

$\qquad\quad x \leq -\dfrac{25}{6} \quad or \qquad x \geq \dfrac{23}{6} \quad$ Dividing by 3

The solution set is $\left\{ x \middle| x \leq -\dfrac{25}{6} \ or \ x \geq \dfrac{23}{6} \right\}$, or

$\left(-\infty, -\dfrac{25}{6} \right] \cup \left[\dfrac{23}{6}, \infty \right)$.

97. $\left| \dfrac{x - 7}{3} \right| < 4$

$\qquad -4 < \dfrac{x - 7}{3} < 4$

$\qquad -12 < x - 7 < 12 \qquad$ Multiplying by 3

$\qquad -5 < x < 19 \qquad\qquad$ Adding 7

The solution set is $\{ x | -5 < x < 19 \}$, or $(-5, 19)$.

99. $\left| \dfrac{2 - 5x}{4} \right| \geq \dfrac{2}{3}$

$\qquad \dfrac{2 - 5x}{4} \leq -\dfrac{2}{3} \quad or \quad \dfrac{2 - 5x}{4} \geq \dfrac{2}{3}$

$\qquad 2 - 5x \leq -\dfrac{8}{3} \quad or \quad 2 - 5x \geq \dfrac{8}{3} \quad$ Multiplying by 4

$\qquad\quad -5x \leq -\dfrac{14}{3} \quad or \quad -5x \geq \dfrac{2}{3} \quad$ Subtracting 2

$\qquad\quad x \geq \dfrac{14}{15} \quad or \qquad x \leq -\dfrac{2}{15} \quad$ Dividing by -5

The solution set is $\left\{ x \middle| x \leq -\dfrac{2}{15} \ or \ x \geq \dfrac{14}{15} \right\}$, or

$\left(-\infty, -\dfrac{2}{15} \right] \cup \left[\dfrac{14}{15}, \infty \right)$.

101. $|m + 5| + 9 \leq 16$

$\qquad |m + 5| \leq 7 \qquad\qquad$ Subtracting 9

$\qquad -7 \leq m + 5 \leq 7$

$\qquad -12 \leq m \leq 2$

The solution set is $\{ m | -12 \leq m \leq 2 \}$, or $[-12, 2]$.

103. $7 - |3 - 2x| \geq 5$

$\qquad -|3 - 2x| \geq -2 \qquad\qquad$ Subtracting 7

$\qquad |3 - 2x| \leq 2 \qquad\qquad$ Multiplying by -1

$\qquad -2 \leq 3 - 2x \leq 2$

$\qquad -5 \leq -2x \leq -1 \qquad$ Subtracting 3

$\qquad \dfrac{5}{2} \geq x \geq \dfrac{1}{2} \qquad\qquad$ Dividing by -2

The solution set is $\left\{ x \middle| \dfrac{5}{2} \geq x \geq \dfrac{1}{2} \right\}$, or $\left\{ x \middle| \dfrac{1}{2} \leq x \leq \dfrac{5}{2} \right\}$, or

$\left[\dfrac{1}{2}, \dfrac{5}{2} \right]$.

105. $\left|\dfrac{2x-1}{0.0059}\right| \leq 1$

$$-1 \leq \frac{2x-1}{0.0059} \leq 1$$

$$-0.0059 \leq 2x - 1 \leq 0.0059$$

$$0.9941 \leq 2x \leq 1.0059$$

$$0.49705 \leq x \leq 0.50295$$

The solution set is $\{x|0.49705 \leq x \leq 0.50295\}$, or $[0.49705, 0.50295]$.

107. Discussion and Writing Exercise

109. The <u>union</u> of two sets A and B is the collection of elements belonging to A and/or B.

111. The expression $x \geq q$ means x is <u>at least</u> q.

113. The <u>absolute value</u> of a number is its distance from zero on the number line.

115. Equations with the same solutions are called <u>equivalent</u> equations.

117. $|d - 6\text{ ft}| \leq \dfrac{1}{2}\text{ ft}$

$$-\frac{1}{2}\text{ ft} \leq d - 6\text{ ft} \leq \frac{1}{2}\text{ ft}$$

$$5\frac{1}{2}\text{ ft} \leq d \leq 6\frac{1}{2}\text{ ft}$$

The solution set is $\left\{d \middle| 5\frac{1}{2}\text{ ft} \leq d \leq 6\frac{1}{2}\text{ ft}\right\}$.

119. $|x + 5| = x + 5$

From the definition of absolute value, $|x+5| = x+5$ only when $x + 5 \geq 0$, or $x \geq -5$. The solution set is $\{x|x \geq -5\}$, or $[-5, \infty)$.

121. $|7x - 2| = x + 4$

From the definition of absolute value, we know $x + 4 \geq 0$, or $x \geq -4$. So we have $x \geq -4$ and

$$7x - 2 = x + 4 \quad or \quad 7x - 2 = -(x + 4)$$
$$6x = 6 \quad or \quad 7x - 2 = -x - 4$$
$$x = 1 \quad or \quad 8x = -2$$
$$x = 1 \quad or \quad x = -\frac{1}{4}$$

The solution set is $\left\{x \middle| x \geq -4 \text{ and } x = 1 \text{ or } x = -\frac{1}{4}\right\}$, or $\left\{1, -\frac{1}{4}\right\}$.

123. $|x - 6| \leq -8$

From the definition of absolute value we know that $|x-6| \geq 0$. Thus $|x - 6| \leq -8$ is false for all x. The solution set is $\emptyset$.

125. $|x + 5| > x$

The inequality is true for all $x < 0$ (because absolute value must be nonnegative). The solution set in this case is $\{x|x < 0\}$. If $x = 0$, we have $|0 + 5| > 0$, which is true.

The solution set in this case is $\{0\}$. If $x > 0$, we have the following:

$$x + 5 < -x \quad or \quad x + 5 > x$$
$$2x < -5 \quad or \quad 5 > 0$$
$$x < -\frac{5}{2} \quad or \quad 5 > 0$$

Although $x > 0$ and $x < -\dfrac{5}{2}$ yields no solution, $x > 0$ and $5 > 0$ (true for all x) yield the solution set $\{x|x > 0\}$ in this case. The solution set for the inequality is $\{x|x < 0\} \cup \{0\} \cup \{x|x > 0\}$, or all real numbers.

127. $-3 < x < 3$ is equivalent to $|x| < 3$.

129. $x \leq -6$ or or $x \geq 6$ is equivalent to $|x| \geq 6$.

131.
$$x < -8 \quad or \qquad x > 2$$
$$x + 3 < -5 \quad or \quad x + 3 > 5 \quad \text{Adding 3}$$
$$|x + 3| > 5$$

Chapter 1 Review Exercises

1.
$$-11 + y = -3$$
$$-11 + y + 11 = -3 + 11$$
$$y = 8$$

The number 8 checks, so it is the solution.

2.
$$-7x = -3$$
$$\frac{-7x}{-7} = \frac{-3}{-7}$$
$$x = \frac{3}{7}$$

The number $\dfrac{3}{7}$ checks, so it is the solution.

3.
$$-\frac{5}{3}x + \frac{7}{3} = -5$$
$$3\left(-\frac{5}{3}x + \frac{7}{3}\right) = 3(-5) \quad \text{Clearing fractions}$$
$$-5x + 7 = -15$$
$$-5x = -22$$
$$x = \frac{22}{5}$$

The number $\dfrac{22}{5}$ checks, so it is the solution.

4.
$$6(2x - 1) = 3 - (x + 10)$$
$$12x - 6 = 3 - x - 10$$
$$12x - 6 = -7 - x$$
$$13x - 6 = -7$$
$$13x = -1$$
$$x = -\frac{1}{13}$$

The number $-\dfrac{1}{13}$ checks, so it is the solution.

5.
$$2.4x + 1.5 = 1.02$$
$$100(2.4x + 1.5) = 100(1.02) \quad \text{Clearing decimals}$$
$$240x + 150 = 102$$
$$240x = -48$$
$$x = -0.2$$

The number -0.2 checks, so it is the solution.

6.
$$2(3 - x) - 4(x + 1) = 7(1 - x)$$
$$6 - 2x - 4x - 4 = 7 - 7x$$
$$2 - 6x = 7 - 7x$$
$$2 + x = 7$$
$$x = 5$$

The number 5 checks, so it is the solution.

7.
$$C = \frac{4}{11}d + 3$$
$$C - 3 = \frac{4}{11}d \qquad \text{Subtracting 3}$$
$$\frac{11}{4}(C - 3) = d \qquad \text{Multiplying by } \frac{11}{4}$$

8.
$$A = 2a - 3b$$
$$A - 2a = -3b$$
$$\frac{A - 2a}{-3} = b, \text{ or}$$
$$\frac{2a - A}{3} = b$$

9. *Familiarize*. Let $x =$ the smaller number. Then $x + 1 =$ the larger number.

***Translate*.**

$$\underbrace{\text{Smaller number}}_{x} \underset{+}{\text{ plus }} \underbrace{\text{larger number}}_{(x+1)} \underset{=}{\text{ is }} \underset{371}{371}.$$

***Solve*.** We solve the equation.
$$x + (x + 1) = 371$$
$$2x + 1 = 371$$
$$2x = 370$$
$$x = 185$$

If $x = 185$, then $x + 1 = 185 + 1 = 186$.

***Check*.** 185 and 186 are consecutive integers and $185 + 186 = 371$. The answer checks.

***State*.** The numbers on the markers are 185 and 186.

10. *Familiarize*. Let $x =$ the length of the longer piece of rope, in meters. Then $\frac{4}{5}x =$ the length of the shorter piece.

***Translate*.**

$$\underbrace{\text{Length of} \atop \text{longer piece}}_{x} \underset{+}{\text{ plus }} \underbrace{\text{Length of} \atop \text{shorter piece}}_{\frac{4}{5}x} \underset{=}{\text{ is }} \underbrace{27 \text{ m}}_{27}.$$

***Solve*.** We solve the equation.
$$x + \frac{4}{5}x = 27$$
$$\frac{9}{5}x = 27$$
$$x = \frac{5}{9} \cdot 27$$
$$x = 15$$

If $x = 15$, then $\frac{4}{5}x = \frac{4}{5} \cdot 15 = 12$.

***Check*.** 12 m is $\frac{4}{5}$ of 15 m and 12 m + 15 m = 27, so the answer checks.

***State*.** The lengths of the pieces are 15 m and 12 m.

11. *Familiarize*. Let $p =$ the former population.

***Translate*.**

$$\underbrace{\text{Former} \atop \text{population}}_{p} \underset{+}{\text{ plus }} \underset{12\%}{\text{12\%}} \underset{\cdot}{\text{ of }} \underbrace{\text{former} \atop \text{population}}_{p} \underset{=}{\text{ is }} \underset{179,200}{179,200}$$

***Solve*.** We solve the equation.
$$p + 12\% \cdot p = 179,200$$
$$p + 0.12p = 179,200$$
$$1.12p = 179,200$$
$$p = 160,000$$

***Check*.** 12% of 160,000 is 0.12(160,000) = 19,200 and 160,000 + 19,200 = 179,200. The answer checks.

***State*.** The former population is 160,000.

12. *Familiarize*. We will use the formula $d = rt$. Arnie's speed on the sidewalk is $3 + 6 = 9$ ft/sec.

***Translate*.**
$$d = rt$$
$$360 = 9t$$

***Solve*.** We solve the equation.
$$360 = 9t$$
$$40 = t$$

***Check*.** If Arnie travels at a speed of 9 ft/sec for 40 sec, he travels $9 \cdot 40 = 360$ ft. The answer checks.

***State*.** It will take Arnie 40 sec to walk the length of the sidewalk.

13. Interval is $[-8, 9)$.

14. Interval notation is $(-\infty, 40]$.

15.
$$x - 2 \leq -4$$
$$x \leq -2$$

The solution set is $(-\infty, -2]$.

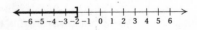

16. $x + 5 > 6$

$x > 1$

The solution set is $(1, \infty)$.

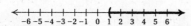

17. $a + 7 \le -14$

$a \le -21$

The solution set is $\{a \mid a \le -21\}$, or $(-\infty, -21]$.

18. $y - 5 \ge -12$

$y \ge -7$

The solution set is $\{y \mid y \ge -7\}$, or $[-7, \infty)$.

19. $4y > -16$

$y > -4$

The solution set is $\{y \mid y > -4\}$, or $(-4, \infty)$.

20. $-0.3y < 9$

$y > -30$ Reversing the inequality symbol

The solution set is $\{y \mid y > -30\}$, or $(-30, \infty)$.

21. $-6x - 5 < 13$

$-6x < 18$

$x > -3$ Reversing the inequality symbol

The solution set is $\{x \mid x > -3\}$, or $(-3, \infty)$.

22. $4y + 3 \le -6y - 9$

$10y + 3 \le -9$

$10y \le -12$

$y \le -\dfrac{6}{5}$

The solution set is $\left\{ y \mid y \le -\dfrac{6}{5} \right\}$, or $\left(-\infty, -\dfrac{6}{5} \right]$.

23. $-\dfrac{1}{2}x - \dfrac{1}{4} > \dfrac{1}{2} - \dfrac{1}{4}x$

$-\dfrac{1}{4}x - \dfrac{1}{4} > \dfrac{1}{2}$

$-\dfrac{1}{4}x > \dfrac{3}{4}$

$x < -3$ Reversing the inequality symbol

The solution set is $\{x \mid x < -3\}$, or $(-\infty, -3)$.

24. $0.3y - 8 < 2.6y + 15$

$-2.3y - 8 < 15$

$-2.3y < 23$

$y > -10$ Reversing the inequality symbol

The solution set is $\{y \mid y > -10\}$, or $(-10, \infty)$.

25. $-2(x - 5) \ge 6(x + 7) - 12$

$-2x + 10 \ge 6x + 42 - 12$

$-2x + 10 \ge 6x + 30$

$-8x + 10 \ge 30$

$-8x \ge 20$

$x \le -\dfrac{5}{2}$ Reversing the inequality symbol

The solution set is $\left\{ x \mid x \le -\dfrac{5}{2} \right\}$, or $\left(-\infty, -\dfrac{5}{2} \right]$.

26. **Familiarize**. Let $t =$ the length of time of the move, in hours. Then Musclebound Movers charges $85 + 40t$ and Champion Moving charges $60t$.

Translate.

Cost of Champion Moving	is more than	Cost of Musclebound Movers
$\downarrow$	$\downarrow$	$\downarrow$
$60t$	$>$	$85 + 40t$

Solve. We solve the inequality.

$60t > 85 + 40t$

$20t > 85$

$t > \dfrac{17}{4}$, or $4\dfrac{1}{4}$

Check. When $t = \dfrac{17}{4}$ hr, Champion Moving charges $60 \cdot \dfrac{17}{4}$, or \$255, and Musclebound Movers charges $85 + 40 \cdot \dfrac{17}{4} = 85 + 170 = \$255$. For a value of t greater than $4\dfrac{1}{4}$, say 5, Champion Moving charges $60 \cdot 5 = \$300$, and Musclebound Movers charges $85 + 40 \cdot 5 = 85 + 200 = \$285$. This partial check tells us that the answer is probably correct.

State. Champion Moving is more expensive for moves taking more than $4\dfrac{1}{4}$ hr. The solution set is $\left\{ t \mid t > 4\dfrac{1}{4} \text{ hr} \right\}$.

27. **Familiarize**. Let $x =$ the amount invested at 13%. Then $30{,}000 - x =$ the amount invested at 15%. The interest earned on the 13% investment is 13%x, or $0.13x$, and the interest earned on the 15% investment is 15%$(30{,}000 - x)$, or $0.15(30{,}000 - x)$.

Translate.

Interest on 13% investment	plus	Interest on 15% investment	is at least	\$4300
$\downarrow$	$\downarrow$	$\downarrow$	$\downarrow$	$\downarrow$
$0.13x$	$+$	$0.15(30{,}000 - x)$	$\ge$	4300

Solve. We solve the inequality.

$0.13x + 0.15(30{,}000 - x) \ge 4300$

$0.13x + 4500 - 0.15x \ge 4300$

$-0.02x + 4500 \ge 4300$

$-0.02x \ge -200$

$x \le 10{,}000$

Check. If $10,000 is invested at 13%, then the amount invested at 15% is $30,000 − $10,000, or $20,000. The interest earned is 0.13($10,000) + 0.15($20,000), or $1300 + $3000, or $4300. Then if less than $10,000 is invested at 13%, the interest earned will be more than $4300. This partial check shows that the answer is probably correct.

State. At most $10,000 can be invested at 13% interest.

28. Interval notation for $-2 \leq x < 5$ is $[-2, 5)$.

29. Interval notation for $x \leq -2$ or $x > 5$ is $(-\infty, -2] \cup (5, \infty)$.

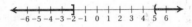

30. $2x - 5 < -7$ and $3x + 8 \geq 14$

$\qquad 2x < -2$ and $\quad 3x \geq 6$

$\qquad \quad x < -1$ and $\qquad x \geq 2$

The intersection of $\{x | x < -1\}$ and $\{x \geq 2\}$ is $\emptyset$, so the solution set is $\emptyset$.

31. $-4 < x + 3 \leq 5$

$\quad -7 < x \leq 2 \qquad$ Subtracting 2

The solution set is $\{x | -7 < x \leq 2\}$, or $(-7, 2]$,

32. $-15 < -4x - 5 < 0$

$\quad -10 < -4x < 5 \qquad$ Adding 5

$\quad \dfrac{5}{2} > x > -\dfrac{5}{4} \qquad$ Dividing by -4 and reversing the inequality symbol

The solution set is $\left\{ x \left| \dfrac{5}{2} > x > -\dfrac{5}{4} \right. \right\}$, or

$\left\{ x \left| -\dfrac{5}{4} < x < \dfrac{5}{2} \right. \right\}$, or $\left(-\dfrac{5}{4}, \dfrac{5}{2} \right)$.

33. $3x < -9$ or $-5x < -5$

$\quad x < -3$ or $\quad x > 1$

The solution set is $\{x | x < -3$ or $x > 1\}$, or $(-\infty, -3) \cup (1, \infty)$.

34. $2x + 5 < -17$ or $-4x + 10 \leq 34$

$\qquad 2x < -22$ or $\qquad -4x \leq 24$

$\qquad \quad x < -11$ or $\qquad \quad x \geq -6$

The solution set is $\{x | x < -11$ or $x \geq -6\}$, or $(-\infty, -11) \cup [-6, \infty)$.

35. $2x + 7 \leq -5$ or $x + 7 \geq 15$

$\qquad 2x \leq -12$ or $\qquad x \geq 8$

$\qquad \quad x \leq -6$ or $\qquad x \geq 8$

The solution set is $\{x | x \leq -6$ or $x \geq 8\}$, or $(-\infty, -6] \cup [8, \infty)$.

36. $\left| -\dfrac{3}{x} \right| = \left| \dfrac{-3}{x} \right| = \dfrac{|-3|}{|x|} = \dfrac{3}{|x|}$

37. $\left| \dfrac{2x}{y^2} \right| = \dfrac{|2x|}{|y^2|} = \dfrac{|2| \cdot |x|}{y^2} = \dfrac{2|x|}{y^2}$

38. $\left| \dfrac{12y}{-3y^2} \right| = \left| \dfrac{-4}{y} \right| = \dfrac{|-4|}{|y|} = \dfrac{4}{|y|}$

39. $|-23 - 39| = |-62| = 62$, or

$|39 - (-23)| = |39 + 23| = |62| = 62$

40. $|x| = 6$

$\quad x = -6$ or $x = 6 \qquad$ Absolute-value principle

The solution set is $\{-6, 6\}$.

41. $|x - 2| = 7$

$\quad x - 2 = -7$ or $\quad x - 2 = 7$

$\qquad \quad x = -5$ or $\qquad \quad x = 9$

The solution set is $\{-5, 9\}$.

42. $|2x + 5| = |x - 9|$

$\quad 2x + 5 = x - 9$ or $\quad 2x + 5 = -(x - 9)$

$\quad x + 5 = -9$ or $\quad 2x + 5 = -x - 9$

$\qquad \quad x = -14$ or $\quad 3x + 5 = -9$

$\qquad \quad x = -14$ or $\qquad \quad 3x = -14$

$\qquad \quad x = -14$ or $\qquad \quad x = -\dfrac{14}{3}$

The solution set is $\left\{ -14, -\dfrac{14}{3} \right\}$.

43. $|5x + 6| = -8$

The absolute value of a number is always nonnegative. Thus, the solution set is $\emptyset$.

44. $|2x + 5| < 12$

$\quad -12 < 2x + 5 < 12$

$\quad -17 < 2x < 7$

$\quad -\dfrac{17}{2} < x < \dfrac{7}{2}$

The solution set is $\left\{ x \left| -\dfrac{17}{2} < x < \dfrac{7}{2} \right. \right\}$, or $\left(-\dfrac{17}{2}, \dfrac{7}{2} \right)$.

45. $|x| \geq 3.5$

$\quad x \leq -3.5$ or $x \geq 3.5$

The solution set is $\{x | x \leq -3.5$ or $x \geq 3.5\}$, or $(-\infty, -3.5] \cup [3.5, \infty)$.

46. $|3x - 4| \geq 15$

$\quad 3x - 4 \leq -15$ or $\quad 3x - 4 \geq 15$

$\qquad 3x \leq -11$ or $\qquad 3x \geq 19$

$\qquad \quad x \leq -\dfrac{11}{3}$ or $\qquad \quad x \geq \dfrac{19}{3}$

The solution set is $\left\{ x \left| x \leq -\dfrac{11}{3} \right. \right.$ or $\left. x \geq \dfrac{19}{3} \right\}$, or

$\left(-\infty, -\dfrac{11}{3} \right] \cup \left[\dfrac{19}{3}, \infty \right)$.

47. $|x| < 0$

The absolute value of a number is always greater than or equal to 0, so the solution set is $\emptyset$.

48. $\{1, 2, 5, 6, 9\} \cap \{1, 3, 5, 9\} = \{1, 5, 9\}$

49. $\{1, 2, 5, 6, 9\} \cup \{1, 3, 5, 9\} = \{1, 2, 3, 5, 6, 9\}$

50. a) In 2008, $t = 2008 - 1988 = 20$.

$$R = -0.0433t + 10.49$$
$$R = -0.0433(20) + 10.49$$
$$R = -0.866 + 10.49$$
$$R \approx 9.62 \text{ sec}$$

b) Substitute 9.5374 for R and solve for t.

$$R = -0.0433t + 10.49$$
$$9.5374 = -0.0433t + 10.49$$
$$-0.9526 = -0.0433t$$
$$22 = t$$

The record will be 9.5374 sec 22 yr after 1988, or in 2010.

c) $\quad 10.15 < -0.0433t + 10.49 < 10.35$

$$-0.34 < -0.0433t < -0.14$$
$$7.85 > t > 3.23$$

The solution set is $\{t | 7.85 > t > 3.23\}$, or $\{t | 3.23 < t < 7.85\}$. This corresponds to about between 3 yr after 1988 to about 8 yr after 1988, or about 1991 to 1996.

51. *Discussion and Writing Exercise.*
(1) $-9(x + 2) = -9x - 18$, not $-9x + 2$. (2) This would be correct if (1) were correct except that the inequality symbol should not have been reversed. (3) If (2) were correct, the right-hand side would be -5, not 8. (4) The inequality symbol should be reversed. The correct solution is

$$7 - 9x + 6x < -9(x + 2) + 10x$$
$$7 - 9x + 6x < -9x - 18 + 10x$$
$$7 - 3x < x - 18$$
$$-4x < -25$$
$$x > \frac{25}{4}.$$

52. *Discussion and Writing Exercise.* "Solve" can mean to find all the replacements that make an equation or inequality true. It can also mean to express a formula as an equivalent equation with a given variable alone on one side.

53. $|2x + 5| \le |x + 3|$

$|2x + 5| \le x + 3 \quad or \quad |2x + 5| \le -(x + 3)$

First we solve $|2x + 5| \le x + 3$.

$$-(x + 3) \le 2x + 5 \quad and \quad 2x + 5 \le x + 3$$
$$-x - 3 \le 2x + 5 \quad and \qquad\quad x \le -2$$
$$-8 \le 3x \qquad and \qquad\quad x \le -2$$
$$-\frac{8}{3} \le x \qquad and \qquad\quad x \le -2$$

The solution set for this portion of the inequality is $\left\{x \,\middle|\, -\frac{8}{3} \le x \le -2\right\}$.

Now we solve $|2x + 5| \le -(x + 3)$.

$$-[-(x + 3)] \le 2x + 5 \quad and \quad 2x + 5 \le -(x + 3)$$
$$x + 3 \le 2x + 5 \quad and \quad 2x + 5 \le -x - 3$$
$$-2 \le x \qquad and \qquad 3x \le -8$$
$$-2 \le x \qquad and \qquad\quad x \le -\frac{8}{3}$$

The solution set for this portion of the inequality is $\emptyset$.

Then the solution set for the original inequality is $\left\{x \,\middle|\, -\frac{8}{3} \le x \le -2\right\} \cup \emptyset$, or $\left\{x \,\middle|\, -\frac{8}{3} \le x \le -2\right\}$. This is expressed in interval notation as $\left[-\frac{8}{3}, -2\right]$.

Chapter 1 Test

1. $\qquad x + 7 = 5$

$$x + 7 - 7 = 5 - 7$$
$$x = -2$$

The number -2 checks, so it is the solution.

2. $\qquad -12x = -8$

$$\frac{-12x}{-12} = \frac{-8}{-12}$$
$$x = \frac{2}{3}$$

The number $\frac{2}{3}$ checks, so it is the solution.

3. $\qquad x - \frac{3}{5} = \frac{2}{3}$

$$x - \frac{3}{5} + \frac{3}{5} = \frac{2}{3} + \frac{3}{5}$$
$$x = \frac{10}{15} + \frac{9}{15}$$
$$x = \frac{19}{15}$$

The number $\frac{19}{15}$ checks, so it is the solution.

4. $3y - 4 = 8$

$$3y = 12 \quad \text{Adding 4}$$
$$y = 4 \quad \text{Dividing by 3}$$

The number 4 checks, so it is the solution.

5. $1.7y - 0.1 = 2.1 - 0.3y$

$$2y - 0.1 = 2.1 \qquad \text{Adding } 0.3y$$
$$2y = 2.2 \qquad \text{Adding } 0.1$$
$$y = 1.1 \qquad \text{Dividing by 2}$$

The number 1.1 checks, so it is the solution.

6. $5(3x + 6) = 6 - (x + 8)$

$$15x + 30 = 6 - x - 8$$
$$15x + 30 = -2 - x$$
$$16x + 30 = -2$$
$$16x = -32$$
$$x = -2$$

The number -2 checks, so it is the solution.

7. $A = 3B - C$

$A + C = 3B$ Adding C

$\dfrac{A + C}{3} = B$ Dividing by 3

8. $m = n - nt$

$m = n(1 - t)$ Factoring out n

$\dfrac{m}{1 - t} = n$ Dividing by $1 - t$

9. *Familiarize*. Let $l =$ the length of the room, in feet. Then $\dfrac{2}{3}l =$ the width. Recall that the formula for the perimeter P of a rectangle with length l and width w is $P = 2l + 2w$.

***Translate*.** We substitute in the formula.

$P = 2l + 2w$

$48 = 2l + 2 \cdot \dfrac{2}{3}l$

***Solve*.** We solve the equation.

$$48 = 2l + 2 \cdot \dfrac{2}{3}l$$

$$48 = 2l + \dfrac{4}{3}l$$

$$48 = \dfrac{10}{3}l$$

$$\dfrac{3}{10} \cdot 48 = l$$

$$\dfrac{72}{5} = l, \text{ or}$$

$$14\dfrac{2}{5} = l$$

If $l = \dfrac{72}{5}$, then $\dfrac{2}{3}l = \dfrac{2}{3} \cdot \dfrac{72}{5} = \dfrac{48}{5}$, or $9\dfrac{3}{5}$.

***Check*.** $9\dfrac{3}{5}$ ft is two-thirds of $14\dfrac{2}{5}$ ft and $2 \cdot 14\dfrac{2}{5} + 2 \cdot 9\dfrac{3}{5} =$ $2 \cdot \dfrac{72}{5} + 2 \cdot \dfrac{48}{5} = \dfrac{144}{5} + \dfrac{96}{5} = \dfrac{240}{5} = 48$. The answer checks.

***State*.** The length of the room is $14\dfrac{2}{5}$ ft and the width is $9\dfrac{3}{5}$ ft.

10. *Familiarize*. Let $c =$ the number of copies the firm can make. The rental cost for 3 months is $3 \cdot \$240$, or $\$720$, and the cost of the copies is $1.8¢ \cdot c$, or $\$0.018c$.

***Translate*.**

$\underbrace{\text{Rental cost}}$ plus $\underbrace{\text{copy cost}}$ $\underbrace{\substack{\text{is no} \\ \text{more than}}}$ $\$1500$

$\quad\downarrow\qquad\quad\downarrow\qquad\quad\downarrow\qquad\qquad\downarrow\qquad\quad\downarrow$

$\quad720\qquad+\quad0.018c\qquad\quad\leq\qquad\quad1500$

***Solve*.** We solve the inequality.

$$720 + 0.018c \leq 1500$$

$$0.018c \leq 780$$

$$c \leq 43,333.\overline{3}$$

***Check*.** Since partial copies cannot be made, we check 43,333. If 43,333 copies are made, the total cost is $\$720 +$ $\$0.018(43,333) \approx \$1499.99$. For more than 43,333 copies,

say 43,334, the total cost is $\$720 + \$0.018(43,334) \approx$ $\$1500.01$. The answer checks.

***State*.** The law firm can make at most 43,333 copies.

11. *Familiarize*. Let $p =$ the former population.

***Translate*.**

$\underbrace{\substack{\text{Former} \\ \text{population}}}$ minus 12% of $\underbrace{\substack{\text{Former} \\ \text{population}}}$ is 158,400.

$\qquad\downarrow\qquad\quad\downarrow\quad\downarrow\quad\downarrow\qquad\quad\downarrow\qquad\quad\downarrow\quad\downarrow$

$\qquad p\qquad\quad-\quad 12\% \cdot\qquad\quad p\qquad\quad = \;158,400$

***Solve*.** We solve the equation.

$$p - 12\% \cdot p = 158,400$$

$$p - 0.12p = 158,400$$

$$0.88p = 158,400$$

$$p = 180,000$$

***Check*.** 12% of 180,000 is $0.12(180,000) = 21,600$ and $180,000 - 21,600 = 158,400$ so the answer checks.

***State*.** The former population of Baytown was 180,000.

12. *Familiarize*. Let $x =$ the measure of the smallest angle. Then $x + 1$ and $x + 2$ represent the measures of the other two angles. Recall that the sum of the measures of the angles in a triangle is $180°$.

***Translate*.**

$\underbrace{\text{The sum of the measures}}$ is $180°$

$\qquad\qquad\downarrow\qquad\qquad\qquad\downarrow\quad\downarrow$

$x + (x + 1) + (x + 2)\quad = \quad 180$

***Solve*.** We solve the equation.

$$x + (x + 1) + (x + 2) = 180$$

$$3x + 3 = 180$$

$$3x = 177$$

$$x = 59$$

If $x = 59$, then $x + 1 = 59 + 1 = 60$ and $x + 2 = 59 + 2 = 61$.

***Check*.** The numbers 59, 60, and 61 are consecutive integers and $59° + 60° + 61° = 180°$. The answer checks.

***State*.** The measures of the angles are $59°$, $60°$, and $61°$.

13. First we will find how long it takes the boat to travel 36 mi downstream.

***Familiarize*.** We will use the formula $d = rt$. Let $t =$ the time, in hours, it will take the boat to travel 36 mi downstream. The speed of the boat traveling downstream is $12 + 3$, or 15 mph.

***Translate*.**

$d = rt$

$36 = 15t$

***Solve*.** We solve the equation.

$$36 = 15t$$

$$\dfrac{12}{5} = t, \text{ or}$$

$$2\dfrac{2}{5} = t$$

Check. If the boat travels at 15 mph for $\frac{12}{5}$ hr, it travels $15 \cdot \frac{12}{5}$, or 36 mi. The answer checks.

State. It will take the boat $2\frac{2}{5}$ hr to travel 36 mi downstream.

Now we find how long it will take the boat to travel 36 mi upstream.

Familiarize. We will use the formula $d = rt$. Let $t =$ the time, in hours, it will take the boat to travel 36 mi upstream. The speed of the boat traveling upstream is $12 - 3$, or 9 mph.

Translate.
$$d = rt$$
$$36 = 9t$$

Solve. We solve the equation.
$$36 = 9t$$
$$4 = t$$

Check. If the boat travels at 9 mph for 4 hr, it travels $9 \cdot 4$, or 36 mi. The answer checks.

State. It will take the boat 4 hr to travel 36 mi upstream.

14. Interval notation for $\{x| -3 < x \le 2\}$ is $(-3, 2]$.

15. Interval notation is $(-4, \infty)$.

16. $\quad x - 2 \le 4$

$\qquad x \le 6 \quad$ Adding 2

The solution set is $\{x|x \le 6\}$, or $(-\infty, 6]$.

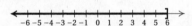

17. $\quad -4y - 3 \ge 5$

$\qquad -4y \ge 8$

$\qquad y \le -2 \quad$ Reversing the inequality symbol

The solution set is $\{y|y \le -2\}$, or $(-\infty, -2]$.

18. $\quad x - 4 \ge 6$

$\qquad x \ge 10 \quad$ Adding 4

The solution set is $\{x|x \ge 10\}$, or $[10, \infty)$.

19. $\quad -0.6y < 30$

$\qquad y > -50 \quad$ Reversing the inequality symbol

The solution set is $\{y|y > -50\}$, or $(-50, \infty)$.

20. $\quad 3a - 5 \le -2a + 6$

$\qquad 5a - 5 \le 6$

$\qquad 5a \le 11$

$\qquad a \le \frac{11}{5}$

The solution set is $\left\{a\middle|a \le \frac{11}{5}\right\}$, or $\left(-\infty, \frac{11}{5}\right]$.

21. $\quad -5y - 1 > -9y + 3$

$\qquad 4y - 1 > 3$

$\qquad 4y > 4$

$\qquad y > 1$

The solution set is $\{y|y > 1\}$, or $(1, \infty)$.

22. $\quad 4(5 - x) < 2x + 5$

$\qquad 20 - 4x < 2x + 5$

$\qquad 20 - 6x < 5$

$\qquad -6x < -15$

$\qquad x > \frac{5}{2}$

The solution set is $\left\{x\middle|x > \frac{5}{2}\right\}$, or $\left(\frac{5}{2}, \infty\right)$.

23. $\quad -8(2x + 3) + 6(4 - 5x) \ge 2(1 - 7x) - 4(4 + 6x)$

$\qquad -16x - 24 + 24 - 30x \ge 2 - 14x - 16 - 24x$

$\qquad -46x \ge -14 - 38x$

$\qquad -8x \ge -14$

$\qquad x \le \frac{7}{4}$

The solution set is $\left\{x\middle|x \le \frac{7}{4}\right\}$, or $\left(-\infty, \frac{7}{4}\right]$.

24. Familiarize. Let $t =$ the length of time of the move, in hours. Then Motivated Movers charges $105 + 30t$ and Quick-Pak Moving charges $80t$.

Translate.

Cost of Quick-Pak	is more than	Cost of Motivated Movers
$80t$	$>$	$105 + 30t$

Solve. We solve the inequality.

$\qquad 80t > 105 + 30t$

$\qquad 50t > 105$

$\qquad t > \frac{21}{10}$, or $2\frac{1}{10}$

Check. When $t = \frac{21}{10}$ hr, Motivated Movers charges $105 + 30 \cdot \frac{21}{10}$, or \$168, and Quick-Pak charges $80 \cdot \frac{21}{10}$, or \$168. For a value of t greater than $2\frac{1}{10}$, say 3, Motivated Movers charges $105 + 30 \cdot 3$, or \$195, and Quick-Pak charges $80 \cdot 3$, or \$240, so Quick-Pak is more expensive. This partial check tells us that the answer is probably correct.

State. Quick-Pak is more expensive for moves more than $2\frac{1}{10}$ hr. The solution set is $\left\{t\middle|t > 2\frac{1}{10} \text{ hr}\right\}$.

25. *Familiarize.* We will use the formula $P = 1 + \dfrac{d}{33}$.

Translate. We want to find those values of P for which

$$2 \le P \le 8$$

or

$$2 \le 1 + \frac{d}{33} \le 8.$$

Solve. We solve the inequality.

$$2 \le 1 + \frac{d}{33} \le 8$$

$$1 \le \frac{d}{33} \le 7$$

$$33 \le d \le 231$$

Check. We could do a partial check by substituting some values for d in the formula. The result checks.

State. The pressure is at least 2 atm and at most 8 atm for depths d in the set $\{d | 33 \text{ ft} \le d \le 231 \text{ ft}\}$.

26. Interval notation for $-3 \le x \le 4$ is $[-3, 4]$.

27. Interval notation for $x < -3 \ or \ x > 4$ is $(-\infty, -3) \cup (4, \infty)$.

28. $5 - 2x \le 1 \quad and \quad 3x + 2 \ge 14$

$\qquad -2x \le -4 \quad and \qquad 3x \ge 12$

$\qquad\quad x \ge 2 \quad and \qquad\quad x \ge 4$

The intersection of $\{x | x \ge 2\}$ *and* $\{x | x \ge 4\}$, is $\{x | x \ge 4\}$, or $[4, \infty)$.

29. $-3 < x - 2 < 4$

$\qquad -1 < x < 6 \qquad$ Adding 2

The solution set is $\{x | -1 < x < 6\}$, or $(-1, 6)$.

30. $-11 \le -5x - 2 < 0$

$\qquad -9 \le -5x < 2$

$\qquad \dfrac{9}{5} \ge x > -\dfrac{2}{5}$

The solution set is $\left\{x \left| \dfrac{9}{5} \ge x > -\dfrac{2}{5}\right.\right\}$, or

$\left\{x \left| -\dfrac{2}{5} < x \le \dfrac{9}{5}\right.\right\}$, or $\left(-\dfrac{2}{5}, \dfrac{9}{5}\right]$.

31. $-3x > 12 \ or \ 4x > -10$

$\qquad x < -4 \ or \quad x > -\dfrac{5}{2}$

The solution set is $\left\{x \left| x < -4 \ or \ x > -\dfrac{5}{2}\right.\right\}$, or

$(-\infty, -4) \cup \left(-\dfrac{5}{2}, \infty\right)$.

32. $x - 7 \le -5 \ or \ x - 7 \ge -10$

$\qquad x \le 2 \quad or \qquad x \ge -3$

The union of $(-\infty, 2]$ and $[-3, \infty)$ is the set of all real numbers, or $(-\infty, \infty)$.

33. $3x - 2 < 7 \ or \ x - 2 > 4$

$\qquad 3x < 9 \ or \qquad x > 6$

$\qquad\ x < 3 \ or \qquad x > 6$

The solution set is $\{x | x < 3 \ or \ x > 6\}$, or $(-\infty, 3) \cup (6, \infty)$.

34. $\left|\dfrac{7}{x}\right| = \dfrac{|7|}{|x|} = \dfrac{7}{|x|}$

35. $\left|\dfrac{-6x^2}{3x}\right| = |-2x| = |-2| \cdot |x| = 2|x|$

36. $|4.8 - (-3.6)| = |4.8 + 3.6| = |8.4| = 8.4$, or

$\qquad |-3.6 - 4.8| = |-8.4| = 8.4$

37. $\{1, 3, 5, 7, 9\} \cap \{3, 5, 11, 13\} = \{3, 5\}$

38. $\{1, 3, 5, 7, 9\} \cup \{3, 5, 11, 13\} = \{1, 3, 5, 7, 9, 11, 13\}$

39. $|x| = 9$

$\qquad x = -9 \ or \ x = 9 \quad$ Absolute-value principle

The solution set is $\{-9, 9\}$.

40. $|x - 3| = 9$

$\qquad x - 3 = -9 \ or \ x - 3 = 9$

$\qquad\quad x = -6 \ or \qquad x = 12$

The solution set is $\{-6, 12\}$.

41. $|x + 10| = |x - 12|$

$\qquad x + 10 = x - 12 \ or \ x + 10 = -(x - 12)$

$\qquad\quad 10 = -12 \quad or \ x + 10 = -x + 12$

$\qquad\quad 10 = -12 \quad or \qquad 2x = 2$

$\qquad\quad 10 = -12 \quad or \qquad\ x = 1$

The first equation has no solution. The solution of the second equation is 1, so the solution set is $\{1\}$.

42. $|2 - 5x| = -10$

The absolute value of a number is always nonnegative. Thus, the solution set is $\emptyset$.

43. $|4x - 1| < 4.5$

$\qquad -4.5 < 4x - 1 < 4.5$

$\qquad -3.5 < 4x < 5.5$

$\qquad -0.875 < x < 1.375$

The solution set is $\{x | -0.875 < x < 1.375\}$, or $(-0.875, 1.375)$. This could also be expressed as

$\left\{x \left| -\dfrac{7}{8} < x < \dfrac{11}{8}\right.\right\}$, or $\left(-\dfrac{7}{8}, \dfrac{11}{8}\right)$.

44. $|x| > 3$

$\qquad x < -3 \ or \ x > 3$

The solution set is $\{x | x < -3 \ or \ x > 3\}$, or $(-\infty, -3) \cup (3, \infty)$.

45.
$$\left|\frac{6-x}{7}\right| \le 15$$

$$-15 \le \frac{6-x}{7} \le 15$$

$$-105 \le 6-x \le 105 \quad \text{Multiplying by 7}$$

$$-111 \le -x \le 99$$

$$111 \ge x \ge -99$$

The solution set is $\{x | 111 \ge x \ge -99\}$, or $\{x | -99 \le x \le 111\}$, or $[-99, 111]$.

46. $|-5x - 3| \ge 10$

$$-5x - 3 \le -10 \quad or \quad -5x - 3 \ge 10$$

$$-5x \le -7 \quad or \quad -5x \ge 13$$

$$x \ge \frac{7}{5} \quad or \quad x \le -\frac{13}{5}$$

The solution set is $\left\{x \middle| x \le -\frac{13}{5} \; or \; x \ge \frac{7}{5}\right\}$, or $\left(-\infty, -\frac{13}{5}\right] \cup \left[\frac{7}{5}, \infty\right)$.

47. $|3x - 4| \le -3$

The absolute value of a number is always nonnegative, so $|3x - 4|$ cannot be less than -3. Thus, the solution set is $\emptyset$.

48. $7x < 8 - 3x < 6 + 7x$

$$7x < 8 - 3x \quad and \quad 8 - 3x < 6 + 7x$$

$$10x < 8 \quad and \quad -10x < -2$$

$$x < \frac{4}{5} \quad and \quad x > \frac{1}{5}$$

The intersection of $\left\{x \middle| x < \frac{4}{5}\right\}$ and $\left\{x \middle| x > \frac{1}{5}\right\}$ is $\left\{x \middle| \frac{1}{5} < x < \frac{4}{5}\right\}$, or $\left(\frac{1}{5}, \frac{4}{5}\right)$.

Chapter 2

Graphs, Functions, and Applications

Exercise Set 2.1

1. $A(4, 1)$ is 4 units right and 1 unit up.

   $B(2, 5)$ is 2 units right and 5 units up.

   $C(0, 3)$ is 0 units left or right and 3 units up.

   $D(0, -5)$ is 0 units left or right and 5 units down.

   $E(6, 0)$ is 6 units right and 0 units up or down.

   $F(-3, 0)$ is 3 units left and 0 units up or down.

   $G(-2, -4)$ is 2 units left and 4 units down.

   $H(-5, 1)$ is 5 units left and 1 unit up.

   $J(-6, 6)$ is 6 units left and 6 units up.

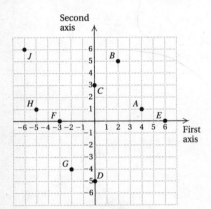

3.

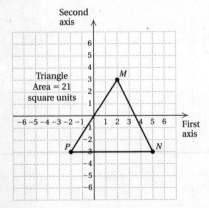

A triangle is formed. The area of a triangle is found by using the formula $A = \frac{1}{2}bh$. In this triangle the base and height are respectively 7 units and 6 units.

$$A = \frac{1}{2}bh = \frac{1}{2} \cdot 7 \cdot 6 = \frac{42}{2} = 21 \text{ square units}$$

5. We substitute 1 for x and -1 for y (alphabetical order of variables).

$$\begin{array}{c|c} y = 2x - 3 \\ \hline -1 \; ? \; 2 \cdot 1 - 3 \\ \; \big| \; -1 & \text{TRUE} \end{array}$$

Thus, $(1, -1)$ is a solution of the equation.

7. We substitute 3 for x and 5 for y (alphabetical order of variables).

$$\begin{array}{c|c} 4x - y = 7 \\ \hline 4 \cdot 3 - 5 \; ? \; 7 \\ 12 - 5 \; \big| \\ 7 \; \big| & \text{TRUE} \end{array}$$

Thus, $(3, 5)$ is a solution of the equation.

9. We substitute 0 for a and $\frac{3}{5}$ for b (alphabetical order of variables).

$$\begin{array}{c|c} 2a + 5b = 7 \\ \hline 2 \cdot 0 + 5 \cdot \frac{3}{5} \; ? \; 7 \\ 0 + 3 \; \big| \\ 3 \; \big| & \text{FALSE} \end{array}$$

Thus, $\left(0, \frac{3}{5}\right)$ is not a solution of the equation.

11. To show that a pair is a solution, we substitute, replacing x with the first coordinate and y with the second coordinate in each pair.

$$\begin{array}{c|c} y = 4 - x \\ \hline 5 \; ? \; 4 - (-1) \\ \big| \; 4 + 1 \\ \big| \; 5 & \text{TRUE} \end{array} \qquad \begin{array}{c|c} y = 4 - x \\ \hline 1 \; ? \; 4 - 3 \\ \big| \; 1 & \text{TRUE} \end{array}$$

In each case the substitution results in a true equation. Thus, $(-1, 5)$ and $(3, 1)$ are both solutions of $y = 4 - x$. We plot these points and sketch the line passing through them.

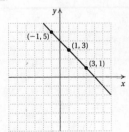

The line appears to pass through $(1, 3)$ also. We check to determine if $(1, 3)$ is a solution.

$$\frac{y = 4 - x}{3 \ ? \ 4 - 1}$$
$$\left. \ \right| \ 3 \qquad \text{TRUE}$$

Thus, $(1, 3)$ is another solution. There are other correct answers, including $(-2, 6)$, $(0, 4)$, $(2, 2)$, $(4, 0)$, and $(5, -1)$.

13. To show that a pair is a solution, we substitute, replacing x with the first coordinate and y with the second coordinate in each pair.

$$\frac{3x + y = 7}{3 \cdot 2 + 1 \ ? \ 7} \qquad \frac{3x + y = 7}{3 \cdot 4 - 5 \ ? \ 7}$$
$$\begin{array}{c|c} 6 + 1 & \\ 7 & \text{TRUE} \end{array} \qquad \begin{array}{c|c} 12 - 5 & \\ 7 & \text{TRUE} \end{array}$$

In each case the substitution results in a true equation. Thus, $(2, 1)$ and $(4, -5)$ are both solutions of $3x + y = 7$. We plot these points and sketch the line passing through them.

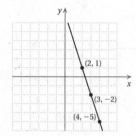

The line appears to pass through $(1, 4)$ also. We check to determine if $(1, 4)$ is a solution of $3x + y = 7$.

$$\frac{3x + y = 7}{3 \cdot 1 + 4 \ ? \ 7}$$
$$\begin{array}{c|c} 3 + 4 & \\ 7 & \text{TRUE} \end{array}$$

Thus, $(1, 4)$ is another solution. There are other correct answers, including $(3, -2)$.

15. To show that a pair is a solution, we substitute, replacing x with the first coordinate and y with the second coordinate in each pair.

$$\frac{6x - 3y = 3}{6 \cdot 1 - 3 \cdot 1 \ ? \ 3}$$
$$\begin{array}{c|c} 6 - 3 & \\ 3 & \text{TRUE} \end{array}$$

$$\frac{6x - 3y = 3}{6(-1) - 3(-3) \ ? \ 3}$$
$$\begin{array}{c|c} -6 + 9 & \\ 3 & \text{TRUE} \end{array}$$

In each case the substitution results in a true equation. Thus, $(1, 1)$ and $(-1, -3)$ are both solutions of $6x - 3y = 3$. We plot these points and sketch the line passing through them.

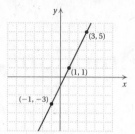

The line appears to pass through $(0, -1)$ also. We check to determine if $(0, -1)$ is a solution of $6x - 3y = 3$.

$$\frac{6x - 3y = 3}{6 \cdot 0 - 3(-1) \ ? \ 3}$$
$$\begin{array}{c|c} 0 + 3 & \\ 3 & \text{TRUE} \end{array}$$

Thus, $(0, -1)$ is another solution. There are other correct answers including $(-2, -5)$, $(2, 3)$, and $(3, 5)$.

17. $y = x - 1$

We find some ordered pairs that are solutions.

When $x = -2$, $y = -2 - 1 = -3$.

When $x = -1$, $y = -1 - 1 = -2$.

When $x = 0$, $y = 0 - 1 = -1$.

When $x = 1$, $y = 1 - 1 = 0$.

When $x = 2$, $y = 2 - 1 = 1$.

When $x = 3$, $y = 3 - 1 = 2$.

x	y
-2	-3
-1	-2
0	-1
1	0
2	1
3	2

Plot these points, draw the line they determine, and label it $y = x - 1$.

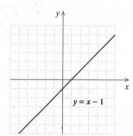

19. $y = x$

We find some ordered pairs that are solutions, plot them, and draw and label the line.

When $x = -2$, $y = -2$.

When $x = -1$, $y = -1$.

When $x = 0$, $y = 0$.

When $x = 1$, $y = 1$.

When $x = 2$, $y = 2$.

When $x = 3$, $y = 3$.

x	y
-2	-2
-1	-1
0	0
1	1
2	2
3	3

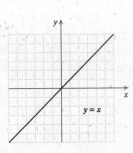

21. $y = \dfrac{1}{4}x$

We find some ordered pairs that are solutions, using multiples of 4 for x to avoid fractions. Then we plot these points and draw and label the line.

When $x = -4$, $y = \dfrac{1}{4}(-4) = -1$.

When $x = 0$, $y = \dfrac{1}{4} \cdot 0 = 0$.

When $x = 4$, $y = \dfrac{1}{4} \cdot 4 = 1$.

x	y	(x, y)
-4	-1	$(-4, -1)$
0	0	$(0, 0)$
4	1	$(4, 1)$

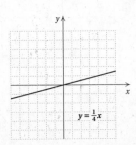

23. $y = 3 - x$

We find some ordered pairs that are solutions, plot them, and draw and label the line.

When $x = -2$, $y = 3 - (-2) = 3 + 2 = 5$.

When $x = 1$, $y = 3 - 1 = 2$.

When $x = 5$, $y = 3 - 5 = -2$.

x	y	(x, y)
-2	5	$(-2, 5)$
1	2	$(1, 2)$
5	-2	$(5, -2)$

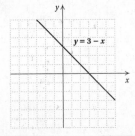

25. $y = 5x - 2$

We find some ordered pairs that are solutions, plot them, and draw and label the line.

When $x = -1$, $y = 5(-1) - 2 = -5 - 2 = -7$.

When $x = 0$, $y = 5 \cdot 0 - 2 = -2$.

When $x = 1$, $y = 5 \cdot 1 - 2 = 5 - 2 = 3$.

x	y	(x, y)
-1	-7	$(-1, -7)$
0	-2	$(0, -2)$
1	3	$(1, 3)$

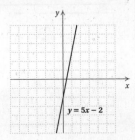

27. $y = \dfrac{1}{2}x + 1$

We find some ordered pairs that are solutions, using even numbers for x to avoid fractions. Then we plot these points and draw and label the line.

When $x = -4$, $y = \dfrac{1}{2}(-4) + 1 = -2 + 1 = -1$.

When $x = 0$, $y = \dfrac{1}{2} \cdot 0 + 1 = 1$.

When $x = 4$, $y = \dfrac{1}{2} \cdot 4 + 1 = 2 + 1 = 3$.

x	y	(x, y)
-4	-1	$(-4, -1)$
0	1	$(0, 1)$
4	3	$(4, 3)$

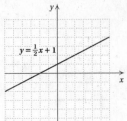

29. $x + y = 5$

First we solve for y.

$$x + y = 5$$
$$y = 5 - x$$

Now we find some ordered pairs that are solutions, plot them, and draw and label the line.

When $x = 0$, $y = 5 - 0 = 5$.

When $x = 2$, $y = 5 - 2 = 3$.

When $x = 5$, $y = 5 - 5 = 0$.

x	y	(x, y)
0	5	$(0, 5)$
2	3	$(2, 3)$
5	0	$(5, 0)$

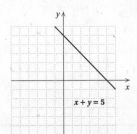

31. $y = -\dfrac{5}{3}x - 2$

We find some ordered pairs that are solutions, using multiples of 3 for x to avoid fractions. Then we plot these points and draw and label the line.

When $x = -3$, $y = -\dfrac{5}{3}(-3) - 2 = 5 - 2 = 3$.

When $x = 0$, $y = -\dfrac{5}{3} \cdot 0 - 2 = -2$.

When $x = 3$, $y = -\dfrac{5}{3} \cdot 3 - 2 = -5 - 2 = -7$.

x	y	(x, y)
-3	3	$(-3, 3)$
0	-2	$(0, -2)$
3	-7	$(3, -7)$

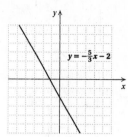

33. $x + 2y = 8$

First we solve for y.

$$x + 2y = 8$$
$$2y = 8 - x$$
$$\frac{1}{2} \cdot 2y = \frac{1}{2}(8 - x)$$
$$y = \frac{1}{2} \cdot 8 - \frac{1}{2} \cdot x$$
$$y = 4 - \frac{1}{2}x$$
$$y = -\frac{1}{2}x + 4$$

Now we find some ordered pairs that are solutions, using even numbers for x to avoid fractions. Then we plot these points and draw and label the line.

When $x = -2$, $y = -\frac{1}{2}(-2) + 4 = 1 + 4 = 5$.

When $x = 2$, $y = -\frac{1}{2} \cdot 2 + 4 = -1 + 4 = 3$.

When $x = 4$, $y = -\frac{1}{2} \cdot 4 + 4 = -2 + 4 = 2$.

x	y	(x, y)
-2	5	$(-2, 5)$
2	3	$(2, 3)$
4	2	$(4, 2)$

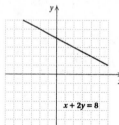

35. $y = \frac{3}{2}x + 1$

We find some ordered pairs that are solutions, using even numbers for x to avoid fractions. Then we plot these points and draw and label the line.

When $x = -4$, $y = \frac{3}{2}(-4) + 1 = -6 + 1 = -5$.

When $x = 0$, $y = \frac{3}{2} \cdot 0 + 1 = 1$.

When $x = 2$, $y = \frac{3}{2} \cdot 2 + 1 = 3 + 1 = 4$.

x	y	(x, y)
-4	-5	$(-4, -5)$
0	1	$(0, 1)$
2	4	$(2, 4)$

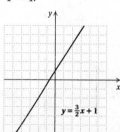

37. $8y + 2x = 4$

First we solve for y.

$$8y + 2x = 4$$
$$8y = 4 - 2x$$
$$\frac{1}{8} \cdot 8y = \frac{1}{8}(4 - 2x)$$
$$y = \frac{1}{8} \cdot 4 - \frac{1}{8} \cdot 2x$$
$$y = \frac{1}{2} - \frac{1}{4}x$$
$$y = -\frac{1}{4}x + \frac{1}{2}$$

Now we find some ordered pairs that are solutions, plot them, and draw and label the line.

When $x = -4$, $y = -\frac{1}{4}(-4) + \frac{1}{2} = 1 + \frac{1}{2} = \frac{3}{2}$.

When $x = 0$, $y = -\frac{1}{4} \cdot 0 + \frac{1}{2} = \frac{1}{2}$.

When $x = 4$, $y = -\frac{1}{4} \cdot 4 + \frac{1}{2} = -1 + \frac{1}{2} = -\frac{1}{2}$.

x	y	(x, y)
-4	$\frac{3}{2}$	$\left(-4, \frac{3}{2}\right)$
0	$\frac{1}{2}$	$\left(0, \frac{1}{2}\right)$
4	$-\frac{1}{2}$	$\left(4, -\frac{1}{2}\right)$

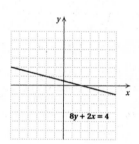

39. $8y + 2x = -4$

First we solve for y.

$$8y + 2x = -4$$
$$8y = -4 - 2x$$
$$\frac{1}{8} \cdot 8y = \frac{1}{8}(-4 - 2x)$$
$$y = \frac{1}{8}(-4) - \frac{1}{8} \cdot 2x$$
$$y = -\frac{1}{2} - \frac{1}{4}x$$
$$y = -\frac{1}{4}x - \frac{1}{2}$$

Now we find some ordered pairs that are solutions, plot them, and draw and label the line.

When $x = -4$, $y = -\frac{1}{4}(-4) - \frac{1}{2} = 1 - \frac{1}{2} = \frac{1}{2}$.

When $x = 0$, $y = -\frac{1}{4} \cdot 0 - \frac{1}{2} = -\frac{1}{2}$.

When $x = 4$, $y = -\frac{1}{4} \cdot 4 - \frac{1}{2} = -1 - \frac{1}{2} = -\frac{3}{2}$.

x	y	(x,y)
-4	$\dfrac{1}{2}$	$\left(-4, \dfrac{1}{2}\right)$
0	$-\dfrac{1}{2}$	$\left(0, -\dfrac{1}{2}\right)$
4	$-\dfrac{3}{2}$	$\left(4, -\dfrac{3}{2}\right)$

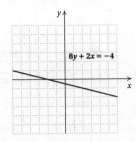

41. $y = x^2$

To find an ordered pair, we choose any number for x and then determine y. For example, if $x = 2$, then $y = 2^2 = 4$. We find several ordered pairs, plot them, and connect them with a smooth curve.

x	y
-2	4
-1	1
0	0
1	1
2	4

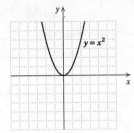

43. $y = x^2 + 2$

To find an ordered pair, we choose any number for x and then determine y. For example, if $x = 2$, then $y = 2^2 + 2 = 4 + 2 = 6$. We find several ordered pairs, plot them, and connect them with a smooth curve.

x	y
-2	6
-1	3
0	2
1	3
2	6

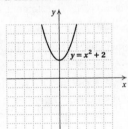

45. $y = x^2 - 3$

To find an ordered pair, we choose any number for x and then determine y. For example, if $x = 2$, then $y = 2^2 - 3 = 4 - 3 = 1$. We find several ordered pairs, plot them, and connect them with a smooth curve.

x	y
-2	1
-1	-2
0	-3
1	-2
2	1
3	6

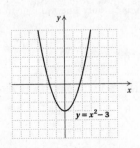

47. $y = -\dfrac{1}{x}$

To find an ordered pair, we choose any number for x and then determine y. For example, if $x = -4$, then $y = \dfrac{1}{4}$. We find several ordered pairs, plot them, and connect them with a smooth curve.

x	y
-4	$\dfrac{1}{4}$
-2	$\dfrac{1}{2}$
$-\dfrac{1}{2}$	2
$\dfrac{1}{2}$	-2
2	$-\dfrac{1}{2}$
4	$-\dfrac{1}{4}$

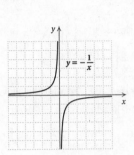

Note that we cannot use 0 as a first-coordinate, since $-1/0$ is undefined. Thus, the graph has two branches, one on each side of the y-axis.

49. $y = |x - 2|$

To find an ordered pair, we choose any number for x and then determine y. For example, if $x = 5$, then $y = |5 - 2| = |3| = 3$. We find several ordered pairs, plot them, and connect them with a smooth curve.

x	y
5	3
3	1
2	0
-1	3
-2	4
-3	5

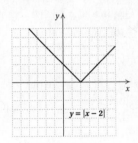

51. $y = x^3$

To find an ordered pair, we choose any number for x and then determine y. For example, if $x = -1$, then $y = (-1)^3 = -1$. We find several ordered pairs, plot them, and connect them with a smooth curve.

x	y
-2	-8
-1	-1
0	0
1	1
2	8

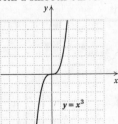

53. Discussion and Writing Exercise

55. $-3 < 2x - 5 \leq 10$

$\qquad 2 < 2x \leq 15 \qquad$ Adding 5

$\qquad 1 < x \leq \dfrac{15}{2} \qquad$ Dividing by 2

The solution set is $\left\{ x \,\middle|\, 1 < x \leq \dfrac{15}{2} \right\}$, or $\left(1, \dfrac{15}{2} \right]$.

57. $3x - 5 \leq -12 \;\; or \;\; 3x - 5 \geq 12$

$\qquad 3x \leq -7 \quad or \qquad 3x \geq 17$

$\qquad x \leq -\dfrac{7}{3} \quad or \qquad x \geq \dfrac{17}{3}$

The solution set is $\left\{ x \,\middle|\, x \leq -\dfrac{7}{3} \;\; or \;\; x \geq \dfrac{17}{3} \right\}$, or

$\left(-\infty, -\dfrac{7}{3} \right] \cup \left[\dfrac{17}{3}, \infty \right)$.

59. ***Familiarize***. The formula for the area of a triangle with base b and height h is $A = \dfrac{1}{2}bh$.

Translate. Substitute 200 for A and 16 for b in the formula.

$$A = \dfrac{1}{2}bh$$

$$200 = \dfrac{1}{2} \cdot 16 \cdot h$$

Carry out. We solve the equation.

$$200 = \dfrac{1}{2} \cdot 16 \cdot h$$

$\qquad 200 = 8h \qquad$ Multiplying

$\qquad 25 = h \qquad$ Dividing by 8 on both sides

Check. The area of a triangle with base 16 ft and height 25 ft is $\dfrac{1}{2} \cdot 16 \cdot 25$, or 200 ft^2. The answer checks.

State. The seed can fill a triangle that is 25 ft tall.

61. ***Familiarize***. Let $m =$ the distance after the first $\dfrac{1}{2}$ mile, in units of $\dfrac{1}{4}$ miles. Then the taxi ride costs $\$1.00 + \$0.30m$ and the total distance is $\dfrac{1}{2} + \dfrac{m}{4}$ miles.

Translate. The taxi ride costs $\$5.20$, so we have

$$1.00 + 0.30m = 5.20.$$

Solve. We solve the equation.

$$1.00 + 0.30m = 5.20$$

$\qquad 10 + 3m = 52 \qquad$ Multiplying by 10

$\qquad 3m = 42$

$\qquad m = 14$

When $m = 14$, the total distance is $\dfrac{1}{2} + \dfrac{14}{4} = \dfrac{16}{4} = 4$ mi.

Check. If the distance is $\dfrac{1}{2}$ mi plus 14 additional $\dfrac{1}{4}$ mi segments, then the fare is $\$1.00 + \$0.30(14) = \$1.00 + \$4.20 = \$5.20$.

State. It is 4 mi from Johnson Street to Elm Street.

63. $y = x^3 - 3x + 2$

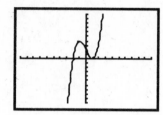

65. We use DOT mode when graphing $y = \dfrac{1}{x-2}$.

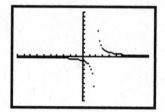

67. Note that the sum of the coordinates of each point on the graph is 4. Thus, we have $x + y = 4$, or $y = -x + 4$.

69. This is the graph of $y = |x|$ (see Example 9) with 3 subtracted from each x-coordinate. Thus, we have $y = |x| - 3$.

Exercise Set 2.2

1. Yes; each member of the domain is matched to only one member of the range.

3. Yes; each member of the domain is matched to only one member of the range.

5. Yes; each member of the domain is matched to only one member of the range.

7. No; a member of the domain is matched to more than one member of the range. In fact, each member of the domain is matched to 3 members of the range.

9. Yes; each member of the domain is matched to only one member of the range.

11. The correspondence is not a function, since a number can be the area of more than one triangle.

13. This correspondence is a function, because each number in the domain, when squared and then increased by 4, corresponds to only one number in the range.

15. $f(x) = x + 5$

a) $f(4) = 4 + 5 = 9$

b) $f(7) = 7 + 5 = 12$

c) $f(-3) = -3 + 5 = 2$

d) $f(0) = 0 + 5 = 5$

e) $f(2.4) = 2.4 + 5 = 7.4$

f) $f\left(\dfrac{2}{3} \right) = \dfrac{2}{3} + 5 = 5\dfrac{2}{3}$

17. $h(p) = 3p$

   a) $h(-7) = 3(-7) = -21$

   b) $h(5) = 3 \cdot 5 = 15$

   c) $h\left(\dfrac{2}{3}\right) = 3 \cdot \dfrac{2}{3} = \dfrac{6}{3} = 2$

   d) $h(0) = 3 \cdot 0 = 0$

   e) $h(6a) = 3 \cdot 6a = 18a$

   f) $h(a+1) = 3(a+1) = 3a + 3$

19. $g(s) = 3s + 4$

   a) $g(1) = 3 \cdot 1 + 4 = 3 + 4 = 7$

   b) $g(-7) = 3(-7) + 4 = -21 + 4 = -17$

   c) $g\left(\dfrac{2}{3}\right) = 3 \cdot \dfrac{2}{3} + 4 = 2 + 4 = 6$

   d) $g(0) = 3 \cdot 0 + 4 = 0 + 4 = 4$

   e) $g(a-2) = 3(a-2) + 4 = 3a - 6 + 4 = 3a - 2$

   f) $g(a+h) = 3(a+h) + 4 = 3a + 3h + 4$

21. $f(x) = 2x^2 - 3x$

   a) $f(0) = 2 \cdot 0^2 - 3 \cdot 0 = 0 - 0 = 0$

   b) $f(-1) = 2(-1)^2 - 3(-1) = 2 + 3 = 5$

   c) $f(2) = 2 \cdot 2^2 - 3 \cdot 2 = 8 - 6 = 2$

   d) $f(10) = 2 \cdot 10^2 - 3 \cdot 10 = 200 - 30 = 170$

   e) $f(-5) = 2(-5)^2 - 3(-5) = 50 + 15 = 65$

   f) $f(4a) = 2(4a)^2 - 3(4a) = 32a^2 - 12a$

23. $f(x) = |x| + 1$

   a) $f(0) = |0| + 1 = 0 + 1 = 1$

   b) $f(-2) = |-2| + 1 = 2 + 1 = 3$

   c) $f(2) = |2| + 1 = 2 + 1 = 3$

   d) $f(-10) = |-10| + 1 = 10 + 1 = 11$

   e) $f(a-1) = |a-1| + 1$

   f) $f(a+h) = |a+h| + 1$

25. $f(x) = x^3$

   a) $f(0) = 0^3 = 0$

   b) $f(-1) = (-1)^3 = -1$

   c) $f(2) = 2^3 = 8$

   d) $f(10) = 10^3 = 1000$

   e) $f(-5) = (-5)^3 = -125$

   f) $f(-3a) = (-3a)^3 = -27a^3$

27. $F(x) = 2.75x + 71.48$

   a) $F(32) = 2.75(32) + 71.48$

   $= 88 + 71.48$

   $= 159.48$ cm

   b) $F(35) = 2.75(35) + 71.48$

   $= 96.25 + 71.48$

   $= 167.73$ cm

29. $P(d) = 1 + \dfrac{d}{33}$

   $P(20) = 1 + \dfrac{20}{33} = 1\dfrac{20}{33}$ atm

   $P(30) = 1 + \dfrac{30}{33} = 1\dfrac{10}{11}$ atm

   $P(100) = 1 + \dfrac{100}{33} = 1 + 3\dfrac{1}{33} = 4\dfrac{1}{33}$ atm

31. $W(d) = 0.112d$

   $W(16) = 0.112(16) = 1.792$ cm

   $W(25) = 0.112(25) = 2.8$ cm

   $W(100) = 0.112(100) = 11.2$ cm

33. Graph $f(x) = -2x$.

Make a list of function values in a table.

$f(-2) = -2(-2) = 4$

$f(-1) = -2(-1) = 2$

$f(0) = -2 \cdot 0 = 0$

$f(2) = -2 \cdot 2 = -4$

x	$f(x)$
-2	4
-1	2
0	0
2	-4

Plot these points and connect them.

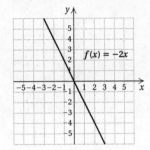

35. Graph $f(x) = 3x - 1$.

Make a list of function values in a table.

$f(-1) = 3(-1) - 1 = -3 - 1 = -4$

$f(0) = 3 \cdot 0 - 1 = 0 - 1 = -1$

$f(1) = 3 \cdot 1 - 1 = 3 - 1 = 2$

$f(2) = 3 \cdot 2 - 1 = 6 - 1 = 5$

x	$f(x)$
-1	-4
0	-1
1	2
2	5

Plot these points and connect them.

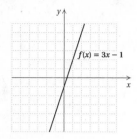

37. Graph $g(x) = -2x + 3$.

Make a list of function values in a table.

$g(-1) = -2(-1) + 3 = 2 + 3 = 5$

$g(0) = -2 \cdot 0 + 3 = 0 + 3 = 3$

$g(3) = -2 \cdot 3 + 3 = -6 + 3 = -3$

x	$g(x)$
-1	5
0	3
3	-3

Plot these points and connect them.

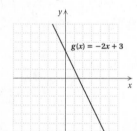

39. Graph $f(x) = \dfrac{1}{2}x + 1$.

Make a list of function values in a table.

$f(-2) = \dfrac{1}{2}(-2) + 1 = -1 + 1 = 0$

$f(0) = \dfrac{1}{2} \cdot 0 + 1 = 0 + 1 = 1$

$f(4) = \dfrac{1}{2} \cdot 4 + 1 = 2 + 1 = 3$

x	$f(x)$
-2	0
0	1
4	3

Plot these points and connect them.

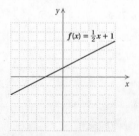

41. Graph $f(x) = 2 - |x|$.

Make a list of function values in a table.

$f(-3) = 2 - |-3| = 2 - 3 = -1$

$f(-2) = 2 - |-2| = 2 - 2 = 0$

$f(-1) = 2 - |-1| = 2 - 1 = 1$

$f(0) = 2 - |0| = 2 - 0 = 2$

$f(1) = 2 - |1| = 2 - 1 = 1$

$f(2) = 2 - |2| = 2 - 2 = 0$

$f(3) = 2 - |3| = 2 - 3 = -1$

x	$f(x)$
-3	-1
-2	0
-1	1
0	2
1	1
2	0
3	-1

Plot these points and connect them.

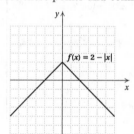

43. Graph $g(x) = |x - 1|$.

Make a list of function values in a table.

$f(-3) = |-3 - 1| = |-4| = 4$

$f(-1) = |-1 - 1| = |-2| = 2$

$f(0) = |0 - 1| = |-1| = 1$

$f(1) = |1 - 1| = |0| = 0$

$f(2) = |2 - 1| = |1| = 1$

$f(4) = |4 - 1| = |3| = 3$

x	$f(x)$
-3	4
-1	2
0	1
1	0
2	1
4	3

Plot these points and connect them.

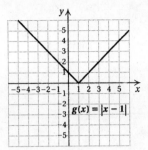

45. Graph $f(x) = x^2$.

Make a list of function values in a table.

$f(-3) = (-3)^2 = 9$

$f(-2) = (-2)^2 = 4$

$f(-1) = (-1)^2 = 1$

$f(0) = 0^2 = 0$

$f(1) = 1^2 = 1$

$f(2) = 2^2 = 4$

$f(3) = 3^2 = 9$

x	$f(x)$
-3	9
-2	4
-1	1
0	0
1	1
2	4
3	9

Plot these points and connect them.

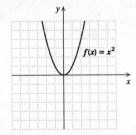

47. Graph $f(x) = x^2 - x - 2$.

Make a list of function values in a table.

$f(-3) = (-3)^2 - (-3) - 2 = 9 + 3 - 2 = 10$

$f(-2) = (-2)^2 - (-2) - 2 = 4 + 2 - 2 = 4$

$f(-1) = (-1)^2 - (-1) - 2 = 1 + 1 - 2 = 0$

$f(0) = 0^2 - 0 - 2 = -2$

$f(1) = 1^2 - 1 - 2 = 1 - 1 - 2 = -2$

$f(2) = 2^2 - 2 - 2 = 4 - 2 - 2 = 0$

$f(3) = 3^2 - 3 - 2 = 9 - 3 - 2 = 4$

x	$f(x)$
-3	10
-2	4
-1	0
0	-2
1	-2
2	0
3	4

Plot these points and connect them.

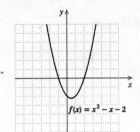

49. Graph $f(x) = 2 - x^2$.

Make a list of function values in a table.

$f(-2) = 2 - (-2)^2 = 2 - 4 = -2$

$f(-1) = 2 - (-1)^2 = 2 - 1 = 1$

$f(0) = 2 - 0^2 = 2$

$f(1) = 2 - 1^2 = 2 - 1 = 1$

$f(2) = 2 - 2^2 = 2 - 4 = -2$

x	$f(x)$
-2	-2
-1	1
0	2
1	1
2	-2

Plot these points and connect them.

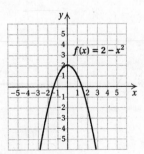

51. Graph $f(x) = x^3 + 1$.

Make a list of function values in a table.

$f(-2) = (-2)^3 + 1 = -8 + 1 = -7$

$f(-1) = (-1)^3 + 1 = -1 + 1 = 0$

$f(0) = 0^3 + 1 = 0 + 1 = 1$

$f(1) = 1^3 + 1 = 1 + 1 = 2$

$f(2) = 2^3 + 1 = 8 + 1 = 9$

x	$f(x)$
-2	-7
-1	0
0	1
1	2
2	9

Plot these points and connect them.

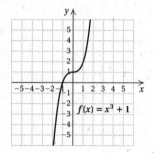

53. We can use the vertical line test:

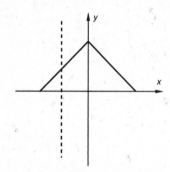

Visualize moving this vertical line across the graph. No vertical line will intersect the graph more than once. Thus, the graph is a graph of a function.

55. We can use the vertical line test:

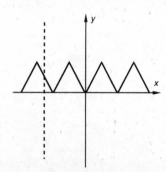

Visualize moving this vertical line across the graph. No vertical line will intersect the graph more than once. Thus, the graph is a graph of a function.

57. We can use the vertical line test.

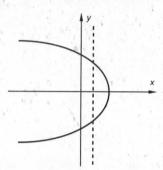

It is possible for a vertical line to intersect the graph more than once. Thus this is not the graph of a function.

59. We can use the vertical line test.

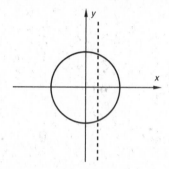

It is possible for a vertical line to intersect the graph more than once. Thus this is not a graph of a function.

61. Locate 1999 on the horizontal axis and then move directly up to the graph. Next move across to the vertical axis. We come to about 1150, so there were about 1150 news/talk radio stations in 1999.

63. Locate 2000 on the horizontal axis and then move directly up to the graph. Next move across to the vertical axis. We come to about 1.9, so about 1.9 billion digital photos that were taken in 2000 were not printed.

65. Discussion and Writing Exercise

67. The axes divide the plane into four regions called quadrants.

69. A function is a correspondence between a first set, called the domain, and a second set, called the range, such that each member of the domain corresponds to exactly one number of the range.

71. Members of the domain of a function are its inputs.

73. The addition principle states that for any real numbers a, b, and c, $a = b$ is equivalent to $a + c = b + c$.

75. To find $f(g(-4))$, we first find $g(-4)$:

$g(-4) = 2(-4) + 5 = -8 + 5 = -3$.

Then $f(g(-4)) = f(-3) = 3(-3)^2 - 1 = 3 \cdot 9 - 1 = 27 - 1 = 26$.

To find $g(f(-4))$, we first find $f(-4)$:

$f(-4) = 3(-4)^2 - 1 = 3 \cdot 16 - 1 = 48 - 1 = 47.$

Then $g(f(-4)) = g(47) = 2 \cdot 47 + 5 = 94 + 5 = 99.$

77. We know that $(-1, -7)$ and $(3, 8)$ are both solutions of $g(x) = mx + b$. Substituting, we have

$$-7 = m(-1) + b, \quad \text{or} \quad -7 = -m + b,$$
$$\text{and} \quad 8 = m(3) + b, \quad \text{or} \quad 8 = 3m + b.$$

Solve the first equation for b and substitute that expression into the second equation.

$-7 = -m + b$	First equation
$m - 7 = b$	Solving for b
$8 = 3m + b$	Second equation
$8 = 3m + (m - 7)$	Substituting
$8 = 3m + m - 7$	
$8 = 4m - 7$	
$15 = 4m$	
$\dfrac{15}{4} = m$	

We know that $m - 7 = b$, so $\dfrac{15}{4} - 7 = b$, or $-\dfrac{13}{4} = b$.

We have $m = \dfrac{15}{4}$ and $b = -\dfrac{13}{4}$, so $g(x) = \dfrac{15}{4}x - \dfrac{13}{4}$.

Exercise Set 2.3

1. a) Locate 1 on the horizontal axis and then find the point on the graph for which 1 is the first coordinate. From that point, look to the vertical axis to find the corresponding y-coordinate, 3. Thus, $f(1) = 3$.

b) The domain is the set of all x-values in the graph. It is $\{-4, -3, -2, -1, 0, 1, 2\}$.

c) To determine which member(s) of the domain are paired with 2, locate 2 on the vertical axis. From there look left and right to the graph to find any points for which 2 is the second coordinate. Two such points exist, $(-2, 2)$ and $(0, 2)$. Thus, the x-values for which $f(x) = 2$ are -2 and 0.

d) The range is the set of all y-values in the graph. It is $\{1, 2, 3, 4\}$.

3. a) Locate 1 on the horizontal axis and then find the point on the graph for which 1 is the first coordinate. From that point, look to the vertical axis to find the corresponding y-coordinate, about $2\frac{1}{2}$. Thus, $f(1) \approx 2\frac{1}{2}$.

b) The set of all x-values in the graph extends from -3 to 5, so the domain is $\{x | -3 \le x \le 5\}$, or $[-3, 5]$.

c) To determine which member(s) of the domain are paired with 2, locate 2 on the vertical axis. From there look left and right to the graph to find any points for which 2 is the second coordinate. One such point exists. Its first coordinate appears to be about $2\frac{1}{4}$. Thus, the x-value for which $f(x) = 2$ is about $2\frac{1}{4}$.

d) The set of all y-values in the graph extends from 1 to 4, so the range is $\{y | 1 \le y \le 4\}$, or $[1, 4]$.

5. a) Locate 1 on the horizontal axis and the find the point on the graph for which 1 is the first coordinate. From that point, look to the vertical axis to find the corresponding y-coordinate. It appears to be about $2\frac{1}{4}$. Thus, $f(1) \approx 2\frac{1}{4}$.

b) The set of all x-values in the graph extends from -4 to 3, so the domain is $\{x | -4 \le x \le 3\}$, or $[-4, 3]$.

c) To determine which member(s) of the domain are paired with 2, locate 2 on the vertical axis. From there look left and right to the graph to find any points for which 2 is the second coordinate. One such point exists. Its first coordinate is about 0, so the x-value for which $f(x) = 2$ is about 0.

d) The set of all y-values in the graph extends from -5 to 4, so the range is $\{y | -5 \le y \le 4\}$, or $[-5, 4]$.

7. a) Locate 1 on the horizontal axis and the find the point on the graph for which 1 is the first coordinate. From that point, look to the vertical axis to find the corresponding y-coordinate. It is 1. Thus, $f(1) = 1$.

b) No endpoints are indicated and we see that the graph extends indefinitely both horizontally and vertically. Thus, the domain is the set of all real numbers.

c) To determine which member(s) of the domain are paired with 2, locate 2 on the vertical axis. From there look left and right to the graph to find any points for which 2 is the second coordinate. One such point exists. Its first coordinate is 3, so the x-value for which $f(x) = 2$ is 3.

d) The range is the set of all real numbers. (See part (b) above.)

9. a) Locate 1 on the horizontal axis and the find the point on the graph for which 1 is the first coordinate. From that point, look to the vertical axis to find the corresponding y-coordinate. It is 1. Thus, $f(1) = 1$.

b) No endpoints are indicated, so we see that the graph extends indefinitely horizontally. Thus, the domain is the set of all real numbers.

c) To determine which member(s) of the domain are paired with 2, locate 2 on the vertical axis. From there look left and right to the graph to find any points for which 2 is the second coordinate. Two such points exist, $(-2, 2)$ and $(2, 2)$. Thus, the x-values for which $f(x) = 2$ are -2 and 2.

d) The smallest y-value is 0. No endpoints are indicated, so we see that the graph extends upward indefinitely from $(0, 0)$. Thus, the range is $\{y | y \ge 0\}$, or $[0, \infty)$.

11. a) Locate 1 on the horizontal axis and then find the point on the graph for which 1 is the first coordinate. From that point, look to the vertical axis to find the corresponding y-coordinate, -1. Thus, $f(1) = -1$.

b) The set of all x-values in the graph extends from -6 to 5, so the domain is $\{x| -6 \leq x \leq 5\}$, or $[-6, 5]$.

c) To determine which member(s) of the domain are paired with 2, locate 2 on the vertical axis. From there look left and right to the graph to find any points for which 2 is the second coordinate. Three such points exist, $(-4, 2)$, $(0, 2)$ and $(3, 2)$. Thus, the x-values for which $f(x) = 2$ are -4, 0, and 3.

d) The set of all y-values in the graph extends from -2 to 2, so the range is $\{y| -2 \leq y \leq 2\}$, or $[-2, 2]$.

13. $f(x) = \dfrac{2}{x+3}$

Since $\dfrac{2}{x+3}$ cannot be calculated when the denominator is 0, we find the x-value that causes $x + 3$ to be 0:

$$x + 3 = 0$$
$$x = -3 \quad \text{Subtracting 3 on both sides}$$

Thus, -3 is not in the domain of f, while all other real numbers are. The domain of f is

$\{x|x \text{ is a real number } and \ x \neq -3\}$, or

$(-\infty, -3) \cup (-3, \infty)$.

15. $f(x) = 2x + 1$

Since we can calculate $2x + 1$ for any real number x, the domain is the set of all real numbers.

17. $f(x) = x^2 + 3$

Since we can calculate $x^2 + 3$ for any real number x, the domain is the set of all real numbers.

19. $f(x) = \dfrac{8}{5x - 14}$

Since $\dfrac{8}{5x - 14}$ cannot be calculated when the denominator is 0, we find the x-value that causes $5x - 14$ to be 0:

$$5x - 14 = 0$$
$$5x = 14$$
$$x = \dfrac{14}{5}$$

Thus, $\dfrac{14}{5}$ is not in the domain of f, while all other real numbers are. The domain of f is

$\left\{x\middle|x \text{ is a real number } and \ x \neq \dfrac{14}{5}\right\}$, or

$\left(\infty, \dfrac{14}{5}\right) \cup \left(\dfrac{14}{5}, \infty\right)$.

21. $f(x) = |x| - 4$

Since we can calculate $|x| - 4$ for any real number x, the domain is the set of all real numbers.

23. $f(x) = \dfrac{4}{|2x - 3|}$

Since $\dfrac{4}{|2x - 3|}$ cannot be calculated when the denominator is 0, we find the x-values that causes $|2x - 3|$ to be 0:

$$|2x - 3| = 0$$
$$2x - 3 = 0$$
$$2x = 3$$
$$x = \dfrac{3}{2}$$

Thus, $\dfrac{3}{2}$ is not in the domain of f, while all other real numbers are. The domain of f is

$\left\{x\middle|x \text{ is a real number } and \ x \neq \dfrac{3}{2}\right\}$, or

$\left(-\infty, \dfrac{3}{2}\right) \cup \left(\dfrac{3}{2}, \infty\right)$.

25. $g(x) = \dfrac{1}{x - 1}$

Since $\dfrac{1}{x - 1}$ cannot be calculated when the denominator is 0, we find the x-value that causes $x - 1$ to be 0:

$$x - 1 = 0$$
$$x = 1$$

Thus, 1 is not in the domain of g, while all other real numbers are. The domain of g is

$\{x|x \text{ is a real number } and \ x \neq 1\}$, or $(-\infty, 1) \cup (1, \infty)$.

27. $g(x) = x^2 - 2x + 1$

Since we can calculate $x^2 - 2x + 1$ for any real number x, the domain is the set of all real numbers.

29. $g(x) = x^3 - 1$

Since we can calculate $x^3 - 1$ for any real number x, the domain is the set of all real numbers.

31. $g(x) = \dfrac{7}{20 - 8x}$

Since $\dfrac{7}{20 - 8x}$ cannot be calculated when the denominator is 0, we find the x-values that cause $20 - 8x$ to be 0:

$$20 - 8x = 0$$
$$-8x = -20$$
$$x = \dfrac{5}{2}$$

Thus, $\dfrac{5}{2}$ is not in the domain of g, while all other real numbers are. The domain of g is

$\left\{x\middle|x \text{ is a real number } and \ x \neq \dfrac{5}{2}\right\}$, or

$\left(-\infty, \dfrac{5}{2}\right) \cup \left(\dfrac{5}{2}, \infty\right)$.

33. $g(x) = |x + 7|$

Since we can calculate $|x + 7|$ for any real number x, the domain is the set of all real numbers.

35. $g(x) = \dfrac{-2}{|4x + 5|}$

Since $\dfrac{-2}{|4x + 5|}$ cannot be calculated when the denominator is 0, we find the x-value that causes $|4x + 5|$ to be 0:

$$|4x + 5| = 0$$
$$4x + 5 = 0$$
$$4x = -5$$
$$x = -\frac{5}{4}$$

Thus, $-\frac{5}{4}$ is not in the domain of g, while all other real numbers are. The domain of g is

$$\left\{ x \middle| x \text{ is a real number } and \ x \neq -\frac{5}{4} \right\}, \text{ or}$$

$$\left(-\infty, -\frac{5}{4} \right) \cup \left(-\frac{5}{4}, \infty \right).$$

37. The input -1 has the output -8, so $f(-1) = -8$;

the input 0 has the output 0, so $f(0) = 0$;

the input 1 has the output -2, so $f(1) = -2$.

39. Discussion and Writing Exercise

41. *Familiarize*. We list the given information in a table.

Plan A: Monthly Income	Plan B: Monthly Income
\$800 salary	\$1000 salary
5% of sales	7% of sales over \$15,000
Total: 800 + 5% of sales	Total: 1000 + 7% of sales over 15,000

Translate. We write an inequality stating that the income from Plan B is greater than the income from Plan A. We let $s =$ sales Then $s - 15,000 =$ sales over 15,000.

$$1000 + 7\%(s - 15,000) > 800 + 5\%s$$

Solve.

$$1000 + 0.07(s - 15,000) > 800 + 0.05s$$
$$1000 + 0.07s - 1050 > 800 + 0.05s$$
$$0.07s - 50 > 800 + 0.05s$$
$$0.02s - 50 > 800$$
$$0.02s > 850$$
$$s > 42,500$$

Check. We can do a partial check by computing the income from each plan for an amount less than 42,500 and for an amount greater than 42,500. We conclude that the answer checks.

State. Plan B is better than Plan A for sales of more than \$42,500. Expressed as an inequality the result is $\{s | s > \$42,500\}$.

43. $|x| = 8$

$x = -8 \ or \ x = 8$

The solution set is $\{-8, 8\}$.

45. $|x - 7| = 11$

$x - 7 = -11 \ or \ x - 7 = 11$

$x = -4 \ or \ x = 18$

The solution set is $\{-4, 18\}$.

47. $|3x - 4| = |x + 2|$

$3x - 4 = x + 2 \ or \ 3x - 4 = -(x + 2)$

$2x - 4 = 2 \ \ or \ 3x - 4 = -x - 2$

$2x = 6 \ \ or \ 4x - 4 = -2$

$x = 3 \ \ or \ \ 4x = 2$

$x = 3 \ \ or \ \ x = \frac{1}{2}$

The solution set is $\left\{ \frac{1}{2}, 3 \right\}$.

49. $|3x - 8| = -11$

Since the absolute value of a number must be nonnegative, the equation has no solution. The solution is $\{ \ \}$, or $\emptyset$.

51. We graph each function and determine the range.

The range of $f(x) = \dfrac{2}{x + 3}$ is $(-\infty, 0) \cup (0, \infty)$; the range of $f(x) = x^2 - 2x + 3$ is $[2, \infty)$; the range of $f(x) = |x| - 4$ is $[-4, \infty)$; the range of $f(x) = |x - 4|$ is $[0, \infty)$.

Exercise Set 2.4

1. $y = 4x + 5$
$y = mx + b$

The slope is 4, and the y-intercept is $(0, 5)$.

3. $f(x) = -2x - 6$
$f(x) = mx + b$

The slope is -2, and the y-intercept is $(0, -6)$.

5. $y = -\dfrac{3}{8}x - \dfrac{1}{5}$
$y = mx + b$

The slope is $-\dfrac{3}{8}$, and the y-intercept is $\left(0, -\dfrac{1}{5} \right)$.

7. $g(x) = 0.5x - 9$
$g(x) = mx + b$

The slope is 0.5, and the y-intercept is $(0, -9)$.

9. First we find the slope-intercept form of the equation by solving for y. This allows us to determine the slope and y-intercept easily.

$$2x - 3y = 8$$
$$-3y = -2x + 8$$
$$\frac{-3y}{-3} = \frac{-2x + 8}{-3}$$
$$y = \frac{2}{3}x - \frac{8}{3}$$

The slope is $\dfrac{2}{3}$, and the y-intercept is $\left(0, -\dfrac{8}{3} \right)$.

11. First we find the slope-intercept form of the equation by solving for y. This allows us to determine the slope and y-intercept easily.

$$9x = 3y + 6$$
$$9x - 6 = 3y$$
$$\frac{9x - 6}{3} = \frac{3y}{y}$$
$$3x - 2 = y, \text{ or}$$
$$y = 3x - 2$$

The slope is 3, and the y-intercept is $(0, -2)$.

13. First we find the slope-intercept form of the equation by solving for y. This allows us to determine the slope and y-intercept easily.

$$3 - \frac{1}{4}y = 2x$$
$$-\frac{1}{4}y = 2x - 3$$
$$-4\left(-\frac{1}{4}y\right) = -4(2x - 3)$$
$$y = -8x + 12$$

The slope is -8, and the y-intercept is $(0, 12)$.

15. First we find the slope-intercept form of the equation by solving for y. This allows us to determine the slope and y-intercept easily.

$$17y + 4x + 3 = 7 + 4x$$
$$17y + 3 = 7$$
$$17y = 4$$
$$y = \frac{4}{17}, \text{ or}$$
$$y = 0 \cdot x + \frac{4}{17}$$

The slope is 0, and the y-intercept is $\left(0, \frac{4}{17}\right)$.

17. We can use any two points on the line, such as $(0, 3)$ and $(4, 1)$.

$$\text{Slope} = \frac{\text{change in } y}{\text{change in } x}$$
$$= \frac{1 - 3}{4 - 0} = \frac{-2}{4} = -\frac{1}{2}$$

19. We can use any two points on the line, such as $(-3, 1)$ and $(3, 3)$.

$$\text{Slope} = \frac{\text{change in } y}{\text{change in } x}$$
$$= \frac{3 - 1}{3 - (-3)} = \frac{2}{6} = \frac{1}{3}$$

21. $\text{Slope} = \dfrac{\text{change in } y}{\text{change in } x} = \dfrac{5 - 9}{4 - 6} = \dfrac{-4}{-2} = 2$

23. $\text{Slope} = \dfrac{\text{change in } y}{\text{change in } x} = \dfrac{-8 - (-4)}{3 - 9} = \dfrac{-4}{-6} = \dfrac{2}{3}$

25. $\text{Slope} = \dfrac{\text{change in } y}{\text{change in } x} = \dfrac{8.7 - 12.4}{-5.2 - (-16.3)} = \dfrac{-3.7}{11.1} =$
$-\dfrac{37}{111} = -\dfrac{1}{3}$

27. $\text{Slope} = \dfrac{0.4}{5} = 0.08 = 8\%$; this can also be expressed as $\dfrac{2}{25}$.

29. $\text{Slope} = \dfrac{43.33}{1238} = 0.035 = 3.5\%$

31. The rate of change can be found using the coordinates of any two points on the line. We use $(2000, 9.7)$ and $(2005, 35)$.

$$\text{Rate} = \frac{\text{change in volume of e-mail}}{\text{corresponding change in time}}$$
$$= \frac{35 - 9.7}{2005 - 2000}$$
$$= \frac{25.3}{5}$$
$$= 5.06 \text{ billion messages daily per year}$$

33. We can use the coordinates of any two points on the line. We'll use $(0, 30)$ and $(3, 3)$.

$$\text{Slope} = \frac{\text{change in } y}{\text{change in } x} = \frac{3 - 30}{3 - 0} = \frac{-27}{3} = -9$$

The rate of change is $-\$900$ per year. That is, the value is decreasing at a rate of $\$900$ per year.

35. We can use the coordinates of any two points on the line. We'll use $(15, 470)$ and $(55, 510)$:

$$\text{Slope} = \frac{\text{change in } y}{\text{change in } x} = \frac{510 - 470}{55 - 15} = \frac{40}{40} = 1$$

The average SAT math score is increasing at a rate of 1 point per thousand dollars of family income.

37. Discussion and Writing Exercise

39. $3^2 - 24 \cdot 56 + 144 \div 12$
$= 9 - 24 \cdot 56 + 144 \div 12$
$= 9 - 1344 + 144 \div 12$
$= 9 - 1344 + 12$
$= -1335 + 12$
$= -1323$

41. $10\{2x + 3[5x - 2(-3x + y^1 - 2)]\}$
$= 10\{2x + 3[5x - 2(-3x + y - 2)]\}$
$= 10\{2x + 3[5x + 6x - 2y + 4]\}$
$= 10\{2x + 3[11x - 2y + 4]\}$
$= 10\{2x + 33x - 6y + 12\}$
$= 10\{35x - 6y + 12\}$
$= 350x - 60y + 120$

43. *Familiarize.* Let t represent the length of a side of the triangle. Then $t - 5$ represents the length of a side of the square.

Translate.

$$\underbrace{\text{Perimeter of the square}}_{4(t-5)} \quad \underbrace{\text{is the same as}}_{=} \quad \underbrace{\text{perimeter of the triangle}}_{3t}$$

Solve.

$$4(t - 5) = 3t$$
$$4t - 20 = 3t$$
$$t - 20 = 0$$
$$t = 20$$

Check. If 20 is the length of a side of the triangle, then the length of a side of the square is $20 - 5$, or 15. The perimeter of the square is $4 \cdot 15$, or 60, and the perimeter of the triangle is $3 \cdot 20$, or 60. The numbers check.

State. The square and triangle have sides of length 15 yd and 20 yd, respectively.

45. $|5x - 8| < 32$

$$-32 < 5x - 8 < 32$$
$$-24 < 5x < 40 \qquad \text{Adding 8}$$
$$-\frac{24}{5} < x < 8 \qquad \text{Dividing by 5}$$

The solution set is $\left\{ x \mid -\dfrac{24}{5} < x < 8 \right\}$, or $\left(-\dfrac{24}{5}, 8 \right)$.

47. $|5x - 8| = -32$

Since the absolute value of a number is nonnegative the equation has no solution. The solution is $\{\ \ \}$, or $\emptyset$.

Exercise Set 2.5

1. $x - 2 = y$

To find the x-intercept we let $y = 0$ and solve for x. We have $x - 2 = 0$, or $x = 2$. The x-intercept is $(2, 0)$.

To find the y-intercept we let $x = 0$ and solve for y.

$$x - 2 = y$$
$$0 - 2 = y$$
$$-2 = y$$

The y-intercept is $(0, -2)$. We plot these points and draw the line.

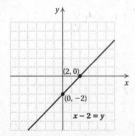

We use a third point as a check. We choose $x = 5$ and solve for y.

$$5 - 2 = y$$
$$3 = y$$

We plot $(5, 3)$ and note that it is on the line.

3. $x + 3y = 6$

To find the x-intercept we let $y = 0$ and solve for x.

$$x + 3y = 6$$
$$x + 3 \cdot 0 = 6$$
$$x = 6$$

The x-intercept is $(6, 0)$.

To find the y-intercept we let $x = 0$ and solve for y.

$$x + 3y = 6$$
$$0 + 3y = 6$$
$$3y = 6$$
$$y = 2$$

The y-intercept is $(0, 2)$.

We plot these points and draw the line.

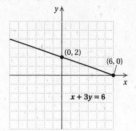

We use a third point as a check. We choose $x = 3$ and solve for y.

$$3 + 3y = 6$$
$$3y = 3$$
$$y = 1$$

We plot $(3, 1)$ and note that it is on the line.

5. $2x + 3y = 6$

To find the x-intercept we let $y = 0$ and solve for x.

$$2x + 3y = 6$$
$$2x + 3 \cdot 0 = 6$$
$$2x = 6$$
$$x = 3$$

The x-intercept is $(3, 0)$.

To find the y-intercept we let $x = 0$ and solve for y.

$$2x + 3y = 6$$
$$2 \cdot 0 + 3y = 6$$
$$3y = 6$$
$$y = 2$$

The y-intercept is $(0, 2)$.

We plot these points and draw the line.

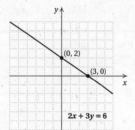

We use a third point as a check. We choose $x = -3$ and solve for y.

$$2(-3) + 3y = 6$$
$$-6 + 3y = 6$$
$$3y = 12$$
$$y = 4$$

We plot $(-3, 4)$ and note that it is on the line.

7. $f(x) = -2 - 2x$

We can think of this equation as $y = -2 - 2x$.

To find the x-intercept we let $f(x) = 0$ and solve for x. We have $0 = -2 - 2x$, or $2x = -2$, or $x = -1$. The x-intercept is $(-1, 0)$.

To find the y-intercept we let $x = 0$ and solve for $f(x)$, or y.

$$y = -2 - 2x$$
$$y = -2 - 2 \cdot 0$$
$$y = -2$$

The y-intercept is $(0, -2)$.

We plot these points and draw the line.

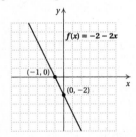

We use a third point as a check. We choose $x = -3$ and calculate y.

$$y = -2 - 2(-3) = -2 + 6 = 4$$

We plot $(-3, 4)$ and note that it is on the line.

9. $5y = -15 + 3x$

To find the x-intercept we let $y = 0$ and solve for x. We have $0 = -15 + 3x$, or $15 = 3x$, or $5 = x$. The x-intercept is $(5, 0)$.

To find the y-intercept we let $x = 0$ and solve for y.

$$5y = -15 + 3x$$
$$5y = -15 + 3 \cdot 0$$
$$5y = -15$$
$$y = -3$$

The y-intercept is $(0, -3)$.

We plot these points and draw the line.

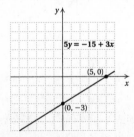

We use a third point as a check. We choose $x = -5$ and solve for y.

$$5y = -15 + 3(-5)$$
$$5y = -15 - 15$$
$$5y = -30$$
$$y = -6$$

We plot $(-5, -6)$ and note that it is on the line.

11. $2x - 3y = 6$

To find the x-intercept we let $y = 0$ and solve for x.

$$2x - 3y = 6$$
$$2x - 3 \cdot 0 = 6$$
$$2x = 6$$
$$x = 3$$

The x-intercept is $(3, 0)$.

To find the y-intercept we let $x = 0$ and solve for y.

$$2x - 3y = 6$$
$$2 \cdot 0 - 3y = 6$$
$$-3y = 6$$
$$y = -2$$

The y-intercept is $(0, -2)$.

We plot these points and draw the line.

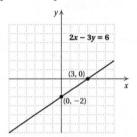

We use a third point as a check. We choose $x = -3$ and solve for y.

$$2(-3) - 3y = 6$$
$$-6 - 3y = 6$$
$$-3y = 12$$
$$y = -4$$

We plot $(-3, -4)$ and note that it is on the line.

13. $2.8y - 3.5x = -9.8$

To find the x-intercept we let $y = 0$ and solve for x.

$$2.8y - 3.5x = -9.8$$
$$2.8(0) - 3.5x = -9.8$$
$$-3.5x = -9.8$$
$$x = 2.8$$

The x-intercept is $(2.8, 0)$.

To find the y-intercept we let $x = 0$ and solve for y.

$$2.8y - 3.5x = -9.8$$
$$2.8y - 3.5(0) = -9.8$$
$$2.8y = -9.8$$
$$y = -3.5$$

The y-intercept is $(0, -3.5)$.

We plot these points and draw the line.

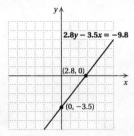

We use a third point as a check. We choose $x = 5$ and solve for y.

$$2.8y - 3.5(5) = -9.8$$
$$2.8y - 17.5 = -9.8$$
$$2.8y = 7.7$$
$$y = 2.75$$

We plot $(5, 2.75)$ and note that it is on the line.

15. $5x + 2y = 7$

To find the x-intercept we let $y = 0$ and solve for x.

$$5x + 2y = 7$$
$$5x + 2 \cdot 0 = 7$$
$$5x = 7$$
$$x = \frac{7}{5}$$

The x-intercept is $\left(\frac{7}{5}, 0\right)$.

To find the y-intercept we let $x = 0$ and solve for y.

$$5x + 2y = 7$$
$$5 \cdot 0 + 2y = 7$$
$$2y = 7$$
$$y = \frac{7}{2}$$

The y-intercept is $\left(0, \frac{7}{2}\right)$.

We plot these points and draw the line.

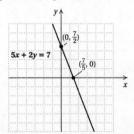

We use a third point as a check. We choose $x = 3$ and solve for y.

$$5 \cdot 3 + 2y = 7$$
$$15 + 2y = 7$$
$$2y = -8$$
$$y = -4$$

We plot $(3, -4)$ and note that it is on the line.

17. $y = \frac{5}{2}x + 1$

First we plot the y-intercept $(0, 1)$. Then we consider the slope $\frac{5}{2}$. Starting at the y-intercept and using the slope, we find another point by moving 5 units up and 2 units to the right. We get to a new point $(2, 6)$.

We can also think of the slope as $\frac{-5}{-2}$. We again start at the y-intercept $(0, 1)$. We move 5 units down and 2 units to the left. We get to another new point $(-2, -4)$. We plot the points and draw the line.

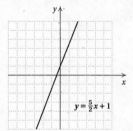

19. $f(x) = -\frac{5}{2}x - 4$

First we plot the y-intercept $(0, -4)$. We can think of the slope as $\frac{-5}{2}$. Starting at the y-intercept and using the slope, we find another point by moving 5 units down and 2 units to the right. We get to a new point $(2, -9)$.

We can also think of the slope as $\frac{5}{-2}$. We again start at the y-intercept $(0, -4)$. We move 5 units up and 2 units to the left. We get to another new point $(-2, 1)$. We plot the points and draw the line.

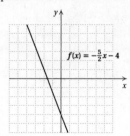

21. $x + 2y = 4$

First we write the equation in slope-intercept form by solving for y.

$$x + 2y = 4$$
$$2y = -x + 4$$
$$\frac{2y}{2} = \frac{-x + 4}{2}$$
$$y = -\frac{1}{2}x + 2$$

Now we plot the y-intercept $(0, 2)$. We can think of the slope as $\frac{-1}{2}$. Starting at the y-intercept and using the slope, we find another point by moving 1 unit down and 2 units to the right. We get to a new point $(2, 1)$.

We can also think of the slope as $\dfrac{1}{-2}$. We again start at the y-intercept $(0, 2)$. We move 1 unit up and 2 units to the left. We get to another new point $(-2, 3)$. We plot the points and draw the line.

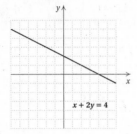

23. $4x - 3y = 12$

First we write the equation in slope-intercept form by solving for y.

$$4x - 3y = 12$$
$$-3y = -4x + 12$$
$$\frac{-3y}{-3} = \frac{-4x + 12}{-3}$$
$$y = \frac{4}{3}x - 4$$

Now we plot the y-intercept $(0, -4)$ and consider the slope $\dfrac{4}{3}$. Starting at the y-intercept and using the slope, we find another point by moving 4 units up and 3 units to the right. We get to a new point $(3, 0)$. In a similar manner we can move from the point $(3, 0)$ to find another point $(6, 4)$. We plot these points and draw the line.

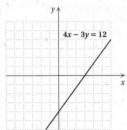

25. $f(x) = \dfrac{1}{3}x - 4$

First we plot the y-intercept $(0, -4)$. Then we consider the slope $\dfrac{1}{3}$. Starting at the y-intercept and using the slope, we find another point by moving 1 unit up and 3 units to the right. We get to a new point $(3, -3)$.

We can also think of the slope as $\dfrac{-1}{-3}$. We again start at the y-intercept $(0, -4)$. We move 1 unit down and 3 units to the left. We get to another new point $(-3, -5)$. We plot these points and draw the line.

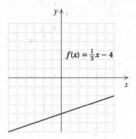

27. $5x + 4 \cdot f(x) = 4$

First we solve for $f(x)$.

$$5x + 4 \cdot f(x) = 4$$
$$4 \cdot f(x) = -5x + 4$$
$$\frac{4 \cdot f(x)}{4} = \frac{-5x + 4}{4}$$
$$f(x) = -\frac{5}{4}x + 1$$

Now we plot the y-intercept $(0, 1)$. We can think of the slope as $\dfrac{-5}{4}$. Starting at the y-intercept and using the slope, we find another point by moving 5 units down and 4 units to the right. We get to a new point $(4, -4)$.

We can also think of the slope as $\dfrac{5}{-4}$. We again start at the y-intercept $(0, 1)$. We move 5 units up and 4 units to the left. We get to another new point $(-4, 6)$. We plot these points and draw the line.

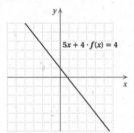

29. $x = 1$

Since y is missing, any number for y will do. Thus all ordered pairs $(1, y)$ are solutions. The graph is parallel to the y-axis.

x	y	
1	-2	
1	0	$\longleftarrow$ x-intercept
1	3	

   ↑   └── Choose any
x must   number for y.
be 1.

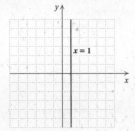

This is a vertical line, so the slope is not defined.

31. $y = -1$

Since x is missing, any number for x will do. Thus all ordered pairs $(x, -1)$ are solutions. The graph is parallel to the x-axis.

x	y
-2	-1
0	-1	$\longleftarrow$ y-intercept
3	-1

$\uparrow$ $\quad$ $\underline{\quad}$ y must be -1.

Choose
any number
for x.

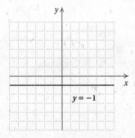

This is a horizontal line, so the slope is 0.

33. $f(x) = -6$

Since x is missing all ordered pairs $(x, 6)$ are solutions. The graph is parallel to the x-axis.

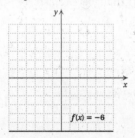

This is a horizontal line, so the slope is 0.

35. $y = 0$

Since x is missing, all ordered pairs $(x, 0)$ are solutions. The graph is the x-axis.

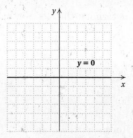

This is a horizontal line, so the slope is 0.

37. $2 \cdot f(x) + 5 = 0$

$$2 \cdot f(x) = -5$$

$$f(x) = -\frac{5}{2}$$

Since x is missing, all ordered pairs $\left(x, -\dfrac{5}{2}\right)$ are solutions. The graph is parallel to the x-axis.

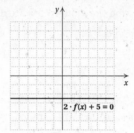

This is a horizontal line, so the slope is 0.

39. $7 - 3x = 4 + 2x$

$$7 - 5x = 4$$

$$-5x = -3$$

$$x = \frac{3}{5}$$

Since y is missing, all ordered pairs $\left(\dfrac{3}{5}, y\right)$ are solutions. The graph is parallel to the y-axis.

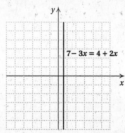

This is a vertical line, so the slope is not defined.

41. We first solve for y and determine the slope of each line.

$$x + 6 = y$$

$$y = x + 6 \quad \text{Reversing the order}$$

The slope of $y = x + 6$ is 1.

$$y - x = -2$$

$$y = x - 2$$

The slope of $y = x - 2$ is 1.

The slopes are the same, and the y-intercepts are different. The lines are parallel.

43. We first solve for y and determine the slope of each line.

$$y + 3 = 5x$$
$$y = 5x - 3$$

The slope of $y = 5x - 3$ is 5.

$$3x - y = -2$$
$$3x + 2 = y$$
$$y = 3x + 2 \quad \text{Reversing the order}$$

The slope of $y = 3x + 2$ is 3.

The slopes are not the same; the lines are not parallel.

45. We determine the slope of each line.

The slope of $y = 3x + 9$ is 3.

$$2y = 6x - 2$$
$$y = 3x - 1$$

The slope of $y = 3x - 1$ is 3.

The slopes are the same, and the y-intercepts are different. The lines are parallel.

47. We solve each equation for x.

$$\begin{array}{cc} 12x = 3 & -7x = 10 \\ x = \dfrac{1}{4} & x = -\dfrac{10}{7} \end{array}$$

We have two vertical lines, so they are parallel.

49. We determine the slope of each line.

The slope of $y = 4x - 5$ is 4.

$$4y = 8 - x$$
$$4y = -x + 8$$
$$y = -\frac{1}{4}x + 2$$

The slope of $4y = 8 - x$ is $-\dfrac{1}{4}$.

The product of their slopes is $4\left(-\dfrac{1}{4}\right)$, or -1; the lines are perpendicular.

51. We determine the slope of each line.

$$x + 2y = 5$$
$$2y = -x + 5$$
$$y = -\frac{1}{2}x + \frac{5}{2}$$

The slope of $x + 2y = 5$ is $-\dfrac{1}{2}$.

$$2x + 4y = 8$$
$$4y = -2x + 8$$
$$y = -\frac{1}{2}x + 2$$

The slope of $2x + 4y = 8$ is $-\dfrac{1}{2}$.

The product of their slopes is $\left(-\dfrac{1}{2}\right)\left(-\dfrac{1}{2}\right)$, or $\dfrac{1}{4}$; the lines are not perpendicular. For the lines to be perpendicular, the product must be -1.

53. We determine the slope of each line.

$$2x - 3y = 7$$
$$-3y = -2x + 7$$
$$y = \frac{2}{3}x - \frac{7}{3}$$

The slope of $2x - 3y = 7$ is $\dfrac{2}{3}$.

$$2y - 3x = 10$$
$$2y = 3x + 10$$
$$y = \frac{3}{2}x + 5$$

The slope of $2y - 3x = 10$ is $\dfrac{3}{2}$.

The product of their slopes is $\dfrac{2}{3} \cdot \dfrac{3}{2} = 1$; the lines are not perpendicular. For the lines to be perpendicular, the product must be -1.

55. Solving the first equation for x and the second for y, we have $x = \dfrac{3}{2}$ and $y = -2$. The graph of $x = \dfrac{3}{2}$ is a vertical line, and the graph of $y = -2$ is a horizontal line. Since one line is vertical and the other is horizontal, the lines are perpendicular.

57. Discussion and Writing Exercise

59.   5.3,000,000,000.

        ⌞————⌟

        10 places

Large number, so the exponent is positive.

$$53,000,000,000 = 5.3 \times 10^{10}$$

61.   0.01. 8

    ⌞⌝

   2 places

Small number, so the exponent is negative.

$$0.018 = 1.8 \times 10^{-2}$$

63. Negative exponent, so the number is small.

   0.00002. 13

    ⌞——⌟

   5 places

$$2.13 \times 10^{-5} = 0.0000213$$

65. Positive exponent, so the number is large.

   2.0000.

   ⌞——⌝

   4 places

$$2 \times 10^4 = 20,000$$

67. $9x - 15y = 3 \cdot 3x - 3 \cdot 5y = 3(3x - 5y)$

69. $21p - 7pq + 14p = 7p \cdot 3 - 7p \cdot q + 7p \cdot 2$
$$= 7p(3 - q + 2)$$

71. The equation will be of the form $y = b$. Since the line passes through $(-2, 3)$, b must be 3. Thus, we have $y = 3$.

73. Find the slope of each line.

$$5y = ax + 5$$

$$y = \frac{a}{5}x + 1$$

The slope of $5y = ax + 5$ is $\frac{a}{5}$.

$$\frac{1}{4}y = \frac{1}{10}x - 1$$

$$4 \cdot \frac{1}{4}y = 4\left(\frac{1}{10}x - 1\right)$$

$$y = \frac{2}{5}x - 4$$

The slope of $\frac{1}{4}y = \frac{1}{10}x - 1$ is $\frac{2}{5}$.

In order for the graphs to be parallel, their slopes must be the same. (Note that the y-intercepts are different.)

$$\frac{a}{5} = \frac{2}{5}$$

$$a = 2 \qquad \text{Multiplying by 5}$$

75. The y-intercept is $\left(0, \frac{2}{5}\right)$, so the equation is of the form $y = mx + \frac{2}{5}$. We substitute -3 for x and 0 for y in this equation to find m.

$$y = mx + \frac{2}{5}$$

$$0 = m(-3) + \frac{2}{5} \qquad \text{Substituting}$$

$$0 = -3m + \frac{2}{5}$$

$$3m = \frac{2}{5} \qquad \text{Adding } 3m$$

$$m = \frac{2}{15} \qquad \text{Multiplying by } \frac{1}{3}$$

The equation is $y = \frac{2}{15}x + \frac{2}{5}$.

(We could also have found the slope as follows:

$$m = \frac{\frac{2}{5} - 0}{0 - (-3)} = \frac{\frac{2}{5}}{3} = \frac{2}{15})$$

77. All points on the x-axis are pairs of the form $(x, 0)$. Thus any number for x will do and y must be 0. The equation is $y = 0$. This equation is a function because its graph passes the vertical-line test.

79. We substitute 4 for x and 0 for y.

$$y = mx + 3$$

$$0 = m(4) + 3$$

$$-3 = 4m$$

$$-\frac{3}{4} = m$$

81. a) Graph II indicates that 200 mL of fluid was dripped in the first 3 hr, a rate of 200/3 mL/hr. It also indicates that 400 mL of fluid was dripped in the next 3 hr, a rate of 400/3 mL/hr, and that this rate continues until the end of the time period shown. Since the rate of 400/3 mL/hr is double the rate of 200/3 mL/hr, this graph is appropriate for the given situation.

b) Graph IV indicates that 300 mL of fluid was dripped in the first 2 hr, a rate of 300/2, or 150 mL/hr. In the next 2 hr, 200 mL was dripped. This is a rate of 200/2, or 100 mL/hr. Then 100 mL was dripped in the next 3 hr, a rate of 100/3, or $33\frac{1}{3}$ mL/hr. Finally, in the remaining 2 hr, 0 mL of fluid was dripped, a rate of 0/2, or 0 mL/hr. Since the rate at which the fluid was given decreased as time progressed and eventually became 0, this graph is appropriate for the given situation.

c) Graph I is the only graph that shows a constant rate for 5 hours, in this case from 3 PM to 8 PM. Thus, it is appropriate for the given situation.

d) Graph III indicates that 100 mL of fluid was dripped in the first 4 hr, a rate of 100/4, or 25 mL/hr. In the next 3 hr, 200 mL was dripped. This is a rate of 200/3, or $66\frac{2}{3}$ mL/hr. Then 100 mL was dripped in the next hour, a rate of 100 mL/hr. In the last hour 200 mL was dripped, a rate of 200 mL/hr. Since the rate at which the fluid was given gradually increased, this graph is appropriate for the given situation.

Exercise Set 2.6

1. We use the slope-intercept equation and substitute -8 for m and 4 for b.

$$y = mx + b$$

$$y = -8x + 4$$

3. We use the slope-intercept equation and substitute 2.3 for m and -1 for b.

$$y = mx + b$$

$$y = 2.3x - 1$$

5. We use the slope-intercept equation and substitute $-\frac{7}{3}$ for m and -5 for b.

$$y = mx + b$$

$$y = -\frac{7}{3}x - 5$$

7. We use the slope-intercept equation and substitute $\frac{2}{3}$ for m and $\frac{5}{8}$ for b.

$$y = mx + b$$

$$y = \frac{2}{3}x + \frac{5}{8}$$

9. Using the point-slope equation:

Substitute 4 for x_1, 3 for y_1, and 5 for m.

$$y - y_1 = m(x - x_1)$$
$$y - 3 = 5(x - 4)$$
$$y - 3 = 5x - 20$$
$$y = 5x - 17$$

Using the slope-intercept equation:

Substitute 4 for x, 3 for y, and 5 for m in $y = mx + b$ and solve for b.

$$y = mx + b$$
$$3 = 5 \cdot 4 + b$$
$$3 = 20 + b$$
$$-17 = b$$

Then we use the equation $y = mx + b$ and substitute 5 for m and -17 for b.

$$y = 5x - 17$$

11. Using the point-slope equation:

Substitute 9 for x_1, 6 for y_1, and -3 for m.

$$y - y_1 = m(x - x_1)$$
$$y - 6 = -3(x - 9)$$
$$y - 6 = -3x + 27$$
$$y = -3x + 33$$

Using the slope-intercept equation:

Substitute 9 for x, 6 for y, and -3 for m in $y = mx + b$ and solve for b.

$$y = mx + b$$
$$6 = -3 \cdot 9 + b$$
$$6 = -27 + b$$
$$33 = b$$

Then we use the equation $y = mx + b$ and substitute -3 for m and 33 for b.

$$y = -3x + 33$$

13. Using the point-slope equation:

Substitute -1 for x_1, -7 for y_1, and 1 for m.

$$y - y_1 = m(x - x_1)$$
$$y - (-7) = 1(x - (-1))$$
$$y + 7 = 1(x + 1)$$
$$y + 7 = x + 1$$
$$y = x - 6$$

Using the slope-intercept equation:

Substitute -1 for x, -7 for y, and 1 for m in $y = mx + b$ and solve for b.

$$y = mx + b$$
$$-7 = 1(-1) + b$$
$$-7 = -1 + b$$
$$-6 = b$$

Then we use the equation $y = mx + b$ and substitute 1 for m and -6 for b.

$$y = 1x - 6, \text{ or } y = x - 6$$

15. Using the point-slope equation:

Substitute 8 for x_1, 0 for y_1, and -2 for m.

$$y - y_1 = m(x - x_1)$$
$$y - 0 = -2(x - 8)$$
$$y = -2x + 16$$

Using the slope-intercept equation:

Substitute 8 for x, 0 for y, and -2 for m in $y = mx + b$ and solve for b.

$$y = mx + b$$
$$0 = -2 \cdot 8 + b$$
$$0 = -16 + b$$
$$16 = b$$

Then we use the equation $y = mx + b$ and substitute -2 for m and 16 for b.

$$y = -2x + 16$$

17. Using the point-slope equation:

Substitute 0 for x_1, -7 for y_1, and 0 for m.

$$y - y_1 = m(x - x_1)$$
$$y - (-7) = 0(x - 0)$$
$$y + 7 = 0$$
$$y = -7$$

Using the slope-intercept equation:

Substitute 0 for x, -7 for y, and 0 for m in $y = mx + b$ and solve for b.

$$y = mx + b$$
$$-7 = 0 \cdot 0 + b$$
$$-7 = b$$

Then we use the equation $y = mx + b$ and substitute 0 for m and -7 for b.

$$y = 0x - 7, \text{ or } y = -7$$

19. Using the point-slope equation:

Substitute 1 for x_1, -2 for y_1, and $\dfrac{2}{3}$ for m.

$$y - y_1 = m(x - x_1)$$
$$y - (-2) = \frac{2}{3}(x - 1)$$
$$y + 2 = \frac{2}{3}x - \frac{2}{3}$$
$$y = \frac{2}{3}x - \frac{8}{3}$$

Using the slope-intercept equation:

Substitute 1 for x, -2 for y and $\dfrac{2}{3}$ for m in $y = mx + b$ and solve for b.

$$y = mx + b$$
$$-2 = \frac{2}{3} \cdot 1 + b$$
$$-2 = \frac{2}{3} + b$$
$$-\frac{8}{3} = b$$

Then we use the equation $y = mx + b$ and substitute $\frac{2}{3}$ for m and $-\frac{8}{3}$ for b.

$$y = \frac{2}{3}x - \frac{8}{3}$$

21. First find the slope of the line:

$$m = \frac{6-4}{5-1} = \frac{2}{4} = \frac{1}{2}$$

Using the point-slope equation:

We choose to use the point $(1, 4)$ and substitute 1 for x_1, 4 for y_1, and $\frac{1}{2}$ for m.

$$y - y_1 = m(x - x_1)$$
$$y - 4 = \frac{1}{2}(x - 1)$$
$$y - 4 = \frac{1}{2}x - \frac{1}{2}$$
$$y = \frac{1}{2}x + \frac{7}{2}$$

Using the slope-intercept equation:

We choose $(1, 4)$ and substitute 1 for x, 4 for y, and $\frac{1}{2}$ for m in $y = mx + b$. Then we solve for b.

$$y = mx + b$$
$$4 = \frac{1}{2} \cdot 1 + b$$
$$4 = \frac{1}{2} + b$$
$$\frac{7}{2} = b$$

Finally, we use the equation $y = mx + b$ and substitute $\frac{1}{2}$ for m and $\frac{7}{2}$ for b.

$$y = \frac{1}{2}x + \frac{7}{2}$$

23. First find the slope of the line:

$$m = \frac{-3-2}{-3-2} = \frac{-5}{-5} = 1$$

Using the point-slope equation:

We choose to use the point $(2, 2)$ and substitute 2 for x_1, 2 for y_1, and 1 for m.

$$y - y_1 = m(x - x_1)$$
$$y - 2 = 1(x - 2)$$
$$y - 2 = x - 2$$
$$y = x$$

Using the slope-intercept equation:

We choose $(2, 2)$ and substitute 2 for x, 2 for y, and 1 for m in $y = mx + b$. Then we solve for b.

$$y = mx + b$$
$$2 = 1 \cdot 2 + b$$
$$2 = 2 + b$$
$$0 = b$$

Finally, we use the equation $y = mx + b$ and substitute 1 for m and 0 for b.

$$y = 1x + 0, \text{ or } y = x$$

25. First find the slope of the line:

$$m = \frac{0-7}{-4-0} = \frac{-7}{-4} = \frac{7}{4}$$

Using the point-slope equation:

We choose $(0, 7)$ and substitute 0 for x_1, 7 for y_1, and $\frac{7}{4}$ for m.

$$y - y_1 = m(x - x_1)$$
$$y - 7 = \frac{7}{4}(x - 0)$$
$$y - 7 = \frac{7}{4}x$$
$$y = \frac{7}{4}x + 7$$

Using the slope-intercept equation:

We choose $(0, 7)$ and substitute 0 for x, 7 for y, and $\frac{7}{4}$ for m in $y = mx + b$. Then we solve for b.

$$y = mx + b$$
$$7 = \frac{7}{4} \cdot 0 + b$$
$$7 = b$$

Finally, we use the equation $y = mx + b$ and substitute $\frac{7}{4}$ for m and 7 for b.

$$y = \frac{7}{4}x + 7$$

27. First find the slope of the line:

$$m = \frac{-6-(-3)}{-4-(-2)} = \frac{-6+3}{-4+2} = \frac{-3}{-2} = \frac{3}{2}$$

Using the point-slope equation:

We choose $(-2, -3)$ and substitute -2 for x_1, -3 for y_1, and $\frac{3}{2}$ for m.

$$y - y_1 = m(x - x_1)$$
$$y - (-3) = \frac{3}{2}(x - (-2))$$
$$y + 3 = \frac{3}{2}(x + 2)$$
$$y + 3 = \frac{3}{2}x + 3$$
$$y = \frac{3}{2}x$$

Using the slope-intercept equation:

We choose $(-2, -3)$ and substitute -2 for x, -3 for y, and $\frac{3}{2}$ for m in $y = mx + b$. Then we solve for b.

$$y = mx + b$$
$$-3 = \frac{3}{2}(-2) + b$$
$$-3 = -3 + b$$
$$0 = b$$

Finally, we use the equation $y = mx + b$ and substitute $\frac{3}{2}$ for m and 0 for b.

$$y = \frac{3}{2}x + 0, \text{ or } y = \frac{3}{2}x$$

29. First find the slope of the line:

$$m = \frac{1 - 0}{6 - 0} = \frac{1}{6}$$

Using the point-slope equation:

We choose $(0, 0)$ and substitute 0 for x_1, 0 for y_1, and $\frac{1}{6}$ for m.

$$y - y_1 = m(x - x_1)$$
$$y - 0 = \frac{1}{6}(x - 0)$$
$$y = \frac{1}{6}x$$

Using the slope-intercept equation:

We choose $(0, 0)$ and substitute 0 for x, 0 for y, and $\frac{1}{6}$ for m in $y = mx + b$. Then we solve for b.

$$y = mx + b$$
$$0 = \frac{1}{6} \cdot 0 + b$$
$$0 = b$$

Finally, we use the equation $y = mx + b$ and substitute $\frac{1}{6}$ for m and 0 for b.

$$y = \frac{1}{6}x + 0, \text{ or } y = \frac{1}{6}x$$

31. First find the slope of the line:

$$m = \frac{-\frac{1}{2} - 6}{\frac{1}{4} - \frac{3}{4}} = \frac{-\frac{13}{2}}{-\frac{1}{2}} = 13$$

Using the point-slope equation:

We choose $\left(\frac{3}{4}, 6\right)$ and substitute $\frac{3}{4}$ for x_1, 6 for y_1, and 13 for m.

$$y - y_1 = m(x - x_1)$$
$$y - 6 = 13\left(x - \frac{3}{4}\right)$$
$$y - 6 = 13x - \frac{39}{4}$$
$$y = 13x - \frac{15}{4}$$

Using the slope-intercept equation:

We choose $\left(\frac{3}{4}, 6\right)$ and substitute $\frac{3}{4}$ for x, 6 for y, and 13 for m in $y = mx + b$. Then we solve for b.

$$y = mx + b$$
$$6 = 13 \cdot \frac{3}{4} + b$$
$$6 = \frac{39}{4} + b$$
$$-\frac{15}{4} = b$$

Finally, we use the equation $y = mx + b$ and substitute 13 for m and $-\frac{15}{4}$ for b.

$$y = 13x - \frac{15}{4}$$

33. First solve the equation for y and determine the slope of the given line.

$$x + 2y = 6 \qquad \text{Given line}$$
$$2y = -x + 6$$
$$y = -\frac{1}{2}x + 3$$

The slope of the given line is $-\frac{1}{2}$. The line through $(3, 7)$ must have slope $-\frac{1}{2}$.

Using the point-slope equation:

Substitute 3 for x_1, 7 for y_1, and $-\frac{1}{2}$ for m.

$$y - y_1 = m(x - x_1)$$
$$y - 7 = -\frac{1}{2}(x - 3)$$
$$y - 7 = -\frac{1}{2}x + \frac{3}{2}$$
$$y = -\frac{1}{2}x + \frac{17}{2}$$

Using the slope-intercept equation:

Substitute 3 for x, 7 for y, and $-\frac{1}{2}$ for m and solve for b.

$$y = mx + b$$
$$7 = -\frac{1}{2} \cdot 3 + b$$
$$7 = -\frac{3}{2} + b$$
$$\frac{17}{2} = b$$

Then we use the equation $y = mx + b$ and substitute $-\frac{1}{2}$ for m and $\frac{17}{2}$ for b.

$$y = -\frac{1}{2}x + \frac{17}{2}$$

35. First solve the equation for y and determine the slope of the given line.

$$5x - 7y = 8 \qquad \text{Given line}$$
$$5x - 8 = 7y$$
$$\frac{5}{7}x - \frac{8}{7} = y$$
$$y = \frac{5}{7}x - \frac{8}{7}$$

The slope of the given line is $\frac{5}{7}$. The line through $(2, -1)$ must have slope $\frac{5}{7}$.

Using the point-slope equation:

Substitute 2 for x_1, -1 for y_1, and $\frac{5}{7}$ for m.

$$y - y_1 = m(x - x_1)$$

$$y - (-1) = \frac{5}{7}(x - 2)$$

$$y + 1 = \frac{5}{7}x - \frac{10}{7}$$

$$y = \frac{5}{7}x - \frac{17}{7}$$

Using the slope-intercept equation:

Substitute 2 for x, -1 for y, and $\frac{5}{7}$ for m and solve for b.

$$y = mx + b$$

$$-1 = \frac{5}{7} \cdot 2 + b$$

$$-1 = \frac{10}{7} + b$$

$$-\frac{17}{7} = b$$

Then we use the equation $y = mx + b$ and substitute $\frac{5}{7}$ for m and $-\frac{17}{7}$ for b.

$$y = \frac{5}{7}x - \frac{17}{7}$$

37. First solve the equation for y and determine the slope of the given line.

$$3x - 9y = 2 \quad \text{Given line}$$

$$3x - 2 = 9y$$

$$\frac{1}{3}x - \frac{2}{9} = y$$

The slope of the given line is $\frac{1}{3}$. The line through $(-6, 2)$ must have slope $\frac{1}{3}$.

Using the point-slope equation:

Substitute -6 for x_1, 2 for y_1, and $\frac{1}{3}$ for m.

$$y - y_1 = m(x - x_1)$$

$$y - 2 = \frac{1}{3}(x - (-6))$$

$$y - 2 = \frac{1}{3}(x + 6)$$

$$y - 2 = \frac{1}{3}x + 2$$

$$y = \frac{1}{3}x + 4$$

Using the slope-intercept equation:

Substitute -6 for x, 2 for y, and $\frac{1}{3}$ for m and solve for b.

$$y = mx + b$$

$$2 = \frac{1}{3}(-6) + b$$

$$2 = -2 + b$$

$$4 = b$$

Then we use the equation $y = mx + b$ and substitute $\frac{1}{3}$ for m and 4 for b.

$$y = \frac{1}{3}x + 4$$

39. First solve the equation for y and determine the slope of the given line.

$$2x + y = -3 \quad \text{Given line}$$

$$y = -2x - 3$$

The slope of the given line is -2. The slope of the perpendicular line is the opposite of the reciprocal of -2. Thus, the line through $(2, 5)$ must have slope $\frac{1}{2}$.

Using the point-slope equation:

Substitute 2 for x_1, 5 for y_1, and $\frac{1}{2}$ for m.

$$y - y_1 = m(x - x_1)$$

$$y - 5 = \frac{1}{2}(x - 2)$$

$$y - 5 = \frac{1}{2}x - 1$$

$$y = \frac{1}{2}x + 4$$

Using the slope-intercept equation:

Substitute 2 for x, 5 for y, and $\frac{1}{2}$ for m and solve for b.

$$y = mx + b$$

$$5 = \frac{1}{2} \cdot 2 + b$$

$$5 = 1 + b$$

$$4 = b$$

Then we use the equation $y = mx + b$ and substitute $\frac{1}{2}$ for m and 4 for b.

$$y = \frac{1}{2}x + 4$$

41. First solve the equation for y and determine the slope of the given line.

$$3x + 4y = 5 \quad \text{Given line}$$

$$4y = -3x + 5$$

$$y = -\frac{3}{4}x + \frac{5}{4}$$

The slope of the given line is $-\frac{3}{4}$. The slope of the perpendicular line is the opposite of the reciprocal of $-\frac{3}{4}$. Thus, the line through $(3, -2)$ must have slope $\frac{4}{3}$.

Using the point-slope equation:

Substitute 3 for x_1, -2 for y_1, and $\frac{4}{3}$ for m.

$$y - y_1 = m(x - x_1)$$

$$y - (-2) = \frac{4}{3}(x - 3)$$

$$y + 2 = \frac{4}{3}x - 4$$

$$y = \frac{4}{3}x - 6$$

Using the slope-intercept equation:

Substitute 3 for x, -2 for y, and $\frac{4}{3}$ for m.

$$y = mx + b$$
$$-2 = \frac{4}{3} \cdot 3 + b$$
$$-2 = 4 + b$$
$$-6 = b$$

Then we use the equation $y = mx + b$ and substitute $\frac{4}{3}$ for m and -6 for b.

$$y = \frac{4}{3}x - 6$$

43. First solve the equation for y and determine the slope of the given line.

$$2x + 5y = 7 \qquad \text{Given line}$$
$$5y = -2x + 7$$
$$y = -\frac{2}{5}x + \frac{7}{5}$$

The slope of the given line is $-\frac{2}{5}$. The slope of the perpendicular line is the opposite of the reciprocal of $-\frac{2}{5}$. Thus, the line through $(0, 9)$ must have slope $\frac{5}{2}$.

Using the point-slope equation:

Substitute 0 for x_1, 9 for y_1, and $\frac{5}{2}$ for m.

$$y - y_1 = m(x - x_1)$$
$$y - 9 = \frac{5}{2}(x - 0)$$
$$y - 9 = \frac{5}{2}x$$
$$y = \frac{5}{2}x + 9$$

Using the slope-intercept equation:

Substitute 0 for x, 9 for y, and $\frac{5}{2}$ for m.

$$y = mx + b$$
$$9 = \frac{5}{2} \cdot 0 + b$$
$$9 = b$$

Then we use the equation $y = mx + b$ and substitute $\frac{5}{2}$ for m and 9 for b.

$$y = \frac{5}{2}x + 9$$

45. a) The problem describes a situation in which an hourly fee is charged after an initial assessment of \$85. After 1 hour, the total cost is $\$85 + \$40 \cdot 1$. After 2 hours, the total cost is $\$85 + \$40 \cdot 2$. Then after t hours, the total cost is $C(t) = 85 + 40t$, or $C(t) = 40t + 85$, where $t > 0$.

b) For $C(t) = 40t + 85$, the y-intercept is $(0, 85)$ and the slope, or rate of change, is \$40 per hour. We plot $(0, 85)$ and from there we count up \$40 and to the right 1 hour. This takes us to $(1, 125)$. Then we draw a line through the points, calculating a third value as a check:

$$C(5) = 40 \cdot 5 + 85 = 285$$

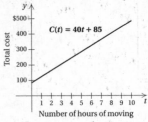

c) To find the cost for $6\frac{1}{2}$ hours of moving service, we determine $C(6.5)$:

$$C(6.5) = 40(6.5) + 85 = 345$$

Thus, it would cost \$345 for $6\frac{1}{2}$ hours of moving service.

47. a) The problem describes a situation in which the value of the fax machine decreases at the rate of \$25 per month from an initial value of \$750. After 1 month, the value is $\$750 - \$25 \cdot 1$. After 2 months, the value is $\$750 - \$25 \cdot 2$. Then after t months, the value is $V(t) = 750 - 25t$, where $t \geq 0$.

b) For $V(t) = 750 - 25t$, or $V(t) = -25t + 750$, the y-intercept is $(0, 750)$ and the slope, or rate of change, is $-\$25$ per month. We think of the slope as $\frac{-25}{1}$. Plot $(0, 750)$ and from there count down \$25 and to the right 1 month. This takes us to $(1, 725)$. We draw the line through the points, calculating a third value as a check:

$$V(6) = 750 - 25 \cdot 6 = 600$$

c) To find the value of the machine after 13 months, we determine $V(13)$:

$$V(13) = 750 - 25 \cdot 13 = 425$$

Thus, after 13 months the value of the machine is \$425.

49. a) One data point is $(0, 2.707)$. In 2003, $x = 2003 - 1999 = 4$, so the other data point is $(4, 3.215)$.

First we find the slope of the line:
$$m = \frac{3.215 - 2.707}{4 - 0} = \frac{0.508}{4} = 0.127.$$

Using the slope, 0.127, and the y-intercept, $(0, 2.707)$, we write the function:
$P(x) = 0.127x + 2.707$.

b) In 2002, $x = 2002 - 1999 = 3$. Substitute 3 for x and do the computation.

$$P(3) = 0.127(3) + 2.707 = 3.088$$

We estimate that 3.088 trillion prescriptions were filled in 2002.

In 2008, $x = 2008 - 1999 = 9$. Substitute 9 for x and do the computation.

$$P(9) = 0.127(9) + 2.707 = 3.85$$

We estimate that the number of prescriptions filled in 2008 is 3.85 trillion.

51. a) We form pairs of the type (t, R) where t is the number of years since 1930 and R is the record. We have two pairs, $(0, 46.8)$ and $(40, 43.8)$. These are two points on the graph of the linear function we are seeking.

First we find the slope:

$$m = \frac{43.8 - 46.8}{40 - 0} = \frac{-3}{40} = -0.075$$

Using the slope and the y-intercept, $(0, 46.8)$ we write the function: $R(t) = -0.075t + 46.8$.

b) 2003 is 73 years since 1930, so to predict the record in 2003, we find $R(73)$:

$$R(73) = -0.075(73) + 46.8$$
$$= 41.325$$

The estimated record is 41.325 seconds in 2003.

2006 is 76 years since 1930, so to predict the record in 2006, we find $R(76)$:

$$R(76) = -0.075(76) + 46.8$$
$$= 41.1$$

The estimated record is 41.1 seconds in 2006.

c) Substitute 40 for $R(t)$ and solve for t:

$$40 = -0.075t + 46.8$$
$$-6.8 = -0.075t$$
$$91 \approx t$$

The record will be 40 seconds about 91 years after 1930, or in 2021.

53. a) We form pairs of the type (t, M) where t is the number of years since 1990 and M is the life expectancy. We have two pairs, $(0, 71.8)$ and $(11, 74.4)$. These are two points on the graph of the linear function we are seeking. First we find the slope.

$$m = \frac{74.4 - 71.8}{11 - 0} = \frac{2.6}{11} \approx 0.236$$

Using the slope and the y-intercept $(0, 71.8)$, we write the function: $M(t) = 0.236t + 71.8$.

b) 2007 is 17 years since 1990, so we find $M(17)$:

$$M(17) = 0.236(17) + 71.8 \approx 75.8$$

We estimate that the life expectancy of males will be about 75.8 years in 2007.

55. Discussion and Writing Exercise

57. $2x + 3 > 51$

$$\quad 2x > 48 \qquad \text{Subtracting 3}$$
$$\quad\;\, x > 24 \qquad \text{Dividing by 2}$$

The solution set is $\{x | x > 24\}$, or $(24, \infty)$.

59. $2x + 3 \leq 51$

$$\quad 2x \leq 48 \qquad \text{Subtracting 3}$$
$$\quad\;\, x \leq 24 \qquad \text{Dividing by 2}$$

The solution set is $\{x | x \leq 24\}$, or $(-\infty, 24]$.

61. $|2x + 3| \leq 13$

$$-13 \leq 2x + 3 \leq 13$$
$$-16 \leq 2x \leq 10 \qquad \text{Subtracting 3}$$
$$\;\;-8 \leq x \leq 5 \qquad \text{Dividing by 2}$$

The solution set is $\{x | -8 \leq x \leq 5\}$, or $[-8, 5]$.

63. $|5x - 4| = -8$

Since the absolute value of a number must be nonnegative, the equation has no solution. The solution set is $\{\ \}$, or $\emptyset$.

65. The value C of the computer, in dollars, after t months can be modeled by a line that contains the points $(6, 900)$ and $(8, 750)$. We find an equation relating C and t.

$$m = \frac{750 - 900}{8 - 6} = \frac{-150}{2} = -75$$
$$C - 900 = -75(t - 6)$$
$$C - 900 = -75t + 450$$
$$\qquad\; C = -75t + 1350$$

Using function notation we have $C(t) = -75t + 1350$. To find how much the computer cost we find $C(0)$:

$$C(0) = -75 \cdot 0 + 1350 = 1350$$

The computer cost \$1350.

Chapter 2 Review Exercises

1. To show that a pair is a solution, we substitute, replacing x with the first coordinate and y with the second coordinate in each pair.

$3x - y = 2$	
$3 \cdot 0 - (-2)$? 2	
$0 + 2$	
2	TRUE

$3x - y = 2$	
$3(-1) - (-5)$? 2	
$-3 + 5$	
2	TRUE

In each case the substitution results in a true equation. Thus, $(0, -2)$ and $(-1, -5)$ are both solutions of $3x - y = 2$. We plot these points and sketch the line passing through them.

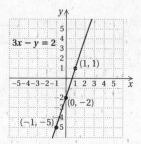

The line appears to pass through $(1, 1)$ also. We check to determine if $(1, 1)$ is a solution of $3x - y = 2$.

$$3x - y = 2$$

$$\frac{}{3 \cdot 1 - 1 \; ? \; 2}$$

$$3 - 1 \quad \Big|$$

$$\qquad \quad 2 \quad \Big| \quad \text{TRUE}$$

Thus, $(1, 1)$ is another solution. There are other correct answers, including $(2, 4)$.

2. $y = 3x + 2$

We find some ordered pairs that are solutions, plot them, and draw and label the line.

When $x = -1$, $y = -3(-1) + 2 = 3 + 2 = 5$.

When $x = 1$, $y = -3 \cdot 1 + 2 = -3 + 2 = -1$

When $x = 2$, $y = -3 \cdot 2 + 2 = -6 + 2 = -4$.

x	y	(x, y)
-1	5	$(-1, 5)$
1	-1	$(1, -1)$
2	-4	$(2, -4)$

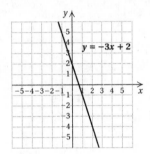

3. $y = \dfrac{5}{2}x - 3$

We find some ordered pairs that are solutions, using multiples of 2 to avoid fractions. Then we plot these points and draw and label the line.

When $x = 0$, $y = \dfrac{5}{2} \cdot 0 - 3 = 0 - 3 = -3$.

When $x = 2$, $y = \dfrac{5}{2} \cdot 2 - 3 = 5 - 3 = 2$.

When $x = 4$, $y = \dfrac{5}{2} \cdot 4 - 3 = 10 - 3 = 7$.

x	y	(x, y)
0	-3	$(0, -3)$
2	2	$(2, 2)$
4	7	$(4, 7)$

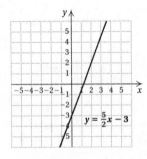

4. $y = |x - 3|$

To find an ordered pair, we choose any number for x and then determine y. For example, if $x = 5$, then $y = |5 - 3| = |2| = 2$. We find several ordered pairs, plot them, and connect them with a smooth curve.

x	y
5	2
3	0
2	1
-1	4
-2	5
-3	6

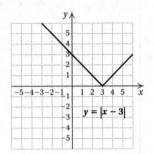

5. $y = 3 - x^2$

To find an ordered pair, we choose any number for x and then determine y. For example, if $x = 2$, then $3 - 2^2 = 3 - 4 = -1$. We find several ordered pairs, plot them, and connect them with a smooth curve.

x	y
-2	-1
-1	2
0	3
1	2
2	-1
3	-6

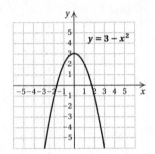

6. No; a member of the domain, 3, is matched to more than one member of the range.

7. Yes; each member of the domain is matched to only one member of the range.

8. $g(x) = -2x + 5$

$g(0) = -2 \cdot 0 + 5 = 0 + 5 = 5$

$g(-1) = -2(-1) + 5 = 2 + 5 = 7$

9. $f(x) = 3x^2 - 2x + 7$

$f(0) = 3 \cdot 0^2 - 2 \cdot 0 + 7 = 0 - 0 + 7 = 7$

$f(-1) = 3(-1)^2 - 2(-1) + 7 = 3 \cdot 1 - 2(-1) + 7 = 3 + 2 + 7 = 12$

10. $C(t) = 309.2t + 3717.7$

$C(10) = 309.2(10) + 3717.7 = 3092 + 3717.7 = 6809.7 \approx 6810$

We estimate that the average cost of tuition and fees will be about \$6810 in 2010.

11. No vertical line will intersect the graph more than once. Thus, the graph is the graph of a function.

12. It is possible for a vertical line to intersect the graph more than once. Thus, this is not the graph of a function.

13. a) Locate 2 on the horizontal axis and then find the point on the graph for which 2 is the first coordinate. From that point, look to the vertical axis to find the corresponding y-coordinate, 3. Thus, $f(2) = 3$.

b) The set of all x-values in the graph extends from -2 to 4, so the domain is $\{x | -2 \le x \le 4\}$, or $[-2, 4]$.

c) To determine which member(s) of the domain are paired with 2, locate 2 on the vertical axis. From there look left and right to the graph to find any points for which 2 is the second coordinate. One such point exists. Its first coordinate appears to be -1. Thus, the x-value for which $f(x) = 2$ is -1.

d) The set of all y-values in the graph extends from 1 to 5, so the range is $\{y|1 \le y \le 5\}$, or $[1, 5]$.

14. $f(x) = \dfrac{5}{x - 4}$

Since $\dfrac{5}{x - 4}$ cannot be calculated when the denominator is 0, we find the x-value that causes $x - 4$ to be 0:

$$x - 4 = 0$$
$$x = 4 \quad \text{Adding 4 on both sides}$$

Thus, 4 is not in the domain of f, while all other real numbers are. The domain of f is

$\{x|x \text{ is a real number } and \ x \ne 4\}$, or $(-\infty, 4) \cup (4, \infty)$.

15. $g(x) = x - x^2$

Since we can calculate $x - x^2$ for any real number x, the domain is the set of all real numbers.

16. $f(x) = -3x + 2$
$\qquad\quad\uparrow\qquad\uparrow$
$f(x) = mx + b$

The slope is -3, and the y-intercept is $(0, 2)$.

17. First we find the slope-intercept form of the equation by solving for y. This allows us to determine the slope and y-intercept easily.

$$4y + 2x = 8$$
$$4y = -2x + 8$$
$$\frac{4y}{4} = \frac{-2x + 8}{4}$$
$$y = -\frac{1}{2}x + 2$$

The slope is $-\dfrac{1}{2}$, and the y-intercept is $(0, 2)$.

18. Slope $= \dfrac{\text{change in } y}{\text{change in } x} = \dfrac{-4 - 7}{10 - 13} = \dfrac{-11}{-3} = \dfrac{11}{3}$

19. $2y + x = 4$

To find the x-intercept we let $y = 0$ and solve for x.

$$2y + x = 4$$
$$2 \cdot 0 + x = 4$$
$$x = 4$$

The x-intercept is $(4, 0)$.

To find the y-intercept we let $x = 0$ and solve for y.

$$2y + x = 4$$
$$2y + 0 = 4$$
$$2y = 4$$
$$y = 2$$

The y-intercept is $(0, 2)$.

We plot these points and draw the line.

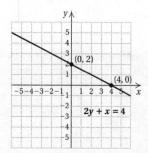

We use a third point as a check. We choose $x = -2$ and solve for y.

$$2y + (-2) = 4$$
$$2y = 6$$
$$y = 3$$

We plot $(-2, 3)$ and note that it is on the line.

20. $2y = 6 - 3x$

To find the x-intercept we let $y = 0$ and solve for x.

$$2y = 6 - 3x$$
$$2 \cdot 0 = 6 - 3x$$
$$0 = 6 - 3x$$
$$3x = 6$$
$$x = 2$$

The x-intercept is $(2, 0)$.

To find the y-intercept we let $x = 0$ and solve for y.

$$2y = 6 - 3x$$
$$2y = 6 - 3 \cdot 0$$
$$2y = 6$$
$$y = 3$$

The y-intercept is $(0, 3)$.

We plot these points and draw the line.

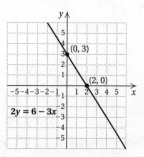

We use a third point as a check. We choose $x = 4$ and solve for y.

$$2y = 6 - 3 \cdot 4$$
$$2y = 6 - 12$$
$$2y = -6$$
$$y = -3$$

We plot $(4, -3)$ and note that it is on the line.

21. $g(x) = -\frac{2}{3}x - 4$

First we plot the y-intercept $(0, -4)$. We can think of the slope as $\frac{-2}{3}$. Starting at the y-intercept and using the slope, we find another point by moving 2 units down and 3 units to the right. We get to a new point $(3, -6)$.

We can also think of the slope as $\frac{2}{-3}$. We again start at the y-intercept $(0, -4)$. We move 2 units up and 3 units to the left. We get to another new point $(-3, -2)$. We plot the points and draw the line.

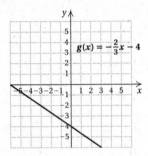

22. $f(x) = \frac{5}{2}x + 3$

First we plot the y-intercept $(0, 3)$. Then we consider the slope $\frac{5}{2}$. Starting at the y-intercept and using the slope, we find another point by moving 5 units up and 2 units to the right. We get to a new point $(2, 8)$.

We can also think of the slope as $\frac{-5}{-2}$. We again start at the y-intercept $(0, 3)$. We move 5 units down and 2 units to the left. We get to another new point $(-2, -2)$. We plot the points and draw the line.

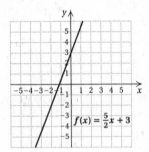

23. $x = -3$

Since y is missing, all ordered pairs $(-3, y)$ are solutions. The graph is parallel to the y-axis.

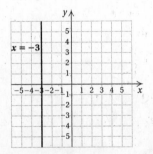

24. $f(x) = 4$

Since x is missing, all ordered pairs $(x, 4)$ are solutions. The graph is parallel to the x-axis.

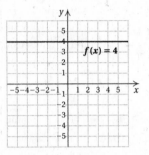

25. We first solve each equation for y and determine the slope of each line.
$$y + 5 = -x$$
$$y = -x - 5$$
The slope of $y + 5 = -x$ is -1.
$$x - y = 2$$
$$x = y + 2$$
$$x - 2 = y$$
The slope of $x - y = 2$ is 1.

The slopes are not the same, so the lines are not parallel. The product of the slopes is $-1 \cdot 1$, or -1, so the lines are perpendicular.

26. We first solve each equation for y and determine the slope of each line.
$$3x - 5 = 7y$$
$$\frac{3}{7}x - \frac{5}{7} = y$$
The slope of $3x - 5 = 7y$ is $\frac{3}{7}$.
$$7y - 3x = 7$$
$$7y = 3x + 7$$
$$y = \frac{3}{7}x + 1$$
The slope of $7y - 3x = 7$ is $\frac{3}{7}$.

The slopes are the same and the y-intercepts are different, so the lines are parallel.

27. We first solve each equation for y and determine the slope of each line.
$$4y + x = 3$$
$$4y = -x + 3$$
$$y = -\frac{1}{4}x + \frac{3}{4}$$
The slope of $4y + x = 3$ is $-\frac{1}{4}$.
$$2x + 8y = 5$$
$$8y = -2x + 5$$
$$y = -\frac{1}{4}x + \frac{5}{8}$$
The slope of $2x + 8y = 5$ is $-\frac{1}{4}$.

The slopes are the same and the y-intercepts are different, so the lines are parallel.

28. $x = 4$ is a vertical line and $y = -3$ is a horizontal line, so the lines are perpendicular.

29. We use the slope-intercept equation and substitute 4.7 for m and -23 for b..
$$y = mx + b$$
$$y = 4.7x - 23$$

30. Using the point-slope equation:

Substitute 3 for x_1, -5 for y_1, and -3 for m.
$$y - y_1 = m(x - x_1)$$
$$y - (-5) = -3(x - 3)$$
$$y + 5 = -3x + 9$$
$$y = -3x + 4$$

Using the slope-intercept equation:

Substitute 3 for x, -5 for y, and -3 for m in $y = mx + b$ and solve for b.
$$y = mx + b$$
$$-5 = -3 \cdot 3 + b$$
$$-5 = -9 + b$$
$$4 = b$$

Then we use the equation $y = mx + b$ and substitute -3 for m and 4 for b.
$$y = -3x + 4$$

31. First find the slope of the line:
$$m = \frac{6 - 3}{-4 - (-2)} = \frac{3}{-2} = -\frac{3}{2}$$

Using the point-slope equation:

We choose to use the point $(-2, 3)$ and substitute -2 for x_1, 3 for y_1, and $-\frac{3}{2}$ for m.
$$y - y_1 = m(x - x_1)$$
$$y - 3 = -\frac{3}{2}(x - (-2))$$
$$y - 3 = -\frac{3}{2}(x + 2)$$
$$y - 3 = -\frac{3}{2}x - 3$$
$$y = -\frac{3}{2}x$$

Using the slope-intercept equation:

We choose $(-2, 3)$ and substitute -2 for x, 3 for y, and $-\frac{3}{2}$ for m in $y = mx + b$. Then we solve for b.
$$3 = -\frac{3}{2}(-2) + b$$
$$3 = 3 + b$$
$$0 = b$$

Finally, we use the equation $y = mx + b$ and substitute $-\frac{3}{2}$ for m and 0 for b.
$$y = -\frac{3}{2}x + 0, \text{ or } y = -\frac{3}{2}x$$

32. First solve the equation for y and determine the slope of the given line.
$$5x + 7y = 8 \qquad \text{Given line}$$
$$7y = -5x + 8$$
$$y = -\frac{5}{7}x + \frac{8}{7}$$

The slope of the given line is $-\frac{5}{7}$. The line through $(14, -1)$ must have slope $-\frac{5}{7}$.

Using the point-slope equation:

Substitute 14 for x_1, -1 for y_1, and $-\frac{5}{7}$ for m.
$$y - y_1 = m(x - x_1)$$
$$y - (-1) = -\frac{5}{7}(x - 14)$$
$$y + 1 = -\frac{5}{7}x + 10$$
$$y = -\frac{5}{7}x + 9$$

Using the slope-intercept equation:

Substitute 14 for x, -1 for y, and $-\frac{5}{7}$ for m and solve for b.
$$y = mx + b$$
$$-1 = -\frac{5}{7} \cdot 14 + b$$
$$-1 = -10 + b$$
$$9 = b$$

Then we use the equation $y = mx + b$ and substitute $-\frac{5}{7}$ for m and 9 for b.
$$y = -\frac{5}{7}x + 9$$

33. First solve the equation for y and determine the slope of the given line.
$$3x + y = 5 \qquad \text{Given line}$$
$$y = -3x + 5$$

The slope of the given line is -3. The slope of the perpendicular line is the opposite of the reciprocal of -3. Thus, the line through $(5, 2)$ must have slope $\frac{1}{3}$.

Using the point-slope equation:

Substitute 5 for x_1, 2 for y_1, and $\frac{1}{3}$ for m.
$$y - y_1 = m(x - x_1)$$
$$y - 2 = \frac{1}{3}(x - 5)$$
$$y - 2 = \frac{1}{3}x - \frac{5}{3}$$
$$y = \frac{1}{3}x + \frac{1}{3}$$

Using the slope-intercept equation:

Substitute 5 for x, 2 for y, and $\frac{1}{3}$ for m and solve for b.

$$y = mx + b$$
$$2 = \frac{1}{3} \cdot 5 + b$$
$$2 = \frac{5}{3} + b$$
$$\frac{1}{3} = b$$

Then we use the equation $y = mx + b$ and substitute $\frac{1}{3}$ for m and $\frac{1}{3}$ for b.

$$y = \frac{1}{3}x + \frac{1}{3}$$

34. a) We form pairs of the type (t, R) where t is the number of years since 1970 and R is the record. We have two pairs, $(0, 46.8)$ and $(34, 44.63)$. These are two points on the graph of the linear function we are seeking.

First we find the slope:

$$m = \frac{44.63 - 46.8}{34 - 0} = \frac{-2.17}{34} \approx -0.064$$

Using the slope and the y-intercept, $(0, 46.8)$ we write the function: $R(t) = -0.064t + 46.8$.

   b) 2008 is 38 years since 1970, so to predict the record in 2008, we find $R(38)$:

$$R(38) = -0.064(38) + 46.8$$
$$\approx 44.37$$

The estimated record is 44.37 seconds in 2008.

2010 is 40 years since 1970, so to predict the record in 2010, we find $R(40)$:

$$R(40) = -0.064(40) + 46.8$$
$$= 44.24$$

The estimated record is 44.24 seconds in 2010.

35. *Discussion and Writing Exercise.* The concept of slope is useful in describing how a line slants. A line with positive slope slants up from left to right. A line with negative slope slants down from left to right. The larger the absolute value of the slope, the steeper the slant.

36. *Discussion and Writing Exercise.* The notation $f(x)$ can be read "f of x" or "f at x" or "the value of f at x." It represents the output of the function f for the input x. The notation $f(a) = b$ provides a concise way to indicate that for the input a, the output of the function f is b.

37. The cost of x jars of preserves is $\$2.49x$, and the shipping charges are $\$3.75 + \$0.60x$. Then the total cost is $\$2.49x + \$3.75 + \$0.60x$, or $\$3.09x + \$3.75$. Thus, a linear function that can be used to determine the cost of buying and shipping x jars of preserves is $f(x) = 3.09x + 3.75$.

Chapter 2 Test

1. To show that a pair is a solution, we substitute, replacing x with the first coordinate and y with the second coordinate.

$$
\begin{array}{c|l}
\multicolumn{2}{c}{y = 5 - 4x} \\
\hline
-3 \; ? \; 5 - 4 \cdot 2 & \\
5 - 8 & \\
-3 & \text{TRUE}
\end{array}
$$

The substitution results in a true equation. Thus, $(2, -3)$ is a solution of $y = 5 - 4x$.

2. To show that a pair is a solution, we substitute, replacing a with the first coordinate and b with the second coordinate.

$$
\begin{array}{c|l}
\multicolumn{2}{c}{5b - 7a = 10} \\
\hline
5(-3) - 7 \cdot 2 \; ? \; 10 & \\
-15 - 14 & \\
-29 & \text{FALSE}
\end{array}
$$

The substitution results in a false equation, so $(2, -3)$ is not a solution of $5b - 7a = 10$.

3. $y = -2x - 5$

We find some ordered pairs that are solutions, plot them, and draw and label the line.

When $x = 0$, $y = -2 \cdot 0 - 5 = 0 - 5 = -5$.

When $x = -2$, $y = -2(-2) - 5 = 4 - 5 = -1$.

When $x = -4$, $y = -2(-4) - 5 = 8 - 5 = 3$.

x	y
0	-5
-2	-1
-4	3

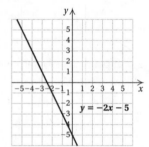

4. $f(x) = -\dfrac{3}{5}x$

We find some function values, plot the corresponding points, and draw the curve.

$$f(-5) = -\frac{3}{5}(-5) = 3$$
$$f(0) = -\frac{3}{5} \cdot 0 = 0$$
$$f(5) = -\frac{3}{5} \cdot 5 = -3$$

x	$f(x)$
-5	3
0	0
5	-3

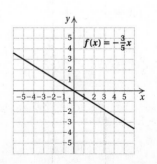

5. $g(x) = 2 - |x|$

We find some function values, plot the corresponding points, and draw the curve.

$g(-4) = 2 - |-4| = 2 - 4 = -2$

$g(-2) = 2 - |-2| = 2 - 2 = 0$

$g(0) = 2 - |0| = 2 - 0 = 2$

$g(3) = 2 - |3| = 2 - 3 = -1$

$g(5) = 2 - |5| = 2 - 5 = -3$

x	$g(x)$
-4	-2
-2	0
0	2
3	-1
5	-3

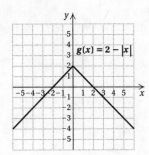

6. $y = \dfrac{4}{x}$

To find an ordered pair, we choose any number for x and then determine y. For example, if $x = -4$, then $y = \dfrac{4}{-4} = -1$. We find several ordered pairs, plot them, and connect them with a smooth curve.

x	y
-4	-1
-2	-2
-1	-4
1	4
2	2
4	1

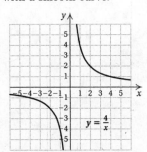

Note that we cannot use 0 as a first-coordinate, since 4/0 is undefined. Thus, the graph has two branches, one on each side of the y-axis.

7. a) In 2002, $x = 2002 - 1990 = 12$. We find $A(12)$.

$A(12) = 0.233(12) + 5.87 = 8.666$

The median age of cars in 2002 was 8.666 yr.

b) Substitute 7.734 for $A(t)$ and solve for t.

$7.734 = 0.233t + 5.87$

$1.864 = 0.233t$

$8 = t$

The median age of cars was 7.734 yr 8 years after 1990, or in 1998.

8. Yes; each member of the domain is matched to only one member of the range.

9. No; a member of the domain, Lake Placid, is matched to more than one member of the range.

10. $f(x) = -3x - 4$

$f(0) = -3 \cdot 0 - 4 = 0 - 4 = -4$

$f(-2) = -3(-2) - 4 = 6 - 4 = 2$

11. $g(x) = x^2 + 7$

$g(0) = 0^2 + 7 = 0 + 7 = 7$

$g(-1) = (-1)^2 + 7 = 1 + 7 = 8$

12. No vertical line will intersect the graph more than once. Thus, the graph is the graph of a function.

13. It is possible for a vertical line to intersect the graph more than once. Thus, this is not the graph of a function.

14. a) Locate 2000 on the x-axis and then move directly up to the graph. The point directly above 2000 lies on the y-axis at about 32. Thus, there were about 32 million persons 65 and older in 2000.

b) Locate 2040 on the x-axis and then move directly up to the graph. Next move across to the y-axis. We come to about 80, so there will be about 80 million persons 65 and older in 2040.

15. a) Locate 2 on the horizontal axis and the find the point on the graph for which 2 is the first coordinate. From that point, look to the vertical axis to find the corresponding y-coordinate. It appears to be about 1.2. Thus, $f(2) \approx 1.2$.

b) The set of all x-values in the graph extends from -3 to 4, so the domain is $\{x | -3 \le x \le 4\}$, or $[-3, 4]$.

c) To determine which member(s) of the domain are paired with 2, locate 2 on the vertical axis. From there look left and right to the graph to find any points for which 2 is the second coordinate. One such point exists. Its first coordinate is -3, so the x-value for which $f(x) = 2$ is -3.

d) The set of all y-values in the graph extends from -1 to 2, so the range is $\{y | -1 \le y \le 2\}$, or $[-1, 2]$.

16. $g(x) = 5 - x^2$

Since we can calculate $5 - x^2$ for any real number x, the domain is the set of all real numbers.

17. $f(x) = \dfrac{8}{2x + 3}$

Since $\dfrac{8}{2x + 3}$ cannot be calculated when the denominator is 0, we find the x-value that causes $2x + 3$ to be 0:

$2x + 3 = 0$

$2x = -3$

$x = -\dfrac{3}{2}$

Thus, $-\dfrac{3}{2}$ is not in the domain of f, while all other real numbers are. The domain of f is

$\left\{ x \middle| x \text{ is a real number } and \ x \ne -\dfrac{3}{2} \right\}$, or

$\left(-\infty, -\dfrac{3}{2} \right) \cup \left(-\dfrac{3}{2}, \infty \right)$.

18. $f(x) = -\dfrac{3}{5}x + 12$

$$f(x) = \underset{\uparrow}{mx} + \underset{\uparrow}{b}$$

The slope is $-\dfrac{3}{5}$, and the y-intercept is $(0, 12)$.

19. First we find the slope-intercept form of the equation by solving for y. This allows us to determine the slope and y-intercept easily.

$$-5y - 2x = 7$$
$$-5y = 2x + 7$$
$$\frac{-5y}{-5} = \frac{2x + 7}{-5}$$
$$y = -\frac{2}{5}x - \frac{7}{5}$$

The slope is $-\dfrac{2}{5}$, and the y-intercept is $\left(0, -\dfrac{7}{5}\right)$.

20. Slope $= \dfrac{\text{change in } y}{\text{change in } x} = \dfrac{-2 - 3}{-2 - 6} = \dfrac{-5}{-8} = \dfrac{5}{8}$

21. Slope $= \dfrac{\text{change in } y}{\text{change in } x} = \dfrac{5.2 - 5.2}{-4.4 - (-3.1)} = \dfrac{0}{-1.3} = 0$

22. We can use the coordinates of any two points on the graph. We'll use $(10, 0)$ and $(25, 12)$.

Slope $= \dfrac{\text{change in } y}{\text{change in } x} = \dfrac{12 - 0}{25 - 10} = \dfrac{12}{15} = \dfrac{4}{5}$

The slope, or rate of change is $\dfrac{4}{5}$ km/min.

23. $2x + 3y = 6$

To find the x-intercept we let $y = 0$ and solve for x.

$$2x + 3y = 6$$
$$2x + 3 \cdot 0 = 6$$
$$2x = 6$$
$$x = 3$$

The x-intercept is $(3, 0)$.

To find the y-intercept we let $x = 0$ and solve for y.

$$2x + 3y = 6$$
$$2 \cdot 0 + 3y = 6$$
$$3y = 6$$
$$y = 2$$

The y-intercept is $(0, 2)$.

We plot these points and draw the line.

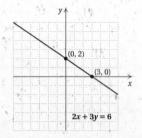

We use a third point as a check. We choose $x = -3$ and solve for y.

$$2(-3) + 3y = 6$$
$$-6 + 3y = 6$$
$$3y = 12$$
$$y = 4$$

We plot $(-3, 4)$ and note that it is on the line.

24. $f(x) = -\dfrac{2}{3}x - 1$

First we plot the y-intercept $(0, -1)$. We can think of the slope as $\dfrac{-2}{3}$. Starting at the y-intercept and using the slope, we find another point by moving 2 units down and 3 units to the right. We get to a new point $(3, -3)$.

We can also think of the slope as $\dfrac{2}{-3}$. We again start at the y-intercept $(0, -1)$. We move 2 units up and 3 units to the left. We get to another new point $(-3, 1)$. We plot the points and draw the line.

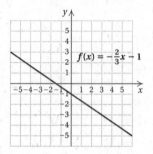

25. $y = f(x) = -3$

Since x is missing, all ordered pairs $(x, -3)$ are solutions. The graph is parallel to the x-axis.

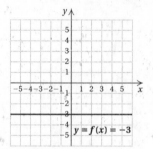

26. $2x = -4$
$$x = -2$$

Since y is missing, all ordered pairs $(-2, y)$ are solutions. The graph is parallel to the y-axis.

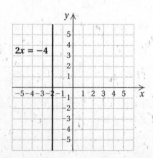

27. We first solve each equation for y and determine the slope of each line.

$$4y + 2 = 3x$$
$$4y = 3x - 2$$
$$y = \frac{3}{4}x - \frac{1}{2}$$

The slope of $4y + 2 = 3x$ is $\frac{3}{4}$.

$$-3x + 4y = -12$$
$$4y = 3x - 12$$
$$y = \frac{3}{4}x - 3$$

The slope of $-3x + 4y = -12$ is $\frac{3}{4}$.

The slopes are the same and the y-intercepts are different, so the lines are parallel.

28. The slope of $y = -2x + 5$ is -2.

We solve the second equation for y and determine the slope.

$$2y - x = 6$$
$$2y = x + 6$$
$$y = \frac{1}{2}x + 3$$

The slopes are not the same, so the lines are not parallel. The product of the slopes is $-2 \cdot \frac{1}{2}$, or -1, so the lines are perpendicular.

29. We use the slope-intercept equation and substitute -3 for m and 4.8 for b.

$$y = mx + b$$
$$y = -3x + 4.8$$

30. $y = f(x) = mx + b$

$$f(x) = 5.2x - \frac{5}{8}$$

31. Using the point-slope equation:

Substitute 1 for x_1, -2 for y_1, and -4 for m.

$$y - y_1 = m(x - x_1)$$
$$y - (-2) = -4(x - 1)$$
$$y + 2 = -4x + 4$$
$$y = -4x + 2$$

Using the slope-intercept equation:

Substitute 1 for x, -2 for y, and -4 for m in $y = mx + b$ and solve for b.

$$y = mx + b$$
$$-2 = -4 \cdot 1 + b$$
$$-2 = -4 + b$$
$$2 = b$$

Then we use the equation $y = mx + b$ and substitute -4 for m and 2 for b.

$$y = -4x + 2$$

32. First find the slope of the line:

$$m = \frac{-6 - 15}{4 - (-10)} = \frac{-21}{14} = -\frac{3}{2}$$

Using the point-slope equation:

We choose to use the point $(4, -6)$ and substitute 4 for x_1, 6 for y_1, and $-\frac{3}{2}$ for m.

$$y - y_1 = m(x - x_1)$$
$$y - (-6) = -\frac{3}{2}(x - 4)$$
$$y + 6 = -\frac{3}{2}x + 6$$
$$y = -\frac{3}{2}x$$

Using the slope-intercept equation:

We choose $(4, -6)$ and substitute 4 for x, -6 for y, and $-\frac{3}{2}$ for m in $y = mx + b$. Then we solve for b.

$$y = mx + b$$
$$-6 = -\frac{3}{2} \cdot 4 + b$$
$$-6 = -6 + b$$
$$0 = b$$

Finally, we use the equation $y = mx + b$ and substitute $-\frac{3}{2}$ for m and 0 for b.

$$y = -\frac{3}{2}x + 0, \text{ or } y = -\frac{3}{2}x$$

33. First solve the equation for y and determine the slope of the given line.

$$x - 2y = 5 \qquad \text{Given line}$$
$$-2y = -x + 5$$
$$y = \frac{1}{2}x - \frac{5}{2}$$

The slope of the given line is $\frac{1}{2}$. The line through $(4, -1)$ must have slope $\frac{1}{2}$.

Using the point-slope equation:

Substitute 4 for x_1, -1 for y_1, and $\frac{1}{2}$ for m.

$$y - y_1 = m(x - x_1)$$
$$y - (-1) = \frac{1}{2}(x - 4)$$
$$y + 1 = \frac{1}{2}x - 2$$
$$y = \frac{1}{2}x - 3$$

Using the slope-intercept equation:

Substitute 4 for x, -1 for y, and $\frac{1}{2}$ for m and solve for b.

$$y = mx + b$$
$$-1 = \frac{1}{2}(4) + b$$
$$-1 = 2 + b$$
$$-3 = b$$

Then we use the equation $y = mx + b$ and substitute $\frac{1}{2}$ for m and -3 for b.

$$y = \frac{1}{2}x - 3$$

34. First solve the equation for y and determine the slope of the given line.

$$\begin{aligned} x + 3y &= 2 \qquad \text{Given line} \\ 3y &= -x + 2 \\ y &= -\frac{1}{3}x + \frac{2}{3} \end{aligned}$$

The slope of the given line is $-\frac{1}{3}$. The slope of the perpendicular line is the opposite of the reciprocal of $-\frac{1}{3}$. Thus, the line through $(2, 5)$ must have slope 3.

Using the point-slope equation:

Substitute 2 for x_1, 5 for y_1, and 3 for m.

$$\begin{aligned} y - y_1 &= m(x - x_1) \\ y - 5 &= 3(x - 2) \\ y - 5 &= 3x - 6 \\ y &= 3x - 1 \end{aligned}$$

Using the slope-intercept equation:

Substitute 2 for x, 5 for y, and 3 for m and solve for b.

$$\begin{aligned} y &= mx + b \\ 5 &= 3 \cdot 2 + b \\ 5 &= 6 + b \\ -1 &= b \end{aligned}$$

Then we use the equation $y = mx + b$ and substitute 3 for m and -1 for b.

$$y = 3x - 1$$

35. a) Note that $2003 - 1970 = 33$. Thus, the data points are $(0, 23.2)$ and $(33, 27.0)$. We find the slope.

$$m = \frac{27.0 - 23.2}{33 - 0} = \frac{3.8}{33} \approx 0.115$$

Using the slope and the y-intercept, $(0, 23.2)$, we write the function: $A(x) = 0.115x + 23.2$

b) In 2008, $x = 2008 - 1970 = 38$.

$A(38) = 0.115(38) + 23.2 = 27.57$ yr

In 2015, $x = 2015 - 1970 = 45$.

$A(45) = 0.115(45) + 23.2 \approx 28.38$ yr

36. First solve each equation for y and determine the slopes.

$$\begin{aligned} 3x + ky &= 17 \\ ky &= -3x + 17 \\ y &= -\frac{3}{k}x + \frac{17}{k} \end{aligned}$$

The slope of $3x + ky = 17$ is $-\frac{3}{k}$.

$$\begin{aligned} 8x - 5y &= 26 \\ -5y &= -8x + 26 \\ y &= \frac{8}{5}x - \frac{26}{5} \end{aligned}$$

The slope of $8x - 5y = 26$ is $\frac{8}{5}$.

If the lines are perpendicular, the product of their slopes is -1.

$$\begin{aligned} -\frac{3}{k} \cdot \frac{8}{5} &= -1 \\ -\frac{24}{5k} &= -1 \\ 24 &= 5k \qquad \text{Multiplying by } -5k \\ \frac{24}{5} &= k \end{aligned}$$

37. Answers may vary. One such function is $f(x) = 3$.

Cumulative Review Chapters R - 2

1. a) Note that $2004 - 1950 = 54$, so 2004 is 54 yr after 1950. Then the data points are $(0, 3.85)$ and $(54, 3.50)$.

First we find the slope.

$$m = \frac{3.50 - 3.85}{54 - 0} = \frac{-0.35}{54} \approx -0.006$$

Using the slope and the y-intercept, $(0, 3.85)$, we write the function: $R(x) = -0.006x + 3.85$.

b) In 2008, $x = 2008 - 1950 = 58$.

$R(58) = -0.006(58) + 3.85 \approx 3.50$ min

In 2010, $x = 2010 - 1950 = 60$.

$R(60) = -0.006(60) + 3.85 \approx 3.49$ min

2. a) Locate 15 on the x-axis and the find the point on the graph for which 15 is the first coordinate. From that point, look to the vertical axis to find the corresponding y-coordinate, 6. Thus, $f(15) = 6$.

b) The set of all x-values in the graph extends from 0 to 30, so the domain is $\{x | 0 \le x \le 30\}$, or $[0, 30]$.

c) To determine which member(s) of the domain are paired with 14, locate 14 on the vertical axis. From there look left and right to the graph to find any points for which 14 is the second coordinate. One such point exists. Its first coordinate is 25. Thus, the x-value for which $f(x) = 14$ is 25.

d) The set of all y-values in the graph extends from 0 to 15, so the range is $\{y | 0 \le y \le 15\}$, or $[0, 15]$.

3. $a^3 + b^0 - c = 3^3 + 7^0 - (-3) = 27 + 1 - (-3) = 27 + 1 + 3 = 31$

4. $|4.3 - 2.1| = |2.2| = 2.2$

5. $\left| -\frac{2}{3} \right| = \frac{2}{3}$

6. $-\frac{1}{3} - \left(-\frac{5}{6} \right) = -\frac{1}{3} + \frac{5}{6} = -\frac{2}{6} + \frac{5}{6} = \frac{3}{6} = \frac{1}{2}$

7. $-3.2(-11.4) = 36.48$

8. $2x - 4(3x - 8) = 2x - 12x + 32 = -10x + 32$

9. $(-16x^3y^{-4})(-2x^5y^3) = -16(-2)x^3 \cdot x^5 \cdot y^{-4} \cdot y^3 =$

$32x^{3+5}y^{-4+3} = 32x^8y^{-1} = \dfrac{32x^8}{y}$

10. $\dfrac{27x^0y^3}{-3x^2y^5} = \dfrac{27 \cdot 1 \cdot y^3}{-3x^2y^5} = \dfrac{27}{-3} \cdot \dfrac{1}{x^2} \cdot \dfrac{y^3}{y^5} =$

$-9 \cdot \dfrac{1}{x^2} \cdot y^{3-5} = -9 \cdot \dfrac{1}{x^2} \cdot y^{-2} = -\dfrac{9}{x^2y^2}$

11. $3(x - 7) - 4[2 - 5(x + 3)] = 3x - 21 - 4[2 - 5x - 15]$

$= 3x - 21 - 4[-13 - 5x]$

$= 3x - 21 + 52 + 20x$

$= 23x + 31$

12. $-128 \div 16 + 32 \cdot (-10) = -8 - 320 = -328$

13. $2^3 - (4 \cdot 2 - 3)^2 + 23^0 \cdot 16^1$

$= 2^3 - (8 - 3)^2 + 23^0 \cdot 16^1$

$= 2^3 - 5^2 + 23^0 \cdot 16^1$

$= 8 - 25 + 1 \cdot 16$

$= 8 - 25 + 16$

$= -17 + 16$

$= -1$

14. $\qquad x + 9.4 = -12.6$

$x + 9.4 - 9.4 = -12.6 - 9.4$

$\qquad x = -22$

The solution is -22.

15. $\qquad \dfrac{2}{3}x - \dfrac{1}{4} = -\dfrac{4}{5}x$

$60\left(\dfrac{2}{3}x - \dfrac{1}{4}\right) = 60\left(-\dfrac{4}{5}x\right) \quad \text{Clearing fractions}$

$60 \cdot \dfrac{2}{3}x - 60 \cdot \dfrac{1}{4} = -48x$

$40x - 15 = -48x$

$40x - 15 - 40x = -48x - 40x$

$-15 = -88x$

$\dfrac{-15}{-88} = \dfrac{-88x}{-88}$

$\dfrac{15}{88} = x$

The solution is $\dfrac{15}{88}$.

16. $\qquad -2.4t = -48$

$\dfrac{-2.4t}{-2.4} = \dfrac{-48}{-2.4}$

$\qquad t = 20$

The solution is 20.

17. $4x + 7 = -14$

$4x = -21 \qquad \text{Subtracting 7}$

$x = -\dfrac{21}{4} \qquad \text{Dividing by 4}$

The solution is $-\dfrac{21}{4}$.

18. $3n - (4n - 2) = 7$

$3n - 4n + 2 = 7$

$-n + 2 = 7$

$-n = 5$

$n = -5 \qquad \text{Multiplying by } -1$

The solution is -5.

19. $\qquad W = Ax + By$

$W - By = Ax \qquad \text{Subtracting } By$

$\dfrac{W - By}{A} = x \qquad \text{Dividing by } A$

20. $\qquad M = A + 4AB$

$\qquad M = A(1 + 4B) \qquad \text{Factoring out } A$

$\dfrac{M}{1 + 4B} = A \qquad \text{Dividing by } 1 + 4B$

21. $y - 12 \le -5$

$y \le 7 \qquad \text{Adding 12}$

The solution set is $\{y | y \le 7\}$, or $(-\infty, 7]$.

22. $6x - 7 < 2x - 13$

$4x - 7 < -13$

$4x < -6$

$x < -\dfrac{3}{2}$

The solution set is $\left\{x \middle| x < -\dfrac{3}{2}\right\}$, or $\left(-\infty, -\dfrac{3}{2}\right)$.

23. $5(1 - 2x) + x < 2(3 + x)$

$5 - 10x + x < 6 + 2x$

$5 - 9x < 6 + 2x$

$5 - 11x < 6$

$-11x < 1$

$x > -\dfrac{1}{11} \qquad \text{Reversing the inequality symbol}$

The solution set is $\left\{x \middle| x > -\dfrac{1}{11}\right\}$, or $\left(-\dfrac{1}{11}, \infty\right)$.

24. $x + 3 < -1 \;\; or \;\; x + 9 \ge 1$

$x < -4 \;\; or \qquad x \ge -8$

The intersection of $\{x | x < -4\}$ and $\{x | x \ge -8\}$ is the set of all real numbers. This is the solution set.

25. $-3 < x + 4 \le 8$

$-7 < x \le 4$

The solution set is $\{x | -7 < x \le 4\}$, or $(-7, 4]$.

26. $-8 \le 2x - 4 \le -1$

$-4 \le 2x \le 3$

$-2 \le x \le \dfrac{3}{2}$

The solution set is $\left\{x \Big| -2 \le x \le \dfrac{3}{2}\right\}$, or $\left[-2, \dfrac{3}{2}\right]$.

27. $|x| = 8$

$x = -8 \quad or \quad x = 8$

The solution set is $\{-8, 8\}$.

28. $|y| > 4$

$x < -4 \quad or \quad y > 4$

The solution set is $\{y | y < -4 \ or \ y > 4\}$, or $(-\infty, -4) \cup (4, \infty)$.

29. $|4x - 1| \le 7$

$-7 \le 4x - 1 \le 7$

$-6 \le 4x \le 8$

$-\dfrac{3}{2} \le x \le 2$

The solution set is $\left\{x \Big| -\dfrac{3}{2} \le x \le 2\right\}$, or $\left[-\dfrac{3}{2}, 2\right]$.

30. First solve the equation for y and determine the slope of the given line.

$4y - x = 3 \qquad$ Given line

$4y = x + 3$

$y = \dfrac{1}{4}x + \dfrac{3}{4}$

The slope of the given line is $\dfrac{1}{4}$. The slope of the perpendicular line is the opposite of the reciprocal of $\dfrac{1}{4}$. Thus, the line through $(-4, -6)$ must have slope -4.

Using the point-slope equation:

Substitute -4 for x_1, -6 for y_1, and -4 for m.

$y - y_1 = m(x - x_1)$

$y - (-6) = -4(x - (-4))$

$y + 6 = -4(x + 4)$

$y + 6 = -4x - 16$

$y = -4x - 22$

Using the slope-intercept equation:

Substitute -4 for x, -6 for y, and -4 for m.

$y = mx + b$

$-6 = -4(-4) + b$

$-6 = 16 + b$

$-22 = b$

Then we use the equation $y = mx + b$ and substitute -4 for m and -22 for b.

$y = -4x - 22$

31. First solve the equation for y and determine the slope of the given line.

$4y - x = 3 \qquad$ Given line

$4y = x + 3$

$y = \dfrac{1}{4}x + \dfrac{3}{4}$

The slope of the given line is $\dfrac{1}{4}$. The line through $(-4, -6)$ must have slope $\dfrac{1}{4}$.

Using the point-slope equation:

Substitute -4 for x_1, -6 for y_1, and $\dfrac{1}{4}$ for m.

$y - y_1 = m(x - x_1)$

$y - (-6) = \dfrac{1}{4}(x - (-4))$

$y + 6 = \dfrac{1}{4}(x + 4)$

$y + 6 = \dfrac{1}{4}x + 1$

$y = \dfrac{1}{4}x - 5$

Using the slope-intercept equation:

Substitute -4 for x, -6 for y, and $\dfrac{1}{4}$ for m and solve for b.

$y = mx + b$

$-6 = \dfrac{1}{4}(-4) + b$

$-6 = -1 + b$

$-5 = b$

Then we use the equation $y = mx + b$ and substitute $\dfrac{1}{4}$ for m and -5 for b.

$y = \dfrac{1}{4}x - 5$

32. $y = -2x + 3$

We find some ordered pairs that are solutions, plot them, and draw and label the graph.

When $x = -1$, $y = -2(-1) + 3 = 2 + 3 = 5$.

When $x = 1$, $y = -2 \cdot 1 + 3 = -2 + 3 = 1$.

When $x = 3$, $y = -2 \cdot 3 + 3 = -6 + 3 = -3$.

x	y
-1	5
1	1
3	-3

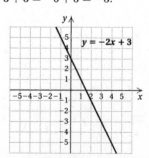

33. $3x = 2y + 6$

To find the x-intercept we let $y = 0$ and solve for x.

$$3x = 2y + 6$$
$$3x = 2 \cdot 0 + 6$$
$$3x = 6$$
$$x = 2$$

The x-intercept is $(2, 0)$.

To find the y-intercept we let $x = 0$ and solve for y.

$$3x = 2y + 6$$
$$3 \cdot 0 = 2y + 6$$
$$0 = 2y + 6$$
$$-2y = 6$$
$$y = -3$$

The y-intercept is $(0, -3)$.

We plot these points and draw the line.

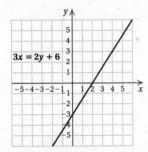

We use a third point as a check. We choose $x = 4$ and solve for y.

$$3 \cdot 4 = 2y + 6$$
$$12 = 2y + 6$$
$$6 = 2y$$
$$3 = y$$

We plot $(4, 3)$ and note that it is on the line.

34. $4x + 16 = 0$

$$4x = -16$$
$$x = -4$$

Since y is missing, all ordered pairs $(-4, y)$ are solutions. The graph is parallel to the y-axis.

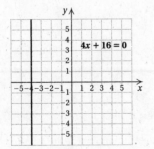

35. $-2y = -6$

$$y = 3$$

Since x is missing, all ordered pairs $(x, 3)$ are solutions. The graph is parallel to the x-axis.

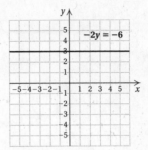

36. $f(x) = \frac{2}{3}x + 1$

We calculate some function values, plot the corresponding points, and connect them.

$$f(-3) = \frac{2}{3}(-3) + 1 = -2 + 1 = -1$$
$$f(0) = \frac{2}{3} \cdot 0 + 1 = 0 + 1 = 1$$
$$f(3) = \frac{2}{3} \cdot 3 + 1 = 2 + 1 = 3$$

x	$f(x)$
-3	-1
0	1
3	3

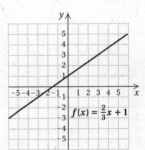

37. $g(x) = 5 - |x|$

We calculate some function values, plot the corresponding points, and connect them.

$$g(-2) = 5 - |-2| = 5 - 2 = 3$$
$$g(-1) = 5 - |-1| = 5 - 1 = 4$$
$$g(0) = 5 - |0| = 5 - 0 = 5$$
$$g(1) = 5 - |1| = 5 - 1 = 4$$
$$g(2) = 5 - |2| = 5 - 2 = 3$$
$$g(3) = 5 - |3| = 5 - 3 = 2$$

x	$g(x)$
-2	3
-1	4
0	5
1	4
2	3
3	2

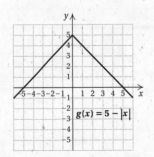

38. First we find the slope-intercept form of the equation by solving for y. This allows us to determine the slope and y-intercept easily.

$$-4y + 9x = 12$$
$$-4y = -9x + 12$$
$$\frac{-4y}{-4} = \frac{-9x + 12}{-4}$$
$$y = \frac{9}{4}x - 3$$

The slope is $\frac{9}{4}$, and the y-intercept is $(0, -3)$.

39. Slope $= \dfrac{\text{change in } y}{\text{change in } x} = \dfrac{3 - 7}{-1 - 2} = \dfrac{-4}{-3} = \dfrac{4}{3}$

40. Using the point-slope equation:

Substitute 2 for x_1, -11 for y_1, and -3 for m.

$$y - y_1 = m(x - x_1)$$
$$y - (-11) = -3(x - 2)$$
$$y + 11 = -3x + 6$$
$$y = -3x - 5$$

Using the slope-intercept equation:

Substitute 2 for x, -11 for y, and -3 for m in $y = mx + b$ and solve for b.

$$y = mx + b$$
$$-11 = -3 \cdot 2 + b$$
$$-11 = -6 + b$$
$$-5 = b$$

Then use the equation $y = mx + b$ and substitute -3 for m and -5 for b.

$$y = -3x - 5$$

41. First find the slope of the line:

$$m = \frac{3 - 2}{-6 - 4} = \frac{1}{-10} = -\frac{1}{10}$$

Using the point-slope equation:

We choose to use the point $(4, 2)$ and substitute 4 for x_1, 2 for y_1, and $-\frac{1}{10}$ for m.

$$y - y_1 = m(x - x_1)$$
$$y - 2 = -\frac{1}{10}(x - 4)$$
$$y - 2 = -\frac{1}{10}x + \frac{2}{5}$$
$$y = -\frac{1}{10}x + \frac{12}{5}$$

Using the slope-intercept equation:

We choose $(4, 2)$ and substitute 4 for x, 2 for y, and $-\frac{1}{10}$ for m in $y = mx + b$. Then we solve for b.

$$y = mx + b$$
$$2 = -\frac{1}{10} \cdot 4 + b$$
$$2 = -\frac{2}{5} + b$$
$$\frac{12}{5} = b$$

Finally, we use the equation $y = mx + b$ and substitute $-\frac{1}{10}$ for m and $\frac{12}{5}$ for b.

$$y = -\frac{1}{10}x + \frac{12}{5}$$

42. Familiarize. Let $w =$ the width, in meters. Then $w + 6 =$ the length. Recall that the formula for the perimeter of a rectangle is $P = 2l + 2w$.

Translate. We use the formula for perimeter.

$$80 - 2(w + 6) + 2w$$

Solve. We solve the equation.

$$80 = 2(w + 6) + 2w$$
$$80 = 2w + 12 + 2w$$
$$80 = 4w + 12$$
$$68 = 4w$$
$$17 = w$$

If $w = 17$, then $w + 6 = 17 + 6 = 23$.

Check. 23 m is 6 m more than 17 m, and $2 \cdot 23 + 2 \cdot 17 = 46 + 34 = 80$ m. The answer checks.

State. The length is 23 m and the width is 17 m.

43. Familiarize. Let $s =$ David's old salary. Then his new salary is $s + 20\%$, or $s + 0.2s$, or $1.2s$.

Translate.

$$\underbrace{\text{New salary}}_{1.2s} \text{ is } \overset{\downarrow}{=} \ \$27{,}000$$

Solve. We solve the equation.

$$1.2s = 27{,}000$$
$$s = 22{,}500$$

Check. 20% of \$22,500 is $0.2(\$22{,}500)$, or \$4500, and $\$22{,}500 + \$4500 = \$27{,}000$. The answer checks.

State. David's old salary was \$22,500.

44.
$$\left(\frac{1}{8}\right)^2 = \frac{1}{8} \cdot \frac{1}{8} = \frac{1}{64}$$
$$\left(\frac{1}{8}\right)^{-2} = \left(\frac{8}{1}\right)^2 = 8^2 = 64$$
$$8^{-2} = \frac{1}{8^2} = \frac{1}{64}$$
$$8^2 = 64$$
$$-8^2 = -64$$
$$(-8)^2 = 64$$
$$\left(-\frac{1}{8}\right)^{-2} = \left(-\frac{8}{1}\right)^2 = (-8)^2 = 64$$
$$\left(-\frac{1}{8}\right)^2 = -\frac{1}{8} \cdot \left(-\frac{1}{8}\right) = \frac{1}{64}$$

45. Using the point-slope equation, $y - y_1 = m(x - x_1)$, with $x_1 = 3$, $y_1 = 1$, and $m = -2$ we have $y - 1 = -2(x - 3)$. Thus, answer (b) is correct.

46. If we let $x =$ the length of the longer piece, in feet, then $\frac{1}{3}x =$ the length of the shorter piece. The problem translates to the equation $x + \frac{1}{3}x = 28$. We solve the equation.

$$x + \frac{1}{3}x = 28$$

$$\frac{4}{3}x = 28$$

$$x = \frac{3}{4} \cdot 28 = 21$$

If $x = 21$, then $\frac{1}{3}x = \frac{1}{3} \cdot 21 = 7$. Thus, answer (b) is correct.

47. $a = a^1$, so answer (d) is correct.

48.
$$[9(2a + 3) - (a - 2)] - [3(2a - 1)]$$
$$= [18a + 27 - a + 2] - [6a - 3]$$
$$= [17a + 29] - [6a - 3]$$
$$= 17a + 29 - 6a + 3$$
$$= 11a + 32$$

Thus, answer (e) is correct.

49. We have two data points, $(1000, 101,000)$ and $(1250, 126,000)$. We find the slope of the line containing these points.

$$m = \frac{126,000 - 101,000}{1250 - 1000} = \frac{25,000}{250} = 100$$

We will use the point-slope equation with $x_1 = 1000$, $y_1 = 101,000$ and $m = 100$.

$$y - y_1 = m(x - x_1)$$
$$y - 101,000 = 100(x - 1000)$$
$$y - 101,000 = 100x - 100,000$$
$$y = 100x + 1000$$

Now we find the value of y when $x = 1500$.

$$y = 100 \cdot 1500 + 1000 = 150,000 + 1000 = 151,000$$

Thus, when \$1500 is spent on advertising, weekly sales increase by \$151,000.

50. First we solve each equation for y and determine the slopes.

a) $7y - 3x = 21$
$$7y = 3x + 21$$
$$y = \frac{3}{7}x + 3$$

The slope is $\frac{3}{7}$.

b) $-3x - 7y = 12$
$$-7y = 3x + 12$$
$$y = -\frac{3}{7}x - \frac{12}{7}$$

The slope is $-\frac{3}{7}$.

c) $7y + 3x = 21$
$$7y = -3x + 21$$
$$y = -\frac{3}{7}x + 3$$

The slope is $-\frac{3}{7}$.

d) $3y + 7x = 12$
$$3y = -7x + 12$$
$$y = -\frac{7}{3}x + 4$$

The slope is $-\frac{7}{3}$.

The only pair of slopes whose product is -1 is $\frac{3}{7}$ and $-\frac{7}{3}$. Thus, equations (1) and (4) represent perpendicular lines.

51. $x + 5 < 3x - 7 \leq x + 13$

$x + 5 < 3x - 7$ *and* $3x - 7 \leq x + 13$

$\quad\; 5 < 2x - 7$ *and* $\quad 2x - 7 \leq 13$

$\quad 12 < 2x$ *and* $\quad 2x \leq 20$

$\quad\; 6 < x$ *and* $\quad x \leq 10$

The solution set is $\{x | 6 < x \text{ and } x \leq 10\}$, or $\{x | 6 < x \leq 10\}$, or $(6, 10]$.

Chapter 3

Systems of Equations

1. Graph both lines on the same set of axes.

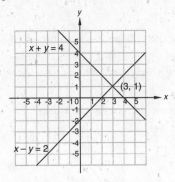

The solution (point of intersection) seems to be the point $(3, 1)$.

Check:

$$\frac{x+y=4}{3+1\ ?\ 4}$$
$$4\ \Big|\ \text{TRUE}$$

$$\frac{x-y=2}{3-1\ ?\ 2}$$
$$2\ \Big|\ \text{TRUE}$$

The solution is $(3, 1)$.

Since the system of equations has a solution it is consistent. Since there is exactly one solution, the equations are independent.

3. Graph both lines on the same set of axes.

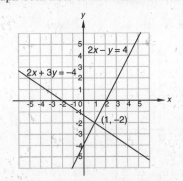

The solution (point of intersection) seems to be the point $(1, -2)$.

Check:

$$\frac{2x-y=4}{2\cdot 1-(-2)\ ?\ 4}$$
$$2+2$$
$$4\ \Big|\ \text{TRUE}$$

$$\frac{2x+3y=-4}{2\cdot 1+3(-2)\ ?\ -4}$$
$$2-6$$
$$-4\ \Big|\ \text{TRUE}$$

The solution is $(1, -2)$.

Since the system of equations has a solution, it is consistent. Since there is exactly one solution, the equations are independent.

5. Graph both lines on the same set of axes.

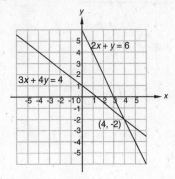

The solution (point of intersection) seems to be the point $(4, -2)$.

Check:

$$\frac{2x+y=6}{2\cdot 4+(-2)\ ?\ 6}$$
$$8-2$$
$$6\ \Big|\ \text{TRUE}$$

$$\frac{3x+4y=4}{3\cdot 4+4(-2)\ ?\ 4}$$
$$12-8$$
$$4\ \Big|\ \text{TRUE}$$

The solution is $(4, -2)$.

Since the system of equations has a solution, it is consistent. Since there is exactly one solution, the equations are independent.

7. Graph both lines on the same set of axes.

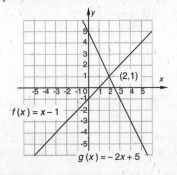

The solution seems to be the point $(2, 1)$.

Check:

$$\frac{f(x)=x-1}{1\ ?\ 2-1}$$
$$1\quad\text{TRUE}$$

$$\frac{g(x)=-2x+5}{1\ ?\ -2\cdot 2+5}$$
$$-4+5$$
$$1\quad\text{TRUE}$$

The solution is $(2, 1)$.

Since the system of equations has a solution, it is consistent. Since there is exactly one solution, the equations are independent.

9. Graph both lines on the same set of axes.

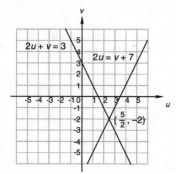

The solution seems to be $\left(\dfrac{5}{2}, -2\right)$.

Check:

$$
\begin{array}{c|c}
2u + v = 3 & 2u = v + 7 \\
\hline
2 \cdot \dfrac{5}{2} + (-2) \ ? \ 3 & 2 \cdot \dfrac{5}{2} \ ? \ -2 + 7 \\
5 - 2 & 5 \ \big| \ 5 \qquad \text{TRUE} \\
3 \ \big| \ \text{TRUE} &
\end{array}
$$

The solution is $\left(\dfrac{5}{2}, -2\right)$.

Since the system of equations has a solution, it is consistent. Since there is exactly one solution, the equations are independent.

11. Graph both lines on the same set of axes.

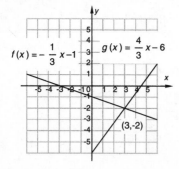

The ordered pair $(3, -2)$ checks in both equations. It is the solution.

Since the system of equations has a solution, it is consistent. Since there is exactly one solution, the equations are independent.

13. Graph both lines on the same set of axes.

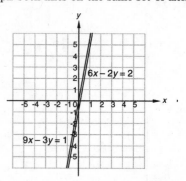

The lines are parallel. There is no solution.

Since the system of equations has no solution, it is inconsistent. Since there is no solution, the equations are independent.

15. Graph both lines on the same set of axes.

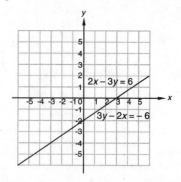

The graphs are the same. Any solution of one of the equations is also a solution of the other. Each equation has an infinite number of solutions. Thus the system of equations has an infinite number of solutions. Since the system of equations has a solution, it is consistent. Since there are infinitely many solutions, the equations are dependent.

17. Graph both lines on the same set of axes.

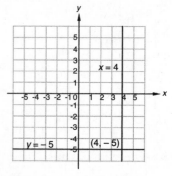

The ordered pair $(4, -5)$ checks in both equations. It is the solution.

Since the system of equations has a solution, it is consistent. Since there is exactly one solution, the equations are independent.

19. Graph both lines on the same set of axes.

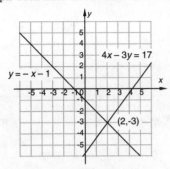

The ordered pair $(2, -3)$ checks in both equations. It is the solution.

Since the system of equations has a solution, it is consistent. Since there is exactly one solution, the equations are independent.

21. Since the system of equations has a solution, it is consistent. Since there is exactly one solution, the equations are independent. The graph of the system consists of a vertical line and a horizontal line, each passing through $(3, 3)$. Thus, system **F** corresponds to this graph.

23. Since the system of equations has a solution, it is consistent. Since there are infinitely many solutions, the equations are dependent. The equations in system B are equivalent, so their graphs are the same. In addition the graph corresponds to the one shown, so system **B** corresponds to this graph.

25. Since the system of equations has no solution, it is inconsistent. Since there is no solution, the equations are independent. The equations in system **D** have the same slope and different y-intercepts and have the graphs shown, so this system corresponds to the given graph.

27. Discussion and Writing Exercise

29. $3x + 4 = x - 2$

$2x + 4 = -2$ Adding $-x$ on both sides

$2x = -6$ Adding -4 on both sides

$x = -3$ Multiplying by $\frac{1}{2}$ on both sides

The solution is -3.

31. $4x - 5x = 8x - 9 + 11x$

$-x = 19x - 9$ Collecting like terms

$-20x = -9$ Adding $-19x$ on both sides

$x = \frac{9}{20}$ Multiplying by $-\frac{1}{20}$ on both sides

The solution is $\frac{9}{20}$.

33. Graph these equations, solving each equation for y first, if necessary. We get $y = \dfrac{13.78 - 2.18x}{7.81}$ and $y = \dfrac{5.79x - 8.94}{3.45}$. Using the INTERSECT feature, we find that the point of intersection is $(2.23, 1.14)$.

35. Graph both lines on the same set of axes.

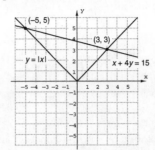

The solutions appear to be $(-5, 5)$ and $(3, 3)$.

Check:

For $(-5, 5)$:

$$\begin{array}{c|c}
y = |x| & \\ \hline
5 \;?\; |-5| & \\
\quad 5 \quad \text{TRUE} &
\end{array}
\qquad
\begin{array}{c|c}
x + 4y = 15 & \\ \hline
-5 + 4 \cdot 5 \;?\; 15 & \\
-5 + 20 & \\
15 & \text{TRUE}
\end{array}$$

For $(3, 3)$:

$$\begin{array}{c|c}
y = |x| & \\ \hline
3 \;?\; |3| & \\
\quad 3 \;\text{TRUE} &
\end{array}
\qquad
\begin{array}{c|c}
x + 4y = 15 & \\ \hline
3 + 4 \cdot 3 \;?\; 15 & \\
3 + 12 & \\
15 & \text{TRUE}
\end{array}$$

Both pairs check. The solutions are $(-5, 5)$ and $(3, 3)$.

Exercise Set 3.2

1. $y = 5 - 4x$,   (1)

$2x - 3y = 13$ (2)

We substitute $5 - 4x$ for y in the second equation and solve for x.

$2x - 3y = 13$ (2)

$2x - 3(5 - 4x) = 13$ Substituting

$2x - 15 + 12x = 13$

$14x - 15 = 13$

$14x = 28$

$x = 2$

Next we substitute 2 for x in either equation of the original system and solve for y.

$y = 5 - 4x$ (1)

$y = 5 - 4 \cdot 2$ Substituting

$y = 5 - 8$

$y = -3$

We check the ordered pair $(2, -3)$.

$$\begin{array}{c|c}
y = 5 - 4x & \\ \hline
-3 \;?\; 5 - 4 \cdot 2 & \\
5 - 8 & \\
-3 & -3 \quad \text{TRUE}
\end{array}$$

$$2x - 3y = 13$$

$$2 \cdot 2 - 3(-3) \ ? \ 13$$
$$4 + 9 \ \big| \$$
$$13 \ \big| \ 13 \quad \text{TRUE}$$

Since $(2, -3)$ checks, it is the solution.

3. $2y + x = 9,$ (1)
$\quad x = 3y - 3$ (2)

We substitute $3y - 3$ for x in the first equation and solve for y.

$$2y + x = 9 \quad (1)$$
$$2y + (3y - 3) = 9 \quad \text{Substituting}$$
$$5y - 3 = 9$$
$$5y = 12$$
$$y = \frac{12}{5}$$

Next we substitute $\frac{12}{5}$ for y in either equation of the original system and solve for x.

$$x = 3y - 3 \qquad\qquad (2)$$
$$x = 3 \cdot \frac{12}{5} - 3 = \frac{36}{5} - \frac{15}{5} = \frac{21}{5}$$

We check the ordered pair $\left(\frac{21}{5}, \frac{12}{5} \right)$.

$$2y + x = 9$$

$$2 \cdot \frac{12}{5} + \frac{21}{15} \ ? \ 9$$
$$\frac{24}{5} + \frac{21}{5} \ \big| \$$
$$\frac{45}{5} \ \big| \$$
$$9 \ \big| \ 9 \quad \text{TRUE}$$

$$x = 3y - 3$$

$$\frac{21}{5} \ ? \ 3 \cdot \frac{12}{5} - 3$$
$$\big| \ \frac{36}{5} - \frac{15}{5}$$
$$\frac{21}{5} \ \big| \ \frac{21}{5} \qquad \text{TRUE}$$

Since $\left(\frac{21}{5}, \frac{12}{5} \right)$ checks, it is the solution.

5. $3s - 4t = 14,$ (1)
$\quad 5s + t = 8$ (2)

We solve the second equation for t.

$$5s + t = 8 \qquad (2)$$
$$t = 8 - 5s \quad (3)$$

We substitute $8 - 5s$ for t in the first equation and solve for s.

$$3s - 4t = 14 \quad (1)$$
$$3s - 4(8 - 5s) = 14 \quad \text{Substituting}$$
$$3s - 32 + 20s = 14$$
$$23s - 32 = 14$$
$$23s = 46$$
$$s = 2$$

Next we substitute 2 for s in Equation (1), (2), or (3). It is easiest to use Equation (3) since it is already solved for t.

$$t = 8 - 5 \cdot 2 = 8 - 10 = -2$$

We check the ordered pair $(2, -2)$.

$$3s - 4t = 14$$

$$3 \cdot 2 - 4(-2) \ ? \ 14$$
$$6 + 8 \ \big| \$$
$$14 \ \big| \ 14 \quad \text{TRUE}$$

$$5s + t = 8$$

$$5 \cdot 2 + (-2) \ ? \ 8$$
$$10 - 2 \ \big| \$$
$$8 \ \big| \ 8 \quad \text{TRUE}$$

Since $(2, -2)$ checks, it is the solution.

7. $9x - 2y = -6,$ (1)
$\quad 7x + 8 = y$ (2)

We substitute $7x + 8$ for y in the first equation and solve for x.

$$9x - 2y = -6 \quad (1)$$
$$9x - 2(7x + 8) = -6 \quad \text{Substituting}$$
$$9x - 14x - 16 = -6$$
$$-5x - 16 = -6$$
$$-5x = 10$$
$$x = -2$$

Next we substitute -2 for x in either equation of the original system and solve for y.

$$7x + 8 = y \quad (2)$$
$$7(-2) + 8 = y$$
$$-14 + 8 = y$$
$$-6 = y$$

We check the ordered pair $(-2, -6)$.

$$9x - 2y = -6$$

$$9(-2) - 2(-6) \ ? \ -6$$
$$-18 + 12 \ \big| \$$
$$-6 \ \big| \qquad \text{TRUE}$$

$$7x + 8 = y$$

$$7(-2) + 8 \ ? \ -6$$
$$-14 + 8 \ \big| \$$
$$-6 \ \big| \qquad \text{TRUE}$$

Since $(-2, -6)$ checks, it is the solution.

9. $-5s + t = 11,$ (1)

$4s + 12t = 4$ (2)

We solve the first equation for t.

$-5s + t = 11$ (1)

$t = 5s + 11$ (3)

We substitute $5s + 11$ for t in the second equation and solve for s.

$4s + 12t = 4$ (2)

$4s + 12(5s + 11) = 4$

$4s + 60s + 132 = 4$

$64s + 132 = 4$

$64s = -128$

$s = -2$

Next we substitute -2 for s in Equation (3).

$t = 5s + 11 = 5(-2) + 11 = -10 + 11 = 1$

We check the ordered pair $(-2, 1)$.

$$\begin{array}{c|c} -5s + t = 11 & \\ \hline -5(-2) + 1 \ ? \ 11 & \\ 10 + 1 & \\ 11 & 11 \quad \text{TRUE} \end{array}$$

$$\begin{array}{c|c} 4s + 12t = 4 & \\ \hline 4(-2) + 12 \cdot 1 \ ? \ 4 & \\ -8 + 12 & \\ 4 & 4 \quad \text{TRUE} \end{array}$$

Since $(-2, 1)$ checks, it is the solution.

11. $2x + 2y = 2,$ (1)

$3x - y = 1$ (2)

We solve the second equation for y.

$3x - y = 1$ (2)

$-y = -3x + 1$

$y = 3x - 1$ (3)

We substitute $3x - 1$ for y in the first equation and solve for x.

$2x + 2y = 2$ (1)

$2x + 2(3x - 1) = 2$

$2x + 6x - 2 = 2$

$8x - 2 = 2$

$8x = 4$

$x = \dfrac{1}{2}$

Next we substitute $\dfrac{1}{2}$ for x in Equation (3).

$y = 3x - 1 = 3 \cdot \dfrac{1}{2} - 1 = \dfrac{3}{2} - 1 = \dfrac{1}{2}$

The ordered pair $\left(\dfrac{1}{2}, \dfrac{1}{2}\right)$ checks in both equations. It is the solution.

13. $3a - b = 7,$ (1)

$2a + 2b = 5$ (2)

We solve the first equation for b.

$3a - b = 7$ (1)

$-b = -3a + 7$

$b = 3a - 7$ (3)

We substitute $3a - 7$ for b in the second equation and solve for a.

$2a + 2b = 5$ (2)

$2a + 2(3a - 7) = 5$

$2a + 6a - 14 = 5$

$8a - 14 = 5$

$8a = 19$

$a = \dfrac{19}{8}$

We substitute $\dfrac{19}{8}$ for a in Equation (3).

$b = 3a - 7 = 3 \cdot \dfrac{19}{8} - 7 = \dfrac{57}{8} - \dfrac{56}{8} = \dfrac{1}{8}$

The ordered pair $\left(\dfrac{19}{8}, \dfrac{1}{8}\right)$ checks in both equations. It is the solution.

15. $2x - 3 = y$ (1)

$y - 2x = 1,$ (2)

We substitute $2x - 3$ for y in the second equation and solve for x.

$y - 2x = 1$ (2)

$2x - 3 - 2x = 1$ Substituting

$-3 = 1$ Collecting like terms

We have a false equation. Therefore, there is no solution.

17. *Familiarize*. Refer to the drawing in the text. Let $l =$ the length of the court and $w =$ the width. Recall that the perimeter P of a rectangle with length l and width w is given by $P = 2l + 2w$.

Translate.

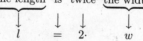

$$\begin{array}{ccc} \underbrace{\text{The perimeter}} & \text{is} & \underbrace{\text{120 ft.}} \\ \downarrow & \downarrow & \downarrow \\ 2l + 2w & = & 120 \end{array}$$

$$\begin{array}{cccc} \underbrace{\text{The length}} & \text{is} & \text{twice} & \underbrace{\text{the width.}} \\ \downarrow & \downarrow & \downarrow & \downarrow \\ l & = & 2 \cdot & w \end{array}$$

We have system of equations.

$2l + 2w = 120,$ (1)

$l = 2w$ (2)

Solve. We substitute $2w$ for l in Equation (1) and solve for w.

$2l + 2w = 120$

$2 \cdot 2w + 2w = 120$

$4w + 2w = 120$

$6w = 120$

$w = 20$

Now substitute 20 for w in Equation (2) and find l.

$$l = 2w = 2 \cdot 20 = 40$$

Check. If the length is 40 ft and the width is 20 ft, then the perimeter is $2 \cdot 40 + 2 \cdot 20$, or 120 ft. Also, the length is twice the width. The answer checks.

State. The length of the court is 40 ft, and the width is 20 ft.

19. Familiarize. Using the drawing in the text, we let x and y represent the measures of the angles.

Translate.

$$\underbrace{\text{The sum of the measures}}_{x + y} \quad \underbrace{\text{is}}_{=} \quad \underbrace{180°.}_{180}$$

$$\underbrace{\text{One angle}}_{x} \quad \underbrace{\text{is}}_{=} \quad \underbrace{3}_{3} \quad \underbrace{\text{times}}_{\cdot} \quad \underbrace{\text{the other}}_{y} \quad \underbrace{\text{less}}_{-} \quad \underbrace{12°.}_{12}$$

We have a system of equations.

$$x + y = 180, \quad (1)$$
$$x = 3y - 12 \quad (2)$$

Solve. Substitute $3y - 12$ for x in Equation (1) and solve for y.

$$x + y = 180$$
$$(3y - 12) + y = 180$$
$$4y - 12 = 180$$
$$4y = 192$$
$$y = 48$$

Now substitute 48 for y in Equation (2) and find x.

$$x = 3y - 12 = 3 \cdot 48 - 12 = 132$$

Check. The sum of the measures is $48° + 132°$, or $180°$. Also, $132°$ is $12°$ less than three times $48°$. The answer checks.

State. The measures of the angles are $48°$ and $132°$.

21. Familiarize. Let x = number of games won and y = number of games tied. The total points earned in x wins is $2x$; the total points earned in y ties is $1 \cdot y$, or y.

Translate.

$$\underbrace{\text{Points from wins}}_{2x} \quad \underbrace{\text{plus}}_{+} \quad \underbrace{\text{points from ties}}_{y} \quad \underbrace{\text{is}}_{=} \quad \underbrace{60.}_{60}$$

$$\underbrace{\text{Number of wins}}_{x} \quad \underbrace{\text{is}}_{=} \quad \underbrace{\text{9 more than the number of ties.}}_{9 + y}$$

We have a system of equations:

$$2x + y = 60,$$
$$x = 9 + y$$

Solve. We solve the system of equations. We use substitution.

$$2(9 + y) + y = 60 \quad \text{Substituting } 9 + y \text{ for } x \text{ in (1)}$$
$$18 + 2y + y = 60$$
$$18 + 3y = 60$$
$$3y = 42$$
$$y = 14$$

$$x = 9 + 14 \quad \text{Substituting 14 for } y \text{ in (2)}$$
$$x = 23$$

Check. The number of wins, 23, is 9 more than the number of ties, 14.

Points from wins:	$23 \times 2 = 46$
Points from ties:	$14 \times 1 = \underline{14}$
	Total $\quad$ 60

The numbers check.

State. The team had 23 wins and 14 ties.

23. Discussion and Writing Exercise

25. $y = 1.3x - 7$

The equation is in slope-intercept form, $y = mx + b$. The slope is 1.3.

27. $\quad A = \dfrac{pq}{7}$

$$7A = pq \quad \text{Multiplying by 7}$$

$$\frac{7A}{q} = p \quad \text{Dividing by } q$$

29. $\quad -4x + 5(x - 7) = 8x - 6(x + 2)$

$$-4x + 5x - 35 = 8x - 6x - 12 \quad \text{Removing parentheses}$$
$$x - 35 = 2x - 12 \qquad \text{Collecting like terms}$$
$$-35 = x - 12 \qquad \text{Subtracting } x$$
$$-23 = x \qquad \text{Adding 12}$$

The solution is -23.

31. $\quad 2 = m + b, \quad (1) \quad \text{Substituting } (1, 2)$

$$4 = -3m + b \quad (2) \quad \text{Substituting } (-3, 4)$$

$$\begin{aligned} 2 &= m + b \quad (1) \\ -4 &= 3m - b \quad \text{Multiplying (2) by } -1 \\ \hline -2 &= 4m \end{aligned}$$

$$-\frac{1}{2} = m$$

Substitute $-\dfrac{1}{2}$ for m in (1).

$$2 = -\frac{1}{2} + b$$

$$\frac{5}{2} = b$$

Thus, $m = -\dfrac{1}{2}$ and $b = \dfrac{5}{2}$.

33. *Familiarize*. Let l = the original length, in inches, and w = the original width, in inches. Then $w - 6$ = the width after 6 in. is cut off.

Translate.

$$\underbrace{\text{The original perimeter}}_{2l + 2w} \;\; \text{is} \;\; \underbrace{156 \text{ in}}_{156}.$$

$$2l + 2w \;\;\;\; = \;\;\;\; 156$$

$$\underbrace{\text{The length}}_{l} \;\; \text{becomes} \;\; \underbrace{4}_{4} \;\; \text{times} \;\; \underbrace{\text{the new width}}_{(w-6)}.$$

$$l \;\; = \;\; 4 \;\; \cdot \;\; (w - 6)$$

We have a system of equations:

$$2l + 2w = 156, \quad (1)$$
$$l = 4 \cdot (w - 6) \quad (2)$$

Solve. Substitute $4(w - 6)$ for l in Equation (1) and solve for w.

$$2l + 2w = 156$$
$$2 \cdot 4(w - 6) + 2w = 156$$
$$8w - 48 + 2w = 156$$
$$10w - 48 = 156$$
$$10w = 204$$
$$w = 20.4$$

Now substitute 20.4 for w in Equation (2) and find l.

$$l = 4 \cdot (w - 6) = 4(20.4 - 6) = 4(14.4) = 57.6$$

Check. The original perimeter is $2(57.6) + 2(20.4)$, or $115.2 + 40.8$, or 156 in. If 6 in. is cut off the width, then the width becomes $20.4 - 6$, or 14.4 in., and the length is 4 times the width, or 57.6 in. The answer checks.

State. The length is 57.6 in., and the width is 20.4 in.

Exercise Set 3.3

1.
$$x + 3y = 7 \quad (1)$$
$$\underline{-x + 4y = 7} \quad (2)$$
$$0 + 7y = 14 \quad \text{Adding}$$
$$7y = 14$$
$$y = 2$$

Substitute 2 for y in one of the original equations and solve for x.

$$x + 3y = 7 \quad \text{Equation (1)}$$
$$x + 3 \cdot 2 = 7 \quad \text{Substituting}$$
$$x + 6 = 7$$
$$x = 1$$

Check:

$x + 3y = 7$	$-x + 4y = 7$
$1 + 3 \cdot 2 \; ? \; 7$	$-1 + 4 \cdot 2 \; ? \; 7$
$1 + 6$	$-1 + 8$
$7 \mid$ TRUE	$7 \mid$ TRUE

Since $(1, 2)$ checks, it is the solution.

3.
$$9x + 5y = 6 \quad (1)$$
$$\underline{2x - 5y = -17} \quad (2)$$
$$11x + 0 = -11 \quad \text{Adding}$$
$$11x = -11$$
$$x = -1$$

Substitute -1 for x in one of the original equations and solve for y.

$$9x + 5y = 6 \quad \text{Equation (1)}$$
$$9(-1) + 5y = 6 \quad \text{Substituting}$$
$$-9 + 5y = 6$$
$$5y = 15$$
$$y = 3$$

We obtain $(-1, 3)$. This checks, so it is the solution.

5.
$$5x + 3y = 19, \quad (1)$$
$$2x - 5y = 11 \quad (2)$$

We multiply twice to make two terms become additive inverses.

From (1):	$25x + 15y = 95$	Multiplying by 5
From (2):	$\underline{6x - 15y = 33}$	Multiplying by 3
	$31x + 0 = 128$	Adding
	$31x = 128$	
	$x = \dfrac{128}{31}$	

Substitute $\dfrac{128}{31}$ for x in one of the original equations and solve for y.

$$5x + 3y = 19 \quad \text{Equation (1)}$$
$$5 \cdot \frac{128}{31} + 3y = 19 \quad \text{Substituting}$$
$$\frac{640}{31} + 3y = \frac{589}{31}$$
$$3y = -\frac{51}{31}$$
$$\frac{1}{3} \cdot 3y = \frac{1}{3} \cdot \left(-\frac{51}{31}\right)$$
$$y = -\frac{17}{31}$$

We obtain $\left(\dfrac{128}{31}, -\dfrac{17}{31}\right)$. This checks, so it is the solution.

7.
$$5r - 3s = 24, \quad (1)$$
$$3r + 5s = 28 \quad (2)$$

We multiply twice to make two terms become additive inverses.

From (1):	$25r - 15s = 120$	Multiplying by 5
From (2):	$\underline{9r + 15s = 84}$	Multiplying by 3
	$34r + 0 = 204$	Adding
	$34r = 204$	
	$r = 6$	

Substitute 6 for r in one of the original equations and solve for s.

$3r + 5s = 28$　　Equation (2)

$3 \cdot 6 + 5s = 28$　　Substituting

$18 + 5s = 28$

$5s = 10$

$s = 2$

We obtain $(6, 2)$. This checks, so it is the solution.

9. $0.3x - 0.2y = 4,$

　　$0.2x + 0.3y = 1$

We first multiply each equation by 10 to clear decimals.

$3x - 2y = 40$　(1)

$2x + 3y = 10$　(2)

We use the multiplication principle with both equations of the resulting system.

From (1):　　$9x - 6y = 120$　　Multiplying by 3

From (2):　　$\underline{4x + 6y = 20}$　　Multiplying by 2

　　　　　$13x + 0 = 140$　　Adding

　　　　　　　$13x = 140$

　　　　　　　　$x = \dfrac{140}{13}$

Substitute $\dfrac{140}{13}$ for x in one of the equations in which the decimals were cleared and solve for y.

$2x + 3y = 10$　　Equation (2)

$2 \cdot \dfrac{140}{13} + 3y = 10$　　Substituting

$\dfrac{280}{13} + 3y = \dfrac{130}{13}$

$3y = -\dfrac{150}{13}$

$y = -\dfrac{50}{13}$

We obtain $\left(\dfrac{140}{13}, -\dfrac{50}{13} \right)$. This checks, so it is the solution.

11. $\dfrac{1}{2}x + \dfrac{1}{3}y = 4,$

　　$\dfrac{1}{4}x + \dfrac{1}{3}y = 3$

We first multiply each equation by the LCM of the denominators to clear fractions.

$3x + 2y = 24$　　Multiplying by 6

$3x + 4y = 36$　　Multiplying by 12

We multiply by -1 on both sides of the first equation and then add.

$-3x - 2y = -24$　　Multiplying by -1

$\underline{3x + 4y = 36}$

$0 + 2y = 12$　　Adding

$2y = 12$

$y = 6$

Substitute 6 for y in one of the equations in which the fractions were cleared and solve for x.

$3x + 2y = 24$

$3x + 2 \cdot 6 = 24$　　Substituting

$3x + 12 = 24$

$3x = 12$

$x = 4$

We obtain $(4, 6)$. This checks, so it is the solution.

13. $\dfrac{2}{5}x + \dfrac{1}{2}y = 2,$

　　$\dfrac{1}{2}x - \dfrac{1}{6}y = 3$

We first multiply each equation by the LCM of the denominators to clear fractions.

$4x + 5y = 20$　　Multiplying by 10

$3x - y = 18$　　Multiplying by 6

We multiply by 5 on both sides of the second equation and then add.

$4x + 5y = 20$

$\underline{15x - 5y = 90}$　　Multiplying by 5

$19x + 0 = 110$　　Adding

$19x = 110$

$x = \dfrac{110}{19}$

Substitute $\dfrac{110}{19}$ for x in one of the equations in which the fractions were cleared and solve for y.

$3x - y = 18$

$3\left(\dfrac{110}{19} \right) - y = 18$　　Substituting

$\dfrac{330}{19} - y = \dfrac{342}{19}$

$-y = \dfrac{12}{19}$

$y = -\dfrac{12}{19}$

We obtain $\left(\dfrac{110}{19}, -\dfrac{12}{19} \right)$. This checks, so it is the solution.

15. $2x + 3y = 1,$

　　$4x + 6y = 2$

Multiply the first equation by -2 and then add.

$-4x - 6y = -2$

$\underline{4x + 6y = 2}$

$0 = 0$　　Adding

We have an equation that is true for all numbers x and y. The system is dependent and has an infinite number of solutions.

17. $2x - 4y = 5,$

$\ 2x - 4y = 6$

Multiply the first equation by -1 and then add.

$-2x + 4y = -5$

$\underline{2x - 4y = 6}$

$\ 0 = 1$

We have a false equation. The system has no solution.

19. $5x - 9y = 7,$

$\ 7y - 3x = -5$

We first write the second equation in the form $Ax + By = C$.

$5x - 9y = 7 \quad (1)$

$-3x + 7y = -5 \quad (2)$

We use the multiplication principle with both equations and then add.

$15x - 27y = 21 \quad$ Multiplying by 3

$\underline{-15x + 35y = -25} \quad$ Multiplying by 5

$\ 0 + \ 8y = \ -4 \quad$ Adding

$\ 8y = -4$

$\ y = -\frac{1}{2}$

Substitute $-\frac{1}{2}$ for y in one of the original equations and solve for x.

$5x - 9y = 7 \quad$ Equation (1)

$5x - 9\left(-\frac{1}{2}\right) = 7 \quad$ Substituting

$5x + \frac{9}{2} = \frac{14}{2}$

$5x = \frac{5}{2}$

$x = \frac{1}{2}$

We obtain $\left(\frac{1}{2}, -\frac{1}{2}\right)$. This checks, so it is the solution.

21. $3(a - b) = 15,$

$\ 4a = b + 1$

We first write each equation in the form $Ax + By = C$.

$3a - 3b = 15 \quad (1)$

$4a - \ b = 1 \quad (2)$

We multiply by -3 on both sides of the second equation and then add.

$3a - \ 3b = \ 15$

$\underline{-12a + \ 3b = -3} \quad$ Multiplying by -3

${-9a} + \ 0 = \ 12$

${-9a} = 12$

$a = -\frac{12}{9}$

$a = -\frac{4}{3}$

Substitute $-\frac{4}{3}$ for a in either Equation (1) or Equation (2) and solve for b.

$4a - b = 1 \quad$ Equation (2)

$4\left(-\frac{4}{3}\right) - b = 1 \quad$ Substituting

$-\frac{16}{3} - b = \frac{3}{3}$

$-b = \frac{19}{3}$

$b = -\frac{19}{3}$

We obtain $\left(-\frac{4}{3}, -\frac{19}{3}\right)$. This checks, so it is the solution.

23. $x - \frac{1}{10}y = 100,$

$\ y - \frac{1}{10}x = -100$

We first write the second equation in the form $Ax + By = C$.

$x - \frac{1}{10}y = 100$

$-\frac{1}{10}x + \ y = -100$

Next we multiply each equation by 10 to clear fractions.

$10x - \ y = 1000 \quad (1)$

$-x + 10y = -1000 \quad (2) \quad$ Equation (1)

We multiply by 10 on both sides of Equation (1) and then add.

$100x - 10y = 10,000 \quad$ Multiplying by 10

$\underline{-x + 10y = -1000}$

$99x + \ 0 = 9000$

$99x = 9000$

$x = \frac{9000}{99}$

$x = \frac{1000}{11}$

Substitute $\frac{1000}{11}$ for x in one of the equations in which the fractions were cleared and solve for y.

$10x - y = 1000 \quad$ Equation (1)

$10\left(\frac{1000}{11}\right) - y = 1000 \quad$ Substituting

$\frac{10,000}{11} - y = \frac{11,000}{11}$

$-y = \frac{1000}{11}$

$y = -\frac{1000}{11}$

We obtain $\left(\frac{1000}{11}, -\frac{1000}{11}\right)$. This checks, so it is the solution.

25. $0.05x + 0.25y = 22,$

$0.15x + 0.05y = 24$

We first multiply each equation by 100 to clear decimals.

$5x + 25y = 2200 \quad (1)$

$15x + 5y = 2400 \quad (2)$

We multiply by -5 on both sides of the second equation and add.

$\underline{\begin{array}{rcl} 5x + 25y = & 2200 \\ -75x - 25y = & -12{,}000 \quad \text{Multiplying by } -5 \end{array}}$

$-70x + \quad 0 = \quad -9800 \quad \text{Adding}$

$-70x = -9800$

$x = \dfrac{-9800}{-70}$

$x = 140$

Substitute 140 for x in one of the equations in which the decimals were cleared and solve for y.

$5x + 25y = 2200 \quad \text{Equation (1)}$

$5 \cdot 140 + 25y = 2200 \quad \text{Substituting}$

$700 + 25y = 2200$

$25y = 1500$

$y = 60$

We obtain $(140, 60)$. This checks, so it is the solution.

27. Familiarize. Let $l =$ the length of the field and $w =$ the width, in meters. Recall that the perimeter P of a rectangle with length l and width w is given by $P = 2l + 2w$.

Translate.

$$\underbrace{\text{The perimeter}}_{2l+2w} \;\; \underbrace{\text{is}}_{=} \;\; \underbrace{340\text{ m}.}_{340}$$

$$\underbrace{\text{The length}}_{l} \;\; \underbrace{\text{is}}_{=} \;\; \underbrace{50\text{ m}}_{50} \;\; \underbrace{\text{more than}}_{+} \;\; \underbrace{\text{the width.}}_{w}$$

We have system of equations.

$2l + 2w = 340, \quad (1)$

$l = 50 + w \quad\quad (2)$

Solve. First we subtract w on both sides of Equation (2).

$l = 50 + w$

$l - w = 50$

Now we have

$2l + 2w = 340, \quad (1)$

$l - \;\; w = 50. \quad (3)$

Multiply Equation (3) by 2 and then add.

$\underline{\begin{array}{rcl} 2l + 2w & = & 340 \\ 2l - 2w & = & 100 \end{array}}$

$4l \quad\quad = 440$

$l = 110$

Now substitute 110 for l in one of the original equations and solve for w.

$l = 50 + w \quad (2)$

$110 = 50 + w$

$60 = w$

Check. The perimeter is $2 \cdot 110 + 2 \cdot 60$, or $220 + 120$, or 340 m. Also, the length, 110 m, exceeds the width, 60 m, by 50 m. The answer checks.

State. The length is 110 m and the width is 60 m.

29. Familiarize. Using the drawing in the text, we let x and y represent the measures of the angles.

Translate.

$$\underbrace{\text{The sum of the measures}}_{x+y} \;\; \underbrace{\text{is}}_{=} \;\; \underbrace{90°.}_{90}$$

$$\underbrace{\text{One angle}}_{y} \; \underbrace{\text{is}}_{=} \; \underbrace{6°}_{6} \; \underbrace{\text{more than}}_{+} \; \underbrace{5}_{5} \; \underbrace{\text{times}}_{\cdot} \; \underbrace{\text{the other.}}_{x}$$

We have system of equations.

$x + y = 90, \quad (1)$

$y = 6 + 5x \quad (2)$

Solve. First we subtract $5x$ on both sides of Equation (2).

$y = 6 + 5x$

$-5x + y = 6 \quad\quad (3)$

Now we have

$x + y = 90, \quad (1)$

$-5x + y = 6. \quad (3)$

Multiply Equation (1) by 5 and then add.

$\underline{\begin{array}{rcl} 5x + 5y & = & 450 \\ -5x + \;\; y & = & 6 \end{array}}$

$6y = 456$

$y = 76$

Substitute 76 for y in one of the original equations and solve for x.

$x + y = 90 \quad (1)$

$x + 76 = 90$

$x = 14$

Check. The sum of the measures is $14° + 76°$, or $90°$. Also, $6°$ more than 5 times the measure of the $14°$ angle is $5 \cdot 14° + 6° = 70° + 6° = 76°$, the measure of the other angle. The answer checks.

State. The angles measure $14°$ and $76°$.

31. Familiarize. Let $c =$ the number of coach seats and $f =$ the number of first-class seats.

Translate.

$$\underbrace{\text{The total number of seats}}_{c+f} \;\; \underbrace{\text{is}}_{=} \;\; \underbrace{152.}_{152}$$

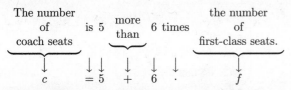

We have a system of equations.

$$c + f = 152, \quad (1)$$
$$c = 5 + 6f \quad (2)$$

Solve. First we subtract $6f$ on both sides of Equation (2).

$$c = 5 + 6f$$
$$c - 6f = 5 \quad (3)$$

Now we have

$$c + \ f = 152, \quad (1)$$
$$c - 6f = 5. \quad (3)$$

Multiply Equation (1) by 6 and then add.

$$6c + 6f = 912$$
$$\underline{c - 6f = 5}$$
$$7c \qquad = 917$$
$$c = 131$$

Substitute 131 for c in one of the original equations and solve for f.

$$c + f = 152 \quad (1)$$
$$131 + f = 152$$
$$f = 21$$

Check. The total number of seats is $131 + 21$, or 152. Five more than six times the number of first-class seats is $5 + 6 \cdot 21$, or $5 + 126$, or 131, the number of coach seats. The answer checks.

State. There are 131 coach-class seats and 21 first-class seats.

33. Discussion and Writing Exercise

35. $f(x) = 3x^2 - x + 1$
$f(0) = 3 \cdot 0^2 - 0 + 1 = 0 - 0 + 1 = 1$

37. $f(x) = 3x^2 - x + 1$
$f(1) = 3 \cdot 1^2 - 1 + 1 = 3 - 1 + 1 = 3$

39. $f(x) = 3x^2 - x + 1$
$f(-2) = 3(-2)^2 - (-2) + 1 = 12 + 2 + 1 = 15$

41. $f(x) = 3x^2 - x + 1$
$f(-4) = 3(-4)^2 - (-4) + 1 = 48 + 4 + 1 = 53$

43. We cannot calculate $f(x)$ when the denominator is 0. We set the denominator equal to 0 and solve for x.

$$x + 7 = 0$$
$$x = -7$$

The domain is $\{x | x \text{ is a real number } and \ x \neq -7\}$, or $(-\infty, -7) \cup (-7, \infty)$.

45. Substitute $-\dfrac{3}{5}$ for m and -7 for b in the slope-intercept equation.

$$y = mx + b$$
$$y = -\frac{3}{5}x - 7$$

47. Graph these equations, solving each equation for y first, if necessary. We get $y = \dfrac{3.5x - 106.2}{2.1}$ and $y = \dfrac{-4.1x - 106.28}{16.7}$. Using the INTERSECT feature, we find that the point of intersection is $(23.12, -12.04)$.

49. Substitute -5 for x and -1 for y in the first equation.

$$A(-5) - 7(-1) = -3$$
$$-5A + 7 = -3$$
$$-5A = -10$$
$$A = 2$$

Then substitute -5 for x and -1 for y in the second equation.

$$-5 - B(-1) = -1$$
$$-5 + B = -1$$
$$B = 4$$

We have $A = 2$, $B = 4$.

51. $(0, -3)$ and $\left(-\dfrac{3}{2}, 6 \right)$ are two solutions of $px - qy = -1$.
Substitute 0 for x and -3 for y.

$$p \cdot 0 - q \cdot (-3) = -1$$
$$3q = -1$$
$$q = -\frac{1}{3}$$

Substitute $-\dfrac{3}{2}$ for x and 6 for y.

$$p \cdot \left(-\frac{3}{2} \right) - q \cdot 6 = -1$$
$$-\frac{3}{2}p - 6q = -1$$

Substitute $-\dfrac{1}{3}$ for q and solve for p.

$$-\frac{3}{2}p - 6 \cdot \left(-\frac{1}{3} \right) = -1$$
$$-\frac{3}{2}p + 2 = -1$$
$$-\frac{3}{2}p = -3$$
$$-\frac{2}{3} \cdot \left(-\frac{3}{2}p \right) = -\frac{2}{3} \cdot (-3)$$
$$p = 2$$

Thus, $p = 2$ and $q = -\dfrac{1}{3}$.

Exercise Set 3.4

1. *Familiarize*. Let x = the number of less expensive brushes sold and y = the number of more expensive brushes sold.

Translate. We organize the information in a table.

Kind of brush	Less expensive	More expensive	Total
Number sold	x	y	45
Price	$8.50	$9.75	
Amount taken in	$8.50x$	$9.75y$	398.75

The "Number sold" row of the table gives us one equation:

$$x + y = 45$$

The "Amount taken in" row gives us a second equation:

$$8.50x + 9.75y = 398.75$$

We have a system of equations:

$$x + y = 45,$$
$$8.50x + 9.75y = 398.75$$

We can multiply the second equation on both sides by 100 to clear the decimals:

$$x + y = 45, \qquad (1)$$
$$850x + 975y = 39{,}875 \qquad (2)$$

Solve. We solve the system of equations using the elimination method. Begin by multiplying Equation (1) by -850.

$$\begin{array}{ll} -850x - 850y = -38{,}250 & \text{Multiplying (1)} \\ \underline{850x + 975y = 39{,}875} & \\ 125y = 1625 & \\ y = 13 & \end{array}$$

Substitute 13 for y in (1) and solve for x.

$$x + 13 = 45$$
$$x = 32$$

Check. The number of brushes sold is $32 + 13$, or 45. The amount taken in was $\$8.50(32) + \$9.75(13) = \$272 + \$126.75 = \$398.75$. The answer checks.

State. 32 of the less expensive brushes were sold, and 13 of the more expensive brushes were sold.

3. *Familiarize*. Let h = the number of vials of Humulin Insulin sold and n = the number of vials of Novolin Velosulin Insulin sold.

Translate. We organize the information in a table.

Brand	Humulin	Novolin	Total
Number sold	h	n	50
Price	$27.06	$34.39	
Amount taken in	$27.06h$	$34.39n$	1567.57

The "Number sold" row of the table gives us one equation:

$$h + n = 50$$

The "Amount taken in" row gives us a second equation:

$$27.06h + 34.39n = 1565.57$$

We have a system of equations:

$$h + n = 50,$$
$$27.06h + 34.39n = 1565.57$$

We can multiply both sides of the second equation by 100 to clear the decimals:

$$h + n = 50,$$
$$2706h + 3439n = 156{,}557$$

Solve. We use the elimination method.

$$\begin{array}{ll} -2706h - 2706n = -135{,}300 & \text{Multiplying (1)} \\ & \text{by } -2095 \\ \underline{2706h + 2706n = 156{,}557} & \\ 733n = 21{,}257 & \\ n = 29 & \end{array}$$

Substitute 29 for n in (1) and solve for h.

$$h + 29 = 50$$
$$h = 21$$

Check. A total of $21 + 29$, or 50 vials, were sold. The amount collected was $\$27.06(21) + \$34.39(29) = \$568.26 + \$997.31 = \$1565.57$. The answer checks.

State. 21 vials of Humulin Insulin and 29 vials of Novolin Velosulin Insulin were sold.

5. *Familiarize*. Let x and y represent the number of 30-sec and 60-sec commercials played, respectively. We will convert 10 min to seconds:

$$10 \text{ min} = 10 \times 1 \text{ min} = 10 \times 60 \text{ sec} = 600 \text{ sec}.$$

Translate. We organize the information in a table.

Type	30-sec	60-sec	Total
Number	x	y	12
Time	$30x$	$60y$	600

The "Number" row of the table gives us one equation:

$$x + y = 12$$

The "Time" row gives us a second equation:

$$30x + 60y = 600$$

We have a system of equations:

$$x + y = 12, \qquad (1)$$
$$30x + 60y = 600 \qquad (2)$$

Solve. We solve the system of equations using the elimination method.

$$\begin{array}{ll} -30x - 30y = -360 & \text{Multiplying (1) by } -30 \\ \underline{30x + 60y = 600} & \\ 30y = 240 & \\ y = 8 & \end{array}$$

Substitute 8 for y in Equation (1) and solve for x.

$$x + 8 = 12$$
$$x = 4$$

Check. If Rudy plays 4 30-sec and 8 60-sec commercials, then the total number of commercials played is $4 + 8$, or 12. Also, the time for 4 30-sec commercials is $4 \cdot 30$, or 120 sec, and the time for 8 60-sec commercials is $8 \cdot 60$, or 480 sec. Then the total commercial time is $120 + 480$, or 600 sec, or 10 min. The answer checks.

State. Rudy plays 4 30-sec commercials and 8 60-sec commercials.

7. *Familiarize*. Let x and y represent the number of pounds of the 40% and the 10% mixture to be used. The final mixture contains 25% (10 lb), or 0.25(10 lb), or 2.5 lb of peanuts.

Translate. We organize the information in a table.

	40% mixture	10% mixture	Wedding mixture
Number of pounds	x	y	10
Percent of peanuts	40%	10%	25%
Pounds of peanuts	$0.4x$	$0.1y$	2.5

The first row of the table gives us one equation:

$$x + y = 10$$

The last row gives us a second equation:

$$0.4x + 0.1y = 2.5$$

After clearing decimals, we have the problem translated to a system of equations:

$$x + y = 10, \quad (1)$$
$$4x + y = 25 \quad (2)$$

Solve. We solve the system of equations using the elimination method.

$$\begin{array}{ll} -x - y = -10 & \text{Multiplying (1) by } -1 \\ \underline{4x + y = 25} & \\ 3x = 15 & \\ x = 5 & \end{array}$$

Now substitute 5 for x in Equation (1) and solve for y.

$$5 + y = 10$$
$$y = 5$$

Check. If 5 lb of each mixture is used, the total wedding mixture is $5 + 5$, or 10 lb. The amount of peanuts in the wedding mixture is $0.4(5) + 0.1(5)$, or $2 + 0.5$, or 2.5 lb. The answer checks.

State. 5 lb of each type of mixture should be used.

9. *Familiarize*. Let $x =$ the number of liters of 25% solution and $y =$ the number of liters of 50% solution to be used. The mixture contains 40%(10 L), or 0.4(10 L) $= 4$ L of acid.

Translate. We organize the information in a table.

	25% solution	50% solution	Mixture
Number of liters	x	y	10
Percent of acid	25%	50%	40%
Amount of acid	$0.25x$	$0.5y$	4 L

We get one equation from the "Number of liters" row of the table.

$$x + y = 10$$

The last row of the table yields a second equation.

$$0.25x + 0.5y = 4$$

After clearing decimals, we have the problem translated to a system of equations:

$$x + y = 10, \quad (1)$$
$$25x + 50y = 400 \quad (2)$$

Solve. We use the elimination method to solve the system of equations.

$$\begin{array}{ll} -25x - 25y = -250 & \text{Multiplying (1) by } -25 \\ \underline{25x + 50y = 400} & \\ 25y = 150 & \\ y = 6 & \end{array}$$

Substitute 6 for y in (1) and solve for x.

$$x + 6 = 10$$
$$x = 4$$

Check. The total amount of the mixture is 4 lb + 6 lb, or 10 lb. The amount of acid in the mixture is $0.25(4 \text{ L}) + 0.5(6 \text{ L}) = 1 \text{ L} + 3 \text{ L} = 4$ L. The answer checks.

State. 4 L of the 25% solution and 6 L of the 50% solution should be mixed.

11. *Familiarize*. Let $n =$ the number of non-silk ties and $s =$ the number of silk ties.

Translate. We organize the information in a table.

	Non-silk	Silk	Total
Number	n	s	17
Price per tie	\$3.25	\$3.60	
Total cost	$3.25n$	$3.6s$	58.75

The first row of the table gives us one equation:

$$n + s = 17$$

The last row of the table yields another equation:

$$3.25n + 3.6s = 58.75$$

After clearing decimals we have the problem translated to a system of equations:

$$n + s = 17 \quad (1)$$
$$325n + 360s = 5875, \quad (2)$$

Solve. We use the elimination method. First we multiply Equation (1) by -325 and then add.

$$-325n - 325s = -5525$$
$$\underline{325n + 360s = 5875}$$
$$35s = 350$$
$$s = 10$$

Although we are asked to find only the number of silk ties, we also find the number of non-silk ties so that we can check the answer. We substitute 10 for s in Equation (1) and solve for n.

$$n + 10 = 17$$
$$n = 7$$

Check. The total number of ties is $7+10$, or 17. The total cleaning charge is $\$3.25(7) + \$3.60(10) = \$22.75 + \$36.00 = \$58.75$. The answer checks.

State. Claudio had 10 silk ties dry cleaned.

13. **Familiarize**. Let $x =$ the amount of the 6% loan and $y =$ the amount of the 9% loan. Recall that the formula for simple interest is

$$\text{Interest} = \text{Principal} \cdot \text{Rate} \cdot \text{Time}.$$

Translate. We organize the information in a table.

	6% loan	9% loan	Total
Principal	x	y	$12,000
Interest Rate	6%	9%	
Time	1 yr	1 yr	
Interest	0.06x	0.09y	$855

The "Principal" row of the table gives us one equation:

$$x + y = 12,000$$

The last row of the table yields another equation:

$$0.06x + 0.09y = 855$$

After clearing decimals, we have the problem translated to a system of equations:

$$x + y = 12,000 \quad (1)$$
$$6x + 9y = 85,500 \quad (2)$$

Solve. We use the elimination method to solve the system of equations.

$$-6x - 6y = -72,000 \quad \text{Multiplying (1) by } -6$$
$$\underline{6x + 9y = 85,500}$$
$$3y = 13,500$$
$$y = 4500$$

Substitute 4500 for y in (1) and solve for x.

$$x + 4500 = 12,000$$
$$x = 7500$$

Check. The loans total $\$7500 + \$4500$, or $\$12,000$. The total interest is $0.06(\$7500) + 0.09(\$4500) = \$450 + \$405 = \$855$. The answer checks.

State. The 6% loan was for $\$7500$, and the 9% loan was for $\$4500$.

15. **Familiarize**. From the bar graph we see that whole milk is 4% milk fat, milk for cream cheese is 8% milk fat, and cream is 30% milk fat. Let $x =$ the number of pounds of whole milk and $y =$ the number of pounds of cream to be used. The mixture contains 8%(200 lb), or 0.08(200 lb) = 16 lb of milk fat.

Translate. We organize the information in a table.

	Whole milk	Cream	Mixture
Number of pounds	x	y	200
Percent of milk fat	4%	30%	8%
Amount of milk fat	0.04x	0.3y	16 lb

We get one equation from the " Number of pounds" row of the table:

$$x + y = 200$$

The last row of the table yields a second equation:

$$0.04x + 0.3y = 16$$

After clearing decimals, we have the problem translated to a system of equations:

$$x + y = 200, \quad (1)$$
$$4x + 30y = 1600 \quad (2)$$

Solve. We use the elimination method to solve the system of equations.

$$-4x - 4y = -800 \qquad \text{Multiplying (1) by } -4$$
$$\underline{4x + 30y = 1600}$$
$$26y = 800$$
$$y = \frac{400}{13}, \text{ or } 30\frac{10}{13}$$

Substitute $\frac{400}{13}$ for y in (1) and solve for x.

$$x + \frac{400}{13} = 200$$
$$x = \frac{2200}{13}, \text{ or } 169\frac{3}{13}$$

Check. The total amount of the mixture is $\frac{2200}{13}$ lb $+ \frac{400}{13}$ lb $= \frac{2600}{13}$ lb $= 200$ lb. The amount of milk fat in the mixture is $0.04\left(\frac{2200}{13} \text{ lb}\right) + 0.3\left(\frac{400}{13} \text{ lb}\right) = \frac{88}{13}$ lb $+ \frac{120}{13}$ lb $= \frac{208}{13}$ lb $= 16$ lb. The answer checks.

State. $169\frac{3}{13}$ lb of whole milk and $30\frac{10}{13}$ lb of cream should be mixed.

17. **Familiarize**. Let $x =$ the number of $5 bills and $y =$ the number of $1 bills. The total value of the $5 bills is $5x$, and the total value of the $1 bills is $1 \cdot y$, or y.

Translate.

The total number of bills is 22.

$$x + y = 22$$

The total value of the bills is $50.

$$5x + y = 50$$

We have a system of equations:

$$x + y = 22, \quad (1)$$
$$5x + y = 50 \quad (2)$$

Solve. We use the elimination method.

$$-x - y = -22 \quad \text{Multiplying (1) by } -1$$

$$\underline{5x + y = 50}$$
$$4x = 28$$
$$x = 7$$

$$7 + y = 22 \quad \text{Substituting 7 for } x \text{ in (1)}$$
$$y = 15$$

Check. Total number of bills: $7 + 15 = 22$

Total value of bills: $\$5 \cdot 7 + \$1 \cdot 15 = \$35 + \$15 =$ $\$50$.

The numbers check.

State. There are 7 $5 bills and 15 $1 bills.

19. Familiarize. Let $x =$ the amount of the 5.5% investment and $y =$ the amount of the 4% investment. Recall that the formula for simple interest is

$$\text{Interest} = \text{Principal} \cdot \text{Rate} \cdot \text{Time}.$$

Translate. We organize the information in a table.

	5.5% investment	4% investment	Total
Amount	x	y	$18,000
Interest Rate	5.5%	4%	
Time	1 yr	1 yr	
Interest	$0.055x$	$0.04y$	$831

The "Amount" row of the table gives us one equation:

$$x + y = 18,000$$

The last row of the table yields another equation:

$$0.055x + 0.04y = 831$$

After clearing decimals, we have the problem translated to a system of equations:

$$x + y = 18,000 \quad (1)$$
$$55x + 40y = 831,000 \quad (2)$$

Solve. We use the elimination method to solve the system of equations.

$$-40x - 40y = -720,000 \quad \text{Multiplying (1) by } -40$$
$$\underline{55x + 40y = 831,000}$$
$$15x = 111,000$$
$$x = 7400$$

Substitute 7400 for y in (1) and solve for x.

$$x + 7400 = 18,000$$
$$x = 10,600$$

Check. The total amount invested was $7400 + \$10,600$, or $18,000. The interest earned was $0.055(\$7400) + 0.04(\$10,600) = \$407 + \$424 = \$831$. The answer checks.

State. $7400 was invested at 5.5% and $10,600 was invested at 4%.

21. Familiarize. We first make a drawing.

Slow train		
d miles	75 mph	$(t+2)$ hr

Fast train		
d miles	125 mph	t hr

From the drawing we see that the distances are the same. Now complete the chart.

$$d \quad = \quad r \quad \cdot \quad t$$

	Distance	Rate	Time	
Slow train	d	75	$t+2$	$\to d = 75(t+2)$
Fast train	d	125	t	$\to d = 125t$

Translate. Using $d = rt$ in each row of the table, we get a system of equations:

$$d = 75(t+2),$$
$$d = 125t$$

Solve. We solve the system of equations.

$$125t = 75(t+2) \quad \text{Using substitution}$$
$$125t = 75t + 150$$
$$50t = 150$$
$$t = 3$$

Then $d = 125t = 125 \cdot 3 = 375$

Check. At 125 mph, in 3 hr the fast train will travel $125 \cdot 3 = 375$ mi. At 75 mph, in $3 + 2$, or 5 hr the slow train will travel $75 \cdot 5 = 375$ mi. The numbers check.

State. The trains will meet 375 mi from the station.

23. Familiarize. We first make a drawing. Let $d =$ the distance and $r =$ the speed of the canoe in still water. Then when the canoe travels downstream its speed is $r + 6$, and its speed upstream is $r - 6$. From the drawing we see that the distances are the same.

Downstream, 6 km/h current

d km, $r + 6$, 4 hr

Upstream, 6 km/h current

d km, $r - 6$, 10 hr

Organize the information in a table.

	Distance	Rate	Time
With current	d	$r+6$	4
Against current	d	$r-6$	10

Translate. Using $d = rt$ in each row of the table, we get a system of equations:

$$d = 4(r+6), \qquad d = 4r + 24,$$
$$\text{or}$$
$$d = 10(r-6) \qquad d = 10r - 60$$

Solve. Solve the system of equations.

$$4r + 24 = 10r - 60 \quad \text{Using substitution}$$
$$24 = 6r - 60$$
$$84 = 6r$$
$$14 = r$$

Check. When $r = 14$, then $r + 6 = 14 + 6 = 20$, and the distance traveled in 4 hr is $4 \cdot 20 = 80$ km. Also, $r - 6 = 14 - 6 = 8$, and the distance traveled in 10 hr is $8 \cdot 10 = 80$ km. The answer checks.

State. The speed of the canoe in still water is 14 km/h.

25. **Familiarize**. We first make a drawing. Let $d =$ the distance and $t =$ the time at 32 mph. At 4 mph faster, the speed is 36 mph.

32 mph	t hr	d mi

36 mph	$\left(t - \dfrac{1}{2}\right)$ hr	d mi

From the drawing, we see that the distances are the same. List the information in a table.

	d	$=$	r	$\cdot$	t
	Distance	Rate	Time		
Slower trip	d	32	t		$\rightarrow d = 32t$
Faster trip	d	36	$t - \dfrac{1}{2}$		$\rightarrow d = 36\left(t - \dfrac{1}{2}\right)$

Translate. Using $d = rt$ in each row of the table, we get a system of equations:

$$d = 32t, \quad (1)$$
$$d = 36\left(t - \frac{1}{2}\right) \quad (2)$$

Solve. We solve the system of equations.

$$32t = 36\left(t - \frac{1}{2}\right) \quad \text{Substituting } 32t \text{ for } d \text{ in (2)}$$
$$32t = 36t - 18$$
$$-4t = -18$$
$$t = \frac{18}{4}, \text{ or } \frac{9}{2}$$

The time at 32 mph is $\dfrac{9}{2}$ hr, and the time at 36 mph is $\dfrac{9}{2} - \dfrac{1}{2}$, or 4 hr.

Check. At 32 mph, in $\dfrac{9}{2}$ hr the salesperson will travel $32 \cdot \dfrac{9}{2}$, or 144 mi. At 36 mph, in 4 hr she will travel $36 \cdot 4$, or 144 mi. Since the distances are the same, the numbers check.

State. The towns are 144 mi apart.

27. **Familiarize**. We first make a drawing. Let $t =$ the time, $d =$ the distance traveled at 190 km/h, and $780 - d =$ the distance traveled at 200 km/h.

190 km/h	t hr	t hr	200 km/h

780 km

We list the information in a table.

	d	$=$	r	$\cdot$	t
	Distance	Rate	Time		
Slower plane	d	190	t		$\rightarrow d = 190t$
Faster plane	$780 - d$	200	t		$\rightarrow 780 - d = 200t$

Translate. Using $d = rt$ in each row of the table, we get a system of equations:

$$d = 190t, \quad (1)$$
$$780 - d = 200t \quad (2)$$

Solve. We solve the system of equations.

$$780 - 190t = 200t \quad \text{Substituting } 190t \text{ for } d \text{ in (2)}$$
$$780 = 390t$$
$$2 = t$$

Check. In 2 hr the slower plane will travel $190 \cdot 2$, or 380 km, and the faster plane will travel $200 \cdot 2$, or 400 km. The sum of the distances is $380 + 400$, or 780 km. The value checks.

State. The planes will meet in 2 hr.

29. **Familiarize**. We first make a drawing. Let $d =$ the distance traveled at 420 km/h and $t =$ the time traveled. Then $1000 - d =$ the distance traveled at 330 km/h.

d km, 420 km/h, t hr	$1000 - d$ km, 330 km/h, t hr

1000 km

We list the information in a table.

	d	$=$	r	$\cdot$	t
	Distance	Rate	Time		
Faster airplane	d	420	t		$\rightarrow d = 420t$
Slower airplane	$1000 - d$	330	t		$\rightarrow 1000 - d = 330t$

Translate. Using $d = rt$ in each row of the table, we get a system of equations:

$$d = 420t, \quad (1)$$
$$1000 - d = 330t \quad (2)$$

Solve. We use substitution.

$$1000 - 420t = 330t \quad \text{Substituting } 420t \text{ for } d \text{ in (2)}$$
$$1000 = 750t$$
$$\frac{4}{3} = t$$

Check. If $t = \dfrac{4}{3}$, then $420 \cdot \dfrac{4}{3} = 560$, the distance traveled by the faster airplane. Also, $330 \cdot \dfrac{4}{3} = 440$, the distance traveled by the slower plane. The sum of the distances is $560 + 440$, or 1000 km. The values check.

State. The airplanes will meet after $\dfrac{4}{3}$ hr, or $1\dfrac{1}{3}$ hr.

31. *Familiarize*. We make a drawing. Note that the plane's speed traveling toward London is $360 + 50$, or 410 mph, and the speed traveling toward New York City is $360 - 50$, or 310 mph. Also, when the plane is d mi from New York City, it is $3458 - d$ mi from London.

New York City   London
310 mph t hours t hours   410 mph

|——————— 3458 mi ———————|

|—— d ——|—— 3458 mi $-d$ ——|

Organize the information in a table.

	Distance	Rate	Time
Toward NYC	d	310	t
Toward London	$3458 - d$	410	t

Translate. Using $d = rt$ in each row of the table, we get a system of equations:

$$d = 310t, \quad (1)$$
$$3458 - d = 410t \quad (2)$$

Solve. We solve the system of equations.

$$3458 - 310t = 410t \quad \text{Using substitution}$$
$$3458 = 720t$$
$$4.8028 \approx t$$

Substitute 4.8028 for t in (1).

$$d \approx 310(4.8028) \approx 1489$$

Check. If the plane is 1489 mi from New York City, it can return to New York City, flying at 310 mph, in $1489/310 \approx$ 4.8 hr. If the plane is $3458 - 1489$, or 1969 mi from London, it can fly to London, traveling at 410 mph, in $1969/410 \approx$ 4.8 hr. Since the times are the same, the answer checks.

State. The point of no return is about 1489 mi from New York City.

33. Discussion and Writing Exercise

35. $f(x) = 4x - 7$
$f(0) = 4 \cdot 0 - 7 = 0 - 7 = -7$

37. $f(x) = 4x - 7$
$f(1) = 4 \cdot 1 - 7 = 4 - 7 = -3$

39. $f(x) = 4x - 7$
$f(-2) = 4(-2) - 7 = -8 - 7 = -15$

41. $f(x) = 4x - 7$
$f(-4) = 4(-4) - 7 = -16 - 7 = -23$

43. $f(x) = 4x - 7$
$f\left(\dfrac{3}{4}\right) = 4 \cdot \dfrac{3}{4} - 7 = 3 - 7 = -4$

45. $f(x) = 4x - 7$
$f(-3h) = 4(-3h) - 7 = -12h - 7$

47. *Familiarize*. Let $x =$ the amount of the original solution that remains after some of the original solution is drained and replaced with pure antifreeze. Let $y =$ the amount of the original solution that is drained and replaced with pure antifreeze.

Translate. We organize the information in a table. Keep in mind that the table contains information regarding the solution *after* some of the original solution is drained and replaced with pure antifreeze.

	Original Solution	Pure Antifreeze	New Mixture
Amount of solution	x	y	16 L
Percent of antifreeze	30%	100%	50%
Amount of antifreeze in solution	$0.3x$	$1 \cdot y$, or y	$0.5(16)$, or 8

The "Amount of solution" row gives us one equation:
$x + y = 16$

The last row gives us a second equation:
$0.3x + y = 8$

After clearing the decimal we have the following system of equations:

$$x + y = 16, \quad (1)$$
$$3x + 10y = 80 \quad (2)$$

Solve. We use the elimination method.

$$-3x - 3y = -48 \quad \text{Multiplying (1) by } -3$$
$$\underline{3x + 10y = 80}$$
$$7y = 32$$
$$y = \frac{32}{7}, \text{ or } 4\frac{4}{7}$$

Although the problem only asks for the amount of pure antifreeze added, we will also find x in order to check.

$$x + 4\frac{4}{7} = 16 \quad \text{Substituting } 4\frac{4}{7} \text{ for } y \text{ in (1)}$$
$$x = 11\frac{3}{7}$$

Check. Total amount of new mixture: $11\frac{3}{7} + 4\frac{4}{7} =$ 16 L

Amount of antifreeze in new mixture:
$$0.3\left(11\frac{3}{7}\right) + 4\frac{4}{7} = \frac{3}{10} \cdot \frac{80}{7} + \frac{32}{7} = \frac{56}{7} = 8 \text{ L}$$
The numbers check.

State. Michelle should drain $4\frac{4}{7}$ L of the original solution and replace it with pure antifreeze.

49. **Familiarize**. Let x and y represent the number of city miles and highway miles that were driven, respectively. Then in city driving, $\dfrac{x}{18}$ gallons of gasoline are used; in highway driving, $\dfrac{y}{24}$ gallons are used.

Translate. We organize the information in a table.

Type of driving	City	Highway	Total
Number of miles	x	y	465
Gallons of gasoline used	$\dfrac{x}{18}$	$\dfrac{y}{24}$	23

The first row of the table gives us one equation:

$$x + y = 465$$

The second row gives us another equation:

$$\frac{x}{18} + \frac{y}{24} = 23$$

After clearing fractions, we have the following system of equations:

$$x + y = 465, \qquad (1)$$
$$24x + 18y = 9936 \qquad (2)$$

Solve. We solve the system of equations using the elimination method.

$$
\begin{array}{ll}
-18x - 18y = -8370 & \text{Multiplying (1) by } -18 \\
\underline{24x + 18y = 9936} & \\
6x = 1566 & \\
x = 261 &
\end{array}
$$

Now substitute 261 for x in Equation (1) and solve for y.

$$261 + y = 465$$
$$y = 204$$

Check. The total mileage is $261 + 204$, or 465. In 216 city miles, 261/18, or 14.5 gal of gasoline are used; in 204 highway miles, 204/24, or 8.5 gal are used. Then a total of $14.5 + 8.5$ or 23 gal of gasoline are used. The answer checks.

State. 261 miles were driven in the city, and 204 miles were driven on the highway.

51. **Familiarize**. Let x = the number of gallons of pure brown and y = the number of gallons of neutral stain that should be added to the original 0.5 gal. Note that a total of 1 gal of stain needs to be added to bring the amount of stain up to 1.5 gal. The original 0.5 gal of stain contains 20%(0.5 gal), or $0.2(0.5 \text{ gal}) = 0.1$ gal of brown stain. The final solution contains 60%(1.5 gal), or $0.6(1.5 \text{ gal}) = 0.9$ gal of brown stain. This is composed of the original 0.1 gal and the x gal that are added.

Translate.

$$\underbrace{\text{The amount of stain added}}_{x+y} \ \underbrace{\text{was}}_{=} \ \underbrace{1 \text{ gal.}}_{1}$$

$$\underbrace{\text{The amount of brown stain in the final solution}}_{0.1 + x} \ \underbrace{\text{is}}_{} \ 0.9 \text{ gal.}$$
$$0.1 + x = 0.9$$

We have a system of equations.

$$x + y = 1, \qquad (1)$$
$$0.1 + x = 0.9 \qquad (2)$$

Carry out. First we solve (2) for x.

$$0.1 + x = 0.9$$
$$x = 0.8$$

Then substitute 0.8 for x in (1) and solve for y.

$$0.8 + y = 1$$
$$y = 0.2$$

Check. Total amount of stain: $0.5 + 0.8 + 0.2 = 1.5$ gal

Total amount of brown stain: $0.1 + 0.8 = 0.9$ gal

Total amount of neutral stain: $0.8(0.5) + 0.2 = 0.4 + 0.2 = 0.6$ gal $= 0.4(1.5$ gal$)$

The answer checks.

State. 0.8 gal of pure brown and 0.2 gal of neutral stain should be added.

Exercise Set 3.5

1.
$$
\begin{array}{rl}
x + y + z = 2, & (1) \\
2x - y + 5z = -5, & (2) \\
-x + 2y + 2z = 1 & (3)
\end{array}
$$

Add Equations (1) and (2) to eliminate y:

$$
\begin{array}{ll}
x + y + z = 2 & (1) \\
\underline{2x - y + 5z = -5} & (2) \\
3x + 6z = -3 & (4) \quad \text{Adding}
\end{array}
$$

Use a different pair of equations and eliminate y:

$$
\begin{array}{ll}
4x - 2y + 10z = -10 & \text{Multiplying (2) by 2} \\
\underline{-x + 2y + 2z = 1} & (3) \\
3x + 12z = -9 & (5) \quad \text{Adding}
\end{array}
$$

Now solve the system of Equations (4) and (5).

$$
\begin{array}{ll}
3x + 6z = -3 & (4) \\
3x + 12z = -9 & (5)
\end{array}
$$

$$
\begin{array}{ll}
-3x - 6z = 3 & \text{Multiplying (4) by } -1 \\
\underline{3x + 12z = -9} & (5) \\
6z = -6 & \text{Adding} \\
z = -1 &
\end{array}
$$

$$
\begin{array}{ll}
3x + 6(-1) = -3 & \text{Substituting } -1 \text{ for } z \text{ in (4)} \\
3x - 6 = -3 & \\
3x = 3 & \\
x = 1 &
\end{array}
$$

$$
\begin{array}{ll}
1 + y + (-1) = 2 & \text{Substituting 1 for } x \text{ and } -1 \\
& \text{for } z \text{ in (1)} \\
y = 2 & \text{Simplifying}
\end{array}
$$

We obtain $(1, 2, -1)$. This checks, so it is the solution.

3. $2x - y + z = 5,$ (1)

$6x + 3y - 2z = 10,$ (2)

$x - 2y + 3z = 5$ (3)

We start by eliminating z from two different pairs of equations.

$4x - 2y + 2z = 10$ Multiplying (1) by 2

$\underline{6x + 3y - 2z = 10}$ (2)

$10x + y \quad = 20$ (4) Adding

$-6x + 3y - 3z = -15$ Multiplying (1) by -3

$\underline{x - 2y + 3z = 5}$ (3)

$-5x + y \quad = -10$ (5) Adding

Now solve the system of Equations (4) and (5).

$10x + y = 20$ (4)

$\underline{5x - y = 10}$ Multiplying (5) by -1

$15x \quad = 30$ Adding

$x = 2$

$10 \cdot 2 + y = 20$ Substituting 2 for x in (4)

$20 + y = 20$

$y = 0$

$2 \cdot 2 - 0 + z = 5$ Substituting 2 for x and 0 for y in (1)

$4 + z = 5$

$z = 1$

We obtain $(2, 0, 1)$. This checks, so it is the solution.

5. $2x - 3y + z = 5,$ (1)

$x + 3y + 8z = 22,$ (2)

$3x - y + 2z = 12$ (3)

We start by eliminating y from two different pairs of equations.

$2x - 3y + z = 5$ (1)

$\underline{x + 3y + 8z = 22}$ (2)

$3x \quad + 9z = 27$ (4) Adding

$x + 3y + 8z = 22$ (2)

$\underline{9x - 3y + 6z = 36}$ Multiplying (3) by 3

$10x \quad + 14z = 58$ (5) Adding

Solve the system of Equations (4) and (5).

$3x + 9z = 27$ (4)

$10x + 14z = 58$ (5)

$30x + 90z = 270$ Multiplying (4) by 10

$\underline{-30x - 42z = -174}$ Multiplying (5) by -3

$48z = 96$ Adding

$z = 2$

$3x + 9 \cdot 2 = 27$ Substituting 2 for z in (4)

$3x + 18 = 27$

$3x = 9$

$x = 3$

$2 \cdot 3 - 3y + 2 = 5$ Substituting 3 for x and 2 for z in (1)

$-3y + 8 = 5$

$-3y = -3$

$y = 1$

We obtain $(3, 1, 2)$. This checks, so it is the solution.

7. $3a - 2b + 7c = 13,$ (1)

$a + 8b - 6c = -47,$ (2)

$7a - 9b - 9c = -3$ (3)

We start by eliminating a from two different pairs of equations.

$3a - 2b + 7c = 13$ (1)

$\underline{-3a - 24b + 18c = 141}$ Multiplying (2) by -3

$- 26b + 25c = 154$ (4) Adding

$-7a - 56b + 42c = 329$ Multiplying (2) by -7

$\underline{7a - 9b - 9c = -3}$ (3)

$- 65b + 33c = 326$ (5) Adding

Now solve the system of Equations (4) and (5).

$-26b + 25c = 154$ (4)

$-65b + 33c = 326$ (5)

$-130b + 125c = 770$ Multiplying (4) by 5

$\underline{130b - 66c = -652}$ Multiplying (5) by -2

$59c = 118$

$c = 2$

$-26b + 25 \cdot 2 = 154$ Substituting 2 for c in (4)

$-26b + 50 = 154$

$-26b = 104$

$b = -4$

$a + 8(-4) - 6(2) = -47$ Substituting -4 for b and 2 for c in (2)

$a - 32 - 12 = -47$

$a - 44 = -47$

$a = -3$

We obtain $(-3, -4, 2)$. This checks, so it is the solution.

9. $2x + 3y + z = 17,$ (1)

$x - 3y + 2z = -8,$ (2)

$5x - 2y + 3z = 5$ (3)

We start by eliminating y from two different pairs of equations.

$2x + 3y + z = 17$ (1)

$\underline{x - 3y + 2z = -8}$ (2)

$3x \quad + 3z = 9$ (4) Adding

$4x + 6y + 2z = 34$ Multiplying (1) by 2

$\underline{15x - 6y + 9z = 15}$ Multiplying (3) by 3

$19x \quad + 11z = 49$ (5) Adding

Now solve the system of Equations (4) and (5).

$$3x + 3z = 9 \quad (4)$$
$$19x + 11z = 49 \quad (5)$$

$$33x + 33z = 99 \quad \text{Multiplying (4) by 11}$$
$$\underline{-57x - 33z = -147} \quad \text{Multiplying (5) by } -3$$
$$-24x \qquad = -48$$
$$x = 2$$

$$3 \cdot 2 + 3z = 9 \quad \text{Substituting 2 for } x \text{ in (4)}$$
$$6 + 3z = 9$$
$$3z = 3$$
$$z = 1$$

$$2 \cdot 2 + 3y + 1 = 17 \quad \text{Substituting 2 for } x \text{ and 1 for}$$
$$\qquad\qquad\qquad\qquad z \text{ in (1)}$$
$$3y + 5 = 17$$
$$3y = 12$$
$$y = 4$$

We obtain $(2, 4, 1)$. This checks, so it is the solution.

11. $2x + y + z = -2, \quad (1)$
$\quad\ \ 2x - y + 3z = 6, \quad (2)$
$\quad\ \ 3x - 5y + 4z = 7 \quad (3)$

We start by eliminating y from two different pairs of equations.

$$2x + y + z = -2 \quad (1)$$
$$\underline{2x - y + 3z = 6} \quad (2)$$
$$4x \quad\ \ + 4z = 4 \quad (4) \quad \text{Adding}$$

$$10x + 5y + 5z = -10 \quad \text{Multiplying (1) by 5}$$
$$\underline{3x - 5y + 4z = 7} \quad (3)$$
$$13x \qquad + 9z = -3 \quad (5) \quad \text{Adding}$$

Now solve the system of Equations (4) and (5).

$$4x + 4z = 4 \quad (4)$$
$$13x + 9z = -3 \quad (5)$$

$$36x + 36z = 36 \quad \text{Multiplying (4) by 9}$$
$$\underline{-52x - 36z = 12} \quad \text{Multiplying (5) by } -4$$
$$-16x \qquad = 48 \quad \text{Adding}$$
$$x = -3$$

$$4(-3) + 4z = 4 \quad \text{Substituting } -3 \text{ for } x \text{ in (4)}$$
$$-12 + 4z = 4$$
$$4z = 16$$
$$z = 4$$

$$2(-3) + y + 4 = -2 \quad \text{Substituting } -3 \text{ for } x \text{ and 4}$$
$$\qquad\qquad\qquad\qquad \text{for } z \text{ in (1)}$$
$$y - 2 = -2$$
$$y = 0$$

We obtain $(-3, 0, 4)$. This checks, so it is the solution.

13. $x - y + z = 4, \quad (1)$
$\quad\ \ 5x + 2y - 3z = 2, \quad (2)$
$\quad\ \ 3x - 7y + 4z = 8 \quad (3)$

We start by eliminating z from two different pairs of equations.

$$3x - 3y + 3z = 12 \quad \text{Multiplying (1) by 3}$$
$$\underline{5x + 2y - 3z = 2} \quad (2)$$
$$8x - y \qquad = 14 \quad (4) \quad \text{Adding}$$

$$-4x + 4y - 4z = -16 \quad \text{Multiplying (1) by } -4$$
$$\underline{3x - 7y + 4z = 8} \quad (3)$$
$$-x - 3y \qquad = -8 \quad (5) \quad \text{Adding}$$

Now solve the system of Equations (4) and (5).

$$8x - y = 14 \quad (4)$$
$$-x - 3y = -8 \quad (5)$$

$$8x - y = 14 \quad (4)$$
$$\underline{-8x - 24y = -64} \quad \text{Multiplying (5) by 8}$$
$$-25y = -50$$
$$y = 2$$

$$8x - 2 = 14 \quad \text{Substituting 2 for } y \text{ in (4)}$$
$$8x = 16$$
$$x = 2$$

$$2 - 2 + z = 4 \quad \text{Substituting 2 for } x \text{ and 2 for}$$
$$\qquad\qquad\qquad y \text{ in (1)}$$
$$z = 4$$

We obtain $(2, 2, 4)$. This checks, so it is the solution.

15. $4x - y - z = 4, \quad (1)$
$\quad\ \ 2x + y + z = -1, \quad (2)$
$\quad\ \ 6x - 3y - 2z = 3 \quad (3)$

We start by eliminating y from two different pairs of equations.

$$4x - y - z = 4 \quad (1)$$
$$\underline{2x + y + z = -1} \quad (2)$$
$$6x \qquad\qquad = 3 \quad (4) \quad \text{Adding}$$

At this point we can either continue by eliminating y from a second pair of equations or we can solve (4) for x and substitute that value in a different pair of the original equations to obtain a system of two equations in two variables. We take the second option.

$$6x = 3 \quad (4)$$
$$x = \frac{1}{2}$$

Substitute $\frac{1}{2}$ for x in (1):

$$4\left(\frac{1}{2}\right) - y - z = 4$$
$$2 - y - z = 4$$
$$-y - z = 2 \quad (5)$$

Substitute $\frac{1}{2}$ for x in (3):

$$6\left(\frac{1}{2}\right) - 3y - 2z = 3$$
$$3 - 3y - 2z = 3$$
$$-3y - 2z = 0 \quad (6)$$

Solve the system of Equations (5) and (6).

$$2y + 2z = -4 \quad \text{Multiplying (5) by } -2$$
$$\underline{-3y - 2z = 0 \quad (6)}$$
$$-y = -4$$
$$y = 4$$

$$-4 - z = 2 \quad \text{Substituting 4 for } y \text{ in (5)}$$
$$-z = 6$$
$$z = -6$$

We obtain $\left(\frac{1}{2}, 4, -6\right)$. This checks, so it is the solution.

17.
$$2r + 3s + 12t = 4, \quad (1)$$
$$4r - 6s + 6t = 1, \quad (2)$$
$$r + s + t = 1 \quad (3)$$

We start by eliminating s from two different pairs of equations.

$$4r + 6s + 24t = 8 \quad \text{Multiplying (1) by 2}$$
$$\underline{4r - 6s + 6t = 1 \quad (2)}$$
$$8r + 30t = 9 \quad (4) \quad \text{Adding}$$

$$4r - 6s + 6t = 1 \quad (2)$$
$$\underline{6r + 6s + 6t = 6 \quad \text{Multiplying (3) by 6}}$$
$$10r + 12t = 7 \quad (5) \quad \text{Adding}$$

Solve the system of Equations (4) and (5).

$$40r + 150t = 45 \quad \text{Multiplying (4) by 5}$$
$$\underline{-40r - 48t = -28 \quad \text{Multiplying (5) by } -4}$$
$$102t = 17$$
$$t = \frac{17}{102}$$
$$t = \frac{1}{6}$$

$$8r + 30\left(\frac{1}{6}\right) = 9 \quad \text{Substituting } \frac{1}{6} \text{ for } t \text{ in (4)}$$
$$8r + 5 = 9$$
$$8r = 4$$
$$r = \frac{1}{2}$$

$$\frac{1}{2} + s + \frac{1}{6} = 1 \quad \text{Substituting } \frac{1}{2} \text{ for } r \text{ and}$$
$$\frac{1}{6} \text{ for } t \text{ in (3)}$$
$$s + \frac{2}{3} = 1$$
$$s = \frac{1}{3}$$

We obtain $\left(\frac{1}{2}, \frac{1}{3}, \frac{1}{6}\right)$. This checks, so it is the solution.

19.
$$4a + 9b = 8, \quad (1)$$
$$8a + 6c = -1, \quad (2)$$
$$ 6b + 6c = -1 \quad (3)$$

We will use the elimination method. Note that there is no c in Equation (1). We will use equations (2) and (3) to obtain another equation with no c terms.

$$8a + 6c = -1 \quad (2)$$
$$\underline{ - 6b - 6c = 1 \quad \text{Multiplying (3) by } -1}$$
$$8a - 6b = 0 \quad (4) \quad \text{Adding}$$

Now solve the system of Equations (1) and (4).

$$-8a - 18b = -16 \quad \text{Multiplying (1) by } -2$$
$$\underline{8a - 6b = 0}$$
$$- 24b = -16$$
$$b = \frac{2}{3}$$

$$8a - 6\left(\frac{2}{3}\right) = 0 \quad \text{Substituting } \frac{2}{3} \text{ for } b \text{ in (4)}$$
$$8a - 4 = 0$$
$$8a = 4$$
$$a = \frac{1}{2}$$

$$8\left(\frac{1}{2}\right) + 6c = -1 \quad \text{Substituting } \frac{1}{2} \text{ for } a \text{ in (2)}$$
$$4 + 6c = -1$$
$$6c = -5$$
$$c = -\frac{5}{6}$$

We obtain $\left(\frac{1}{2}, \frac{2}{3}, -\frac{5}{6}\right)$. This checks, so it is the solution.

21.
$$x + y + z = 57, \quad (1)$$
$$-2x + y = 3, \quad (2)$$
$$x - z = 6 \quad (3)$$

We will use the substitution method. Solve Equations (2) and (3) for y and z, respectively. Then substitute in Equation (1) to solve for x.

$$-2x + y = 3 \quad \text{Solving (2) for } y$$
$$y = 2x + 3$$
$$x - z = 6 \quad \text{Solving (3) for } z$$
$$-z = -x + 6$$
$$z = x - 6$$

$$x + (2x + 3) + (x - 6) = 57 \quad \text{Substituting in (1)}$$
$$4x - 3 = 57$$
$$4x = 60$$
$$x = 15$$

To find y, substitute 15 for x in $y = 2x + 3$:

$$y = 2 \cdot 15 + 3 = 33$$

To find z, substitute 15 for x in $z = x - 6$:

$$z = 15 - 6 = 9$$

We obtain $(15, 33, 9)$. This checks, so it is the solution.

23. $\quad r + s \quad\quad = \quad 5, \quad (1)$

$\quad\quad\quad 3s + 2t = -1, \quad (2)$

$\quad 4r \quad\quad + \ t = \ 14 \quad (3)$

We will use the elimination method. Note that there is no t in Equation (1). We will use Equations (2) and (3) to obtain another equation with no t terms.

$$3s + 2t = -1 \quad (2)$$

$$\underline{-8r \quad\quad -2t = -28} \quad \text{Multiplying (3) by } -2$$

$$-8r + 3s \quad\quad = -29 \quad (4) \quad \text{Adding}$$

Now solve the system of Equations (1) and (4).

$$r + \ s = \quad 5 \quad (1)$$

$$-8r + 3s = -29 \quad (4)$$

$$8r + \ 8s = \quad 40 \quad \text{Multiplying (1) by 8}$$

$$\underline{-8r + \ 3s = -29} \quad (4)$$

$$11s = \quad 11 \quad \text{Adding}$$

$$s = 1$$

$$r + 1 = 5 \quad \text{Substituting 1 for } s \text{ in (1)}$$

$$r = 4$$

$$4 \cdot 4 + t = 14 \quad \text{Substituting 4 for } r \text{ in (3)}$$

$$16 + t = 14$$

$$t = -2$$

We obtain $(4, 1, -2)$. This checks, so it is the solution.

25. Discussion and Writing Exercise

27. $F = 3ab$

$\quad \dfrac{F}{3b} = a \quad \text{Dividing by } 3b$

29. $\quad F = \dfrac{1}{2}t(c - d)$

$\quad 2F = t(c - d) \quad \text{Multiplying by 2}$

$\quad 2F = tc - td \quad \text{Removing parentheses}$

$2F + td = tc \quad\quad \text{Adding } td$

$\dfrac{2F + td}{t} = c, \text{ or} \quad \text{Dividing by } t$

$\dfrac{2F}{t} + d = c$

31. $Ax - By = c$

$\quad Ax = By + c \quad \text{Adding } By$

$Ax - c = By \quad\quad \text{Subtracting } c$

$\dfrac{Ax - c}{B} = y \quad\quad \text{Dividing by } B$

33. $y = -\dfrac{2}{3}x - \dfrac{5}{4}$

The equation is in slope-intercept form, $y = mx + b$. The slope is $-\dfrac{2}{3}$, and the y-intercept is $\left(0, -\dfrac{5}{4}\right)$.

35. $2x - 5y = 10$

$\quad -5y = -2x + 10$

$-\dfrac{1}{5}(-5y) = -\dfrac{1}{5}(-2x + 10)$

$\quad y = \dfrac{2}{5}x - 2$

The equation is now in slope-intercept form, $y = mx + b$. The slope is $\dfrac{2}{5}$, and the y-intercept is $(0, -2)$.

37. $w + \ x + \ y + \ z = 2, \quad (1)$

$w + 2x + 2y + 4z = 1, \quad (3)$

$w - \ x + \ y + \ z = 6, \quad (3)$

$w - 3x - \ y + \ z = 2 \quad (4)$

Start by eliminating w from three different pairs of equations.

$$w + \ x + \ y + \ z = \quad 2 \quad (1)$$

$$\underline{-w - 2x - 2y - 4z = -1} \quad \text{Multiplying (2) by } -1$$

$$-x - \ y - 3z = \quad 1 \quad (5) \quad \text{Adding}$$

$$w + \ x + y + z = \quad 2 \quad (1)$$

$$\underline{-w + \ x - y - z = -6} \quad \text{Multiplying (3) by } -1$$

$$2x \quad\quad\quad = -4 \quad (6) \quad \text{Adding}$$

$$w + \ x + \ y + z = \quad 2 \quad (1)$$

$$\underline{-w + 3x + \ y - z = -2} \quad \text{Multiplying (4) by } -1$$

$$4x + 2y \quad\quad = \quad 0 \quad (7) \quad \text{Adding}$$

We can solve (6) for x:

$$2x = -4$$

$$x = -2$$

Substitute -2 for x in (7):

$$4(-2) + 2y = 0$$

$$-8 + 2y = 0$$

$$2y = 8$$

$$y = 4$$

Substitute -2 for x and 4 for y in (5):

$$-(-2) - 4 - 3z = 1$$

$$-2 - 3z = 1$$

$$-3z = 3$$

$$z = -1$$

Substitute -2 for x, 4 for y, and -1 for z in (1):

$$w - 2 + 4 - 1 = 2$$

$$w + 1 = 2$$

$$w = 1$$

We obtain $(1, -2, 4, -1)$. This checks, so it is the solution.

Exercise Set 3.6

1. Familiarize. Let x = the number of 20-oz drinks, y = the number of 32-oz drinks, and z = the number of 40-oz drinks sold. Then the amounts collected from selling each size of drink are $3.25x$, $4.75y$, and $5.75z$, respectively.

Translate.

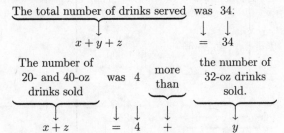

The total number of drinks served was 34.

$$x + y + z = 34$$

The number of 20- and 40-oz drinks sold was 4 more than the number of 32-oz drinks sold.

$$x + z = 4 + y$$

The total amount collected was $153.

$$1.09x + 1.29y + 1.49z = 153$$

Now we have a system of equations.

$$x + y + z = 34,$$
$$x + z = 4 + y,$$
$$3.25x + 4.75y + 5.75z = 153$$

Solve. Solving the system we get $(11, 15, 8)$.

Check. The total number of drinks served was $11+15+8$, or 34. The number of 20-oz and 40-oz drinks combined was $11 + 8$, or 19 drinks. This is 4 more than 15, the number of 32-oz drinks sold. The total amount collected was $\$3.25(11) + \$4.75(15) + \$5.75(8) = \$35.75 + \$71.25 + \$46 = \$153$. The answer checks.

State. Jake sold 11 20-oz drinks, 15 32-oz drinks, and 8 40-oz drinks.

3. Familiarize. We first make a drawing.

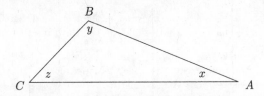

We let x, y, and z represent the measures of angles A, B, and C, respectively. The measures of the angles of a triangle add up to $180°$.

Translate.

The sum of the measures is $180°$.

$$x + y + z = 180$$

The measure of angle B is three times the measure of angle A.

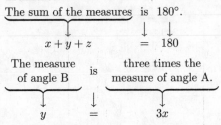

$$y = 3x$$

The measure of angle C is $20°$ more than the measure of angle A.

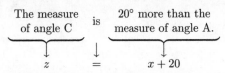

$$z = x + 20$$

We now have a system of equations.

$$x + y + z = 180,$$
$$y = 3x,$$
$$z = x + 20$$

Solve. Solving the system we get $(32, 96, 52)$.

Check. The sum of the measures is $32° + 96° + 52°$, or $180°$. Three times the measure of angle A is $3 \cdot 32°$, or $96°$, the measure of angle B. $20°$ more than the measure of angle A is $32° + 20°$, or $52°$, the measure of angle C. The numbers check.

State. The measures of angles A, B, and C are $32°$, $96°$, and $52°$, respectively.

5. Familiarize. Let x, y, and z represent the smallest, middle, and largest numbers, respectively.

Translate.

Sum of the numbers is 55.

$$x + y + z = 55$$

Largest minus smallest is 49.

$$z - x = 49$$

Smallest plus middle is 13.

$$x + y = 13$$

We now have a system of equations.

$$x + y + z = 55,$$
$$z - x = 49,$$
$$x + y = 13$$

Solve. Solving the system we get $(-7, 20, 42)$.

Check. $-7 + 20 + 42 = 55$; $42 - (-7) = 49$; and $-7 + 20 = 13$. The answer checks.

State. The numbers are -7, 20, and 42.

7. Familiarize. Let x = the cost of automatic transmission, y = the cost of power door locks, and z = the cost of air conditioning. The prices of the options are added to the basic price of $12,685.

Translate.

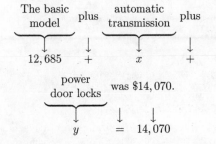

The basic model plus automatic transmission plus

$$12,685 + x +$$

power door locks was $14,070.

$$y = 14,070$$

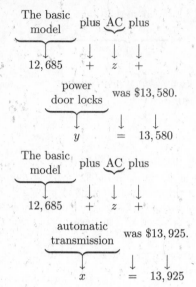

The basic model plus AC plus

$12,685 \quad + \quad z \quad +$

power door locks   was $13,580.

$y \quad = \quad 13,580$

The basic model plus AC plus

$12,685 \quad + \quad z \quad +$

automatic transmission   was $13,925.

$x \quad = \quad 13,925$

We now have a system of equations.

$$12,685 + x + y = 14,070,$$
$$12,685 + z + y = 13,580,$$
$$12,685 + z + x = 13,925$$

Solve. Solving the system we get $(865, 520, 375)$.

Check. The basic model with automatic transmission and power door locks costs $12,685 + \$865 + \$520$, or $14,070. The basic model with AC and power door locks costs $12,685 + \$375 + \$520$, or $13,580. The basic model with AC and automatic transmission costs $12,685 + \$375 + \$865$, or $13,925. The numbers check.

State. Automatic transmission costs $865, power door locks cost $520, and AC costs $375.

9. **Familiarize.** It helps to organize the information in a table. We let x, y, and z represent the weekly productions of the individual machines.

Machines Working	A	B	C
Weekly Production	x	y	z

Machines Working	A & B	B & C	A, B, & C
Weekly Production	3400	4200	5700

Translate. From the table, we obtain three equations.

$x + y + z = 5700$ (All three machines working)

$x + y \quad\quad = 3400$ (A and B working)

$\quad\; y + z = 4200$ (B and C working)

Solve. Solving the system we get $(1500, 1900, 2300)$.

Check. The sum of the weekly productions of machines A, B & C is $1500 + 1900 + 2300$, or 5700. The sum of the weekly productions of machines A and B is $1500 + 1900$,

or 3400. The sum of the weekly productions of machines B and C is $1900 + 2300$, or 4200. The numbers check.

State. In a week Machine A can polish 1500 lenses, Machine B can polish 1900 lenses, and Machine C can polish 2300 lenses.

11. **Familiarize.** Let $x =$ the amount invested in the first fund, $y =$ the amount invested in the second fund, and $z =$ the amount invested in the third fund. Then the earnings from the investments were $0.1x$, $0.06y$, and $0.15z$.

Translate.

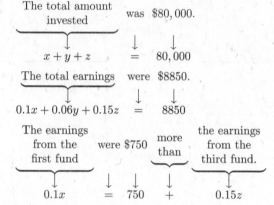

The total amount invested   was $80,000.

$x + y + z \quad = \quad 80,000$

The total earnings   were $8850.

$0.1x + 0.06y + 0.15z \quad = \quad 8850$

The earnings from the first fund   were $750 more than   the earnings from the third fund.

$0.1x \quad = \quad 750 \quad + \quad 0.15z$

Now we have a system of equations.

$$x + y + z = 80,000$$
$$0.1x + 0.06y + 0.15z = 8850,$$
$$0.1x = 750 + 0.15z$$

Solve. Solving the system we get $(45,000,\ 10,000,\ 25,000)$.

Check. The total investment was $45,000 + \$10,000 + \$25,000$, or $80,000. The total earnings were $0.1(\$45,000) + 0.06(10,000) + 0.15(25,000) = \$4500 + \$600 + \$3750 = \$8850$. The earnings from the first fund, $4500, were $750 more than the earnings from the second fund, $3750.

State. $45,000 was invested in the first fund, $10,000 in the second fund, and $25,000 in the third fund.

13. **Familiarize.** Let x, y, and z represent the number of fraternal twin births for Asian-Americans, African-Americans, and Caucasians in the U.S., respectively, out of every 15,400 births.

Translate. Out of every 15,400 births, we have the following statistics:

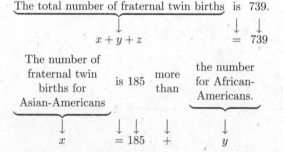

The total number of fraternal twin births   is 739.

$x + y + z \quad = \quad 739$

The number of fraternal twin births for Asian-Americans   is 185 more than   the number for African-Americans.

$x \quad = \quad 185 \quad + \quad y$

The number of fraternal twin births for Asian-Americans $\underbrace{}$ is 231 more than the number for Caucasians $\underbrace{}$.

$$x \qquad = 231 \quad + \quad z$$

We have a system of equations.

$$x + y + z = 739,$$
$$x = 185 + y,$$
$$x = 231 + y$$

Solve. Solving the system we get $(385, 200, 154)$.

Check. The total of the numbers is 739. Also 385 is 185 more than 200, and it is 231 more than 154.

State. Out of every 15,400 births, there are 385 births of fraternal twins for Asian-Americans, 200 for African-Americans, and 154 for Caucasians.

15. Familiarize. Let $r =$ the number of servings of roast beef, $p =$ the number of baked potatoes, and $b =$ the number of servings of broccoli. Then r servings of roast beef contain $300r$ Calories, $20r$ g of protein, and no vitamin C. In p baked potatoes there are $100p$ Calories, $5p$ g of protein, and $20p$ mg of vitamin C. And b servings of broccoli contain $50b$ Calories, $5b$ g of protein, and $100b$ mg of vitamin C. The patient requires 800 Calories, 55 g of protein, and 220 mg of vitamin C.

Translate. Write equations for the total number of calories, the total amount of protein, and the total amount of vitamin C.

$$300r + 100p + 50b = 800 \quad \text{(Calories)}$$
$$20r + 5p + 5b = 55 \quad \text{(protein)}$$
$$20p + 100b = 220 \quad \text{(vitamin C)}$$

We now have a system of equations.

Solve. Solving the system we get $(2, 1, 2)$.

Check. Two servings of roast beef provide 600 Calories, 40 g of protein, and no vitamin C. One baked potato provides 100 Calories, 5 g of protein, and 20 mg of vitamin C. And 2 servings of broccoli provide 100 Calories, 10 g of protein, and 200 mg of vitamin C. Together, then, they provide 800 Calories, 55 g of protein, and 220 mg of vitamin C. The values check.

State. The dietician should prepare 2 servings of roast beef, 1 baked potato, and 2 servings of broccoli.

17. Familiarize. Let x, y, and z represent the number of par-3, par-4, and par-5 holes, respectively. Then a par golfer shoots $3x$ on the par-3 holes, $4x$ on the par-4 holes, and $5x$ on the par-5 holes.

Translate.

The total number of holes $\underbrace{}$ is 18.

$$x + y + z \qquad = 18$$

A par golfer's score $\underbrace{}$ is 70.

$$3x + 4y + 5z \qquad = 70$$

The number of par-4 holes $\underbrace{}$ is 2 times the number of par-5 holes $\underbrace{}$.

$$y \qquad = 2 \cdot z$$

We have a system of equations.

$$x + y + z = 18,$$
$$3x + 4y + 5z = 70,$$
$$y = 2z$$

Solve. Solving the system we get $(6, 8, 4)$.

Check. The numbers add up to 18. A par golfer would shoot $3 \cdot 6 + 4 \cdot 8 + 5 \cdot 4$, or 70. The number of par-4 holes, 8, is twice the number of par-5 holes, 4. The numbers check.

State. There are 6 par-3 holes, 8 par-4 holes, and 4 par-5 holes.

19. Familiarize. Let x, y, and z represent the number of 2-point field goals, 3-point field goals, and 1-point foul shots made, respectively. The total number of points scored from each of these types of goals is $2x$, $3y$, and z.

Translate.

The total number of points $\underbrace{}$ was 92.

$$2x + 3y + z \qquad = 92$$

The total number of baskets $\underbrace{}$ was 50.

$$x + y + z \qquad = 50$$

The number of 2-pointers $\underbrace{}$ was 19 more than the number of foul shots $\underbrace{}$.

$$x \qquad = 19 \quad + \quad z$$

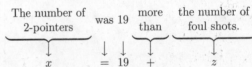

Now we have a system of equations.

$$2x + 3y + z = 92,$$
$$x + y + z = 50,$$
$$x = 19 + z$$

Solve. Solving the system we get $(32, 5, 13)$.

Check. The total number of points was $2 \cdot 32 + 3 \cdot 5 + 13 = 64 + 15 + 13 = 92$. The number of baskets was $32 + 5 + 13$, or 50. The number of 2-pointers, 32, was 19 more than the number of foul shots, 13. The numbers check.

State. The Knicks made 32 two-point field goals, 5 three-point field goals, and 13 foul shots.

21. Discussion and Writing Exercise

23. Discussion and Writing Exercise

25. The expression $x \le q$ means x is <u>at most</u> q.

27. The graph of a <u>linear</u> equation is a line.

29. A <u>consistent</u> system of equations has at least one solution.

31. The y-intercept of the graph of $f(x) = mx + b$ is the point $(0, b)$.

33. *Familiarize*. We first make a drawing with additional labels.

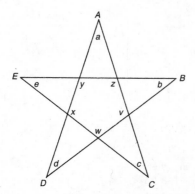

We let a, b, c, d, and e represent the angle measures at the tips of the star. We also label the interior angles of the pentagon v, w, x, y, and z. We recall the following geometric fact:

The sum of the measures of the interior angles of a polygon of n sides is given by $(n - 2)180°$.

Using this fact we know:

1. The sum of the angle measures of a triangle is $(3 - 2)180°$, or $180°$.

2. The sum of the angle measures of a pentagon is $(5 - 2)180°$, or $3(180°)$.

Translate. Using fact (1) listed above we obtain a system of 5 equations.

$$a + v + d = 180$$
$$b + w + e = 180$$
$$c + x + a = 180$$
$$d + y + b = 180$$
$$e + z + c = 180$$

Solve. Adding we obtain

$$2a + 2b + 2c + 2d + 2e + v + w + x + y + z =$$
$$5(180)$$
$$2(a + b + c + d + e) + (v + w + x + y + z) =$$
$$5(180)$$

Using fact (2) listed above we substitute $3(180)$ for $(v + w + x + y + z)$ and solve for $(a + b + c + d + e)$.

$$2(a + b + c + d + e) + 3(180) = 5(180)$$
$$2(a + b + c + d + e) = 2(180)$$
$$a + b + c + d + e = 180$$

Check. We should repeat the above calculations.

State. The sum of the angle measures at the tips of the star is $180°$.

35. *Familiarize*. Let $x =$ the one's digit, $y =$ the ten's digit, and $z =$ the hundred's digit. Then the number is represented by $100z + 10y + x$. When the digits are reversed, the resulting number is represented by $100x + 10y + z$.

Translate.

The sum of the digits is 14.

$$x + y + z = 14$$

The ten's digit is 2 more than the one's digit.

$$y = 2 + x$$

The number is the same as the number with the digits reversed.

$$100z + 10y + x = 100x + 10y + z$$

Now we have a system of equations.

$$x + y + z = 14,$$
$$y = 2 + x,$$
$$100z + 10y + x = 100x + 10y + z$$

Solve. Solving the system we get $(4, 6, 4)$.

Check. If the number is 464, then the sum of the digits is $4 + 6 + 4$, or 14. The ten's digit, 6, is 2 more than the one's digit, 4. If the digits are reversed the number is unchanged. The result checks.

State. The number is 464.

Exercise Set 3.7

1. We use alphabetical order to replace x by -3 and y by 3.

$$\frac{3x + y < -5}{3(-3) + 3 \ ? \ -5}$$
$$-9 + 3$$
$$-6 \ \bigg| \quad \text{TRUE}$$

Since $-6 < -5$ is true, $(-3, 3)$ is a solution.

3. We use alphabetical order to replace x by 5 and y by 9.

$$\frac{2x - y > -1}{2 \cdot 5 - 9 \ ? \ -1}$$
$$10 - 9$$
$$1 \ \bigg| \quad \text{TRUE}$$

Since $1 > -1$ is true, $(5, 9)$ is a solution.

5. Graph: $y > 2x$

We first graph the line $y = 2x$. We draw the line dashed since the inequality symbol is $>$. To determine which half-plane to shade, test a point not on the line. We try $(1, 1)$ and substitute:

$$\frac{y > 2x}{1 \ ? \ 2 \cdot 1}$$
$$\bigg| \ 2 \quad \text{FALSE}$$

Since $1 > 2$ is false, $(1, 1)$ is not a solution, nor are any points in the half-plane containing $(1, 1)$. The points in the opposite half-plane are solutions, so we shade that half-plane and obtain the graph.

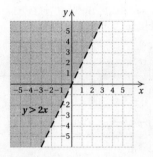

7. Graph: $y < x + 1$

First graph the line $y = x + 1$. Draw it dashed since the inequality symbol is $<$. Test the point $(0, 0)$ to determine if it is a solution.

$$\begin{array}{c|c} y < x + 1 \\ \hline 0 \;?\; 0 + 1 \\ \mid 1 & \text{TRUE} \end{array}$$

Since $0 < 1$ is true, we shade the half-plane containing $(0, 0)$ and obtain the graph.

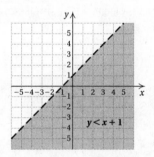

9. Graph: $y > x - 2$

First graph the line $y = x - 2$. Draw a dashed line since the inequality symbol is $>$. Test the point $(0, 0)$ to determine if it is a solution.

$$\begin{array}{c|c} y > x - 2 \\ \hline 0 \;?\; 0 - 2 \\ \mid -2 & \text{TRUE} \end{array}$$

Since $0 > -2$ is true, we shade the half-plane containing $(0, 0)$ and obtain the graph.

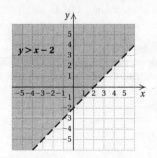

11. Graph: $x + y < 4$

First graph $x + y = 4$. Draw the line dashed since the inequality symbol is $<$. Test the point $(0, 0)$ to determine if it is a solution.

$$\begin{array}{c|c} x + y < 4 \\ \hline 0 + 0 \;?\; 4 \\ 0 \mid & \text{TRUE} \end{array}$$

Since $0 < 4$ is true, we shade the half-plane containing $(0, 0)$ and obtain the graph.

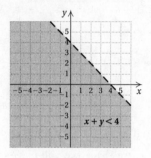

13. Graph: $3x + 4y \leq 12$

We first graph $3x + 4y = 12$. Draw the line solid since the inequality symbol is $\leq$. Test the point $(0, 0)$ to determine if it is a solution.

$$\begin{array}{c|c} 3x + 4y \leq 12 \\ \hline 3 \cdot 0 + 4 \cdot 0 \;?\; 12 \\ 0 \mid & \text{TRUE} \end{array}$$

Since $0 \leq 12$ is true, we shade the half-plane containing $(0, 0)$ and obtain the graph.

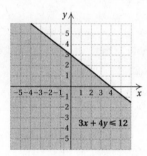

15. Graph: $2y - 3x > 6$

We first graph $2y - 3x = 6$. Draw the line dashed since the inequality symbol is $>$. Test the point $(0, 0)$ to determine if it is a solution.

$$\begin{array}{c|c} 2y - 3x > 6 \\ \hline 2 \cdot 0 - 3 \cdot 0 \;?\; 6 \\ 0 \mid & \text{FALSE} \end{array}$$

Since $0 > 6$ is false, we shade the half-plane that does not contain $(0, 0)$ and obtain the graph.

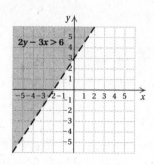

17. Graph: $3x - 2 \leq 5x + y$

$$-2 \leq 2x + y$$

We first graph $-2 = 2x + y$. Draw the line solid since the inequality symbol is $\leq$. Test the point $(0,0)$ to determine if it is a solution.

$$\frac{-2 \leq 2x + y}{-2 \;?\; 2 \cdot 0 + 0}$$
$$\begin{array}{c|c} & 0 & \text{TRUE} \end{array}$$

Since $-2 \leq 0$ is true, we shade the half-plane containing $(0,0)$ and obtain the graph.

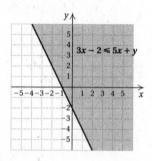

19. Graph: $x < 5$

We first graph $x = 5$. Draw the line dashed since the inequality symbol is $<$. Test the point $(0,0)$ to determine if it is a solution.

$$\frac{x < 5}{0 \;?\; 5 \;\; \text{TRUE}}$$

Since $0 < 5$ is true, we shade the half-plane containing $(0,0)$ and obtain the graph.

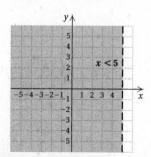

21. Graph: $y > 2$

We first graph $y = 2$. We draw the line dashed since the inequality symbol is $>$. Test the point $(0,0)$ to determine if it is a solution.

$$\frac{y > 2}{0 \;?\; 2 \;\; \text{FALSE}}$$

Since $0 > 2$ is false, we shade the half-plane that does not contain $(0,0)$ and obtain the graph.

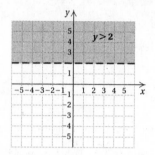

23. Graph: $2x + 3y \leq 6$

We first graph $2x + 3y = 6$. We draw the line solid since the inequality symbol is $\leq$. Test the point $(0,0)$ to determine if it is a solution.

$$\frac{2x + 3y \leq 6}{2 \cdot 0 + 3 \cdot 0 \;?\; 6}$$
$$\begin{array}{c|c} 0 & \text{TRUE} \end{array}$$

Since $0 \leq 6$ is true, we shade the half-plane containing $(0,0)$ and obtain the graph.

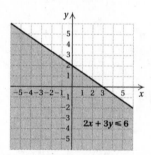

25. The intercepts of the graph of the related equation are $(0, -2)$ and $(3, 0)$, so inequality **F** could be the correct one. Since $(0,0)$ is in the solution set of this inequality and the half-plane containing $(0,0)$ is shaded, we know that inequality **F** corresponds to this graph.

27. The intercepts of the graph of the related equation are $(-5, 0)$ and $(0, 3)$, so inequality **B** could be the correct one. Since $(0,0)$ is in the solution set of this inequality and the half-plane containing $(0,0)$ is shaded, we know that inequality **B** corresponds to this graph.

29. The intercepts of the graph of the related equation are $(-3, 0)$ and $(0, -3)$, so inequality **C** could be the correct one. Since $(0,0)$ is not in the solution set of the inequality and the half-plane that does not contain $(0,0)$ is shaded, we know that inequality **C** corresponds to this graph.

31. Graph: $y \geq x$,

$$y \leq -x + 2$$

We graph the lines $y = x$ and $y = -x + 2$, using solid lines. We indicate the region for each inequality by arrows at the ends of the lines. Note where the regions overlap, and shade the region of solutions.

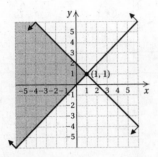

To find the vertex we solve the system of related equations:

$$y = x,$$
$$y = -x + 2$$

Solving, we obtain the vertex $(1, 1)$.

33. Graph: $y > x,$

$\qquad\qquad y < -x + 1$

We graph the lines $y = x$ and $y = -x + 1$, using dashed lines. We indicate the region for each inequality by arrows at the ends of the lines. Note where the regions overlap, and shade the region of solutions.

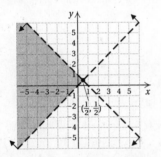

To find the vertex we solve the system of related equations:

$$y = x,$$
$$y = -x + 1$$

Solving, we obtain the vertex $\left(\dfrac{1}{2}, \dfrac{1}{2}\right)$.

35. Graph: $y \geq -2,$

$\qquad\qquad x \geq 1$

We graph the lines $y = -2$ and $x = 1$, using solid lines. We indicate the region for each inequality by arrows. Shade the region where they overlap.

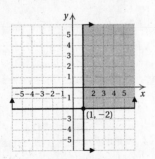

To find the vertex, we solve the system of related equations:

$$y = -2,$$
$$x = 1$$

Solving, we obtain the vertex $(1, -2)$.

37. Graph: $x \leq 3,$

$\qquad\qquad y \geq -3x + 2$

Graph the lines $x = 3$ and $y = -3x + 2$, using solid lines. Indicate the region for each inequality by arrows, and shade the region where they overlap.

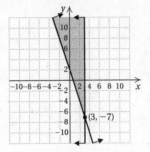

To find the vertex we solve the system of related equations:

$$x = 3,$$
$$y = -3x + 2$$

Solving, we obtain the vertex $(3, -7)$.

39. Graph: $y \leq 2x + 1,$ (1)

$\qquad\qquad y \geq -2x + 1,$ (2)

$\qquad\qquad x \leq 2$ (3)

Shade the intersection of the graphs of $y \leq 2x + 1$, $y \geq -2x + 1$, and $x \leq 2$.

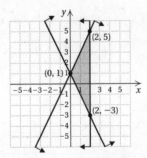

To find the vertices we solve three different systems of equations. From (1) and (2) we obtain the vertex $(0, 1)$. From (1) and (3) we obtain the vertex $(2, 5)$. From (2) and (3) we obtain the vertex $(2, -3)$.

41. Graph: $x + y \leq 1,$

$\qquad\qquad x - y \leq 2$

Graph the lines $x + y = 1$ and $x - y = 2$, using solid lines. Indicate the region for each inequality by arrows, and shade the region where they overlap.

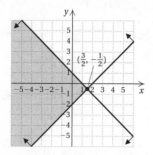

To find the vertex we solve the system of related equations:

$$x + y = 1,$$
$$x - y = 2$$

The vertex is $\left(\dfrac{3}{2}, -\dfrac{1}{2}\right)$.

43. Graph: $x + 2y \le 12$, (1)
   $2x + y \le 12$, (2)
   $x \ge 0$, (3)
   $y \ge 0$ (4)

Shade the intersection of the graphs of the four inequalities above.

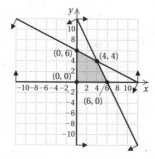

To find the vertices we solve four different systems of equations, as follows:

System of equations	Vertex
From (1) and (2)	$(4, 4)$
From (1) and (3)	$(0, 6)$
From (2) and (4)	$(6, 0)$
From (3) and (4)	$(0, 0)$

45. Graph: $8x + 5y \le 40$, (1)
   $x + 2y \le 8$, (2)
   $x \ge 0$, (3)
   $y \ge 0$ (4)

Shade the intersection of the graphs of the four inequalities above.

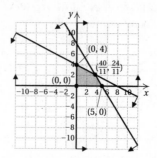

To find the vertices we solve four different systems of equations, as follows:

System of equations	Vertex
From (1) and (2)	$\left(\dfrac{40}{11}, \dfrac{24}{11}\right)$
From (1) and (4)	$(5, 0)$
From (2) and (3)	$(0, 4)$
From (3) and (4)	$(0, 0)$

47. Discussion and Writing Exercise

49. $5(3x - 4) = -2(x + 5)$
   $15x - 20 = -2x - 10$
   $17x - 20 = -10$
   $17x = 10$
   $x = \dfrac{10}{17}$

The solution is $\dfrac{10}{17}$.

51. $2(x - 1) + 3(x - 2) - 4(x - 5) = 10$
   $2x - 2 + 3x - 6 - 4x + 20 = 10$
   $x + 12 = 10$
   $x = -2$

The solution is -2.

53. $5x + 7x = -144$
   $12x = -144$
   $x = -12$

The solution is -12.

55. $f(x) = |2 - x|$
   $f(0) = |2 - 0| = |2| = 2$

57. $f(x) = |2 - x|$
   $f(1) = |2 - 1| = |1| = 1$

59. $f(x) = |2 - x|$
   $f(-2) = |2 - (-2)| = |4| = 4$

61. $f(x) = |2 - x|$
   $f(-4) = |2 - (-4)| = |6| = 6$

63. Both the width and the height must be positive, so we have

$$w > 0,$$
$$h > 0.$$

To be checked as luggage, the sum of the length, width, and height cannot exceed 62 in., so we have

$$30 + w + h \leq 62, \text{ or}$$
$$w + h \leq 32.$$

To be mailed the sum of the length and girth cannot exceed 130 in., so we have

$$30 + 2w + 2h \leq 130, \text{ or}$$
$$2w + 2h \leq 100, \text{ or}$$
$$w + h \leq 50.$$

Thus, have a system of inequalities:

$$w > 0,$$
$$h > 0,$$
$$w + h \leq 32,$$
$$w + h \leq 50$$

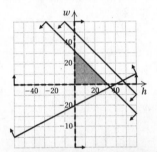

65.

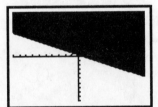

67.

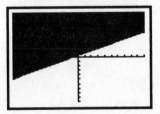

69. Left to the student

Exercise Set 3.8

1. $C(x) = 25x + 270,000 \qquad R(x) = 70x$

a) $P(x) = R(x) - C(x)$

$\qquad = 70x - (25x + 270,000)$

$\qquad = 70x - 25x - 270,000$

$\qquad = 45x - 270,000$

b) To find the break-even point we solve the system

$$R(x) = 70x,$$
$$C(x) = 25x + 270,000.$$

Since both $R(x)$ and $C(x)$ are in dollars and they are equal at the break-even point, we can rewrite the system:

$$d = 70x, \qquad (1)$$
$$d = 25x + 270,000 \quad (2)$$

We solve using substitution.

$70x = 25x + 270,000 \quad$ Substituting $65x$
$\qquad\qquad\qquad\qquad$ for d in (2)

$45x = 270,000$

$x = 6000$

Thus, 6000 units must be produced and sold in order to break even.

The amount taken in is $R(6000) = 70 \cdot 6000 = \$420,000$. Thus, the break-even point is $(6000, \$420,000)$.

3. $C(x) = 10x + 120,000 \qquad R(x) = 60x$

a) $P(x) = R(x) - C(x)$

$\qquad = 60x - (10x + 120,000)$

$\qquad = 60x - 10x - 120,000$

$\qquad = 50x - 120,000$

b) Solve the system

$$R(x) = 60x,$$
$$C(x) = 10x + 120,000.$$

Since both $R(x)$ and $C(x)$ are in dollars and they are equal at the break-even point, we can rewrite the system:

$$d = 60x, \qquad (1)$$
$$d = 10x + 120,000 \quad (2)$$

We solve using substitution.

$60x = 10x + 120,000 \quad$ Substituting $60x$
$\qquad\qquad\qquad\qquad$ for d in (2)

$50x = 120,000$

$x = 2400$

Thus, 2400 units must be produced and sold in order to break even.

The amount taken in is $R(2400) = 60 \cdot 2400 = \$144,000$. Thus, the break-even point is $(2400, \$144,000)$.

5. $C(x) = 20x + 10,000 \qquad R(x) = 100x$

a) $P(x) = R(x) - C(x)$

$\qquad = 100x - (20x + 10,000)$

$\qquad = 100x - 20x - 10,000$

$\qquad = 80x - 10,000$

b) Solve the system

$$R(x) = 100x,$$
$$C(x) = 20x + 10,000.$$

Since both $R(x)$ and $C(x)$ are in dollars and they are equal at the break-even point, we can rewrite the system:

$$d = 100x, \qquad (1)$$
$$d = 20x + 10,000 \quad (2)$$

We solve using substitution.

$$100x = 20x + 10,000 \quad \text{Substituting } 100x$$
$$\text{for } d \text{ in (2)}$$
$$80x = 10,000$$
$$x = 125$$

Thus, 125 units must be produced and sold in order to break even.

The amount taken in is $R(125) = 100 \cdot 125 = \$12,500$. Thus, the break-even point is $(125, \$12,500)$.

7. $C(x) = 22x + 16,000 \qquad R(x) = 40x$

a) $P(x) = R(x) - C(x)$
$$= 40x - (22x + 16,000)$$
$$= 40x - 22x - 16,000$$
$$= 18x - 16,000$$

b) Solve the system
$$R(x) = 40x,$$
$$C(x) = 22x + 16,000.$$

Since both $R(x)$ and $C(x)$ are in dollars and they are equal at the break-even point, we can rewrite the system:

$$d = 40x, \qquad (1)$$
$$d = 22x + 16,000 \quad (2)$$

We solve using substitution.

$$40x = 22x + 16,000 \quad \text{Substituting } 40x \text{ for}$$
$$d \text{ in (2)}$$
$$18x = 16,000$$
$$x \approx 889 \text{ units}$$

Thus, 889 units must be produced and sold in order to break even.

The amount taken in is $R(889) = 40 \cdot 889 = \$35,560$. Thus, the break-even point is $(889, \$35,560)$.

9. $C(x) = 50x + 195,000 \qquad R(x) = 125x$

a) $P(x) = R(x) - C(x)$
$$= 125x - (50x + 195,000)$$
$$= 125x - 50x - 195,000$$
$$= 75x - 195,000$$

b) Solve the system
$$R(x) = 125x,$$
$$C(x) = 50x + 195,000.$$

Since $R(x) = C(x)$ at the break-even point, we can rewrite the system:

$$R(x) = 125x, \qquad (1)$$
$$R(x) = 50x + 195,000 \quad (2)$$

We solve using substitution.

$$125x = 50x + 195,000 \quad \text{Substituting } 125x$$
$$\text{for } R(x) \text{ in (2)}$$
$$75x = 195,000$$
$$x = 2600$$

To break even 2600 units must be produced and sold.

The amount taken in is $R(2600) = 125 \cdot 2600 = \$325,000$. Thus, the break-even point is $(2600, \$325,000)$.

11. a) $C(x) = \text{Fixed costs} + \text{Variable costs}$
$$C(x) = 22,500 + 40x,$$

where x is the number of lamps produced.

b) Each lamp sells for \$85. The total revenue is 85 times the number of lamps sold. We assume that all lamps produced are sold.

$$R(x) = 85x$$

c) $P(x) = R(x) - C(x)$
$$P(x) = 85x - (22,500 + 40x)$$
$$= 85x - 22,500 - 40x$$
$$= 45x - 22,500$$

d) $P(3000) = 45(3000) - 22,500$
$$= 135,000 - 22,500$$
$$= 112,500$$

The company will realize a profit of \$112,500 when 3000 lamps are produced and sold.

$$P(400) = 45(400) - 22,500$$
$$= 18,000 - 22,500$$
$$= -4500$$

The company will realize a \$4500 loss when 400 lamps are produced and sold.

e) Solve the system
$$R(x) = 85x,$$
$$C(x) = 22,500 + 40x.$$

Since both $R(x)$ and $C(x)$ are in dollars and they are equal at the break-even point, we can rewrite the system:

$$d = 85x, \qquad (1)$$
$$d = 22,500 + 40x \quad (2)$$

We solve using substitution.

$$85x = 22,500 + 40x \quad \text{Substituting } 85x \text{ for } d$$
$$\text{in (2)}$$
$$45x = 22,500$$
$$x = 500$$

The firm will break even if it produces and sells 500 lamps and takes in a total of $R(500) = 85 \cdot 500 = \$42,500$ in revenue. Thus, the break-even point is $(500, \$42,500)$.

13. a) $C(x) = \text{Fixed costs} + \text{Variable costs}$
$$C(x) = 16,404 + 6x,$$

where x is the number of caps produced, in dozens.

b) Each dozen caps sell for \$18. The total revenue is 18 times the number of caps sold, in dozens. We assume that all caps produced are sold.

$$R(x) = 18x$$

c) $P(x) = R(x) - C(x)$

$$P(x) = 18x - (16,404 + 6x)$$
$$= 18x - 16,404 - 6x$$
$$= 12x - 16,404$$

d) $P(3000) = 12(3000) - 16,404$
$$= 36,000 - 16,404$$
$$= 19,596$$

The company will realize a profit of \$19,596 when 3000 dozen caps are produced and sold.

$$P(1000) = 12(1000) - 16,404$$
$$= 12,000 - 16,404$$
$$= -4404$$

The company will realize a \$4404 loss when 1000 dozen caps are produced and sold.

e) Solve the system

$$R(x) = 18x,$$
$$C(x) = 16,404 + 6x.$$

Since both $R(x)$ and $C(x)$ are in dollars and they are equal at the break-even point, we can rewrite the system:

$$d = 18x, \qquad (1)$$
$$d = 16,404 + 6x \qquad (2)$$

We solve using substitution.

$18x = 16,404 + 6x$ Substituting $18x$ for d in (2)

$$12x = 16,404$$
$$x = 1367$$

The firm will break even if it produces and sells 1367 dozen caps and takes in a total of $R(1367) = 18 \cdot 1367 = \$24,606$ in revenue. Thus, the break-even point is (1367, \$24,606).

15. $D(p) = 1000 - 10p,$

$\quad S(p) = 230 + p$

Since both demand and supply are quantities, the system can be rewritten:

$$q = 1000 - 10p, \quad (1)$$
$$q = 230 + p \qquad (2)$$

Substitute $1000 - 10p$ for q in (2) and solve.

$$1000 - 10p = 230 + p$$
$$770 = 11p$$
$$70 = p$$

The equilibrium price is \$70 per unit. To find the equilibrium quantity we substitute \$70 into either $D(p)$ or $S(p)$.

$D(70) = 1000 - 10 \cdot 70 = 1000 - 700 = 300$

The equilibrium quantity is 300 units.

The equilibrium point is (\$70, 300).

17. $D(p) = 760 - 13p,$

$\quad S(p) = 430 + 2p$

Rewrite the system:

$$q = 760 - 13p, \quad (1)$$
$$q = 430 + 2p \qquad (2)$$

Substitute $760 - 13p$ for q in (2) and solve.

$$760 - 13p = 430 + 2p$$
$$330 = 15p$$
$$22 = p$$

The equilibrium price is \$22 per unit.

To find the equilibrium quantity we substitute \$22 into either $D(p)$ or $S(p)$.

$S(22) = 430 + 2(22) = 430 + 44 = 474$

The equilibrium quantity is 474 units.

The equilibrium point is (\$22, 474).

19. $D(p) = 7500 - 25p,$

$\quad S(p) = 6000 + 5p$

Rewrite the system:

$$q = 7500 - 25p, \quad (1)$$
$$q = 6000 + 5p \qquad (2)$$

Substitute $7500 - 25p$ for q in (2) and solve.

$$7500 - 25p = 6000 + 5p$$
$$1500 = 30p$$
$$50 = p$$

The equilibrium price is \$50 per unit.

To find the equilibrium quantity we substitute \$50 into either $D(p)$ or $S(p)$.

$D(50) = 7500 - 25(50) = 7500 - 1250 = 6250$

The equilibrium quantity is 6250 units.

The equilibrium point is (\$50, 6250).

21. $D(p) = 1600 - 53p,$

$\quad S(p) = 320 + 75p$

Rewrite the system:

$$q = 1600 - 53p, \quad (1)$$
$$q = 320 + 75p \qquad (2)$$

Substitute $1600 - 53p$ for q in (2) and solve.

$$1600 - 53p = 320 + 75p$$
$$1280 = 128p$$
$$10 = p$$

The equilibrium price is \$10 per unit.

To find the equilibrium quantity we substitute \$10 into either $D(p)$ or $S(p)$.

$S(10) = 320 + 75(10) = 320 + 750 = 1070$

The equilibrium quantity is 1070 units.

The equilibrium point is (\$10, 1070).

23. Discussion and Writing Exercise

25. $5y - 3x = 8$

$$5y = 3x + 8$$

$$\frac{1}{5} \cdot 5y = \frac{1}{5}(3x + 8)$$

$$y = \frac{3}{5}x + \frac{8}{5}$$

The equation is now in slope-intercept form, $y = mx + b$. The slope is $\frac{3}{5}$, and the y-intercept is $\left(0, \frac{8}{5}\right)$.

27. $2y = 3.4x + 98$

$$\frac{1}{2} \cdot 2y = \frac{1}{2}(3.4x + 98)$$

$$y = 1.7x + 49$$

The equation is now in slope-intercept form, $y = mx + b$. The slope is 1.7, and the y-intercept is $(0, 49)$.

Chapter 3 Review Exercises

1. Graph both lines on the same set of axes.

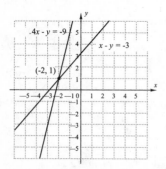

The solution (point of intersection) seems to be the point $(-2, 1)$.

Check:

$4x - y = -9$		$x - y = -3$	
$4(-2) - 1 \; ? \; -9$		$-2 - 1 \; ? \; -3$	
$-8 - 1$		-3	TRUE
-9	TRUE		

The solution is $(-2, 1)$.

Since the system of equations has a solution it is consistent. Since there is exactly one solution, the equations are independent.

2. Graph both lines on the same set of axes.

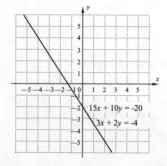

The graphs are the same. Any solution of one of the equations is also a solution of the other. Each equation has an infinite number of solutions. Thus the system of equations has an infinite number of solutions.

Since the system of equations has a solution, it is consistent. Since there are infinitely many solutions, the equations are dependent.

3. Graph both lines on the same set of axes.

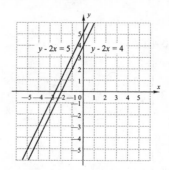

The lines are parallel. There is no solution.

Since the system of equations has no solution, it is inconsistent. Since there is no solution, the equations are independent.

4. $7x - 4y = 6$,   (1)

$y - 3x = -2$ (2)

First solve the second equation for y.

$$y - 3x = -2 \qquad (2)$$

$$y = 3x - 2 \quad (3)$$

Now substitute $3x - 2$ for y in the first equation and solve for x.

$$7x - 4y = 6 \qquad (1)$$

$$7x - 4(3x - 2) = 6$$

$$7x - 12x + 8 = 6$$

$$-5x + 8 = 6$$

$$-5x = -2$$

$$x = \frac{2}{5}$$

Next substitute $\frac{2}{5}$ for x in Equation (3) and find y.

$$y = 3 \cdot \frac{2}{5} - 2 = \frac{6}{5} - \frac{10}{5} = -\frac{4}{5}$$

Since $\left(\frac{2}{5}, -\frac{4}{5}\right)$ checks, it is the solution.

5. $y = x + 2$,   (1)

$y - x = 8$ (2)

Substitute $x + 2$ for y in the second equation and solve for x.

$$y - x = 8 \quad (2)$$

$$x + 2 - x = 8$$

$$2 = 8$$

We get a false equation. There is no solution.

6. $9x - 6y = 2$, (1)

$x = 4y + 5$ (2)

Substitute $4y + 5$ for x in the first equation and solve for y.

$$9x - 6y = 2 \qquad (1)$$
$$9(4y + 5) - 6y = 2$$
$$36y + 45 - 6y = 2$$
$$30y + 45 = 2$$
$$30y = -43$$
$$y = -\frac{43}{30}$$

Now substitute $-\dfrac{43}{30}$ for y in the second equation and find x.

$$x = 4\left(-\frac{43}{30}\right) + 5 = -\frac{86}{15} + \frac{75}{15} = -\frac{11}{15}$$

Since $\left(-\dfrac{11}{15}, -\dfrac{43}{30}\right)$ checks, it is the solution.

7. $8x - 2y = 10$,

$-4y - 3x = -17$

We rewrite the second equation in the form $Ax + By = C$.

$$8x - 2y = 10 \qquad (1)$$
$$-3x - 4y = -17 \qquad (2)$$

Now we multiply Equation (1) by -2 and add.

$$-16x + 4y = -20$$
$$\underline{-3x - 4y = -17}$$
$$-19x \qquad = -37$$
$$x = \frac{37}{19}$$

Substitute $\dfrac{37}{19}$ for x in one of the equations and solve for y.

$$8x - 2y = 10 \qquad (1)$$
$$8 \cdot \frac{37}{19} - 2y = 10$$
$$\frac{296}{19} - 2y = 10$$
$$-2y = -\frac{106}{19}$$
$$y = \frac{53}{19}$$

Since $\left(\dfrac{37}{19}, \dfrac{53}{19}\right)$ checks, it is the solution.

8. $4x - 7y = 18$, (1)

$9x + 14y = 40$ (2)

First multiply Equation (1) by 2 and then add.

$$8x - 14y = 36$$
$$\underline{9x + 14y = 40}$$
$$17x \qquad = 76$$
$$x = \frac{76}{17}$$

Now substitute $\dfrac{76}{17}$ for x in one of the equations and solve for y.

$$4x - 7y = 18 \qquad (1)$$
$$4 \cdot \frac{76}{17} - 7y = 18$$
$$\frac{304}{17} - 7y = 18$$
$$-7y = \frac{2}{17}$$
$$y = -\frac{2}{119}$$

Since $\left(\dfrac{76}{17}, -\dfrac{2}{119}\right)$ checks, it is the solution.

9. $3x - 5y = -4$, (1)

$5x - 3y = 4$ (2)

We multiply Equation (1) by 3 and Equation (2) by -5 and then add.

$$9x - 15y = -12$$
$$\underline{-25x + 15y = -20}$$
$$-16x \qquad = -32$$
$$x = 2$$

Now substitute 2 for x in one of the equations and solve for y.

$$3x - 5y = -4 \qquad (1)$$
$$3 \cdot 2 - 5y = -4$$
$$6 - 5y = -4$$
$$-5y = -10$$
$$y = 2$$

Since $(2, 2)$ checks, it is the solution.

10. $1.5x - 3 = -2y$,

$3x + 4y = 6$

First we rewrite the first equation in the form $Ax + By = C$.

$$1.5x + 2y = 3, \ (1)$$
$$3x + 4y = 6 \ (2)$$

Now multiply Equation (1) by -2 and then add.

$$-3x - 4y = -6$$
$$\underline{3x + 4y = 6}$$
$$0 = 0$$

We get an equation that is true for all numbers x and y. The system has an infinite number of solutions.

11. *Familiarize*. Let $x =$ the price of one CD and $y =$ the cost of one cassette.

Translate. Two CDs and one cassette cost $37, so we have

$$2x + y = 37.$$

One CD and two cassettes cost $5 less than $37, or $32, so we have

$$x + 2y = 32.$$

We have a system of equations.

$$2x + y = 37, \quad (1)$$
$$x + 2y = 32 \quad (2)$$

Solve. We use the elimination method. First we multiply Equation (2) by -2 and then add.

$$2x + y = 37$$
$$\underline{-2x - 4y = -64}$$
$$-3y = -27$$
$$y = 9$$

Now substitute 9 for y in one of the equations and solve for x.

$$x + 2y = 32$$
$$x + 2 \cdot 9 = 32$$
$$x + 18 = 32$$
$$x = 14$$

Check. If a CD costs \$14 and a cassette costs \$9, then two CDs and one cassette cost $2 \cdot \$14 + \$9$, or $\$28 + \$9$, or \$37. Also, one CD and two cassettes cost $\$14 + 2 \cdot \$9$, or $\$14 + \$18$, or \$32 and this is \$5 less than \$37. The answer checks.

State. A CD costs \$14 and a cassette costs \$9.

12. Familiarize. Let x and y represent the number of liters of Orange Thirst and Quencho that should be used, respectively. The amount of orange juice in the mixture is $10\%(10 \text{ L})$, or $0.1(10 \text{ L})$, or 1 L.

Translate. We organize the information in a table.

	Orange Thirst	Quencho	Mixture
Number of liters	x	y	10
Percent of juice	15%	5%	10%
Liters of juice	$0.15x$	$0.05y$	1

The first row of the table gives us one equation.

$$x + y = 10$$

The last row yields a second equation.

$$0.15x + 0.05y = 1$$

After clearing decimals we have the following system of equations.

$$x + y = 10, \quad (1)$$
$$15x + 5y = 100 \quad (2)$$

Solve. We use the elimination method. First we multiply Equation (1) by -5 and then add.

$$-5x - 5y = -50$$
$$\underline{15x + 5y = 100}$$
$$10x = 50$$
$$x = 5$$

Now substitute 5 for x in one of the equations and solve for y.

$$x + y = 10 \quad (1)$$
$$5 + y = 10$$
$$y = 5$$

Check. If 5 L of each type of juice are used, then the mixture contains $5 + 5$, or 10 L. The amount of orange juice in the mixture is $0.15(5) + 0.05(5)$, or $0.75 + 0.25$, or 1 L. The answer checks.

State. 5 L of Orange Thirst and 5 L of Quencho should be used.

13. Familiarize. We first make a drawing.

Slow train d miles	44 mph	$(t+1)$ hr
Fast train d miles	52 mph	t hr

From the drawing we see that the distances are the same. We organize the information in a table.

	Distance	Rate	Time	
Slow train	d	44	$t+1$	$\rightarrow d = 44(t+1)$
Fast train	d	52	t	$\rightarrow d = 52t$

$$d = r \cdot t$$

Translate. Using $d = rt$ in each row of the table, we get a system of equations:

$$d = 44(t+1),$$
$$d = 52t$$

Solve. We solve the system of equations.

$$52t = 44(t+1) \quad \text{Using substitution}$$
$$52t = 44t + 44$$
$$8t = 44$$
$$t = \frac{11}{2}, \text{ or } 5\frac{1}{2}$$

Check. At 52 mph, in $5\frac{1}{2}$ hr the fast train will travel $52 \cdot \frac{11}{2}$, or 286 mi. At 44 mph, in $5\frac{1}{2} + 1$, or $6\frac{1}{2}$ hr, the slow train travels $44 \cdot \frac{13}{2}$, or 286 mi. Since the distances are the same, the answer checks.

State. The second train will travel $5\frac{1}{2}$ hr before it overtakes the first train.

14.
$$x + 2y + z = 10, \quad (1)$$
$$2x - y + z = 8, \quad (2)$$
$$3x + y + 4z = 2 \quad (3)$$

We start by eliminating y from two different pairs of equations.

$$x + 2y + z = 10 \quad (1)$$
$$\underline{4x - 2y + 2z = 16} \quad \text{Multiplying (2) by 2}$$
$$5x + 3z = 26 \quad (4)$$

$$2x - y + z = 8 \quad (2)$$
$$\underline{3x + y + 4z = 2} \quad (3)$$
$$5x \qquad + 5z = 10 \quad (5)$$

Now solve the system of Equations (4) and (5). We multiply Equation (5) by -1 and then add.

$$5x + 3z = 26 \qquad (4)$$
$$\underline{-5x - 5z = -10}$$
$$-2z = 16$$
$$z = -8$$

$$5x + 3(-8) = 26 \quad \text{Substituting } -8 \text{ for } y \text{ in } (4)$$
$$5x - 24 = 26$$
$$5x = 50$$
$$x = 10$$

$$3 \cdot 10 + y + 4(-8) = 2 \quad \text{Substituting } 10 \text{ for } x \text{ and}$$
$$\qquad\qquad\qquad\qquad -8 \text{ for } y \text{ in } (3)$$
$$30 + y - 32 = 2$$
$$y - 2 = 2$$
$$y = 4$$

We obtain $(10, 4, -8)$. This checks, so it is the solution.

15.
$$3x + 2y + z = 3, \quad (1)$$
$$6x - 4y - 2z = -34, \quad (2)$$
$$-x + 3y - 3z = 14 \quad (3)$$

We start by eliminating z from two different pairs of equations.

$$6x + 4y + 2z = 6 \qquad \text{Multiplying } (1) \text{ by } 2$$
$$\underline{6x - 4y - 2z = -34}$$
$$12x \qquad\qquad = -28 \quad (4)$$

$$9x + 6y + 3z = 9 \qquad \text{Multiplying } (1) \text{ by } 3$$
$$\underline{-x + 3y - 3z = 14}$$
$$8x + 9y \qquad = 23 \quad (5)$$

From Equation (4), we have $12x = -28$, or $x = -\dfrac{7}{3}$. Substitute $-\dfrac{7}{3}$ for x in Equation (5) and solve for y.

$$8\left(-\frac{7}{3}\right) + 9y = 23$$
$$-\frac{56}{3} + 9y = 23$$
$$9y = \frac{125}{3}$$
$$y = \frac{125}{27}$$

Now substitute $-\dfrac{7}{3}$ for x and $\dfrac{125}{27}$ for y in one of the original equations and solve for z. We use Equation (1).

$$3\left(-\frac{7}{3}\right) + 2\left(\frac{125}{27}\right) + z = 3$$
$$-7 + \frac{250}{27} + z = 3$$
$$\frac{61}{27} + z = 3$$
$$z = \frac{20}{27}$$

We obtain $\left(-\dfrac{7}{3}, \dfrac{125}{27}, \dfrac{20}{27}\right)$. This checks, so it is the solution.

16.
$$2x - 5y - 2z = -4, \quad (1)$$
$$7x + 2y - 5z = -6, \quad (2)$$
$$-2x + 3y + 2z = 4 \quad (3)$$

We start by eliminating x from two different pairs of equations.

$$14x - 35y - 14z = -28 \quad \text{Multiplying } (1) \text{ by } 7$$
$$\underline{-14x - 4y + 10z = 12} \quad \text{Multiplying } (2) \text{ by } -2$$
$$-39y - 4z = -16 \quad (4)$$

$$2x - 5y - 2z = -4 \quad (1)$$
$$\underline{-2x + 3y + 2z = 4} \quad (3)$$
$$-2y \qquad = 0$$
$$y = 0$$

Substitute 0 for y in Equation (4) and solve for z.

$$-39 \cdot 0 - 4x = -16$$
$$-4x = -16$$
$$z = 4$$

Now substitute 0 for y and 4 for z in one of the original equations and solve for x. We use Equation (1).

$$2x - 5 \cdot 0 - 2 \cdot 4 = -4$$
$$2x - 8 = -4$$
$$2x = 4$$
$$x = 2$$

We obtain $(2, 0, 4)$. This checks, so it is the solution.

17.
$$x + y + 2z = 1, \quad (1)$$
$$x - y + z = 1, \quad (2)$$
$$x + 2y + z = 2 \quad (3)$$

We start by eliminating y from two different pairs of equations.

$$x + y + 2z = 1 \quad (1)$$
$$\underline{x - y + z = 1} \quad (2)$$
$$2x \qquad + 3z = 2 \quad (4)$$

$$2x - 2y + 2z = 2 \quad \text{Multiplying } (2) \text{ by } 2$$
$$\underline{x + 2y + z = 2}$$
$$3x \qquad + 3z = 4 \quad (5)$$

Now solve the system of Equations (4) and (5).

$$2x + 3z = 2 \qquad (4)$$

$$\underline{-3x - 3z = -4} \qquad \text{Multiplying (5) by } -1$$

$$-x \qquad = -2$$

$$x = 2$$

$$2 \cdot 2 + 3z = 2 \qquad \text{Substituting 2 for } x \text{ in (4)}$$

$$4 + 3z = 2$$

$$3z = -2$$

$$z = -\frac{2}{3}$$

$$2 + y + 2\left(-\frac{2}{3}\right) = 1 \qquad \text{Substituting 2 for } x \text{ and}$$

$$-\frac{2}{3} \text{ for } y \text{ in (1)}$$

$$2 + y - \frac{4}{3} = 1$$

$$y + \frac{2}{3} = 1$$

$$y = \frac{1}{3}$$

We obtain $\left(2, \frac{1}{3}, -\frac{2}{3}\right)$. This checks, so it is the solution.

18. Familiarize. Let a, b, and c represent the measures of angles A, B, and C, respectively. Recall that the sum of the measures of the angles of a triangle is $180°$.

Translate.

The sum of the measures is $180°$.

$$a + b + c \qquad = \qquad 180$$

The measure of angle A is 4 times the measure of angle C.

$$a \qquad = \quad 4 \quad \cdot \qquad c$$

The measure of angle B is $45°$ more than the measure of angle C.

$$b \qquad = \quad 45 \quad + \qquad c$$

We have a system of equations.

$$a + b + c = 180,$$

$$a = 4c,$$

$$b = 45 + c$$

Solve. Solving the system we get $\left(90, 67\frac{1}{2}, 22\frac{1}{2}\right)$.

Check. The sum of the measures is $90° + 67\frac{1}{2}° + 22\frac{1}{2}°$, or $180°$. Four times the measure of angle C is $4\left(22\frac{1}{2}°\right)$, or $90°$, the measure of angle B; $45°$ more than the measure of angle C is $22\frac{1}{2}° + 45°$, or $67\frac{1}{2}°$, the measure of angle B. The answer checks.

State. The measures of angles A, B, and C are $90°$, $67\frac{1}{2}°$, and $22\frac{1}{2}°$, respectively.

19. Familiarize. Let x, y, and z represent the number of $20 bills, $5 bills, and $1 bills, respectively. The total values of each type of bill are $20x$, $5y$, and z.

Translate.

Total value is $194.

$$20x + 5y + z \quad = \quad 194$$

Number of $1 bills is total number of $20 and $5 bills less 1.

$$z \qquad = \qquad x + y \qquad - \quad 1$$

Total number of bills is 39.

$$x + y + z \qquad = \quad 39$$

We have a system of equations.

$$20x + 5y + z = 194,$$

$$z = x + y - 1,$$

$$x + y + z = 39$$

Solve. Solving the system we get $(5, 15, 19)$.

Check. The total value of the bills is $5 \cdot \$20 + 15 \cdot \$5 + \$19$, or $\$100 + \$75 + \$19$, or $\$194$. The total number of $20 and $5 bills is $5 + 15$, or 20. This is 1 less than 19, the number of $1 bills. The total number of bills is $5 + 15 + 19$, or 39. The answer checks.

State. Elaine has 5 $20 bills, 15 $5 bills, and 19 $1 bills.

20. Graph: $2x + 3y < 12$

First graph the line $2x + 3y = 12$. Draw it dashed since the inequality symbol is $<$. Test the point $(0, 0)$ to determine if it is a solution.

$$\frac{2x + 3y < 12}{2 \cdot 0 + 3 \cdot 0 \;?\; 12}$$

$$0 \;\;\Big|\;\; \text{TRUE}$$

Since $0 < 12$ is true, we shade the half-plane containing $(0, 0)$ and draw the graph.

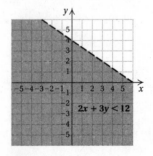

21. Graph: $y \leq 0$

First graph the line $y = 0$ (the x-axis). Draw it solid since the inequality symbol is $\leq$. Test the point $(1, 2)$ to determine if it is a solution.

$$\frac{y \leq 0}{2 \;?\; 0 \;\; \text{FALSE}}$$

Since $2 \le 0$ is false, we shade the half-plane that does not contain $(1, 2)$ and obtain the graph.

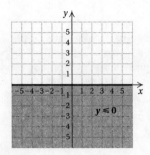

22. Graph: $x + y \ge 1$

First graph the line $x + y = 1$. Draw it solid since the inequality symbol is $\le$. Test the point $(0, 0)$ to determine if it is a solution.

$$\frac{x + y \ge 1}{0 + 0 \; ? \; 1}$$
$$0 \quad | \quad \text{FALSE}$$

Since $0 \ge 1$ is false, we shade the half-plane that does not contain $(0, 0)$ and obtain the graph.

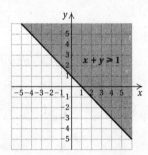

23. Graph: $y \ge -3$,
$\quad\quad\quad\quad x \ge 2$

We graph the lines $y = -3$ and $x = 2$ using solid lines. We indicate the region for each inequality by arrows at the ends of the lines. Note where the regions overlap, and shade the region of solutions.

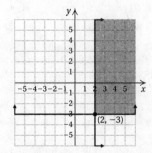

To find the vertex we solve the system of related equations:

$$y = -3,$$
$$x = 2$$

We see that the vertex is $(2, -3)$.

24. Graph: $x + 3y \ge -1$,
$\quad\quad\quad\quad x + 3y \le 4$

We graph the lines $x + 3y = -1$ and $x + 3y = 4$ using solid lines. We indicate the region for each inequality by arrows at the ends of the lines. Note where the regions overlap, and shade the region of solutions.

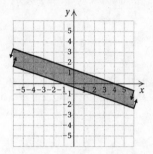

25. Graph: $x - y \le 3$,   (1)
$\quad\quad\quad\quad x + y \ge -1$,   (2)
$\quad\quad\quad\quad\quad\quad y \le 2$ (3)

We graph the lines $x - y = 3$, $x + y = -1$, and $y = 2$ using solid lines. We indicate the region for each inequality by arrows at the ends of the lines. Note where the regions overlap, and shade the region of solutions.

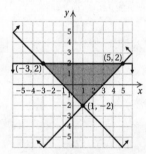

To find the vertices we solve three different systems of equations. From (1) and (2) we obtain the vertex $(1, -2)$. From (1) and (3) we obtain the vertex $(5, 2)$. From (2) and (3) we obtain the vertex $(-3, 2)$.

26. a) $C(x) = $ Fixed costs $+$ Variable costs
$\quad\quad\;\; C(x) = 35,000 + 175x$,

where x is the number of beds produced.

b) Each bed sells for \$300. The total revenue is 300 times the number of beds sold. We assume that all beds produced are sold.

$\quad R(x) = 300x$

c) $P(x) = R(x) - C(x)$
$\quad P(x) = 300x - (35,000 + 175x)$
$\quad\quad\quad\;\; = 300x - 35,000 - 175x$
$\quad\quad\quad\;\; = 125x - 35,000$

d) $P(1200) = 125(1200) - 35,000$
$\quad\quad\quad\quad\quad = 150,000 - 35,000$
$\quad\quad\quad\quad\quad = 115,000$

The company will realize a profit of \$115,000 when 1200 beds are produced and sold.

$$P(200) = 125(200) - 35,000$$
$$= 25,000 - 35,000$$
$$= -10,000$$

The company will realize a \$10,000 loss when 200 beds are produced and sold.

e) Solve the system

$$R(x) = 300x,$$
$$C(x) = 35,000 + 125x.$$

Since both $R(x)$ and $C(x)$ are in dollars and they are equal at the break-even point, we can rewrite the system:

$$d = 300x, \qquad (1)$$
$$d = 35,000 + 125x \qquad (2)$$

We solve using substitution.

$$300x = 35,000 + 175x \quad \text{Substituting } 175x \text{ for } d$$
$$\text{in (2)}$$
$$125x = 35,000$$
$$x = 280$$

The firm will break even if it produces and sells 280 beds and takes in a total of $R(280) = 300 \cdot 280 = \$84,000$ in revenue. Thus, the break-even point is (200, \$84,000).

27. $D(p) = 120 - 13p,$
$S(p) = 60 + 7p$

Since both demand and supply are quantities, the system can be rewritten:

$$q = 120 - 13p, \quad (1)$$
$$q = 60 + 7p \qquad (2)$$

Substitute $120 - 13p$ for q in (2) and solve.

$$120 - 13p = 60 + 7p$$
$$60 = 20p$$
$$3 = p$$

The equilibrium price is \$3 per unit. To find the equilibrium quantity we substitute \$3 into either $D(p)$ or $S(p)$.

$$D(3) = 120 - 13 \cdot 3 = 120 - 39 = 81$$

The equilibrium quantity is 81 units.

The equilibrium point is (\$3, 81).

28. *Discussion and Writing Exercise.* The comparison is summarized in the table in Section 3.3.

29. *Discussion and Writing Exercise.* Many problems that deal with more than one unknown quantity are easier to translate to a system of equations than to a single equation. Problems involving complementary or supplementary angles, the dimensions of a geometric figure, mixtures, and the measures of the angles of a triangle are examples.

30. Let $d =$ the number of dimes and $q =$ the number of quarters. Then we translate. The total number of coins is 20 so we have one equation:

$$d + q = 20$$

The value of d dimes and q quarters, in cents, is $10d + 25q$. The value of q dimes and d quarters is $10q + 25d$. Since the value of q dimes and d quarters is 90¢ more than the value of d dimes and q quarters, we have a second equation:

$$10q + 25d = 10d + 25q + 90, \text{ or}$$
$$15d - 15q = 90, \text{ or } d - q = 6 \text{ (dividing by 15)}$$

We have a system of equations.

$$d + q = 20, \quad (1)$$
$$d - q = 6 \qquad (2)$$

We can solve the system using the elimination method.

$$\begin{array}{rl} d + q = 20 \\ \underline{d - q = 6} \\ 2d \phantom{{}+q} = 26 \\ d = 13 \end{array}$$

Now substitute 13 for d in one of the equations and solve for q. We use Equation (1).

$$13 + q = 20$$
$$q = 7$$

There were 13 dimes and 7 quarters.

31. We graph the equations and find the points of intersection.

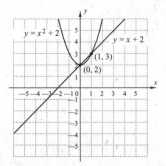

The solutions are $(0, 2)$ and $(1, 3)$.

32. a) $M(45) = 1.88(45) + 81.31 = 165.91$ cm

b) $F(45) = 1.95(45) + 72.85 = 160.60$ cm

c) We can use a graphing calculator to graph the equations and then use the INTERSECT feature to find the point of intersection. It is approximately $(120.857, 308.521)$.

d) The length, or value of x, for which the heights are the same is the first coordinate of the point of intersection of the graphs of the functions. Then from part (c) we know that this height is about 120.857 cm.

Chapter 3 Test

1. Graph both lines on the same set of axes.

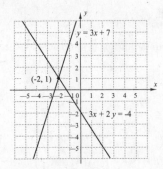

The solution (point of intersection) seems to be the point $(-2, 1)$.

Check:

$$
\begin{array}{c|c}
y = 3x + 7 & 3x + 2y = -4 \\
\hline
1 \ ? \ 3(-2) + 7 & 3(-2) + 2 \cdot 1 \ ? \ -4 \\
\quad -6 + 7 & -6 + 2 \\
\quad 1 \qquad \text{TRUE} & \qquad -4 \quad \text{TRUE}
\end{array}
$$

The solution is $(-2, 1)$.

Since the system of equations has a solution, it is consistent. Since there is exactly one solution, the equations are independent.

2. Graph both lines on the same set of axes.

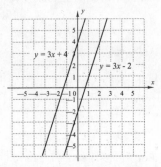

The lines are parallel. There is no solution.

Since the system of equations has no solution, it is inconsistent. Since there is no solution, the equations are independent.

3. Graph both lines on the same set of axes.

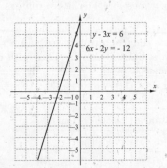

The graphs are the same. Any solution of one of the equations is also a solution of the other. Each equation has an infinite number of solutions. Thus the system of equations has an infinite number of solutions.

Since the system of equations has a solution, it is consistent. Since there are infinitely many solutions, the equations are dependent.

4. $x + 3y = -8, \quad (1)$

$4x - 3y = 23 \quad (2)$

Solve the first equation for x.

$$x + 3y = -8$$
$$x = -3y - 8 \quad (3)$$

We substitute $-3y - 8$ for x in the second equation and solve for y.

$$4x - 3y = 23 \qquad (2)$$
$$4(-3y - 8) - 3y = 23$$
$$-12y - 32 - 3y = 23$$
$$-15y - 32 = 23$$
$$-15y = 55$$
$$y = -\frac{11}{3}$$

Next we substitute $-\dfrac{11}{3}$ for y in Equation (3).

$$x = -3y - 8 = -3\left(-\frac{11}{3}\right) - 8 = 11 - 8 = 3$$

The ordered pair $\left(3, -\dfrac{11}{3}\right)$ checks in both equations. It is the solution.

5. $2x + 4y = -6, \quad (1)$

$y = 3x - 9 \qquad (2)$

We substitute $3x - 9$ for y in the first equation and solve for x.

$$2x + 4y = -6 \quad (1)$$
$$2x + 4(3x - 9) = -6$$
$$2x + 12x - 36 = -6$$
$$14x - 36 = -6$$
$$14x = 30$$
$$x = \frac{15}{7}$$

Next we substitute $\dfrac{15}{7}$ for x in Equation (2).

$$y = 3 \cdot \frac{15}{7} - 9 = \frac{45}{7} - 9 = \frac{45}{7} - \frac{63}{7} = -\frac{18}{7}$$

The ordered pair $\left(\dfrac{15}{7}, -\dfrac{18}{7}\right)$ checks in both equations. It is the solution.

6. $4x - 6y = 3, \quad (1)$

$6x - 4y = -3 \quad (2)$

We multiply Equation (1) by 2 and Equation (2) by -3 to make two terms additive inverses.

$$8x - 12y = 6$$
$$-18x + 12y = 9$$
$$\overline{-10x = 15} \quad \text{Adding}$$
$$x = -\frac{3}{2}$$

Now substitute $-\frac{3}{2}$ for x in one of the original equations and solve for y.

$$4x - 6y = 3 \qquad \text{Equation (1)}$$
$$4\left(-\frac{3}{2}\right) - 6y = 3$$
$$-6 - 6y = 3$$
$$-6y = 9$$
$$y = -\frac{3}{2}$$

The ordered pair $\left(-\frac{3}{2}, -\frac{3}{2}\right)$ checks in both equations. It is the solution.

7. $4y + 2x = 18$,
$$3x + 6y = 26$$

We rewrite the first equation in the form $Ax + By = C$.

$$2x + 4y = 18, \quad (1)$$
$$3x + 6y = 26 \quad (2)$$

Now multiply Equation (1) by 3 and Equation (2) by -2 and then add.

$$6x + 12y = 54$$
$$-6x - 12y = -52$$
$$\overline{0 = 2}$$

We have a false equation. There is no solution.

8. Familiarize. Let $x =$ the number of milliliters of 34% solution and $y =$ the number of milliliters of 61% solution to be used. We organize the information in a table.

	34% solution	61% solution	Mixture
Number of milliliters	x	y	120
Percent of salt	34%	61%	50%
Amount of salt	$0.34x$	$0.61y$	0.5(120) or 60

Translate. We get one equation from the first row of the table.

$$x + y = 120$$

The last row of the table yields a second equation.

$$0.34x + 0.61y = 60$$

After clearing decimals, we have the following system of equations.

$$x + y = 120, \quad (1)$$
$$34x + 61y = 6000 \quad (2)$$

Solve. We use the elimination method.

$$-34x - 34y = -4080 \quad \text{Multiplying (1) by } -34$$
$$34x + 61y = 6000$$
$$\overline{27y = 1920}$$
$$y = \frac{640}{9}, \text{ or } 71\frac{1}{9}$$

Substitute $71\frac{1}{9}$ for y in Equation (1) and solve for x.

$$x + 71\frac{1}{9} = 120$$
$$x = 48\frac{8}{9}$$

Check. $48\frac{8}{9}$ mL $+ 71\frac{1}{9}$ mL $= 120$ mL. The amount of salt in the mixture is $0.34\left(48\frac{8}{9}\right) + 0.61\left(71\frac{1}{9}\right)$, or 60 mL. The answer checks.

State. $48\frac{8}{9}$ mL of 34% solution and $71\frac{1}{9}$ mL of 61% solution should be used.

9. Familiarize. Let $b =$ the number of buckets of wings and $d =$ the number of chicken dinners sold. We organize the information in a table.

	Buckets	Dinners	Total
Number sold	b	d	28
Price	\$12	\$7	
Amount collected	$12b$	$7d$	281

Translate. The first and last rows of the table give us two equations.

$$b + d = 28, \qquad (1)$$
$$12b + 7d = 281 \quad (2)$$

Solve. We use the elimination method.

$$-7b - 7d = -196 \quad \text{Multiplying (1) by } -7$$
$$12b + 7d = 281$$
$$\overline{5b = 85}$$
$$b = 17$$

Substitute 17 for b in Equation (1) and solve for d.

$$17 + d = 28$$
$$d = 11$$

Check. The number of orders filled was $17 + 11$, or 28. The amount taken in was $\$12 \cdot 17 + \$7 \cdot 11$, or \$281. The answer checks.

State. 17 buckets of wings and 11 chicken dinners were sold.

10. Familiarize. Let $d =$ the distance traveled and $r =$ the speed of the plane in still air, in km/h. Then with a 20-km/h tailwind the plane's speed is $r + 20$. The speed of the plane traveling against the wind is $r - 20$. We organize the information in a table.

	Distance	Rate	Time
With wind	d	$r + 20$	5
Against wind	d	$r - 20$	7

Translate. Using $d = rt$ in each row of the table, we get a system of equations.

$$d = (r + 20)5,$$
$$d = (r - 20)7, \text{ or}$$

$$d = 5r + 100,$$
$$d = 7r - 140$$

Solve. We use the substitution method.

$$5r + 100 = 7r - 140$$
$$100 = 2r - 140$$
$$240 = 2r$$
$$120 = r$$

Check. If $r = 120$, then $r + 20 = 120 + 20 = 140$, and the distance traveled in 5 hr is $140 \cdot 5$, or 700 km. Also, $r - 20 = 120 - 20 = 100$, and the distance traveled in 7 hr is $100 \cdot 7$, or 700 km. The distances are the same, so the answer checks.

State. The speed of the plane in still air is 120 km/h.

11. *Familiarize*. Let l = the length and w = the width, in feet.

Translate. The width is 42 ft less than the length, so we have one equation:

$$w = l - 42.$$

We use the formula $P = 2l + 2w$ to get a second equation:

$$288 = 2l + 2w.$$

We have a system of equations.

$$w = l - 42, \qquad (1)$$
$$288 = 2l + 2w \qquad (2)$$

Solve. We use the substitution method.

$$288 = 2l + 2(l - 42)$$
$$288 = 2l + 2l - 84$$
$$288 = 4l - 84$$
$$372 = 4l$$
$$93 = l$$

Substitute 93 for l in Equation (1) and find w.

$$w = 93 - 42 = 51$$

Check. The length, 93 ft, is 42 ft more than the width, 51 ft. Also, $2 \cdot 93 + 2 \cdot 51 = 288$ ft. The answer checks.

State. The length is 93 ft, and the width is 51 ft.

12.
$$6x + 2y - 4z = 15, \quad (1)$$
$$-3x - 4y + 2z = -6, \quad (2)$$
$$4x - 6y + 3z = 8 \quad (3)$$

We start by eliminating y from two different pairs of equations.

$$\begin{array}{ll} 12x + 4y - 8z = 30 & \text{Multiplying (1) by 2} \\ \underline{-3x - 4y + 2z = -6} & (2) \\ 9x \quad\;\; - 6z = 24 & (4) \end{array}$$

$$\begin{array}{ll} 18x + 6y - 12z = 45 & \text{Multiplying (1) by 3} \\ \underline{4x - 6y + 3z = 8} & (3) \\ 22x \quad\quad - 9z = 53 & (5) \end{array}$$

Now solve the system of Equations (4) and (5).

$$9x - 6z = 24 \quad (4)$$
$$22x - 9z = 53 \quad (5)$$

$$\begin{array}{ll} 27x - 18z = 72 & \text{Multiplying (4) by 3} \\ \underline{-44x + 18z = -106} & \text{Multiplying (5) by } -2 \\ -17x \quad\quad = -34 & \\ x = 2 & \end{array}$$

$$9 \cdot 2 - 6z = 24 \quad \text{Substituting 2 for } x \text{ in (4)}$$
$$18 - 6z = 24$$
$$-6z = 6$$
$$z = -1$$

$$6 \cdot 2 + 2y - 4(-1) = 15 \qquad \begin{array}{l}\text{Substituting 2 for } x \\ \text{and } -1 \text{ for } z \text{ in (1)}\end{array}$$
$$12 + 2y + 4 = 15$$
$$16 + 2y = 15$$
$$2y = -1$$
$$y = -\frac{1}{2}$$

We obtain $\left(2, -\dfrac{1}{2}, -1\right)$. This checks, so it is the solution.

13.
$$D(p) = 79 - 8p,$$
$$S(p) = 37 + 6p$$

Since both $D(p)$ and $S(p)$ represent a quantity, we can rewrite the system:

$$q = 79 - 8p, \quad (1)$$
$$q = 37 + 6p \quad (2)$$

Substitute $79 - 8p$ for q in (2) and solve.

$$79 - 8p = 37 + 6p$$
$$79 = 37 + 14p$$
$$42 = 14p$$
$$3 = p$$

The equilibrium price is $3 per unit.

To find the equilibrium quantity we substitute $3 into either $D(p)$ or $S(p)$.

$$D(3) = 79 - 8 \cdot 3 = 79 - 24 = 55$$

The equilibrium quantity is 55 units.

The equilibrium point is $(\$3, 55)$.

14. *Familiarize*. Let x, y, and z represent the number of hours worked by the electrician, the carpenter, and the plumber, respectively. In x hours the carpenter earns $21x$; in y hours the carpenter earns $19.50y$; and in z hours the plumber earns $24z$.

Translate.

$$\underbrace{\text{Total time worked}}_{x + y + z} \;\; \underbrace{\text{is}}_{=} \;\; \underbrace{21.5 \text{ hr.}}_{21.5}$$

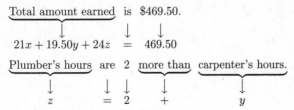

Total amount earned is $469.50.

$$21x + 19.50y + 24z = 469.50$$

Plumber's hours are 2 more than carpenter's hours.

$$z = 2 + y$$

We have a system of equations.

$$x + y + z = 21.5,$$
$$21x + 19.50y + 24z = 469.50,$$
$$z = 2 + y$$

Solve. Solving the system, we get $(3.5, 8, 10)$.

Check. The total time worked was $3.5 + 8 + 10$, or 21.5 hr. The amount earned was $\$21(3.5) + \$19.50(8) + \$24(10)$, or $\$469.50$. The time worked by the plumber, 10 hr, is 2 more than the time worked by the carpenter, 8 hr. The answer checks.

State. The electrician worked 3.5 hr.

15. a) The total cost is the fixed cost plus the variable costs.

$$C(x) = 40,000 + 30x$$

b) The total revenue is $80 times the number of rackets sold.

$$R(x) = 80x$$

c) $P(x) = R(x) - C(x)$
$$= 80x - (40,000 + 30x)$$
$$= 80x - 40,000 - 30x$$
$$= 50x - 40,000$$

d) $P(1200) = 50 \cdot 1200 - 40,000$
$$= 60,000 - 40,000$$
$$= 20,000$$

There is a $20,000 profit when 1200 rackets are produced and sold.

$P(200) = 50 \cdot 200 - 40,000$
$$= 10,000 - 40,000$$
$$= -30,000$$

There is a $30,000 loss when 200 rackets are produced and sold.

e) Solve the system
$$R(x) = 80x,$$
$$C(x) = 40,000 + 30x.$$

Since both $R(x)$ and $C(x)$ are in dollars and they are equal at the break-even point, we can rewrite the system:

$$d = 80x, \qquad (1)$$
$$d = 40,000 + 30x \quad (2)$$

We solve using substitution.

$$80x = 40,000 + 30x$$
$$50x = 40,000$$
$$x = 800$$

The company will break even if it produces and sells 800 rackets and takes in a total of $R(800) = 80 \cdot 800 = \$64,000$ in revenue. Thus, the break-even point is $(800, \$64,000)$.

16. Graph: $x - 6y < -6$

We first graph the line $x - 6y = 6$. We draw the line dashed since the inequality symbol is $<$. Test the point $(0,0)$ to determine if it is a solution.

$$\frac{x - 6y < -6}{0 - 6 \cdot 0 \; ? \; -6}$$
$$0 \quad | \quad \text{FALSE}$$

Since $0 < -6$ is false, we shade the half-plane that does not contain $(0,0)$.

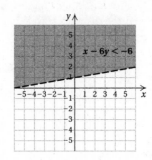

17. Graph: $x + y \geq 3,$
$$x - y \geq 5$$

Graph the lines $x + y = 3$ and $x - y = 5$ using solid lines. Indicate the region for each inequality by arrows and shade the region where they overlap.

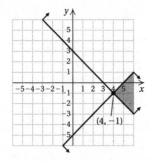

To find the vertex we solve the system of related equations:

$$x + y = 3,$$
$$x - y = 5$$

Solving, we obtain the vertex $(4, -1)$.

18. Graph: $2y - x \geq -4, \quad (1)$
$$2y + 3x \leq -6, \quad (2)$$
$$y \leq 0, \quad (3)$$
$$x \leq 0 \quad (4)$$

Shade the intersection of the graphs of the four inequalities above.

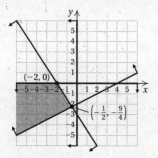

To find the two vertices we solve two systems of equations, as follows:

System of equations	Vertex
From (1) and (2)	$\left(-\dfrac{1}{2}, -\dfrac{9}{4}\right)$
From (2) and (3)	$(-2, 0)$

19. Substituting -1 for x and 3 for $f(x)$, we have

$3 = m(-1) + b$, or $3 = -m + b$.

Substituting -2 for x and -4 for $f(x)$, we have

$-4 = m(-2) + b$, or $-4 = -2m + b$.

Then we have a system of equations:

$$3 = -m + b, \quad (1)$$
$$-4 = -2m + b \quad (2)$$

We solve the system of equations.

$$\begin{aligned} -3 = m - b & \quad \text{Multiplying (1) by } -1 \\ \underline{-4 = -2m + b} & \quad (2) \\ -7 = -m & \\ 7 = m & \end{aligned}$$

Substitute 7 for m in (1) and solve for b.

$$3 = -7 + b$$
$$10 = b$$

Thus, we have $m = 7$ and $b = 10$.

Chapter 4

Polynomials and Polynomial Functions

Exercise Set 4.1

1. $-9x^4 - x^3 + 7x^2 + 6x - 8$

Term	$-9x^4$	$-x^3$	$7x^2$	$6x$	-8
Degree	4	3	2	1	0
Degree of polynomial	4				
Leading term	$-9x^4$				
Leading coefficient	-9				
Constant term	-8				

3. $t^3 + 4t^7 + s^2t^4 - 2$

Term	t^3	$4t^7$	s^2t^4	-2
Degree	3	7	6	0
Degree of polynomial	7			
Leading term	$4t^7$			
Leading coefficient	4			
Constant term	-2			

5. $u^7 + 8u^2v^6 + 3uv + 4u - 1$

Term	u^7	$8u^2v^6$	$3uv$	$4u$	-1
Degree	7	8	2	1	0
Degree of polynomial	8				
Leading term	$8u^2v^6$				
Leading coefficient	8				
Constant term	-1				

7. $-4y^3 - 6y^2 + 7y + 23$

9. $-xy^3 + x^2y^2 + x^3y + 1$

11. $-9b^5y^5 - 8b^2y^3 + 2by$

13. $5 + 12x - 4x^3 + 8x^5$

15. $3xy^3 + x^2y^2 - 9x^3y + 2x^4$

17. $-7ab + 4ax - 7ax^2 + 4x^6$

19. $P(x) = 3x^2 - 2x + 5$
$$P(4) = 3 \cdot 4^2 - 2 \cdot 4 + 5$$
$$= 48 - 8 + 5$$
$$= 45$$
$$P(-2) = 3(-2)^2 - 2(-2) + 5$$
$$= 12 + 4 + 5$$
$$= 21$$
$$P(0) = 3 \cdot 0^2 - 2 \cdot 0 + 5$$
$$= 0 - 0 + 5$$
$$= 5$$

21. $p(x) = 9x^3 + 8x^2 - 4x - 9$
$$p(-3) = 9(-3)^3 + 8(-3)^2 - 4(-3) - 9$$
$$= -243 + 72 + 12 - 9$$
$$= -168$$
$$p(0) = 9 \cdot 0^3 + 8 \cdot 0^2 - 4 \cdot 0 - 9$$
$$= 0 + 0 - 0 - 9$$
$$= -9$$
$$p(1) = 9 \cdot 1^3 + 8 \cdot 1^2 - 4 \cdot 1 - 9$$
$$= 9 + 8 - 4 - 9$$
$$= 4$$
$$p\left(\frac{1}{2}\right) = 9\left(\frac{1}{2}\right)^3 + 8\left(\frac{1}{2}\right)^2 - 4 \cdot \frac{1}{2} - 9$$
$$= \frac{9}{8} + 2 - 2 - 9$$
$$= -\frac{63}{8}, \text{ or } -7\frac{7}{8}$$

23. $N(x) = \frac{1}{3}x^3 + \frac{1}{2}x^2 + \frac{1}{6}x$
$$N(8) = \frac{1}{3} \cdot 8^3 + \frac{1}{2} \cdot 8^2 + \frac{1}{6} \cdot 8$$
$$= \frac{512}{3} + \frac{64}{2} + \frac{8}{6}$$
$$= \frac{1024}{6} + \frac{192}{6} + \frac{8}{6}$$
$$= \frac{1224}{6} = 204$$

There are 204 golf balls in the first stack.

$N(7) = \frac{1}{3} \cdot 7^3 + \frac{1}{2} \cdot 7^2 + \frac{1}{6} \cdot 7$

$= \frac{343}{3} + \frac{49}{2} + \frac{7}{6}$

$= \frac{686}{6} + \frac{147}{6} + \frac{7}{6}$

$= \frac{840}{6} = 140$

There are 140 golf balls in the second stack.

$N(6) = \frac{1}{3} \cdot 6^3 + \frac{1}{2} \cdot 6^2 + \frac{1}{6} \cdot 6$

$= 72 + 18 + 1 = 91$

There are 91 golf balls in the third stack.

$N(5) = \frac{1}{3} \cdot 5^3 + \frac{1}{2} \cdot 5^2 + \frac{1}{6} \cdot 5$

$= \frac{125}{3} + \frac{25}{2} + \frac{5}{6}$

$= \frac{250}{6} + \frac{75}{6} + \frac{5}{6}$

$= \frac{330}{6} = 55$

There are 55 golf balls in the fourth stack.

25. a) Locate 2 on the horizontal axis. From there move vertically to the graph and then horizontally to the $M(t)$-axis. This locates a value of about 340. Thus, about 340 mg of ibuprofen is in the the bloodstream 2 hr after 400 mg have been swallowed.

b) Locate 4 on the horizontal axis. From there move vertically to the graph and then horizontally to the $M(t)$-axis. This locates a value of about 190. Thus, about 190 mg of ibuprofen is in the the bloodstream 4 hr after 400 mg have been swallowed.

c) Locate 5 on the horizontal axis. From there move vertically to the graph and then horizontally to the $M(t)$-axis. This locates a value of about 65. Thus, $M(5) \approx 65$.

d) Locate 3 on the horizontal axis. From there move vertically to the graph and then horizontally to the $M(t)$-axis. This locates a value of about 300. Thus, $M(3) \approx 300$.

27. $R(x) = 280x - 0.4x^2$

a) $R(75) = 280 \cdot 75 - 0.4(75)^2$

$= 21,000 - 0.4(5625)$

$= 21,000 - 2250 = 18,750$

The total revenue is \$18,750.

b) $R(100) = 280(100) - 0.4(100)^2$

$= 28,000 - 4000$

$= 24,000$

The total revenue is \$24,000.

29. We subtract:

$P(x) = R(x) - C(x)$

$= 280x - 0.4x^2 - (7000 + 0.6x^2)$

$= 280x - 0.4x^2 + (-7000 - 0.6x^2)$ Adding the opposite

$= -x^2 + 280x - 7000$

The total profit is given by $P(x) = -x^2 + 280x - 7000$.

31. Substitute 162 for G, 78 for W_1, and 68 for L_2.

$M = G - W_1 - L_2 + 1$

$= 162 - 78 - 68 + 1$

$= 17$

The magic number is 17.

33. Substitute 162 for G, 86 for W_1, and 69 for L_2.

$M = G - W_1 - L_2 + 1$

$= 162 - 86 - 69 + 1$

$= 8$

The magic number is 8.

35. $6x^2 - 7x^2 + 3x^2 = (6 - 7 + 3)x^2 = 2x^2$

37. $7x - 2y - 4x + 6y$

$= (7 - 4)x + (-2 + 6)y$

$= 3x + 4y$

39. $3a + 9 - 2 + 8a - 4a + 7$

$= (3 + 8 - 4)a + (9 - 2 + 7)$

$= 7a + 14$

41. $3a^2b + 4b^2 - 9a^2b - 6b^2$

$= (3 - 9)a^2b + (4 - 6)b^2$

$= -6a^2b - 2b^2$

43. $8x^2 - 3xy + 12y^2 + x^2 - y^2 + 5xy + 4y^2$

$= (8 + 1)x^2 + (-3 + 5)xy + (12 - 1 + 4)y^2$

$= 9x^2 + 2xy + 15y^2$

45. $4x^2y - 3y + 2xy^2 - 5x^2y + 7y + 7xy^2$

$= (4 - 5)x^2y + (-3 + 7)y + (2 + 7)xy^2$

$= -x^2y + 4y + 9xy^2$

47. $(3x^2 + 5y^2 + 6) + (2x^2 - 3y^2 - 1)$

$= (3 + 2)x^2 + (5 - 3)y^2 + (6 - 1)$

$= 5x^2 + 2y^2 + 5$

49. $(2a + 3b - c) + (4a - 2b + 2c)$

$= (2 + 4)a + (3 - 2)b + (-1 + 2)c$

$= 6a + b + c$

51. $(a^2 - 3b^2 + 4c^2) + (-5a^2 + 2b^2 - c^2)$

$= (1 - 5)a^2 + (-3 + 2)b^2 + (4 - 1)c^2$

$= -4a^2 - b^2 + 3c^2$

53. $\quad (x^2 + 3x - 2xy - 3) + (-4x^2 - x + 3xy + 2)$

$\quad = (1 - 4)x^2 + (3 - 1)x + (-2 + 3)xy + (-3 + 2)$

$\quad = -3x^2 + 2x + xy - 1$

55. $\quad (7x^2y - 3xy^2 + 4xy) + (-2x^2y - xy^2 + xy)$

$\quad = (7 - 2)x^2y + (-3 - 1)xy^2 + (4 + 1)xy$

$\quad = 5x^2y - 4xy^2 + 5xy$

57. $\quad (2r^2 + 12r - 11) + (6r^2 - 2r + 4) + (r^2 - r - 2)$

$\quad = (2 + 6 + 1)r^2 + (12 - 2 - 1)r + (-11 + 4 - 2)$

$\quad = 9r^2 + 9r - 9$

59. $\quad \left(\dfrac{2}{3}xy + \dfrac{5}{6}xy^2 + 5.1x^2y\right) + \left(-\dfrac{4}{5}xy + \dfrac{3}{4}xy^2 - 3.4x^2y\right)$

$\quad = \left(\dfrac{2}{3} - \dfrac{4}{5}\right)xy + \left(\dfrac{5}{6} + \dfrac{3}{4}\right)xy^2 + (5.1 - 3.4)x^2y$

$\quad = \left(\dfrac{10}{15} - \dfrac{12}{15}\right)xy + \left(\dfrac{10}{12} + \dfrac{9}{12}\right)xy^2 + 1.7x^2y$

$\quad = -\dfrac{2}{15}xy + \dfrac{19}{12}xy^2 + 1.7x^2y$

61. $5x^3 - 7x^2 + 3x - 6$

   a) $-(5x^3 - 7x^2 + 3x - 6)$ Writing an inverse sign in front

   b) $-5x^3 + 7x^2 - 3x + 6$ Writing the opposite of each term

63. $-13y^2 + 6ay^4 - 5by^2$

   a) $-(-13y^2 + 6ay^4 - 5by^2)$

   b) $13y^2 - 6ay^4 + 5by^2$

65. $\quad (7x - 2) - (-4x + 5)$

$\quad = (7x - 2) + (4x - 5)$ Adding the opposite

$\quad = 11x - 7$

67. $\quad (-3x^2 + 2x + 9) - (x^2 + 5x - 4)$

$\quad = (-3x^2 + 2x + 9) + (-x^2 - 5x + 4)$ Adding the opposite

$\quad = -4x^2 - 3x + 13$

69. $\quad (5a - 2b + c) - (3a + 2b - 2c)$

$\quad = (5a - 2b + c) + (-3a - 2b + 2c)$

$\quad = 2a - 4b + 3c$

71. $\quad (3x^2 - 2x - x^3) - (5x^2 - 8x - x^3)$

$\quad = (3x^2 - 2x - x^3) + (-5x^2 + 8x + x^3)$

$\quad = -2x^2 + 6x$

73. $\quad (5a^2 + 4ab - 3b^2) - (9a^2 - 4ab + 2b^2)$

$\quad = (5a^2 + 4ab - 3b^2) + (-9a^2 + 4ab - 2b^2)$

$\quad = -4a^2 + 8ab - 5b^2$

75. $\quad (6ab - 4a^2b + 6ab^2) - (3ab^2 - 10ab - 12a^2b)$

$\quad = (6ab - 4a^2b + 6ab^2) + (-3ab^2 + 10ab + 12a^2b)$

$\quad = 16ab + 8a^2b + 3ab^2$

77. $\quad (0.09y^4 - 0.052y^3 + 0.93) -$
$\qquad\qquad (0.03y^4 - 0.084y^3 + 0.94y^2)$

$\quad = (0.09y^4 - 0.052y^3 + 0.93) +$
$\qquad\qquad (-0.03y^4 + 0.084y^3 - 0.94y^2)$

$\quad = 0.06y^4 + 0.032y^3 - 0.94y^2 + 0.93$

79. $\quad \left(\dfrac{5}{8}x^4 - \dfrac{1}{4}x^2 - \dfrac{1}{2}\right) - \left(-\dfrac{3}{8}x^4 + \dfrac{3}{4}x^2 + \dfrac{1}{2}\right)$

$\quad = \left(\dfrac{5}{8}x^4 - \dfrac{1}{4}x^2 - \dfrac{1}{2}\right) + \left(\dfrac{3}{8}x^4 - \dfrac{3}{4}x^2 - \dfrac{1}{2}\right)$

$\quad = x^4 - x^2 - 1$

81. Discussion and Writing Exercise

83. Graph: $f(x) = \dfrac{2}{3}x - 1$.

We find some ordered pairs that are solutions. By choosing multiples of 3 for x we can avoid fractional values when calculating $f(x)$.

For $x = -3$, $f(-3) = \dfrac{2}{3}(-3) - 1 = -2 - 1 = -3$.

For $x = 0$, $f(0) = \dfrac{2}{3} \cdot 0 - 1 = 0 - 1 = -1$.

For $x = 3$, $f(3) = \dfrac{2}{3} \cdot 3 - 1 = 2 - 1 = 1$.

x	$f(x)$	$(x, f(x))$
-3	-3	$(-3, -3)$
0	-1	$(0, -1)$
3	1	$(3, 1)$

We plot these points and draw the graph.

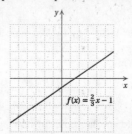

$f(x) = \frac{2}{3}x - 1$

85. Graph: $g(x) = \dfrac{4}{x - 3}$.

We choose x-values and find the corresponding values of $g(x)$. We list the results in a table.

x	$f(x)$	$(x, f(x))$
-5	$-\dfrac{1}{2}$	$\left(-5, -\dfrac{1}{2}\right)$
-3	$-\dfrac{2}{3}$	$\left(-3, -\dfrac{2}{3}\right)$
-1	-1	$(-1, -1)$
0	$-\dfrac{4}{3}$	$\left(0, -\dfrac{4}{3}\right)$
1	-2	$(1, -2)$
2	-4	$(2, -4)$
4	4	$(4, 4)$
5	2	$(5, 2)$
6	$\dfrac{4}{3}$	$\left(6, \dfrac{4}{3}\right)$
7	1	$(7, 1)$
9	$\dfrac{2}{3}$	$\left(9, \dfrac{2}{3}\right)$

Since $4/0$ is undefined, we cannot use 3 as a first coordinate. The graph has two branches, one on each side of the line $x = 3$.

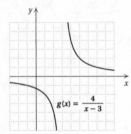

87. $3(y - 2) = 3 \cdot y - 3 \cdot 2 = 3y - 6$

89. $-14(3p - 2q - 10) = -14 \cdot 3p - (-14)(2q) - (-14)(10)$
$$= -42p + 28q + 140$$

91. Graph: $y = \dfrac{4}{3}x + 2$.

First we plot the y-intercept $(0, 2)$. Starting at $(0, 2)$ and using the slope, $\dfrac{4}{3}$, we find another point by moving 4 units up and 3 units to the right. We get to the point $(3, 6)$. We can also think of the slope as $\dfrac{-4}{-3}$. Starting again at $(0, 2)$, move 4 units down and 3 units to the left to the point $(-3, -2)$. Plot the points and draw the line.

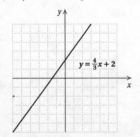

93. Graph: $y = 0.4x - 3$

We write the equation as $y = \dfrac{4}{10}x - 3$, or $y = \dfrac{2}{5}x - 3$. First we plot the y-intercept $(0, -3)$. Starting at $(0, -3)$ and using the slope, $\dfrac{2}{5}$, we find another point by moving 2 units up and 5 units to the right. We get to the point $(5, -1)$. We can also think of the slope as $\dfrac{-2}{-5}$. Starting again at $(0, -3)$, move 2 units down and 5 units to the left to the point $(-5, -5)$. Plot the points and draw the line.

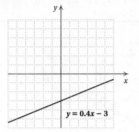

95. First we find the number of truffles in the display.
$$N(x) = \frac{1}{6}x^3 + \frac{1}{2}x^2 + \frac{1}{3}x$$
$$N(5) = \frac{1}{6} \cdot 5^3 + \frac{1}{2} \cdot 5^2 + \frac{1}{3} \cdot 5$$
$$= \frac{1}{6} \cdot 125 + \frac{1}{2} \cdot 25 + \frac{5}{3}$$
$$= \frac{125}{6} + \frac{25}{2} + \frac{5}{3}$$
$$= \frac{125}{6} + \frac{75}{6} + \frac{10}{6}$$
$$= \frac{210}{6} = 35$$

There are 35 truffles in the display. Now find the volume of one truffle. Each truffle's diameter is 3 cm, so the radius is $\dfrac{3}{2}$, or 1.5 cm.
$$V(r) = \frac{4}{3}\pi r^3$$
$$V(1.5) \approx \frac{4}{3}(3.14)(1.5)^3 \approx 14.13 \text{ cm}^3$$

Finally, multiply the number of truffles and the volume of a truffle to find the total volume of chocolate.
$$35(14.13 \text{ cm}^3) = 494.55 \text{ cm}^3$$
The display contains about 494.55 cm^3 of chocolate.

97. $(47x^{4a} + 3x^{3a} + 22x^{2a} + x^a + 1) +$
$$(37x^{3a} + 8x^{2a} + 3)$$
$$= 47x^{4a} + (3 + 37)x^{3a} + (22 + 8)x^{2a} + x^a + (1 + 3)$$
$$= 47x^{4a} + 40x^{3a} + 30x^{2a} + x^a + 4$$

99. Left to the student

Exercise Set 4.2

1. $8y^2 \cdot 3y = (8 \cdot 3)(y^2 \cdot y) = 24y^3$

3. $2x(-10x^2y) = [2(-10)](x \cdot x^2)(y) = -20x^3y$

5. $(5x^5y^4)(-2xy^3) = [5(-2)](x^5 \cdot x)(y^4 \cdot y^3) = -10x^6y^7$

7. $\quad 2z(7 - x)$

$= 2z \cdot 7 - 2z \cdot x$ Using a distributive law

$= 14z - 2zx$ Multiplying monomials

9. $\quad 6ab(a + b)$

$= 6ab \cdot a + 6ab \cdot b$ Using a distributive law

$= 6a^2b + 6ab^2$ Multiplying monomials

11. $\quad 5cd(3c^2d - 5cd^2)$

$= 5cd \cdot 3c^2d - 5cd \cdot 5cd^2$

$= 15c^3d^2 - 25c^2d^3$

13. $\quad (5x + 2)(3x - 1)$

$= 15x^2 - 5x + 6x - 2$ FOIL

$= 15x^2 + x - 2$

15. $\quad (s + 3t)(s - 3t)$

$= s^2 - (3t)^2$ $(A + B)(A - B) = A^2 - B^2$

$= s^2 - 9t^2$

17. $\quad (x - y)(x - y)$

$= x^2 - 2xy + y^2$ $(A - B)^2 = A^2 - 2AB + B^2$

19. $\quad (x^3 + 8)(x^3 - 5)$

$= x^6 - 5x^3 + 8x^3 - 40$ FOIL

$= x^6 + 3x^3 - 40$

21. $\quad (a^2 - 2b^2)(a^2 - 3b^2)$

$= a^4 - 3a^2b^2 - 2a^2b^2 + 6b^4$ FOIL

$= a^4 - 5a^2b^2 + 6b^4$

23. $\quad (x - 4)(x^2 + 4x + 16)$

$= (x - 4)(x^2) + (x - 4)(4x) + (x - 4)(16)$
           Using a distributive law

$= x(x^2) - 4(x^2) + x(4x) - 4(4x) + x(16) - 4(16)$
           Using a distributive law

$= x^3 - 4x^2 + 4x^2 - 16x + 16x - 64$
           Multiplying monomials

$= x^3 - 64$ Collecting like terms

25. $\quad (x + y)(x^2 - xy + y^2)$

$= (x + y)x^2 + (x + y)(-xy) + (x + y)(y^2)$

$= x(x^2) + y(x^2) + x(-xy) + y(-xy) + x(y^2) + y(y^2)$

$= x^3 + x^2y - x^2y - xy^2 + xy^2 + y^3$

$= x^3 + y^3$

27.

$$
\begin{array}{r}
a^2 + a - 1 \\
a^2 + 4a - 5 \\
\hline
-5a^2 - 5a + 5 \quad \text{Multiplying by } -5 \\
4a^3 + 4a^2 - 4a \quad \text{Multiplying by } 4a \\
a^4 + a^3 - a^2 \quad \text{Multiplying by } a^2 \\
\hline
a^4 + 5a^3 - 2a^2 - 9a + 5 \quad \text{Adding}
\end{array}
$$

29.

$$
\begin{array}{lr}
4a^2b - 2ab + 3b^2 & \\
ab - 2b + a & \\
\hline
4a^3b - 2a^2b + 3ab^2 & (1) \\
-6b^3 \qquad\quad + 4ab^2 - 8a^2b^2 & (2) \\
3ab^3 \qquad\qquad\quad - 2a^2b^2 + 4a^3b^2 & (3) \\
\hline
3ab^3 - 6b^3 + 4a^3b - 2a^2b + 7ab^2 - 10a^2b^2 + 4a^3b^2 & (4)
\end{array}
$$

(1) Multiplying by a

(2) Multiplying by $-2b$

(3) Multiplying by ab

(4) Adding

31. $\quad \left(x + \dfrac{1}{4}\right)\left(x + \dfrac{1}{4}\right)$

$= x^2 + \dfrac{1}{4}x + \dfrac{1}{4}x + \dfrac{1}{16}$ FOIL

$= x^2 + \dfrac{1}{2}x + \dfrac{1}{16}$

33. $\quad \left(\dfrac{1}{2}x - \dfrac{2}{3}\right)\left(\dfrac{1}{4}x + \dfrac{1}{3}\right)$

$= \dfrac{1}{8}x^2 + \dfrac{1}{6}x - \dfrac{1}{6}x - \dfrac{2}{9}$ FOIL

$= \dfrac{1}{8}x^2 - \dfrac{2}{9}$

35. $\quad (1.3x - 4y)(2.5x + 7y)$

$= 3.25x^2 + 9.1xy - 10xy - 28y^2$ FOIL

$= 3.25x^2 - 0.9xy - 28y^2$

37. $\quad (a + 8)(a + 5)$

$= a^2 + 5a + 8a + 40$ FOIL

$= a^2 + 13a + 40$

39. $\quad (y + 7)(y - 4)$

$= y^2 - 4y + 7y - 28$ FOIL

$= y^2 + 3y - 28$

41. $\quad \left(3a + \dfrac{1}{2}\right)^2$

$= (3a)^2 + 2(3a)\left(\dfrac{1}{2}\right) + \left(\dfrac{1}{2}\right)^2$
         $(A + B)^2 = A^2 + 2AB + B^2$

$= 9a^2 + 3a + \dfrac{1}{4}$

43. $\quad (x - 2y)^2$

$= x^2 - 2(x)(2y) + (2y)^2$
         $(A - B)^2 = A^2 - 2AB + B^2$

$= x^2 - 4xy + 4y^2$

45. $\left(b - \dfrac{1}{3}\right)\left(b - \dfrac{1}{2}\right)$

$= b^2 - \dfrac{1}{2}b - \dfrac{1}{3}b + \dfrac{1}{6}$ FOIL

$= b^2 - \dfrac{3}{6}b - \dfrac{2}{6}b + \dfrac{1}{6}$

$= b^2 - \dfrac{5}{6}b + \dfrac{1}{6}$

47. $(2x + 9)(x + 2)$

$= 2x^2 + 4x + 9x + 18$ FOIL

$= 2x^2 + 13x + 18$

49. $(20a - 0.16b)^2$

$= (20a)^2 - 2(20a)(0.16b) + (0.16b)^2$

$\qquad\qquad (A - B)^2 = A^2 - 2AB + B^2$

$= 400a^2 - 6.4ab + 0.0256b^2$

51. $(2x - 3y)(2x + y)$

$= 4x^2 + 2xy - 6xy - 3y^2$ FOIL

$= 4x^2 - 4xy - 3y^2$

53. $(x^3 + 2)^2$

$= (x^3)^2 + 2 \cdot x^3 \cdot 2 + 2^2$ $(A + B)^2 = A^2 + 2AB + B^2$

$= x^6 + 4x^3 + 4$

55. $(2x^2 - 3y^2)^2$

$= (2x^2)^2 - 2(2x^2)(3y^2) + (3y^2)^2$

$\qquad\qquad (A - B)^2 = A^2 - 2AB + B^2$

$= 4x^4 - 12x^2y^2 + 9y^4$

57. $(a^3b^2 + 1)^2$

$= (a^3b^2)^2 + 2 \cdot a^3b^2 \cdot 1 + 1^2$

$\qquad\qquad (A + B)^2 = A^2 + 2AB + B^2$

$= a^6b^4 + 2a^3b^2 + 1$

59. $(0.1a^2 - 5b)^2$

$= (0.1a^2)^2 - 2(0.1a^2)(5b) + (5b)^2$

$\qquad\qquad (A - B)^2 = A^2 - 2AB + B^2$

$= 0.01a^4 - a^2b + 25b^2$

61. $A = P(1 + i)^2$

$A = P(1 + 2i + i^2)$ FOIL

$A = P + 2Pi + Pi^2$ Multiplying by P

63. $(d + 8)(d - 8)$

$= d^2 - 8^2$ $(A + B)(A - B) = A^2 - B^2$

$= d^2 - 64$

65. $(2c + 3)(2c - 3)$

$= (2c)^2 - 3^2$ $(A + B)(A - B) = A^2 - B^2$

$= 4c^2 - 9$

67. $(6m - 5n)(6m + 5n)$

$= (6m)^2 - (5n)^2$ $(A + B)(A - B) = A^2 - B^2$

$= 36m^2 - 25n^2$

69. $(x^2 + yz)(x^2 - yz)$

$= (x^2)^2 - (yz)^2$ $(A + B)(A - B) = A^2 - B^2$

$= x^4 - y^2z^2$

71. $(-mn + m^2)(mn + m^2)$

$= (m^2 - mn)(m^2 + mn)$

$= (m^2)^2 - (mn)^2$ $(A + B)(A - B) = A^2 - B^2$

$= m^4 - m^2n^2$

73. $(-3pq + 4p^2)(4p^2 - 3pq)$

$= (4p^2 - 3pq)(4p^2 + 3pq)$

$= (4p^2)^2 - (3pq)^2$

$= 16p^4 - 9p^2q^2$

75. $\left(\dfrac{1}{2}p - \dfrac{2}{3}q\right)\left(\dfrac{1}{2}p + \dfrac{2}{3}q\right)$

$= \left(\dfrac{1}{2}p\right)^2 - \left(\dfrac{2}{3}q\right)^2$

$= \dfrac{1}{4}p^2 - \dfrac{4}{9}q^2$

77. $(x + 1)(x - 1)(x^2 + 1)$

$= (x^2 - 1^2)(x^2 + 1)$

$= (x^2 - 1)(x^2 + 1)$

$= (x^2)^2 - 1^2$

$= x^4 - 1$

79. $(a - b)(a + b)(a^2 - b^2)$

$= (a^2 - b^2)(a^2 - b^2)$

$= (a^2 - b^2)^2$

$= (a^2)^2 - 2(a^2)(b^2) + (b^2)^2$

$= a^4 - 2a^2b^2 + b^4$

81. $(a + b + 1)(a + b - 1)$

$= [(a + b) + 1][(a + b) - 1]$

$= (a + b)^2 - 1^2$

$= a^2 + 2ab + b^2 - 1$

83. $(2x + 3y + 4)(2x + 3y - 4)$

$= [(2x + 3y) + 4][(2x + 3y) - 4]$

$= (2x + 3y)^2 - 4^2$

$= 4x^2 + 12xy + 9y^2 - 16$

85. $f(x) = 5x + x^2$

$f(t - 1) = 5(t - 1) + (t - 1)^2$

$\qquad = 5t - 5 + t^2 - 2t + 1$

$\qquad = t^2 + 3t - 4$

$f(p + 1) = 5(p + 1) + (p + 1)^2$

$\qquad = 5p + 5 + p^2 + 2p + 1$

$\qquad = p^2 + 7p + 6$

$f(a + h) - f(a)$

$= [5(a + h) + (a + h)^2] - [5a + a^2]$

$= 5a + 5h + a^2 + 2ah + h^2 - 5a - a^2$

$= h^2 + 5h + 2ah$

$$f(t-2) + c = 5(t-2) + (t-2)^2 + c$$
$$= 5t - 10 + t^2 - 4t + 4 + c$$
$$= t^2 + t - 6 + c$$
$$f(a) + 5 = 5a + a^2 + 5$$

87. $f(x) = 3x^2 - 7x + 8$

$$f(t-1)$$
$$= 3(t-1)^2 - 7(t-1) + 8$$
$$= 3(t^2 - 2t + 1) - 7t + 7 + 8$$
$$= 3t^2 - 6t + 3 - 7t + 7 + 8$$
$$= 3t^2 - 13t + 18$$

$$f(p+1)$$
$$= 3(p+1)^2 - 7(p+1) + 8$$
$$= 3(p^2 + 2p + 1) - 7p - 7 + 8$$
$$= 3p^2 + 6p + 3 - 7p - 7 + 8$$
$$= 3p^2 - p + 4$$

$$f(a+h) - f(a)$$
$$= [3(a+h)^2 - 7(a+h) + 8] - [3a^2 - 7a + 8]$$
$$= 3(a^2 + 2ah + h^2) - 7a - 7h + 8 - 3a^2 + 7a - 8$$
$$= 3a^2 + 6ah + 3h^2 - 7a - 7h + 8 - 3a^2 + 7a - 8$$
$$= 3h^2 + 6ah - 7h$$

$$f(t-2) + c$$
$$= 3(t-2)^2 - 7(t-2) + 8 + c$$
$$= 3(t^2 - 4t + 4) - 7t + 14 + 8 + c$$
$$= 3t^2 - 12t + 12 - 7t + 14 + 8 + c$$
$$= 3t^2 - 19t + 34 + c$$

$$f(a) + 5 = 3a^2 - 7a + 8 + 5$$
$$= 3a^2 - 7a + 13$$

89. $f(x) = 5x - x^2$

$$f(t-1) = 5(t-1) - (t-1)^2$$
$$= 5t - 5 - (t^2 - 2t + 1)$$
$$= 5t - 5 - t^2 + 2t - 1$$
$$= -t^2 + 7t - 6$$

$$f(p+1) = 5(p+1) - (p+1)^2$$
$$= 5p + 5 - (p^2 + 2p + 1)$$
$$= 5p + 5 - p^2 - 2p - 1$$
$$= -p^2 + 3p + 4$$

$$f(a+h) - f(a)$$
$$= [5(a+h) - (a+h)^2] - [5a - a^2]$$
$$= 5a + 5h - (a^2 + 2ah + h^2) - 5a + a^2$$
$$= 5a + 5h - a^2 - 2ah - h^2 - 5a + a^2$$
$$= -h^2 + 5h - 2ah$$

$$f(t-2) + c = 5(t-2) - (t-2)^2 + c$$
$$= 5t - 10 - (t^2 - 4t + 4) + c$$
$$= 5t - 10 - t^2 + 4t - 4 + c$$
$$= -t^2 + 9t - 14 + c$$

$$f(a) + 5 = 5a - a^2 + 5$$

91. $f(x) = 4 + 3x - x^2$

$$f(t-1) = 4 + 3(t-1) - (t-1)^2$$
$$= 4 + 3t - 3 - (t^2 - 2t + 1)$$
$$= 4 + 3t - 3 - t^2 + 2t - 1$$
$$= -t^2 + 5t$$

$$f(p+1) = 4 + 3(p+1) - (p+1)^2$$
$$= 4 + 3p + 3 - (p^2 + 2p + 1)$$
$$= 4 + 3p + 3 - p^2 - 2p - 1$$
$$= -p^2 + p + 6$$

$$f(a+h) - f(a)$$
$$= [4 + 3(a+h) - (a+h)^2] - [4 + 3a - a^2]$$
$$= 4 + 3a + 3h - (a^2 + 2ah + h^2) - 4 - 3a + a^2$$
$$= 4 + 3a + 3h - a^2 - 2ah - h^2 - 4 - 3a + a^2$$
$$= -h^2 + 3h - 2ah$$

$$f(t-2) + c = 4 + 3(t-2) - (t-2)^2 + c$$
$$= 4 + 3t - 6 - (t^2 - 4t + 4) + c$$
$$= 4 + 3t - 6 - t^2 + 4t - 4 + c$$
$$= -t^2 + 7t - 6 + c$$

$$f(a) + 5 = 4 + 3a - a^2 + 5$$
$$= -a^2 + 3a + 9$$

93. Discussion and Writing Exercise

95. *Familiarize*. When Rachel's sister catches up with her, their distances traveled are the same. Let $d =$ this distance. Also let $t =$ the time for Rachel's sister to catch up with her. Then $t + 2 =$ the time Rachel travels before her sister catches up with her. Organize the information in a table.

	d	=	r	·	t
	Distance		Rate		Time
Rachel	d		55		$t+2$
Sister	d		75		t

Translate. Using $d = r \cdot t$ in each row of the table, we get a system of equations.

$$d = 55(t+2), \quad (1)$$
$$d = 75t \qquad (2)$$

Solve. We use the substitution method. First, substitute $75t$ for d in Equation (1) and solve for t.

$$75t = 55(t+2)$$
$$75t = 55t + 110$$
$$20t = 110$$
$$t = 5.5$$

If the sister's time is 5.5 hr, then Rachel's time is $5.5 + 2$, or 7.5 hr.

Check. At 55 mph, in 7.5 hr Rachel travels 55(7.5), or 412.5 mi. At 75 mph, in 5.5 hr her sister travels 75(5.5), or 412.5 mi. Since the distances are the same, the answer checks.

State. Rachel's sister will catch up with her 5.5 hr after the sister leaves (or 7.5 hr after Rachel leaves).

97.
$$5x + 9y = 2 \quad (1)$$
$$\underline{4x - 9y = 10 \quad (2)}$$
$$9x \quad\quad = 12 \quad \text{Adding}$$
$$x = \frac{4}{3}$$

Substitute $\frac{4}{3}$ for x in one of the equations and solve for y.

$$5x + 9y = 2 \quad\quad (1)$$
$$5 \cdot \frac{4}{3} + 9y = 2$$
$$\frac{20}{3} + 9y = 2$$
$$9y = -\frac{14}{3}$$
$$y = -\frac{14}{27}$$

The ordered pair $\left(\frac{4}{3}, -\frac{14}{27}\right)$ checks. It is the solution.

99. $2x - 3y = 1, \quad (1)$
$4x - 6y = 2 \quad (2)$

We multiply Equation (1) by -2 and add.

$$-4x + 6y = -2$$
$$\underline{4x - 6y = \quad 2}$$
$$0 = \quad 0$$

We get an equation that is true for all values of x and y. There are infinitely many solutions.

101. Left to the student

103. $(z^{n^2})^{n^3}(z^{4n^3})^{n^2} = z^{n^5} \cdot z^{4n^5} = z^{n^5+4n^5} = z^{5n^5}$

105.
$$(r^2 + s^2)^2(r^2 + 2rs + s^2)(r^2 - 2rs + s^2)$$
$$= (r^2 + s^2)^2(r + s)^2(r - s)^2$$
$$= (r^2 + s^2)^2[(r + s)(r - s)]^2$$
$$= (r^2 + s^2)^2(r^2 - s^2)^2$$
$$= [(r^2 + s^2)(r^2 - s^2)]^2$$
$$= (r^4 - s^4)^2$$
$$= r^8 - 2r^4s^4 + s^8$$

107.
$$\left(3x^5 - \frac{5}{11}\right)^2$$
$$= (3x^5)^2 - 2(3x^5)\left(\frac{5}{11}\right) + \left(\frac{5}{11}\right)^2$$
$$= 9x^{10} - \frac{30}{11}x^5 + \frac{25}{121}$$

109.
$$(x^a + y^b)(x^a - y^b)(x^{2a} + y^{2b})$$
$$= (x^{2a} - y^{2b})(x^{2a} + y^{2b})$$
$$= x^{4a} - y^{4b}$$

111.
$$(x - 1)(x^2 + x + 1)(x^3 + 1)$$
$$= (x^3 + x^2 + x - x^2 - x - 1)(x^3 + 1)$$
$$= (x^3 - 1)(x^3 + 1)$$
$$= x^6 - 1$$

Exercise Set 4.3

1. $6a^2 + 3a$
$$= 3a \cdot 2a + 3a \cdot 1$$
$$= 3a(2a + 1)$$

3. $x^3 + 9x^2$
$$= x^2 \cdot x + x^2 \cdot 9$$
$$= x^2(x + 9)$$

5. $8x^2 - 4x^4$
$$= 4x^2 \cdot 2 - 4x^2 \cdot x^2$$
$$= 4x^2(2 - x^2)$$

7. $4x^2y - 12xy^2$
$$= 4xy \cdot x - 4xy \cdot 3y$$
$$= 4xy(x - 3y)$$

9. $3y^2 - 3y - 9$
$$= 3 \cdot y^2 - 3 \cdot y - 3 \cdot 3$$
$$= 3(y^2 - y - 3)$$

11. $4ab - 6ac + 12ad$
$$= 2a \cdot 2b - 2a \cdot 3c + 2a \cdot 6d$$
$$= 2a(2b - 3c + 6d)$$

13. $10a^4 + 15a^2 - 25a - 30$
$$= 5 \cdot 2a^4 + 5 \cdot 3a^2 - 5 \cdot 5a - 5 \cdot 6$$
$$= 5(2a^4 + 3a^2 - 5a - 6)$$

15. $15x^2y^5z^3 - 12x^4y^4z^7$
$$= 3x^2y^4z^3 \cdot 5y - 3x^2y^4z^3 \cdot 4x^2z^4$$
$$= 3x^2y^4z^3(5y - 4x^2z^4)$$

17. $14a^4b^3c^5 + 21a^3b^5c^4 - 35a^4b^4c^3$
$$= 7a^3b^3c^3 \cdot 2ac^2 + 7a^3b^3c^3 \cdot 3b^2c - 7a^3b^3c^3 \cdot 5ab$$
$$= 7a^3b^3c^3(2ac^2 + 3b^2c - 5ab)$$

19. $-5x - 45 = -5(x + 9)$

21. $-6a - 84 = -6(a + 14)$

23. $-2x^2 + 2x - 24 = -2(x^2 - x + 12)$

25. $-3y^2 + 24y = -3y(y - 8)$

27. $-a^4 + 2a^3 - 13a^2 - 1$
$$= -1(a^4 - 2a^3 + 13a^2 + 1), \text{ or}$$
$$= -(a^4 - 2a^3 + 13a^2 + 1)$$

29. $-3y^3 + 12y^2 - 15y + 24 = -3(y^3 - 4y^2 + 5y - 8)$

31. $N(x) = \frac{1}{3}x^3 + \frac{1}{2}x^2 + \frac{1}{6}x$
$$N(x) = x\left(\frac{1}{3}x^2 + \frac{1}{2}x + \frac{1}{6}\right), \quad \text{Factoring out } x$$
$$\text{or } \frac{1}{6}x(2x^3 + 3x^2 + 1) \text{ Factoring out } \frac{1}{6}x$$

33. a) $\quad h(t) = -16t^2 + 72t$

$\qquad h(t) = -8t(2t - 9)$

b) $\quad h(2) = -16 \cdot 2^2 + 72 \cdot 2 = 80$

$\qquad h(2) = -8(2)(2 \cdot 2 - 9) = -8(2)(-5) = 80$

The expressions have the same value for $t = 2$, so the factorization is probably correct.

35. $\quad R(x) = 280x - 0.4x^2$

$\qquad R(x) = 0.4x(700 - x)$

37. $a(b-2) + c(b-2) = (a+c)(b-2)$

39. $\quad (x-2)(x+5) + (x-2)(x+8)$

$\quad = (x-2)[(x+5) + (x+8)]$

$\quad = (x-2)(2x+13)$

41. $a^2(x-y) + a^2(x-y) = 2a^2(x-y)$

43. $\quad ac + ad + bc + bd$

$\quad = a(c+d) + b(c+d)$

$\quad = (a+b)(c+d)$

45. $\quad b^3 - b^2 + 2b - 2$

$\quad = b^2(b-1) + 2(b-1)$

$\quad = (b^2+2)(b-1)$

47. $\quad y^3 - 8y^2 + y - 8$

$\quad = y^2(y-8) + 1(y-8)$

$\quad = (y^2+1)(y-8)$

49. $\quad 24x^3 - 36x^2 + 72x - 108$

$\quad = 12(2x^3 - 3x^2 + 6x - 9)$

$\quad = 12[x^2(2x-3) + 3(2x-3)]$

$\quad = 12(x^2+3)(2x-3)$

51. $a^4 - a^3 + a^2 + a = a(a^3 - a^2 + a + 1)$

53. $\quad 2y^4 + 6y^2 + 5y^2 + 15$

$\quad = 2y^2(y^2+3) + 5(y^2+3)$

$\quad = (2y^2+5)(y^2+3)$

55. Discussion and Writing Exercise

57. The equation $y = mx + b$ is called the slope-intercept equation of the line with slope m and y-intercept $(0, b)$.

59. If the slope of a line is less than 0, the graph slants down from left to right.

61. The equation $y - y_1 = m(x - x_1)$, where m is the slope of the line and (x_1, y_1) is a point on the line, is called the point-slope equation.

63. When the terms of a polynomial are written such that the exponents increase from left to right, we say the polynomial is written in ascending order.

65. $x^5y^4 + \underline{\quad} = x^3y(\underline{\quad} + xy^5)$

The term that goes in the first blank is the product of x^3y and xy^5, or x^4y^6.

The term that goes in the second blank is the expression that is multiplied with x^3y to obtain x^5y^4, or x^2y^3. Thus, we have

$\qquad x^5y^4 + x^4y^6 = x^3y(x^2y^3 + xy^5).$

67. $\quad rx^2 - rx + 5r + sx^2 - sx + 5s$

$\quad = r(x^2 - x + 5) + s(x^2 - x + 5)$

$\quad = (x^2 - x + 5)(r + s)$

69. $\quad a^4x^4 + a^4x^2 + 5a^4 + a^2x^4 + a^2x^2 + 5a^2 +$

$\qquad\qquad\qquad\qquad 5x^4 + 5x^2 + 25$

$\quad = a^4(x^4 + x^2 + 5) + a^2(x^4 + x^2 + 5) + 5(x^4 + x^2 + 5)$

$\quad = (x^4 + x^2 + 5)(a^4 + a^2 + 5)$

71. $x^{1/3} - 7x^{4/3} = x^{1/3} \cdot 1 - 7x \cdot x^{1/3} = x^{1/3}(1 - 7x)$

73. $\quad x^{1/3} - 5x^{1/2} + 3x^{3/4}$

$\quad = x^{4/12} - 5x^{6/12} + 3x^{9/12}$

$\quad = x^{4/12}(1 - 5x^{2/12} + 3x^{5/12})$

$\quad = x^{1/3}(1 - 5x^{1/6} + 3x^{5/12})$

75. $\quad 3a^{n+1} + 6a^n - 15a^{n+2}$

$\quad = 3a^n \cdot a + 3a^n \cdot 2 - 3a^n(5a^2)$

$\quad = 3a^n(a + 2 - 5a^2)$

77. $\quad 7y^{2a+b} - 5y^{a+b} + 3y^{a+2b}$

$\quad = y^{a+b} \cdot 7y^a - y^{a+b}(5) + y^{a+b} \cdot 3y^b$

$\quad = y^{a+b}(7y^a - 5 + 3y^b)$

Exercise Set 4.4

1. $x^2 + 13x + 36$

We look for two numbers whose product is 36 and whose sum is 13. Since both 36 and 13 are positive, we need consider only positive factors.

Pairs of Factors	Sums of Factors
1, 36	37
2, 18	20
3, 12	15
4, 9	13
6, 6	12

The numbers we need are 4 and 9. The factorization is $(x+4)(x+9)$.

3. $t^2 - 8t + 15$

Since the constant term, 15, is positive and the coefficient of the middle term, -8, is negative, we look for a factorization of 15 in which both factors are negative. Their sum must be -8.

Pairs of Factors	Sums of Factors
$-1, -15$	-16
$-3, -5$	-8

The numbers we need are -3 and -5. The factorization is $(t-3)(t-5)$.

5. $x^2 - 8x - 33$

Since the constant term, -33, is negative, we look for a factorization of -33 in which one factor is positive and one factor is negative. The sum of the factors must be -8, so the negative factor must have the larger absolute value. Thus we consider only pairs of factors in which the negative factor has the larger absolute value.

Pairs of Factors	Sums of Factors
1, -33	-32
3, -11	-8

The numbers we want are 3 and -11. The factorization is $(x + 3)(x - 11)$.

7. $2y^2 - 16y + 32$

$= 2(y^2 - 8y + 16)$ Removing the common factor

We now factor $y^2 - 8y + 16$. We look for two numbers whose product is 16 and whose sum is -8. Since the constant term is positive and the coefficient of the middle term is negative, we look for a factorization of 16 in which both factors are negative.

Pairs of Factors	Sums of Factors
$-1, -16$	-17
$-2, \ -8$	-10
$-4, \ -4$	-8

The numbers we need are -4 and -4.

$y^2 - 8y + 16 = (y - 4)(y - 4)$

We must not forget to include the common factor 2.

$2y^2 - 16y + 32 = 2(y - 4)(y - 4)$.

9. $p^2 + 3p - 54$

Since the constant term is negative, we look for a factorization of -54 in which one factor is positive and one factor is negative. We consider only pairs of factors in which the positive factor has the larger absolute value, since the sum of the factors, 3, is positive.

Pairs of Factors	Sums of Factors
54, -1	53
27, -2	25
18, -3	15
9, -6	3

The numbers we need are 9 and -6. The factorization is $(p + 9)(p - 6)$.

11. $12x + x^2 + 27 = x^2 + 12x + 27$

We look for two numbers whose product is 27 and whose sum is 12. Since both 27 and 12 are positive, we need consider only positive factors.

Pairs of Factors	Sums of Factors
1, 27	28
3, 9	12

The numbers we want are 3 and 9. The factorization is $(x + 3)(x + 9)$.

13. $y^2 - \frac{2}{3}y + \frac{1}{9}$

Since the constant term, $\frac{1}{9}$, is positive and the coefficient of the middle term, $-\frac{2}{3}$, is negative, we look for a factorization of $\frac{1}{9}$ in which both factors are negative. Their sum must be $-\frac{2}{3}$.

Pairs of Factors	Sums of Factors
$-1, -\frac{1}{9}$	$-\frac{10}{9}$
$-\frac{1}{3}, -\frac{1}{3}$	$-\frac{2}{3}$

The numbers we need are $-\frac{1}{3}$ and $-\frac{1}{3}$. The factorization is $\left(y - \frac{1}{3}\right)\left(y - \frac{1}{3}\right)$.

15. $t^2 - 4t + 3$

Since the constant term, 3, is positive and the coefficient of the middle term, -4, is negative, we look for a factorization of 3 in which both factors are negative. Their sum must be -4. The only possibility is $-1, -3$. These are the numbers we need. The factorization is $(t - 1)(t - 3)$.

17. $5x + x^2 - 14 = x^2 + 5x - 14$

Since the constant term, -14, is negative, we look for a factorization of -14 in which one factor is positive and one factor is negative. Their sum must be 5, so the positive factor must have the larger absolute value. We consider only pairs of factors in which the positive factor has the larger absolute value.

Pairs of Factors	Sums of Factors
$-1, 14$	13
$-2, \ 7$	5

The numbers we need are -2 and 7. The factorization is $(x - 2)(x + 7)$.

19. $x^2 + 5x + 6$

We look for two numbers whose product is 6 and whose sum is 5. Since 6 and 5 are both positive, we need consider only positive factors.

Pairs of Factors	Sums of Factors
1, 6	7
2, 3	5

The numbers we need are 2 and 3. The factorization is $(x + 2)(x + 3)$.

21. $56 + x - x^2 = -x^2 + x + 56 = -1(x^2 - x - 56)$

We now factor $x^2 - x - 56$. Since the constant term, -56, is negative, we look for a factorization of -56 in which one factor is positive and one factor is negative. We consider only pairs of factors in which the negative factor has the larger absolute value, since the sum of the factors, -1, is negative.

Pairs of Factors	Sums of Factors
−56, 1	−55
−28, 2	−26
−14, 4	−10
−8, 7	−1

The numbers we need are −8 and 7. Thus, $x^2 - x - 56 = (x - 8)(x + 7)$. We must not forget to include the factor that was factored out earlier:

$$-x^2 + x + 56$$
$$= -1(x^2 - x - 56)$$
$$= -1(x - 8)(x + 7)$$
$$= (-x + 8)(x + 7) \qquad \text{Multiplying } x - 8 \text{ by } -1$$
$$= (x - 8)(-x - 7) \qquad \text{Multiplying } x + 7 \text{ by } -1$$

23. $32y + 4y^2 - y^3$

There is a common factor, y. We also factor out −1 in order to make the leading coefficient positive.

$$32y + 4y^2 - y^3 = -y(-32 - 4y + y^2)$$
$$= -y(y^2 - 4y - 32)$$

Now we factor $y^2 - 4y - 32$. Since the constant term, −32, is negative, we look for a factorization of −32 in which one factor is positive and one factor is negative. We consider only pairs of factors in which the negative factor has the larger absolute value, since the sum of the factors, −4, is negative.

Pairs of Factors	Sums of Factors
−32, 1	−31
−16, 2	−14
−8, 4	−4

The numbers we need are −8 and 4. Thus, $y^2 - 4y - 32 = (y - 8)(y + 4)$. We must not forget to include the common factor:

$$32y + 4y^2 - y^3$$
$$= -y(y^2 - 4y - 32)$$
$$= -y(y - 8)(y + 4)$$
$$= -1 \cdot y(y - 8)(y + 4)$$
$$= y(-y + 8)(y + 4) \qquad \text{Multiplying } y - 8 \text{ by } -1$$
$$= y(y - 8)(-y - 4) \qquad \text{Multiplying } y + 4 \text{ by } -1$$

25. $x^4 + 11x^2 - 80$

First make a substitution. We let $u = x^2$, so $u^2 = x^4$. Then we consider $u^2 + 11u - 80$. We look for pairs of factors of −80, one positive and one negative, such that the positive factor has the larger absolute value and the sum of the factors is 11.

Pairs of Factors	Sums of Factors
80, −1	79
40, −2	38
20, −4	16
16, −5	11
10, −8	2

The numbers we need are 16 and −5. Then $u^2 + 11u - 80 = (u + 16)(u - 5)$. Replacing u by x^2 we

obtain the factorization of the original trinomial:
$(x^2 + 16)(x^2 - 5)$.

27. $x^2 - 3x + 7$

There are no factors of 7 whose sum is −3. This trinomial is not factorable into binomials with integer coefficients.

29. $x^2 + 12xy + 27y^2$

We look for numbers p and q such that $x^2 + 12xy + 27y^2 = (x + py)(x + qy)$. Our thinking is much the same as if we were factoring $x^2 + 12x + 27$. We look for factors of 27 whose sum is 12. Those factors are 9 and 3. Then

$$x^2 + 12xy + 27y^2 = (x + 9y)(x + 3y).$$

31. $45 + 4x - x^2 = -x^2 + 4x + 45 = -1(x^2 - 4x - 45)$

Now we factor $x^2 - 4x - 45$. We look for two numbers whose product is −45 and whose sum is −4. The numbers we need are −9 and 5. We have:

$$-x^2 + 4x + 45$$
$$= -1(x^2 - 4x - 45)$$
$$= -1(x - 9)(x + 5)$$
$$= (-x + 9)(x + 5) \qquad \text{Multiplying } x - 9 \text{ by } -1$$
$$= (x - 9)(-x - 5) \qquad \text{Multiplying } x + 5 \text{ by } -1$$

33. $-z^2 + 36 - 9z = -z^2 - 9z + 36 = -1(z^2 + 9z - 36)$

Now we factor $z^2 + 9z - 36$. We look for two numbers whose product is −36 and whose sum is 9. The numbers we need are 12 and −3. We have:

$$-z^2 - 9z + 36$$
$$= -1(z^2 + 9z - 36)$$
$$= -1(z + 12)(z - 3)$$
$$= (-z - 12)(z - 3) \qquad \text{Multiplying } z + 12 \text{ by } -1$$
$$= (z + 12)(-z + 3) \qquad \text{Multiplying } z - 3 \text{ by } -1$$

35. $x^4 + 50x^2 + 49$

Substitute u for x^2 (and hence u^2 for x^4). Consider $u^2 + 50u + 49$. We look for a pair of positive factors of 49 whose sum is 50.

Pairs of Factors	Sums of Factors
7, 7	14
1, 49	50

The numbers we need are 1 and 49. Then $u^2 + 50u + 49 = (u + 1)(u + 49)$. Replacing u by x^2 we have

$$x^4 + 50x^2 + 49 = (x^2 + 1)(x^2 + 49).$$

37. $x^6 + 11x^3 + 18$

Substitute u for x^3 (and hence u^2 for x^6). Consider $u^2 + 11u + 18$. We look for two numbers whose product is 18 and whose sum is 11. Since both 18 and 11 are positive, we need consider only positive factors.

Pairs of Factors	Sums of Factors
1, 18	19
2, 9	11
3, 6	9

The numbers we need are 2 and 9. Then $u^2 + 11u + 18 = (u + 2)(u + 9)$. Replacing u by x^3 we obtain the factorization of the original trinomial: $(x^3 + 2)(x^3 + 9)$.

39. $x^8 - 11x^4 + 24$

Substitute u for x^4 (and hence u^2 for x^8). Consider $u^2 - 11u + 24$. Since the constant term, 24, is positive and the coefficient of the middle term, -11, is negative, we look for a factorization of 24 in which both factors are negative. Their sum must be -11.

Pairs of Factors	Sums of Factors
$-1, -24$	-25
$-2, -12$	-14
$-3, \;\; -8$	-11
$-4, \;\; -6$	-10

The numbers we need are -3 and -8. Then $u^2 - 11u + 24 = (u-3)(u-8)$. Replacing u by x^4 we obtain the factorization of the original trinomial: $(x^4 - 3)(x^4 - 8)$.

41. Discussion and Writing Exercise

43. *Familiarize.* Let $x =$ the number of pounds of Countryside rice and $y =$ the number of pounds of Mystic rice to be used in the mixture.

Translate. We organize the information in a table.

	Countryside	Mystic	Mixture
Number of pounds	x	y	25
Percent of wild rice	10%	50%	35%
Pounds of wild rice	$0.1x$	$0.5x$	$0.35(25)$

From the "Number of pounds" row of the table we get one equation.

$$x + y = 25$$

We get a second equation from the last row of the table.

$$0.1x + 0.5x = 0.35(25), \text{ or}$$
$$0.1x + 0.5x = 8.75$$

Clearing decimals, we have the following system of equations:

$$x + y = 25, \quad (1)$$
$$10x + 50y = 875 \quad (2)$$

Solve. We use the elimination method to solve the system of equations. We multiply Equation (1) by -10 and add.

$$\begin{array}{r} -10x - 10y = -250 \\ \underline{10x + 50y = 875} \\ 40y = 625 \\ y = 15.625 \end{array}$$

Now substitute 15.625 for y in Equation (1) and solve for x.

$$x + 15.625 = 25$$
$$x = 9.375$$

Check. The total weight of the mixture is $9.375 + 15.625$, or 25 lb. The amount of wild rice in the mixture is $0.1(9.375) + 0.5(15.625) = 0.9375 + 7.8125 = 8.75$ lb. This is 35% of 25 lb. The numbers check.

State. 9.375 lb, or $9\frac{3}{8}$ lb, of Countryside Rice and 15.625 lb, or $15\frac{5}{8}$ lb of Mystic Rice should be used.

45. The graph is that of a function, because no vertical line can cross the graph at more than one point.

47. The graph is not that of a function, because a vertical line can cross the graph at more than one point.

49. $f(x) = x^2 - 2$

$f(x)$ can be calculated for any number x, so the domain is the set of all real numbers.

51. $f(x) = \dfrac{3}{4x - 7}$

We cannot calculate $\dfrac{3}{4x - 7}$ when the denominator is 0.

$$4x - 7 = 0$$
$$4x = 7$$
$$x = \frac{7}{4}$$

The domain is $\left\{ x \,\middle|\, x \text{ is a real number and } x \neq \dfrac{7}{4} \right\}$, or $\left(-\infty, \dfrac{7}{4} \right) \cup \left(\dfrac{7}{4}, \infty \right)$.

53. All such m are the sums of the factors of 75.

Pair of Factors	Sum of Factors
$75, \;\; 1$	76
$-75, -1$	-76
$25, \;\; 3$	28
$-25, -3$	-28
$15, \;\; 5$	20
$-15, -5$	-20

m can be 76, -76, 28, -28, 20, or -20.

55. $20(-365) = -7300$ and $20 + (-365) = -345$ so the other factor is $(x - 365)$.

Exercise Set 4.5

1. $3x^2 - 14x - 5$

We will use the FOIL method.

1) There is no common factor (other than 1 or -1).

2) We factor the first term, $3x^2$. The factors are $3x$ and x. We have this possibility:

$$(3x +)(x +)$$

3) Next we factor the last term, -5. The possibilities are $-1 \cdot 5$ and $1(-5)$.

4) We look for combinations of factors from steps (2) and (3) such that the sum of their products is the middle term, $-14x$. We try the possibilities:

$$(3x - 1)(x + 5) = 3x^2 + 14x - 5$$
$$(3x + 1)(x - 5) = 3x^2 - 14x - 5$$

The factorization is $(3x + 1)(x - 5)$.

3. $10y^3 + y^2 - 21y$

We will use the *ac*-method.

1) Look for a common factor. We factor out y:

$$y(10y^2 + y - 21)$$

2) Factor the trinomial $10y^2 + y - 21$. Multiply the leading coefficient, 10, and the constant, -21.

$$10(-21) = -210$$

3) Look for a factorization of -210 in which the sum of the factors is the coefficient of the middle term, 1.

Pairs of Factors	Sums of Factors
-1, 210	209
1, -210	-209
-2, 105	103
2, -105	-103
-3, 70	67
3, -70	-67
-5, 42	37
5, -42	-37
-6, 35	29
6, -35	-29
-7, 30	23
7, -30	-23
-10, 21	11
10, -21	-11
-14, 15	1
14, -15	-1

$\leftarrow -14 + 15 = 1$

4) Next, split the middle term, y, as follows:

$$y = -14y + 15y$$

5) Factor by grouping:

$$10y^2 + y - 21 = 10y^2 - 14y + 15y - 21$$
$$= 2y(5y - 7) + 3(5y - 7)$$
$$= (2y + 3)(5y - 7)$$

We must include the common factor to get a factorization of the original trinomial:

$$10y^3 + y^2 - 21y = y(2y + 3)(5y - 7)$$

5. $3c^2 - 20c + 32$

We will use the FOIL method.

1) There is no common factor(other than 1 or -1).

2) Factor the first term, $3c^2$. The factors are $3c$ and c. We have this possibility:

$$(3c+ \quad)(c+ \quad)$$

3) Next we factor the last term, 32. The possibilities are $1 \cdot 32$, $-1(-32)$, $2 \cdot 16$, $(-2)(-16)$, $4 \cdot 8$, and $-4(-8)$.

4) We look for a combination of factors from steps (2) and (3) such that the sum of their products is the middle term, $-20c$. Trial and error leads us to the correct factorization, $(3c - 8)(c - 4)$.

7. $35y^2 + 34y + 8$

We will use the *ac*-method.

1) There is no common factor (other than 1 or -1).

2) Multiply the leading coefficient, 35, and the constant, 8: $35(8) = 280$

3) Try to factor 280 so the sum of the factors is 34. We need only consider pairs of positive factors since 280 and 34 are both positive.

Pairs of Factors	Sums of Factors
280, 1	281
140, 2	142
70, 4	74
56, 5	61
40, 7	47
28, 10	38
20, 14	34

4) Split $34y$ as follows:

$$34y = 20y + 14y$$

5) Factor by grouping:

$$35y^2 + 34y + 8 = 35y^2 + 20y + 14y + 8$$
$$= 5y(7y + 4) + 2(7y + 4)$$
$$= (7y + 4)(5y + 2)$$

9. $4t + 10t^2 - 6 = 10t^2 + 4t - 6$

We will use the FOIL method.

1) Factor out the common factor, 2:

$$2(5t^2 + 2t - 3)$$

2) Now we factor out the trinomial $5t^2 + 2t - 3$.

Factor the first term, $5t^2$. The factors are $5t$ and t. We have this possibility:

$$(5t+ \quad)(t+ \quad)$$

3) Factor the last term, -3. The possibilities are $1(-3)$ and $-1 \cdot 3$.

4) Look for factors in steps (2) and (3) such that the sum of the products is the middle term, $2t$. Trial and error leads us to the correct factorization:
$5t^2 + 2t - 3 = (5t - 3)(t + 1)$

We must include the common factor to get a factorization of the original trinomial:

$$4t + 10t^2 - 6 = 2(5t - 3)(t + 1)$$

11. $8x^2 - 16 - 28x = 8x^2 - 28x - 16$

We will use the *ac*-method.

1) Factor out the common factor, 4:

$$4(2x^2 - 7x - 4)$$

2) Now we factor the trinomial $2x^2 - 7x - 4$. Multiply the leading coefficient, 2, and the constant, -4: $2(-4) = -8$

3) Factor -8 so the sum of the factors is -7. We need only consider parts of factors in which the negative factor has the larger absolute value, since their sum is negative.

Pairs of Factors	Sums of Factors
-4, 2	-2
-8, 1	-7

4) Split $-7x$ as follows:
$$-7x = -8x + x$$

5) Factor by grouping:
$$2x^2 - 7x - 4 = 2x^2 - 8x + x - 4$$
$$= 2x(x - 4) + (x - 4)$$
$$= (x - 4)(2x + 1)$$

We must include the common factor to get a factorization of the original trinomial:
$$8x^2 - 16 - 28x = 4(x - 4)(2x + 1)$$

13. $18a^2 - 51a + 15$

We will use the FOIL method.

1) Factor out the common factor, 3:
$$3(6a^2 - 17a + 5)$$

2) We now factor the trinomial $6a^2 - 17a + 5$.

Factor the first term, $6a^2$. The factors are $6a$, a and $3a$, $2a$. We have these possibilities:
$$(6a+ \quad)(a+ \quad) \text{ and } (3a+ \quad)(2a+ \quad)$$

3) Factor the last term, 5. The possibilities are $5 \cdot 1$ and $-5(-1)$.

4) Look for factors in steps (2) and (3) such that the sum of the products is the middle term, $-17a$. Trial and error leads us to the correct factorization:
$6a^2 - 17a + 5 = (3a - 1)(2a - 5)$

We must include the common factor to get a factorization of the original trinomial:
$$18a^2 - 51a + 15 = 3(3a - 1)(2a - 5)$$

15. $30t^2 + 85t + 25$

We will use the ac-method.

1) Factor out the common factor, 5:
$$5(6t^2 + 17t + 5)$$

2) Now we factor the trinomial $6t^2 + 17t + 5$. Multiply the leading coefficient, 6, and the constant, 5:
$6 \cdot 5 = 30$

3) Factor 30 so the sum of the factors is 17. We need to consider only positive pairs of factors since the middle term and the constant are both positive.

Pairs of Factors	Sums of Factors
30, 1	31
15, 2	17
10, 3	13
6, 5	11

4) Split $17t$ as follows:
$$17t = 15t + 2t$$

5) Factor by grouping:
$$6t^2 + 17t + 5 = 6t^2 + 15t + 2t + 5$$
$$= 3t(2t + 5) + (2t + 5)$$
$$= (3t + 1)(2t + 5)$$

We must include the common factor to get a factorization of the original trinomial:
$$30t^2 + 85t + 25 = 5(3t + 1)(2t + 5)$$

17. $12x^3 - 31x^2 + 20x$

We will use the FOIL method.

1) Factor out the common factor, x:
$$x(12x^2 - 31x + 20)$$

2) We now factor the trinomial $12x^2 - 31x + 20$. Factor the first term, $12x^2$. The factors are $12x$, x and $6x$, $2x$ and $4x$, $3x$. We have these possibilities:
$(12x+ \quad)(x+ \quad)$, $(6x+ \quad)(2x+ \quad)$, $(4x+ \quad)(3x+ \quad)$

3) Factor the last term, 20. The possibilities are $20 \cdot 1$, $-20(-1)$, $10 \cdot 2$, $-10(-2)$, $5 \cdot 4$, and $-5(-4)$.

4) Look for factors in steps (2) and (3) such that the sum of the products is the middle term, $-31x$. Trial and error leads us to the correct factorization:
$12x^2 - 31x + 20 = (4x - 5)(3x - 4)$

We must include the common factor to get a factorization of the original trinomial:
$$12x^3 - 31x^2 + 20x = x(4x - 5)(3x - 4)$$

19. $14x^4 - 19x^3 - 3x^2$

We will use the ac-method.

1) Factor out the common factor, x^2:
$$x^2(14x^2 - 19x - 3)$$

2) Now we factor the trinomial $14x^2 - 19x - 3$. Multiply the leading coefficient, 14, and the constant, -3:
$14(-3) = -42$

3) Factor -42 so the sum of the factors is -19. We need only consider pairs of factors in which the negative factor has the larger absolute value, since the sum is negative.

Pairs of Factors	Sums of Factors
-42, 1	-41
-21, 2	-19
-14, 3	-11
-7, 6	-1

4) Split $-19x$ as follows:
$$-19x = -21x + 2x$$

5) Factor by grouping:
$$14x^2 - 19x - 3 = 14x^2 - 21x + 2x - 3$$
$$= 7x(2x - 3) + 2x - 3$$
$$= (2x - 3)(7x + 1)$$

We must include the common factor to get a factorization of the original trinomial:
$$14x^4 - 19x^3 - 3x^2 = x^2(2x - 3)(7x + 1)$$

21. $3a^2 - a - 4$

We will use the FOIL method.

1) There is no common factor (other than 1 or -1).

2) Factor the first term, $3a^2$. The factors are $3a$ and a. We have this possibility: $(3a+\ \ )(a+\ \ )$

3) Factor the last term, -4. The possibilities are $4(-1)$, $-4 \cdot 1$, and $2(-2)$.

4) Look for factors in steps (2) and (3) such that the sum of the products is the middle term, $-a$. Trial and error leads us to the correct factorization: $(3a - 4)(a + 1)$

23. $9x^2 + 15x + 4$

We will use the ac-method.

1) There is no common factor (other than 1 or -1).

2) Multiply the leading coefficient and the constant: $9(4) = 36$

3) Factor 36 so the sum of the factors is 15. We need only consider pairs of positive factors since 36 and 15 are both positive.

Pairs of Factors	Sums of Factors
36, 1	37
18, 2	20
12, 3	15
9, 4	13
6, 6	12

4) Split $15x$ as follows:
$$15x = 12x + 3x$$

5) Factor by grouping:
$$9x^2 + 15x + 4 = 9x^2 + 12x + 3x + 4$$
$$= 3x(3x + 4) + 3x + 4$$
$$= (3x + 4)(3x + 1)$$

25. $3 + 35z - 12z^2 = -12z^2 + 35z + 3$

We will use the FOIL method.

1) Factor out -1 so the leading coefficient is positive: $-1(12z^2 - 35z - 3)$

2) Now we factor the trinomial $12z^2 - 35z - 3$. Factor the first term, $12z^2$. The factors are $12z$, z and $6z$, $2z$ and $4z$, $3z$. We have these possibilities: $(12z+\ \ )(z+\ \ )$, $(6z+\ \ )(2z+\ \ )$, $(4z+\ \ )(3z+\ \ )$

3) Factor the last term, -3. The possibilities are $3(-1)$ and $-3 \cdot 1$.

4) Look for factors in steps (2) and (3) such that the sum of the products is the middle term, $-35z$. Trial and error leads us to the correct factorization: $(12z + 1)(z - 3)$

We must include the common factor to get a factorization of the original trinomial:

$$3 + 35z - 12z^2 = -1(12z + 1)(z - 3), \text{ or}$$
$$(-12z - 1)(z - 3), \text{ or } (12z + 1)(-z + 3)$$

27. $-4t^2 - 4t + 15$

We will use the ac-method.

1) Factor out -1 so the leading coefficient is positive: $-1(4t^2 + 4t - 15)$

2) Now we factor the trinomial $4t^2 + 4t - 15$. Multiply the leading coefficient and the constant: $4(-15) = -60$

3) Factor -60 so the sum of the factors is 4. The desired factorization is $10(-6)$.

4) Split $4t$ as follows:
$$4t = 10t - 6t$$

5) Factor by grouping:
$$4t^2 + 4t - 15 = 4t^2 + 10t - 6t - 15$$
$$= 2t(2t + 5) - 3(2t + 5)$$
$$= (2t + 5)(2t - 3)$$

We must include the common factor to get a factorization of the original trinomial:

$$-4t^2 - 4t + 15 = -1(2t + 5)(2t - 3), \text{ or}$$
$$(-2t - 5)(2t - 3), \text{ or } (2t + 5)(-2t + 3)$$

29. $3x^3 - 5x^2 - 2x$

We will use the FOIL method.

1) Factor out the common factor, x:
$x(3x^2 - 5x - 2)$

2) Now we factor the trinomial $3x^2 - 5x - 2$. Factor the first term, $3x^2$. The factors are $3x$ and x. We have this possibility: $(3x+\ \ )(x+\ \ )$

3) Factor the last term, -2. The possibilities are $2(-1)$ and $-2 \cdot 1$.

4) Look for factors in steps (2) and (3) such that the sum of the products is the middle term, $-5x$. Trial and error leads us to the correct factorization: $(3x + 1)(x - 2)$

We must include the common factor to get a factorization of the original trinomial:

$$3x^3 - 5x^2 - 2x = x(3x + 1)(x - 2)$$

31. $24x^2 - 2 - 47x = 24x^2 - 47x - 2$

We will use the ac-method.

1) There is no common factor (other than 1 or -1).

2) Multiply the leading coefficient and the constant: $24(-2) = -48$

3) Factor -48 so the sum of the factors is -47. The desired factorization is $-48 \cdot 1$.

4) Split $-47x$ as follows:
$$-47x = -48x + x$$

5) Factor by grouping:
$$24x^2 - 47x - 2 = 24x^2 - 48x + x - 2$$
$$= 24x(x - 2) + (x - 2)$$
$$= (x - 2)(24x + 1)$$

33. $-8t^3 - 8t^2 + 30t$

We will use the FOIL method.

1) Factor out the common factor, $-2t$:
$$-2t(4t^2 + 4t - 15)$$

2) Now factor the trinomial $4t^2 + 4t - 15$. Factor the first term, $4t^2$. The possibilities are $4t \cdot t$ and $2t \cdot 2t$. The possible factorizations are of the form:
$$(4t+\quad)(t+\quad) \text{ and } (2t+\quad)(2t+\quad)$$

3) Factor the last term, -15. The possibilities are $-15 \cdot 1$, $15(-1)$, $-5 \cdot 3$, and $5(-3)$.

4) Look for factors in steps (2) and (3) such that the sum of the products is the middle term, $4t$. Trial and error leads us to the correct factorization:
$$(2t+5)(2t-3)$$

We must include the common factor to get a factorization of the original trinomial:
$$-8t^3 - 8t^2 + 30t = -2t(2t+5)(2t-3)$$

35. $-24x^3 + 2x + 47x^2$, or $-24x^3 + 47x^2 + 2x$

We will use the ac-method.

1) Factor out the common factor, $-x$:
$$-x(24x^2 - 47x - 2)$$

Now factor the trinomial $24x^2 - 47x - 2$. From Exercise 31 we know that the factorization is $(x-2)(24x+1)$. We must include the common factor to get a factorization of the original trinomial:
$$-24x^3 + 2x + 47x^2 = -x(x-2)(24x+1)$$

37. $21x^2 + 37x + 12$

We will use the FOIL method.

1) There is no common factor (other than 1 or -1).

2) Factor the first term $21x^2$. The factors are $21x$, x and $7x$, $3x$. We have these possibilities: $(21x+\quad)(x+\quad)$ and $(7x+\quad)(3x+\quad)$.

3) Factor the last term, 12. The possibilities are $12 \cdot 1$, $-12 \cdot (-1)$, $6 \cdot 2$, $-6 \cdot (-2)$, $4 \cdot 3$, and $-4 \cdot (-3)$.

4) Look for factors in steps (2) and (3) such that the sum of the products is the middle term, $37x$. Trial and error leads us to the correct factorization:
$$(7x+3)(3x+4)$$

39. $40x^4 + 16x^2 - 12$

We will use the ac-method.

1) Factor out the common factor, 4.
$$4(10x^4 + 4x^2 - 3)$$

Now we will factor the trinomial $10x^4 + 4x^2 - 3$. Substitute u for x^2 (and u^2 for x^4), and factor $10u^2 + 4u - 3$.

2) Multiply the leading coefficient and the constant: $10(-3) = -30$

3) Factor -30 so the sum of the factors is 4. This cannot be done. The trinomial $10u^2 + 4u - 3$ cannot be factored into binomials with integer coefficients. We have
$$40x^4 + 16x^2 - 12 = 4(10x^4 + 4x^2 - 3)$$

41. $12a^2 - 17ab + 6b^2$

We will use the FOIL method. (Our thinking is much the same as if we were factoring $12a^2 - 17a + 6$.)

1) There is no common factor (other than 1 or -1).

2) Factor the first term, $12a^2$. The factors are $12a$, a and $6a$, $2a$ and $4a$, $3a$. We have these possibilities: $(12a+\quad)(a+\quad)$ and $(6a+\quad)(2a+\quad)$ and $(4a+\quad)(3a+\quad)$.

3) Factor the last term, $6b^2$. The possibilities are $6b \cdot b$, $-6b \cdot (-b)$, $3b \cdot 2b$, and $-3b \cdot (-2b)$.

4) Look for factors in steps (2) and (3) such that the sum of the products is the middle term, $-17ab$. Trial and error leads us to the correct factorization:
$$(4a-3b)(3a-2b)$$

43. $2x^2 + xy - 6y^2$

We will use the ac-method.

1) There is no common factor (other than 1 or -1).

2) Multiply the coefficients of the first and last terms: $2(-6) = -12$

3) Factor -12 so the sum of the factors is 1. The desired factorization is $4(-3)$.

4) Split xy as follows:
$$xy = 4xy - 3xy$$

5) Factor by grouping:
$$\begin{aligned} 2x^2 + xy - 6y^2 &= 2x^2 + 4xy - 3xy - 6y^2 \\ &= 2x(x+2y) - 3y(x+2y) \\ &= (x+2y)(2x-3y) \end{aligned}$$

45. $12x^2 - 58xy + 56y^2$

We will use the FOIL method.

1) Factor out the common factor, 2:
$$2(6x^2 - 29xy + 28y^2)$$

2) Now we factor the trinomial $6x^2 - 29xy + 28y^2$. Factor the first term, $6x^2$. The factors are $6x$, x and $3x$, $2x$. We have these possibilities: $(6x+\quad)(x+\quad)$ and $(3x+\quad)(2x+\quad)$.

3) Factor the last term, $28y^2$. The possibilities are $28y \cdot y$, $-28y \cdot (-y)$, $14y \cdot 2y$, $-14 \cdot (-2y)$, $7y \cdot 4y$, and $-7y \cdot (-4y)$.

4) Look for factors in steps (2) and (3) such that the sum of the products is the middle term, $-29xy$. Trial and error leads us to the correct factorization:
$$(3x-4y)(2x-7y)$$

We must include the common factor to get a factorization of the original trinomial:
$$12x^2 - 58xy + 56y^2 = 2(3x-4y)(2x-7y)$$

47. $9x^2 - 30xy + 25y^2$

We will use the ac-method.

1) There is no common factor (other than 1 or -1).

2) Multiply the coefficients of the first and last terms: $9(25) = 225$

3) Factor 225 so the sum of the factors is -30. The desired factorization is $-15(-15)$.

4) Split $-30xy$ as follows:
$$-30xy = -15xy - 15xy$$

5) Factor by grouping:
$$\begin{aligned} 9x^2 - 30xy + 25y^2 &= 9x^2 - 15xy - 15xy + 25y^2 \\ &= 3x(3x - 5y) - 5y(3x - 5y) \\ &= (3x - 5y)(3x - 5y) \end{aligned}$$

49. $3x^6 + 4x^3 - 4$

We will use the FOIL method.

1) There is no common factor (other than 1 or -1). Substitute u for x^3 (and hence u^2 for x^6). We factor $3u^2 + 4u - 4$.

2) Factor the first term, $3u^2$. The factors are $3u$ and u. We have this possibility:
$(3u+\quad)(u+\quad)$

3) Factor the last term, -4. The possibilities are $-1 \cdot 4$, $1 \cdot (-4)$, and $-2 \cdot 2$.

4) Look for factors in steps (2) and (3) such that the sum of the products is the middle term, $4u$. Trial and error leads us to the correct factorization of $3u^2 + 4u - 4$: $(3u - 2)(u + 2)$. Replacing u with x^3 we have the factorization of the original trinomial: $(3x^3 - 2)(x^3 + 2)$.

51. a) $h(10) = -16(0)^2 + 80(0) + 224 = 224$ ft

$h(1) = -16(1)^2 + 80(1) + 224 = 288$ ft

$h(3) = -16(3)^2 + 80(3) + 224 = 320$ ft

$h(4) = -16(4)^2 + 80(4) + 224 = 288$ ft

$h(6) = -16(6)^2 + 80(6) + 224 = 128$ ft

b) $h(t) = -16t^2 + 80t + 224$

We will use the grouping method.

1) Factor out -16 so the leading coefficient is positive: $-16(t^2 - 5t - 14)$

2) Factor the trinomial $t^2 - 5t - 14$. Multiply the leading coefficient and the constant: $1(-14) = -14$

3) Factor -14 so the sum of the factors is -5. The desired factorization is $-7 \cdot 2$.

4) Split $-5t$ as follows:
$$-5t = -7t + 2t$$

5) Factor by grouping:
$$\begin{aligned} t^2 - 5t - 14 &= t^2 - 7t + 2t - 14 \\ &= t(t - 7) + 2(t - 7) \\ &= (t - 7)(t + 2) \end{aligned}$$

We must include the common factor to get a factorization of the original trinomial.
$$h(t) = -16(t - 7)(t + 2)$$

53. Discussion and Writing Exercise

55.
$$\begin{aligned} x + 2y - z &= 0, \quad (1) \\ 4x + 2y + 5z &= 6, \quad (2) \\ 2x - y + z &= 5 \quad (3) \end{aligned}$$

First we will eliminate z from Equations (1) and (2).
$$\begin{array}{ll} 5x + 10y - 5z = 0 & \text{Multiplying (1) by 5} \\ 4x + 2y + 5z = 6 & (2) \\ \hline 9x + 12y \quad\quad = 6 & (4) \end{array}$$

Now add Equations (1) and (3) to eliminate z.
$$\begin{array}{l} x + 2y - z = 0 \\ 2x - y + z = 5 \\ \hline 3x + y \quad\quad = 5 \quad (5) \end{array}$$

Now solve the system composed of Equations (4) and (5). We multiply Equation (5) by -3 and add.
$$\begin{array}{l} 9x + 12y = \quad 6 \\ -9x - 3y = -15 \\ \hline 9y = -9 \\ \quad\; y = -1 \end{array}$$

Substitute -1 for y in Equation (4) or (5) and solve for x. We use Equation (5) here.
$$\begin{aligned} 3x - 1 &= 5 \\ 3x &= 6 \\ x &= 2 \end{aligned}$$

Now substitute 2 for x and -1 for y in one of the original equations and solve for z. We use Equation (3).
$$\begin{aligned} 2 \cdot 2 - (-1) + z &= 5 \\ 4 + 1 + z &= 5 \\ 5 + z &= 5 \\ z &= 0 \end{aligned}$$

The triple $(2, -1, 0)$ checks, so it is the solution.

57.
$$\begin{aligned} 2x + 9y + 6z &= 5, \quad (1) \\ x - y + z &= 4, \quad (2) \\ 3x + 2y + 3z &= 7 \quad (3) \end{aligned}$$

First we will eliminate x from Equations (1) and (2).
$$\begin{array}{ll} 2x + 9y + 6z = \quad 5 & (1) \\ -2x + 2y - 2z = -8 & \text{Multiplying (2) by } -2 \\ \hline 11y + 4z = -3 & (4) \end{array}$$

Now eliminate x from Equations (2) and (3).
$$\begin{array}{ll} -3x + 3y - 3z = -12 & \text{Multiplying (2) by } -3 \\ 3x + 2y + 3z = \quad 7 & (3) \\ \hline 5y \quad\quad\quad = -5 \\ y = -1 \end{array}$$

Note that the last step eliminated z as well as x, allowing us to solve for y. Now substitute -1 for y in Equation (4) and solve for z.

$$11(-1) + 4z = -3$$
$$-11 + 4z = -3$$
$$4z = 8$$
$$z = 2$$

Substitute -1 for y and 2 for z in one of the original equations and solve for x. We use Equation (2).

$$x - (-1) + 2 = 4$$
$$x + 1 + 2 = 4$$
$$x + 3 = 4$$
$$x = 1$$

The triple $(1, -1, 2)$ checks, so it is the solution.

59. Write the first equation in slope-intercept form.

$$y - 2x = 18$$
$$y = 2x + 18$$

The second equation, $2x - 7 = y$, or $y = 2x - 7$, is in slope-intercept form.

Since the equations have the same slope, 2, and different y-intercepts, $(0, 8)$ and $(0, -7)$, the graphs of the equations are parallel lines.

61. Write each equation in slope-intercept form.

$$2x + 5y = 4 \qquad\qquad 2x - 5y = -3$$
$$5y = -2x + 4 \qquad\qquad -5y = -2x - 3$$
$$y = -\frac{2}{5}x + \frac{4}{5} \qquad\qquad y = \frac{2}{5}x + \frac{3}{5}$$

The slope of the first line is $-\frac{2}{5}$, and the slope of the second is $\frac{2}{5}$. The slopes are not the same nor is their product -1, so the graphs of the equations are neither parallel nor perpendicular.

63. First we find the slope.

$$m = \frac{-4 - (-3)}{5 - (-2)} = \frac{-1}{7} = -\frac{1}{7}$$

Now substitute $-\frac{1}{7}$ for m and the coordinates of either point in the slope-intercept equation $y = mx + b$ and solve for b. We will use the point $(5, -4)$.

$$y = mx + b$$
$$-4 = -\frac{1}{7} \cdot 5 + b$$
$$-4 = -\frac{5}{7} + b$$
$$-\frac{23}{7} = b$$

Now use the equation $y = mx + b$ again, substituting $-\frac{1}{7}$ for m and $-\frac{23}{7}$ for b.

$$y = -\frac{1}{7}x - \frac{23}{7}$$

65. First we find the slope.

$$m = \frac{-4 - 3}{7 - (-10)} = \frac{-7}{17} = -\frac{7}{17}$$

Now substitute $-\frac{7}{17}$ for m and the coordinates of either point in the slope-intercept equation $y = mx + b$ and solve for b. We will use the point $(-10, 3)$.

$$y = mx + b$$
$$3 = -\frac{7}{17}(-10) + b$$
$$3 = \frac{70}{17} + b$$
$$-\frac{19}{17} = b$$

Now use the equation $y = mx + b$ again, substituting $-\frac{7}{17}$ for m and $-\frac{19}{17}$ for b.

$$y = -\frac{7}{17}x - \frac{19}{17}$$

67. Left to the student

69. $p^2 q^2 + 7pq + 12$

The factorization will be of the form $(pq + \quad)(pq + \quad)$. We look for factors of 12 whose sum is 7. The factors we need are 4 and 3. The factorization is $(pq + 4)(pq + 3)$.

71. $x^2 - \frac{4}{25} + \frac{3}{5}x = x^2 + \frac{3}{5}x - \frac{4}{25}$

We look for factors of $-\frac{4}{25}$ whose sum is $\frac{3}{5}$. The factors are $\frac{4}{5}$ and $-\frac{1}{5}$. The factorization is $\left(x + \frac{4}{5}\right)\left(x - \frac{1}{5}\right)$.

73. $y^2 + 0.4y - 0.05$

We look for factors of -0.05 whose sum is 0.4. The factors are -0.1 and 0.5. The factorization is $(y - 0.1)(y + 0.5)$.

75. $7a^2 b^2 + 6 + 13ab = 7a^2 b^2 + 13ab + 6$

We will use the grouping method. There is no common factor (other than 1 or -1). Multiply the leading coefficient and the constant: $7(6) = 42$. Factor 42 so the sum of the factors is 13. The desired factorization is $6 \cdot 7$. Split the middle term and factor by grouping.

$$7a^2 b^2 + 13ab + 6 = 7a^2 b^2 + 6ab + 7ab + 6$$
$$= ab(7ab + 6) + 7ab + 6$$
$$= (7ab + 6)(ab + 1)$$

77. $3x^2 + 12x - 495$

Factor out the common factor, 3.

$$3(x^2 + 4x - 165)$$

Now factor $x^2 + 4x - 165$. Find factors of -165 whose sum is 4. The factors are -11 and 15. Then $x^2 + 4x - 165 = (x - 11)(x + 15)$, and $3x^2 + 12x - 495 = 3(x - 11)(x + 15)$.

79. $216x + 78x^2 + 6x^3 = 6x^3 + 78x^2 + 216x$

Factor out the common factor, $6x$.

$6x(x^2 + 13x + 36)$

Now factor $x^2 + 13x + 36$. Look for factors of 36 whose sum is 13. The factors are 9 and 4. Then $x^2 + 13x + 36 = (x + 9)(x + 4)$, and $6x^3 + 78x^2 + 216x = 6x(x + 9)(x + 4)$.

81. $x^{2a} + 5x^a - 24$

$x^{2a} = (x^a)^2$, so the factorization is of the form $(x^a + \quad)(x^a + \quad)$.

Look for factors of -24 whose sum is 5. The factors are 8 and -3. Then the factorization is $(x^a + 8)(x^a - 3)$.

Exercise Set 4.6

1. $x^2 - 4x + 4 = (x - 2)^2$ Find the square terms and write their square roots with a minus sign between them.

3. $y^2 + 18y + 81 = (y + 9)^2$ Find the square terms and write their square roots with a minus sign between them.

5. $x^2 + 1 + 2x = x^2 + 2x + 1$ Writing in descending order
$\quad = (x + 1)^2$ Factoring the trinomial square

7. $9y^2 + 12y + 4 = (3y + 2)^2$ Find the square terms and write their square roots with a minus sign between them.

9. $-18y^2 + y^3 + 81y = y^3 - 18y^2 + 81y$ Writing in descending order
$\quad = y(y^2 - 18y + 81)$ Removing the common factor
$\quad = y(y - 9)^2$ Factoring the trinomial square

11. $12a^2 + 36a + 27 = 3(4a^2 + 12a + 9)$ Removing the common factor
$\quad = 3(2a + 3)^2$ Factoring the trinomial square

13. $2x^2 - 40x + 200 = 2(x^2 - 20x + 100)$
$\quad = 2(x - 10)^2$

15. $1 - 8d + 16d^2 = (1 - 4d)^2$,   Find the square terms
or $(4d - 1)^2$ and write their square roots with a minus sign between them.

17. $y^4 - 8y^2 + 16 = (y^2 - 4)^2$ Find the square terms and write their square roots with a minus sign between them.

$\quad = [(y + 2)(y - 2)]^2$ Factoring the difference of squares
$\quad = (y + 2)^2(y - 2)^2$

19. $0.25x^2 + 0.30x + 0.09 = (0.5x + 0.3)^2$ Find the square terms and write their square roots with a plus sign between them.

21. $p^2 - 2pq + q^2 = (p - q)^2$

23. $a^2 + 4ab + 4b^2 = (a + 2b)^2$

25. $25a^2 - 30ab + 9b^2 = (5a - 3b)^2$

27. $y^6 + 26y^3 + 169 = (y^3 + 13)^2$ Find the square terms and write their square roots with a plus sign between them

29. $16x^{10} - 8x^5 + 1 = (4x^5 - 1)^2$ $[16x^{10} = (4x^5)^2]$

31. $x^4 + 2y^2y^2 + y^4 = (x^2 + y^2)^2$

33. $x^2 - 16 = x^2 - 4^2 = (x + 4)(x - 4)$

35. $p^2 - 49 = p^2 - 7^2 = (p + 7)(p - 7)$

37. $p^2q^2 - 25 = (pq)^2 - 5^2 = (pq + 5)(pq - 5)$

39. $\quad 6x^2 - 6y^2$
$\quad = 6(x^2 - y^2)$ Removing the common factor
$\quad = 6(x + y)(x - y)$ Factoring the difference of squares

41. $\quad 4xy^4 - 4xz^4$
$\quad = 4x(y^4 - z^4)$ Removing the common factor
$\quad = 4x[(y^2)^2 - (z^2)^2]$
$\quad = 4x(y^2 + z^2)(y^2 - z^2)$ Factoring the difference of squares
$\quad = 4x(y^2 + z^2)(y + z)(y - z)$ Factoring $y^2 - z^2$

43. $4a^3 - 49a = a(4a^2 - 49)$
$\quad = a[(2a)^2 - 7^2]$
$\quad = a(2a + 7)(2a - 7)$

45. $3x^8 - 3y^8 = 3(x^8 - y^8)$
$\quad = 3[(x^4)^2 - (y^4)^2)]$
$\quad = 3(x^4 + y^4)(x^4 - y^4)$
$\quad = 3(x^4 + y^4)[(x^2)^2 - (y^2)^2]$
$\quad = 3(x^4 + y^4)(x^2 + y^2)(x^2 - y^2)$
$\quad = 3(x^4 + y^4)(x^2 + y^2)(x + y)(x - y)$

47. $9a^4 - 25a^2b^4 = a^2(9a^2 - 25b^4)$
$\quad = a^2[(3a)^2 - (5b^2)^2]$
$\quad = a^2(3a + 5b^2)(3a - 5b^2)$

49. $\dfrac{1}{36} - z^2 = \left(\dfrac{1}{6}\right)^2 - z^2 = \left(\dfrac{1}{6} + z\right)\left(\dfrac{1}{6} - z\right)$

51. $0.04x^2 - 0.09y^2 = (0.2x)^2 - (0.3y)^2$
$\quad = (0.2x + 0.3y)(0.2x - 0.3y)$

53. $m^3 - 7m^2 - 4m + 28$

$= m^2(m-7) - 4(m-7)$ Factoring by grouping

$= (m^2 - 4)(m-7)$

$= (m+2)(m-2)(m-7)$ Factoring the difference of squares

55. $a^3 - ab^2 - 2a^2 + 2b^2$

$= a(a^2 - b^2) - 2(a^2 - b^2)$ Factoring by grouping

$= (a-2)(a^2 - b^2)$

$= (a-2)(a+b)(a-b)$ Factoring the difference of squares

57. $(a+b)^2 - 100 = (a+b)^2 - 10^2$

$= (a+b+10)(a+b-10)$

59. $144 - (p-8)^2$ Difference of squares

$= [12 + (p-8)][12 - (p-8)]$

$= (12 + p - 8)(12 - p + 8)$

$= (4+p)(20-p)$

61. $a^2 + 2ab + b^2 - 9$

$= (a^2 + 2ab + b^2) - 9$ Grouping as a difference of squares

$= (a+b)^2 - 3^2$

$= (a+b+3)(a+b-3)$

63. $r^2 - 2r + 1 - 4s^2$

$= (r^2 - 2r + 1) - 4s^2$ Grouping as a difference of squares

$= (r-1)^2 - (2s)^2$

$= (r-1+2s)(r-1-2s)$

65. $2m^2 + 4mn + 2n^2 - 50b^2$

$= 2(m^2 + 2mn + n^2 - 25b^2)$ Removing the common factor

$= 2[(m^2 + 2mn + n^2) - 25b^2]$ Grouping as a difference of squares

$= 2[(m+n)^2 - (5b)^2]$

$= 2(m+n+5b)(m+n-5b)$

67. $9 - (a^2 + 2ab + b^2) = 9 - (a+b)^2$

$= [3 + (a+b)][3 - (a+b)]$, or

$(3 + a + b)(3 - a - b)$

69. $z^3 + 27 = z^3 + 3^3$

$= (z+3)(z^2 - 3z + 9)$

$A^3 + B^3 = (A+B)(A^2 - AB + B^2)$

71. $x^3 - 1 = x^3 - 1^3$

$= (x-1)(x^2 + x + 1)$

$A^3 - B^3 = (A-B)(A^2 + AB + B^2)$

73. $y^3 + 125 = y^3 + 5^3$

$= (y+5)(y^2 - 5y + 25)$

$A^3 + B^3 = (A+B)(A^2 - AB + B^2)$

75. $8a^3 + 1 = (2a)^3 + 1^3$

$= (2a+1)(4a^2 - 2a + 1)$

$A^3 + B^3 = (A+B)(A^2 - AB + B^2)$

77. $y^3 - 8 = y^3 - 2^3$

$= (y-2)(y^2 + 2y + 4)$

$A^3 - B^3 = (A-B)(A^2 + AB + B^2)$

79. $8 - 27b^3 = 2^3 - (3b)^3$

$= (2 - 3b)(4 + 6b + 9b^2)$

81. $64y^3 + 1 = (4y)^3 + 1^3$

$= (4y+1)(16y^2 - 4y + 1)$

83. $8x^3 + 27 = (2x)^3 + 3^3$

$= (2x+3)(4x^2 - 6x + 9)$

85. $a^3 - b^3 = (a-b)(a^2 + ab + b^2)$

87. $a^3 + \dfrac{1}{8} = a^3 + \left(\dfrac{1}{2}\right)^3$

$= \left(a + \dfrac{1}{2}\right)\left(a^2 - \dfrac{1}{2}a + \dfrac{1}{4}\right)$

89. $2y^3 - 128 = 2(y^3 - 64)$

$= 2(y^3 - 4^3)$

$= 2(y-4)(y^2 + 4y + 16)$

91. $24a^3 + 3 = 3(8a^3 + 1)$

$= 3[(2a)^3 + 1^3]$

$= 3(2a+1)(4a^2 - 2a + 1)$

93. $rs^3 + 64r = r(s^3 + 64)$

$= r(s^3 + 4^3)$

$= r(s+4)(s^2 - 4s + 16)$

95. $5x^3 - 40z^3 = 5(x^3 - 8z^3)$

$= 5[x^3 - (2z)^3]$

$= 5(x - 2z)(x^2 + 2xz + 4z^2)$

97. $x^3 + 0.001 = x^3 + (0.1)^3$

$= (x + 0.1)(x^2 - 0.1x + 0.01)$

99. $64x^6 - 8t^6 = 8(8x^6 - t^6)$

$= 8[(2x^2)^3 - (t^2)^3]$

$= 8(2x^2 - t^2)(4x^4 + 2x^2t^2 + t^4)$

101. $2y^4 - 128y = 2y(y^3 - 64)$

$= 2y(y^3 - 4^3)$

$= 2y(y-4)(y^2 + 4y + 16)$

103. $z^6 - 1$

$= (z^3)^2 - 1^2$ Writing as a difference of squares

$= (z^3 + 1)(z^3 - 1)$ Factoring a difference of squares

$= (z+1)(z^2 - z + 1)(z-1)(z^2 + z + 1)$

Factoring a sum and a difference of cubes

105. $t^6 + 64y^6 = (t^2)^3 + (4y^2)^3$

$$= (t^2 + 4y^2)(t^4 - 4t^2y^2 + 16y^4)$$

107. Discussion and Writing Exercise

109. $7x - 2y = -11$,   (1)

      $2x + 7y = 18$ (2)

To eliminate y, multiply Equation (1) by 7 and Equation (2) by 2 and add.

$$49x - 14y = -77$$

$$\underline{4x + 14y = 36}$$

$$53x = -41$$

$$x = -\frac{41}{53}$$

Now substitute $-\dfrac{41}{53}$ for x in one of the original equations and solve for y.

$$2x + 7y = 18 \qquad (2)$$

$$2\left(-\frac{41}{53}\right) + 7y = 18$$

$$-\frac{82}{53} + 7y = 18$$

$$7y = \frac{1036}{53}$$

$$y = \frac{148}{53}$$

The pair $\left(-\dfrac{41}{53}, \dfrac{148}{53}\right)$ checks, so it is the solution.

111. $x - y = -12$,   (1)

    $\underline{x + y = 14}$ (2)

  $2x = 2$ Adding

    $x = 1$

Now substitute 1 for x in one of the original equations and solve for y.

$$x + y = 14 \quad (2)$$

$$1 + y = 14$$

$$y = 13$$

The pair $(1, 13)$ checks, so it is the solution.

113. $x - y \leq 5$,

    $x + y \geq 3$

Graph the lines $x - y = 5$ and $x + y = 3$ using solid lines. We indicate the region for each inequality by arrows. Shade the region where they overlap.

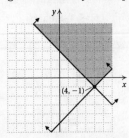

To find the vertex, solve the system of related equations:

$x - y = 5$

$x + y = 3$

Solving, we obtain the vertex $(4, -1)$.

115. $x - y \geq 5$,   (1)

    $x + y \leq 3$,   (2)

     $x \geq 1$ (3)

Shade the intersection of the graphs of $x - y \geq 5$, $x + y \leq 3$, and $x \geq 1$.

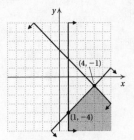

To find the vertices we solve two different systems of equations. From (1) and (2) we obtain the vertex $(4, -1)$. From (1) and (3) we obtain the vertex $(1, -4)$.

117. To find an equation of the line through $(-2, -4)$ parallel to $x - y = 5$, we first write $x - y = 5$ in slope-intercept form.

$$x - y = 5$$

$$-y = -x + 5$$

$$y = x - 5$$

The slope of $y = x - 5$ is 1, so we want to find an equation with slope 1 containing the point $(-2, -4)$. First substitute 1 for m, -2 for x, and -4 for y in $y = mx + b$.

$$y = mx + b$$

$$-4 = 1(-2) + b$$

$$-4 = -2 + b$$

$$-2 = b$$

Now use $y = mx + b$ again, substituting 1 for m and -2 for b.

$$y = 1 \cdot x - 2, \text{ or } y = x - 2$$

We have seen above that the slope of $x - y = 5$ is 1. Then the slope of a line perpendicular to this line is the opposite of the reciprocal of 1, or -1. Thus, we want to find an equation of the line with slope -1 containing $(-2, -4)$. First substitute -1 for m, -2 for x, and -4 for y in $y = mx + b$.

$$y = mx + b$$

$$-4 = -1(-2) + b$$

$$-4 = 2 + b$$

$$-6 = b$$

Now use $y = mx + b$ again, substituting -1 for m and -6 for b.

$$y = -1 \cdot x - 6, \text{ or } y = -x - 6$$

119. To find an equation of the line through $(4,5)$ parallel to $y = -\dfrac{1}{2}x + 3$, we first observe that the slope of $y = -\dfrac{1}{2}x + 3$ is $-\dfrac{1}{2}$. Thus, we want to find an equation of the line with slope $-\dfrac{1}{2}$ containing $(4,5)$. Begin by substituting $-\dfrac{1}{2}$ for m, 4 for x, and 5 for y in $y = mx + b$.

$$y = mx + b$$
$$5 = -\frac{1}{2} \cdot 4 + b$$
$$5 = -2 + b$$
$$7 = b$$

Now use $y = mx + b$ again, substituting $-\dfrac{1}{2}$ for m and 7 for b.

$$y = -\frac{1}{2}x + 7$$

The slope of a line perpendicular to $y = -\dfrac{1}{2}x + 3$ is the opposite of the reciprocal of $-\dfrac{1}{2}$, or 2. Thus, we want to find an equation of the line with slope 2 containing $(4,5)$. First substitute 2 for m, 4 for x, and 5 for y in $y = mx + b$.

$$y = mx + b$$
$$5 = 2 \cdot 4 + b$$
$$5 = 8 + b$$
$$-3 = b$$

Now use $y = mx + b$ again, substituting 2 for m and -3 for b.

$$y = 2x - 3$$

121. If $P(x) = x^3$, then

$$P(a + h) - P(a)$$
$$= (a + h)^3 - a^3$$
$$= (a + h - a)[(a + h)^2 + (a + h)(a) + a^2]$$
$$= h(a^2 + 2ah + h^2 + a^2 + ah + a^2)$$
$$= h(3a^2 + 3ah + h^2)$$

123. The model shows a cube with volume a^3 from which a portion whose volume is b^3 has been removed. This leaves a remaining volume which can be expressed as $a^2(a-b) + ab(a-b) + b^2(a-b)$, or $(a-b)(a^2+ab+b^2)$. Thus, $a^3 - b^3 = (a - b)(a^2 + ab + b^2)$.

125.
$$5c^{100} - 80d^{100}$$
$$= 5(c^{100} - 16d^{100})$$
$$= 5(c^{50} + 4d^{50})(c^{50} - 4d^{50})$$
$$= 5(c^{50} + 4d^{50})(c^{25} + 2d^{25})(c^{25} - 2d^{25})$$

127.
$$x^{6a} + y^{3b} = (x^{2a})^3 + (y^b)^3$$
$$= (x^{2a} + y^b)(x^{4a} - x^{2a}y^b + y^{2b})$$

129.
$$3x^{3a} + 24y^{3b} = 3(x^{3a} + 8y^{3b})$$
$$= 3[(x^a)^3 + (2y^b)^3]$$
$$= 3(x^a + 2y^b)(x^{2a} - 2x^ay^b + 4y^{2b})$$

131.
$$\frac{1}{24}x^3y^3 + \frac{1}{3}z^3 = \frac{1}{3}\left(\frac{1}{8}x^3y^3 + z^3\right)$$
$$= \frac{1}{3}\left[\left(\frac{1}{2}xy\right)^3 + z^3\right]$$
$$= \frac{1}{3}\left(\frac{1}{2}xy + z\right)\left(\frac{1}{4}x^2y^2 - \frac{1}{2}xyz + z^2\right)$$

133.
$$(x + y)^3 - x^3$$
$$= [(x + y) - x][(x + y)^2 + x(x + y) + x^2]$$
$$= (x + y - x)(x^2 + 2xy + y^2 + x^2 + xy + x^2)$$
$$= y(3x^2 + 3xy + y^2)$$

135.
$$(a + 2)^3 - (a - 2)^3$$
$$= [(a+2) - (a-2)][(a+2)^2 + (a+2)(a-2) + (a-2)^2]$$
$$= (a + 2 - a + 2)(a^2 + 4a + 4 + a^2 - 4 + a^2 - 4a + 4)$$
$$= 4(3a^2 + 4)$$

Exercise Set 4.7

1.
$$y^2 - 225$$
$$= y^2 - 15^2 \quad \text{Difference of squares}$$
$$= (y + 15)(y - 15)$$

3.
$$2x^2 + 11x + 12$$
$$= (2x + 3)(x + 4) \qquad \text{FOIL or } ac\text{-method}$$

5.
$$5x^4 - 20$$
$$= 5(x^4 - 4) \quad \text{Removing the common factor}$$
$$= 5[(x^2)^2 - 2^2] \quad \text{Difference of squares}$$
$$= 5(x^2 + 2)(x^2 - 2)$$

7.
$$p^2 + 36 + 12p$$
$$= p^2 + 12p + 36 \quad \text{Trinomial square}$$
$$= (p + 6)^2$$

9.
$$2x^2 - 10x - 132$$
$$= 2(x^2 - 5x - 66)$$
$$= 2(x - 11)(x + 6) \quad \text{Trial and error}$$

11.
$$9x^2 - 25y^2$$
$$= (3x)^2 - (5y)^2 \quad \text{Difference of squares}$$
$$= (3x + 5y)(3x - 5y)$$

13.
$$4m^4 - 100$$
$$= 4(m^4 - 25) \quad \text{Removing the common factor}$$
$$= 4[(m^2)^2 - 5^2] \quad \text{Difference of squares}$$
$$= 4(m^2 + 5)(m^2 - 5)$$

15.
$$3x^2 + 15x - 252$$
$$= 3(x^2 + 5x - 84)$$
$$= 3(x + 12)(x - 7) \quad \text{FOIL or } ac\text{-method}$$

17.
$$2xy^2 - 50x$$
$$= 2x(y^2 - 25)$$
$$= 2x(y + 5)(y - 5)$$

19. $225 - (a-3)^2$ Difference of squares
$= [15 + (a-3)][15 - (a-3)]$
$= [15 + a - 3][15 - a + 3]$
$= (12 + a)(18 - a)$

21. $m^6 - 1$
$= (m^3)^2 - 1^2$ Difference of squares
$= (m^3 + 1)(m^3 - 1)$ Sum and difference of cubes
$= (m+1)(m^2 - m + 1)(m-1)(m^2 + m + 1)$

23. $x^2 + 6x - y^2 + 9$
$= x^2 + 6x + 9 - y^2$
$= (x+3)^2 - y^2$ Difference of squares
$= [(x+3) + y][(x+3) - y]$
$= (x + 3 + y)(x + 3 - y)$

25. $250x^3 - 128y^3$
$= 2(125x^3 - 64y^3)$
$= 2[(5x)^3 - (4y)^3]$ Difference of cubes
$= 2(5x - 4y)(25x^2 + 20xy + 16y^2)$

27. $8m^3 + m^6 - 20$
$= m^6 + 8m^3 - 20$
$= (m^3)^2 + 8m^3 - 20$
$= (m^3 - 2)(m^3 + 10)$ Trial and error

29. $ac + cd - ab - bd$
$= c(a + d) - b(a + d)$ Factoring by grouping
$= (c - b)(a + d)$

31. $50b^2 - 5ab - a^2$
$= (5b - a)(10b + a)$ FOIL or ac-method

33. $-7x^2 + 2x^3 + 4x - 14$
$= 2x^3 - 7x^2 + 4x - 14$
$= x^2(2x - 7) + 2(2x - 7)$ Factoring by grouping
$= (x^2 + 2)(2x - 7)$

35. $2x^3 + 6x^2 - 8x - 24$
$= 2(x^3 + 3x^2 - 4x - 12)$
$= 2[x^2(x + 3) - 4(x + 3)]$ Factoring by grouping
$= 2(x^2 - 4)(x + 3)$ Difference of squares
$= 2(x + 2)(x - 2)(x + 3)$

37. $16x^3 + 54y^3$
$= 2(8x^3 + 27y^3)$
$= 2[(2x)^3 + (3y)^3]$ Sum of cubes
$= 2(2x + 3y)(4x^2 - 6xy + 9y^2)$

39. $36y^2 - 35 + 12y$
$= 36y^2 + 12y - 35$
$= (6y - 5)(6y + 7)$ FOIL or ac-method

41. $a^8 - b^8$ Difference of squares
$= (a^4 + b^4)(a^4 - b^4)$ Difference of squares
$= (a^4 + b^4)(a^2 + b^2)(a^2 - b^2)$ Difference of squares
$= (a^4 + b^4)(a^2 + b^2)(a + b)(a - b)$

43. $a^3b - 16ab^3$
$= ab(a^2 - 16b^2)$ Difference of squares
$= ab(a + 4b)(a - 4b)$

45. $\frac{1}{16}x^2 - \frac{1}{6}xy^2 + \frac{1}{9}y^4$
$= \frac{1}{16}x^2 - \frac{1}{6}xy^2 + \frac{1}{9}(y^2)^2$ Trinomial square
$= \left(\frac{1}{4}x - \frac{1}{3}y^2\right)^2$

47. $5x^3 - 5x^2y - 5xy^2 + 5y^3$
$= 5(x^3 - x^2y - xy^2 + y^3)$
$= 5[x^2(x - y) - y^2(x - y)]$ Factoring by grouping
$= 5(x^2 - y^2)(x - y)$
$= 5(x + y)(x - y)(x - y)$ Factoring the difference of squares
$= 5(x + y)(x - y)^2$

49. $42ab + 27a^2b^2 + 8$
$= 27a^2b^2 + 42ab + 8$
$= (9ab + 2)(3ab + 4)$ FOIL or ac-method

51. $8y^4 - 125y$
$= y(8y^3 - 125)$
$= y[(2y)^3 - 5^3]$ Difference of cubes
$= y(2y - 5)(4y^2 + 10y + 25)$

53. $a^2 - b^2 - 6b - 9$
$= a^2 - (b^2 + 6b + 9)$ Factoring out -1
$= a^2 - (b + 3)^2$ Difference of squares
$= [a + (b + 3)][a - (b + 3)]$
$= (a + b + 3)(a - b - 3)$

55. Discussion and Writing Exercise

57. *Familiarize.* Let $x =$ the number of correct answers and $y =$ the number of incorrect answers. Then the total number of points awarded for the correct answers is $2x$, the total number of points deducted for the incorrect answers is $\frac{1}{2}y$, and the total score is $2x - \frac{1}{2}y$. Assume that all the questions were answered.

Translate.

$\underbrace{\text{The number of questions}}_{x + y}$ is 75.
$= 75$

$\underbrace{\text{The total score}}_{2x - \frac{1}{2}y}$ is 100.
$= 100$

After clearing the fraction in the second equation, we have a system of equations.

$$x + y = 75,$$
$$4x - y = 200$$

Solve. We use the elimination method. Begin by adding the equations.

$$x + y = 75$$
$$\underline{4x - y = 200}$$
$$5x = 275$$
$$x = 55$$

Now substitute 55 for x in the first equation and solve for y.

$$55 + y = 75$$
$$y = 20$$

Check. If there are 55 correct answers and 20 incorrect answers, the total number of questions is $55+20$, or 75. For the 55 correct answers, $2 \cdot 55$ or 110 points are awarded; for the 20 incorrect answers, $\frac{1}{2} \cdot 20$ or 10, points are deducted. Then the total score is $110-10$ or 100. The answer checks.

State. There were 55 correct answers and 20 incorrect answers.

59. $\quad 30y^4 - 97xy^2 + 60x^2$

$= (5y^2 - 12x)(6y^2 - 5x)$ FOIL or ac-method

61. $\quad 5x^3 - \dfrac{5}{27}$

$= 5\left(x^3 - \dfrac{1}{27}\right)$

$= 5\left[x^3 - \left(\dfrac{1}{3}\right)^3\right]$ Difference of cubes

$= 5\left(x - \dfrac{1}{3}\right)\left(x^2 + \dfrac{1}{3}x + \dfrac{1}{9}\right)$

63. $\quad (x - p)^2 - p^2$ Difference of squares

$= (x - p + p)(x - p - p)$

$= x(x - 2p)$

65. $\quad (y - 1)^4 - (y - 1)^2$

$= (y - 1)^2[(y - 1)^2 - 1]$
    Removing the common factor

$= (y - 1)^2[(y - 1) + 1][(y - 1) - 1]$
    Factoring the difference of squares

$= (y - 1)^2(y)(y - 2)$, or $y(y - 1)^2(y - 2)$

67. $\quad 4x^2 + 4xy + y^2 - r^2 + 6rs - 9s^2$

$= (4x^2 + 4xy + y^2) - 1(r^2 - 6rs + 9s^2)$ Grouping

$= (2x + y)^2 - (r - 3s)^2$ Difference of squares

$= [(2x + y) + (r - 3s)][(2x + y) - (r - 3s)]$

$= (2x + y + r - 3s)(2x + y - r + 3s)$

69. $\quad c^{2w+1} + 2c^{w+1} + c$

$= c^{2w} \cdot c + 2c^w \cdot c + c$

$= c(c^{2w} + 2c^w + 1)$

$= c[(c^w)^2 + 2(c^w) + 1]$ Trinomial square

$= c(c^w + 1)^2$

71. $\quad 3(x + 1)^2 + 9(x + 1) - 12$

$= 3[(x + 1)^2 + 3(x + 1) - 4]$

$= 3[(x + 1) + 4][(x + 1) - 1]$ Factor u^2+3u-4
                                    where $u=x+1$

$= 3(x + 5)(x)$, or

$ 3x(x + 5)$

73. $\quad x^6 - 2x^5 + x^4 - x^2 + 2x - 1$

$= x^4(x^2 - 2x + 1) - (x^2 - 2x + 1)$

$= (x^2 - 2x + 1)(x^4 - 1)$

$= (x - 1)^2(x^2 + 1)(x + 1)(x - 1)$

$= (x - 1)^3(x^2 + 1)(x + 1)$

75. $\quad y^9 - y$

$= y(y^8 - 1)$

$= y(y^4 + 1)(y^4 - 1)$

$= y(y^4 + 1)(y^2 + 1)(y^2 - 1)$

$= y(y^4 + 1)(y^2 + 1)(y + 1)(y - 1)$

Exercise Set 4.8

1. $\qquad x^2 + 3x = 28$

$x^2 + 3x - 28 = 0$ Getting 0 on one side

$(x + 7)(x - 4) = 0$ Factoring

$x + 7 = 0 \quad or \quad x - 4 = 0$ Principle of zero
                                        products

$x = -7 \ or \qquad x = 4$

The solutions are -7 and 4.

3. $\qquad y^2 + 9 = 6y$

$y^2 - 6y + 9 = 0$ Getting 0 on one side

$(y - 3)(y - 3) = 0$ Factoring

$y - 3 = 0 \ or \ y - 3 = 0$ Principle of zero
                                products

$y = 3 \ or \qquad y = 3$

There is only one solution, 3.

5. $\quad x^2 + 20x + 100 = 0$

$(x + 10)(x + 10) = 0$ Factoring

$x + 10 = 0 \quad or \quad x + 10 = 0$ Principle of zero
                                            products

$x = -10 \ or \qquad x = -10$

There is only one solution, -10.

7. $9x + x^2 + 20 = 0$

$x^2 + 9x + 20 = 0$ Changing order

$(x + 5)(x + 4) = 0$ Factoring

$x + 5 = 0$ *or* $x + 4 = 0$ Principle of zero products

    $x = -5$ *or* $x = -4$

The solutions are -5 and -4.

9. $x^2 + 8x = 0$

$x(x + 8) = 0$ Factoring

$x = 0$ *or* $x + 8 = 0$ Principle of zero products

$x = 0$ *or* $x = -8$

The solutions are 0 and -8.

11. $x^2 - 25 = 0$

$(x + 5)(x - 5) = 0$ Factoring

$x + 5 = 0$ *or* $x - 5 = 0$ Principle of zero products

    $x = -5$ *or* $x = 5$

The solutions are -5 and 5.

13. $z^2 = 144$

$z^2 - 144 = 0$ Getting 0 on one side

$(z + 12)(z - 12) = 0$ Factoring

$z + 12 = 0$ *or* $z - 12 = 0$ Principle of zero products

    $z = -12$ *or* $z = 12$

The solutions are -12 and 12.

15. $y^2 + 2y = 63$

$y^2 + 2y - 63 = 0$

$(y + 9)(y - 7) = 0$

$y + 9 = 0$ *or* $y - 7 = 0$

   $y = -9$ *or* $y = 7$

The solutions are -9 and 7.

17. $32 + 4x - x^2 = 0$

     $0 = x^2 - 4x - 32$

     $0 = (x - 8)(x + 4)$

$x - 8 = 0$ *or* $x + 4 = 0$

  $x = 8$ *or* $x = -4$

The solutions are 8 and -4.

19. $3b^2 + 8b + 4 = 0$

$(3b + 2)(b + 2) = 0$

$3b + 2 = 0$ *or* $b + 2 = 0$

  $3b = -2$ *or* $b = -2$

   $b = -\dfrac{2}{3}$ *or* $b = -2$

The solutions are $-\dfrac{2}{3}$ and -2.

21. $8y^2 - 10y + 3 = 0$

$(4y - 3)(2y - 1) = 0$

$4y - 3 = 0$ *or* $2y - 1 = 0$

  $4y = 3$ *or* $2y = 1$

   $y = \dfrac{3}{4}$ *or* $y = \dfrac{1}{2}$

The solutions are $\dfrac{3}{4}$ and $\dfrac{1}{2}$.

23. $6z - z^2 = 0$

     $0 = z^2 - 6z$

     $0 = z(z - 6)$

$z = 0$ *or* $z - 6 = 0$

$z = 0$ *or* $z = 6$

The solutions are 0 and 6.

25. $12z^2 + z = 6$

$12z^2 + z - 6 = 0$

$(4z + 3)(3z - 2) = 0$

$4z + 3 = 0$ *or* $3z - 2 = 0$

  $4z = -3$ *or* $3z = 2$

   $z = -\dfrac{3}{4}$ *or* $z = \dfrac{2}{3}$

The solutions are $-\dfrac{3}{4}$ and $\dfrac{2}{3}$.

27. $7x^2 - 7 = 0$

$7(x^2 - 1) = 0$

$7(x + 1)(x - 1) = 0$

$x + 1 = 0$ *or* $x - 1 = 0$

  $x = -1$ *or* $x = 1$

The solutions are -1 and 1.

29. $21r^2 + r - 10 = 0$

$(3r - 2)(7r + 5) = 0$

$3r - 2 = 0$ *or* $7r + 5 = 0$

  $3r = 2$ *or* $7r = -5$

   $r = \dfrac{2}{3}$ *or* $r = -\dfrac{5}{7}$

The solutions are $\dfrac{2}{3}$ and $-\dfrac{5}{7}$.

31. $15y^2 = 3y$

$15y^2 - 3y = 0$

$3y(5y - 1) = 0$

$3y = 0$ *or* $5y - 1 = 0$

  $y = 0$ *or* $5y = 1$

  $y = 0$ *or* $y = \dfrac{1}{5}$

The solutions are 0 and $\dfrac{1}{5}$.

33. $14 = x(x - 5)$

$14 = x^2 - 5x$

$0 = x^2 - 5x - 14$ Getting 0 on one side

$0 = (x - 7)(x + 2)$

$x - 7 = 0$ *or* $x + 2 = 0$

$x = 7$ *or* $x = -2$

The solutions are 7 and -2.

35. $2x^3 - 2x^2 = 12x$

$2x^3 - 2x^2 - 12x = 0$

$2x(x^2 - x - 6) = 0$

$2x(x - 3)(x + 2) = 0$

$2x = 0$ *or* $x - 3 = 0$ *or* $x + 2 = 0$

$x = 0$ *or* $x = 3$ *or* $x = -2$

The solutions are 0, 3, and -2.

37. $2x^3 = 128x$

$2x^3 - 128x = 0$

$2x(x^2 - 64) = 0$

$2x(x + 8)(x - 8) = 0$

$x = 0$ *or* $x + 8 = 0$ *or* $x - 8 = 0$

$x = 0$ *or* $x = -8$ *or* $x = 8$

The solutions are 0, -8, and 8.

39. $t^4 - 26t^2 + 25 = 0$

$(t^2 - 1)(t^2 - 25) = 0$

$(t + 1)(t - 1)(t + 5)(t - 5) = 0$

$t + 1 = 0$ *or* $t - 1 = 0$ *or* $t + 5 = 0$ *or* $t - 5 = 0$

$t = -1$ *or* $t = 1$ *or* $t = -5$ *or* $t = 5$

The solutions are -1, 1, -5, and 5.

41. $(a - 4)(a + 4) = 20$

$a^2 - 16 = 20$

$a^2 - 36 = 0$

$(a + 6)(a - 6) = 0$

$a + 6 = 0$ *or* $a - 6 = 0$

$a = -6$ *or* $a = 6$

The solutions are -6 and 6.

43. $x(5 + 12x) = 28$

$5x + 12x^2 = 28$

$12x^2 + 5x - 28 = 0$

$(4x + 7)(3x - 4) = 0$

$4x + 7 = 0$ *or* $3x - 4 = 0$

$4x = -7$ *or* $3x = 4$

$x = -\dfrac{7}{4}$ *or* $x = \dfrac{4}{3}$

The solutions are $-\dfrac{7}{4}$ and $\dfrac{4}{3}$.

45. We set $f(x)$ equal to 8.

$x^2 + 12x + 40 = 8$

$x^2 + 12x + 32 = 0$

$(x + 8)(x + 4) = 0$

$x + 8 = 0$ *or* $x + 4 = 0$

$x = -8$ *or* $x = -4$

The values of x for which $f(x) = 8$ are -8 and -4.

47. We set $g(x)$ equal to 12.

$2x^2 + 5x = 12$

$2x^2 + 5x - 12 = 0$

$(2x - 3)(x + 4) = 0$

$2x - 3 = 0$ *or* $x + 4 = 0$

$2x = 3$ *or* $x = -4$

$x = \dfrac{3}{2}$ *or* $x = -4$

The values of x for which $g(x) = 12$ are $\dfrac{3}{2}$ and -4.

49. We set $h(x)$ equal to -27.

$12x + x^2 = -27$

$12x + x^2 + 27 = 0$

$x^2 + 12x + 27 = 0$ Rearranging

$(x + 3)(x + 9) = 0$

$x + 3 = 0$ *or* $x + 9 = 0$

$x = -3$ *or* $x = -9$

The values of x for which $h(x) = -27$ are -3 and -9.

51. $f(x) = \dfrac{3}{x^2 - 4x - 5}$

$f(x)$ cannot be calculated for any x-value for which the denominator, $x^2 - 4x - 5$, is 0. To find the excluded values, we solve:

$x^2 - 4x - 5 = 0$

$(x - 5)(x + 1) = 0$

$x - 5 = 0$ *or* $x + 1 = 0$

$x = 5$ *or* $x = -1$

The domain of f is $\{x | x$ is a real number *and* $x \neq 5$ *and* $x \neq -1\}$.

53. $f(x) = \dfrac{x}{6x^2 - 54}$

$f(x)$ cannot be calculated for any x-value for which the denominator, $6x^2 - 54$, is 0. To find the excluded values, we solve:

$6x^2 - 54 = 0$

$6(x^2 - 9) = 0$

$6(x + 3)(x - 3) = 0$

$x + 3 = 0$ *or* $x - 3 = 0$

$x = -3$ *or* $x = 3$

The domain of f is $\{x | x$ is a real number *and* $x \neq -3$ *and* $x \neq 3\}$.

55. $f(x) = \dfrac{x-5}{25x^2 - 10x + 1}$

$f(x)$ cannot be calculated for any x-value for which the denominator, $25x^2 - 10x + 1$, is 0. To find the excluded values, we solve:

$$25x^2 - 10x + 1 = 0$$
$$(5x - 1)(5x - 1) = 0$$
$$5x - 1 = 0 \quad or \quad 5x - 1 = 0$$
$$5x = 1 \quad or \qquad 5x = 1$$
$$x = \frac{1}{5} \quad or \qquad x = \frac{1}{5}$$

The domain of f is $\left\{ x \middle| x \text{ is a real number } and \ x \neq \dfrac{1}{5} \right\}$.

57. $f(x) = \dfrac{7}{5x^3 - 35x^2 + 50x}$

$f(x)$ cannot be calculated for any x-value for which the denominator, $5x^3 - 35x^2 + 50x$, is 0. To find the excluded values, we solve:

$$5x^3 - 35x^2 + 50x = 0$$
$$5x(x^2 - 7x + 10) = 0$$
$$5x(x - 2)(x - 5) = 0$$
$$5x = 0 \quad or \quad x - 2 = 0 \quad or \quad x - 5 = 0$$
$$x = 0 \quad or \qquad x = 2 \quad or \qquad x = 5$$

The domain of f is $\{x|x \text{ is a real number } and$
$x \neq 0 \ and \ x \neq 2 \ and \ x \neq 5\}$.

59. From the graph we see that the x-intercepts are $(-5, 0)$ and $(9, 0)$. The solutions of the equation are the first coordinates of the x-intercepts, -5 and 9.

61. From the graph we see that the x-intercepts are $(-4, 0)$ and $(8, 0)$. The solutions of the equation are the first coordinates of the x-intercepts, -4 and 8.

63. *Familiarize*. Let w = the width of the envelope. Then $w + 4$ = the length. Recall that the area of a rectangle is given by length $\times$ width.

Translate.

$$\underbrace{\text{The area}}_{w(w+4)} \ \underbrace{\text{is}}_{=} \ \underbrace{96 \text{ cm}^2}_{96}.$$

Solve. We solve the equation.

$$w(w + 4) = 96$$
$$w^2 + 4w = 96$$
$$w^2 + 4w - 96 = 0$$
$$(w + 12)(w - 8) = 0$$
$$w + 12 = 0 \quad or \quad w - 8 = 0$$
$$w = -12 \quad or \qquad w = 8$$

Check. The width cannot be negative so we check only 8. If the width is 8 cm, then the length is $8 + 4$, or 12 cm, and the area is $8 \cdot 12$, or 96 cm^2. The answer checks.

State. The length is 12 cm, and the width is 8 cm.

65. *Familiarize*. Using the labels on the drawing in the text, we let x represent the base of the triangle and $x + 2$ represent the height. Recall that the formula for the area of the triangle with base b and height h is $\dfrac{1}{2}bh$.

Translate.

$$\underbrace{\text{The area}}_{\frac{1}{2}x(x+2)} \ \underbrace{\text{is}}_{=} \ \underbrace{12 \text{ ft}^2}_{12}.$$

Solve. We solve the equation:

$$\frac{1}{2}x(x + 2) = 12$$
$$x(x + 2) = 24 \quad \text{Multiplying by 2}$$
$$x^2 + 2x = 24$$
$$x^2 + 2x - 24 = 0$$
$$(x + 6)(x - 4) = 0$$
$$x + 6 = 0 \quad or \quad x - 4 = 0$$
$$x = -6 \quad or \qquad x = 4$$

Check. We check only 4 since the length of the base cannot be negative. If the base is 4 ft, then the height is $4 + 2$, or 6 ft, and the area is $\dfrac{1}{2} \cdot 4 \cdot 6$, or 12 ft^2. The answer checks.

State. The height is 6 ft, and the base is 4 ft.

67. *Familiarize*. We make a drawing. We let l = the length of the flower bed and w = the width.

Recall the formula for the area of a rectangle: $A = lw$.

Translate. We translate to a system of equations.

$$108 = lw, \quad \text{Substituting 108 for } A$$
$$l = w + 3$$

Solve. We will use the system of equations to find an equation in one variable. We substitute $w + 3$ for l in the first equation.

$$108 = (w + 3)w$$
$$108 = w^2 + 3w$$
$$0 = w^2 + 3w - 108$$
$$0 = (w + 12)(w - 9)$$
$$w + 12 = 0 \quad or \quad w - 9 = 0$$
$$w = -12 \quad or \qquad w = 9$$

Check. Width cannot be negative, so -12 cannot be a solution. If the width is 9 and the length is 3 m longer, or 12, then the area will be $12 \cdot 9$, or 108 m^2. We have a solution.

State. The length is 12 m, and the width is 9 m.

69. Familiarize. If $d =$ the distance from the base of the tower to the end of the wire, then $d + 4 =$ the height of the tower.

Translate. We use the Pythagorean theorem.
$$d^2 + (d + 4)^2 = 20^2$$

Solve. We solve the equation.
$$d^2 + (d^2 + 8d + 16) = 400$$
$$2d^2 + 8d + 16 = 400$$
$$2d^2 + 8d - 384 = 0$$
$$2(d + 16)(d - 12) = 0$$
$$d + 16 = 0 \quad or \quad d - 12 = 0$$
$$d = -16 \quad or \quad \quad d = 12$$

Check. The distance cannot be negative, so we check only 12. If $d = 12$, then $d + 4 = 16$ and $12^2 + 16^2 = 400 = 20^2$. The answer checks.

State. The distance d is 12 ft, and the height of the tower is 16 ft.

71. Familiarize. Let x represent the first integer, $x + 2$ the second, and $x + 4$ the third.

Translate.

$$\underbrace{\text{Square of the third}}_{(x+4)^2} \underbrace{\text{is}}_{=} \underbrace{76}_{76} \underbrace{\begin{array}{c}\text{more} \\ \text{than}\end{array}}_{+} \underbrace{\text{square of the second.}}_{(x+2)^2}$$

Solve. We solve the equation:
$$(x + 4)^2 = 76 + (x + 2)^2$$
$$x^2 + 8x + 16 = 76 + x^2 + 4x + 4$$
$$x^2 + 8x + 16 = x^2 + 4x + 80$$
$$4x = 64$$
$$x = 16$$

Check. We check the integers 16, 18, and 20. The square of 20, or 400, is 76 more than 324, the square of 18. The answer checks.

State. The integers are 16, 18, and 20.

73. Familiarize. We make a drawing and label it. We let x represent the length of a side of the original square.

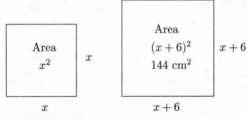

Translate.

$$\underbrace{\text{Area of new square}}_{(x+6)^2} \underbrace{\text{is}}_{=} \underbrace{144 \text{ cm}^2}_{144}.$$

Solve. We solve the equation:
$$(x + 6)^2 = 144$$
$$x^2 + 12x + 36 = 144$$
$$x^2 + 12x - 108 = 0$$
$$(x - 6)(x + 18) = 0$$
$$x - 6 = 0 \quad or \quad x + 18 = 0$$
$$x = 6 \quad or \quad \quad x = -18$$

Check. We only check 6 since the length of a side cannot be negative. If we increase the length by 6, the new length is $6+6$, or 12 cm. Then the new area is $12 \cdot 12$, or 144 cm^2. We have a solution.

State. The length of a side of the original square is 6 cm.

75. Familiarize. Using the labels in the text, we let $x =$ the width of the frame, in centimeters. Then we see that $20 - 2x =$ the length of the picture and $12 - 2x =$ the width of the picture. Recall that the area A of a rectangle with length l and width w is given by $A = lw$.

Translate.

$$\underbrace{\text{The area of the picture}}_{(20 - 2x)(12 - 2x)} \underbrace{\text{is}}_{=} \underbrace{84 \text{ cm}^2}_{84}.$$

Solve. We solve the equation.
$$(20 - 2x)(12 - 2x) = 84$$
$$240 - 64x + 4x^2 = 84$$
$$4x^2 - 64x + 156 = 0$$
$$4(x^2 - 16x + 39) = 0$$
$$x^2 - 16x + 39 = 0$$
$$(x - 13)(x - 3) = 0$$
$$x - 13 = 0 \quad or \quad x - 3 = 0$$
$$x = 13 \quad or \quad \quad x = 3$$

Check. If the width of the frame were 13 cm, then the width of the picture would be $12 - 2 \cdot 13$, or -14 cm. This is not possible, so 13 cannot be a solution. If the width of the frame is 3 cm, then the length and width of the picture are $20 - 2 \cdot 3$, or 14 cm, and $12 - 2 \cdot 3$, or 6 cm, respectively. The area of the picture is 14 cm $\cdot$ 6 cm, or 84 cm^2. The number 3 checks.

State. The frame is 3 cm wide.

77. Familiarize. Let $w =$ the width of the parking lot. Then $w + 50 =$ the length.

Translate. We use the Pythagorean theorem.
$$a^2 + b^2 = c^2$$
$$w^2 + (w + 50)^2 = 250^2$$

Solve. We solve the equation:
$$w^2 + (w + 50)^2 = 250^2$$
$$w^2 + w^2 + 100w + 2500 = 62,500$$
$$2w^2 + 100w + 2500 = 62,500$$
$$2w^2 + 100w - 60,000 = 0$$
$$w^2 + 50w - 30,000 = 0 \qquad \text{Dividing by 2}$$
$$(w - 150)(w + 200) = 0$$

$w - 150 = 0 \quad or \quad w + 200 = 0$

$w = 150 \quad or \qquad w = -200$

Check. Since the width cannot be negative, -200 cannot be a solution. We check 150. When $w = 150$, then $w + 150 = 200$, and $150^2 + 200^2 = 250^2$. The answer checks.

State. The parking lot is 150 ft by 200 ft.

79. Familiarize. Let $x =$ the other leg of the triangle. Then $x + 1 =$ the hypotenuse.

Translate. We use the Pythagorean theorem.

$a^2 + b^2 = c^2$

$9^2 + x^2 = (x + 1)^2$

Solve. We solve the equation:

$9^2 + x^2 = (x + 1)^2$

$81 + x^2 = x^2 + 2x + 1$

$80 = 2x$

$40 = x$

Check. When $x = 40$, then $x + 1 = 41$, and $9^2 + 40^2 = 1681 = 41^2$. The numbers check.

State. The other sides have lengths of 40 m and 41 m.

81. Familiarize. Let $x =$ the length of one leg, in ft. Then $x + 1 =$ the length of the hypotenuse.

Translate. We use the Pythagorean theorem.

$a^2 + b^2 = c^2$

$7^2 + x^2 = (x + 1)^2$

Solve. We solve the equation:

$7^2 + x^2 = (x + 1)^2$

$49 + x^2 = x^2 + 2x + 1$

$48 = 2x$

$24 = x$

Check. When $x = 24$, then $x + 1 = 25$, and $7^2 + 24^2 = 625 = 25^2$. The numbers check.

State. The length of one leg is 24 ft and the length of the hypotenuse is 25 ft.

83. Familiarize. We will use the equation
$h(t) = -16t^2 + 80t + 224$.

Translate.

$$\underbrace{\text{Height}} \quad \text{is} \quad \underbrace{\text{0 ft.}}$$
$$\qquad \downarrow \qquad \downarrow$$
$$-16t^2 + 80t + 224 = \quad 0$$

Solve. We solve the equation:

$-16(t^2 - 5t - 14) = 0$

$-16(t - 7)(t + 2) = 0$

$t - 7 = 0 \quad or \quad t + 2 = 0$

$t = 7 \quad or \qquad t = -2$

Check. The number -2 is not a solution, since time cannot be negative in this application. When $t = 7$, $h(t) = -16 \cdot 7^2 + 80 \cdot 7 + 224 = 0$. We have a solution.

State. The object reaches the ground after 7 sec.

85. Discussion and Writing Exercise

87. $|4 - (-3)| = |7| = 7$, or $|-3 - 4| = |-7| = 7$

89. $|3 - (-4)| = |7| = 7$, or $|-4 - 3| = |-7| = 7$

91. $|3.6 - 4.9| = |-1.3| = 1.3$, or $|4.9 - 3.6| = |1.3| = 1.3$

93. $|-123 - 568| = |-691| = 691$, or
$|568 - (-123)| = |691| = 691$

95. First find the slope.

$$m = \frac{-4 - 7}{-8 - (-2)} = \frac{-11}{-6} = \frac{11}{6}$$

Now substitute $\frac{11}{6}$ for m and the coordinates of one of the given points for x and y in the equation $y = mx + b$ and solve for b. We will use the point $(-2, 7)$.

$y = mx + b$

$7 = \frac{11}{6}(-2) + b$

$7 = -\frac{11}{3} + b$

$\frac{32}{3} = b$

Now use $y = mx + b$ again, substituting $\frac{11}{6}$ for m and $\frac{32}{3}$ for b.

$$y = \frac{11}{6}x + \frac{32}{3}$$

97. First find the slope.

$$m = \frac{4 - 7}{8 - (-2)} = \frac{-3}{10} = -\frac{3}{10}$$

Now substitute $-\frac{3}{10}$ for m and the coordinates of one of the given points for x and y in the equation $y = mx + b$ and solve for b. We will use the point $(-2, 7)$.

$y = mx + b$

$7 = -\frac{3}{10}(-2) + b$

$7 = \frac{3}{5} + b$

$\frac{32}{5} = b$

Now we use $y = mx + b$ again, substituting $-\frac{3}{10}$ for m and $\frac{32}{5}$ for b.

$$y = -\frac{3}{10}x + \frac{32}{5}$$

99.

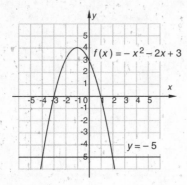

The solutions of $-x^2 - 2x + 3 = 0$ are the first coordinates of the x-intercepts. From the graph we see that these are -3 and 1. The solution set is $\{-3, 1\}$.

To solve $-x^2 - 2x + 3 \geq -5$ we find the x-values for which $f(x) \geq -5$. From the graph we see that these are the values in the interval $[-4, 2]$. The solution set can also be expressed as $\{x| -4 \leq x \leq 2\}$.

101. Left to the student

103. a) One method is to graph $y = x^4 - 3x^3 - x^2 + 5$ and then use the ZERO feature. The solutions are 1.2522305 and 3.1578935.

   b) One method is to graph $y_1 = x^4 - 3x^3 - x^2 + 5$ and $y_2 = 5$. Then use the INTERSECT feature to find the first coordinates of the points of intersection of the graphs. The solutions are -0.3027756, 0, and 3.3027756.

   c) One method is to graph $y_1 = x^4 - 3x^3 - x^2 + 5$ and $y_2 = -8$. Then use the INTERSECT feature to find the first coordinates of the points of intersection of the graphs. The solutions are 2.1387475 and 2.7238657.

   d) One method is to graph $y_1 = x^4$ and $y_2 = 1 + 3x^3 + x^2$. Then use the INTERSECT feature to find the first coordinates of the points of intersection of the graphs. The solutions are -0.7462555 and 3.3276509.

Chapter 4 Review Exercises

1. $3x^6y - 7x^8y^3 + 2x^3 - 3x^2$

   a) The degrees of the terms are $6 + 1$, or 7; $8 + 3$, or 11; 3; and 2. The degree of the polynomial is 11.

   b) The leading term is the term of highest degree, $-7x^8y^3$. The leading coefficient is -7.

   c) $-3x^2 + 2x^3 + 3x^6y - 7x^8y^3$

   d) $-7x^8y^3 + 3x^6y + 2x^3 - 3x^2$, or
   $-7x^8y^3 + 3x^6y - 3x^2 + 2x^3$

2. $P(x) = x^3 - x^2 + 4x$

   $P(0) = 0^3 - 0^2 + 4 \cdot 0 = 0$

   $P(-1) = (-1)^3 - (-1)^2 + 4(-1)$

   $\quad = -1 - 1 - 4$

   $\quad = -6$

3. $P(x) = 4 - 2x - x^2$

   $P(-2) = 4 - 2(-2) - (-2)^2$

   $\quad = 4 + 4 - 4$

   $\quad = 4$

   $P(5) = 4 - 2 \cdot 5 - 5^2$

   $\quad = 4 - 10 - 25$

   $\quad = -31$

4. a) $2006 - 1996 = 10$, so we locate 10 on the t-axis. From there move vertically to the graph and then horizontally to the $E(t)$-axis. This locates a value of about 230, so we predict that there will be about $230{,}000$ visits in 2006.

   b) $2010 - 1996 = 14$, so we find $E(14)$.

   $E(14) = 1.55(14)^2 + 2.71(14) + 47.04 = 388.78 \approx 389$

   Thus, we predict that there will be about $389{,}000$ visits in 2010.

5. $4x^2y - 3xy^2 - 5x^2y + xy^2$

   $= (4 - 5)x^2y + (-3 + 1)xy^2$

   $= -x^2y - 2xy^2$

6. $3ab - 10 + 5ab^2 - 2ab + 7ab^2 + 14$

   $= (3 - 2)ab + (5 + 7)ab^2 + (-10 + 14)$

   $= ab + 12ab^2 + 4$

7. $(-6x^3 - 4x^2 + 3x + 1) + (5x^3 + 2x + 6x^2 + 1)$

   $= (-6 + 5)x^3 + (-4 + 6)x^2 + (3 + 2)x + (1 + 1)$

   $= -x^3 + 2x^2 + 5x + 2$

8. $(4x^3 - 2x^2 - 7x + 5) + (8x^2 - 3x^3 - 9 + 6x)$

   $= (4 - 3)x^3 + (-2 + 8)x^2 + (-7 + 6)x + (5 - 9)$

   $= x^3 + 6x^2 - x - 4$

9. $-9xy^2 - \ xy + \ 6x^2y$

   $\quad 4xy^2 - \ xy - \ 5x^2y$

   $\quad \underline{-3xy^2 + 6xy + 12x^2y}$

   $\quad -8xy^2 + 4xy + 13x^2y$

10. $(3x - 5) - (-6x + 2)$

   $= (3x - 5) + (6x - 2)$

   $= 9x - 7$

11. $(4a - b + 3c) - (6a - 7b - 4c)$

   $= (4a - b + 3c) + (-6a + 7b + 4c)$

   $= -2a + 6b + 7c$

12. $(9p^2 - 4p + 4) - (-7p^2 + 4p + 4)$

   $= (9p^2 - 4p + 4) + (7p^2 - 4p - 4)$

   $= 16p^2 - 8p$

13. $(6x^2 - 4xy + y^2) - (2y^2 + 3xy - 2y^2)$

$= (6x^2 - 4xy + y^2) + (-2y^2 - 3xy + 2y^2)$

$= 6x^2 - 7xy + y^2$

14. $(3x^2y)(-6xy^3) = [3(-6)](x^2 \cdot x)(y \cdot y^3) = -18x^3y^4$

15.
$$x^4 - 2x^2 + 3$$
$$x^4 + x^2 - 1$$
$$\underline{- x^4 + 2x^2 - 3}$$
$$x^6 - 2x^4 + 3x^2$$
$$\underline{x^8 - 2x^6 + 3x^4}$$
$$x^8 - x^6 + 5x^2 - 3$$

16. $(4ab + 3c)(2ab - c)$

$= 8a^2b^2 - 4abc + 6abc - 3c^2$ FOIL

$= 8a^2b^2 + 2abc - 3c^2$

17. $(2x + 5y)(2x - 5y) = (2x)^2 - (5y)^2 = 4x^2 - 25y^2$

18. $(2x - 5y)^2 = (2x)^2 - 2 \cdot 2x \cdot 5y + (5y)^2 = 4x^2 - 20xy + 25y^2$

19.
$$5x^2 - 7x + 3$$
$$\underline{4x^2 + 2x - 9}$$
$$-45x^2 + 63x - 27$$
$$10x^3 - 14x^2 + 6x$$
$$\underline{20x^4 - 28x^3 + 12x^2}$$
$$20x^4 - 18x^3 - 47x^2 + 69x - 27$$

20. $(x^2 + 4y^3)^2 = (x^2)^2 + 2 \cdot x^2 \cdot 4y^3 + (4y^3)^2 = x^4 + 8x^2y^3 + 16y^6$

21. $(x - 5)(x^2 + 5x + 25)$

$= (x - 5)(x^2) + (x - 5)(5x) + (x - 5)(25)$

$= x^3 - 5x^2 + 5x^2 - 25x + 25x - 125$

$= x^3 - 125$

22. $\left(x - \dfrac{1}{3}\right)\left(x - \dfrac{1}{6}\right)$

$= x^2 - \dfrac{1}{6}x - \dfrac{1}{3}x + \dfrac{1}{18}$ FOIL

$= x^2 - \dfrac{1}{2}x + \dfrac{1}{18}$

23. $f(x) = x^2 - 2x - 7$

$f(a - 1) = (a - 1)^2 - 2(a - 1) - 7$

$ = a^2 - 2a + 1 - 2a + 2 - 7$

$ = a^2 - 4a - 4$

$f(a + h) - f(a)$

$= (a + h)^2 - 2(a + h) - 7 - (a^2 - 2a - 7)$

$= a^2 + 2ah + h^2 - 2a - 2h - 7 - a^2 + 2a + 7$

$= 2ah + h^2 - 2h$

24. $9y^4 - 3y^2 = 3y^2 \cdot 3y^2 - 3y^2 \cdot 1$

$ = 3y^2(3y^2 - 1)$

25. $15x^4 - 18x^3 + 21x^2 - 9x = 3x(5x^3 - 6x^2 + 7x - 3)$

26. $a^2 - 12a + 27$

$= (a - 3)(a - 9)$ Trial and error

27. $3m^2 + 14m + 8$

$= (3m + 2)(m + 4)$ FOIL or ac-method

28. $25x^2 + 20x + 4$

$= (5x)^2 + 2 \cdot 5x \cdot 2 + 2^2$ Trinomial square

$= (5x + 2)^2$

29. $4y^2 - 16$

$= 4(y^2 - 4)$ Difference of squares

$= 4(y + 2)(y - 2)$

30. $ax + 2bx - ay - 2by$

$= x(a + 2b) - y(a + 2b)$ Factoring by grouping

$= (x - y)(a + 2b)$

31. $4x^4 + 4x^2 + 20 = 4(x^4 + x^2 + 5)$

32. $27x^3 - 8$

$= (3x)^3 - 2^3$ Difference of cubes

$= (3x - 2)(9x^2 + 6x + 4)$

33. $0.064b^3 - 0.125c^3$

$= (0.4b)^3 - (0.5c)^3$ Difference of cubes

$= (0.4b - 0.5c)(0.16b^2 + 0.2bc + 0.25c^2)$

34. $y^5 - y$

$= y(y^4 - 1)$

$= y(y^2 + 1)(y^2 - 1)$

$= y(y^2 + 1)(y + 1)(y - 1)$

35. $2z^8 - 16z^6$

$= 2z^6(z^2 - 8)$

36. $54x^6y - 2y$

$= 2y(27z^6 - 1)$ Difference of cubes

$= 2y(3x^2 - 1)(9x^4 + 3x^2 + 1)$

37. $1 + a^3$ Sum of cubes

$= (1 + a)(1 - a + a^2)$

38. $36x^2 - 120x + 100$

$= 4(9x^2 - 30x + 25)$ Trinomial square

$= 4(3x - 5)^2$

39. $6t^2 + 17pt + 5p^2$

$= (3t + p)(2t + 5p)$ FOIL or ac-method

40. $x^3 + 2x^2 - 9x - 18$

$= x^2(x + 2) - 9(x + 2)$

$= (x^2 - 9)(x + 2)$

$= (x + 3)(x - 3)(x + 2)$

41. $a^2 - 2ab + b^2 - 4t^2$

$= (a - b)^2 - 4t^2$

$= (a - b + 2t)(a - b + 2t)$

42.
$$x^2 - 20x = -100$$
$$x^2 - 20x + 100 = 0$$
$$(x - 10)(x - 10) = 0$$
$$x - 10 = 0 \quad or \quad x - 10 = 0$$
$$x = 10 \quad or \qquad x = 10$$
The solution is 10.

43.
$$6b^2 - 13b + 6 = 0$$
$$(2b - 3)(3b - 2) = 0$$
$$2b - 3 = 0 \quad or \quad 3b - 2 = 0$$
$$2b = 3 \quad or \qquad 3b = 2$$
$$b = \frac{3}{2} \quad or \qquad b = \frac{2}{3}$$
The solutions are $\frac{3}{2}$ and $\frac{2}{3}$.

44.
$$8y^2 = 14y$$
$$8y^2 - 14y = 0$$
$$2y(4y - 7) = 0$$
$$2y = 0 \quad or \quad 4y - 7 = 0$$
$$y = 0 \quad or \qquad 4y = 7$$
$$y = 0 \quad or \qquad y = \frac{7}{4}$$
The solutions are 0 and $\frac{7}{4}$.

45.
$$r^2 = 16$$
$$r^2 - 16 = 0$$
$$(r + 4)(r - 4) = 0$$
$$r + 4 = 0 \quad or \quad r - 4 = 0$$
$$r = -4 \quad or \qquad r = 4$$
The solutions are -4 and 4.

46. We set $f(x)$ equal to 4.
$$x^2 - 7x - 40 = 4$$
$$x^2 - 7x - 44 = 0$$
$$(x + 4)(x - 11) = 0$$
$$x + 4 = 0 \quad or \quad x - 11 = 0$$
$$x = -4 \quad or \qquad x = 11$$
The values of x for which $f(x) = 4$ are -4 and 11.

47. $f(x) = \dfrac{x - 3}{3x^2 + 19x - 14}$

$f(x)$ cannot be calculated for any x-value for which the denominator, $3x^2 + 19x - 14$, is 0. To find the excluded values, we solve:
$$3x^2 + 19x - 14 = 0$$
$$(3x - 2)(x + 7) = 0$$
$$3x - 2 = 0 \quad or \quad x + 7 = 0$$
$$3x = 2 \quad or \qquad x = -7$$
$$x = \frac{2}{3} \quad or \qquad x = -7$$

The domain of f is $\left\{ x \middle| x \text{ is a real number } and \; x \neq \frac{2}{3} \text{ and } x \neq -7 \right\}$.

48. *Familiarize*. Using the labels on the drawing in the text, we let $w =$ the width of the photograph and $w + 3 =$ the length, in inches. Then the dimensions with the border added are $w + 2 + 2$ and $w + 3 + 2 + 2$, or $w + 4$ and $w + 7$.

***Translate*.** We use the formula for the area of a rectangle, $A = lw$.
$$(w + 7)(w + 4) = 108$$

***Solve*.** We solve the equation.
$$(w + 7)(w + 4) = 108$$
$$w^2 + 11w + 28 = 108$$
$$w^2 + 11w - 80 = 0$$
$$(w + 16)(w - 5) = 0$$
$$w + 16 = 0 \quad or \quad w - 5 = 0$$
$$w = -16 \quad or \qquad w = 5$$

***Check*.** The width cannot be negative, so we check only 5. If the width of the photograph is 5 in., then the length is $5 + 3$, or 8 in. With the border added, the dimensions are $5 + 2 + 2$ and $8 + 2 + 2$, or 9 in. and 12 in. The area is $9 \cdot 12$, or 108 in². The answer checks.

***State*.** The length of the photograph is 8 in., and the width is 5 in.

49. *Familiarize*. Let x, $x + 2$, and $x + 4$ represent the integers.

***Translate*.** The sum of the squares of the integers is 83, so we have
$$x^2 + (x + 2)^2 + (x + 4)^2 = 83.$$

***Solve*.** We solve the equation.
$$x^2 + (x + 2)^2 + (x + 4)^2 = 83$$
$$x^2 + x^2 + 4x + 4 + x^2 + 8x + 16 = 83$$
$$3x^2 + 12x + 20 = 83$$
$$3x^2 + 12x - 63 = 0$$
$$3(x^2 + 4x - 21) = 0$$
$$3(x + 7)(x - 3) = 0$$
$$x + 7 = 0 \quad or \quad x - 3 = 0$$
$$x = -7 \quad or \qquad x = 3$$
If $x = -7$, then $x + 2 = -7 + 2$, or -5, and $x + 4 = -7 + 4 = -3$. If $x = 3$, then $x + 2 = 3 + 2$, or 5, and $x + 4 = 3 + 4$, or 7.

***Check*.** -7, -5, and -3 are consecutive odd integers and $(-7)^2 + (-5)^2 + (-3)^2 = 49 + 25 + 9 = 83$. Also, 3, 5, and 7 are consecutive odd integers and $3^2 + 5^2 + 7^2 = 9 + 25 + 49 = 83$. Both answers check.

***State*.** The integers are -7, -5, and -3 or 3, 5, and 7.

50. *Familiarize*. Let $s =$ the length of a side of the square. Then the area is $s \cdot s$, or s^2.

Translate.

Area is 7 $\underbrace{\text{more}}_{\text{than}}$ six times $\underbrace{\text{the length}}_{\text{of a side.}}$

$\downarrow \quad \downarrow \quad \downarrow \qquad \downarrow \qquad \downarrow \qquad \downarrow$
$s^2 \quad = \quad 7 \quad + \quad 6 \quad \cdot \qquad s$

Solve. We solve the equation.

$$s^2 = 7 + 6s$$
$$s^2 - 6s - 7 = 0$$
$$(s - 7)(s + 1) = 0$$
$$s - 7 = 0 \ \ or \ \ s + 1 = 0$$
$$s = 7 \ \ or \qquad s = -1$$

Check. The length of a side cannot be negative, so we check only 7. The area is 7^2, or 49, and 7 more than six times 7 is $7 + 6 \cdot 7$, or $7 + 42$, or 49. The answer checks.

State. The length of a side of the square is 7.

51. *Discussion and Writing Exercise.* The discussion could include the following points:

a) We can now solve certain polynomial equations.

b) Whereas most linear equations have exactly one solution, polynomial equations can have more than one solution.

c) We used factoring and the principle of zero products to solve polynomial equations.

52. *Discussion and Writing Exercise.*

a) The middle term, $2 \cdot a \cdot 3$, is missing.
$$(a + 3)^2 = a^2 + 6a + 9$$

b) The middle term of the trinomial factor should be $-ab$.
$$a^3 + b^3 = (a + b)(a^2 - ab + b^2)$$

c) The middle term, $-2ab$, is missing and the sign preceding b^2 is incorrect.
$$(a - b)(a - b) = a^2 - 2ab + b^2$$

d) The product of the outside terms and the product of the inside terms are missing.
$$(x + 3)(x - 4) = x^2 - x - 12$$

e) There should be a minus sign between the terms of the product.
$$(p + 7)(p - 7) = p^2 - 49$$

f) The middle term, $-2 \cdot t \cdot 3$, is missing and the sign preceding 9 is incorrect.
$$(t - 3)^2 = t^2 - 6t + 9$$

53. $\quad 128x^6 - 2y^6$
$$= 2(64x^6 - y^6) \qquad \text{Difference of squares}$$
$$= 2(8x^3 + y^3)(8x^3 - y^3) \quad \text{Sum of cubes and}$$
$$\qquad\qquad\qquad\qquad\qquad \text{difference of cubes}$$
$$= 2(2x + y)(4x^2 - 2xy + y^2)(2x - y)(4x^2 + 2xy + y^2)$$

54. $\quad (x + 1)^3 - (x - 1)^3 \qquad \text{Difference of cubes}$
$$= [(x + 1) - (x - 1)][(x + 1)^2 + (x + 1)(x - 1) + (x - 1)^2]$$
$$= (x + 1 - x + 1)(x^2 + 2x + 1 + x^2 - 1 + x^2 - 2x + 1)$$
$$= 2(3x^2 + 1)$$

55. $[a - (b - 1)][(b - 1)^2 + a(b - 1) + a^2] =$
$$[a - (b - 1)][a^2 + a(b - 1) + (b - 1)^2]$$

This product is of the form $(A - B)(A^2 + AB + B^2)$ where $A = a$ and $B = b - 1$. We know that this product is the factorization of $A^3 - B^2$, so we have $a^3 - (b - 1)^3$.

56. $\qquad\qquad 64x^3 = x$
$$64x^3 - x = 0$$
$$x(64x^2 - 1) = 0$$
$$x(8x + 1)(8x - 1) = 0$$
$$x = 0 \ \ or \ \ 8x + 1 = 0 \quad or \ \ 8x - 1 = 0$$
$$x = 0 \ \ or \qquad 8x = -1 \ \ or \qquad 8x = 1$$
$$x = 0 \ \ or \qquad x = -\frac{1}{8} \ \ or \qquad x = \frac{1}{8}$$

The solutions are 0, $-\frac{1}{8}$, and $\frac{1}{8}$.

Chapter 4 Test

1. $3xy^3 - 4x^2y + 5x^5y^4 - 2x^4y$

a), b)

Term	$3xy^3$	$-4x^2y$	$5x^5y^4$	$-2x^4y$
Degree	4	3	9	5
Degree of polynomial	9			
Leading term	$5x^5y^4$			
Leading coefficient	5			

c) $3xy^3 - 4x^2y - 2x^4y + 5x^5y^4$

d) $5x^5y^4 + 3xy^3 - 4x^2y - 2x^4y$ or
$5x^5y^4 + 3xy^3 - 2x^4y - 4x^2y$

2. $P(x) = 2x^3 + 3x^2 - x + 4$
$$P(0) = 2 \cdot 0^3 + 3 \cdot 0^2 - 0 + 4$$
$$= 0 + 0 - 0 + 4$$
$$= 0$$
$$P(-2) = 2(-2)^3 + 3(-2)^2 - (-2) + 4$$
$$= 2(-8) + 3(4) - (-2) + 4$$
$$= -16 + 12 + 2 + 4$$
$$= 2$$

3. a) Locate 6 on the horizontal axis. $(2010 - 2004 = 6)$ From there move vertically to the graph and then horizontally to the $S(t)$-axis. This locates a value of about 1.66. Thus, we estimate that sales of video games will be about \$1.66 billion in 2010.

b) In 2009, $t = 2009 - 2004 = 5$.
$$S(5) = 0.0496 \cdot 5^4 - 0.6705 \cdot 5^3 + 2.6367 \cdot 5^2 - 2.3880 \cdot 5 + 1.6123 \approx 2.8$$

We estimate that sales of video games will be about \$2.8 billion in 2009.

4. $\quad 5xy - 2xy^2 - 2xy + 5xy^2$
$= (5-2)xy + (-2+5)xy^2$
$= 3xy + 3xy^2$

5. $\quad (-6x^3 + 3x^2 - 4y) + (3x^3 - 2y - 7y^2)$
$= (-6+3)x^3 + 3x^2 + (-4-2)y - 7y^2$
$= -3x^3 + 3x^2 - 6y - 7y^2$

6. $\quad (4a^3 - 2a^2 + 6a - 5) + (3a^3 - 3a + 2 - 4a^2)$
$= (4+3)a^3 + (-2-4)a^2 + (6-3)a + (-5+2)$
$= 7a^3 - 6a^2 + 3a - 3$

7. $\quad (5m^3 - 4m^2n - 6mn^2 - 3n^3) +$
$\qquad (9mn^2 - 4n^3 + 2m^3 + 6m^2n)$
$= (5+2)m^3 + (-4+6)m^2n + (-6+9)mn^2 + (-3-4)n^3$
$= 7m^3 + 2m^2n + 3mn^2 - 7n^3$

8. $\quad (9a - 4b) - (3a + 4b) = (9a - 4b) + (-3a - 4b)$
$\qquad\qquad\qquad\qquad = 6a - 8b$

9. $\quad (4x^2 - 3x + 7) - (-3x^2 + 4x - 6)$
$= (4x^2 - 3x + 7) + (3x^2 - 4x + 6)$
$= 7x^2 - 7x + 13$

10. $\quad (6y^2 - 2y - 5y^3) - (4y^2 - 7y - 6y^3)$
$= (6y^2 - 2y - 5y^3) + (-4y^2 + 7y + 6y^3)$
$= 2y^2 + 5y + y^3$

11. $(-4x^2y)(-16xy^2) = [-4(-16)](x^2 \cdot x)(y \cdot y^2) = 64x^3y^3$

12. $\quad (6a - 5b)(2a + b)$
$= 12a^2 + 6ab - 10ab - 5b^2 \qquad$ FOIL
$= 12a^2 - 4ab - 5b^2$

13. $\quad (x - y)(x^2 - xy - y^2)$
$= (x-y)(x^2) - (x-y)(xy) - (x-y)(y^2)$
$= x \cdot x^2 - y \cdot x^2 - x \cdot xy - (-y)(xy) - x \cdot y^2 - (-y)(y^2)$
$= x^3 - x^2y - x^2y + xy^2 - xy^2 + y^3$
$= x^3 - 2x^2y + y^3$

14.
$$
\begin{array}{r}
3m^2 + 4m - 2 \\
\underline{-\ \ m^2 - 3m + 5} \\
15m^2 + 20m - 10 \\
-\ 9m^3 - 12m^2 + 6m \\
\underline{-3m^4 - 4m^3 + 2m^2 \qquad} \\
-3m^4 - 13m^3 + 5m^2 + 26m - 10
\end{array}
$$

15. $\quad (4y - 9)^2$
$= (4y)^2 - 2 \cdot 4y \cdot 9 + 9^2$
$\qquad\qquad\qquad (A - B)^2 = A^2 - 2AB + B^2$
$= 16y^2 - 72y + 81$

16. $\quad (x - 2y)(x + 2y)$
$= x^2 - (2y)^2 \quad (A + B)(A - B) = A^2 - B^2$
$= x^2 - 4y^2$

17. $f(x) = x^2 - 5x$
$f(a + 10) = (a + 10)^2 - 5(a + 10)$
$\qquad\qquad = a^2 + 20a + 100 - 5a - 50$
$\qquad\qquad = a^2 + 15a + 50$
$f(a + h) - f(a) = (a + h)^2 - 5(a + h) - (a^2 - 5a)$
$\qquad\qquad = a^2 + 2ah + h^2 - 5a - 5h - a^2 + 5a$
$\qquad\qquad = 2ah + h^2 - 5h$

18. $9x^2 + 7x = x \cdot 9x + x \cdot 7 = x(9x + 7)$

19. $24y^3 + 16y^2 = 8y^2 \cdot 3y + 8y^2 \cdot 2 = 8y^2(3y + 2)$

20. $\quad y^3 + 5y^2 - 4y - 20 = y^2(y + 5) - 4(y + 5)$
$\qquad\qquad\qquad\qquad = (y^2 - 4)(y + 5)$
$\qquad\qquad\qquad\qquad = (y + 2)(y - 2)(y + 5)$

21. $p^2 - 12p - 28$

We look for a pair of factors of -28 whose sum is -12. The numbers we need are -14 and 2.

$p^2 - 12p - 28 = (p - 14)(p + 2)$

22. $12m^2 + 20m + 3$

We will use the FOIL method.

1) There are no common factors (other than 1 or -1).

2) Factor the first term, $12m^2$. The possibilities are $(12m+\quad)(m+\quad)$ and $(6m+\quad)(2m+\quad)$ and $(4m+\quad)(3m+\quad)$.

3) Factor the last term, 3. We need to consider only positive factors because both the middle term and the last term are positive. The factors are 3 and 1.

4) Look for factors in steps (2) and (3) such that the sum of the products is the middle term, $20m$. Trial and error leads us to the correct factorization: $(6m + 1)(2m + 3)$.

23. $9y^2 - 25 = (3y)^2 - 5^2 = (3y + 5)(3y - 5)$

24. $\quad 3r^3 - 3 = 3(r^3 - 1)$
$\qquad\qquad = 3(r - 1)(r^2 + r + 1)$
$\qquad\qquad\quad A^3 - B^3 = (A - B)(A^2 + AB + B^2)$

25. $9x^2 + 25 - 30x = 9x^2 - 30x + 25$
$\qquad\qquad\qquad = (3x)^2 - 2 \cdot 3x \cdot 5 + 5^2$
$\qquad\qquad\qquad = (3x - 5)^2$

26. $(z + 1)^2 - b^2 = (z + 1 + b)(z + 1 - b)$

27. $\quad x^8 - y^8 = (x^4)^2 - (y^4)^2$
$\qquad\qquad = (x^4 + y^4)(x^4 - y^4)$
$\qquad\qquad = (x^4 + y^4)[(x^2)^2 - (y^2)^2]$
$\qquad\qquad = (x^4 + y^4)(x^2 + y^2)(x^2 - y^2)$
$\qquad\qquad = (x^4 + y^4)(x^2 + y^2)(x + y)(x - y)$

28. $\quad y^2 + 8y + 16 - 100t^2$
$= (y + 4)^2 - (10t)^2$
$= (y + 4 + 10t)(y + 4 - 10t)$

29. $20a^2 - 5b^2 = 5(4a^2 - b^2) = 5(2a + b)(2a - b)$

30. $24x^2 - 46x + 10$

We will use the *ac*-method.

1) We factor out the common factor, 2.
$$2(12x^2 - 23x + 5)$$

2) Now we factor the trinomial $12x^2 - 23x + 5$. Multiply the leading coefficient, 12, and the constant, 5.
$$12 \cdot 5 = 60$$

3) Look for a factorization of 60 in which the sum of the factors is the coefficient of the middle term, -23. The factors we need are -20 and -3.

4) Split the middle term as follows:
$$-23x = -20x - 3x$$

5) Factor by grouping.
$$12x^2 - 23x + 5 = 12x^2 - 20x - 3x + 5$$
$$= 4x(3x - 5) - (3x - 5)$$
$$= (4x - 1)(3x - 5)$$

We must include the common factor to get a factorization of the original trinomial.
$$24x^2 - 46x + 10 = 2(4x - 1)(3x - 5)$$

31. $16a^7b + 54ab^7$
$$= 2ab(8a^6 + 27b^6)$$
$$= 2ab[(2a^2)^3 + (3b^2)^3]$$
$$= 2ab(2a^2 + 3b^2)[(2a^2)^2 - 2a^2 \cdot 3b^2 + (3b^2)^2]$$
$$A^3 + B^3 = (A + B)(A^2 - AB + B^2)$$
$$= 2ab(2a^2 + 3b^2)(4a^4 - 6a^2b^2 + 9b^4)$$

32. $x^2 - 18 = 3x$
$$x^2 - 3x - 18 = 0$$
$$(x - 6)(x + 3) = 0$$
$$x - 6 = 0 \quad or \quad x + 3 = 0$$
$$x = 6 \quad or \qquad x = -3$$
The solutions are 6 and -3.

33. $5y^2 - 125 = 0$
$$5(y^2 - 25) = 0$$
$$5(y + 5)(y - 5) = 0$$
$$y + 5 = 0 \quad or \quad y - 5 = 0$$
$$y = -5 \quad or \quad y = 5$$
The solutions are -5 and 5.

34. $2x^2 + 21 = -17x$
$$2x^2 + 17x + 21 = 0$$
$$(2x + 3)(x + 7) = 0$$
$$2x + 3 = 0 \quad or \quad x + 7 = 0$$
$$2x = -3 \quad or \qquad x = -7$$
$$y = -\frac{3}{2} \quad or \qquad x = -7$$
The solutions are $-\frac{3}{2}$ and -7.

35. We set $f(x)$ equal to 11.
$$3x^2 - 15x + 11 = 11$$
$$3x^2 - 15x = 0$$
$$3x(x - 5) = 0$$
$$3x = 0 \quad or \quad x - 5 = 0$$
$$x = 0 \quad or \qquad x = 5$$
The values of x for which $f(x) = 11$ are 0 and 5.

36. $f(x) = \dfrac{3 - x}{x^2 + 2x + 1}$

$f(x)$ cannot be calculated for any x-value for which the denominator is 0. To find the excluded values, we solve:
$$x^2 + 2x + 1 = 0$$
$$(x + 1)(x + 1) = 0$$
$$x + 1 = 0 \quad or \quad x + 1 = 0$$
$$x = -1 \quad or \qquad x = -1$$
The domain of f is $\{x | x$ is a real number *and* $x \neq -1\}$, or $(-\infty, -1) \cup (-1, \infty)$.

37. *Familiarize.* Let $w =$ the width, in cm. Then $w + 3 =$ the length.

Translate. We use the formula for the area of a rectangle, $A = l \cdot w$.
$$40 = (w + 3)w$$

Solve. We solve the equation.
$$40 = (w + 3)w$$
$$40 = w^2 + 3w$$
$$0 = w^2 + 3w - 40$$
$$0 = (w + 8)(w - 5)$$
$$w + 8 = 0 \quad or \quad w - 5 = 0$$
$$w = -8 \quad or \qquad w = 5$$

Check. The width cannot be negative, so we check only 5. When $w = 5$, then $w + 3 = 5 + 3 = 8$. If the length is 8 cm and the width is 5 cm, then the length is 3 cm more than the width and the area is $8 \cdot 5$, or 40 cm^2. The answer checks.

State. The length is 8 cm, and the width is 5 cm.

38. *Familiarize.* Let $d =$ the distance the ladder reaches up the wall, in feet. Then $d + 2 =$ the length of the ladder. We make a drawing.

Translate. We use the Pythagorean theorem.
$$a^2 + b^2 = c^2$$
$$10^2 + d^2 = (d + 2)^2$$

Solve. We solve the equation.

$$10^2 + d^2 = (d+2)^2$$
$$100 + d^2 = d^2 + 4d + 4$$
$$100 = 4d + 4$$
$$96 = 4d$$
$$24 = d$$

If $d = 24$, then $d + 2 = 24 + 2 = 26$.

Check. 26 ft is 2 ft more than 24 ft. Also, $10^2 + 24^2 = 100 + 576 = 676 = 26^2$, so the answer checks.

State. The ladder reaches 24 ft up the wall.

39. *Familiarize*. Let $s =$ the length of a side of the square.

Translate.

Area	is	5	more than	4	times	length.
↓	↓	↓	↓	↓	↓	↓
s^2	$=$	5	$+$	4	$\cdot$	s

Solve. We solve the equation.

$$s^2 = 5 + 4 \cdot s$$
$$s^2 - 4s - 5 = 0$$
$$(s-5)(s+1) = 0$$
$$s - 5 = 0 \ \ or \ \ s + 1 = 0$$
$$s = 5 \ \ or \ \ \ \ \ \ \ s = -1$$

Check. Since the length of a side cannot be negative, we check only 5. The area of a square with sides of length 5 is 5^2, or 25, and 4 times the length of a side is $4 \cdot 5$, or 20. Since 25 is 5 more than 20, the answer checks.

State. The length of a side of the square is 5.

40. $f(n) = \dfrac{1}{2}n^2 - \dfrac{1}{2}n$

$$f(n) = \frac{1}{2}n \cdot n - \frac{1}{2}n \cdot 1$$

$$f(n) = \frac{1}{2}n(n-1)$$

41. $6x^{2n} - 7x^n - 20 = 6(x^n)^2 - 7x^n - 20 = (3x^n + 4)(2x^n - 5)$

42. $(p+q)^2 = p^2 + 2pq + q^2$

$\ \ \ \ \ \ 29 = p^2 + 2 \cdot 5 + q^2$ Substituting 29 for $(p+q)^2$

$\ $ and 5 for pq

$\ \ \ \ \ \ 29 = p^2 + 10 + q^2$

$\ \ \ \ \ \ 19 = p^2 + q^2$

Cumulative Review Chapters R - 4

1. 4.000.

  └──↑

   3 places

Positive exponent, so the number is large.

$4 \times 10^3 = 4000$

4. 0,000.

 ↑──┘

 4 places

Large number, so the exponent is positive.

$40,000 = 4 \times 10^4$

3. 00000.

 └──↑

 5 places

Positive exponent, so the number is large.

$3 \times 10^5 = 300,000$

5. 77,000.

 ↑──┘

 5 places

Large number, so the exponent is positive.

$577,000 = 5.77 \times 10^5$

2. $\dfrac{2m-n}{4} = \dfrac{2 \cdot 3 - 2}{4} = \dfrac{6-2}{4} = \dfrac{4}{4} = 1$

3. The distance of 0 from 0 is 0, so $|0| = 0$.

4. $-8 - (-4) = -8 + 4 = -4$

5. $\ \ \ \ \ -3a + (6a - 1) - 4(a+1)$

$\ \ = -3a + 6a - 1 - 4a - 4$

$\ \ = -a - 5$

6. $\ \ \ \ \ 2[5(3x+4) - 2x] - [7(3x+4) - 8]$

$\ \ = 2[15x + 20 - 2x] - [21x + 28 - 8]$

$\ \ = 2[13x + 20] - [21x + 20]$

$\ \ = 26x + 40 - 21x - 20$

$\ \ = 5x + 20$

7. $[(-3a^{-6}b^2)^5]^{-2} = (-3a^{-6}b^2)^{5(-2)}$

$\ = (-3a^{-6}b^2)^{-10}$

$\ = (-3)^{-10}(a^{-6})^{-10}(b^2)^{-10}$

$\ = (-3)^{-10}a^{60}b^{-20}$

$\ = \dfrac{a^{60}}{(-3)^{-10}b^{20}}$

8. $\ \ \ \ \ (x^2 + 4x - xy - 9) + (-3x^2 - 3x + 8)$

$\ \ = (1-3)x^2 + (4-3)x - xy + (-9+8)$

$\ \ = -2x^2 + x - xy - 1$

9. $\ \ \ \ \ (6x^2 - 3x + 2x^3) - (8x^2 - 9x + 2x^3)$

$\ \ = (6x^2 - 3x + 2x^3) + (-8x^2 + 9x - 2x^3)$

$\ \ = (6-8)x^2 + (-3+9)x + (2-2)x^3$

$\ \ = -2x^2 + 6x$

10.

$$\begin{array}{r}
a^2 - a - 3 \\
a^2 + 2a - 3 \\
\hline
-3a^2 + 3a + 9 \\
2a^3 - 2a^2 - 6a \\
a^4 - a^3 - 3a^2 \\
\hline
a^4 + a^3 - 8a^2 - 3a + 9
\end{array}$$

11. $(x + 4)(x + 9)$

$= x^2 + 9x + 4x + 36$ FOIL

$= x^2 + 13x + 36$

12. $8 - 3x = 6x - 10$

$8 - 9x = -10$ Subtracting $6x$

$-9x = -18$ Subtracting 8

$x = 2$ Dividing by -9

The solution is 2.

13. $\dfrac{1}{2}x - 3 = \dfrac{7}{2}$

$2\left(\dfrac{1}{2}x - 3\right) = 2 \cdot \dfrac{7}{2}$ Clearing fractions

$x - 6 = 7$

$x = 13$

The solution is 13.

14. $A = \dfrac{1}{2}h(a + b)$

$\dfrac{2A}{h} = a + b$ Multiplying by $\dfrac{2}{h}$

$\dfrac{2A}{h} - a = b$, or

$\dfrac{2A - ah}{h} = b$

15. $6x - 1 \le 3(5x + 2)$

$6x - 1 \le 15x + 6$

$-9x - 1 \le 6$

$-9x \le 7$

$x \ge -\dfrac{7}{9}$ Dividing by -9 and reversing the inequality symbol

The solution set is $\left\{x \middle| x \ge -\dfrac{7}{9}\right\}$, or $\left[-\dfrac{7}{9}, \infty\right)$.

16. $4x - 3 < 2$ *or* $x - 3 > 1$

$4x < 5$ *or* $x > 4$

$x < \dfrac{5}{4}$ *or* $x > 4$

The solution set is $\left\{x \middle| x < \dfrac{5}{4} \ or \ x > 4\right\}$, or

$\left(-\infty, \dfrac{5}{4}\right) \cup (4, \infty)$.

17. $|2x - 3| < 7$

$-7 < 2x - 3 < 7$

$-4 < 2x < 10$

$-2 < x < 5$

The solution set is $\{x| -2 < x < 5\}$, or $(-2, 5)$.

18. $x + y + z = -5$,   (1)

$x - z = 10$,   (2)

$y - z = 12$ (3)

Since Equation (2) does not have a y-term, we eliminate y from a pair of equations.

$\begin{array}{l} x + y + z = -5 \quad (1) \\ \underline{ - y + z = -12} \quad \text{Multiplying (3) by } -1 \\ x \phantom{{}+ y} + 2z = -17 \quad (5) \end{array}$

Now we solve the system of Equations (2) and (5).

$x - z = 10$ (2)

$x + 2z = -17$ (5)

We multiply Equation (2) by 2 and then add.

$\begin{array}{l} 2x - 2z = 20 \\ \underline{x + 2z = -17} \\ 3x \phantom{{}+ 2z} = 3 \\ x = 1 \end{array}$

$1 - z = 10$ Substituting in (2)

$-z = 9$

$z = -9$

$1 + y - 9 = -5$ Substituting in (1)

$y - 8 = -5$

$y = 3$

The solution is $(1, 3, -9)$.

19. $2x + 5y = -2$,   (1)

$5x + 3y = 14$ (2)

We multiply Equation (1) by 3 and Equation (2) by -5 and then add.

$\begin{array}{l} 6x + 15y = -6 \\ \underline{-25x - 15y = -70} \\ -19x \phantom{{}+ 15y} = -76 \\ x = 4 \end{array}$

$2 \cdot 4 + 5y = -2$ Substituting in (1)

$8 + 5y = -2$

$5y = -10$

$y = -2$

The solution is $(4, -2)$.

20. $3x - y = 7$,   (1)

$2x + 2y = 5$ (2)

We multiply Equation (1) by 2 and then add.

$\begin{array}{l} 6x - 2y = 14 \\ \underline{2x + 2y = 5} \\ 8x \phantom{{}- 2y} = 19 \\ x = \dfrac{19}{8} \end{array}$

$2 \cdot \dfrac{19}{8} + 2y = 5$ Substituting in (2)

$\dfrac{19}{4} + 2y = 5$

$2y = \dfrac{1}{4}$

$y = \dfrac{1}{8}$

The solution is $\left(\dfrac{19}{8}, \dfrac{1}{8}\right)$.

21. $x + 2y - \ z = \ 0,$ (1)

$3x + \ y - 2z = -1,$ (2)

$x - 4y + \ z = -2$ (3)

First we eliminate z from two different pairs of equations.

$-2x - 4y + 2z = 0$ Multiplying (1) by -2

$\underline{\ 3x + \ y - 2z = -1}$ (2)

$x - 3y \qquad\quad = -1$ (4)

$x + 2y - z = 0$ (1)

$\underline{x - 4y + z = -2}$ (3)

$2x - 2y \qquad = -2$ (5)

Now we solve the system of equations (4) and (5). We multiply Equation (4) by -2 and then add.

$-2x + 6y = 2$

$\underline{\ 2x - 2y = -2}$

$4y = 0$

$y = 0$

$x - 3 \cdot 0 = -1$ Substituting in (4)

$x = -1$

$-1 - 4 \cdot 0 + z = -2$ Substituting in (1)

$-1 + z = -2$

$z = -1$

The solution is $(-1, 0, -1)$.

22. $2x + 3y = \ 2,$ (1)

$-4x + 3y = -7$ (2)

We multiply Equation (2) by -1 and then add.

$2x + 3y = 2$

$\underline{4x - 3y = 7}$

$6x \qquad = 9$

$x = \dfrac{3}{2}$

$2 \cdot \dfrac{3}{2} + 3y = 2$ Substituting in (1)

$3 + 3y = 2$

$3y = -1$

$y = -\dfrac{1}{3}$

The solution is $\left(\dfrac{3}{2}, -\dfrac{1}{3}\right)$.

23. $-3a + 2b = 0,$ (1)

$\underline{\ 3a - 4b = -1}$ (2)

$-2b = -1$ Adding

$b = \dfrac{1}{2}$

$3a - 4 \cdot \dfrac{1}{2} = -1$ Substituting in (2)

$3a - 2 = -1$

$3a = 1$

$a = \dfrac{1}{3}$

The solution is $\left(\dfrac{1}{3}, \dfrac{1}{2}\right)$.

24. $2x + 2y - 4z = \ 1,$ (1)

$-2x - 4y + 8z = -1,$ (2)

$4x + 4y + 4z = \ 5$ (3)

First we add Equations (1) and (2).

$2x + 2y - 4z = 1$

$\underline{-2x - 4y + 8z = -1}$

$-2y + 4z = 0$ (4)

Next we multiply Equation (2) by 2 and then add the resulting equation to Equation (3).

$-4x - 8y + 16z = -2$

$\underline{\ 4x + 4y + \ 4z = 5}$

$-4y + 20z = 3$ (5)

Now we solve the system of Equations (4) and (5). We multiply Equation (4) by -2 and then add.

$4y - \ 8z = 0$

$\underline{-4y + 20z = 3}$

$12z = 3$

$z = \dfrac{1}{4}$

$-2y + 4 \cdot \dfrac{1}{4} = 0$ Substituting in (4)

$-2y + 1 = 0$

$-2y = -1$

$y = \dfrac{1}{2}$

$2x + 2 \cdot \dfrac{1}{2} - 4 \cdot \dfrac{1}{4} = 1$

$2x + 1 - 1 = 1$

$2x = 1$

$x = \dfrac{1}{2}$

The solution is $\left(\dfrac{1}{2}, \dfrac{1}{2}, \dfrac{1}{4}\right)$.

25. $11x + x^2 + 24 = 0$

$x^2 + 11x + 24 = 0$ Rearranging

$(x + 3)(x + 8) = 0$

$x + 3 = 0$ *or* $x + 8 = 0$

$x = -3$ *or* $x = -8$

The solutions are -3 and -8.

26. $2x^2 - 15x = -7$

$2x^2 - 15x + 7 = 0$

$(2x - 1)(x - 7) = 0$

$2x - 1 = 0$ *or* $x - 7 = 0$

$2x = 1$ *or* $x = 7$

$x = \dfrac{1}{2}$ *or* $x = 7$

The solutions are $\dfrac{1}{2}$ and 7.

27. We set $f(x)$ equal to 4.

$3x^2 + 4x = 4$

$3x^2 + 4x - 4 = 0$

$(3x - 2)(x + 2) = 0$

$3x + 2 = 0$ *or* $x + 2 = 0$

$3x = -2$ *or* $x = -2$

$x = -\dfrac{2}{3}$ *or* $x = -2$

The values of x for which $f(x) = 4$ are $-\dfrac{2}{3}$ and -2.

28. $F(x) = \dfrac{x + 7}{x^2 - 2x - 15}$

The values of x excluded from the domain are those for which the denominator is 0. We find those values.

$x^2 - 2x - 15 = 0$

$(x - 5)(x + 3) = 0$

$x - 5 = 0$ *or* $x + 3 = 0$

$x = 5$ *or* $x = -3$

The domain of F is $\{x | x$ is a real number *and* $x \neq 5$ *and* $x \neq -3\}$.

29. $3x^3 - 12x^2 = 3x^2 \cdot x - 3x^2 \cdot 4 = 3x^2(x - 4)$

30. $2x^4 + x^3 + 2x + 1$

$= x^3(2x + 1) + (2x + 1)$ Factoring by grouping

$= (x^3 + 1)(2x + 1)$ Sum of cubes

$= (x + 1)(x^2 - x + 1)(2x + 1)$

31. $x^2 + 5x - 14$

$= (x + 7)(x - 2)$ Trial and error

32. $20a^2 - 23a + 6$

$= (4a - 3)(5a - 2)$ FOIL or *ac*-method

33. $4x^2 - 25$ Difference of squares

$= (2x + 5)(2x - 5)$

34. $2x^2 - 28x + 98$

$= 2(x^2 - 14x + 49)$ Trinomial square

$= 2(x - 7)^2$

35. $a^3 + 64$ Sum of cubes

$= (a + 4)(a^2 - 4a + 16)$

36. $8x^3 - 1$ Difference of cubes

$= (2x - 1)(4x^2 + 2x + 1)$

37. $4a^3 + a^6 - 12$

$= a^6 + 4a^3 - 12$ Rearranging

$= (a^3 + 6)(a^3 - 2)$ Trial and error

38. $4x^4y^2 - x^2y^4$

$= x^2y^2(4x^2 - y^2)$

$= x^2y^2(2x + y)(2x - y)$

39. $x < 1$ *or* $x \geq 2$

We shade all points to the left of 1 and use a parentheses at 1 to show that it is not a solution. We also shade all points to the right of 2 and use a bracket at 2 to show that 2 is a solution.

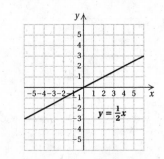

40. $y = -2x$

We find some ordered pairs that are solutions and draw and label the line.

When $x = -2$, $y = -2(-2) = 4$.

When $x = 0$, $y = -2 \cdot 0 = 0$.

When $x = 2$, $y = -2 \cdot 2 = -4$.

x	y
-2	4
0	0
2	-4

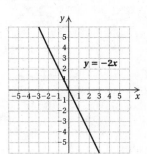

41. $y = \dfrac{1}{2}x$

We find some ordered pairs that are solutions and draw and label the line.

When $x = -4$, $y = \dfrac{1}{2}(-4) = -2$.

When $x = 0$, $y = \dfrac{1}{2} \cdot 0 = 0$.

When $x = 2$, $y = \dfrac{1}{2} \cdot 2 = 1$.

x	y
-4	-2
0	0
2	1

42. $4y + 3x = 12 + 3x$

$\qquad 4y = 12 \qquad$ Subtracting $3x$

$\qquad y = 3$

Since x is missing, all ordered pairs $(x, 3)$ are solutions. The graph is parallel to the x-axis.

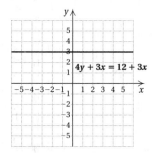

43. $6y + 24 = 0$

$\qquad 6y = -24$

$\qquad y = -4$

Since x is missing, all ordered pairs $(x, -4)$ are solutions. The graph is parallel to the x-axis.

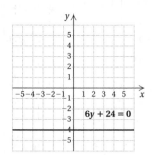

44. $y > x + 6$

First graph $y = x + 6$. Draw the line dashed since the inequality symbol is $>$. Test the point $(0, 0)$ to determine if it is a solution.

$$\frac{y > x + 6}{0 \ ? \ 0 + 6}$$

$\qquad \Big| \ 6 \qquad$ FALSE

Since $0 > 6$ is false, we shade the half-plane that does not contain $(0, 0)$.

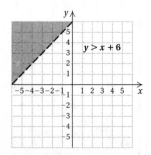

45. $2x + y \leq 2$

First graph $2x + y = 2$. Draw the line solid since the inequality symbol is $\leq$. Test the point $(0, 0)$ to determine if it is a solution.

$$\frac{2x + y \leq 2}{2 \cdot 0 + 0 \ ? \ 2}$$

$\qquad 0 \ \Big| \qquad$ TRUE

Since $0 \leq 2$ is true, we shade the half-plane that contains $(0, 0)$.

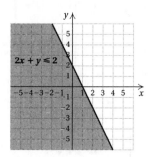

46. $f(x) = x^2 - 3$

Make a list of function values in a table.

$f(-2) = (-2)^2 - 3 = 4 - 3 = 1$

$f(-1) = (-1)^2 - 3 = 1 - 3 = -2$

$f(0) = 0^2 - 3 = -3$

$f(1) = 1^2 - 3 = 1 - 3 = -2$

$f(2) = 2^2 - 3 = 4 - 3 = 1$

x	$f(x)$
-2	1
-1	-2
0	-3
1	-2
2	1

Plot these points and connect them.

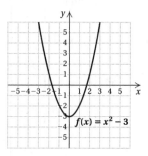

47. $g(x) = 4 - |x|$

Make a list of function values in a table.

$g(-5) = 4 - |-5| = 4 - 5 = -1$

$g(-3) = 4 - |-3| = 4 - 3 = 1$

$g(-1) = 4 - |-1| = 4 - 1 = 3$

$g(0) = 4 - |0| = 4 - 0 = 4$

$g(2) = 4 - |2| = 4 - 2 = 2$

$g(4) = 4 - |4| = 4 - 4 = 0$

x	$g(x)$
-5	-1
-3	1
-1	3
0	4
2	2
4	0

Plot these points and connect them.

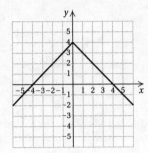

48. $y \geq -x$,

$y \leq 2x + 1$

Graph the lines $y = -x$ and $y = 2x+1$ using solid lines. Indicate the region for each inequality by arrows, and shade the region where they overlap.

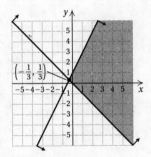

To find the vertex we solve the system of related equations:

$y = -x$,

$y = 2x + 1$

The vertex is $\left(-\dfrac{1}{3}, \dfrac{1}{3} \right)$.

49. $2x + 3y \leq 6$,   (1)

$5x - 5y \leq 15$,   (2)

$x \geq 0$ (3)

Shade the intersection of the graphs of the three inequalities above.

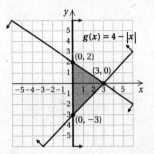

To find the vertices we solve four systems of equations, as follows:

System of equations	Vertex
From (1) and (2)	$(3, 0)$
From (1) and (3)	$(0, 2)$
From (2) and (3)	$(0, -3)$

50. First solve the equation for y and determine the slope of the given line.

$x + 2y = 6$ Given line

$2y = -x + 6$

$y = -\dfrac{1}{2}x + 3$

The slope of the given line is $-\dfrac{1}{2}$. The line through $(3, 7)$ must have slope $-\dfrac{1}{2}$.

Using the point-slope equation:

Substitute 3 for x_1, 7 for y_1, and $-\dfrac{1}{2}$ for m.

$y - y_1 = m(x - x_1)$

$y - 7 = -\dfrac{1}{2}(x - 3)$

$y - 7 = -\dfrac{1}{2}x + \dfrac{3}{2}$

$y = -\dfrac{1}{2}x + \dfrac{17}{2}$.

Using the slope-intercept equation:

Substitute 3 for x, 7 for y, and $-\dfrac{1}{2}$ for m and solve for b.

$y = mx + b$

$7 = -\dfrac{1}{2} \cdot 3 + b$

$7 = -\dfrac{3}{2} + b$

$\dfrac{17}{2} = b$

Then we use the equation $y = mx + b$ and substitute $-\dfrac{1}{2}$ for m and $\dfrac{17}{2}$ for b.

$y = -\dfrac{1}{2}x + \dfrac{17}{2}$

51. First solve the equation for y and determine the slope of the given line.

$$3x + 4y = 5 \qquad \text{Given line}$$
$$4y = -3x + 5$$
$$y = -\frac{3}{4}x + \frac{5}{4}$$

The slope of the given line is $-\frac{3}{4}$. The slope of the perpendicular line is the opposite of the reciprocal of $-\frac{3}{4}$. Thus, the line through $(3, -2)$ must have slope $\frac{4}{3}$.

Using the point-slope equation:

Substitute 3 for x_1, -2 for y_1, and $\frac{4}{3}$ for m.

$$y - y_1 = m(x - x_1)$$
$$y - (-2) = \frac{4}{3}(x - 3)$$
$$y + 2 = \frac{4}{3}x - 4$$
$$y = \frac{4}{3}x - 6$$

Using the slope-intercept equation:

Substitute 3 for x, -2 for y, and $\frac{4}{3}$ for m.

$$y = mx + b$$
$$-2 = \frac{4}{3} \cdot 3 + b$$
$$-2 = 4 + b$$
$$-6 = b$$

Then we use the equation $y = mx + b$ and substitute $\frac{4}{3}$ for m and -6 for b.

$$y = \frac{4}{3}x - 6$$

52. First find the slope of the line.

$$m = \frac{0 - 4}{-2 - (-1)} = \frac{-4}{-1} = 4$$

Using the point-slope equation:

We choose $(-2, 0)$ and substitute -2 for x_1, 0 for y_1, and 4 for m.

$$y - y_1 = m(x - x_1)$$
$$y - 0 = 4(x - (-2))$$
$$y = 4(x + 2)$$
$$y = 4x + 8$$

Using the slope-intercept equation:

We choose $(-2, 0)$ and substitute -2 for x, 0 for y, and 4 for m. Then solve for b.

$$y = mx + b$$
$$0 = 4(-2) + b$$
$$0 = -8 + b$$
$$8 = b$$

Finally, we use the equation $y = mx + b$ and substitute 4 for m and 8 for b.

$$y = 4x + 8$$

53. Using the point-slope equation:

Substitute 2 for x_1, 1 for y_1, and -3 for m.

$$y - y_1 = m(x - x_1)$$
$$y - 1 = -3(x - 2)$$
$$y - 1 = -3x + 6$$
$$y = -3x + 7$$

Using the slope-intercept equation:

Substitute 2 for x, 1 for y, and -3 for b. Then solve for b.

$$y = mx + b$$
$$1 = -3 \cdot 2 + b$$
$$1 = -6 + b$$
$$7 = b$$

Finally, we use the equation $y = mx + b$ and substitute -3 for m and 7 for b.

$$y = -3x + 7$$

54. *Familiarize.* It helps to organize the information in a table. We let x, y, and z represent the weekly productions of the individual machines.

Machines Working	A	B	C
Weekly Production	x	y	z

Machines Working	A & B	B & C	A, B, & C
Weekly Production	3400	4200	5700

Translate. From the table, we obtain three equations.

$x + y + z = 5700$ (All three machines working)

$x + y \quad\; = 3400$ (A and B working)

$\quad\; y + z = 4200$ (B and C working)

Solve. Solving the system we get $(1500, 1900, 2300)$.

Check. The sum of the weekly productions of machines A, B & C is $1500 + 1900 + 2300$, or 5700. The sum of the weekly productions of machines A and B is $1500 + 1900$, or 3400. The sum of the weekly productions of machines B and C is $1900 + 2300$, or 4200. The numbers check.

State. In a week Machine A can polish 1500 lenses, Machine B can polish 1900 lenses, and Machine C can polish 2300 lenses.

55. a) In 2010, $t = 2010 - 2000 = 10$.

$$P(10) = 0.505(10) + 46.84 = 51.89\%$$

b) Substitute 60 for $P(t)$ and solve for t.

$$60 = 0.505t + 46.84$$
$$13.16 = 0.505t$$
$$26 \approx t$$

60% of food dollars will be spent eating away from home about 26 yr after 2000, or in 2026.

c) $50 < 0.505t + 46.84 < 70$

    $3.16 < 0.505t < 23.16$

    $6.3 < t < 45.9$

The percentage of food dollars spent eating away from home will be between 50% and 70% from about 6 yr after 2000 to about 45 yr after 2000, or from 2006 to 2045.

56. a) $N(6) = 6^2 - 6 = 36 - 6 = 30$ games

   b) $72 = n^2 - n$

      $0 = n^2 - n - 72$

      $0 = (n - 9)(n + 8)$

      $n - 9 = 0 \;\; or \;\; n + 8 = 0$

      $n = 9 \;\; or \;\;\;\;\;\; n = -8$

The number of games cannot be negative, so there are 9 teams in the league.

57. *Familiarize*. Referring to the drawing in the text, we see that the dimensions of the Lucite are $5 + 2x$ cm by $4 + 2x$ cm.

Translate.

$$\underbrace{\text{Area of 1 piece of Lucite}} \;\; \text{is} \;\; 5\tfrac{1}{2} \;\; \text{times} \;\; \underbrace{\text{area of card.}}$$

$$(5 + 2x)(4 + 2x) \;\; = \;\; 5\tfrac{1}{2} \;\; \cdot \;\; 5 \cdot 4$$

Solve. We solve the equation.

$$(5 + 2x)(4 + 2x) = 5\tfrac{1}{2} \cdot 5 \cdot 4$$

$$20 + 18x + 4x^2 = 110$$

$$4x^2 + 18x - 90 = 0$$

$$2(2x^2 + 9x - 45) = 0$$

$$2(2x + 15)(x - 3) = 0$$

$$2x + 15 = 0 \;\;\;\; or \;\; x - 3 = 0$$

$$2x = -15 \;\; or \;\;\;\;\;\; x = 3$$

$$x = -\frac{15}{2} \;\; or \;\;\;\;\;\; x = 3$$

Check. Since the width of the border cannot be negative, we check only 3. If $x = 6$, then $5 + 2x = 5 + 2 \cdot 3 = 5 + 6 = 11$ and $4 + 2x = 4 + 2 \cdot 3 = 4 + 6 = 10$.

Then the area of a piece of the Lucite is $11 \cdot 10$, or 110 cm, and the area of the card is $5 \cdot 4$, or 20 cm. Since $5\tfrac{1}{2} \cdot 20 = 110$, the answer checks.

State. The dimensions of the Lucite are 11 cm by 10 cm.

58. $m = \dfrac{1 - (-5)}{-7 - (-8)} = \dfrac{6}{1} = 6$

Answer (d) is correct.

59. First we find the slope of the given line.

$$-2x + y = -5$$

$$y = 2x - 5$$

The slope of the given line is 2. Now we find the equation of the line with slope 2 and containing $(3, 5)$. We will use the point-slope equation.

$$y - y_1 = m(x - x_1)$$

$$y - 5 = 2(x - 3)$$

$$y - 5 = 2x - 6$$

$$y = 2x - 1$$

Answer (b) is correct.

60. If we substitute -2 for x and 3 for y in each pair of equations, we find that $(-2, 3)$ is a solution of only system (b), so (b) is the correct answer.

61. $t^3 - 64 = (t - 4)(t^2 + 4t + 16)$

Answer (e) is correct.

62. $|x + 1| \le |x - 3|$

$|x + 1| \le x - 3 \;\; or \;\; |x + 1| \le -(x - 3)$

First we solve $|x + 1| \le x - 3$.

$$-(x - 3) \le x + 1 \;\; and \;\; x + 1 \le x - 3$$

$$-x + 3 \le x + 1 \;\; and \;\;\;\;\;\; 1 \le -3$$

$$2 \le 2x \;\;\;\;\; and \;\;\;\;\;\; 1 \le -3$$

$$1 \le x \;\;\;\;\;\; and \;\;\;\;\;\; 1 \le -3$$

Since $1 \le -3$ is false for all values of x, the solution set for this portion of the inequality is $\emptyset$.

Now we solve $|x + 1| \le -(x - 3)$.

$$-[-(x - 3)] \le x + 1 \;\; and \;\; x + 1 \le -(x - 3)$$

$$x - 3 \le x + 1 \;\; and \;\; x + 1 \le -x + 3$$

$$-3 \le 1 \;\;\;\;\;\; and \;\;\;\;\; 2x \le 2$$

$$-3 \le 1 \;\;\;\;\;\; and \;\;\;\;\;\; x \le 1$$

Since $-3 \le 1$ is true for all values of x, the solution set for this portion of the inequality is $\{x | x \le 1\}$, or $(-\infty, 1]$.

Then the solution set for the original inequality is $\emptyset \cup \{x | x \le 1\}$, or $\{x \le 1\}$, or $(-\infty, 1]$.

63. *Familiarize*. Let b = the number of years Bert has taught and s = the number of years Sally has taught. Then two years ago Bert and Sally had taught $b - 2$ and $s - 2$ years, respectively.

Translate. Together Bert and Sally have taught 46 years, so we have one equation.

$$b + s = 46$$

Two years ago Bert had taught 2.5 times as long as Sally, so we have

$$b - 2 = 2.5(s - 2), \;\; or$$

$$b - 2 = 2.5s - 5, \;\; or$$

$$b - 2.5s = -3.$$

Thus, we have a system of equations.

$$b + s = 46, \;\; (1)$$

$$b - 2.5s = -3 \;\; (2)$$

Solve. We will use the elimination method. Multiply Equation (2) by -1 and then add.

$$b + s = 46$$
$$\underline{-b + 2.5s = 3}$$
$$3.5s = 49$$
$$s = 14$$

Now substitute 14 for s in Equation (1) and solve for b.

$$b + 14 = 46$$
$$b = 32$$

Check. If Bert has taught 32 yr and Sally has taught 14 yr, then together they have taught $32 + 14$, or 46 yr. Two years ago Bert had taught $32 - 2$, or 30 yr, and Sally had taught $14 - 2$, or 12 yr, and $2.5(12) = 30$, so the answer checks.

State. Bert has taught 32 yr, and Sally has taught 14 yr.

Chapter 5

Rational Expressions, Equations, and Functions

Exercise Set 5.1

1. $\dfrac{5t^2 - 64}{3t + 17}$

We set the denominator equal to 0 and solve.

$$3t + 17 = 0$$
$$3t = -17$$
$$t = -\dfrac{17}{3}$$

The expression is not defined for the number $-\dfrac{17}{3}$.

3. $\dfrac{x^3 - x^2 + x + 2}{x^2 + 12x + 35}$

We set the denominator equal to 0 and solve.

$$x^2 + 12x + 35 = 0$$
$$(x + 5)(x + 7) = 0$$
$$x + 5 = 0 \quad or \quad x + 7 = 0$$
$$x = -5 \quad or \quad x = -7$$

The expression is not defined for the numbers -5 and -7.

5. In Exercise 1 we found that the only replacement for which the rational expression $\dfrac{5t^2 - 64}{3t + 17}$ is not defined is $-\dfrac{17}{3}$. Then the domain of $f(t) = \dfrac{5t^2 - 64}{3t + 17}$ is $\left\{ x \,\middle|\, x \text{ is a real number } and \ x \neq -\dfrac{17}{3} \right\}$, or $\left(-\infty, -\dfrac{17}{3} \right) \cup \left(-\dfrac{17}{3}, \infty \right)$.

7. In Exercise 3 we found that the replacements for which the rational expression $\dfrac{x^3 - x^2 + x + 2}{x^2 + 12x + 35}$ is not defined are -7 and -5. Then the domain of $f(x) = \dfrac{x^3 - x^2 + x + 2}{x^2 + 12x + 35}$ is $\{x | x \text{ is a real number } and \ x \neq -7 \ and \ x \neq -5\}$, or $(-\infty, -7) \cup (-7, -5) \cup (-5, \infty)$.

9. $\dfrac{7x}{7x} \cdot \dfrac{x + 2}{x + 8} = \dfrac{7x(x + 2)}{7x(x + 8)}$ Multiplying numerators and multiplying denominators

11. $\dfrac{q - 5}{q + 3} \cdot \dfrac{q + 5}{q + 5} = \dfrac{(q - 5)(q + 5)}{(q + 3)(q + 5)}$ Multiplying numerators and multiplying denominators

13. $\dfrac{15y^5}{5y^4} = \dfrac{3 \cdot 5 \cdot y^4 \cdot y}{5 \cdot y^4 \cdot 1}$ Factoring the numerator and the denominator

$= \dfrac{5y^4}{5y^4} \cdot \dfrac{3y}{1}$ Factoring the rational expression

$= 1 \cdot 3y$ $\dfrac{5y^4}{5y^4} = 1$

$= 3y$ Removing a factor of 1

15. $\dfrac{16p^3}{24p^7} = \dfrac{8p^3 \cdot 2}{8p^3 \cdot 3p^4}$ Factoring the numerator and the denominator

$= \dfrac{8p^3}{8p^3} \cdot \dfrac{2}{3p^4}$ Factoring the rational expression

$= 1 \cdot \dfrac{2}{3p^4}$ $\dfrac{8p^3}{8p^3} = 1$

$= \dfrac{2}{3p^4}$ Removing a factor of 1

17. $\dfrac{9a - 27}{9} = \dfrac{9(a - 3)}{9 \cdot 1}$ Factoring the numerator and the denominator

$= \dfrac{9}{9} \cdot \dfrac{a - 3}{1}$

$= \dfrac{a - 3}{1}$ Removing a factor of 1

$= a - 3$

19. $\dfrac{12x - 15}{21} = \dfrac{3(4x - 5)}{3 \cdot 7}$ Factoring the numerator and the denominator

$= \dfrac{3}{3} \cdot \dfrac{4x - 5}{7}$

$= \dfrac{4x - 5}{7}$ Removing a factor of 1

21. $\dfrac{4y - 12}{4y + 12} = \dfrac{4(y - 3)}{4(y + 3)} = \dfrac{4}{4} \cdot \dfrac{y - 3}{y + 3} = \dfrac{y - 3}{y + 3}$

23. $\dfrac{t^2 - 16}{t^2 - 8t + 16} = \dfrac{(t + 4)(t - 4)}{(t - 4)(t - 4)} = \dfrac{t + 4}{t - 4} \cdot \dfrac{t - 4}{t - 4} = \dfrac{t + 4}{t - 4}$

25. $\dfrac{x^2 - 9x + 8}{x^2 + 3x - 4} = \dfrac{(x - 8)(x - 1)}{(x + 4)(x - 1)} = \dfrac{x - 8}{x + 4} \cdot \dfrac{x - 1}{x - 1} = \dfrac{x - 8}{x + 4}$

27. $\dfrac{w^3 - z^3}{w^2 - z^2} = \dfrac{(w - z)(w^2 + wz + z^2)}{(w + z)(w - z)} =$

$\dfrac{w - z}{w - z} \cdot \dfrac{w^2 + wz + z^2}{w + z} = \dfrac{w^2 + wz + z^2}{w + z}$

29. $\dfrac{x^4}{3x+6} \cdot \dfrac{5x+10}{5x^7}$

$= \dfrac{x^4(5x+10)}{(3x+6)(5x^7)}$ — Multiplying the numerators and the denominators

$= \dfrac{x^4(5)(x+2)}{3(x+2)(5)(x^4)(x^3)}$ — Factoring the numerator and the denominator

$= \dfrac{\cancel{x^4}(\cancel{5})(\cancel{x+2})(1)}{3(\cancel{x+2})(\cancel{5})(\cancel{x^4})(x^3)}$ — Removing a factor of 1: $\dfrac{(x^4)(5)(x+2)}{(x+2)(5)(x^4)} = 1$

$= \dfrac{1}{3x^3}$ — Simplifying

31. $\dfrac{x^2-16}{x^2} \cdot \dfrac{x^2-4x}{x^2-x-12}$

$= \dfrac{(x^2-16)(x^2-4x)}{x^2(x^2-x-12)}$ — Multiplying the numerators and the denominators

$= \dfrac{(x+4)(x-4)(x)(x-4)}{x \cdot x(x-4)(x+3)}$ — Factoring the numerator and the denominator

$= \dfrac{(x+4)(\cancel{x-4})(\cancel{x})(x-4)}{\cancel{x} \cdot x(\cancel{x-4})(x+3)}$ — Removing a factor of 1

$= \dfrac{(x+4)(x-4)}{x(x+3)}$

33. $\dfrac{y^2-16}{2y+6} \cdot \dfrac{y+3}{y-4} = \dfrac{(y^2-16)(y+3)}{(2y+6)(y-4)}$

$= \dfrac{(y+4)(y-4)(y+3)}{2(y+3)(y-4)}$

$= \dfrac{(y+4)(\cancel{y-4})(\cancel{y+3})}{2(\cancel{y+3})(\cancel{y-4})}$

$= \dfrac{y+4}{2}$

35. $\dfrac{x^2-2x-35}{2x^3-3x^2} \cdot \dfrac{4x^3-9x}{7x-49}$

$= \dfrac{(x^2-2x-35)(4x^3-9x)}{(2x^3-3x^2)(7x-49)}$

$= \dfrac{(x-7)(x+5)(x)(2x+3)(2x-3)}{x \cdot x(2x-3)(7)(x-7)}$

$= \dfrac{(\cancel{x-7})(x+5)(\cancel{x})(2x+3)(\cancel{2x-3})}{\cancel{x} \cdot x(\cancel{2x-3})(7)(\cancel{x-7})}$

$= \dfrac{(x+5)(2x+3)}{7x}$

37. $\dfrac{c^3+8}{c^2-4} \cdot \dfrac{c^2-4c+4}{c^2-2c+4}$

$= \dfrac{(c^3+8)(c^2-4c+4)}{(c^2-4)(c^2-2c+4)}$

$= \dfrac{(c+2)(c^2-2c+4)(c-2)(c-2)}{(c+2)(c-2)(c^2-2c+4) \cdot 1}$

$= \dfrac{(c+2)(c^2-2c+4)(c-2)}{(c+2)(c^2-2c+4)(c-2)} \cdot \dfrac{c-2}{1}$

$= \dfrac{c-2}{1}$

$= c-2$

39. $\dfrac{x^2-y^2}{x^3-y^3} \cdot \dfrac{x^2+xy+y^2}{x^2+2xy+y^2}$

$= \dfrac{(x^2-y^2)(x^2+xy+y^2)}{(x^3-y^3)(x^2+2xy+y^2)}$

$= \dfrac{(x+y)(x-y)(x^2+xy+y^2) \cdot 1}{(x-y)(x^2+xy+y^2)(x+y)(x+y)}$

$= \dfrac{(x+y)(x-y)(x^2+xy+y^2)}{(x+y)(x-y)(x^2+xy+y^2)} \cdot \dfrac{1}{x+y}$

$= \dfrac{1}{x+y}$

41. $\dfrac{12x^8}{3y^4} \div \dfrac{16x^3}{6y}$

$= \dfrac{12x^8}{3y^4} \cdot \dfrac{6y}{16x^3}$ — Multiplying by the reciprocal of the divisor

$= \dfrac{12x^8(6y)}{3y^4(16x^3)}$ — Multiplying the numerators and the denominators

$= \dfrac{3 \cdot 4 \cdot x^3 \cdot x^5 \cdot 2 \cdot 3 \cdot y}{3 \cdot y \cdot y^3 \cdot 4 \cdot 2 \cdot 2 \cdot x^3}$ — Factoring the numerator and the denominator

$= \dfrac{\cancel{3} \cdot \cancel{4} \cdot \cancel{x^3} \cdot x^5 \cdot \cancel{2} \cdot 3 \cdot \cancel{y}}{\cancel{3} \cdot \cancel{y} \cdot y^3 \cdot \cancel{4} \cdot \cancel{2} \cdot 2 \cdot \cancel{x^3}}$ — Removing a factor of 1

$= \dfrac{3x^5}{2y^3}$

43. $\dfrac{3y+15}{y} \div \dfrac{y+5}{y} = \dfrac{3y+15}{y} \cdot \dfrac{y}{y+5}$

$= \dfrac{(3y+15)(y)}{y(y+5)}$

$= \dfrac{3(y+5)(y)}{y(y+5) \cdot 1}$

$= \dfrac{3(\cancel{y+5})(\cancel{y})}{\cancel{y}(\cancel{y+5}) \cdot 1}$

$= \dfrac{3}{1}$

$= 3$

45. $\dfrac{y^2-9}{y} \div \dfrac{y+3}{y+2} = \dfrac{y^2-9}{y} \cdot \dfrac{y+2}{y+3}$

$$= \dfrac{(y^2-9)(y+2)}{y(y+3)}$$

$$= \dfrac{(y+3)(y-3)(y+2)}{y(y+3)}$$

$$= \dfrac{(y+3)(y-3)(y+2)}{y(y+3)}$$

$$= \dfrac{(y-3)(y+2)}{y}$$

47. $\dfrac{4a^2-1}{a^2-4} \div \dfrac{2a-1}{a-2} = \dfrac{4a^2-1}{a^2-4} \cdot \dfrac{a-2}{2a-1}$

$$= \dfrac{(4a^2-1)(a-2)}{(a^2-4)(2a-1)}$$

$$= \dfrac{(2a+1)(2a-1)(a-2)}{(a+2)(a-2)(2a-1)}$$

$$= \dfrac{(2a+1)(2a-1)(a-2)}{(a+2)(a-2)(2a-1)}$$

$$= \dfrac{2a+1}{a+2}$$

49. $\dfrac{x^2-16}{x^2-10x+25} \div \dfrac{3x-12}{x^2-3x-10}$

$$= \dfrac{x^2-16}{x^2-10x+25} \cdot \dfrac{x^2-3x-10}{3x-12}$$

$$= \dfrac{(x^2-16)(x^2-3x-10)}{(x^2-10x+25)(3x-12)}$$

$$= \dfrac{(x+4)(x-4)(x-5)(x+2)}{(x-5)(x-5)(3)(x-4)}$$

$$= \dfrac{(x+4)(x-4)(x-5)(x+2)}{(x-5)(x-5)(3)(x-4)}$$

$$= \dfrac{(x+4)(x+2)}{3(x-5)}$$

51. $\dfrac{y^3+3y}{y^2-9} \div \dfrac{y^2+5y-14}{y^2+4y-21}$

$$= \dfrac{y^3+3y}{y^2-9} \cdot \dfrac{y^2+4y-21}{y^2+5y-14}$$

$$= \dfrac{(y^3+3y)(y^2+4y-21)}{(y^2-9)(y^2+5y-14)}$$

$$= \dfrac{y(y^2+3)(y+7)(y-3)}{(y+3)(y-3)(y+7)(y-2)}$$

$$= \dfrac{y(y^2+3)(y+7)(y-3)}{(y+3)(y-3)(y+7)(y-2)}$$

$$= \dfrac{y(y^2+3)}{(y+3)(y-2)}$$

53. $\dfrac{x^3-64}{x^3+64} \div \dfrac{x^2-16}{x^2-4x+16}$

$$= \dfrac{x^3-64}{x^3+64} \cdot \dfrac{x^2-4x+16}{x^2-16}$$

$$= \dfrac{(x^3-64)(x^2-4x+16)}{(x^3+64)(x^2-16)}$$

$$= \dfrac{(x-4)(x^2+4x+16)(x^2-4x+16)}{(x+4)(x^2-4x+16)(x+4)(x-4)}$$

$$= \dfrac{(x-4)(x^2-4x+16)}{(x-4)(x^2-4x+16)} \cdot \dfrac{x^2+4x+16}{(x+4)(x+4)}$$

$$= \dfrac{x^2+4x+16}{(x+4)(x+4)}, \text{ or } \dfrac{x^2+4x+16}{(x+4)^2}$$

55. $\dfrac{8x^3y^3+27x^3}{64x^3y^3-x^3} \div \dfrac{4x^2y^2-9x^2}{16x^2y^2+4x^2y+x^2}$

$$= \dfrac{8x^3y^3+27x^3}{64x^3y^3-x^3} \cdot \dfrac{16x^2y^2+4x^2y+x^2}{4x^2y^2-9x^2}$$

$$= \dfrac{(8x^3y^3+27x^3)(16x^2y^2+4x^2y+x^2)}{(64x^3y^3-x^3)(4x^2y^2-9x^2)}$$

$$= \dfrac{x^3(8y^3+27)(x^2)(16y^2+4y+1)}{x^3(64y^3-1)(x^2)(4y^2-9)}$$

$$= \dfrac{x^3(2y+3)(4y^2-6y+9)(x^2)(16y^2+4y+1)}{x^3(4y-1)(16y^2+4y+1)(x^2)(2y+3)(2y-3)}$$

$$= \dfrac{x^3(2y+3)(x^2)(16y^2+4y+1)}{x^3(16y^2+4y+1)(x^2)(2y+3)} \cdot \dfrac{4y^2-6y+9}{(4y-1)(2y-3)}$$

$$= \dfrac{4y^2-6y+9}{(4y-1)(2y-3)}$$

57. $\left[\dfrac{r^2-4s^2}{r+2s} \div (r+2s)\right] \cdot \dfrac{2s}{r-2s}$

$$= \left[\dfrac{r^2-4s^2}{r+2s} \cdot \dfrac{1}{r+2s}\right] \cdot \dfrac{2s}{r-2s}$$

$$= \dfrac{(r^2-4s^2)(1)(2s)}{(r+2s)(r+2s)(r-2s)}$$

$$= \dfrac{(r+2s)(r-2s)(2s)}{(r+2s)(r+2s)(r-2s)}$$

$$= \dfrac{(r+2s)(r-2s)(2s)}{(r+2s)(r+2s)(r-2s)}$$

$$= \dfrac{2s}{r+2s}$$

59. Discussion and Writing Exercise

61. The function can be written as a set of six ordered pairs, $\{(-4,3),(-2,1),(0,-3),(2,-2),(4,0),(6,4)\}$.

The domain is the set of all first coordinates, $\{-4,-2,0,2,4,6\}$.

The range is the set of all second coordinates, $\{-3,-2,0,1,3,4\}$.

63. The domain is the set of all x-values on the graph, $[-5,5]$.

The range is the set of all y-values on the graph, $[-4,4]$.

65. $6a^2 + 5ab - 25b^2$

We can factor this using the FOIL method or the grouping method. The factorization is $(3a - 5b)(2a + 5b)$.

67. $10x^2 - 80x + 70 = 10(x^2 - 8x + 7)$

To factor $x^2 - 8x + 7$ we find factors of 7 whose sum is -8. The numbers we need are -1 and -7, so $x^2 - 8x + 7 = (x - 1)(x - 7)$. Then $10x^2 - 80x + 70 = 10(x - 1)(x - 7)$.

69. $21p^2 + p - 10$

We can factor this using the FOIL method or the grouping method. The factorization is $(7p + 5)(3p - 2)$.

71. $2x^3 - 16x^2 - 66x = 2x(x^2 - 8x - 33)$

To factor $x^2 - 8x - 33$ we find factors of -33 whose sum is -8. The numbers we need are -11 and 3, so $x^2 - 8x - 33 = (x - 11)(x + 3)$. Then $2x^3 - 16x^2 - 66x = 2x(x - 11)(x + 3)$.

73. Substitute $-\dfrac{2}{3}$ for m and -5 for b in the slope-intercept

equation, $y = mx + b$. The equation is $y = -\dfrac{2}{3}x - 5$.

75.
$$\frac{x(x+1) - 2(x+3)}{(x+1)(x+2)(x+3)} = \frac{x^2 + x - 2x - 6}{(x+1)(x+2)(x+3)}$$
$$= \frac{x^2 - x - 6}{(x+1)(x+2)(x+3)}$$
$$= \frac{(x-3)(x+2)}{(x+1)(x+2)(x+3)}$$
$$= \frac{(x-3)\cancel{(x+2)}}{(x+1)\cancel{(x+2)}(x+3)}$$
$$= \frac{x-3}{(x+1)(x+3)}$$

77.
$$\frac{m^2 - t^2}{m^2 + t^2 + m + t + 2mt} = \frac{m^2 - t^2}{(m^2 + 2mt + t^2) + (m+t)}$$
$$= \frac{(m+t)(m-t)}{(m+t)^2 + (m+t)}$$
$$= \frac{(m+t)(m-t)}{(m+t)[(m+t)+1]}$$
$$= \frac{\cancel{(m+t)}(m-t)}{\cancel{(m+t)}(m+t+1)}$$
$$= \frac{m-t}{m+t+1}$$

79. $g(x) = \dfrac{2x+3}{4x-1}$

$g(5) = \dfrac{2 \cdot 5 + 3}{4 \cdot 5 - 1} = \dfrac{10+3}{20-1} = \dfrac{13}{19}$

$g(0) = \dfrac{2 \cdot 0 + 3}{4 \cdot 0 - 1} = \dfrac{3}{-1} = -3$

$g\left(\dfrac{1}{4}\right) = \dfrac{2 \cdot \dfrac{1}{4} + 3}{4 \cdot \dfrac{1}{4} - 1} = \dfrac{\dfrac{1}{2} + 3}{1 - 1} = \dfrac{\dfrac{7}{2}}{0}$; since division by 0 is not

defined, $g(0)$ is not defined.

$g(a+h) = \dfrac{2(a+h) + 3}{4(a+h) - 1} = \dfrac{2a + 2h + 3}{4a + 4h - 1}$

Exercise Set 5.2

1. $15 = 3 \cdot 5$

$40 = 2 \cdot 2 \cdot 2 \cdot 5$

$\text{LCM} = 2 \cdot 2 \cdot 2 \cdot 3 \cdot 5$, or 120

(We used each factor the greatest number of times that it occurs in any one prime factorization.)

3. $18 = 2 \cdot 3 \cdot 3$

$48 = 2 \cdot 2 \cdot 2 \cdot 2 \cdot 3$

$\text{LCM} = 2 \cdot 2 \cdot 2 \cdot 2 \cdot 3 \cdot 3$, or 144

5. $30 = 2 \cdot 3 \cdot 5$

$105 = 3 \cdot 5 \cdot 7$

$\text{LCM} = 2 \cdot 3 \cdot 5 \cdot 7$, or 210

7. $9 = 3 \cdot 3$

$15 = 3 \cdot 5$

$5 = 5$

$\text{LCM} = 3 \cdot 3 \cdot 5$, or 45

9. $\dfrac{5}{6} + \dfrac{4}{15} = \dfrac{5}{2 \cdot 3} + \dfrac{4}{3 \cdot 5}$, $\text{LCD} = 2 \cdot 3 \cdot 5$, or 30

$= \dfrac{5}{2 \cdot 3} \cdot \dfrac{5}{5} + \dfrac{4}{3 \cdot 5} \cdot \dfrac{2}{2}$

$= \dfrac{25}{2 \cdot 3 \cdot 5} + \dfrac{8}{2 \cdot 3 \cdot 5}$

$= \dfrac{33}{2 \cdot 3 \cdot 5} = \dfrac{\cancel{3} \cdot 11}{2 \cdot \cancel{3} \cdot 5}$

$= \dfrac{11}{10}$

11. $\dfrac{7}{36} + \dfrac{1}{24}$

$= \dfrac{7}{2 \cdot 2 \cdot 3 \cdot 3} + \dfrac{1}{2 \cdot 2 \cdot 2 \cdot 3}$, $\text{LCD} = 2 \cdot 2 \cdot 2 \cdot 3 \cdot 3$, or 72

$= \dfrac{7}{2 \cdot 2 \cdot 3 \cdot 3} \cdot \dfrac{2}{2} + \dfrac{1}{2 \cdot 2 \cdot 2 \cdot 3} \cdot \dfrac{3}{3}$

$= \dfrac{14}{2 \cdot 2 \cdot 2 \cdot 3 \cdot 3} + \dfrac{3}{2 \cdot 2 \cdot 2 \cdot 3 \cdot 3}$

$= \dfrac{17}{2 \cdot 2 \cdot 2 \cdot 3 \cdot 3}$

$= \dfrac{17}{72}$

13. $\dfrac{3}{4} + \dfrac{7}{30} + \dfrac{1}{16}$

$= \dfrac{3}{2 \cdot 2} + \dfrac{7}{2 \cdot 3 \cdot 5} + \dfrac{1}{2 \cdot 2 \cdot 2 \cdot 2}$, LCD $= 2 \cdot 2 \cdot 2 \cdot 2 \cdot 3 \cdot 5$

$= \dfrac{3}{2 \cdot 2} \cdot \dfrac{2 \cdot 2 \cdot 3 \cdot 5}{2 \cdot 2 \cdot 3 \cdot 5} + \dfrac{7}{2 \cdot 3 \cdot 5} \cdot \dfrac{2 \cdot 2 \cdot 2}{2 \cdot 2 \cdot 2} +$

$\dfrac{1}{2 \cdot 2 \cdot 2 \cdot 2} \cdot \dfrac{3 \cdot 5}{3 \cdot 5}$

$= \dfrac{180}{2 \cdot 2 \cdot 2 \cdot 2 \cdot 3 \cdot 5} + \dfrac{56}{2 \cdot 2 \cdot 2 \cdot 2 \cdot 3 \cdot 5} + \dfrac{15}{2 \cdot 2 \cdot 2 \cdot 2 \cdot 3 \cdot 5}$

$= \dfrac{251}{2 \cdot 2 \cdot 2 \cdot 2 \cdot 3 \cdot 5}$

$= \dfrac{251}{240}$

15. $21x^2 y = 3 \cdot 7 \cdot x \cdot x \cdot y$

$7xy = 7 \cdot x \cdot y$

LCM $= 3 \cdot 7 \cdot x \cdot x \cdot y$, or $21x^2 y$

17. $y^2 - 100 = (y+10)(y-10)$

$10y + 100 = 10(y+10)$

LCM $= 10(y+10)(y-10)$

19. $15ab^2 = 3 \cdot 5 \cdot a \cdot b \cdot b$

$3ab = 3 \cdot a \cdot b$

$10a^3 b = 2 \cdot 5 \cdot a \cdot a \cdot a \cdot b$

LCM $= 2 \cdot 3 \cdot 5 \cdot a \cdot a \cdot a \cdot b \cdot b$, or $30a^3 b^2$

21. $5y - 15 = 5(y-3)$

$y^2 - 6y + 9 = (y-3)(y-3)$

LCM $= 5(y-3)(y-3)$, or $5(y-3)^2$

23. $y^2 - 25 = (y+5)(y-5)$

$5 - y$

We can use $y-5$ from the prime factorization of $y^2 - 25$ or $5 - y$ from the second expression, but not both.

LCM $= (y+5)(y-5)$, or $(y+5)(5-y)$

25. $2r^2 - 5r - 12 = (2r+3)(r-4)$

$3r^2 - 13r + 4 = (3r-1)(r-4)$

$r^2 - 16 = (r+4)(r-4)$

LCM $= (2r+3)(3r-1)(r+4)(r-4)$

27. $x^5 + 4x^3 = x^3(x^2+4) = x \cdot x \cdot x(x^2+4)$

$x^3 - 4x^2 + 4x = x(x^2 - 4x + 4) = x(x-2)(x-2)$

LCM $= x \cdot x \cdot x(x-2)(x-2)(x^2+4)$, or $x^3(x-2)^2(x^2+4)$

29. $x^5 - 2x^4 + x^3 = x^3(x^2 - 2x + 1) = x \cdot x \cdot x(x-1)(x-1)$

$2x^3 + 2x = 2x(x^2+1)$

$5x + 5 = 5(x+1)$

LCM $= 2 \cdot 5 \cdot x \cdot x \cdot x(x-1)(x-1)(x+1)(x^2+1)$, or $10x^3(x-1)^2(x+1)(x^2+1)$

31. $\dfrac{x-2y}{x+y} + \dfrac{x+9y}{x+y}$

$= \dfrac{x - 2y + x + 9y}{x+y}$ Adding the numerators

$= \dfrac{2x + 7y}{x+y}$

33. $\dfrac{4y+2}{y-2} - \dfrac{y-3}{y-2}$

$= \dfrac{4y + 2 - (y-3)}{y-2}$ Subtracting numerators

$= \dfrac{4y + 2 - y + 3}{y-2}$

$= \dfrac{3y + 5}{y-2}$

35. $\dfrac{a^2}{a-b} + \dfrac{b^2}{b-a}$

$= \dfrac{a^2}{a-b} + \dfrac{b^2}{b-a} \cdot \dfrac{-1}{-1}$ Multiplying by 1, using $\dfrac{-1}{-1}$

$= \dfrac{a^2}{a-b} + \dfrac{-b^2}{a-b}$

$= \dfrac{a^2 - b^2}{a-b}$ Adding numerators

$= \dfrac{(a+b)(a-b)}{a-b}$ Factoring the numerator

$= \dfrac{(a+b)\cancel{(a-b)}}{1 \cdot \cancel{(a-b)}}$ Removing a factor of 1

$= \dfrac{a+b}{1}$

$= a+b$

37. $\dfrac{6}{y} - \dfrac{7}{-y}$

$= \dfrac{6}{y} - \dfrac{7}{-y} \cdot \dfrac{-1}{-1}$ Multiplying by 1, using $\dfrac{-1}{-1}$

$= \dfrac{6}{y} - \dfrac{-7}{y}$

$= \dfrac{6 - (-7)}{y}$ Subtracting numerators

$= \dfrac{13}{y}$

39. $\dfrac{4a-2}{a^2-49}+\dfrac{5+3a}{49-a^2}$

$=\dfrac{4a-2}{a^2-49}+\dfrac{5+3a}{49-a^2}\cdot\dfrac{-1}{-1}$ Multiplying by 1, using $\dfrac{-1}{-1}$

$=\dfrac{4a-2}{a^2-49}+\dfrac{-5-3a}{a^2-49}$

$=\dfrac{4a-2-5-3a}{a^2-49}$ Adding numerators

$=\dfrac{a-7}{a^2-49}$

$=\dfrac{a-7}{(a+7)(a-7)}$ Factoring

$=\dfrac{(a-7)\cdot 1}{(a+7)(a-7)}$ Removing a factor of 1

$=\dfrac{1}{a+7}$

41. $\dfrac{a^3}{a-b}+\dfrac{b^3}{b-a}$

$=\dfrac{a^3}{a-b}+\dfrac{b^3}{b-a}\cdot\dfrac{-1}{-1}$

$=\dfrac{a^3}{a-b}+\dfrac{-b^3}{a-b}$

$=\dfrac{a^3-b^3}{a-b}$

$=\dfrac{(a-b)(a^2+ab+b^2)}{a-b}$

$=\dfrac{(a-b)(a^2+ab+b^2)}{(a-b)\cdot 1}$

$=a^2+ab+b^2$

43. $\dfrac{y-2}{y+4}+\dfrac{y+3}{y-5}$ LCD $=(y+4)(y-5)$

$=\dfrac{y-2}{y+4}\cdot\dfrac{y-5}{y-5}+\dfrac{y+3}{y-5}\cdot\dfrac{y+4}{y+4}$

$=\dfrac{(y^2-7y+10)+(y^2+7y+12)}{(y+4)(y-5)}$

$=\dfrac{2y^2+22}{(y+4)(y-5)}$, or $\dfrac{2y^2+22}{y^2-y-20}$

45. $\dfrac{4xy}{x^2-y^2}+\dfrac{x-y}{x+y}$

$=\dfrac{4xy}{(x+y)(x-y)}+\dfrac{x-y}{x+y}$ LCD $=(x+y)(x-y)$

$=\dfrac{4xy}{(x+y)(x-y)}+\dfrac{x-y}{x+y}\cdot\dfrac{x-y}{x-y}$

$=\dfrac{4xy+x^2-2xy+y^2}{(x+y)(x-y)}$

$=\dfrac{x^2+2xy+y^2}{(x+y)(x-y)}=\dfrac{(x+y)(x+y)}{(x+y)(x-y)}$

$=\dfrac{(x+y)(x+y)}{(x+y)(x-y)}=\dfrac{x+y}{x-y}$

47. $\dfrac{9x+2}{3x^2-2x-8}+\dfrac{7}{3x^2+x-4}$

$=\dfrac{9x+2}{(3x+4)(x-2)}+\dfrac{7}{(3x+4)(x-1)}$

$\qquad\qquad$ LCD $=(3x+4)(x-2)(x-1)$

$=\dfrac{9x+2}{(3x+4)(x-2)}\cdot\dfrac{x-1}{x-1}+\dfrac{7}{(3x+4)(x-1)}\cdot\dfrac{x-2}{x-2}$

$=\dfrac{9x^2-7x-2+7x-14}{(3x+4)(x-2)(x-1)}$

$=\dfrac{9x^2-16}{(3x+4)(x-2)(x-1)}=\dfrac{(3x+4)(3x-4)}{(3x+4)(x-2)(x-1)}$

$=\dfrac{(3x+4)(3x-4)}{(3x+4)(x-2)(x-1)}$

$=\dfrac{3x-4}{(x-2)(x-1)}$, or $\dfrac{3x-4}{x^2-3x+2}$

49. $\dfrac{4}{x+1}+\dfrac{x+2}{x^2-1}+\dfrac{3}{x-1}$

$=\dfrac{4}{x+1}+\dfrac{x+2}{(x+1)(x-1)}+\dfrac{3}{x-1}$

$\qquad\qquad$ LCD $=(x+1)(x-1)$

$=\dfrac{4}{x+1}\cdot\dfrac{x-1}{x-1}+\dfrac{x+2}{(x+1)(x-1)}+\dfrac{3}{x-1}\cdot\dfrac{x+1}{x+1}$

$=\dfrac{4x-4+x+2+3x+3}{(x+1)(x-1)}$

$=\dfrac{8x+1}{(x+1)(x-1)}$, or $\dfrac{8x+1}{x^2-1}$

51. $\dfrac{x-1}{3x+15}-\dfrac{x+3}{5x+25}$

$=\dfrac{x-1}{3(x+5)}-\dfrac{x+3}{5(x+5)}$

$\qquad\qquad$ LCD $=3\cdot 5(x+5)$, or $15(x+5)$

$=\dfrac{x-1}{3(x+5)}\cdot\dfrac{5}{5}-\dfrac{x+3}{5(x+5)}\cdot\dfrac{3}{3}$

$=\dfrac{5x-5-(3x+9)}{15(x+5)}$

$=\dfrac{5x-5-3x-9}{15(x+5)}$

$=\dfrac{2x-14}{15(x+5)}$, or $\dfrac{2x-14}{15x+75}$

53. $\dfrac{5ab}{a^2-b^2}-\dfrac{a-b}{a+b}$

$=\dfrac{5ab}{(a+b)(a-b)}-\dfrac{a-b}{a+b}$ LCD $=(a+b)(a-b)$

$=\dfrac{5ab}{(a+b)(a-b)}-\dfrac{a-b}{a+b}\cdot\dfrac{a-b}{a-b}$

$=\dfrac{5ab-(a^2-2ab+b^2)}{(a+b)(a-b)}$

$=\dfrac{5ab-a^2+2ab-b^2}{(a+b)(a-b)}$

$=\dfrac{-a^2+7ab-b^2}{(a+b)(a-b)}$, or $\dfrac{-a^2+7ab-b^2}{a^2-b^2}$

55.
$$\frac{3y}{y^2 - 7y + 10} - \frac{2y}{y^2 - 8y + 15}$$

$$= \frac{3y}{(y-5)(y-2)} - \frac{2y}{(y-5)(y-3)}$$

$$\text{LCD} = (y-5)(y-2)(y-3)$$

$$= \frac{3y}{(y-5)(y-2)} \cdot \frac{y-3}{y-3} - \frac{2y}{(y-5)(y-3)} \cdot \frac{y-2}{y-2}$$

$$= \frac{3y^2 - 9y - (2y^2 - 4y)}{(y-5)(y-2)(y-3)}$$

$$= \frac{3y^2 - 9y - 2y^2 + 4y}{(y-5)(y-2)(y-3)}$$

$$= \frac{y^2 - 5y}{(y-5)(y-2)(y-3)} = \frac{y(y-5)}{(y-5)(y-2)(y-3)}$$

$$= \frac{y(y-5)}{(y-5)(y-2)(y-3)}$$

$$= \frac{y}{(y-2)(y-3)}, \text{ or } \frac{y}{y^2 - 5y + 6}$$

57.
$$\frac{y}{y^2 - y - 20} + \frac{2}{y+4}$$

$$= \frac{y}{(y-5)(y+4)} + \frac{2}{y+4} \qquad \text{LCD} = (y-5)(y+4)$$

$$= \frac{y}{(y-5)(y+4)} + \frac{2}{y+4} \cdot \frac{y-5}{y-5}$$

$$= \frac{y + 2y - 10}{(y-5)(y+4)}$$

$$= \frac{3y - 10}{(y-5)(y+4)}, \text{ or } \frac{3y - 10}{y^2 - y - 20}$$

59.
$$\frac{3y+2}{y^2 + 5y - 24} + \frac{7}{y^2 + 4y - 32}$$

$$= \frac{3y+2}{(y+8)(y-3)} + \frac{7}{(y+8)(y-4)}$$

$$\text{LCD} = (y+8)(y-3)(y-4)$$

$$= \frac{3y+2}{(y+8)(y-3)} \cdot \frac{y-4}{y-4} + \frac{7}{(y+8)(y-4)} \cdot \frac{y-3}{y-3}$$

$$= \frac{3y^2 - 10y - 8 + 7y - 21}{(y+8)(y-3)(y-4)}$$

$$= \frac{3y^2 - 3y - 29}{(y+8)(y-3)(y-4)}$$

61.
$$\frac{3x-1}{x^2 + 2x - 3} - \frac{x+4}{x^2 - 9}$$

$$= \frac{3x-1}{(x+3)(x-1)} - \frac{x+4}{(x+3)(x-3)}$$

$$\text{LCD} = (x+3)(x-1)(x-3)$$

$$= \frac{3x-1}{(x+3)(x-1)} \cdot \frac{x-3}{x-3} - \frac{x+4}{(x+3)(x-3)} \cdot \frac{x-1}{x-1}$$

$$= \frac{3x^2 - 10x + 3 - (x^2 + 3x - 4)}{(x+3)(x-1)(x-3)}$$

$$= \frac{3x^2 - 10x + 3 - x^2 - 3x + 4}{(x+3)(x-1)(x-3)}$$

$$= \frac{2x^2 - 13x + 7}{(x+3)(x-1)(x-3)}$$

63.
$$\frac{1}{x+1} - \frac{x}{x-2} + \frac{x^2 + 2}{x^2 - x - 2}$$

$$= \frac{1}{x+1} - \frac{x}{x-2} + \frac{x^2 + 2}{(x-2)(x+1)}$$

$$\text{LCD} = (x+1)(x-2)$$

$$= \frac{1}{x+1} \cdot \frac{x-2}{x-2} - \frac{x}{x-2} \cdot \frac{x+1}{x+1} + \frac{x^2 + 2}{(x-2)(x+1)}$$

$$= \frac{x - 2 - (x^2 + x) + x^2 + 2}{(x+1)(x-2)}$$

$$= \frac{x - 2 - x^2 - x + x^2 + 2}{(x+1)(x-2)}$$

$$= \frac{0}{(x+1)(x-2)}$$

$$= 0$$

65.
$$\frac{x-1}{x-2} - \frac{x+1}{x+2} + \frac{x-6}{x^2 - 4}$$

$$= \frac{x-1}{x-2} - \frac{x+1}{x+2} + \frac{x-6}{(x+2)(x-2)}$$

$$\text{LCD} = (x-2)(x+2)$$

$$= \frac{x-1}{x-2} \cdot \frac{x+2}{x+2} - \frac{x+1}{x+2} \cdot \frac{x-2}{x-2} + \frac{x-6}{(x+2)(x-2)}$$

$$= \frac{(x^2 + x - 2) - (x^2 - x - 2) + (x-6)}{(x-2)(x+2)}$$

$$= \frac{x^2 + x - 2 - x^2 + x + 2 + x - 6}{(x-2)(x+2)}$$

$$= \frac{3x - 6}{(x-2)(x+2)}$$

$$= \frac{3(x-2)}{(x-2)(x+2)}$$

$$= \frac{3(x-2)}{(x-2)(x+2)}$$

$$= \frac{3}{x+2}$$

67.
$$\frac{y+2}{y+4} + \frac{y-7}{y^2 - 16} - \frac{y-3}{y-4}$$

$$= \frac{y+2}{y+4} + \frac{y-7}{(y+4)(y-4)} - \frac{y-3}{y-4}$$

$$\text{LCD} = (y+4)(y-4)$$

$$= \frac{y+2}{y+4} \cdot \frac{y-4}{y-4} + \frac{y-7}{(y+4)(y-4)} - \frac{y-3}{y-4} \cdot \frac{y+4}{y+4}$$

$$= \frac{(y^2 - 2y - 8) + (y-7) - (y^2 + y - 12)}{(y+4)(y-4)}$$

$$= \frac{y^2 - 2y - 8 + y - 7 - y^2 - y + 12}{(y+4)(y-4)}$$

$$= \frac{-2y - 3}{(y+4)(y-4)}$$

69. $\dfrac{4x}{x^2-1}+\dfrac{3x}{1-x}-\dfrac{4}{x-1}$

$=\dfrac{4x}{x^2-1}+\dfrac{3x}{1-x}\cdot\dfrac{-1}{-1}-\dfrac{4}{x-1}$

$=\dfrac{4x}{(x+1)(x-1)}+\dfrac{-3x}{x-1}-\dfrac{4}{x-1}$

$\qquad\qquad \text{LCD} = (x+1)(x-1)$

$=\dfrac{4x}{(x+1)(x-1)}+\dfrac{-3x}{x-1}\cdot\dfrac{x+1}{x+1}-\dfrac{4}{x-1}\cdot\dfrac{x+1}{x+1}$

$=\dfrac{4x-3x^2-3x-4x-4}{(x+1)(x-1)}$

$=\dfrac{-3x^2-3x-4}{x^2-1}$

71. $\dfrac{1}{x+y}+\dfrac{1}{y-x}-\dfrac{2x}{x^2-y^2}$

$=\dfrac{1}{x+y}+\dfrac{1}{y-x}\cdot\dfrac{-1}{-1}-\dfrac{2x}{x^2-y^2}$

$=\dfrac{1}{x+y}+\dfrac{-1}{x-y}-\dfrac{2x}{(x+y)(x-y)}$

$\qquad\qquad \text{LCD} = (x+y)(x-y)$

$=\dfrac{1}{x+y}\cdot\dfrac{x-y}{x-y}+\dfrac{-1}{x-y}\cdot\dfrac{x+y}{x+y}-\dfrac{2x}{(x+y)(x-y)}$

$=\dfrac{x-y-x-y-2x}{(x+y)(x-y)}$

$=\dfrac{-2x-2y}{(x+y)(x-y)}=\dfrac{-2(x+y)}{(x+y)(x-y)}$

$=\dfrac{-2(\cancel{x+y})}{(\cancel{x+y})(x-y)}$

$=\dfrac{-2}{x-y},\text{ or }\dfrac{2}{y-x}$

73. Discussion and Writing Exercise

75. Graph: $2x-3y>6$

We first graph the line $2x-3y=6$. The intercepts are $(0,-2)$ and $(3,0)$. We draw the line dashed since the inequality symbol is $>$. To determine which half-plane to shade, we consider a test point not on the line. We try $(0,0)$:

$$\dfrac{2x-3y>6}{2\cdot 0-3\cdot 0 \ ? \ 6}$$
$$\qquad 0 \ \Big| \qquad \text{FALSE}$$

Since $0>6$ is false, we shade the half-plane that does not contain $(0,0)$.

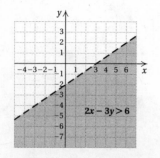

77. Graph: $5x+3y\le 15$

We first graph the line $5x+3y=15$. The intercepts are $(3,0)$ and $(0,5)$. We draw the line solid since the inequality symbol is $\le$. To determine which half-plane to shade, we consider a test point not on the line. We try $(0,0)$:

$$\dfrac{5x+3y\le 15}{5\cdot 0+3\cdot 0 \ ? \ 15}$$
$$\qquad 0 \ \Big| \qquad \text{TRUE}$$

Since $0\le 15$ is true, we shade the half-plane that contains $(0,0)$.

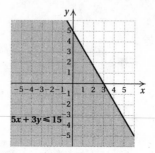

79. $t^3-8=t^3-2^3$

$\qquad = (t-2)(t^2+2t+4)$

81. $\quad 23x^4+23x$

$= 23x(x^3+1)$

$= 23x(x+1)(x^2-x+1)$

83. $\dfrac{15x^{-7}y^{12}z^4}{35x^{-2}y^6z^{-3}}=\dfrac{15}{35}x^{-7-(-2)}y^{12-6}z^{4-(-3)}$

$\qquad = \dfrac{3}{7}x^{-5}y^6z^7,\text{ or}$

$\qquad = \dfrac{3y^6z^7}{7x^5}$

85. From the graph or from the equation of the function we see that the domain is $(-\infty,2)\cup(2,\infty)$. From the graph we see that the range is $(-\infty,0)\cup(0,\infty)$.

87. $x^8-x^4=x^4(x^2+1)(x+1)(x-1)$

$x^5-x^2=x^2(x-1)(x^2+x+1)$

$x^5-x^3=x^3(x+1)(x-1)$

$x^5+x^2=x^2(x+1)(x^2-x+1)$

LCM =
$x^4(x^2+1)(x+1)(x-1)(x^2+x+1)(x^2-x+1)$

89.

$$\frac{x+y+1}{y-(x+1)} + \frac{x+y-1}{x-(y-1)} - \frac{x-y-1}{1-(y-x)}$$

$$= \frac{x+y+1}{-x+y-1} + \frac{x+y-1}{x-y+1} - \frac{x-y-1}{x-y+1}$$

$$= \frac{x+y+1}{-x+y-1} \cdot \frac{-1}{-1} + \frac{x+y-1}{x-y+1} - \frac{x-y-1}{x-y+1}$$

$$= \frac{-x-y-1}{x-y+1} + \frac{x+y-1}{x-y+1} - \frac{x-y-1}{x-y+1}$$

$$= \frac{-x-y-1+x+y-1-(x-y-1)}{x-y+1}$$

$$= \frac{-x-y-1+x+y-1-x+y+1}{x-y+1}$$

$$= \frac{-x+y-1}{x-y+1}$$

$$= \frac{-(x-y+1)}{x-y+1}$$

$$= \frac{-1 \cdot (x-y+1)}{1 \cdot (x-y+1)}$$

$$= \frac{-1}{1} \cdot \frac{x-y+1}{x-y+1}$$

$$= -1$$

91.

$$\frac{x}{x^4-y^4} - \frac{1}{x^2+2xy+y^2}$$

$$= \frac{x}{(x^2+y^2)(x+y)(x-y)} - \frac{1}{(x+y)(x+y)}$$

$$= \frac{x}{(x^2+y^2)(x+y)(x-y)} \cdot \frac{x+y}{x+y} -$$

$$\frac{1}{(x+y)(x+y)} \cdot \frac{(x^2+y^2)(x-y)}{(x^2+y^2)(x-y)}$$

$$= \frac{x^2+xy}{(x^2+y^2)(x+y)(x-y)(x+y)} -$$

$$\frac{x^3-x^2y+xy^2-y^3}{(x^2+y^2)(x+y)(x-y)(x+y)}$$

$$= \frac{x^2+xy-(x^3-x^2y+xy^2-y^3)}{(x^2+y^2)(x+y)(x-y)(x+y)}$$

$$= \frac{x^2+xy-x^3+x^2y-xy^2+y^3}{(x^2+y^2)(x+y)(x-y)(x+y)}$$

Exercise Set 5.3

1.

$$\frac{24x^6+18x^5-36x^2}{6x^2}$$

$$= \frac{24x^6}{6x^2} + \frac{18x^5}{6x^2} - \frac{36x^2}{6x^2}$$

$$= 4x^4+3x^3-6$$

3.

$$\frac{45y^7-20y^4+15y^2}{5y^2}$$

$$= \frac{45y^7}{5y^2} - \frac{20y^4}{5y^2} + \frac{15y^2}{5y^2}$$

$$= 9y^5-4y^2+3$$

5.

$$(32a^4b^3+14a^3b^2-22a^2b) \div 2a^2b$$

$$= \frac{32a^4b^3+14a^3b^2-22a^2b}{2a^2b}$$

$$= \frac{32a^4b^3}{2a^2b} + \frac{14a^3b^2}{2a^2b} - \frac{22a^2b}{2a^2b}$$

$$= 16a^2b^2+7ab-11$$

7.

$$\begin{array}{r} x+7 \\ x+3 \overline{\smash{\big)}\, x^2+10x+21} \\ \underline{x^2+3x} \\ 7x+21 \\ \underline{7x+21} \\ 0 \end{array}$$

$(x^2+10x)-(x^2+3x)=7x$

The answer is $x+7$.

9.

$$\begin{array}{r} a-12 \\ a+4 \overline{\smash{\big)}\, a^2-8a-16} \\ \underline{a^2+4a} \\ -12a-16 \\ \underline{-12a-48} \\ 32 \end{array}$$

$(a^2-8a)-(a^2+4a)=-12a$

$(-12a-16)-(-12a-48)=32$

The answer is $a-12$, R 32, or $a-12+\dfrac{32}{a+4}$.

11.

$$\begin{array}{r} x+2 \\ x+5 \overline{\smash{\big)}\, x^2+7x+14} \\ \underline{x^2+5x} \\ 2x+14 \\ \underline{2x+10} \\ 4 \end{array}$$

$(x^2+7x)-(x^2+5x)=2x$

$(2x+14)-(2x+10)=4$

The answer is $x+2$, R 4, or $x+2+\dfrac{4}{x+5}$.

13.

$$\begin{array}{r} 2y^2-y+2 \\ 2y+4 \overline{\smash{\big)}\, 4y^3+6y^2+0y+14} \\ \underline{4y^3+8y^2} \\ -2y^2+0y \\ \underline{-2y^2-4y} \\ 4y+14 \\ \underline{4y+8} \\ 6 \end{array}$$

The answer is $2y^2-y+2$, R 6, or $2y^2-y+2+\dfrac{6}{2y+4}$.

15.

$$\begin{array}{r} 2y^2+2y-1 \\ 5y-2 \overline{\smash{\big)}\, 10y^3+6y^2-9y+10} \\ \underline{10y^3-4y^2} \\ 10y^2-9y \\ \underline{10y^2-4y} \\ -5y+10 \\ \underline{-5y+2} \\ 8 \end{array}$$

The answer is $2y^2+2y-1$, R 8, or $2y^2+2y-1+\dfrac{8}{5y-2}$.

17.

$$
\require{enclose}
\begin{array}{r}
2x^2 - x - 9 \\
x^2 + 2 \enclose{longdiv}{2x^4 - x^3 - 5x^2 + x - 6} \\
\underline{2x^4 \phantom{{}- x^3} + 4x^2 \phantom{{}+ x - 6}} \\
-x^3 - 9x^2 + x \phantom{{}- 6} \\
\underline{-x^3 \phantom{{}- 9x^2} - 2x \phantom{{}- 6}} \\
-9x^2 + 3x - 6 \\
\underline{-9x^2 \phantom{{}+ 3x} - 18} \\
3x + 12
\end{array}
$$

The answer is $2x^2 - x - 9$, R $(3x + 12)$, or
$2x^2 - x - 9 + \dfrac{3x + 12}{x^2 + 2}$.

19.

$$
\require{enclose}
\begin{array}{r}
2x^3 + 5x^2 + 17x + 51 \\
x^2 - 3x \enclose{longdiv}{2x^5 - x^4 + 2x^3 + 0x^2 - x} \\
\underline{2x^5 - 6x^4 \phantom{{}+ 2x^3 + 0x^2 - x}} \\
5x^4 + 2x^3 \\
\underline{5x^4 - 15x^3 \phantom{{}+ 0x^2}} \\
17x^3 + 0x^2 \\
\underline{17x^3 - 51x^2 \phantom{{}- x}} \\
51x^2 - x \\
\underline{51x^2 - 153x} \\
152x
\end{array}
$$

The answer is $2x^3 + 5x^2 + 17x + 51$, R $152x$, or
$2x^3 + 5x^2 + 17x + 51 + \dfrac{152x}{x^2 - 3x}$.

21. $(x^3 - 2x^2 + 2x - 5) \div (x - 1)$

$$
\begin{array}{r|rrrr}
1 & 1 & -2 & 2 & -5 \\
  &   & 1 & -1 & 1 \\
\hline
  & 1 & -1 & 1 & -4
\end{array}
$$

The answer is $x^2 - x + 1$, R -4, or $x^2 - x + 1 + \dfrac{-4}{x - 1}$.

23. $(a^2 + 11a - 19) \div (a + 4) =$
$(a^2 + 11a - 19) \div [a - (-4)]$

$$
\begin{array}{r|rrr}
-4 & 1 & 11 & -19 \\
   &   & -4 & -28 \\
\hline
   & 1 & 7 & -47
\end{array}
$$

The answer is $a + 7$, R -47, or $a + 7 + \dfrac{-47}{a + 4}$.

25. $(x^3 - 7x^2 - 13x + 3) \div (x - 2)$

$$
\begin{array}{r|rrrr}
2 & 1 & -7 & -13 & 3 \\
  &   & 2 & -10 & -46 \\
\hline
  & 1 & -5 & -23 & -43
\end{array}
$$

The answer is $x^2 - 5x - 23$, R -43, or $x^2 - 5x - 23 + \dfrac{-43}{x - 2}$.

27. $(3x^3 + 7x^2 - 4x + 3) \div (x + 3) =$
$(3x^3 + 7x^2 - 4x + 3) \div [x - (-3)]$

$$
\begin{array}{r|rrrr}
-3 & 3 & 7 & -4 & 3 \\
   &   & -9 & 6 & -6 \\
\hline
   & 3 & -2 & 2 & -3
\end{array}
$$

The answer is $3x^2 - 2x + 2$, R -3, or $3x^2 - 2x + 2 + \dfrac{-3}{x + 3}$.

29. $(y^3 - 3y + 10) \div (y - 2) =$
$(y^3 + 0y^2 - 3y + 10) \div (y - 2)$

$$
\begin{array}{r|rrrr}
2 & 1 & 0 & -3 & 10 \\
  &   & 2 & 4 & 2 \\
\hline
  & 1 & 2 & 1 & 12
\end{array}
$$

The answer is $y^2 + 2y + 1$, R 12, or $y^2 + 2y + 1 + \dfrac{12}{y - 2}$.

31. $(3x^4 - 25x^2 - 18) \div (x - 3) =$
$(3x^4 + 0x^3 - 25x^2 + 0x - 18) \div (x - 3)$

$$
\begin{array}{r|rrrrr}
3 & 3 & 0 & -25 & 0 & -18 \\
  &   & 9 & 27 & 6 & 18 \\
\hline
  & 3 & 9 & 2 & 6 & 0
\end{array}
$$

The answer is $3x^3 + 9x^2 + 2x + 6$.

33. $(x^3 - 8) \div (x - 2) = (x^3 + 0x^2 + 0x - 8) \div (x - 2)$

$$
\begin{array}{r|rrrr}
2 & 1 & 0 & 0 & -8 \\
  &   & 2 & 4 & 8 \\
\hline
  & 1 & 2 & 4 & 0
\end{array}
$$

The answer is $x^2 + 2x + 4$.

35. $(y^4 - 16) \div (y - 2) =$
$(y^4 + 0y^3 + 0y^2 + 0y - 16) \div (y - 2)$

$$
\begin{array}{r|rrrrr}
2 & 1 & 0 & 0 & 0 & -16 \\
  &   & 2 & 4 & 8 & 16 \\
\hline
  & 1 & 2 & 4 & 8 & 0
\end{array}
$$

The answer is $y^3 + 2y^2 + 4y + 8$.

37. Discussion and Writing Exercise

39. Graph: $2x - 3y < 6$

We first graph the line $2x - 3y = 6$. The intercepts are $(0, -2)$ and $(3, 0)$. We draw the line dashed since the inequality symbol is $<$. To determine which half-plane to shade, we consider a test point not on the line. We try $(0, 0)$:

$$
\begin{array}{c}
2x - 3y < 6 \\
\hline
2 \cdot 0 - 3 \cdot 0 \; ? \; 6 \\
0 \; \Big| \qquad \text{TRUE}
\end{array}
$$

Since $0 < 6$ is true, we shade the half-plane that contains $(0, 0)$.

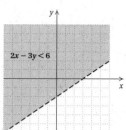

41. Graph $y > 4$.

We first graph the line $y = 4$. This is a horizontal line 4 units above the x-axis. We draw the line dashed since the inequality symbol is $>$. To determine which half-plane to shade, we write $y > 4$ as $0x + y > 4$ and consider a test point not on the line. We try $(0, 0)$:

$$\begin{array}{c} 0x + y > 4 \\ \hline 0 \cdot 0 + 0 \; ? \; 4 \\ 0 \;\mid\; \quad \text{FALSE} \end{array}$$

Since $0 > 4$ is false, we shade the half-plane that does not contain $(0, 0)$.

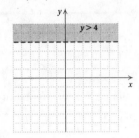

43. Graph: $f(x) = x^2$.

We select x-values and find the corresponding y-values. We plot these ordered pairs and draw the graph.

x	y
-2	4
-1	1
0	0
1	1
2	4

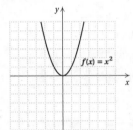

45. Graph: $f(x) = 3 - x^2$.

We select x-values and find the corresponding y-values. We plot these ordered pairs and draw the graph.

x	y
-3	-6
-2	-1
-1	2
0	3
1	2
2	-1
3	-6

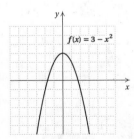

47. $\quad x^2 - 5x = 0$

$x(x - 5) = 0$

$x = 0 \quad \text{or} \quad x - 5 = 0$

$x = 0 \quad \text{or} \qquad x = 5$

The solutions are 0 and 5.

49. $\qquad 12x^2 = 17x + 5$

$12x^2 - 17x - 5 = 0$

$(4x + 1)(3x - 5) = 0$

$4x + 1 = 0 \quad \text{or} \quad 3x - 5 = 0$

$4x = -1 \quad \text{or} \qquad 3x = 5$

$x = -\dfrac{1}{4} \quad \text{or} \qquad x = \dfrac{5}{3}$

The solutions are $-\dfrac{1}{4}$ and $\dfrac{5}{3}$.

51. $\quad f(x) = 4x^3 + 16x^2 - 3x - 45$

$f(-3) = 4(-3)^3 + 16(-3)^2 - 3(-3) - 45$

$\qquad = -108 + 144 + 9 - 45$

$\qquad = 0$

Since $f(-3) = 0$, we know that -3 is a solution of $f(x) = 0$. That is, $f(x) = 0$ when $x = -3$, or when $x + 3 = 0$. We find $f(x) \div (x + 3)$.

$$\begin{array}{r} 4x^2 + 4x - 15 \\ x + 3 \enclose{longdiv}{4x^3 + 16x^2 - 3x - 45} \\ \underline{4x^3 + 12x^2} \\ 4x^2 - 3x \\ \underline{4x^2 + 12x} \\ -15x - 45 \\ \underline{-15x - 45} \\ 0 \end{array}$$

Now we know that $(x + 3)(4x^2 + 4x - 15) = 0$. To find the other solutions of $f(x) = 0$ we solve $4x^2 + 4x - 15 = 0$.

$4x^2 + 4x - 15 = 0$

$(2x - 3)(2x + 5) = 0$

$2x - 3 = 0 \quad \text{or} \quad 2x + 5 = 0$

$2x = 3 \quad \text{or} \qquad 2x = -5$

$x = \dfrac{3}{2} \quad \text{or} \qquad x = -\dfrac{5}{2}$

Then the solutions of $f(x) = 0$ are -3, $\dfrac{3}{2}$, and $-\dfrac{5}{2}$.

53.

$$\begin{array}{r} x - 5 \\ x + 2 \enclose{longdiv}{x^2 - 3x + 2k} \\ \underline{x^2 + 2x} \\ -5x + 2k \\ \underline{-5x - 10} \\ 2k + 10 \end{array}$$

The remainder is 7. Thus, we solve the following equation for k.

$2k + 10 = 7$

$2k = -3$

$k = -\dfrac{3}{2}$

55.

$$\begin{array}{r} a^2 + ab \\ a^2 + 3ab + 2b^2 \enclose{longdiv}{a^4 + 4a^3b + 5a^2b^2 + 2ab^3} \\ \underline{a^4 + 3a^3b + 2a^2b^2} \\ a^3b + 3a^2b^2 + 2ab^3 \\ \underline{a^3b + 3a^2b^2 + 2ab^3} \\ 0 \end{array}$$

The answer is $a^2 + ab$.

Exercise Set 5.4

1. $\dfrac{2 + \frac{3}{5}}{4 - \frac{1}{2}}$

The LCM of the denominators is $5 \cdot 2$, or 10. We multiply by 1 using 10/10.

$$\frac{2 + \frac{3}{5}}{4 - \frac{1}{2}} = \frac{2 + \frac{3}{5}}{4 - \frac{1}{2}} \cdot \frac{10}{10}$$

$$= \frac{\left(2 + \frac{3}{5}\right) \cdot 10}{\left(4 - \frac{1}{2}\right) \cdot 10}$$

$$= \frac{2 \cdot 10 + \frac{3}{5} \cdot 10}{4 \cdot 10 - \frac{1}{2} \cdot 10}$$

$$= \frac{20 + 6}{40 - 5}$$

$$= \frac{26}{35}$$

3. $\dfrac{\frac{2}{3} + \frac{4}{5}}{\frac{3}{4} - \frac{1}{2}}$

The LCM of the denominators is $3 \cdot 5 \cdot 2 \cdot 2$, or 60. We multiply by 1 using 60/60.

$$\frac{\frac{2}{3} + \frac{4}{5}}{\frac{3}{4} - \frac{1}{2}} = \frac{\frac{2}{3} + \frac{4}{5}}{\frac{3}{4} - \frac{1}{2}} \cdot \frac{60}{60}$$

$$= \frac{\left(\frac{2}{3} + \frac{4}{5}\right) \cdot 60}{\left(\frac{3}{4} - \frac{1}{2}\right) \cdot 60}$$

$$= \frac{\frac{2}{3} \cdot 60 + \frac{4}{5} \cdot 60}{\frac{3}{4} \cdot 60 - \frac{1}{2} \cdot 60}$$

$$= \frac{40 + 48}{45 - 30}$$

$$= \frac{88}{15}$$

5. $\dfrac{\frac{x}{y^2}}{\frac{y^3}{x^2}} = \dfrac{x}{y^2} \cdot \dfrac{x^2}{y^3}$ Multiplying by the reciprocal of the divisor

$$= \frac{x^3}{y^5}$$

7. $= \dfrac{\frac{9x^2 - y^2}{xy}}{\frac{3x - y}{y}}$

$$= \frac{9x^2 - y^2}{xy} \cdot \frac{y}{3x - y} \quad \text{Multiplying by the reciprocal of the divisor}$$

$$= \frac{(9x^2 - y^2)(y)}{xy(3x - y)}$$

$$= \frac{(3x + y)(3x - y)(y)}{xy\,(3x - y)}$$

$$= \frac{3x + y}{x}$$

9. $\dfrac{\frac{1}{a} + 2}{\frac{1}{a} - 1}$

The LCM of all the denominators is a. We multiply by 1 using a/a.

$$\frac{\frac{1}{a} + 2}{\frac{1}{a} - 1} = \frac{\frac{1}{a} + 2}{\frac{1}{a} - 1} \cdot \frac{a}{a}$$

$$= \frac{\left(\frac{1}{a} + 2\right) \cdot a}{\left(\frac{1}{a} - 1\right) \cdot a}$$

$$= \frac{\frac{1}{a} \cdot a + 2a}{\frac{1}{a} \cdot a - a}$$

$$= \frac{1 + 2a}{1 - a}$$

11. $\dfrac{x - \frac{1}{x}}{x + \frac{1}{x}}$

The LCM of all the denominators is x. We multiply by 1 using x/x.

$$\frac{x - \frac{1}{x}}{x + \frac{1}{x}} = \frac{x - \frac{1}{x}}{x + \frac{1}{x}} \cdot \frac{x}{x}$$

$$= \frac{\left(x - \frac{1}{x}\right) \cdot x}{\left(x + \frac{1}{x}\right) \cdot x}$$

$$= \frac{x \cdot x - \frac{1}{x} \cdot x}{x \cdot x + \frac{1}{x} \cdot x}$$

$$= \frac{x^2 - 1}{x^2 + 1}$$

13. $\dfrac{\dfrac{3}{x}+\dfrac{4}{y}}{\dfrac{4}{x}-\dfrac{3}{y}}$

The LCM of all the denominators is xy. We multiply by 1 using $\dfrac{xy}{xy}$.

$$\dfrac{\dfrac{3}{x}+\dfrac{4}{y}}{\dfrac{4}{x}-\dfrac{3}{y}}=\dfrac{\dfrac{3}{x}+\dfrac{4}{y}}{\dfrac{4}{x}-\dfrac{3}{y}}\cdot\dfrac{xy}{xy}$$

$$=\dfrac{\left(\dfrac{3}{x}+\dfrac{4}{y}\right)\cdot xy}{\left(\dfrac{4}{x}-\dfrac{3}{y}\right)\cdot xy}$$

$$=\dfrac{\dfrac{3}{x}\cdot xy+\dfrac{4}{y}\cdot xy}{\dfrac{4}{x}\cdot xy-\dfrac{3}{y}\cdot xy}$$

$$=\dfrac{3y+4x}{4y-3x}$$

15. $\dfrac{a-\dfrac{3a}{b}}{b-\dfrac{b}{a}}$

$$=\dfrac{a-\dfrac{3a}{b}}{b-\dfrac{b}{a}}\cdot\dfrac{ab}{ab}\qquad\text{Using the LCM of the denominators}$$

$$=\dfrac{\left(a-\dfrac{3a}{b}\right)\cdot ab}{\left(b-\dfrac{b}{a}\right)\cdot ab}$$

$$=\dfrac{a(ab)-\dfrac{3a}{b}\cdot ab}{b(ab)-\dfrac{b}{a}\cdot ab}$$

$$=\dfrac{a^2b-3a^2}{ab^2-b^2}$$

$$=\dfrac{a^2(b-3)}{b^2(a-1)}$$

17. $\dfrac{\dfrac{1}{a}+\dfrac{1}{b}}{\dfrac{a^2-b^2}{ab}}$

$$=\dfrac{\dfrac{1}{a}+\dfrac{1}{b}}{\dfrac{a^2-b^2}{ab}}\cdot\dfrac{ab}{ab}\qquad\text{Using the LCM of the denominators}$$

$$=\dfrac{\left(\dfrac{1}{a}+\dfrac{1}{b}\right)\cdot ab}{\left(\dfrac{a^2-b^2}{ab}\right)\cdot ab}$$

$$=\dfrac{\dfrac{1}{a}\cdot ab+\dfrac{1}{b}\cdot ab}{\dfrac{a^2-b^2}{ab}\cdot ab}$$

$$=\dfrac{b+a}{a^2-b^2}=\dfrac{b+a}{(a+b)(a-b)}$$

$$=\dfrac{a+b}{a+b}\cdot\dfrac{1}{a-b}\qquad(b+a=a+b)$$

$$=\dfrac{1}{a-b}$$

19. $\dfrac{\dfrac{1}{x+h}-\dfrac{1}{x}}{h}$

$$=\dfrac{\dfrac{1}{x+h}\cdot\dfrac{x}{x}-\dfrac{1}{x}\cdot\dfrac{x+h}{x+h}}{h}\qquad\text{Adding in the numerator}$$

$$=\dfrac{\dfrac{x-x-h}{x(x+h)}}{h}=\dfrac{\dfrac{-h}{x(x+h)}}{h}$$

$$=\dfrac{-h}{x(x+h)}\cdot\dfrac{1}{h}\qquad\begin{array}{l}\text{Multiplying by the reciprocal}\\\text{of the divisor}\end{array}$$

$$=\dfrac{-h}{hx(x+h)}$$

$$=\dfrac{h}{h}\cdot\dfrac{-1}{x(x+h)}\qquad\text{Removing a factor of 1}$$

$$=\dfrac{-1}{x(x+h)},\ \text{or}\ -\dfrac{1}{x(x+h)}$$

21. $\dfrac{\dfrac{x^2-x-12}{x^2-2x-15}}{\dfrac{x^2+8x+12}{x^2-5x-14}}$

$$=\dfrac{x^2-x-12}{x^2-2x-15}\cdot\dfrac{x^2-5x-14}{x^2+8x+12}\qquad\begin{array}{l}\text{Multiplying by the}\\\text{reciprocal of}\\\text{the divisor}\end{array}$$

$$=\dfrac{(x^2-x-12)(x^2-5x-14)}{(x^2-2x-15)(x^2+8x+12)}$$

$$=\dfrac{(x+3)(x-4)(x-7)(x+2)}{(x-5)(x+3)(x+6)(x+2)}\qquad\begin{array}{l}\text{Removing a factor}\\\text{of 1}\end{array}$$

$$=\dfrac{(x-4)(x-7)}{(x-5)(x+6)}$$

23. $\dfrac{\dfrac{1}{x+2}+\dfrac{4}{x-3}}{\dfrac{2}{x-3}-\dfrac{7}{x+2}}$

$=\dfrac{\dfrac{1}{x+2}+\dfrac{4}{x-3}}{\dfrac{2}{x-3}-\dfrac{7}{x+2}}\cdot\dfrac{(x+2)(x-3)}{(x+2)(x-3)}$ Using the LCM of the denominators

$=\dfrac{\dfrac{1}{x+2}\cdot(x+2)(x-3)+\dfrac{4}{x-3}\cdot(x+2)(x-3)}{\dfrac{2}{x-3}\cdot(x+2)(x-3)-\dfrac{7}{x+2}\cdot(x+2)(x-3)}$

$=\dfrac{x-3+4(x+2)}{2(x+2)-7(x-3)}$

$=\dfrac{x-3+4x+8}{2x+4-7x+21}$

$=\dfrac{5x+5}{-5x+25}$

$=\dfrac{\cancel{5}\,(x+1)}{\cancel{5}\,(-x+5)}$ Removing a factor of 1

$=\dfrac{x+1}{-x+5}$, or $\dfrac{x+1}{5-x}$

25. $\dfrac{\dfrac{6}{x^2-4}-\dfrac{5}{x+2}}{\dfrac{7}{x^2-4}-\dfrac{4}{x-2}}$

$=\dfrac{\dfrac{6}{(x+2)(x-2)}-\dfrac{5}{x+2}\cdot\dfrac{x-2}{x-2}}{\dfrac{7}{(x+2)(x-2)}-\dfrac{4}{x-2}\cdot\dfrac{(x+2)}{(x+2)}}$

Finding the LCM of the denominators and multiplying by 1

$=\dfrac{\dfrac{6}{(x+2)(x-2)}-\dfrac{5x-10}{(x+2)(x-2)}}{\dfrac{7}{(x+2)(x-2)}-\dfrac{4x+8}{(x+2)(x-2)}}$

$=\dfrac{\dfrac{6-5x+10}{(x+2)(x-2)}}{\dfrac{7-4x-8}{(x+2)(x-2)}}$ Subtracting in the numerator and in the denominator

$=\dfrac{\dfrac{16-5x}{(x+2)(x-2)}}{\dfrac{-1-4x}{(x+2)(x-2)}}$

$=\dfrac{16-5x}{(x+2)(x-2)}\cdot\dfrac{(x+2)(x-2)}{-1-4x}$

Multiplying by the reciprocal of the denominator

$=\dfrac{(16-5x)\cancel{(x+2)}\cancel{(x-2)}}{\cancel{(x+2)}\cancel{(x-2)}(-1-4x)}$ Removing a factor of 1

$=\dfrac{16-5x}{-1-4x}$

$=\dfrac{\cancel{-1}\cdot(5x-16)}{\cancel{-1}(4x+1)}=\dfrac{5x-16}{4x+1}$

27. $\dfrac{\dfrac{1}{z^2}-\dfrac{1}{w^2}}{\dfrac{1}{z^3}+\dfrac{1}{w^3}}$

$=\dfrac{\dfrac{1}{z^2}-\dfrac{1}{w^2}}{\dfrac{1}{z^3}+\dfrac{1}{w^3}}\cdot\dfrac{z^3w^3}{z^3w^3}$ Multiplying by the LCM of the denominators

$=\dfrac{\left(\dfrac{1}{z^2}-\dfrac{1}{w^2}\right)\cdot z^3w^3}{\left(\dfrac{1}{z^3}+\dfrac{1}{w^3}\right)\cdot z^3w^3}$

$=\dfrac{zw^3-z^3w}{w^3+z^3}$

$=\dfrac{zw(w^2-z^2)}{(w+z)(w^2-wz+z^2)}$

$=\dfrac{zw(w\cancel{+}z)(w-z)}{(w\cancel{+}z)(w^2-wz+z^2)}$

$=\dfrac{zw(w-z)}{w^2-wz+z^2}$

29. $\dfrac{\dfrac{3}{x^2+2x-3}-\dfrac{1}{x^2-3x-10}}{\dfrac{3}{x^2-6x+5}-\dfrac{1}{x^2+5x+6}}$

$=\dfrac{\dfrac{3}{(x+3)(x-1)}-\dfrac{1}{(x-5)(x+2)}}{\dfrac{3}{(x-5)(x-1)}-\dfrac{1}{(x+3)(x+2)}}$

$=\dfrac{\dfrac{3}{(x+3)(x-1)}-\dfrac{1}{(x-5)(x+2)}}{\dfrac{3}{(x-5)(x-1)}-\dfrac{1}{(x+3)(x+2)}}\cdot$

$\dfrac{(x+3)(x-1)(x-5)(x+2)}{(x+3)(x-1)(x-5)(x+2)}$

Multiplying by 1, using the LCD

$=\dfrac{3(x-5)(x+2)-(x+3)(x-1)}{3(x+3)(x+2)-(x-1)(x-5)}$

$=\dfrac{3(x^2-3x-10)-(x^2+2x-3)}{3(x^2+5x+6)-(x^2-6x+5)}$

$=\dfrac{3x^2-9x-30-x^2-2x+3}{3x^2+15x+18-x^2+6x-5}$

$=\dfrac{2x^2-11x-27}{2x^2+21x+13}$

31. Discussion and Writing Exercise

33. $4x^3+20x^2+6x$

$=2x(2x^2+10x+3)$ Factoring out the largest common factor

The trinomial $2x^2+10x+3$ cannot be factored as a product of binomials so we have the complete factorization.

35. $y^3-8=y^3-2^3=(y-2)(y^2+2y+4)$

37. $1000x^3+1=(10x)^3+1^3=(10x+1)(100x^2-10x+1)$

39. $y^3 - 64x^3 = y^3 - (4x)^3 = (y - 4x)(y^2 + 4xy + 16x^2)$

41.
$$T = \frac{r + s}{3}$$
$$3T = r + s \qquad \text{Multiplying by 3}$$
$$3T - r = s \qquad \text{Subtracting } r$$

43.
$$f(x) = x^2 - 3$$
$$f(-5) = (-5)^2 - 3 = 25 - 3 = 22$$

45. $f(a) = \dfrac{3}{a^2}, \ f(a + h) = \dfrac{3}{(a + h)^2}$

$$\frac{f(a + h) - f(a)}{h} = \frac{\dfrac{3}{(a + h)^2} - \dfrac{3}{a^2}}{h}$$

$$= \frac{3a^2 - 3(a + h)^2}{a^2(a + h)^2} \cdot \frac{1}{h}$$

$$= \frac{3a^2 - 3a^2 - 6ah - 3h^2}{a^2(a + h)^2} \cdot \frac{1}{h}$$

$$= \frac{-6ah - 3h^2}{a^2(a + h)^2 h}$$

$$= \frac{-3h(2a + h)}{a^2(a + h)^2 h}$$

$$= \frac{-3\cancel{h}(2a + h)}{a^2(a + h)^2 \cancel{h}}$$

$$= \frac{-3(2a + h)}{x^2(a + h)^2}, \text{ or}$$

$$= \frac{-6a - 3h}{a^2(a + h)^2}$$

47. $f(a) = \dfrac{1}{1 - a}, \ f(a + h) = \dfrac{1}{1 - (a + h)} = \dfrac{1}{1 - a - h}$

$$\frac{f(a + h) - f(a)}{h}$$

$$= \frac{\dfrac{1}{1 - a - h} - \dfrac{1}{1 - a}}{h}$$

$$= \frac{1 - a - (1 - a - h)}{(1 - a - h)(1 - a)} \cdot \frac{1}{h}$$

$$= \frac{1 - a - 1 + a + h}{(1 - a - h)(1 - a)} \cdot \frac{1}{h}$$

$$= \frac{h}{(1 - a - h)(1 - a)h}$$

$$= \frac{\cancel{h} \cdot 1}{(1 - a - h)(1 - a)\cancel{h}}$$

$$= \frac{1}{(1 - a - h)(1 - a)}$$

49.
$$\frac{5x^{-1} - 5y^{-1} + 10x^{-1}y^{-1}}{6x^{-1} - 6y^{-1} + 12x^{-1}y^{-1}}$$

$$= \frac{\dfrac{5}{x} - \dfrac{5}{y} + \dfrac{10}{xy}}{\dfrac{6}{x} - \dfrac{6}{y} + \dfrac{12}{xy}}$$

$$= \frac{\dfrac{5}{x} - \dfrac{5}{y} + \dfrac{10}{xy}}{\dfrac{6}{x} - \dfrac{6}{y} + \dfrac{12}{xy}} \cdot \frac{xy}{xy}$$

$$= \frac{5y - 5x + 10}{6y - 6x + 12}$$

$$= \frac{5(y - x + 2)}{6(y - x + 2)}$$

$$= \frac{5}{6} \cdot \frac{y - x + 2}{y - x + 2}$$

$$= \frac{5}{6}$$

51.
$$\frac{1}{x^2 - \dfrac{1}{x}} = \frac{1}{x^2 \cdot \dfrac{x}{x} - \dfrac{1}{x}} = \frac{1}{\dfrac{x^3}{x} - \dfrac{1}{x}} =$$

$$\frac{1}{\dfrac{x^3 - 1}{x}} = 1 \cdot \frac{x}{x^3 - 1} = \frac{x}{x^3 - 1}$$

53.
$$\frac{\dfrac{1}{a^3 + b^3}}{a + b} = 1 \cdot \frac{a + b}{a^3 + b^3}$$

$$= \frac{1 \cdot (\cancel{a + b})}{(\cancel{a + b})(a^2 - ab + b^2)}$$

$$= \frac{1}{a^2 - ab + b^2}$$

Exercise Set 5.5

1. $\dfrac{y}{10} = \dfrac{2}{5} + \dfrac{3}{8}$, LCM is 40

$$40 \cdot \frac{y}{10} = 40 \cdot \left(\frac{2}{5} + \frac{3}{8}\right) \qquad \text{Multiplying by the LCM}$$

$$4y = 40 \cdot \frac{2}{5} + 40 \cdot \frac{3}{8} \qquad \text{Removing parentheses}$$

$$4y = 16 + 15$$

$$4y = 31$$

$$y = \frac{31}{4}$$

Check:

$$\frac{y}{10} = \frac{2}{5} + \frac{3}{8}$$

$$\frac{\dfrac{31}{4}}{10} \; ? \; \frac{2}{5} + \frac{3}{8}$$

$$\frac{31}{4} \cdot \frac{1}{10} \; \Big| \; \frac{16}{40} + \frac{15}{40}$$

$$\frac{31}{40} \; \Big| \; \frac{31}{40} \qquad \text{TRUE}$$

The solution is $\dfrac{31}{4}$.

3. $\dfrac{1}{4} - \dfrac{5}{6} = \dfrac{1}{a}$, LCM is $12a$

$12a \cdot \left(\dfrac{1}{4} - \dfrac{5}{6}\right) = 12a \cdot \dfrac{1}{a}$ Multiplying by the LCM

$12a \cdot \dfrac{1}{4} - 12a \cdot \dfrac{5}{6} = 12$

$3a - 10a = 12$

$-7a = 12$

$a = -\dfrac{12}{7}$

Check: $\dfrac{\dfrac{1}{4} - \dfrac{5}{6} = \dfrac{1}{a}}{}$

$\dfrac{1}{4} - \dfrac{5}{6} \;?\; \dfrac{1}{-\dfrac{12}{7}}$

$\dfrac{3}{12} - \dfrac{10}{12} \;\Big|\; 1 \cdot \left(-\dfrac{7}{12}\right)$

$-\dfrac{7}{12} \;\Big|\; -\dfrac{7}{12}$ TRUE

The solution is $-\dfrac{12}{7}$.

5. $\dfrac{x}{3} - \dfrac{x}{4} = 12$, LCM is 12

$12 \cdot \left(\dfrac{x}{3} - \dfrac{x}{4}\right) = 12 \cdot 12$

$12 \cdot \dfrac{x}{3} - 12 \cdot \dfrac{x}{4} = 144$

$4x - 3x = 144$

$x = 144$

Check: $\dfrac{x}{3} - \dfrac{x}{4} = 12$

$\dfrac{\dfrac{144}{3} - \dfrac{144}{4} \;?\; 12}{}$

$48 - 36 \;\Big|$

$12 \;\Big|$ TRUE

The solution is 144.

7. $x + \dfrac{8}{x} = -9$, LCM is x

$x\left(x + \dfrac{8}{x}\right) = x(-9)$

$x \cdot x + x \cdot \dfrac{8}{x} = -9x$

$x^2 + 8 = -9x$

$x^2 + 9x + 8 = 0$

$(x+1)(x+8) = 0$

$x + 1 = 0 \quad or \quad x + 8 = 0$ Principle of zero products

$x = -1 \quad or \qquad x = -8$

Check:

For -1:

$\dfrac{x + \dfrac{8}{x} = -9}{}$

$-1 + \dfrac{8}{-1} \;?\; -9$

$-1 - 8 \;\Big|$

$-9 \;\Big|$ TRUE

For -8:

$\dfrac{x + \dfrac{8}{x} = -9}{}$

$-8 + \dfrac{8}{-8} \;?\; -9$

$-8 - 1 \;\Big|$

$-9 \;\Big|$ TRUE

The solutions are -1 and -8.

9. $\dfrac{3}{y} + \dfrac{7}{y} = 5$, LCM is y

$y\left(\dfrac{3}{y} + \dfrac{7}{y}\right) = y \cdot 5$

$y \cdot \dfrac{3}{y} + y \cdot \dfrac{7}{y} = 5y$

$3 + 7 = 5y$

$10 = 5y$

$2 = y$

Check: $\dfrac{3}{y} + \dfrac{7}{y} = 5$

$\dfrac{\dfrac{3}{2} + \dfrac{7}{2} \;?\; 5}{}$

$\dfrac{10}{2} \;\Big|$

$5 \;\Big|$ TRUE

The solution is 2.

11. $\dfrac{1}{2} = \dfrac{z-5}{z+1}$, LCM is $2(z+1)$

$2(z+1) \cdot \dfrac{1}{2} = 2(z+1) \cdot \dfrac{z-5}{z+1}$

$z + 1 = 2(z - 5)$

$z + 1 = 2z - 10$

$11 = z$

Check: $\dfrac{1}{2} = \dfrac{z-5}{z+1}$

$\dfrac{\dfrac{1}{2} \;?\; \dfrac{11-5}{11+1}}{}$

$\Big|\; \dfrac{6}{12}$

$\Big|\; \dfrac{1}{2}$ TRUE

The solution is 11.

13.
$$\frac{3}{y+1} = \frac{2}{y-3}, \text{ LCM is } (y+1)(y-3)$$

$$(y+1)(y-3) \cdot \frac{3}{y+1} = (y+1)(y-3) \cdot \frac{2}{y-3}$$

$$3(y-3) = 2(y+1)$$

$$3y - 9 = 2y + 2$$

$$y = 11$$

Check:
$$\frac{3}{y+1} = \frac{2}{y-3}$$

$$\frac{3}{11+1} \; ? \; \frac{2}{11-3}$$

$$\frac{3}{12} \; \bigg| \; \frac{2}{8}$$

$$\frac{1}{4} \; \bigg| \; \frac{1}{4} \qquad \text{TRUE}$$

The solution is 11.

15.
$$\frac{y-1}{y-3} = \frac{2}{y-3}, \text{ LCM is } y-3$$

$$(y-3) \cdot \frac{y-1}{y-3} = (y-3) \cdot \frac{2}{y-3}$$

$$y - 1 = 2$$

$$y = 3$$

Check:
$$\frac{y-1}{y-3} = \frac{2}{y-3}$$

$$\frac{3-1}{3-3} \; ? \; \frac{2}{3-3}$$

$$\frac{2}{0} \; \bigg| \; \frac{2}{0} \qquad \text{UNDEFINED}$$

We know that 3 is not a solution of the original equation, because it results in division by 0. The equation has no solution.

17.
$$\frac{x+1}{x} = \frac{3}{2}, \text{ LCM is } 2x$$

$$2x \cdot \frac{x+1}{x} = 2x \cdot \frac{3}{2}$$

$$2(x+1) = x \cdot 3$$

$$2x + 2 = 3x$$

$$2 = x$$

Check:
$$\frac{x+1}{x} = \frac{3}{2}$$

$$\frac{2+1}{2} \; ? \; \frac{3}{2}$$

$$\frac{3}{2} \; \bigg| \qquad \text{TRUE}$$

The solution is 2.

19.
$$\frac{1}{2} - \frac{4}{9x} = \frac{4}{9} - \frac{1}{6x}, \text{ LCM is } 18x$$

$$18x\left(\frac{1}{2} - \frac{4}{9x}\right) = 18x\left(\frac{4}{9} - \frac{1}{6x}\right)$$

$$18x \cdot \frac{1}{2} - 18x \cdot \frac{4}{9x} = 18x \cdot \frac{4}{9} - 18x \cdot \frac{1}{6x}$$

$$9x - 8 = 8x - 3$$

$$x = 5$$

Since 5 checks, it is the solution.

21.
$$\frac{60}{x} - \frac{60}{x-5} = \frac{2}{x}, \text{ LCM is } x(x-5)$$

$$x(x-5)\left(\frac{60}{x} - \frac{60}{x-5}\right) = x(x-5) \cdot \frac{2}{x}$$

$$60(x-5) - 60x = 2(x-5)$$

$$60x - 300 - 60x = 2x - 10$$

$$-300 = 2x - 10$$

$$-290 = 2x$$

$$-145 = x$$

Since -145 checks, it is the solution.

23.
$$\frac{7}{5x-2} = \frac{5}{4x}, \text{ LCM is } 4x(5x-2)$$

$$4x(5x-2) \cdot \frac{7}{5x-2} = 4x(5x-2) \cdot \frac{5}{4x}$$

$$4x \cdot 7 = 5(5x-2)$$

$$28x = 25x - 10$$

$$3x = -10$$

$$x = -\frac{10}{3}$$

Since $-\dfrac{10}{3}$ checks, it is the solution.

25.
$$\frac{x}{x-2} + \frac{x}{x^2-4} = \frac{x+3}{x+2}$$

$$\frac{x}{x-2} + \frac{x}{(x+2)(x-2)} = \frac{x+3}{x+2}$$

$$\text{LCM is } (x+2)(x-2)$$

$$(x+2)(x-2)\left(\frac{x}{x-2} + \frac{x}{(x+2)(x-2)}\right) =$$

$$(x+2)(x-2) \cdot \frac{x+3}{x+2}$$

$$x(x+2) + x =$$

$$(x-2)(x+3)$$

$$x^2 + 2x + x = x^2 + x - 6$$

$$3x = x - 6$$

$$2x = -6$$

$$x = -3$$

Since -3 checks, it is the solution.

27.
$$\frac{6}{x^2 - 4x + 3} - \frac{1}{x - 3} = \frac{1}{4x - 4}$$

$$\frac{6}{(x-3)(x-1)} - \frac{1}{x-3} = \frac{1}{4(x-1)}$$

LCM is $4(x-3)(x-1)$

$$4(x-3)(x-1)\left(\frac{6}{(x-3)(x-1)} - \frac{1}{x-3}\right) =$$

$$4(x-3)(x-1) \cdot \frac{1}{4(x-1)}$$

$$4 \cdot 6 - 4(x-1) = x - 3$$

$$24 - 4x + 4 \quad x - 3$$

$$-5x = -31$$

$$x = \frac{31}{5}$$

Since $\frac{31}{5}$ checks, it is the solution.

29.
$$\frac{5}{y + 3} = \frac{1}{4y^2 - 36} + \frac{2}{y - 3}$$

$$\frac{5}{y + 3} = \frac{1}{4(y+3)(y-3)} + \frac{2}{y - 3}$$

LCM is $4(y+3)(y-3)$

$$4(y+3)(y-3) \cdot \frac{5}{y+3} =$$

$$4(y+3)(y-3)\left(\frac{1}{4(y+3)(y-3)} + \frac{2}{y-3}\right)$$

$$4 \cdot 5(y-3) = 1 + 4 \cdot 2(y+3)$$

$$20y - 60 = 1 + 8y + 24$$

$$12y = 85$$

$$y = \frac{85}{12}$$

Since $\frac{85}{12}$ checks, it is the solution.

31.
$$\frac{a}{2a - 6} - \frac{3}{a^2 - 6a + 9} = \frac{a - 2}{3a - 9}$$

$$\frac{a}{2(a-3)} - \frac{3}{(a-3)(a-3)} = \frac{a - 2}{3(a-3)}$$

LCM is $2 \cdot 3(a-3)(a-3)$

$$6(a-3)(a-3)\left(\frac{a}{2(a-3)} - \frac{3}{(a-3)(a-3)}\right) =$$

$$6(a-3)(a-3) \cdot \frac{a - 2}{3(a-3)}$$

$$3a(a-3) - 6 \cdot 3 = 2(a-3)(a-2)$$

$$3a^2 - 9a - 18 = 2(a^2 - 5a + 6)$$

$$3a^2 - 9a - 18 = 2a^2 - 10a + 12$$

$$a^2 + a - 30 = 0$$

$$(a+6)(a-5) = 0$$

$$a + 6 = 0 \quad or \quad a - 5 = 0$$

$$a = -6 \quad or \qquad a = 5$$

Both -6 and 5 check. The solutions are -6 and 5.

33.
$$\frac{2x + 3}{x - 1} = \frac{10}{x^2 - 1} + \frac{2x - 3}{x + 1}$$

$$\frac{2x + 3}{x - 1} = \frac{10}{(x+1)(x-1)} + \frac{2x - 3}{x + 1}$$

LCM is $(x+1)(x-1)$

$$(x+1)(x-1) \cdot \frac{2x + 3}{x - 1} =$$

$$(x+1)(x-1)\left(\frac{10}{(x+1)(x-1)} + \frac{2x - 3}{x + 1}\right)$$

$$(x+1)(2x+3) = 10 + (x-1)(2x-3)$$

$$2x^2 + 5x + 3 = 10 + 2x^2 - 5x + 3$$

$$5x + 3 = 13 - 5x$$

$$10x = 10$$

$$x = 1$$

We know that 1 is not a solution of the original equation, because it results in division by 0. The equation has no solution.

35. $\dfrac{4}{x + 3} + \dfrac{7}{x^2 - 3x + 9} = \dfrac{108}{x^3 + 27}$

Note: $x^3 + 27 = (x+3)(x^2 - 3x + 9)$

Thus the LCM is $(x+3)(x^2 - 3x + 9)$.

$$(x+3)(x^2-3x+9)\left(\frac{4}{x+3} + \frac{7}{x^2 - 3x + 9}\right) =$$

$$(x+3)(x^2-3x+9)\left(\frac{108}{(x+3)(x^2 - 3x + 9)}\right)$$

$$4(x^2 - 3x + 9) + 7(x+3) = 108$$

$$4x^2 - 12x + 36 + 7x + 21 = 108$$

$$4x^2 - 5x - 51 = 0$$

$$(4x - 17)(x + 3) = 0$$

$$4x - 17 = 0 \quad or \quad x + 3 = 0$$

$$4x = 17 \quad or \qquad x = -3$$

$$x = \frac{17}{4} \quad or \qquad x = -3$$

We know that -3 is not a solution of the original equation, because it results in division by 0. Since $\dfrac{17}{4}$ checks, it is the solution.

37.
$$\frac{5x}{x-7} - \frac{35}{x+7} = \frac{490}{x^2-49}$$

$$\frac{5x}{x-7} - \frac{35}{x+7} = \frac{490}{(x+7)(x-7)}$$

LCM is $(x+7)(x-7)$

$$(x+7)(x-7)\left(\frac{5x}{x-7} - \frac{35}{x+7}\right) =$$

$$(x+7)(x-7)\cdot\frac{490}{(x+7)(x-7)}$$

$$5x(x+7) - 35(x-7) = 490$$
$$5x^2 + 35x - 35x + 245 = 490$$
$$5x^2 + 245 = 490$$
$$5x^2 - 245 = 0$$
$$5(x^2 - 49) = 0$$
$$5(x+7)(x-7) = 0$$

$$x+7 = 0 \quad or \quad x-7 = 0$$
$$x = -7 \quad or \quad \quad x = 7$$

The numbers -7 and 7 are possible solutions. Each makes a denominator 0, so the equation has no solution.

39.
$$\frac{x-1}{3} + \frac{6x+1}{15} + \frac{2(x-2)}{13-7x} = \frac{2(x+2)}{5},$$

LCM is $15(13-7x)$

$$15(13-7x)\left(\frac{x-1}{3} + \frac{6x+1}{15} + \frac{2(x-2)}{13-7x}\right) =$$

$$15(13-7x)\cdot\frac{2(x+2)}{5}$$

$$5(13-7x)(x-1) + (13-7x)(6x+1) + 15\cdot 2(x-2) =$$
$$3(13-7x)(2)(x+2)$$
$$5(-7x^2+20x-13) + (-42x^2+71x+13) + 30x-60 =$$
$$6(-7x^2-x+26)$$
$$-35x^2+100x-65 - 42x^2+71x+13+30x-60 =$$
$$-42x^2-6x+156$$
$$-77x^2+201x-112 = -42x^2-6x+156$$
$$0 = 35x^2 - 207x + 268$$
$$0 = (35x-67)(x-4)$$

$$35x-67 = 0 \quad or \quad x-4 = 0$$
$$35x = 67 \quad or \quad \quad x = 4$$
$$x = \frac{67}{35} \quad or \quad \quad x = 4$$

Both numbers check. The solutions are $\frac{67}{35}$ and 4.

41. We find all values of x for which $2x - \frac{6}{x} = 1$. First note that $x \neq 0$. Then multiply on both sides by the LCD, x.

$$x\left(2x - \frac{6}{x}\right) = x\cdot 1$$
$$x\cdot 2x - x\cdot\frac{6}{x} = x$$
$$2x^2 - 6 = x$$
$$2x^2 - x - 6 = 0$$
$$(2x+3)(x-2) = 0$$
$$x = -\frac{3}{2} \quad or \quad x = 2$$

Both values check. The solutions are $-\frac{3}{2}$ and 2.

43. We find all values of x for which $\frac{x-3}{x+2} = \frac{1}{5}$. First note that $x \neq -2$. Then multiply on both sides by the LCD, $5(x+2)$.

$$5(x+2)\cdot\frac{x-3}{x+2} = 5(x+2)\cdot\frac{1}{5}$$
$$5(x-3) = x+2$$
$$5x-15 = x+2$$
$$4x = 17$$
$$x = \frac{17}{4}$$

This value checks. The solution is $\frac{17}{4}$.

45. We find all values of x for which $\frac{6}{x} - \frac{6}{2x} = 5$. First note that $x \neq 0$. Then multiply on both sides by the LCD, $2x$.

$$2x\left(\frac{6}{x} - \frac{6}{2x}\right) = 2x\cdot 5$$
$$2x\cdot\frac{6}{x} - 2x\cdot\frac{6}{2x} = 10x$$
$$12 - 6 = 10x$$
$$6 = 10x$$
$$\frac{3}{5} = x$$

This value checks. The solution is $\frac{3}{5}$.

47. Discussion and Writing Exercise

49.
$$1 - t^6$$
$$= 1^2 - (t^3)^2$$
$$= (1+t^3)(1-t^3)$$
$$= (1+t)(1-t+t^2)(1-t)(1+t+t^2)$$

51.
$$a^3 - 8b^3$$
$$= a^3 - (2b)^3$$
$$= (a-2b)(a^2+2ab+4b^2)$$

53. $(x-3)(x+4) = 0$
$$x-3 = 0 \quad or \quad x+4 = 0$$
$$x = 3 \quad or \quad \quad x = -4$$

The solutions are 3 and -4.

55. $12x^2 - 11x + 2 = 0$

$(4x - 1)(3x - 2) = 0$

$4x - 1 = 0 \quad or \quad 3x - 2 = 0$

$4x = 1 \quad or \qquad 3x = 2$

$x = \dfrac{1}{4} \quad or \qquad x = \dfrac{2}{3}$

The solutions are $\dfrac{1}{4}$ and $\dfrac{2}{3}$.

57. a) Graph $y_1 = \dfrac{x+3}{x+2} - \dfrac{x+4}{x+3}$ and $y_2 = \dfrac{x+5}{x+4} -$ $\dfrac{x+6}{x+5}$ and use the INTERSECT feature to find the point of intersection of the graphs. It is $(-3.5, 1.\overline{3})$.

b) Set $f(x)$ equal to $g(x)$ and solve for x.

$\dfrac{x+3}{x+2} - \dfrac{x+4}{x+3} = \dfrac{x+5}{x+4} - \dfrac{x+6}{x+5}$

Note that $x \neq -2$ and $x \neq -3$ and $x \neq -4$ and $x \neq -5$.

$(x+2)(x+3)(x+4)(x+5)\left(\dfrac{x+3}{x+2} - \dfrac{x+4}{x+3}\right) =$

$(x+2)(x+3)(x+4)(x+5)\left(\dfrac{x+5}{x+4} - \dfrac{x+6}{x+5}\right)$

$(x+3)(x+4)(x+5)(x+3) - (x+2)(x+4)(x+5)(x+4) =$
$(x+2)(x+3)(x+5)(x+5) - (x+2)(x+3)(x+4)(x+6)$

$x^4 + 15x^3 + 83x^2 + 201x + 180 -$
$(x^4 + 15x^3 + 82x^2 + 192x + 160) =$
$x^4 + 15x^3 + 81x^2 + 185x + 150 -$
$(x^4 + 15x^3 + 80x^2 + 180x + 144)$

$x^2 + 9x + 20 = x^2 + 5x + 6$

$4x = -14$

$x = -\dfrac{7}{2}$

This value checks. When $x = -\dfrac{7}{2}$, $f(x) = g(x)$. This confirms that the graphs of $f(x)$ and $g(x)$ intersect when $x = -\dfrac{7}{2}$, or -3.5.

c) Answers will vary.

Exercise Set 5.6

1. Familiarize. Let $t =$ the time it takes them, working together, to fill the order.

Translate. Using the work principle, we get the following equation:

$$\frac{t}{5} + \frac{t}{9} = 1$$

Solve. We solve the equation.

$$\frac{t}{5} + \frac{t}{9} = 1, \text{ LCM is } 45$$

$$45\left(\frac{t}{5} + \frac{t}{9}\right) = 45 \cdot 1$$

$$45 \cdot \frac{t}{5} + 45 \cdot \frac{t}{9} = 45$$

$$9t + 5t = 45$$

$$14t = 45$$

$$t = \frac{45}{14}, \text{ or } 3\frac{3}{14}$$

Check. We verify the work principle.

$$\frac{\frac{45}{14}}{5} + \frac{\frac{45}{14}}{9} = \frac{45}{14} \cdot \frac{1}{5} + \frac{45}{14} \cdot \frac{1}{9} = \frac{9}{14} + \frac{5}{14} = \frac{14}{14} = 1$$

State. It will take them $3\dfrac{3}{14}$ hr, working together.

3. Familiarize. Let $t =$ the time it will take to fill the pool using both the pipe and the hose.

Translate. Using the work principle, we get the following equation:

$$\frac{t}{12} + \frac{t}{30} = 1$$

Solve. We solve the equation.

$$\frac{t}{12} + \frac{t}{30} = 1, \text{ LCM is } 60$$

$$60\left(\frac{t}{12} + \frac{t}{30}\right) = 60 \cdot 1$$

$$\frac{60t}{12} + \frac{60t}{30} = 60$$

$$5t + 2t = 60$$

$$7t = 60$$

$$t = \frac{60}{7}, \text{ or } 8\frac{4}{7}$$

Check. We verify the work principle.

$$\frac{\frac{60}{7}}{12} + \frac{\frac{60}{7}}{30} = \frac{60}{7} \cdot \frac{1}{12} + \frac{60}{7} \cdot \frac{1}{30} = \frac{5}{7} + \frac{2}{7} = 1$$

State. It will take them $8\dfrac{4}{7}$ hr to fill the pool using both the pipe and the hose.

5. Familiarize. Let $t =$ the time it will take to print the order if both presses are used.

Translate. Using the work principle, we get the following equation:

$$\frac{t}{4.5} + \frac{t}{5.5} = 1$$

Solve. We solve the equation.

$$\frac{t}{4.5} + \frac{t}{5.5} = 1, \text{ LCM is } (4.5)(5.5)$$

$$(4.5)(5.5)\left(\frac{t}{4.5} + \frac{t}{5.5}\right) = (4.5)(5.5)(1)$$

$$(4.5)(5.5) \cdot \frac{t}{4.5} + (4.5)(5.5) \cdot \frac{t}{5.5} = 24.75$$

$$5.5t + 4.5t = 24.75$$

$$10t = 24.75$$

$$t = 2.475, \text{ or } 2\frac{19}{40}$$

Check. We verify the work principle.

$$\frac{2.475}{4.5} + \frac{2.475}{5.5} = 0.55 + 0.45 = 1$$

State. It will take 2.475 hr, or $2\frac{19}{40}$ hr, to print the order if both presses are used.

7. Familiarize. Let a = the number of days it takes Juan to paint the house. Then $4a$ = the number of days it takes Ariel to paint the house.

Translate. Using the work principle, we get the following equation:

$$\frac{8}{a} + \frac{8}{4a} = 1, \text{ or } \frac{8}{a} + \frac{2}{a} = 1$$

Solve. We solve the equation.

$$\frac{8}{a} + \frac{2}{a} = 1$$

$$\frac{10}{a} = 1 \quad \text{Adding}$$

$$10 = a \quad \text{Multiplying by } a$$

Check. If it takes Juan 10 days to paint the house, then it takes Ariel $4 \cdot 10$, or 40 days. We verify the work principle.

$$\frac{8}{10} + \frac{8}{40} = \frac{4}{5} + \frac{1}{5} = \frac{5}{5} = 1$$

State. It would take Juan 10 days to paint the house alone, and it would take Ariel 40 days.

9. Familiarize. Let t = the number of hours it would take Skyler to split the firewood working alone. Then $t - 6$ = the time it would take Jake to split the firewood working alone.

Translate. Using the work principle, we get the following equation:

$$\frac{4}{t} + \frac{4}{t-6} = 1$$

Solve. We solve the equation.

$$\frac{4}{t} + \frac{4}{t-6} = 1, \text{ LCM is } t(t-6)$$

$$t(t-6)\left(\frac{4}{t} + \frac{4}{t-6}\right) = t(t-6) \cdot 1$$

$$t(t-6) \cdot \frac{4}{t} + t(t-6) \cdot \frac{4}{t-6} = t(t-6)$$

$$4(t-6) + 4t = t^2 - 6t$$

$$4t - 24t + 4t = t^2 - 6t$$

$$8t - 24 = t^2 - 6t$$

$$0 = t^2 - 14t + 24$$

$$0 = (t-2)(t-12)$$

$$t - 2 = 0 \quad or \quad t - 12 = 0$$

$$t = 2 \quad or \quad t = 12$$

Check. If it takes Skyler 2 hr to split the firewood, then it would take Jack $2 - 6$, or -4 hr. Since negative values of time have no meaning in this application, 2 is not a solution.

If it takes Skyler 12 hr to split the firewood, then it would take Jake $12 - 6$, or 6 hr. We verify the work principle.

$$\frac{4}{12} + \frac{4}{6} = \frac{1}{3} + \frac{2}{3} = 1$$

State. It would take Skyler 12 hr to split the firewood working alone, and it would take Jake 6 hr.

11. Familiarize. Let w = the waist measurement that allows Marta to meet the recommendation.

Translate. We translate to a proportion. We write 0.85 as $\frac{85}{100}$.

$$\begin{array}{l} \text{Waist} \rightarrow \\ \text{Hip} \rightarrow \end{array} \frac{85}{100} = \frac{w}{40} \begin{array}{l} \leftarrow \text{Waist} \\ \leftarrow \text{Hip} \end{array}$$

Solve. We solve the proportion.

$$\frac{85}{100} = \frac{w}{40}$$

$$85 \cdot 40 = 100w \quad \text{Equating cross products}$$

$$\frac{85 \cdot 40}{100} = w$$

$$34 = w$$

Check. We substitute into the proportion and check cross products.

$$\frac{85}{100} = \frac{34}{40}; \; 85 \cdot 40 = 3400; \; 100 \cdot 34 = 3400$$

Since the cross products are the same, the answer checks.

State. Marta's waist measurement should be 34 in. or less to meet the recommendation.

13. Familiarize. Let h = the number of home runs Alex Rodriguez would hit in the 162-game season.

Translate. We translate to a proportion.

$$\begin{array}{l} \text{Home runs} \rightarrow \\ \text{Games} \rightarrow \end{array} \frac{14}{45} = \frac{h}{162} \begin{array}{l} \leftarrow \text{Home runs} \\ \leftarrow \text{Games} \end{array}$$

Solve. We solve the proportion.

$$\frac{14}{45} = \frac{h}{162}$$

$$14 \cdot 162 = 45h \quad \text{Equating cross products}$$

$$\frac{14 \cdot 162}{45} = h$$

$$50 \approx h$$

Check. We substitute into the proportion and check cross products.

$$\frac{14}{45} = \frac{50}{162}; \; 14 \cdot 162 = 2268; \; 45 \cdot 50 = 2250$$

The cross products are approximately the same. Remember that we rounded the value of h. The answer checks.

State. If he continued to bat at the same rate, Rodriguez would hit about 50 home runs in the 162-game season.

15. *Familiarize*. Let t = the number of trees required to produce beans for 638 kg of coffee.

Translate. We translate to a proportion.

$$\text{Trees} \rightarrow \frac{14}{7.7} = \frac{t}{638} \leftarrow \text{Trees}$$
$$\text{Coffee} \rightarrow \qquad\qquad \leftarrow \text{Coffee}$$

Solve. We solve the proportion.

$$\frac{14}{7.7} = \frac{t}{638}$$
$$14 \cdot 638 = 7.7t \quad \text{Equating cross products}$$
$$\frac{14 \cdot 638}{7.7} = t$$
$$1160 = t$$

Check. We substitute into the proportion and check cross products.

$$\frac{14}{7.7} = \frac{1160}{638}; \; 14 \cdot 638 = 8932; \; 7.7(1160) = 8932$$

Since the cross products are the same, the answer checks.

State. The beans from 1160 trees are required to produce 638 kg of coffee.

17. *Familiarize*. Let D = the number of deer in the game preserve.

Translate. We translate to a proportion.

$$\begin{array}{l}\text{Deer tagged} \\ \text{originally} \\ \text{Deer in} \\ \text{game preserve}\end{array} \rightarrow \frac{318}{D} = \frac{56}{168} \begin{array}{l}\leftarrow \text{Tagged deer} \\ \leftarrow \text{caught later} \\ \leftarrow \text{Deer caught} \\ \text{later}\end{array}$$

Solve. We solve the proportion.

$$\frac{318}{D} = \frac{56}{168}$$
$$318 \cdot 168 = D \cdot 56 \quad \text{Equating cross products}$$
$$\frac{318 \cdot 168}{56} = D$$
$$954 = D$$

Check. We substitute into the proportion and check cross products.

$$\frac{318}{954} = \frac{56}{168}; \; 318 \cdot 168 = 53,424; \; 954 \cdot 56 = 53,424$$

Since the cross products are the same, the answer checks.

State. There are 954 deer in the game preserve.

19. a) *Familiarize*. Let w = the number of tons the rocket will weigh on Mars.

Translate. We translate to a proportion.

$$\text{Weight on Mars} \rightarrow \frac{0.4}{1} = \frac{w}{12} \leftarrow \text{Weight on Mars}$$
$$\text{Weight on earth} \rightarrow \qquad\qquad \leftarrow \text{Weight on earth}$$

Solve. We solve the proportion.

$$\frac{0.4}{1} = \frac{w}{12}$$
$$0.4 \cdot 12 = 1 \cdot w \quad \text{Equating cross products}$$
$$4.8 = w$$

Check. We substitute into the proportion and check cross products.

$$\frac{0.4}{1} = \frac{4.8}{12}; \; 0.4(12) = 4.8; \; 1 \cdot (4.8) = 4.8$$

Since the cross products are the same, the answer checks.

State. A 12-T rocket will weigh 4.8 T on Mars.

b) *Familiarize*. Let w = the number of pounds the astronaut will weigh on Mars.

Translate. We translate to a proportion.

$$\text{Weight on Mars} \rightarrow \frac{0.4}{1} = \frac{w}{120} \leftarrow \text{Weight on Mars}$$
$$\text{Weight on earth} \rightarrow \qquad\qquad \leftarrow \text{Weight on earth}$$

Solve. We solve the proportion.

$$\frac{0.4}{1} = \frac{w}{120}$$
$$0.4 \cdot 120 = 1 \cdot w \quad \text{Equating cross products}$$
$$48 = w$$

Check. We substitute into the proportion and check cross products.

$$\frac{0.4}{1} = \frac{48}{120}; \; 0.4(120) = 48; \; 1 \cdot 48 = 48$$

Since the cross products are the same, the answer checks.

State. A 120-lb astronaut will weigh 48 lb on Mars.

21. *Familiarize*. Let x = the number that is added to each of the given numbers.

Translate. We translate to a proportion.

$$\frac{1+x}{2+x} = \frac{3+x}{5+x}$$

Solve. We solve the proportion.

$$\frac{1+x}{2+x} = \frac{3+x}{5+x}$$
$$(1+x)(5+x) = (2+x)(3+x) \quad \begin{array}{l}\text{Equating cross} \\ \text{products}\end{array}$$
$$5 + 6x + x^2 = 6 + 5x + x^2$$
$$x = 1$$

Check. We substitute into the proportion and check cross products.

$$\frac{1+1}{2+1} = \frac{3+1}{5+1}, \text{ or } \frac{2}{3} = \frac{4}{6}; \; 2 \cdot 6 = 12; \; 3 \cdot 4 = 12$$

Since the cross products are the same, the answer checks.

State. The number is 1.

23. *Familiarize*. We first make a drawing. Let r = the boat's speed in still water in mph. Then $r - 4$ = the speed upstream and $r + 4$ = the speed downstream.

$$\text{Upstream} \quad 6 \text{ miles} \quad r - 4 \text{ mph} \longrightarrow$$
$$\text{12 miles} \quad r + 4 \text{ mph} \quad \text{Downstream}$$

We organize the information in a table. The time is the same both upstream and downstream so we use t for each time.

	Distance	Speed	Time
Upstream	6	$r-4$	t
Downstream	12	$r+4$	t

Translate. Using the formula Time = Distance/Rate in each row of the table and the fact that the times are the same, we can write an equation.
$$\frac{6}{r-4}=\frac{12}{r+4}$$
Solve. We solve the equation.
$$\frac{6}{r-4}=\frac{12}{r+4},$$
LCD is $(r-4)(r+4)$
$$(r-4)(r+4)\cdot\frac{6}{r-4}=(r-4)(r+4)\cdot\frac{12}{r+4}$$
$$6(r+4)=12(r-4)$$
$$6r+24=12r-48$$
$$72=6r$$
$$12=r$$

Check. If the boat's speed in still water is 12 mph, then its speed upstream is $12-4$, or 8 mph, and its speed downstream is $12+4$, or 16 mph. Traveling 6 mi at 8 mph takes the boat $\frac{6}{8}=\frac{3}{4}$ hr. Traveling 12 mi at 16 mph takes the boat $\frac{12}{16}=\frac{3}{4}$ hr. Since the times are the same, the answer checks.

State. The boat's speed in still water is 12 mph.

25. Familiarize. Using the labels on the drawing in the text, we let $c=$ the speed of the current. Then $10-c=$ the speed of the tugboat upstream and $10+c=$ the speed downstream. We organize the information in a table.

	Distance	Speed	Time
Upstream	24	$10-c$	$\frac{24}{10-c}$
Downstream	24	$10+c$	$\frac{24}{10+c}$

Translate. Since the total time is 5 hr, we use the last column of the table to write an equation.
$$\frac{24}{10-c}+\frac{24}{10+c}=5$$
Solve. We solve the equation.
$$\frac{24}{10-c}+\frac{24}{10+c}=5,$$
LCD is $(10-c)(10+c)$
$$(10-c)(10+c)\left(\frac{24}{10-c}+\frac{24}{10+c}\right)=(10-c)(10+c)\cdot5$$
$$24(10+c)+24(10-c)=5(100-c^2)$$
$$240+24c+240-24c=500-5c^2$$
$$480=500-5c^2$$
$$5c^2-20=0$$
$$5(c^2-4)=0$$
$$5(c+2)(c-2)=0$$

$$c+2=0 \quad or \quad c-2=0$$
$$c=-2 \quad or \quad c=2$$

Check. Since negative time has no meaning in this application, we check only 2. If the current is 2 mph, then the boat's speed upstream is $10-2$, or 8 mph, and speed downstream is $10+2$, or 12 mph. Traveling 24 mi at 8 mph takes 24/8, or 3 hr; traveling 24 mi at 12 mph takes 24/12, or 2 hr. The total time is $3+2$, or 5 hr, so the answer checks.

State. The speed of the current is 2 mph.

27. Familiarize. We first make a drawing. Let $r=$ Camille's speed on a nonmoving sidewalk in ft/sec. Then her speed moving forward on the moving sidewalk is $r+1.8$, and her speed in the opposite direction is $r-1.8$.

Forward $r+1.8$ 105 ft

51 ft $r-1.8$ Opposite direction

We organize the information in a table. The time is the same both forward and in the opposite direction so we use t for each time.

	Distance	Speed	Time
Forward	105	$r+1.8$	t
Opposite direction	51	$r-1.8$	t

Translate. Using the formula Time = Distance/Rate in each row of the table and the fact that the times are the same, we can write an equation.
$$\frac{105}{r+1.8}=\frac{51}{r-1.8}$$
Solve. We solve the equation.
$$\frac{105}{r+1.8}=\frac{51}{r-1.8},$$
LCD is $(r+1.8)(r-1.8)$
$$(r+1.8)(r-1.8)\cdot\frac{105}{r+1.8}=(r+1.8)(r-1.8)\cdot\frac{51}{r-1.8}$$
$$105(r-1.8)=51(r+1.8)$$
$$105r-189=51r+91.8$$
$$54r=280.8$$
$$r=5.2$$

Check. If Camille's speed on a nonmoving sidewalk is 5.2 ft/sec, then her speed moving forward on the moving sidewalk is $5.2+1.8$, or 7 ft/sec, and her speed moving in the opposite direction on the sidewalk is $5.2-1.8$, or 3.4 ft/sec. Moving 105 ft at 7 ft/sec takes $\frac{105}{7}=15$ sec. Moving 51 ft at 3.4 ft/sec takes 15 sec. Since the times are the same, the answer checks.

State. Camille would be walking 5.2 ft/sec on a nonmoving sidewalk.

29. Familiarize. Let $r=$ Simone's walking speed, in mph. Then $r-2=$ Rosanna's speed. The times are the same, so we let $t=$ the time. We organize the information in a table.

	Distance	Speed	Time
Simone	8	r	t
Rosanna	5	$r-2$	t

Translate. Using the formula Time = Distance/Rate in each row of the table and the fact that the times are the same, we can write an equation.

$$\frac{8}{r} = \frac{5}{r-2}$$

Solve. We solve the equation.

$$\frac{8}{r} = \frac{5}{r-2}, \text{ LCD is } r(r-2)$$

$$r(r-2) \cdot \frac{8}{r} = r(r-2) \cdot \frac{5}{r-2}$$

$$8(r-2) = 5r$$

$$8r - 16 = 5r$$

$$3r = 16$$

$$r = \frac{16}{3}$$

Check. If Simone's speed is $\frac{16}{3}$ mph, then Rosanna's speed is $\frac{16}{3} - 2$, or $\frac{10}{3}$ mph. Walking 8 mi at $\frac{16}{3}$ mph takes Simone 8/(16/3), or $\frac{3}{2}$ hr. Walking 5 mi at $\frac{10}{3}$ mph takes Rosanna 5/(10/3), or $\frac{3}{2}$ hr. Since the times are the same, the answer checks.

State. Simone's speed is $\frac{16}{3}$, or $5\frac{1}{3}$ mph, and Rosanna's speed is $\frac{10}{3}$, or $3\frac{1}{3}$ mph.

31. Discussion and Writing Exercise

33. The domain is the set of all x-values on the graph, $[-5, 5]$. The range is the set of all y-values on the graph, $[-4, 3]$.

35. The domain is the set of all x-values on the graph, $[-5, 5]$. The range is the set of all y-values on the graph, $[-5, 3]$.

37. Graph: $x - 4y \geq 4$

We first graph the line $x - 4y = 4$. The intercepts are $(0, -1)$ and $(4, 0)$. We draw the line solid since the inequality symbol is $\geq$. To determine which half-plane to shade, we consider a test point not on the line. We try $(0, 0)$:

$$\frac{x - 4y \geq 4}{0 - 4 \cdot 0 \ ? \ 4}$$
$$0 \ \Big| \quad \text{FALSE}$$

Since $0 \geq 4$ is false, we shade the half-plane that does not contain $(0, 0)$.

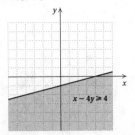

39. Graph: $f(x) = |x + 3|$

We select x-values and find the corresponding y-values. We plot these ordered pairs and draw the graph.

x	y
-5	2
-3	0
-1	2
0	3
2	5

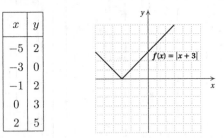

41. Familiarize. Let d = the distance, in miles, Melissa lives from work. Also let t = the travel time in hours, when Melissa arrives on time. Note that 1 min = $\frac{1}{60}$ hr and 5 min = $\frac{5}{60}$, or $\frac{1}{12}$ hr.

Translate. Melissa's travel time at 50 mph is $\frac{d}{50}$. This is $\frac{1}{60}$ hr more than t, so we write an equation using this information:

$$\frac{d}{50} = t + \frac{1}{60}$$

Her travel time at 60 mph, $\frac{d}{60}$, is $\frac{1}{12}$ hr less than t, so we write a second equation:

$$\frac{d}{60} = t - \frac{1}{12}$$

We have a system of equations:

$$\frac{d}{50} = t + \frac{1}{60},$$
$$\frac{d}{60} = t - \frac{1}{12}$$

Solve. Solving the system of equations, we get $\left(30, \frac{7}{12}\right)$.

Check. Traveling 30 mi at 50 mph takes $\frac{30}{50}$, or $\frac{3}{5}$ hr. Since $\frac{7}{12} + \frac{1}{60} = \frac{36}{60} = \frac{3}{5}$, this time makes Melissa $\frac{1}{60}$ hr, or 1 min late. Traveling 30 mi at 60 mph takes $\frac{30}{60}$, or $\frac{1}{2}$ hr. Since $\frac{7}{12} - \frac{1}{12} = \frac{6}{12} = \frac{1}{2}$, this time makes Melissa $\frac{1}{12}$ hr, or 5 min early. The answer checks.

State. Melissa lives 30 mi from work.

43. Familiarize. Let x = the number of miles driven in the city and y = the number of miles driven on the highway. Then $\frac{x}{22.5}$ = the number of gallons of gasoline used in city driving and $\frac{y}{30}$ = the number of gallons used in highway driving.

Translate. The total number of miles is 465, so we have one equation:

$$x + y = 465$$

The total amount of gasoline used is 18.4 gal, so we have a second equation.

$$\frac{x}{22.5} + \frac{y}{30} = 18.4$$

After clearing fractions, we have the following system of equations:

$$x + y = 465,$$
$$30x + 22.5y = 12,420$$

Solve. The solution of the system of equations is $(261, 204)$.

Check. If 261 city miles and 204 highway miles are driven, the total number of miles is $261 + 204$, or 465. In 261 city miles $261/22.5$ or 11.6 gal, of gasoline are used; in 204 highway miles $204/30$, or 6.8 gal, are used. Then the total amount of gasoline used is $11.6 + 6.8$, or 18.4 gal. The answer checks.

State. 261 mi were driven in the city, and 204 mi were driven on the highway.

45. Familiarize. It helps to first make a drawing.

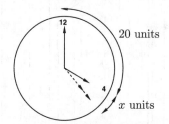

20 units

x units

The minute hand moves 60 units per hour while the hour hand moves 5 units per hour, where one unit represents one minute on the face of the clock. When the hands are in the same position the first time, the hour hand will have moved x units and the minute hand will have moved $x + 20$ units. The times are the same.

We organize the information in a table.

	Distance	Speed	Time
Minute	$x + 20$	60	t
Hour	x	5	t

Translate. Since Time = Distance/Speed and since the times are the same, we can write an equation:

$$\frac{x + 20}{60} = \frac{x}{5}$$

Solve. We solve the equation.

$$\frac{x + 20}{60} = \frac{x}{5}, \text{ LCD is } 60$$
$$60 \cdot \frac{x + 20}{60} = 60 \cdot \frac{x}{5}$$
$$x + 20 = 12x$$
$$20 = 11x$$
$$\frac{20}{11} = x, \text{ or}$$
$$1\frac{9}{11} = x$$

When the hour hand has moved $1\frac{9}{11}$ units, the minute hand has moved $20 + 1\frac{9}{11}$, or $21\frac{9}{11}$ units, so the time will be $21\frac{9}{11}$ minutes after 4:00.

Check. At 60 units per hour the minute hand moves $21\frac{9}{11}$ units in $60/(240/11)$, or $11/4$ min. At 5 units per hour the hour hand moves $1\frac{9}{11}$ units in $5/(20/11)$, or $11/4$ min. Since the times are the same, the answer checks.

State. The hands will first be in the same position $21\frac{9}{11}$ min after 4:00.

Exercise Set 5.7

1.
$$\frac{W_1}{W_2} = \frac{d_1}{d_2}$$

$$W_2 d_2 \cdot \frac{W_1}{W_2} = W_2 d_2 \cdot \frac{d_1}{d_2} \quad \text{Multiplying by the LCM}$$

$$d_2 W_1 = W_2 d_1$$

$$\frac{d_2 W_1}{d_1} = W_2 \quad \text{Dividing by } d_1$$

3.
$$\frac{1}{R} = \frac{1}{r_1} + \frac{1}{r_2}$$

$$R r_1 r_2 \cdot \frac{1}{R} = R r_1 r_2 \left(\frac{1}{r_1} + \frac{1}{r_2} \right) \quad \begin{array}{l}\text{Multiplying by} \\ \text{the LCM}\end{array}$$

$$r_1 r_2 = R r_2 + R r_1$$

$$r_1 r_2 - R r_2 = R r_1 \quad \text{Subtracting } R r_2$$

$$r_2 (r_1 - R) = R r_1 \quad \text{Factoring}$$

$$r_2 = \frac{R r_1}{r_1 - R} \quad \text{Dividing by } r_1 - R$$

5.
$$s = \frac{(v_1 + v_2)t}{2}$$

$$2 \cdot s = 2 \cdot \frac{(v_1 + v_2)t}{2} \quad \text{Multiplying by 2}$$

$$2s = (v_1 + v_2)t$$

$$\frac{2s}{v_1 + v_2} = t \quad \text{Dividing by } v_1 + v_2$$

7.
$$R = \frac{gs}{g + s}$$

$$(g + s) \cdot R = (g + s) \cdot \frac{gs}{g + s} \quad \text{Multiplying by } g + s$$

$$R(g + s) = gs$$

$$Rg + Rs = gs \quad \text{Removing parentheses}$$

$$Rg = gs - Rs \quad \text{Subtracting } Rs$$

$$Rg = s(g - R) \quad \text{Factoring}$$

$$\frac{Rg}{g - R} = s \quad \text{Dividing by } g - R$$

9.
$$\frac{1}{p} + \frac{1}{q} = \frac{1}{f}$$

$$pqf\left(\frac{1}{p} + \frac{1}{q}\right) = pqf \cdot \frac{1}{f} \qquad \text{Multiplying by } pqf$$

$$pqf \cdot \frac{1}{p} + pqf \cdot \frac{1}{q} = pq$$

$$qf + pf = pq$$

$$qf = pq - pf \qquad \text{Subtracting } pf$$

$$qf = p(q - f) \qquad \text{Factoring}$$

$$\frac{qf}{q - f} = p \qquad \text{Dividing by } q - f$$

11.
$$\frac{t}{a} + \frac{t}{b} = 1$$

$$ab\left(\frac{t}{a} + \frac{t}{b}\right) = ab \cdot 1 \qquad \text{Multiplying by } ab$$

$$ab \cdot \frac{t}{a} + ab \cdot \frac{t}{b} = ab \qquad \text{Removing parentheses}$$

$$bt + at = ab$$

$$bt = ab - at \qquad \text{Subtracting } at$$

$$bt = a(b - t) \qquad \text{Factoring}$$

$$\frac{bt}{b - t} = a \qquad \text{Dividing by } b - t$$

13.
$$I = \frac{nE}{E + nr}$$

$$(E + nr)I = (E + nr) \cdot \frac{nE}{E + nr} \qquad \text{Multiplying by } E + nr$$

$$EI + nrI = nE$$

$$nrI = nE - EI \qquad \text{Subtracting } EI$$

$$nrI = E(n - I) \qquad \text{Factoring}$$

$$\frac{nrI}{n - I} = E \qquad \text{Dividing by } n - I$$

15.
$$I = \frac{704.5W}{H^2}$$

$$H^2 I = 704.5W \qquad \text{Multiplying by } H^2$$

$$H^2 = \frac{704.5W}{I} \qquad \text{Dividing by } I$$

17.
$$\frac{E}{e} = \frac{R + r}{r}$$

$$er \cdot \frac{E}{e} = er \cdot \frac{R + r}{r} \qquad \text{Multiplying by } er$$

$$Er = eR + er$$

$$Er - er = eR \qquad \text{Subtracting } er$$

$$r(E - e) = eR \qquad \text{Factoring}$$

$$r = \frac{eR}{E - e} \qquad \text{Dividing by } E - e$$

19.
$$V = \frac{1}{3}\pi h^2(3R - h)$$

$$\frac{3}{\pi h^2} \cdot V = \frac{3}{\pi h^2} \cdot \frac{1}{3}\pi h^2(3R - h) \qquad \text{Multiplying by } \frac{3}{\pi h^2}$$

$$\frac{3V}{\pi h^2} = 3R - h$$

$$\frac{3V}{\pi h^2} + h = 3R \qquad \text{Adding } h$$

$$\frac{1}{3}\left(\frac{3V}{\pi h^2} + h\right) = \frac{1}{3} \cdot 3R \qquad \text{Multiplying by } \frac{1}{3}$$

$$\frac{V}{\pi h^2} + \frac{h}{3} = R, \text{ or}$$

$$\frac{3V + \pi h^3}{3\pi h^2} = R$$

21.
$$P = \frac{A}{1 + r}$$

$$P(1 + r) = \frac{A}{1 + r} \cdot (1 + r) \qquad \text{Multiplying by the LCD}$$

$$P(1 + r) = A$$

$$1 + r = \frac{A}{P} \qquad \text{Dividing by } P$$

$$r = \frac{A}{P} - 1, \text{ or } \frac{A - P}{P}$$

23.
$$\frac{V^2}{R^2} = \frac{2g}{R + h}$$

$$(R + h) \cdot \frac{V^2}{R^2} = (R + h) \cdot \frac{2g}{R + h} \qquad \text{Multiplying by } (R + h)$$

$$\frac{(R + h)V^2}{R^2} = 2g$$

$$R + h = \frac{2gR^2}{V^2} \qquad \text{Multiplying by } \frac{R^2}{V^2}$$

$$h = \frac{2gR^2}{V^2} - R \qquad \text{Adding } -R$$

The result can also be expressed as
$$h = \frac{2gR^2 - RV^2}{V^2}.$$

25.
$$A = \frac{2Tt + Qq}{2T + Q}$$

$$(2T + Q) \cdot A = (2T + Q) \cdot \frac{2Tt + Qq}{2T + Q}$$

$$2AT + AQ = 2Tt + Qq$$

$$AQ - Qq = 2Tt - 2AT \qquad \text{Adding } -2AT \text{ and } -Qq$$

$$Q(A - q) = 2Tt - 2AT$$

$$Q = \frac{2Tt - 2AT}{A - q}$$

27. Discussion and Writing Exercise

29. *Familiarize.* Let x = the number of 30-min tapes bought and y = the number of 60-min tapes. We will express all times in minutes so we convert 10 hr to minutes: 10 hr = 10×60 min = 600 min. The recording time for the 30-min tapes is $30x$, and the recording time for the 60-min tapes is $60y$.

Translate. Esther buys 12 tapes, so we have one equation:

$$x + y = 12$$

The total recording time is 600 min, so we have a second equation:

$$30x + 60y = 600$$

We have a system of equations:

$$x + y = 12,$$
$$30x + 60y = 600$$

Solve. The solution of the system of equations is $(4, 8)$, so 4 30-min tapes and 8 60-min tapes were bought.

Check. The total number of tapes is $4 + 8$, or 12. The total recording time is $30 \cdot 4 + 8 \cdot 60$, or 600 min. The answer checks.

State. Esther bought 4 30-min tapes and 8 60-min tapes.

31. $f(x) = x^3 - x$

$\quad f(2) = 2^3 - 2 = 8 - 2 = 6$

33. $f(x) = x^3 - x$

$\quad f(2a) = (2a)^3 - 2a = 8a^3 - 2a$

35. First we find the slope.

$$m = \frac{5 - (-3)}{-2 - 8} = \frac{8}{-10} = -\frac{4}{5}$$

Now we use the slope and one of the given points to find b. We will use $(-2, 5)$. Substitute -2 for x, 5 for y, and $-\frac{4}{5}$ for m in $y = mx + b$ and solve for b.

$$y = mx + b$$
$$5 = -\frac{4}{5}(-2) + b$$
$$5 = \frac{8}{5} + b$$
$$\frac{17}{5} = b$$

Use $y = mx + b$ again, substituting $-\frac{4}{5}$ for m and $\frac{17}{5}$ for b. The equation is

$$y = -\frac{4}{5}x + \frac{17}{5}.$$

Exercise Set 5.8

1. $\quad y = kx$

$\quad 40 = k \cdot 8 \quad$ Substituting

$\quad \ 5 = k \quad$ Solving for k

The variation constant is 5.

The equation of variation is $y = 5x$.

3. $\quad y = kx$

$\quad \ 4 = k \cdot 30 \quad$ Substituting

$\quad \dfrac{4}{30} = k, \text{ or } \quad$ Solving for k

$\quad \dfrac{2}{15} = k \quad$ Simplifying

The variation constant is $\dfrac{2}{15}$.

The equation of variation is $y = \dfrac{2}{15}x$.

5. $\quad y = kx$

$\quad 0.9 = k \cdot 0.4 \quad$ Substituting

$\quad \dfrac{0.9}{0.4} = k, \text{ or}$

$\quad \dfrac{9}{4} = k$

The variation constant is $\dfrac{9}{4}$.

The equation of variation is $y = \dfrac{9}{4}x$.

7. Let $p =$ the number of people using the cans.

$\quad N = kp \qquad N$ varies directly as p.

$\quad 60,000 = k \cdot 250 \qquad$ Substituting

$\quad \dfrac{60,000}{250} = k \qquad$ Solving for k

$\quad \ \ 240 = k \qquad$ Variation constant

$N = 240p \qquad$ Equation of variation

$N = 240(1,189,000) \quad$ Substituting

$N = 285,360,000$

In Dallas 285,360,000 cans are used each year.

9. $\quad d = kw \qquad d$ varies directly as w.

$\quad 40 = k \cdot 3 \quad$ Substituting

$\quad \dfrac{40}{3} = k \qquad$ Variation constant

$\quad d = \dfrac{40}{3}w \qquad$ Equation of variation

$\quad d = \dfrac{40}{3} \cdot 5 \qquad$ Substituting

$\quad d = \dfrac{200}{3}, \text{ or } 66\dfrac{2}{3}$

The spring is stretched $66\dfrac{2}{3}$ cm by a 5-kg barbell.

11. Let $F =$ the number of grams of fat and $w =$ the weight.

$\quad F = kw \qquad F$ varies directly as w.

$\quad 60 = k \cdot 120 \quad$ Substituting

$\quad \dfrac{60}{120} = k, \text{ or} \qquad$ Solving for k

$\quad \dfrac{1}{2} = k \qquad$ Variation constant

$\quad F = \dfrac{1}{2}w \qquad$ Equation of variation

$\quad F = \dfrac{1}{2} \cdot 180 \qquad$ Substituting

$\quad F = 90$

The maximum daily fat intake for a person weighing 180 lb is 90 g.

13. Let $m =$ the mass of the body.

$W = km$ W varies directly as m.

$64 = k \cdot 96$ Substituting

$\dfrac{64}{96} = k$ Solving for k

$\dfrac{2}{3} = k$ Variation constant

$W = \dfrac{2}{3}m$ Equation of variation

$W = \dfrac{2}{3} \cdot 60$ Substituting

$W = 40$

There are 40 kg of water in a 60-kg person.

15. $y = \dfrac{k}{x}$

$14 = \dfrac{k}{7}$ Substituting

$7 \cdot 14 = k$ Solving for k

$98 = k$

The variation constant is 98.

The equation of variation is $y = \dfrac{98}{x}$.

17. $y = \dfrac{k}{x}$

$3 = \dfrac{k}{12}$ Substituting

$12 \cdot 3 = k$ Solving for k

$36 = k$

The variation constant is 36.

The equation of variation is $y = \dfrac{36}{x}$.

19. $y = \dfrac{k}{x}$

$0.1 = \dfrac{k}{0.5}$ Substituting

$0.5(0.1) = k$ Solving for k

$0.05 = k$

The variation constant is 0.05.

The equation of variation is $y = \dfrac{0.05}{x}$.

21. $T = \dfrac{k}{P}$ T varies inversely as P.

$5 = \dfrac{k}{7}$ Substituting

$35 = k$ Variation constant

$T = \dfrac{35}{P}$ Equation of variation

$T = \dfrac{35}{10}$ Substituting

$T = 3.5$

It will take 10 bricklayers 3.5 hr to complete the job.

23. $I = \dfrac{k}{R}$ I varies inversely as R.

$\dfrac{1}{2} = \dfrac{k}{240}$ Substituting

$240 \cdot \dfrac{1}{2} = k$

$120 = k$ Variation constant

$I = \dfrac{120}{R}$ Equation of variation

$I = \dfrac{120}{540}$ Substituting

$I = \dfrac{2}{9}$

When the resistance is 540 ohms, the current is $\dfrac{2}{9}$ ampere.

25. $P = \dfrac{k}{W}$ P varies inversely as W.

$330 = \dfrac{k}{3.2}$ Substituting

$1056 = k$ Variation constant

$P = \dfrac{1056}{W}$ Equation of variation

$550 = \dfrac{1056}{W}$ Substituting

$550W = 1056$ Multiplying by W

$W = \dfrac{1056}{550}$ Dividing by 550

$W = 1.92$ Simplifying

A tone with a pitch of 550 vibrations per second has a wavelength of 1.92 ft.

27. $V = \dfrac{k}{P}$ V varies inversely as P.

$200 = \dfrac{k}{32}$ Substituting

$6400 = k$ Variation constant

$V = \dfrac{6400}{P}$ Equation of variation

$V = \dfrac{6400}{40}$ Substituting

$V = 160$

The volume is 160 cm^3 under a pressure of 40 kg/cm^2.

29. $y = kx^2$

$0.15 = k(0.1)^2$ Substituting

$0.15 = 0.01k$

$\dfrac{0.15}{0.01} = k$

$15 = k$

The equation of variation is $y = 15x^2$.

31.
$$y = \frac{k}{x^2}$$
$$0.15 = \frac{k}{(0.1)^2} \quad \text{Substituting}$$
$$0.15 = \frac{k}{0.01}$$
$$0.15(0.01) = k$$
$$0.0015 = k$$

The equation of variation is $y = \dfrac{0.0015}{x^2}$.

33. $y = kxz$
$$56 = k \cdot 7 \cdot 8 \quad \text{Substituting}$$
$$56 = 56k$$
$$1 = k$$

The equation of variation is $y = xz$.

35. $y = kxz^2$
$$105 = k \cdot 14 \cdot 5^2 \quad \text{Substituting}$$
$$105 = 350k$$
$$\frac{105}{350} = k$$
$$\frac{3}{10} = k$$

The equation of variation is $y = \dfrac{3}{10}xz^2$.

37. $y = k\dfrac{xz}{wp}$
$$\frac{3}{28} = k\frac{3 \cdot 10}{7 \cdot 8} \quad \text{Substituting}$$
$$\frac{3}{28} = k \cdot \frac{30}{56}$$
$$\frac{3}{28} \cdot \frac{56}{30} = k$$
$$\frac{1}{5} = k$$

The equation of variation is $y = \dfrac{xz}{5wp}$.

39. $d = kr^2$
$$200 = k \cdot 60^2 \quad \text{Substituting}$$
$$200 = 3600k$$
$$\frac{200}{3600} = k$$
$$\frac{1}{18} = k$$

The equation of variation is $d = \dfrac{1}{18}r^2$.

Substitute 72 for d and find r.
$$72 = \frac{1}{18}r^2$$
$$1296 = r^2$$
$$36 = r$$

A car can travel 36 mph and still stop in 72 ft.

41.
$$I = \frac{k}{d^2}$$
$$90 = \frac{k}{5^2} \quad \text{Substituting}$$
$$90 = \frac{k}{25}$$
$$2250 = k$$

The equation of variation is $I = \dfrac{2250}{d^2}$.

Substitute 40 for I and find d.
$$40 = \frac{2250}{d^2}$$
$$40d^2 = 2250$$
$$d^2 = 56.25$$
$$d = 7.5$$

The distance from 5 m to 7.5 m is $7.5 - 5$, or 2.5 m, so it is 2.5 m further to a point where the intensity is 40 W/m².

43. $E = \dfrac{kR}{I}$

We first find k.
$$1.87 = \frac{k \cdot 44}{211\frac{1}{3}}$$
$$1.87 = \frac{k \cdot 44}{634/3} \quad \left(211\frac{1}{3} = \frac{634}{3}\right)$$
$$\frac{634}{3}(1.87) = k \cdot 44$$
$$\frac{\frac{634}{3}(1.87)}{44} = k$$
$$8.98 \approx k$$

The equation of variation is $E = \dfrac{8.98R}{I}$.

Substitute 1.87 for E and 240 for I and solve for R.
$$1.87 = \frac{8.98R}{240}$$
$$\frac{240}{8.98}(1.87) = R$$
$$50 \approx R$$

Roger Clemens would have given up about 50 earned runs if he had pitched 240 innings.

45. $Q = kd^2$

We first find k.
$$225 = k \cdot 5^2$$
$$225 = 25k$$
$$9 = k$$

The equation of variation is $Q = 9d^2$.

Substitute 9 for d and compute Q.
$$Q = 9 \cdot 9^2$$
$$Q = 9 \cdot 81$$
$$Q = 729$$

729 gallons of water are emptied by a pipe that is 9 in. in diameter.

47. Discussion and Writing Exercise

49. When two terms have the same variable(s) raised to the same power(s), they are called <u>like</u> terms.

51. If the sum of two polynomials is 0, they are called <u>opposites</u>, or <u>additive</u> inverses of each other.

53. The <u>intersection</u> of two sets A and B is the set of all members that are common to A and B.

55. The <u>multiplication principle</u> states that for any real number a, b, and c, $c \neq 0$, $a = b$ is equivalent to $a \cdot c = b \cdot c$.

57. We are told $A = kd^2$, and we know $A = \pi r^2$ so we have:

$$kd^2 = \pi r^2$$

$$kd^2 = \pi \left(\frac{d}{2}\right)^2 \qquad r = \frac{d}{2}$$

$$kd^2 = \frac{\pi d^2}{4}$$

$$k = \frac{\pi}{4} \qquad \text{Variation constant}$$

59. $Q = \dfrac{kp^2}{q^3}$

Q varies directly as the square of p and inversely as the cube of q.

61. Let V represent the volume and p represent the price of a jar of peanut butter. The volume of a right circular cylinder with radius r and height h is $\pi r^2 h$. If the diameter is 3 in., then the radius is $\dfrac{3}{2}$ in.

$$V = kp$$

$$\pi r^2 h = kp$$

$$\pi \left(\frac{3}{2}\right)^2 (4) = k(1.2) \qquad \text{Substituting}$$

$$7.5\pi = k$$

$$V = 7.5\pi p \qquad \text{Equation of variation}$$

Now we substitute $\dfrac{6}{2}$, or 3, for r and 6 for h and solve for p.

$$V = 7.5\pi p$$

$$\pi r^2 h = 7.5\pi p$$

$$\pi \cdot 3^2 \cdot 6 = 7.5\pi p \qquad \text{Substituting}$$

$$\pi \cdot 9 \cdot 6 = 7.5\pi p$$

$$54\pi = 7.5\pi p$$

$$7.2 = p \qquad \text{Dividing by } 7.5\pi$$

The bigger jar should cost $7.20.

Chapter 5 Review Exercises

1. $\dfrac{x^2 - 3x + 2}{x^2 - 9}$

We set the denominator equal to 0 and solve.

$$x^2 - 9 = 0$$

$$(x + 3)(x - 3) = 0$$

$$x + 3 = 0 \quad or \quad x - 3 = 0$$

$$x = -3 \quad or \qquad x = 3$$

The expression is not defined for the numbers -3 and 3.

2. In Exercise 1 we found that the replacements for which $\dfrac{x^2 - 3x + 2}{x^2 - 9}$ is not defined are -3 and 3. Then the domain of $f(x)$ is $\{x | x$ is a real number $and\ x \neq -3\ and\ x \neq 3\}$, or $(-\infty, -3) \cup (-3, 3) \cup (3, \infty)$.

3.
$$\frac{4x^2 - 7x - 2}{12x^2 + 11x + 2}$$

$$= \frac{(4x + 1)(x - 2)}{(4x + 1)(3x + 2)}$$

$$= \frac{(4x + 1)(x - 2)}{(4x + 1)(3x + 2)}$$

$$= \frac{x - 2}{3x + 2}$$

4. $\dfrac{a^2 + 2a + 4}{a^3 - 8} = \dfrac{a^2 + 2a + 4}{(a - 2)(a^2 + 2a + 4)}$

$$= \frac{a^2 + 2a + 4}{a^2 + 2a + 4} \cdot \frac{1}{a - 2}$$

$$= \frac{1}{a - 2}$$

5. $6x^3 = 2 \cdot 3 \cdot x \cdot x \cdot x$

$16x^2 = 2 \cdot 2 \cdot 2 \cdot 2 \cdot x \cdot x$

LCM $= 2 \cdot 2 \cdot 2 \cdot 2 \cdot 3 \cdot x \cdot x \cdot x$, or $48x^3$

6. $x^2 - 49 = (x + 7)(x - 7)$

$3x + 1 = 3x + 1$

LCM $= (x + 7)(x - 7)(3x + 1)$

7. $x^2 + x - 20 = (x + 5)(x - 4)$

$x^2 + 3x - 10 = (x + 5)(x - 2)$

LCM $= (x + 5)(x - 4)(x - 2)$

8. $\dfrac{y^2 - 64}{2y + 10} \cdot \dfrac{y + 5}{y + 8} = \dfrac{(y^2 - 64)(y + 5)}{(2y + 10)(y + 8)}$

$$= \frac{(y + 8)(y - 8)(y + 5)}{2(y + 5)(y + 8)}$$

$$= \frac{(y + 8)(y - 8)(y + 5)}{2(y + 5)(y + 8)}$$

$$= \frac{y - 8}{2}$$

9.
$$\frac{x^3 - 8}{x^2 - 25} \cdot \frac{x^2 + 10x + 25}{x^2 + 2x + 4}$$

$$= \frac{(x^3 - 8)(x^2 + 10x + 25)}{(x^2 - 25)(x^2 + 2x + 4)}$$

$$= \frac{(x - 2)(x^2 + 2x + 4)(x + 5)(x + 5)}{(x + 5)(x - 5)(x^2 + 2x + 4)}$$

$$= \frac{(x^2 + 2x + 4)(x + 5)}{(x^2 + 2x + 4)(x + 5)} \cdot \frac{(x - 2)(x + 5)}{x - 5}$$

$$= \frac{(x - 2)(x + 5)}{x - 5}$$

10. $\dfrac{9a^2-1}{a^2-9} \div \dfrac{3a+1}{a+3} = \dfrac{9a^2-1}{a^2-9} \cdot \dfrac{a+3}{3a+1}$

$= \dfrac{(9a^2-1)(a+3)}{(a^2-9)(3a+1)}$

$= \dfrac{(3a+1)(3a-1)(a+3)}{(a+3)(a-3)(3a+1)}$

$= \dfrac{(\cancel{3a+1})(3a-1)(\cancel{a+3})}{(\cancel{a+3})(a-3)(\cancel{3a+1})}$

$= \dfrac{3a-1}{a-3}$

11. $\dfrac{x^3-64}{x^2-16} \div \dfrac{x^2+5x+6}{x^2-3x-18}$

$= \dfrac{x^3-64}{x^2-16} \cdot \dfrac{x^2-3x-18}{x^2+5x+6}$

$= \dfrac{(x^3-64)(x^2-3x-18)}{(x^2-16)(x^2+5x+6)}$

$= \dfrac{(x-4)(x^2+4x+16)(x-6)(x+3)}{(x+4)(x-4)(x+2)(x+3)}$

$= \dfrac{(\cancel{x-4})(x^2+4x+16)(x-6)(\cancel{x+3})}{(x+4)(\cancel{x-4})(x+2)(\cancel{x+3})}$

$= \dfrac{(x^2+4x+16)(x-6)}{(x+4)(x+2)}$

12. $\dfrac{x}{x^2+5x+6} - \dfrac{2}{x^2+3x+2}$

$= \dfrac{x}{(x+2)(x+3)} - \dfrac{2}{(x+1)(x+2)}$

$\qquad\qquad$ LCD $= (x+2)(x+3)(x+1)$

$= \dfrac{x}{(x+2)(x+3)} \cdot \dfrac{x+1}{x+1} - \dfrac{2}{(x+1)(x+2)} \cdot \dfrac{x+3}{x+3}$

$= \dfrac{x^2+x-(2x+6)}{(x+2)(x+3)(x+1)}$

$= \dfrac{x^2+x-2x-6}{(x+2)(x+3)(x+1)}$

$= \dfrac{x^2-x-6}{(x+2)(x+3)(x+1)}$

$= \dfrac{(x+2)(x-3)}{(x+2)(x+3)(x+1)}$

$= \dfrac{(\cancel{x+2})(x-3)}{(\cancel{x+2})(x+3)(x+1)}$

$= \dfrac{x-3}{(x+3)(x+1)}$

13. $\dfrac{2x^2}{x-y} + \dfrac{2y^2}{x+y}$ $\quad$ LCD $= (x-y)(x+y)$

$= \dfrac{2x^2}{x-y} \cdot \dfrac{x+y}{x+y} + \dfrac{2y^2}{x+y} \cdot \dfrac{x-y}{x-y}$

$= \dfrac{2x^3+2x^2y+2xy^2-2y^3}{(x-y)(x+y)}$

14. $\dfrac{3}{y+4} - \dfrac{y}{y-1} + \dfrac{y^2+3}{y^2+3y-4}$

$= \dfrac{3}{y+4} - \dfrac{y}{y-1} + \dfrac{y^2+3}{(y-1)(y+4)}$

$\qquad\qquad$ LCD $= (y+4)(y-1)$

$= \dfrac{3}{y+4} \cdot \dfrac{y-1}{y-1} - \dfrac{y}{y-1} \cdot \dfrac{y+4}{y+4} + \dfrac{y^2+3}{(y-1)(y+4)}$

$= \dfrac{3y-3-(y^2+4y)+y^2+3}{(y+4)(y-1)}$

$= \dfrac{3y-3-y^2-4y+y^2+3}{(y+4)(y-1)}$

$= \dfrac{-y}{(y+4)(y-1)}$

15. $(16ab^3c - 10ab^2c^2 + 12a^2b^2c) \div (4ab)$

$= \dfrac{16ab^3c - 10ab^2c^2 + 12a^2b^2c}{4ab}$

$= \dfrac{16ab^3c}{4ab} - \dfrac{10ab^2c^2}{4ab} + \dfrac{12a^2b^2c}{4ab}$

$= 4b^2c - \dfrac{5}{2}bc^2 + 3abc$

16.

$$
\begin{array}{r}
y-14 \\
y-6 \enclose{longdiv}{y^2-20y+64} \\
\underline{y^2-6y} \\
-14y+64 \\
\underline{-14y+84} \\
-20
\end{array}
$$

The answer is $y-14$, R -20, or $y-14+\dfrac{-20}{y-6}$.

17.

$$
\begin{array}{r}
6x^2 - 9 \\
x^2+2 \enclose{longdiv}{6x^4+0x^3+3x^2+5x+4} \\
\underline{6x^4+12x^2} \\
-9x^2 \\
\underline{-9x^2-18} \\
5x+22
\end{array}
$$

The answer is $6x^2-9$, R $(5x+22)$, or $6x^2-9+\dfrac{5x+22}{x^2+2}$.

18.

$$
\begin{array}{r|rrrr}
4 & 1 & 5 & 4 & -7 \\
& & 4 & 36 & 160 \\
\hline
& 1 & 9 & 40 & 153
\end{array}
$$

The answer is $x^2+9x+40$, R 153, or $x^2+9x+40+\dfrac{153}{x-4}$.

19. $(3x^4-5x^3+2x-7) \div (x+1) =$

$(3x^4-5x^3+0x^2+2x-7) \div [x-(-1)]$

$$
\begin{array}{r|rrrrr}
-1 & 3 & -5 & 0 & 2 & -7 \\
& & -3 & 8 & -8 & 6 \\
\hline
& 3 & -8 & 8 & -6 & -1
\end{array}
$$

The answer is $3x^3-8x^2+8x-6$, R -1, or

$3x^3-8x^2+8x-6+\dfrac{-1}{x+1}$.

20. $\dfrac{3+\dfrac{3}{y}}{4+\dfrac{4}{y}} = \dfrac{3+\dfrac{3}{y}}{4+\dfrac{4}{y}} \cdot \dfrac{y}{y}$

$= \dfrac{\left(3+\dfrac{3}{y}\right)\cdot y}{\left(4+\dfrac{4}{y}\right)\cdot y}$

$= \dfrac{3\cdot y + \dfrac{3}{y}\cdot y}{4\cdot y + \dfrac{4}{y}\cdot y}$

$= \dfrac{3y+3}{4y+4}$

$= \dfrac{3(y+1)}{4(y+1)}$

$= \dfrac{3(\cancel{y+1})}{4(\cancel{y+1})}$

$= \dfrac{3}{4}$

21. $\dfrac{\dfrac{2}{a}+\dfrac{2}{b}}{\dfrac{4}{a^3}+\dfrac{4}{b^3}} = \dfrac{\dfrac{2}{a}\cdot\dfrac{b}{b}+\dfrac{2}{b}\cdot\dfrac{a}{a}}{\dfrac{4}{a^3}\cdot\dfrac{b^3}{b^3}+\dfrac{4}{b^3}\cdot\dfrac{a^3}{a^3}}$

$= \dfrac{\dfrac{2b}{ab}+\dfrac{2a}{ab}}{\dfrac{4b^3}{a^3b^3}+\dfrac{4a^3}{a^3b^3}}$

$= \dfrac{\dfrac{2b+2a}{ab}}{\dfrac{4b^3+4a^3}{a^3b^3}}$

$= \dfrac{2b+2a}{ab}\cdot\dfrac{a^3b^3}{4b^3+4a^3}$

$= \dfrac{(2b+2a)\cdot a^3b^3}{ab(4b^3+4a^3)}$

$= \dfrac{2(b+a)\cdot ab \cdot a^2b^2}{ab\cdot 2\cdot 2(b+a)(b^2-ab+a^2)}$

$= \dfrac{2(b+a)\cdot ab}{2(b+a)\cdot ab}\cdot\dfrac{a^2b^2}{2(b^2-ab+a^2)}$

$= \dfrac{a^2b^2}{2(b^2-ab+a^2)}$

22. $\dfrac{\dfrac{x^2-5x-36}{x^2-36}}{\dfrac{x^2+x-12}{x^2-12x+36}}$

$= \dfrac{x^2-5x-36}{x^2-36}\cdot\dfrac{x^2-12x+36}{x^2+x-12}$

$= \dfrac{(x^2-5x-36)(x^2-12x+36)}{(x^2-36)(x^2+x-12)}$

$= \dfrac{(x-9)(x+4)(x-6)(x-6)}{(x+6)(x-6)(x+4)(x-3)}$

$= \dfrac{(x-9)(\cancel{x+4})(\cancel{x-6})(x-6)}{(x+6)(\cancel{x-6})(\cancel{x+4})(x-3)}$

$= \dfrac{(x-9)(x-6)}{(x+6)(x-3)}$

23. $\dfrac{\dfrac{4}{x+3}-\dfrac{2}{x^2-3x+2}}{\dfrac{3}{x-2}+\dfrac{1}{x^2+2x-3}}$

$= \dfrac{\dfrac{4}{x+3}-\dfrac{2}{(x-1)(x-2)}}{\dfrac{3}{x-2}+\dfrac{1}{(x+3)(x-1)}}$

$= \dfrac{\dfrac{4}{x+3}-\dfrac{2}{(x-1)(x-2)}}{\dfrac{3}{x-2}+\dfrac{1}{(x+3)(x-1)}}\cdot\dfrac{(x+3)(x-1)(x-2)}{(x+3)(x-1)(x-2)}$

$= \dfrac{\dfrac{4}{x+3}\cdot(x+3)(x-1)(x-2)-\dfrac{2}{(x-1)(x-2)}\cdot(x+3)(x-1)(x-2)}{\dfrac{3}{x-2}\cdot(x+3)(x-1)(x-2)+\dfrac{1}{(x+3)(x-1)}\cdot(x+3)(x-1)(x-2)}$

$= \dfrac{4(x-1)(x-2)-2(x+3)}{3(x+3)(x-1)+(x-2)}$

$= \dfrac{4(x^2-3x+2)-2x-6}{3(x^2+2x-3)+x-2}$

$= \dfrac{4x^2-12x+8-2x-6}{3x^2+6x-9+x-2}$

$= \dfrac{4x^2-14x+2}{3x^2+7x-11}$

24. $\dfrac{x}{4}+\dfrac{x}{7}=1,$ LCM is 28

$28\left(\dfrac{x}{4}+\dfrac{x}{7}\right)=28\cdot 1$

$28\cdot\dfrac{x}{4}+28\cdot\dfrac{x}{7}=28$

$7x+4x=28$

$11x=28$

$x=\dfrac{28}{11}$

Since $\dfrac{28}{11}$ checks, it is the solution.

25. $\dfrac{5}{3x+2} = \dfrac{3}{2x}$, LCM is $2x(3x+2)$

$$2x(3x+2) \cdot \dfrac{5}{3x+2} = 2x(3x+2) \cdot \dfrac{3}{2x}$$

$$2x \cdot 5 = 3(3x+2)$$

$$10x = 9x + 6$$

$$x = 6$$

Since 6 checks, it is the solution.

26. $\dfrac{4x}{x+1} + \dfrac{4}{x} + 9 = \dfrac{4}{x^2 + x}$

$$\dfrac{4x}{x+1} + \dfrac{4}{x} + 9 = \dfrac{4}{x(x+1)},$$

$$\text{LCM is } x(x+1)$$

$$x(x+1)\left(\dfrac{4x}{x+1} + \dfrac{4}{x} + 9\right) = x(x+1) \cdot \dfrac{4}{x(x+1)}$$

$$x \cdot 4x + 4(x+1) + 9x(x+1) = 4$$

$$4x^2 + 4x + 4 + 9x^2 + 9x = 4$$

$$13x^2 + 13x + 4 = 4$$

$$13x^2 + 13x = 0$$

$$13x(x+1) = 0$$

$$13x = 0 \ \ or \ \ x + 1 = 0$$

$$x = 0 \ \ or \ \ \ \ \ x = -1$$

The numbers 0 and -1 each make a denominator 0, so there are no solutions.

27. $\dfrac{90}{x^2 - 3x + 9} - \dfrac{5x}{x+3} = \dfrac{405}{x^3 + 27}$

$$\dfrac{90}{x^2 - 3x + 9} - \dfrac{5x}{x+3} = \dfrac{405}{(x+3)(x^2 - 3x + 9)}$$

$$\text{LCM is } (x+3)(x^2 - 3x + 9)$$

$$(x+3)(x^2 - 3x + 9) \cdot \left(\dfrac{90}{x^2 - 3x + 9} - \dfrac{5x}{x+3}\right) =$$

$$(x+3)(x^2 - 3x + 9) \cdot \dfrac{405}{(x+3)(x^2 + 3x + 9)}$$

$$90(x+3) - 5x(x^2 - 3x + 9) = 405$$

$$90x + 270 - 5x^3 + 15x^2 - 45x = 405$$

$$-5x^3 + 15x^2 + 45x + 270 = 405$$

$$-5x^3 + 15x^2 + 45x - 135 = 0$$

$$5x^3 - 15x^2 - 45x + 135 = 0 \quad \text{Multiplying by } -1$$

$$5(x^3 - 3x^2 - 9x + 27) = 0$$

$$5[x^2(x-3) - 9(x-3)] = 0$$

$$5(x^2 - 9)(x - 3) = 0$$

$$5(x+3)(x-3)(x-3) = 0$$

$$x + 3 = 0 \ \ or \ \ x - 3 = 0 \ or \ \ x - 3 = 0$$

$$x = -3 \ or \ \ \ \ x = 3 \ or \ \ \ \ \ \ x = 3$$

We know that -3 is not a solution of the original equation, because it results in division by 0. Since 3 checks, it is the solution.

28. $\dfrac{2}{x-3} + \dfrac{1}{4x+20} = \dfrac{1}{x^2 + 2x - 15}$

$$\dfrac{2}{x-3} + \dfrac{1}{4(x+5)} = \dfrac{1}{(x-3)(x+5)}$$

$$\text{LCM is } 4(x-3)(x+5)$$

$$4(x-3)(x+5)\left(\dfrac{2}{x-3} + \dfrac{1}{4(x+5)}\right) = 4(x-3)(x+5) \cdot \dfrac{1}{(x-3)(x+5)}$$

$$4(x+5) \cdot 2 + (x-3) = 4$$

$$8x + 40 + x - 3 = 4$$

$$9x + 37 = 4$$

$$9x = -33$$

$$x = -\dfrac{11}{3}$$

Since $-\dfrac{11}{3}$ checks, it is the solution.

29. $\dfrac{6}{x} + \dfrac{4}{x} = 5$

$$\dfrac{10}{x} = 5 \quad \text{Adding}$$

$$x \cdot \dfrac{10}{x} = x \cdot 5$$

$$10 = 5x$$

$$2 = x$$

Since 2 checks, the value of x for which $f(x) = 5$ is 2.

30. *Familiarize.* Let $t =$ the time it takes them to paint the house, working together.

Translate. Using the work principle, we get the following equation:

$$\dfrac{t}{12} + \dfrac{t}{9} = 1$$

Solve. We solve the equation.

$$\dfrac{t}{12} + \dfrac{t}{9} = 1, \text{ LCM is } 36$$

$$36\left(\dfrac{t}{12} + \dfrac{t}{9}\right) = 36 \cdot 1$$

$$36 \cdot \dfrac{t}{12} + 36 \cdot \dfrac{t}{9} = 36$$

$$3t + 4t = 36$$

$$7t = 36$$

$$t = \dfrac{36}{7}, \text{ or } 5\dfrac{1}{7}$$

Check. We verify the work principle.

$$\dfrac{\frac{36}{7}}{12} + \dfrac{\frac{36}{7}}{9} = \dfrac{36}{7} \cdot \dfrac{1}{12} + \dfrac{36}{7} \cdot \dfrac{1}{9} = \dfrac{3}{7} + \dfrac{4}{7} = 1$$

State. It will take them $5\dfrac{1}{7}$ hr to paint the house, working together.

31. *Familiarize.* Let $r =$ the speed of the boat in still water. Then the boat's speed traveling downstream is $r + 6$ and the speed upstream is $r - 6$. Since the time is the same downstream and upstream, we let t represent each time.

	Distance	Speed	Time
Downstream	50	$r+6$	t
Upstream	30	$r-6$	t

Translate. Using the formula Time = Distance/Rate in each row of the table and the fact that the times are the same, we can write an equation.

$$\frac{50}{r+6} = \frac{30}{r-6}$$

Solve.

$$\frac{50}{r+6} = \frac{30}{r-6},$$

$$\text{LCD is } (r+6)(r-6)$$

$$(r+6)(r-6) \cdot \frac{50}{r+6} = (r+6)(r-6) \cdot \frac{30}{r-6}$$

$$50(r-6) = 30(r+6)$$

$$50r - 300 = 30r + 180$$

$$20r = 480$$

$$r = 24$$

Check. If the speed of the boat in still water is 24 mph, then the speed downstream is $24+6$, or 30 mph. It would take 50/30, or 5/3 hr, to travel 50 mi downstream. The speed upstream is $24-6$, or 18 mph. It would take 30/18, or 5/3 hr, to travel 30 mi upstream. The times are the same, so the answer checks.

State. The speed of the boat in still water is 24 mph.

32. Familiarize. Let d = the number of miles Fred will travel in 15 days.

Translate. We translate to a proportion.

$$\begin{array}{l} \text{Miles} \rightarrow \\ \text{Days} \rightarrow \end{array} \frac{800}{3} = \frac{d}{15} \begin{array}{l} \leftarrow \text{Miles} \\ \leftarrow \text{Days} \end{array}$$

Solve.

$$\frac{800}{3} = \frac{d}{15}$$

$$800 \cdot 15 = 3 \cdot d$$

$$\frac{800 \cdot 15}{3} = d$$

$$4000 = d$$

Check. We substitute in the proportion and check cross products.

$$\frac{800}{3} = \frac{4000}{15}; \; 800 \cdot 15 = 12,000; \; 3 \cdot 4000 = 12,000$$

Since the cross products are the same the answer checks.

State. Fred will travel 4000 mi in 15 days.

33. First we solve for d.

$$W = \frac{cd}{c+d}$$

$$W(c+d) = cd \qquad \text{Multiplying by } c+d$$

$$Wc + Wd = cd$$

$$Wc = cd - Wd$$

$$Wc = d(c - W)$$

$$\frac{Wc}{c-W} = d$$

Now we solve for c.

$$W = \frac{cd}{c+d}$$

$$W(c+d) = cd \qquad \text{Multiplying by } c+d$$

$$Wc + Wd = cd$$

$$Wd = cd - Wc$$

$$Wd = c(d - W)$$

$$\frac{Wd}{d-W} = c$$

34. First we solve for b.

$$S = \frac{p}{a} + \frac{t}{b}$$

$$ab \cdot S = ab\left(\frac{p}{a} + \frac{t}{b}\right)$$

$$abS = ab \cdot \frac{p}{a} + ab \cdot \frac{t}{b}$$

$$abS = bp + at$$

$$abS - bp = at$$

$$b(aS - p) = at$$

$$b = \frac{at}{aS - p}$$

Now we solve for t.

$$S = \frac{p}{a} + \frac{t}{b}$$

$$ab \cdot S = ab\left(\frac{p}{a} + \frac{t}{b}\right)$$

$$abS = ab \cdot \frac{p}{a} + ab \cdot \frac{t}{b}$$

$$abS = bp + at$$

$$abS - bp = at$$

$$\frac{abS - bp}{a} = t$$

35. $\quad y = kx$

$$100 = k \cdot 25$$

$$4 = k$$

$$y = 4x \quad \text{Equation of variation}$$

36. $\quad y = \frac{k}{x}$

$$100 = \frac{k}{25}$$

$$2500 = k$$

$$y = \frac{2500}{x} \quad \text{Equation of variation}$$

37. $\quad t = \frac{k}{r} \quad t$ varies inversely as r.

$$35 = \frac{k}{800} \quad \text{Substituting}$$

$$28,000 = k$$

$$t = \frac{28,000}{r} \quad \text{Equation of variation}$$

$$t = \frac{28,000}{1400}$$

$$t = 20$$

It will take the pump 20 min to empty the tank at the rate of 1400 kL per minute.

38. $N = ka$ N varies directly as a.

$87 = k \cdot 28$ Substituting

$\dfrac{87}{28} = k$

$N = \dfrac{87}{28} a$

$N = \dfrac{87}{28} \cdot 25$

$N \approx 77.7$

Ellen's score would have been about 77.7 if she had answered 25 questions correctly.

39. $P = kC^2$ P varies directly as the square of C.

$180 = k \cdot 6^2$ Substituting

$180 = 36k$

$5 = k$

$P = 5C^2$ Equation of variation

$P = 5 \cdot 10^2$

$P = 5 \cdot 100$

$P = 500$

The circuit expends 500 watts of heat when the current is 10 amperes.

40. *Discussion and Writing Exercise.* When adding or subtracting rational expressions, we use the LCM of the denominators (the LCD). When solving a rational equation or when solving a formula for a given letter, we multiply by the LCM of all the denominators to clear fractions. When simplifying a complex rational expression, we can use the LCM in either of two ways. We can multiply by a/a, where a is the LCM of all the denominators occurring in the expression. Or we can use the LCM to add or subtract as necessary in the numerator and in the denominator.

41. *Discussion and Writing Exercise.* Rational equations differ from those previously studied because they contain variables in denominators. Because of this, possible solutions must be checked in the original equation to avoid division by 0.

42. The reciprocal of $\dfrac{a-b}{a^3-b^3}$ is $\dfrac{a^3-b^3}{a-b}$.

$\dfrac{a^3-b^3}{a-b} = \dfrac{(a-b)(a^2+ab+b^2)}{(a-b)\cdot 1}$

$\qquad = \dfrac{(a-b)(a^2+ab+b^2)}{(a-b)\cdot 1}$

$\qquad = a^2+ab+b^2$

43.

$$\dfrac{5}{x-13} - \dfrac{5}{x} = \dfrac{65}{x^2-13x}$$

$$\dfrac{5}{x-13} - \dfrac{5}{x} = \dfrac{65}{x(x-13)}$$

$$\text{LCD is } x(x-13)$$

$$x(x-13)\left(\dfrac{5}{x-13} - \dfrac{5}{x}\right) = x(x-13)\cdot\dfrac{65}{x(x-13)}$$

$$5x - 5(x-13) = 65$$

$$5x - 5x + 65 = 65$$

$$65 = 65$$

We get a true equation. Thus, all real numbers except those that make a denominator 0 are solutions of the equation. The numbers that make a denominator 0 are 13 and 0, so all real numbers except 13 and 0 are solutions.

Chapter 5 Test

1. $\dfrac{x^2-16}{x^2-3x+2}$

We set the denominator equal to 0 and solve.

$$x^2 - 3x + 2 = 0$$

$$(x-1)(x-2) = 0$$

$$x - 1 = 0 \ \text{ or } \ x - 2 = 0$$

$$x = 1 \ \text{ or } \qquad x = 2$$

The expression is not defined for the numbers 1 and 2.

2. In Exercise 1 we found that the replacements for which $\dfrac{x^2-16}{x^2-3x+2}$ is not defined are 1 and 2. Then the domain of $f(x) = \dfrac{x^2-16}{x^2-3x+2}$ is $(-\infty, 1) \cup (1, 2) \cup (2, \infty)$.

3. $\dfrac{12x^2+11x+2}{4x^2-7x-2} = \dfrac{(4x+1)(3x+2)}{(4x+1)(x-2)}$

$\qquad = \dfrac{(4x+1)(3x+2)}{(4x+1)(x-2)}$

$\qquad = \dfrac{3x+2}{x-2}$

4. $\dfrac{p^3+1}{p^2-p-2} = \dfrac{(p+1)(p^2-p+1)}{(p+1)(p-2)}$

$\qquad = \dfrac{(p+1)(p^2-p+1)}{(p+1)(p-2)}$

$\qquad = \dfrac{p^2-p+1}{p-2}$

5. $x^2 + x - 6 = (x+3)(x-2)$

$x^2 + 8x + 15 = (x+3)(x+5)$

$\text{LCM} = (x+3)(x-2)(x+5)$

6. $\dfrac{2x^2 + 20x + 50}{x^2 - 4} \cdot \dfrac{x + 2}{x + 5}$

$= \dfrac{(2x^2 + 20x + 50)(x + 2)}{(x^2 - 4)(x + 5)}$

$= \dfrac{2(x^2 + 10x + 25)(x + 2)}{(x + 2)(x - 2)(x + 5)}$

$= \dfrac{2(x + 5)(x + 5)(x + 2)}{(x + 2)(x - 2)(x + 5)}$

$= \dfrac{2(\cancel{x+5})(x + 5)(\cancel{x+2})}{(\cancel{x+2})(x - 2)(\cancel{x+5})}$

$= \dfrac{2(x + 5)}{x - 2}$

7. $\dfrac{x}{x^2 + 11x + 30} - \dfrac{5}{x^2 + 9x + 20}$

$= \dfrac{x}{(x + 5)(x + 6)} - \dfrac{5}{(x + 4)(x + 5)}$

$\qquad \text{LCD is} (x + 5)(x + 6)(x + 4)$

$= \dfrac{x}{(x + 5)(x + 6)} \cdot \dfrac{x + 4}{x + 4} - \dfrac{5}{(x + 4)(x + 5)} \cdot \dfrac{x + 6}{x + 6}$

$= \dfrac{x^2 + 4x - (5x + 30)}{(x + 5)(x + 6)(x + 4)}$

$= \dfrac{x^2 + 4x - 5x - 30}{(x + 5)(x + 6)(x + 4)}$

$= \dfrac{x^2 - x - 30}{(x + 5)(x + 6)(x + 4)}$

$= \dfrac{(x + 5)(x - 6)}{(x + 5)(x + 6)(x + 4)}$

$= \dfrac{(\cancel{x+5})(x - 6)}{(\cancel{x+5})(x + 6)(x + 4)}$

$= \dfrac{x - 6}{(x + 6)(x + 4)}$

8. $\dfrac{y^2 - 16}{2y + 6} \div \dfrac{y - 4}{y + 3} = \dfrac{y^2 - 16}{2y + 6} \cdot \dfrac{y + 3}{y - 4}$

$\qquad\qquad = \dfrac{(y^2 - 16)(y + 3)}{(2y + 6)(y - 4)}$

$\qquad\qquad = \dfrac{(y + 4)(y - 4)(y + 3)}{2(y + 3)(y - 4)}$

$\qquad\qquad = \dfrac{(y + 4)(\cancel{y-4})(\cancel{y+3})}{2(\cancel{y+3})(\cancel{y-4})}$

$\qquad\qquad = \dfrac{y + 4}{2}$

9. $\dfrac{x^2}{x - y} + \dfrac{y^2}{y - x} = \dfrac{x^2}{x - y} + \dfrac{y^2}{y - x} \cdot \dfrac{-1}{-1}$

$\qquad\qquad = \dfrac{x^2}{x - y} + \dfrac{-y^2}{x - y}$

$\qquad\qquad = \dfrac{x^2 - y^2}{x - y}$

$\qquad\qquad = \dfrac{(x + y)(x - y)}{(x - y)}$

$\qquad\qquad = \dfrac{(x + y)(\cancel{x-y})}{(\cancel{x-y}) \cdot 1}$

$\qquad\qquad = x + y$

10. $\dfrac{1}{x + 1} - \dfrac{x + 2}{x^2 - 1} + \dfrac{3}{x - 1}$

$= \dfrac{1}{x + 1} - \dfrac{x + 2}{(x + 1)(x - 1)} + \dfrac{3}{x - 1} \quad \text{LCD is } (x + 1)(x - 1)$

$= \dfrac{1}{x + 1} \cdot \dfrac{x - 1}{x - 1} - \dfrac{x + 2}{(x + 1)(x - 1)} + \dfrac{3}{x - 1} \cdot \dfrac{x + 1}{x + 1}$

$= \dfrac{x - 1 - (x + 2) + 3x + 3}{(x + 1)(x - 1)}$

$= \dfrac{x - 1 - x - 2 + 3x + 3}{(x + 1)(x - 1)}$

$= \dfrac{3x}{(x + 1)(x - 1)}$

11. $\dfrac{a}{a - b} + \dfrac{b}{a^2 + ab + b^2} - \dfrac{2}{a^3 - b^3}$

$= \dfrac{a}{a - b} + \dfrac{b}{a^2 + ab + b^2} - \dfrac{2}{(a - b)(a^2 + ab + b^2)}$

$\qquad\qquad \text{LCD is } (a - b)(a^2 + ab + b^2)$

$= \dfrac{a}{a - b} \cdot \dfrac{a^2 + ab + b^2}{a^2 + ab + b^2} + \dfrac{b}{a^2 + ab + b^2} \cdot \dfrac{a - b}{a - b} -$

$\qquad\qquad \dfrac{2}{(a - b)(a^2 + ab + b^2)}$

$= \dfrac{a^3 + a^2 b + ab^2 + ab - b^2 - 2}{(a - b)(a^2 + ab + b^2)}$

12. $(20r^2 s^3 + 15r^2 s^2 - 10r^3 s^3) \div (5r^2 s)$

$= \dfrac{20r^2 s^3 + 15r^2 s^2 - 10r^3 s^3}{5r^2 s}$

$= \dfrac{20r^2 s^3}{5r^2 s} + \dfrac{15r^2 s^2}{5r^2 s} - \dfrac{10r^3 s^3}{5r^2 s}$

$= 4s^2 + 3s - 2rs^2$

13.

$$
\begin{array}{r}
y^2 - 5y + 25 \\
y + 5 \overline{\smash{\big)}\ y^3 + 0y^2 + 0y + 125} \\
\underline{y^3 + 5y^2} \\
-5y^2 + 0y \\
\underline{-5y^2 - 25y} \\
25y + 125 \\
\underline{25y + 125} \\
0
\end{array}
$$

The answer is $y^2 - 5y + 25$.

14.

$$
\begin{array}{r}
4x^2 + 3x - 4 \\
x^2 + 1 \overline{\smash{\big)}\ 4x^4 + 3x^3 + 0x^2 - 5x - 2} \\
\underline{4x^4 \qquad\quad + 4x^2} \\
3x^3 - 4x^2 \\
\underline{3x^3 \qquad\quad + 3x} \\
-4x^2 - 8x \\
\underline{-4x^2 \qquad\quad - 4} \\
-8x + 2
\end{array}
$$

The answer is $4x^2 + 3x - 4$, R $(-8x + 2)$, or

$4x^2 + 3x - 4 + \dfrac{-8x + 2}{x^2 + 1}$.

15. $(x^3 + 3x^2 + 2x - 6) \div (x - 3)$

$$\underline{3}\,|\ \begin{array}{rrrr} 1 & 3 & 2 & -6 \\ & 3 & 18 & 60 \\ \hline 1 & 6 & 20\,| & 54 \end{array}$$

The answer is $x^2 + 6x + 20$, R 54, or $x^2 + 6x + 20 + \dfrac{54}{x-3}$.

16. $(4x^3 - 6x^2 - 9) \div (x+5) = (4x^3 - 6x^2 + 0x - 9) \div [x - (-5)]$

$$\underline{-5}\,|\ \begin{array}{rrrr} 4 & -6 & 0 & -9 \\ & -20 & 130 & -650 \\ \hline 4 & -26 & 130\,| & -659 \end{array}$$

The answer is $4x^2 - 26x + 130$, R -659, or
$4x^2 - 26x + 130 + \dfrac{-659}{x+5}$.

17. $\dfrac{1 - \dfrac{1}{x^2}}{1 - \dfrac{1}{x}}$

The LCM of the denominators is x^2. We multiply by 1 using x^2/x^2.

$$\dfrac{1 - \dfrac{1}{x^2}}{1 - \dfrac{1}{x}} = \dfrac{1 - \dfrac{1}{x^2}}{1 - \dfrac{1}{x}} \cdot \dfrac{x^2}{x^2}$$

$$= \dfrac{\left(1 - \dfrac{1}{x^2}\right) \cdot x^2}{\left(1 - \dfrac{1}{x}\right) \cdot x^2}$$

$$= \dfrac{1 \cdot x^2 - \dfrac{1}{x^2} \cdot x^2}{1 \cdot x^2 - \dfrac{1}{x} \cdot x^2}$$

$$= \dfrac{x^2 - 1}{x^2 - x}$$

$$= \dfrac{(x+1)(x-1)}{x(x-1)}$$

$$= \dfrac{(x+1)(x-1)}{x(x-1)}$$

$$= \dfrac{x+1}{x}$$

18. $\dfrac{\dfrac{1}{a^3} + \dfrac{1}{b^3}}{\dfrac{1}{a} + \dfrac{1}{b}}$

The LCM of the denominators is $a^3 b^3$. We multiply by 1 using $(a^3 b^3)/(a^3 b^3)$.

$$\dfrac{\dfrac{1}{a^3} + \dfrac{1}{b^3}}{\dfrac{1}{a} + \dfrac{1}{b}} = \dfrac{\dfrac{1}{a^3} + \dfrac{1}{b^3}}{\dfrac{1}{a} + \dfrac{1}{b}} \cdot \dfrac{a^3 b^3}{a^3 b^3}$$

$$= \dfrac{\left(\dfrac{1}{a^3} + \dfrac{1}{b^3}\right) \cdot a^3 b^3}{\left(\dfrac{1}{a} + \dfrac{1}{b}\right) \cdot a^3 b^3}$$

$$= \dfrac{\dfrac{1}{a^3} \cdot a^3 b^3 + \dfrac{1}{b^3} \cdot a^3 b^3}{\dfrac{1}{a} \cdot a^3 b^3 + \dfrac{1}{b} \cdot a^3 b^3}$$

$$= \dfrac{b^3 + a^3}{a^2 b^3 + a^3 b^2}$$

$$= \dfrac{(b+a)(b^2 - ba + a^2)}{a^2 b^2 (b+a)}$$

$$= \dfrac{(b+a)(b^2 - ba + a^2)}{a^2 b^2 (b+a)}$$

$$= \dfrac{b^2 - ba + a^2}{a^2 b^2}$$

19.
$$\dfrac{2}{x-1} + \dfrac{2}{x+2} = 1 \quad \text{LCM is } (x-1)(x+2)$$

$$(x-1)(x+2)\left(\dfrac{2}{x-1} + \dfrac{2}{x+2}\right) = (x-1)(x+2) \cdot 1$$

$$2(x+2) + 2(x-1) = x^2 + x - 2$$

$$2x + 4 + 2x - 2 = x^2 + x - 2$$

$$4x + 2 = x^2 + x - 2$$

$$0 = x^2 - 3x - 4$$

$$0 = (x-4)(x+1)$$

$$x - 4 = 0 \quad or \quad x + 1 = 0$$

$$x = 4 \quad or \quad x = -1$$

Both values check. The values of x for which $f(x) = 1$ are 4 and -1.

20.
$$\dfrac{2}{x-1} = \dfrac{3}{x+3} \quad \text{LCM is } (x-1)(x+3)$$

$$(x-1)(x+3) \cdot \dfrac{2}{x-1} = (x-1)(x+3) \cdot \dfrac{3}{x+3}$$

$$2(x+3) = 3(x-1)$$

$$2x + 6 = 3x - 3$$

$$9 = x$$

This value checks. The solution is 9.

21.
$$\frac{7x}{x+3} + \frac{21}{x-3} = \frac{126}{x^2-9}$$
$$\frac{7x}{x+3} + \frac{21}{x-3} = \frac{126}{(x+3)(x-3)}$$

LCM is $(x+3)(x-3)$

$$(x+3)(x-3)\left(\frac{7x}{x+3} + \frac{21}{x-3}\right) =$$
$$(x+3)(x-3)\cdot\frac{126}{(x+3)(x-3)}$$
$$7x(x-3) + 21(x+3) = 126$$
$$7x^2 - 21x + 21x + 63 = 126$$
$$7x^2 + 63 = 126$$
$$7x^2 - 63 = 0$$
$$7(x^2-9) = 0$$
$$7(x+3)(x-3) = 0$$
$$x+3 = 0 \quad or \quad x-3 = 0$$
$$x = -3 \quad or \quad x = 3$$

We know that neither number can be a solution of the original equation because each one results in division by 0. Thus, the equation has no solution.

22.
$$\frac{2x}{x+7} = \frac{5}{x+1} \quad \text{LCM is } (x+7)(x+1)$$
$$(x+7)(x+1)\cdot\frac{2x}{x+7} = (x+7)(x+1)\cdot\frac{5}{x+1}$$
$$2x(x+1) = 5(x+7)$$
$$2x^2 + 2x = 5x + 35$$
$$2x^2 - 3x - 35 = 0$$
$$(2x+7)(x-5) = 0$$
$$2x+7 = 0 \quad or \quad x-5 = 0$$
$$2x = -7 \quad or \quad x = 5$$
$$x = -\frac{7}{2} \quad or \quad x = 5$$

Both values check. The solutions are $-\frac{7}{2}$ and 5.

23.
$$\frac{1}{3x-6} - \frac{1}{x^2-4} = \frac{3}{x+2}$$
$$\frac{1}{3(x-2)} - \frac{1}{(x+2)(x-2)} = \frac{3}{x+2}$$

LCM is $3(x-2)(x+2)$

$$3(x-2)(x+2)\left(\frac{1}{3(x-2)} - \frac{1}{(x+2)(x-2)}\right) =$$
$$3(x-2)(x+2)\cdot\frac{3}{x+2}$$
$$x+2-3 = 9(x-2)$$
$$x-1 = 9x-18$$
$$-8x = -17$$
$$x = \frac{17}{8}$$

This value checks. The solution is $\frac{17}{8}$.

24. Familiarize. Let s = the number of hours it takes Sam to complete the puzzle working alone. Then $s+4$ = Bella's time.

Translate. We use the work principle.
$$\frac{t}{a} + \frac{t}{b} = 1$$
$$\frac{1.5}{s} + \frac{1.5}{s+4} = 1$$

Solve. We solve the equation.
$$\frac{1.5}{s} + \frac{1.5}{s+4} = 1 \quad \text{LCM is } s(s+4)$$
$$s(s+4)\left(\frac{1.5}{s} + \frac{1.5}{s+4}\right) = s(s+4)\cdot 1$$
$$1.5(s+4) + 1.5s = s^2 + 4s$$
$$1.5s + 6 + 1.5s = s^2 + 4s$$
$$3s + 6 = s^2 + 4s$$
$$0 = s^2 + s - 6$$
$$0 = (s+3)(s-2)$$
$$s+3 = 0 \quad or \quad s-2 = 0$$
$$s = -3 \quad or \quad s = 2$$

Check. The time cannot be negative, so we check only 2. If $s = 2$, then $s+4 = 2+4 = 6$. In 1.5 hr Sam does $\frac{1.5}{2}$, or 0.75, of the job and Bella does $\frac{1.5}{6}$, or 0.25, of the job. Together they do $0.75 + 0.25$, or 1 entire job. The answer checks.

State. It would take Sam 2 hr to complete the puzzle working alone.

25. Familiarize. Let w = the speed of the wind. Then Jody's speed against the wind is $12 - w$, and the speed with the wind is $12 + w$. We organize the information in a table.

	Distance	Speed	Time
Against wind	8	$12-w$	t
With wind	14	$12+w$	t

Translate. Using the formula Time = Distance/Rate in each row of the table and the fact that the times are the same, we can write an equation.
$$\frac{8}{12-w} = \frac{14}{12+w}$$

Solve. We solve the equation.
$$\frac{8}{12-w} = \frac{14}{12+w}$$

LCM is $(12-w)(12+w)$

$$(12-w)(12+w)\cdot\frac{8}{12-w} = (12-w)(12+w)\cdot\frac{14}{12+w}$$
$$8(12+w) = 14(12-w)$$
$$96 + 8w = 168 - 14w$$
$$22w = 72$$
$$w = \frac{36}{11}, \text{ or } 3\frac{3}{11}$$

Check. Jody's speed against a $3\frac{3}{11}$ mph wind is $12 - 3\frac{3}{11}$, or $8\frac{8}{11}$ mph. At this speed Jody travels 8 mi in $8 \div \left(8\frac{8}{11}\right)$, or $\frac{11}{12}$ hr. The speed with the wind is $12 + 3\frac{3}{11}$, or $15\frac{3}{11}$ mph. At this speed Jody travels 14 mi in $14 \div \left(15\frac{3}{11}\right)$, or $\frac{11}{12}$ hr. The times are the same, so the answer checks.

State. The speed of the wind is $3\frac{3}{11}$ mph.

26. Familiarize. Let $p =$ the number of gallons of paint needed to paint 6000 ft^2 of clapboard.

Translate. We translate to a proportion.

$$\text{Paint} \rightarrow \frac{4}{1700} = \frac{p}{6000} \leftarrow \text{Paint}$$
$$\text{Area} \rightarrow \phantom{\frac{4}{1700}} \phantom{\frac{p}{6000}} \leftarrow \text{Area}$$

Solve. We solve the proportion.

$$\frac{4}{1700} = \frac{p}{6000}$$
$$4 \cdot 6000 = 1700 \cdot p \quad \text{Equating cross products}$$
$$\frac{4 \cdot 6000}{1700} = p$$
$$\frac{240}{17} = p, \text{ or}$$
$$14\frac{2}{17} = p$$

Check. We substitute into the proportion and check cross products.

$$\frac{4}{1700} = \frac{240/17}{6000}; \ 4 \cdot 6000 = 24,000; \ 1700 \cdot \frac{240}{17} = 24,000$$

Since the cross products are the same, the answer checks.

State. $14\frac{2}{17}$ gal of paint would be needed.

27. Solve for a:

$$T = \frac{ab}{a - b}$$
$$(a - b)T = (a - b) \cdot \frac{ab}{a - b}$$
$$aT - bT = ab$$
$$aT - ab = bT$$
$$a(T - b) = bT$$
$$a = \frac{bT}{T - b}$$

Solve for b:

$$T = \frac{ab}{a - b}$$
$$(a - b)T = (a - b) \cdot \frac{ab}{a - b}$$
$$aT - bT = ab$$
$$aT = ab + bT$$
$$aT = b(a + T)$$
$$\frac{aT}{a + T} = b$$

28.

$$Q = \frac{2}{a} - \frac{t}{b}$$
$$ab \cdot Q = ab\left(\frac{2}{a} - \frac{t}{b}\right)$$
$$abQ = ab \cdot \frac{2}{a} - ab \cdot \frac{t}{b}$$
$$abQ = 2b - at$$
$$abQ + at = 2b$$
$$a(bQ + t) = 2b$$
$$a = \frac{2b}{bQ + t}$$

29.

$$Q = kxy$$
$$25 = k \cdot 2 \cdot 5$$
$$25 = 10k$$
$$\frac{5}{2} = k$$
$$Q = \frac{5}{2}xy \quad \text{Equation of variation}$$

30.

$$y = \frac{k}{x}$$
$$10 = \frac{k}{25}$$
$$250 = k$$
$$y = \frac{250}{x} \quad \text{Equation of variation}$$

31. We first find an equation of variation.

$$I = kt$$
$$275 = k \cdot 40$$
$$6.875 = k$$
$$I = 6.875t \quad \text{Equation of variation}$$

Now we use the equation to find the pay for working 72 hr.

$$I = 6.875t$$
$$I = 6.875(72)$$
$$I = 495$$

Dean is paid \$495 for working 72 hr.

32. We first find an equation of variation.

$$t = \frac{k}{r}$$
$$5 = \frac{k}{60}$$
$$300 = k$$
$$t = \frac{300}{r} \quad \text{Equation of variation}$$

Now we use the equation to find how long it would take to drive the same distance at 40 km/h.

$$t = \frac{300}{r}$$
$$t = \frac{300}{40}$$
$$t = \frac{15}{2}, \text{ or } 7\frac{1}{2}$$

It would take $7\frac{1}{2}$ hr at a speed of 40 km/h.

33. First we find an equation of variation.

$$A = kr^2$$

$$314 = k \cdot 5^2$$

$$314 = 25k$$

$$12.56 = k$$

$$A = 12.56r^2 \quad \text{Equation of variation}$$

Now we use the equation to find the area when the radius is 7 cm.

$$A = 12.56r^2$$

$$A = 12.56 \cdot 7^2$$

$$A = 12.56(49)$$

$$A = 615.44$$

The area is 615.44 cm^2.

34.
$$\frac{6}{x-15} - \frac{6}{x} = \frac{90}{x^2 - 15x}$$

$$\frac{6}{x-15} - \frac{6}{x} = \frac{90}{x(x-15)} \quad \text{LCM is } x(x-15)$$

$$x(x-15)\left(\frac{6}{x-15} - \frac{6}{x}\right) = x(x-15) \cdot \frac{90}{x(x-15)}$$

$$6x - 6(x-15) = 90$$

$$6x - 6x + 90 = 90$$

$$90 = 90$$

We get an equation that is true for all values of x. Thus, all real numbers except those that result in division by 0 in the original equation are solutions. We see that division by 0 results when $x = 0$ or $x = 15$, so all real numbers except 0 and 15 are solutions.

35. $1 - t^6 = (1 + t^3)(1 - t^3)$

$$= (1+t)(1-t+t^2)(1-t)(1+t+t^2)$$

$1 + t^6 = (1 + t^2)(1 - t^2 + t^4)$

The LCM is

$(1+t)(1-t+t^2)(1-t)(1+t+t^2)(1+t^2)(1-t^2+t^4)$, or $(1-t^6)(1+t^6)$.

36. To find the x-intercept we set $f(x)$ equal to 0 and solve for x.

$$\frac{\dfrac{5}{x+4} - \dfrac{3}{x-2}}{\dfrac{2}{x-3} + \dfrac{1}{x+4}} = 0$$

$$\frac{5}{x+4} - \frac{3}{x-2} = 0 \quad \text{Multiplying by}$$
$$\frac{2}{x-3} + \frac{1}{x+4}$$

$$(x+4)(x-2)\left(\frac{5}{x+4} - \frac{3}{x-2}\right) = (x+4)(x-2) \cdot 0$$

$$5(x-2) - 3(x+4) = 0$$

$$5x - 10 - 3x - 12 = 0$$

$$2x - 22 = 0$$

$$2x = 22$$

$$x = 11$$

The x-intercept is $(11, 0)$.

To find the y-intercept we find $f(0)$.

$$\frac{\dfrac{5}{0+4} - \dfrac{3}{0-2}}{\dfrac{2}{0-3} + \dfrac{1}{0+4}} = \frac{\dfrac{5}{4} + \dfrac{3}{2}}{-\dfrac{2}{3} + \dfrac{1}{4}}$$

$$= \frac{\dfrac{5}{4} + \dfrac{6}{4}}{-\dfrac{8}{12} + \dfrac{3}{12}}$$

$$= \frac{\dfrac{11}{4}}{-\dfrac{5}{12}}$$

$$= \frac{11}{4} \cdot \left(-\frac{12}{5}\right)$$

$$= -\frac{11 \cdot 12}{4 \cdot 5}$$

$$= -\frac{11 \cdot 3 \cdot 4}{4 \cdot 5}$$

$$= -\frac{11 \cdot 3 \cdot \cancel{4}}{\cancel{4} \cdot 5}$$

$$= -\frac{33}{5}$$

The y-intercept is $\left(0, -\dfrac{33}{5}\right)$.

Chapter 6

Radical Expressions, Equations, and Functions

Exercise Set 6.1

1. The square roots of 16 are 4 and -4, because $4^2 = 16$ and $(-4)^2 = 16$.

3. The square roots of 144 are 12 and -12, because $12^2 = 144$ and $(-12)^2 = 144$.

5. The square roots of 400 are 20 and -20, because $20^2 = 400$ and $(-20)^2 = 400$.

7. $-\sqrt{\dfrac{49}{36}} = -\dfrac{7}{6}$ Since $\sqrt{\dfrac{49}{36}} = \dfrac{7}{6}$, $-\sqrt{\dfrac{49}{36}} = -\dfrac{7}{6}$.

9. $\sqrt{196} = 14$ Remember, $\sqrt{}$ indicates the principle square root.

11. $\sqrt{0.0036} = 0.06$

13. $\sqrt{-225}$ does not exist as a real number because negative numbers do not have real-number square roots.

15. $\sqrt{347} \approx 18.628$

17. $\sqrt{\dfrac{285}{74}} \approx 1.962$

19. $9\sqrt{y^2 + 16}$

The radicand is the expression written under the radical sign, $y^2 + 16$.

21. $x^4 y^5 \sqrt{\dfrac{x}{y-1}}$

The radicand is the expression written under the radical sign, $\dfrac{x}{y-1}$.

23. $f(x) = \sqrt{5x - 10}$

$f(6) = \sqrt{5 \cdot 6 - 10} = \sqrt{20} \approx 4.472$

$f(2) = \sqrt{5 \cdot 2 - 10} = \sqrt{0} = 0$

$f(1) = \sqrt{5 \cdot 1 - 10} = \sqrt{-5}$

Since negative numbers do not have real-number square roots, $f(1)$ does not exist as a real number.

$f(-1) = \sqrt{5(-1) - 10} = \sqrt{-15}$

Since negative numbers do not have real-number square roots, $f(-1)$ does not exist as a real number.

25. $g(x) = \sqrt{x^2 - 25}$

$g(-6) = \sqrt{(-6)^2 - 25} = \sqrt{11} \approx 3.317$

$g(3) = \sqrt{3^2 - 25} = \sqrt{-16}$

Since negative numbers do not have real-number square roots, $g(3)$ does not exist as a real number.

$g(6) = \sqrt{6^2 - 25} = \sqrt{11} \approx 3.317$

$g(13) = \sqrt{13^2 - 25} = \sqrt{144} = 12$

27. The domain of $f(x) = \sqrt{5x - 10}$ is the set of all x-values for which $5x - 10 \geq 0$.

$5x - 10 \geq 0$

$5x \geq 10$

$x \geq 2$

The domain is $\{x | x \geq 2\}$, or $[2, \infty)$.

29. $S(x) = 2\sqrt{5x}$

$S(30) = 2\sqrt{5 \cdot 30} = 2\sqrt{150} \approx 24.5$

The speed of a car that left skid marks of length 30 ft was about 24.5 mph.

$S(150) = 2\sqrt{5 \cdot 150} = 2\sqrt{750} \approx 54.8$

The speed of a car that left skid marks of length 150 ft was about 54.8 mph.

31. Graph: $f(x) = 2\sqrt{x}$.

We find some ordered pairs, plot points, and draw the curve.

x	$f(x)$	$(x, f(x))$
0	0	$(0, 0)$
1	2	$(1, 2)$
2	2.8	$(2, 2.8)$
3	3.5	$(3, 3.5)$
4	4	$(4, 4)$
5	4.5	$(5, 4.5)$

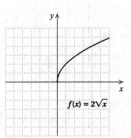

$f(x) = 2\sqrt{x}$

33. Graph: $F(x) = -3\sqrt{x}$.

We find some ordered pairs, plot points, and draw the curve.

x	$f(x)$	$(x, f(x))$
0	0	$(0, 0)$
1	-3	$(1, -3)$
2	-4.2	$(2, -4.2)$
3	-5.2	$(3, -5.2)$
4	-6	$(4, -6)$
5	-6.7	$(5, -6.7)$

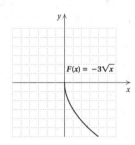

35. Graph: $f(x) = \sqrt{x}$.

We find some ordered pairs, plot points, and draw the curve.

x	$f(x)$	$(x, f(x))$
0	0	$(0, 0)$
1	1	$(1, 1)$
2	1.4	$(2, 1.4)$
3	1.7	$(3, 1.7)$
4	2	$(4, 2)$
5	2.2	$(5, 2.2)$

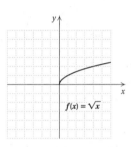

37. Graph: $f(x) = \sqrt{x - 2}$.

We find some ordered pairs, plot points, and draw the curve.

x	$f(x)$	$(x, f(x))$
2	0	$(2, 0)$
3	1	$(3, 1)$
4	1.4	$(4, 1.4)$
5	1.7	$(5, 1.7)$
7	2.2	$(7, 2.2)$
9	2.6	$(9, 2.6)$

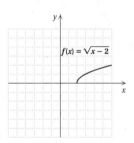

39. Graph: $f(x) = \sqrt{12 - 3x}$.

We find some ordered pairs, plot points, and draw the curve.

x	$f(x)$	$(x, f(x))$
-5	5.2	$(-5, 5.2)$
-3	4.6	$(-3, 4.6)$
-1	3.9	$(-1, 3.9)$
0	3.5	$(0, 3.5)$
2	2.4	$(2, 2.4)$
4	0	$(4, 0)$

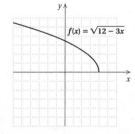

41. Graph: $g(x) = \sqrt{3x + 9}$.

We find some ordered pairs, plot points, and draw the curve.

x	$f(x)$	$(x, f(x))$
-3	0	$(-3, 0)$
-1	2.4	$(-1, 2.4)$
0	3	$(0, 3)$
1	3.5	$(1, 3.5)$
3	4.2	$(3, 4.2)$
5	4.9	$(5, 4.9)$

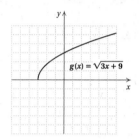

43. $\sqrt{16x^2} = \sqrt{(4x)^2} = |4x| = 4|x|$

(The absolute value is used to ensure that the principal square root is nonnegative.)

45. $\sqrt{(-12c)^2} = |-12c| = |-12| \cdot |c| = 12|c|$

(The absolute value is used to ensure that the principal square root is nonnegative.)

47. $\sqrt{(p + 3)^2} = |p + 3|$

(The absolute value is used to ensure that the principal square root is nonnegative.)

49. $\sqrt{x^2 - 4x + 4} = \sqrt{(x - 2)^2} = |x - 2|$

(The absolute value is used to ensure that the principal square root is nonnegative.)

51. $\sqrt[3]{27} = 3 \quad [3^3 = 27]$

53. $\sqrt[3]{-64x^3} = -4x \quad [(-4x)^3 = -64x^3]$

55. $\sqrt[3]{-216} = -6 \quad [(-6)^3 = -216]$

57. $\sqrt[3]{0.343(x + 1)^3} = 0.7(x + 1)$
$$[(0.7(x + 1))^3 = 0.343(x + 1)^3]$$

59. $\quad f(x) = \sqrt[3]{x + 1}$
$$f(7) = \sqrt[3]{7 + 1} = \sqrt[3]{8} = 2$$
$$f(26) = \sqrt[3]{26 + 1} = \sqrt[3]{27} = 3$$
$$f(-9) = \sqrt[3]{-9 + 1} = \sqrt[3]{-8} = -2$$
$$f(-65) = \sqrt[3]{-65 + 1} = \sqrt[3]{-64} = -4$$

61. $\quad f(x) = -\sqrt[3]{3x + 1}$
$$f(0) = -\sqrt[3]{3 \cdot 0 + 1} = -\sqrt[3]{1} = -1$$
$$f(-7) = -\sqrt[3]{3(-7) + 1} = -\sqrt[3]{-20}, \text{ or } \sqrt[3]{20} \approx 2.7144$$
$$f(21) = -\sqrt[3]{3 \cdot 21 + 1} = -\sqrt[3]{64} = -4$$
$$f(333) = -\sqrt[3]{3 \cdot 333 + 1} = -\sqrt[3]{1000} = -10$$

63. $-\sqrt[4]{625} = -5 \quad$ Since $5^4 = 625$, then $\sqrt[4]{625} = 5$ and $-\sqrt[4]{625} = -5$.

65. $\sqrt[5]{-1} = -1 \quad$ Since $(-1)^5 = -1$

67. $\sqrt[5]{-\dfrac{32}{243}} = -\dfrac{2}{3} \quad$ Since $\left(-\dfrac{2}{3}\right)^5 = -\dfrac{32}{243}$

69. $\sqrt[6]{x^6} = |x|$

The index is even so we use absolute-value notation.

71. $\sqrt[4]{(5a)^4} = |5a| = 5|a|$

The index is even so we use absolute-value notation.

73. $\sqrt[10]{(-6)^{10}} = |-6| = 6$

75. $\sqrt[414]{(a+b)^{414}} = |a+b|$

The index is even so we use absolute-value notation.

77. $\sqrt[7]{y^7} = y$

We do not use absolute-value notation when the index is odd.

79. $\sqrt[5]{(x-2)^5} = x-2$

We do not use absolute-value notation when the index is odd.

81. Discussion and Writing Exercise

83.
$$x^2 + x - 2 = 0$$
$$(x+2)(x-1) = 0 \quad \text{Factoring}$$
$$x+2 = 0 \quad or \quad x-1 = 0 \quad \text{Principle of zero}$$
$$\text{products}$$
$$x = -2 \quad or \quad x = 1$$

The solutions are -2 and 1.

85.
$$4x^2 - 49 = 0$$
$$(2x+7)(2x-7) = 0 \quad \text{Factoring}$$
$$2x+7 = 0 \quad or \quad 2x-7 = 0 \quad \text{Principle of zero}$$
$$\text{products}$$
$$2x = -7 \quad or \quad 2x = 7$$
$$x = -\frac{7}{2} \quad or \quad x = \frac{7}{2}$$

The solutions are $-\frac{7}{2}$ and $\frac{7}{2}$.

87.
$$3x^2 + x = 10$$
$$3x^2 + x - 10 = 0$$
$$(3x-5)(x+2) = 0$$
$$3x-5 = 0 \quad or \quad x+2 = 0$$
$$3x = 5 \quad or \quad x = -2$$
$$x = \frac{5}{3} \quad or \quad x = -2$$

The solutions are $\frac{5}{3}$ and -2.

89.
$$4x^3 - 20x^2 + 25x = 0$$
$$x(4x^2 - 20x + 25) = 0$$
$$x(2x-5)(2x-5) = 0$$
$$x = 0 \quad or \quad 2x-5 = 0 \quad or \quad 2x-5 = 0$$
$$x = 0 \quad or \quad 2x = 5 \quad or \quad 2x = 5$$
$$x = 0 \quad or \quad x = \frac{5}{2} \quad or \quad x = \frac{5}{2}$$

The solutions are 0 and $\frac{5}{2}$.

91. $(a^3b^2c^5)^3 = a^{3\cdot3}b^{2\cdot3}c^{5\cdot3} = a^9b^6c^{15}$

93. $f(x) = \dfrac{\sqrt{x+3}}{\sqrt{2-x}}$

In the numerator we must have $x+3 \geq 0$, or $x \geq -3$, and in the denominator we must have $2-x > 0$, or $x < 2$. Thus, we have $x \geq -3$ *and* $x < 2$, so

Domain of $f = \{x| -3 \leq x < 2\}$, or $[-3, 2)$.

95. From 3 on the x-axis, go up to the graph and across to the y-axis to find $f(3) = \sqrt{3} \approx 1.7$.

From 5 on the x-axis, go up to the graph and across to the y-axis to find $f(5) = \sqrt{5} \approx 2.2$.

From 10 on the x-axis, go up to the graph and across to the y-axis to find $f(10) = \sqrt{10} \approx 3.2$.

97. a) $f(x) = \sqrt[3]{x}$

Domain $= (-\infty, \infty)$; range $= (-\infty, \infty)$

b) $g(x) = \sqrt[3]{4x-5}$

Domain $= (-\infty, \infty)$; range $= (-\infty, \infty)$

c) $q(x) = 2 - \sqrt{x+3}$

Domain $= [-3, \infty)$; range $= (-\infty, 2]$

d) $h(x) = \sqrt[4]{x}$

Domain $= [0, \infty)$; range $= [0, \infty)$

e) $t(x) = \sqrt[4]{x-3}$

Domain $= [3, \infty)$; range $= [0, \infty)$

Exercise Set 6.2

1. $y^{1/7} = \sqrt[7]{y}$

3. $(8)^{1/3} = \sqrt[3]{8} = 2$

5. $(a^3b^3)^{1/5} = \sqrt[5]{a^3b^3}$

7. $16^{3/4} = \sqrt[4]{16^3} = (\sqrt[4]{16})^3 = 2^3 = 8$

9. $49^{3/2} = \sqrt{49^3} = (\sqrt{49})^3 = 7^3 = 343$

11. $\sqrt{17} = 17^{1/2}$

13. $\sqrt[3]{18} = 18^{1/3}$

15. $\sqrt[5]{xy^2z} = (xy^2z)^{1/5}$

17. $(\sqrt{3mn})^3 = (3mn)^{3/2}$

19. $(\sqrt[7]{8x^2y})^5 = (8x^2y)^{5/7}$

21. $27^{-1/3} = \dfrac{1}{27^{1/3}} = \dfrac{1}{\sqrt[3]{27}} = \dfrac{1}{3}$

23. $100^{-3/2} = \dfrac{1}{100^{3/2}} = \dfrac{1}{(\sqrt{100})^3} = \dfrac{1}{10^3} = \dfrac{1}{1000}$

25. $3x^{-1/4} = 3 \cdot \dfrac{1}{x^{1/4}} = \dfrac{3}{x^{1/4}}$

27. $(2rs)^{-3/4} = \dfrac{1}{(2rs)^{3/4}}$

29. $2a^{3/4}b^{-1/2}c^{2/3} = 2 \cdot a^{3/4} \cdot \dfrac{1}{b^{1/2}} \cdot c^{2/3} = \dfrac{2a^{3/4}c^{2/3}}{b^{1/2}}$

31. $\left(\dfrac{7x}{8yz}\right)^{-3/5} = \left(\dfrac{8yz}{7x}\right)^{3/5}$ $\left(\text{Since } \left(\dfrac{a}{b}\right)^{-n} = \left(\dfrac{b}{a}\right)^{n}\right)$

33. $\dfrac{1}{x^{-2/3}} = x^{2/3}$

35. $2^{-1/3}x^4 y^{-2/7} = \dfrac{1}{2^{1/3}} \cdot x^4 \cdot \dfrac{1}{y^{2/7}} = \dfrac{x^4}{2^{1/3}y^{2/7}}$

37. $\dfrac{7x}{\sqrt[3]{z}} = \dfrac{7x}{z^{1/3}}$

39. $\dfrac{5a}{3c^{-1/2}} = \dfrac{5a}{3} \cdot c^{1/2} = \dfrac{5ac^{1/2}}{3}$

41. $5^{3/4} \cdot 5^{1/8} = 5^{3/4+1/8} = 5^{6/8+1/8} = 5^{7/8}$

43. $\dfrac{7^{5/8}}{7^{3/8}} = 7^{5/8-3/8} = 7^{2/8} = 7^{1/4}$

45. $\dfrac{4.9^{-1/6}}{4.9^{-2/3}} = 4.9^{-1/6-(-2/3)} = 4.9^{-1/6+4/6} = 4.9^{3/6} = 4.9^{1/2}$

47. $(6^{3/8})^{2/7} = 6^{3/8 \cdot 2/7} = 6^{6/56} = 6^{3/28}$

49. $a^{2/3} \cdot a^{5/4} = a^{2/3+5/4} = a^{8/12+15/12} = a^{23/12}$

51. $(a^{2/3} \cdot b^{5/8})^4 = (a^{2/3})^4 (b^{5/8})^4 = a^{8/3}b^{20/8} = a^{8/3}b^{5/2}$

53. $(x^{2/3})^{-3/7} = x^{2/3(-3/7)} = x^{-2/7} = \dfrac{1}{x^{2/7}}$

55. $\sqrt[6]{a^2} = a^{2/6}$ Converting to exponential notation

$= a^{1/3}$ Simplifying the exponent

$= \sqrt[3]{a}$ Returning to radical notation

57. $\sqrt[3]{x^{15}} = x^{15/3}$ Converting to exponential notation

$= x^5$ Simplifying

59. $\sqrt[6]{x^{-18}} = x^{-18/6}$ Converting to exponential notation

$= x^{-3}$ Simplifying

$= \dfrac{1}{x^3}$

61. $(\sqrt[3]{ab})^{15} = (ab)^{15/3}$ Converting to exponential notation

$= (ab)^5$ Simplifying the exponent

$= a^5 b^5$ Using the law of exponents

63. $\sqrt[14]{128} = \sqrt[14]{2^7} = 2^{7/14} = 2^{1/2} = \sqrt{2}$

65. $\sqrt[6]{4x^2} = (2^2 x^2)^{1/6} = 2^{2/6}x^{2/6}$

$= 2^{1/3}x^{1/3} = (2x)^{1/3} = \sqrt[3]{2x}$

67. $\sqrt{x^4 y^6} = (x^4 y^6)^{1/2} = x^{4/2}y^{6/2} = x^2 y^3$

69. $\sqrt[5]{32c^{10}d^{15}} = (2^5 c^{10}d^{15})^{1/5} = 2^{5/5}c^{10/5}d^{15/5}$

$= 2c^2 d^3$

71. $\sqrt[3]{7} \cdot \sqrt[4]{5} = 7^{1/3} \cdot 5^{1/4} = 7^{4/12} \cdot 5^{3/12} =$

$(7^4 \cdot 5^3)^{1/12} = \sqrt[12]{7^4 \cdot 5^3}$

73. $\sqrt[4]{5} \cdot \sqrt[5]{7} = 5^{1/4} \cdot 7^{1/5} = 5^{5/20} \cdot 7^{4/20} = (5^5 \cdot 7^4)^{1/20} =$

$\sqrt[20]{5^5 \cdot 7^4}$

75. $\sqrt{x}\sqrt[3]{2x} = x^{1/2} \cdot (2x)^{1/3} = x^{3/6} \cdot (2x)^{2/6} =$

$[x^3(2x)^2]^{1/6} = (x^3 \cdot 4x^2)^{1/6} = (4x^5)^{1/6} = \sqrt[6]{4x^5}$

77. $(\sqrt[5]{a^2 b^4})^{15} = (a^2 b^4)^{15/5} = (a^2 b^4)^3 = a^6 b^{12}$

79. $\sqrt[3]{\sqrt[6]{m}} = \sqrt[3]{m^{1/6}} = (m^{1/6})^{1/3} = m^{1/18} = \sqrt[18]{m}$

81. $x^{1/3} \cdot y^{1/4} \cdot z^{1/6} = x^{4/12} \cdot y^{3/12} \cdot z^{2/12} =$

$(x^4 y^3 z^2)^{1/12} = \sqrt[12]{x^4 y^3 z^2}$

83. $\left(\dfrac{c^{-4/5}d^{5/9}}{c^{3/10}d^{1/6}}\right)^3 = (c^{-4/5-3/10}d^{5/9-1/6})^3 =$

$(c^{-8/10-3/10}d^{10/18-3/18})^3 = (c^{-11/10}d^{7/18})^3 =$

$c^{-33/10}d^{7/6} = c^{-99/30}d^{35/30} = (c^{-99}d^{35})^{1/30} =$

$\left(\dfrac{d^{35}}{c^{99}}\right)^{1/30} = \sqrt[30]{\dfrac{d^{35}}{c^{99}}}$

85. Discussion and Writing Exercise

87. $A = \dfrac{ab}{a+b}$

$A(a+b) = ab$ Multiplying by $a+b$

$Aa + Ab = ab$

$Ab = ab - Aa$ Subtracting Aa

$Ab = a(b-A)$ Factoring

$\dfrac{Ab}{b-A} = a$ Dividing by $b-A$

89. $Q = \dfrac{st}{s-t}$

$Q(s-t) = st$ Multiplying by $s-t$

$Qs - Qt = st$

$Qs = st + Qt$ Adding Qt

$Qs = t(s+Q)$ Factoring

$\dfrac{Qs}{s+Q} = t$ Dividing by $s+Q$

91.

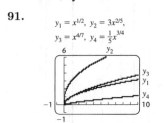

Exercise Set 6.3

1. $\sqrt{24} = \sqrt{4 \cdot 6} = \sqrt{4}\,\sqrt{6} = 2\sqrt{6}$

3. $\sqrt{90} = \sqrt{9 \cdot 10} = \sqrt{9}\,\sqrt{10} = 3\sqrt{10}$

5. $\sqrt[3]{250} = \sqrt[3]{125 \cdot 2} = \sqrt[3]{125}\,\sqrt[3]{2} = 5\sqrt[3]{2}$

7. $\sqrt{180x^4} = \sqrt{36 \cdot 5 \cdot x^4} = \sqrt{36x^4}\,\sqrt{5} = 6x^2\sqrt{5}$

9. $\sqrt[3]{54x^8} = \sqrt[3]{27 \cdot 2 \cdot x^6 \cdot x^2} = \sqrt[3]{27x^6}\sqrt[3]{2x^2} = 3x^2\sqrt[3]{2x^2}$

11. $\sqrt[3]{80t^8} = \sqrt[3]{8 \cdot 10 \cdot t^6 \cdot t^2} = \sqrt[3]{8t^6}\sqrt[3]{10t^2} = 2t^2\sqrt[3]{10t^2}$

13. $\sqrt[4]{80} = \sqrt[4]{16 \cdot 5} = \sqrt[4]{16}\sqrt[4]{5} = 2\sqrt[4]{5}$

15. $\sqrt{32a^2b} = \sqrt{16a^2 \cdot 2b} = \sqrt{16a^2} \cdot \sqrt{2b} = 4a\sqrt{2b}$

17. $\sqrt[4]{243x^8y^{10}} = \sqrt[4]{81x^8y^8 \cdot 3y^2} = \sqrt[4]{81x^8y^8}\sqrt[4]{3y^2} =$
$3x^2y^2\sqrt[4]{3y^2}$

19. $\sqrt[5]{96x^7y^{15}} = \sqrt[5]{32x^5y^{15} \cdot 3x^2} = \sqrt[5]{32x^5y^{15}}\sqrt[5]{3x^2} =$
$2xy^3\sqrt[5]{3x^2}$

21. $\sqrt{10}\sqrt{5} = \sqrt{10 \cdot 5} = \sqrt{50} = \sqrt{25 \cdot 2} = 5\sqrt{2}$

23. $\sqrt{15}\sqrt{6} = \sqrt{15 \cdot 6} = \sqrt{90}$
$\qquad = \sqrt{9 \cdot 10} = \sqrt{9}\sqrt{10} = 3\sqrt{10}$

25. $\sqrt[3]{2}\sqrt[3]{4} = \sqrt[3]{2 \cdot 4} = \sqrt[3]{8} = 2$

27. $\sqrt{45}\sqrt{60} = \sqrt{45 \cdot 60} = \sqrt{2700}$
$\qquad = \sqrt{900 \cdot 3} = \sqrt{900}\sqrt{3} = 30\sqrt{3}$

29. $\sqrt{3x^3}\sqrt{6x^5} = \sqrt{18x^8} = \sqrt{9x^8 \cdot 2} = 3x^4\sqrt{2}$

31. $\sqrt{5b^3}\sqrt{10c^4} = \sqrt{5b^3 \cdot 10c^4}$
$\qquad = \sqrt{50b^3c^4}$
$\qquad = \sqrt{25 \cdot 2 \cdot b^2 \cdot b \cdot c^4}$
$\qquad = \sqrt{25b^2c^4}\sqrt{2b}$
$\qquad = 5bc^2\sqrt{2b}$

33. $\sqrt[3]{5a^2}\sqrt[3]{2a} = \sqrt[3]{5a^2 \cdot 2a} = \sqrt[3]{10a^3} = \sqrt[3]{a^3 \cdot 10} = a\sqrt[3]{10}$

35. $\sqrt[3]{y^4}\sqrt[3]{16y^5} = \sqrt[3]{y^4 \cdot 16y^5}$
$\qquad = \sqrt[3]{16y^9}$
$\qquad = \sqrt[3]{8 \cdot 2 \cdot y^9}$
$\qquad = \sqrt[3]{8y^9}\sqrt[3]{2}$
$\qquad = 2y^3\sqrt[3]{2}$

37. $\sqrt[4]{16}\sqrt[4]{64} = \sqrt[4]{16 \cdot 64} = \sqrt[4]{1024} = \sqrt[4]{256 \cdot 4} =$
$\sqrt[4]{256}\sqrt[4]{4} = 4\sqrt[4]{4}$

39. $\sqrt{12a^3b}\sqrt{8a^4b^2} = \sqrt{12a^3b \cdot 8a^4b^2} =$
$\sqrt{96a^7b^3} = \sqrt{16a^6b^2 \cdot 6ab} = \sqrt{16a^6b^2}\sqrt{6ab} =$
$4a^3b\sqrt{6ab}$

41. $\quad \sqrt{2}\sqrt[3]{5}$
$= 2^{1/2} \cdot 5^{1/3}$ Converting to exponential notation
$= 2^{3/6} \cdot 5^{2/6}$ Rewriting so that exponents have a common denominator
$= (2^3 \cdot 5^2)^{1/6}$ Using $a^n b^n = (ab)^n$
$= \sqrt[6]{2^3 \cdot 5^2}$ Converting to radical notation
$= \sqrt[6]{8 \cdot 25}$ Simplifying
$= \sqrt[6]{200}$ Multiplying

43. $\quad \sqrt[4]{3}\sqrt{2}$
$= 3^{1/4} \cdot 2^{1/2}$ Converting to exponential notation
$= 3^{1/4} \cdot 2^{2/4}$ Rewriting so that exponents have a common denominator
$= (3 \cdot 2^2)^{1/4}$ Using $a^n b^n = (ab)^n$
$= \sqrt[4]{3 \cdot 2^2}$ Converting to radical notation
$= \sqrt[4]{3 \cdot 4}$ Squaring 2
$= \sqrt[4]{12}$ Multiplying

45. $\quad \sqrt{a}\sqrt[4]{a^3}$
$= a^{1/2} \cdot a^{3/4}$ Converting to exponential notation
$= a^{5/4}$ Adding exponents
$= a^{1+1/4}$ Writing 5/4 as a mixed number
$= a \cdot a^{1/4}$ Factoring
$= a\sqrt[4]{a}$ Returning to radical notation

47. $\quad \sqrt[5]{b^2}\sqrt{b^3}$
$= b^{2/5} \cdot b^{3/2}$ Converting to exponential notation
$= b^{19/10}$ Adding exponents
$= b^{1+9/10}$ Writing 19/10 as a mixed number
$= b \cdot b^{9/10}$ Factoring
$= b\sqrt[10]{b^9}$ Returning to radical notation

49. $\sqrt{xy^3}\sqrt[3]{x^2y} = (xy^3)^{1/2}(x^2y)^{1/3}$
$\qquad = (xy^3)^{3/6}(x^2y)^{2/6}$
$\qquad = [(xy^3)^3(x^2y)^2]^{1/6}$
$\qquad = \sqrt[6]{x^3y^9 \cdot x^4y^2}$
$\qquad = \sqrt[6]{x^7y^{11}}$
$\qquad = \sqrt[6]{x^6y^6 \cdot xy^5}$
$\qquad = xy\sqrt[6]{xy^5}$

51. $\dfrac{\sqrt{90}}{\sqrt{5}} = \sqrt{\dfrac{90}{5}} = \sqrt{18} = \sqrt{9 \cdot 2} = \sqrt{9}\sqrt{2} = 3\sqrt{2}$

53. $\dfrac{\sqrt{35q}}{\sqrt{7q}} = \sqrt{\dfrac{35q}{7q}} = \sqrt{5}$

55. $\dfrac{\sqrt[3]{54}}{\sqrt[3]{2}} = \sqrt[3]{\dfrac{54}{2}} = \sqrt[3]{27} = 3$

57. $\dfrac{\sqrt{56xy^3}}{\sqrt{8x}} = \sqrt{\dfrac{56xy^3}{8x}} = \sqrt{7y^3} = \sqrt{y^2 \cdot 7y} =$
$\sqrt{y^2}\sqrt{7y} = y\sqrt{7y}$

59. $\dfrac{\sqrt[3]{96a^4b^2}}{\sqrt[3]{12a^2b}} = \sqrt[3]{\dfrac{96a^4b^2}{12a^2b}} = \sqrt[3]{8a^2b} = \sqrt[3]{8}\sqrt[3]{a^2b} = 2\sqrt[3]{a^2b}$

61. $\dfrac{\sqrt{128xy}}{2\sqrt{2}} = \dfrac{1}{2}\dfrac{\sqrt{128xy}}{\sqrt{2}} = \dfrac{1}{2}\sqrt{\dfrac{128xy}{2}} = \dfrac{1}{2}\sqrt{64xy} =$
$\dfrac{1}{2}\sqrt{64}\sqrt{xy} = \dfrac{1}{2} \cdot 8\sqrt{xy} = 4\sqrt{xy}$

63. $\dfrac{\sqrt[4]{48x^9y^{13}}}{\sqrt[4]{3xy^5}} = \sqrt[4]{\dfrac{48x^9y^{13}}{3xy^5}} = \sqrt[4]{16x^8y^8} = 2x^2y^2$

65. $\dfrac{\sqrt[3]{a}}{\sqrt{a}}$

$= \dfrac{a^{1/3}}{a^{1/2}}$ Converting to exponential notation

$= a^{1/3-1/2}$ Subtracting exponents

$= a^{2/6-3/6}$

$= a^{-1/6}$

$= \dfrac{1}{a^{1/6}}$

$= \dfrac{1}{\sqrt[6]{a}}$ Converting to radical notation

67. $\dfrac{\sqrt[3]{a^2}}{\sqrt[4]{a}}$

$= \dfrac{a^{2/3}}{a^{1/4}}$ Converting to exponential notation

$= a^{2/3-1/4}$ Subtracting exponents

$= a^{5/12}$ Converting back

$= \sqrt[12]{a^5}$ to radical notation

69. $\dfrac{\sqrt[4]{x^2y^3}}{\sqrt[3]{xy}}$

$= \dfrac{(x^2y^3)^{1/4}}{(xy)^{1/3}}$ Converting to exponential notation

$= \dfrac{x^{2/4}y^{3/4}}{x^{1/3}y^{1/3}}$ Using the power and product rules

$= x^{2/4-1/3}y^{3/4-1/3}$ Subtracting exponents

$= x^{2/12}y^{5/12}$

$= (x^2y^5)^{1/2}$ Converting back to

$= \sqrt[12]{x^2y^5}$ radical notation

71. $\sqrt{\dfrac{25}{36}} = \dfrac{\sqrt{25}}{\sqrt{36}} = \dfrac{5}{6}$

73. $\sqrt{\dfrac{16}{49}} = \dfrac{\sqrt{16}}{\sqrt{49}} = \dfrac{4}{7}$

75. $\sqrt[3]{\dfrac{125}{27}} = \dfrac{\sqrt[3]{125}}{\sqrt[3]{27}} = \dfrac{5}{3}$

77. $\sqrt{\dfrac{49}{y^2}} = \dfrac{\sqrt{49}}{\sqrt{y^2}} = \dfrac{7}{y}$

79. $\sqrt{\dfrac{25y^3}{x^4}} = \dfrac{\sqrt{25y^3}}{\sqrt{x^4}} = \dfrac{\sqrt{25y^2 \cdot y}}{\sqrt{x^4}} = \dfrac{\sqrt{25y^2}\,\sqrt{y}}{\sqrt{x^4}} =$

$\dfrac{5y\sqrt{y}}{x^2}$

81. $\sqrt[3]{\dfrac{27a^4}{8b^3}} = \dfrac{\sqrt[3]{27a^4}}{\sqrt[3]{8b^3}} = \dfrac{\sqrt[3]{27a^3 \cdot a}}{\sqrt[3]{8b^3}} = \dfrac{\sqrt[3]{27a^3}\,\sqrt[3]{a}}{\sqrt[3]{8b^3}} =$

$\dfrac{3a\sqrt[3]{a}}{2b}$

83. $\sqrt[4]{\dfrac{81x^4}{16}} = \dfrac{\sqrt[4]{81x^4}}{\sqrt[4]{16}} = \dfrac{3x}{2}$

85. $\sqrt[5]{\dfrac{32x^8}{y^{10}}} = \dfrac{\sqrt[5]{32x^8}}{\sqrt[5]{y^{10}}} = \dfrac{\sqrt[5]{32 \cdot x^5 \cdot x^3}}{\sqrt[5]{y^{10}}} = \dfrac{\sqrt[5]{32x^5}\,\sqrt[5]{x^3}}{\sqrt[5]{y^{10}}} =$

$\dfrac{2x\sqrt[5]{x^3}}{y^2}$

87. $\sqrt[6]{\dfrac{x^{13}}{y^6z^{12}}} = \dfrac{\sqrt[6]{x^{13}}}{\sqrt[6]{y^6z^{12}}} = \dfrac{\sqrt[6]{x^{12} \cdot x}}{\sqrt[6]{y^6z^{12}}} = \dfrac{\sqrt[6]{x^{12}}\,\sqrt[6]{x}}{\sqrt[6]{y^6z^{12}}} = \dfrac{x^2\sqrt[6]{x}}{yz^2}$

89. Discussion and Writing Exercise

91. *Familiarize.* We will use the formula $d = rt$. When the boat travels downstream, its rate is $14 + 7$, or 21 mph. Its rate traveling upstream is $14 - 7$ or 7 mph.

Translate. We substitute in the formula.

Downstream: $56 = 21t$

Upstream: $56 = 7t$

Solve. We solve the equation.

Downstream: $56 = 21t$

$\dfrac{56}{21} = t$

$\dfrac{8}{3} = t$, or

$2\dfrac{2}{3} = t$

Upstream: $56 = 7t$

$8 = t$

Check. At a rate of 21 mph, in $\dfrac{8}{3}$ hr the boat would travel $21 \cdot \dfrac{8}{3}$, or 56 mi. At a rate of 7 mph, in 8 hr the boat would travel $7 \cdot 8$, or 56 mi. The answer checks.

State. It will take the boat $2\dfrac{2}{3}$ hr to travel 56 mi downstream and 8 hr to travel 56 mi upstream.

93. $\dfrac{12x}{x-4} - \dfrac{3x^2}{x+4} = \dfrac{384}{x^2-16}$

$\dfrac{12x}{x-4} - \dfrac{3x^2}{x+4} = \dfrac{384}{(x+4)(x-4)}$,

         LCM is $(x+4)(x-4)$.

$(x+4)(x-4)\left[\dfrac{12x}{x-4} - \dfrac{3x^2}{x+4}\right] = (x+4)(x-4) \cdot \dfrac{384}{(x+4)(x-4)}$

$12x(x+4) - 3x^2(x-4) = 384$

$12x^2 + 48x - 3x^3 + 12x^2 = 384$

$-3x^3 + 24x^2 + 48x - 384 = 0$

$-3(x^3 - 8x^2 - 16x + 128) = 0$

$-3[x^2(x-8) - 16(x-8)] = 0$

$-3(x-8)(x^2-16) = 0$

$-3(x-8)(x+4)(x-4) = 0$

$x - 8 = 0$ *or* $x + 4 = 0$ *or* $x - 4 = 0$

$x = 8$ *or* $x = -4$ *or* $x = 4$

Check: For 8:

$$\frac{12x}{x-4} - \frac{3x^2}{x+4} = \frac{384}{x^2-16}$$

$$\frac{12 \cdot 8}{8-4} - \frac{3 \cdot 8^2}{8+4} \ ? \ \frac{384}{8^2-16}$$

$$\frac{96}{4} - \frac{192}{12} \ \bigg| \ \frac{384}{48}$$

$$24 - 16 \ \bigg| \ 8$$

$$8 \ \bigg| \qquad \text{TRUE}$$

8 is a solution.

For -4:

$$\frac{12x}{x-4} - \frac{3x^2}{x+4} = \frac{384}{x^2-16}$$

$$\frac{12(-4)}{-4-4} - \frac{3(-4)^2}{-4+4} \ ? \ \frac{384}{(-4)^2-16}$$

$$\frac{-48}{-8} - \frac{48}{0} \ \bigg| \ \frac{384}{16-16} \qquad \text{UNDEFINED}$$

-4 is not a solution.

For 4:

$$\frac{12x}{x-4} - \frac{3x^2}{x+4} = \frac{384}{x^2-16}$$

$$\frac{12 \cdot 4}{4-4} - \frac{3 \cdot 4^2}{4+4} \ ? \ \frac{384}{4^2-16}$$

$$\frac{48}{0} - \frac{48}{8} \ \bigg| \ \frac{384}{16-16} \qquad \text{UNDEFINED}$$

4 is not a solution.

The solution is 8.

95.
$$\frac{18}{x^2-3x} = \frac{2x}{x-3} - \frac{6}{x}$$

$$\frac{18}{x(x-3)} = \frac{2x}{x-3} - \frac{6}{x},$$

$$\text{LCM is } x(x-3)$$

$$x(x-3) \cdot \frac{18}{x(x-3)} = x(x-3)\left(\frac{2x}{x-3} - \frac{6}{x}\right)$$

$$18 = x(x-3) \cdot \frac{2x}{x-3} - x(x-3) \cdot \frac{6}{x}$$

$$18 = 2x^2 - 6x + 18$$

$$0 = 2x^2 - 6x$$

$$0 = 2x(x-3)$$

$$2x = 0 \ \ or \ \ x - 3 = 0$$

$$x = 0 \ \ or \qquad x = 3$$

Each value makes a denominator 0. There is no solution.

97. a) $T = 2\pi\sqrt{\dfrac{65}{980}} \approx 1.62$ sec

b) $T = 2\pi\sqrt{\dfrac{98}{980}} \approx 1.99$ sec

c) $T = 2\pi\sqrt{\dfrac{120}{980}} \approx 2.20$ sec

99. $\dfrac{\sqrt{44x^2y^9z}\sqrt{22y^9z^6}}{(\sqrt{11xy^8z^2})^2} = \dfrac{\sqrt{44 \cdot 22x^2y^{18}z^7}}{\sqrt{11 \cdot 11x^2y^{16}z^4}} =$

$$\sqrt{\frac{44 \cdot 22x^2y^{18}z^7}{11 \cdot 11x^2y^{16}z^4}} = \sqrt{4 \cdot 2y^2z^3} = \sqrt{4y^2z^2 \cdot 2z} = 2yz\sqrt{2z}$$

Exercise Set 6.4

1. $7\sqrt{5} + 4\sqrt{5} = (7+4)\sqrt{5} \qquad$ Factoring out $\sqrt{5}$
$$= 11\sqrt{5}$$

3. $6\sqrt[3]{7} - 5\sqrt[3]{7} = (6-5)\sqrt[3]{7} \qquad$ Factoring out $\sqrt[3]{7}$
$$= \sqrt[3]{7}$$

5. $4\sqrt[3]{y} + 9\sqrt[3]{y} = (4+9)\sqrt[3]{y} = 13\sqrt[3]{y}$

7. $5\sqrt{6} - 9\sqrt{6} - 4\sqrt{6} = (5-9-4)\sqrt{6} = -8\sqrt{6}$

9. $4\sqrt[3]{3} - \sqrt{5} + 2\sqrt[3]{3} + \sqrt{5} =$
$$(4+2)\sqrt[3]{3} + (-1+1)\sqrt{5} = 6\sqrt[3]{3}$$

11. $8\sqrt{27} - 3\sqrt{3} = 8\sqrt{9 \cdot 3} - 3\sqrt{3} \ \left.\right\}$ Factoring the
$$= 8\sqrt{9} \cdot \sqrt{3} - 3\sqrt{3} \ \left.\right\} \text{ first radical}$$
$$= 8 \cdot 3\sqrt{3} - 3\sqrt{3} \quad \text{Taking the square root}$$
$$= 24\sqrt{3} - 3\sqrt{3}$$
$$= (24-3)\sqrt{3} \quad \text{Factoring out } \sqrt{3}$$
$$= 21\sqrt{3}$$

13. $8\sqrt{45} + 7\sqrt{20} = 8\sqrt{9 \cdot 5} + 7\sqrt{4 \cdot 5} \ \left.\right\}$ Factoring the radicals
$$= 8\sqrt{9} \cdot \sqrt{5} + 7\sqrt{4} \cdot \sqrt{5}$$
$$= 8 \cdot 3\sqrt{5} + 7 \cdot 2\sqrt{5} \quad \text{Taking the square roots}$$
$$= 24\sqrt{5} + 14\sqrt{5}$$
$$= (24+14)\sqrt{5} \quad \text{Factoring out } \sqrt{5}$$
$$= 38\sqrt{5}$$

15. $18\sqrt{72} + 2\sqrt{98} = 18\sqrt{36 \cdot 2} + 2\sqrt{49 \cdot 2} =$
$$18\sqrt{36} \cdot \sqrt{2} + 2\sqrt{49} \cdot \sqrt{2} = 18 \cdot 6\sqrt{2} + 2 \cdot 7\sqrt{2} =$$
$$108\sqrt{2} + 14\sqrt{2} = (108+14)\sqrt{2} = 122\sqrt{2}$$

17. $3\sqrt[3]{16} + \sqrt[3]{54} = 3\sqrt[3]{8 \cdot 2} + \sqrt[3]{27 \cdot 2} =$
$$3\sqrt[3]{8} \cdot \sqrt[3]{2} + \sqrt[3]{27} \cdot \sqrt[3]{2} = 3 \cdot 2\sqrt[3]{2} + 3\sqrt[3]{2} =$$
$$6\sqrt[3]{2} + 3\sqrt[3]{2} = (6+3)\sqrt[3]{2} = 9\sqrt[3]{2}$$

19. $2\sqrt{128} - \sqrt{18} + 4\sqrt{32} =$
$$2\sqrt{64 \cdot 2} - \sqrt{9 \cdot 2} + 4\sqrt{16 \cdot 2} =$$
$$2\sqrt{64} \cdot \sqrt{2} - \sqrt{9} \cdot \sqrt{2} + 4\sqrt{16} \cdot \sqrt{2} =$$
$$2 \cdot 8\sqrt{2} - 3\sqrt{2} + 4 \cdot 4\sqrt{2} = 16\sqrt{2} - 3\sqrt{2} + 16\sqrt{2} =$$
$$(16 - 3 + 16)\sqrt{2} = 29\sqrt{2}$$

21. $\sqrt{5a} + 2\sqrt{45a^3} = \sqrt{5a} + 2\sqrt{9a^2 \cdot 5a} =$
$$\sqrt{5a} + 2\sqrt{9a^2} \cdot \sqrt{5a} = \sqrt{5a} + 2 \cdot 3a\sqrt{5a} =$$
$$\sqrt{5a} + 6a\sqrt{5a} = (1 + 6a)\sqrt{5a}$$

23. $\sqrt[3]{24x} - \sqrt[3]{3x^4} = \sqrt[3]{8 \cdot 3x} - \sqrt[3]{x^3 \cdot 3x} =$
$\sqrt[3]{8} \cdot \sqrt[3]{3x} - \sqrt[3]{x^3} \cdot \sqrt[3]{3x} = 2\sqrt[3]{3x} - x\sqrt[3]{3x} =$
$(2 - x)\sqrt[3]{3x}$

25. $7\sqrt{27x^3} + \sqrt{3x} = 7\sqrt{9x^2 \cdot 3x} + \sqrt{3x} =$
$7 \cdot \sqrt{9x^2} \cdot \sqrt{3x} + \sqrt{3x} = 7 \cdot 3x \cdot \sqrt{3x} + \sqrt{3x} =$
$21x\sqrt{3x} + \sqrt{3x} = (21x + 1)\sqrt{3x}$

27. $\sqrt{4} + \sqrt{18} = 2 + \sqrt{9 \cdot 2} = 2 + \sqrt{9} \cdot \sqrt{2} = 2 + 3\sqrt{2}$

29. $5\sqrt[3]{32} - \sqrt[3]{108} + 2\sqrt[3]{256} =$
$5\sqrt[3]{8 \cdot 4} - \sqrt[3]{27 \cdot 4} + 2\sqrt[3]{64 \cdot 4} =$
$5\sqrt[3]{8} \cdot \sqrt[3]{4} - \sqrt[3]{27} \cdot \sqrt[3]{4} + 2\sqrt[3]{64} \cdot \sqrt[3]{4} =$
$5 \cdot 2\sqrt[3]{4} - 3\sqrt[3]{4} + 2 \cdot 4\sqrt[3]{4} = 10\sqrt[3]{4} - 3\sqrt[3]{4} + 8\sqrt[3]{4} =$
$(10 - 3 + 8)\sqrt[3]{4} = 15\sqrt[3]{4}$

31. $\quad \sqrt[3]{6x^4} + \sqrt[3]{48x} - \sqrt[3]{6x}$
$= \sqrt[3]{x^3 \cdot 6x} + \sqrt[3]{8 \cdot 6x} - \sqrt[3]{6x}$
$= \sqrt[3]{x^3} \cdot \sqrt[3]{6x} + \sqrt[3]{8} \cdot \sqrt[3]{6x} - \sqrt[3]{6x}$
$= x\sqrt[3]{6x} + 2\sqrt[3]{6x} - \sqrt[3]{6x}$
$= (x + 2 - 1)\sqrt[3]{6x}$
$= (x + 1)\sqrt[3]{6x}$

33. $\sqrt{4a - 4} + \sqrt{a - 1} = \sqrt{4(a - 1)} + \sqrt{a - 1}$
$= \sqrt{4} \cdot \sqrt{a - 1} + \sqrt{a - 1}$
$= 2\sqrt{a - 1} + \sqrt{a - 1}$
$= (2 + 1)\sqrt{a - 1}$
$= 3\sqrt{a - 1}$

35. $\sqrt{x^3 - x^2} + \sqrt{9x - 9} = \sqrt{x^2(x - 1)} + \sqrt{9(x - 1)}$
$= \sqrt{x^2}\sqrt{x - 1} + \sqrt{9}\sqrt{x - 1}$
$= x\sqrt{x - 1} + 3\sqrt{x - 1}$
$= (x + 3)\sqrt{x - 1}$

37. $\sqrt{5}(4 - 2\sqrt{5}) = \sqrt{5} \cdot 4 - 2(\sqrt{5})^2$ Distributive law
$= 4\sqrt{5} - 2 \cdot 5$
$= 4\sqrt{5} - 10$

39. $\sqrt{3}(\sqrt{2} - \sqrt{7}) = \sqrt{3}\sqrt{2} - \sqrt{3}\sqrt{7}$ Distributive law
$= \sqrt{6} - \sqrt{21}$

41. $\sqrt{3}(-4\sqrt{3} + 6) = \sqrt{3} \cdot (-4\sqrt{3}) + \sqrt{3} \cdot 6 =$
$-4 \cdot 3 + 6\sqrt{3} = -12 + 6\sqrt{3}$

43. $\sqrt{3}(2\sqrt{5} - 3\sqrt{4}) = \sqrt{3}(2\sqrt{5} - 3 \cdot 2) =$
$\sqrt{3} \cdot 2\sqrt{5} - \sqrt{3} \cdot 6 = 2\sqrt{15} - 6\sqrt{3}$

45. $\sqrt[3]{2}(\sqrt[3]{4} - 2\sqrt[3]{32}) = \sqrt[3]{2} \cdot \sqrt[3]{4} - \sqrt[3]{2} \cdot 2\sqrt[3]{32} =$
$\sqrt[3]{8} - 2\sqrt[3]{64} = 2 - 2 \cdot 4 = 2 - 8 = -6$

47. $\sqrt[3]{a}(\sqrt[3]{2a^2} + \sqrt[3]{16a^2}) = \sqrt[3]{a} \cdot \sqrt[3]{2a^2} + \sqrt[3]{a} \cdot \sqrt[3]{16a^2} =$
$\sqrt[3]{2a^3} + \sqrt[3]{16a^3} = \sqrt[3]{a^3 \cdot 2} + \sqrt[3]{8a^3 \cdot 2} = a\sqrt[3]{2} + 2a\sqrt[3]{2} =$
$3a\sqrt[3]{2}$

49. $(\sqrt{3} - \sqrt{2})(\sqrt{3} + \sqrt{2}) = (\sqrt{3})^2 - (\sqrt{2})^2 = 3 - 2 = 1$

51. $(\sqrt{8} + 2\sqrt{5})(\sqrt{8} - 2\sqrt{5}) = (\sqrt{8})^2 - (2\sqrt{5})^2 =$
$8 - 4 \cdot 5 = 8 - 20 = -12$

53. $(7 + \sqrt{5})(7 - \sqrt{5}) = 7^2 - (\sqrt{5})^2 = 49 - 5 = 44$

55. $(2 - \sqrt{3})(2 + \sqrt{3}) = 2^2 - (\sqrt{3})^2 = 4 - 3 = 1$

57. $(\sqrt{8} + \sqrt{5})(\sqrt{8} - \sqrt{5}) = (\sqrt{8})^2 - (\sqrt{5})^2 = 8 - 5 = 3$

59. $(3 + 2\sqrt{7})(3 - 2\sqrt{7}) = 3^2 - (2\sqrt{7})^2 =$
$9 - 4 \cdot 7 = 9 - 28 = -19$

61. $(\sqrt{a} + \sqrt{b})(\sqrt{a} - \sqrt{b}) = (\sqrt{a})^2 - (\sqrt{b})^2 = a - b$

63. $\quad (3 - \sqrt{5})(2 + \sqrt{5})$
$= 3 \cdot 2 + 3\sqrt{5} - 2\sqrt{5} - (\sqrt{5})^2$ Using FOIL
$= 6 + 3\sqrt{5} - 2\sqrt{5} - 5$
$= 1 + \sqrt{5}$ Simplifying

65. $\quad (\sqrt{3} + 1)(2\sqrt{3} + 1)$
$= \sqrt{3} \cdot 2\sqrt{3} + \sqrt{3} \cdot 1 + 1 \cdot 2\sqrt{3} + 1^2$ Using FOIL
$= 2 \cdot 3 + \sqrt{3} + 2\sqrt{3} + 1$
$= 7 + 3\sqrt{3}$ Simplifying

67. $(2\sqrt{7} - 4\sqrt{2})(3\sqrt{7} + 6\sqrt{2}) =$
$2\sqrt{7} \cdot 3\sqrt{7} + 2\sqrt{7} \cdot 6\sqrt{2} - 4\sqrt{2} \cdot 3\sqrt{7} - 4\sqrt{2} \cdot 6\sqrt{2} =$
$6 \cdot 7 + 12\sqrt{14} - 12\sqrt{14} - 24 \cdot 2 =$
$42 + 12\sqrt{14} - 12\sqrt{14} - 48 = -6$

69. $(\sqrt{a} + \sqrt{2})(\sqrt{a} + \sqrt{3}) =$
$(\sqrt{a})^2 + \sqrt{a} \cdot \sqrt{3} + \sqrt{2} \cdot \sqrt{a} + \sqrt{2} \cdot \sqrt{3} =$
$a + \sqrt{3a} + \sqrt{2a} + \sqrt{6}$

71. $(2\sqrt[3]{3} + \sqrt[3]{2})(\sqrt[3]{3} - 2\sqrt[3]{2}) =$
$2\sqrt[3]{3} \cdot \sqrt[3]{3} - 2\sqrt[3]{3} \cdot 2\sqrt[3]{2} + \sqrt[3]{2} \cdot \sqrt[3]{3} - \sqrt[3]{2} \cdot 2\sqrt[3]{2} =$
$2\sqrt[3]{9} - 4\sqrt[3]{6} + \sqrt[3]{6} - 2\sqrt[3]{4} = 2\sqrt[3]{9} - 3\sqrt[3]{6} - 2\sqrt[3]{4}$

73. $(2 + \sqrt{3})^2 = 2^2 + 4\sqrt{3} + (\sqrt{3})^2$ Squaring a binomial
$= 4 + 4\sqrt{3} + 3$
$= 7 + 4\sqrt{3}$

75. $\quad (\sqrt[5]{9} - \sqrt[5]{3})(\sqrt[5]{8} + \sqrt[5]{27})$
$= \sqrt[5]{9} \cdot \sqrt[5]{8} + \sqrt[5]{9} \cdot \sqrt[5]{27} - \sqrt[5]{3} \cdot \sqrt[5]{8} - \sqrt[5]{3} \cdot \sqrt[5]{27}$
                                        Using FOIL
$= \sqrt[5]{72} + \sqrt[5]{243} - \sqrt[5]{24} - \sqrt[5]{81}$
$= \sqrt[5]{72} + 3 - \sqrt[5]{24} - \sqrt[5]{81}$

77. Discussion and Writing Exercise

79.
$$\frac{x^3 + 4x}{x^2 - 16} \div \frac{x^2 + 8x + 15}{x^2 + x - 20}$$

$$= \frac{x^3 + 4x}{x^2 - 16} \cdot \frac{x^2 + x - 20}{x^2 + 8x + 15}$$

$$= \frac{(x^3 + 4x)(x^2 + x - 20)}{(x^2 - 16)(x^2 + 8x + 15)}$$

$$= \frac{x(x^2 + 4)(x + 5)(x - 4)}{(x + 4)(x - 4)(x + 3)(x + 5)}$$

$$= \frac{x(x^2 + 4)(x + 5)(x - 4)}{(x + 4)(x - 4)(x + 3)(x + 5)}$$

$$= \frac{x(x^2 + 4)}{(x + 4)(x + 3)}$$

81.
$$\frac{a^3 + 8}{a^2 - 4} \cdot \frac{a^2 - 4a + 4}{a^2 - 2a + 4}$$

$$= \frac{(a^3 + 8)(a^2 - 4a + 4)}{(a^2 - 4)(a^2 - 2a + 4)}$$

$$= \frac{(a + 2)(a^2 - 2a + 4)(a - 2)(a - 2)}{(a + 2)(a - 2)(a^2 - 2a + 4)(1)}$$

$$= \frac{(a + 2)(a^2 - 2a + 4)(a - 2)}{(a + 2)(a^2 - 2a + 4)(a - 2)} \cdot \frac{a - 2}{1}$$

$$= a - 2$$

83.
$$\frac{x - \dfrac{1}{3}}{x + \dfrac{1}{4}} = \frac{x - \dfrac{1}{3}}{x + \dfrac{1}{4}} \cdot \frac{12}{12}$$

$$= \frac{\left(x - \dfrac{1}{3}\right)(12)}{\left(x + \dfrac{1}{4}\right)(12)}$$

$$= \frac{12x - 4}{12x + 3}, \text{ or } \frac{4(3x - 1)}{3(4x + 1)}$$

85.
$$\frac{\dfrac{1}{p} - \dfrac{1}{q}}{\dfrac{1}{p^2} - \dfrac{1}{q^2}} = \frac{\dfrac{1}{p} - \dfrac{1}{q}}{\dfrac{1}{p^2} - \dfrac{1}{q^2}} \cdot \frac{p^2 q^2}{p^2 q^2}$$

$$= \frac{\left(\dfrac{1}{p} - \dfrac{1}{q}\right)(p^2 q^2)}{\left(\dfrac{1}{p^2} - \dfrac{1}{q^2}\right)(p^2 q^2)}$$

$$= \frac{p^2 q^2 \cdot \dfrac{1}{p} - p^2 q^2 \cdot \dfrac{1}{q}}{p^2 q^2 \cdot \dfrac{1}{p^2} - p^2 q^2 \cdot \dfrac{1}{q^2}}$$

$$= \frac{pq^2 - p^2 q}{q^2 - p^2}$$

$$= \frac{pq(q - p)}{(q + p)(q - p)}$$

$$= \frac{pq(q - p)}{(q + p)(q - p)}$$

$$= \frac{pq}{q + p}$$

87. $|3x + 7| = 22$

$$3x + 7 = -22 \quad or \quad 3x + 7 = 22$$

$$3x = -29 \quad or \quad 3x = 15$$

$$x = -\frac{29}{3} \quad or \quad x = 5$$

The solutions are $-\dfrac{29}{3}$ and 5.

89. $|3x + 7| \geq 22$

$$3x + 7 \leq -22 \quad or \quad 3x + 7 \geq 22$$

$$3x \leq -29 \quad or \quad 3x \geq 15$$

$$x \leq -\frac{29}{3} \quad or \quad x \geq 5$$

The solution set is $\left\{x \middle| x \leq -\dfrac{29}{3} \text{ or } x \geq 5\right\}$, or $\left(-\infty, -\dfrac{29}{3}\right] \cup [5, \infty)$.

91.

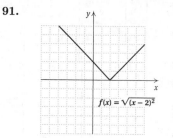

$$f(x) = \sqrt{(x - 2)^2}$$

Since $(x - 2)^2$ is nonnegative for all values of x, the domain of f is $\{x | x \text{ is a real number}\}$, or $(-\infty, \infty)$.

93. $\sqrt{9 + 3\sqrt{5}}\sqrt{9 - 3\sqrt{5}} = \sqrt{(9 + 3\sqrt{5})(9 - 3\sqrt{5})} = $
$\sqrt{9^2 - (3\sqrt{5})^2} = \sqrt{81 - 9 \cdot 5} = \sqrt{81 - 45} = \sqrt{36} = 6$

95. $(\sqrt{3} + \sqrt{5} - \sqrt{6})^2 = [(\sqrt{3} + \sqrt{5}) - \sqrt{6}]^2 = $
$(\sqrt{3} + \sqrt{5})^2 - 2(\sqrt{3} + \sqrt{5})(\sqrt{6}) + (\sqrt{6})^2 = $
$3 + 2\sqrt{15} + 5 - 2\sqrt{18} - 2\sqrt{30} + 6 = $
$14 + 2\sqrt{15} - 2\sqrt{9 \cdot 2} - 2\sqrt{30} = $
$14 + 2\sqrt{15} - 6\sqrt{2} - 2\sqrt{30}$

97. $(\sqrt[3]{9} - 2)(\sqrt[3]{9} + 4)$
$= \sqrt[3]{9}\sqrt[3]{9} + 4\sqrt[3]{9} - 2\sqrt[3]{9} - 2 \cdot 4$
$= \sqrt[3]{81} + 2\sqrt[3]{9} - 8$
$= \sqrt[3]{27 \cdot 3} + 2\sqrt[3]{9} - 8$
$= 3\sqrt[3]{3} + 2\sqrt[3]{9} - 8$

Exercise Set 6.5

1. $\sqrt{\dfrac{5}{3}} = \sqrt{\dfrac{5}{3} \cdot \dfrac{3}{3}} = \sqrt{\dfrac{15}{9}} = \dfrac{\sqrt{15}}{\sqrt{9}} = \dfrac{\sqrt{15}}{3}$

3. $\sqrt{\dfrac{11}{2}} = \sqrt{\dfrac{11}{2} \cdot \dfrac{2}{2}} = \sqrt{\dfrac{22}{4}} = \dfrac{\sqrt{22}}{\sqrt{4}} = \dfrac{\sqrt{22}}{2}$

5. $\dfrac{2\sqrt{3}}{7\sqrt{5}} = \dfrac{2\sqrt{3}}{7\sqrt{5}} \cdot \dfrac{\sqrt{5}}{\sqrt{5}} = \dfrac{2\sqrt{15}}{7\sqrt{5^2}} = \dfrac{2\sqrt{15}}{7 \cdot 5} = \dfrac{2\sqrt{15}}{35}$

7. $\sqrt[3]{\dfrac{16}{9}} = \sqrt[3]{\dfrac{16}{9} \cdot \dfrac{3}{3}} = \sqrt[3]{\dfrac{48}{27}} = \dfrac{\sqrt[3]{8 \cdot 6}}{\sqrt[3]{27}} = \dfrac{2\sqrt[3]{6}}{3}$

9. $\dfrac{\sqrt[3]{3a}}{\sqrt[3]{5c}} = \dfrac{\sqrt[3]{3a}}{\sqrt[3]{5c}} \cdot \dfrac{\sqrt[3]{5^2 c^2}}{\sqrt[3]{5^2 c^2}} = \dfrac{\sqrt[3]{75ac^2}}{\sqrt[3]{5^3 c^3}} = \dfrac{\sqrt[3]{75ac^2}}{5c}$

11. $\dfrac{\sqrt[3]{2y^4}}{\sqrt[3]{6x^4}} = \dfrac{\sqrt[3]{2y^4}}{\sqrt[3]{6x^4}} \cdot \dfrac{\sqrt[3]{6^2 x^2}}{\sqrt[3]{6^2 x^2}} = \dfrac{\sqrt[3]{72x^2 y^4}}{\sqrt[3]{6^3 x^6}} = \dfrac{\sqrt[3]{8y^3 \cdot 9x^2 y}}{6x^2} =$

$\dfrac{2y\sqrt[3]{9x^2 y}}{6x^2} = \dfrac{y\sqrt[3]{9x^2 y}}{3x^2}$

13. $\dfrac{1}{\sqrt[4]{st}} = \dfrac{1}{\sqrt[4]{st}} \cdot \dfrac{\sqrt[4]{s^3 t^3}}{\sqrt[4]{s^3 t^3}} = \dfrac{\sqrt[4]{s^3 t^3}}{\sqrt[4]{s^4 t^4}} = \dfrac{\sqrt[4]{s^3 t^3}}{st}$

15. $\sqrt{\dfrac{3x}{20}} = \sqrt{\dfrac{3x}{20} \cdot \dfrac{5}{5}} = \sqrt{\dfrac{15x}{100}} = \dfrac{\sqrt{15x}}{\sqrt{100}} = \dfrac{\sqrt{15x}}{10}$

17. $\sqrt[3]{\dfrac{4}{5x^5 y^2}} = \sqrt[3]{\dfrac{4}{5x^5 y^2} \cdot \dfrac{25xy}{5^2 xy}} = \sqrt[3]{\dfrac{100xy}{5^3 x^6 y^3}} =$

$\dfrac{\sqrt[3]{100xy}}{\sqrt[3]{5^3 x^6 y^3}} = \dfrac{\sqrt[3]{100xy}}{5x^2 y}$

19. $\sqrt[4]{\dfrac{1}{8x^7 y^3}} = \sqrt[4]{\dfrac{1}{2^3 x^7 y^3} \cdot \dfrac{2xy}{2xy}} = \sqrt[4]{\dfrac{2xy}{2^4 x^8 y^4}} = \dfrac{\sqrt[4]{2xy}}{\sqrt[4]{2^4 x^8 y^4}} =$

$\dfrac{\sqrt[4]{2xy}}{2x^2 y}$

21. $\dfrac{9}{6 - \sqrt{10}} = \dfrac{9}{6 - \sqrt{10}} \cdot \dfrac{6 + \sqrt{10}}{6 + \sqrt{10}} = \dfrac{9(6 + \sqrt{10})}{6^2 - (\sqrt{10})^2} =$

$\dfrac{9(6 + \sqrt{10})}{36 - 10} = \dfrac{54 + 9\sqrt{10}}{26}$

23. $\dfrac{-4\sqrt{7}}{\sqrt{5} - \sqrt{3}} = \dfrac{-4\sqrt{7}}{\sqrt{5} - \sqrt{3}} \cdot \dfrac{\sqrt{5} + \sqrt{3}}{\sqrt{5} + \sqrt{3}} =$

$\dfrac{-4\sqrt{7}(\sqrt{5} + \sqrt{3})}{(\sqrt{5})^2 - (\sqrt{3})^2} = \dfrac{-4\sqrt{7}(\sqrt{5} + \sqrt{3})}{5 - 3} =$

$\dfrac{-4\sqrt{7}(\sqrt{5} + \sqrt{3})}{2} = -2\sqrt{7}(\sqrt{5} + \sqrt{3}) = -2\sqrt{35} - 2\sqrt{21}$

25. $\dfrac{\sqrt{5} - 2\sqrt{6}}{\sqrt{3} - 4\sqrt{5}} = \dfrac{\sqrt{5} - 2\sqrt{6}}{\sqrt{3} - 4\sqrt{5}} \cdot \dfrac{\sqrt{3} + 4\sqrt{5}}{\sqrt{3} + 4\sqrt{5}} =$

$\dfrac{\sqrt{15} + 4 \cdot 5 - 2\sqrt{18} - 8\sqrt{30}}{(\sqrt{3})^2 - (4\sqrt{5})^2} =$

$\dfrac{\sqrt{15} + 20 - 2\sqrt{9 \cdot 2} - 8\sqrt{30}}{3 - 16 \cdot 5} =$

$\dfrac{\sqrt{15} + 20 - 6\sqrt{2} - 8\sqrt{30}}{-77}, \text{ or } -\dfrac{\sqrt{15} + 20 - 6\sqrt{2} - 8\sqrt{30}}{77}$

27. $\dfrac{2 - \sqrt{a}}{3 + \sqrt{a}} = \dfrac{2 - \sqrt{a}}{3 + \sqrt{a}} \cdot \dfrac{3 - \sqrt{a}}{3 - \sqrt{a}} = \dfrac{6 - 2\sqrt{a} - 3\sqrt{a} + a}{9 - a} =$

$\dfrac{6 - 5\sqrt{a} + a}{9 - a}$

29. $\dfrac{5\sqrt{3} - 3\sqrt{2}}{3\sqrt{2} - 2\sqrt{3}} = \dfrac{5\sqrt{3} - 3\sqrt{2}}{3\sqrt{2} - 2\sqrt{3}} \cdot \dfrac{3\sqrt{2} + 2\sqrt{3}}{3\sqrt{2} + 2\sqrt{3}} =$

$\dfrac{15\sqrt{6} + 10 \cdot 3 - 9 \cdot 2 - 6\sqrt{6}}{9 \cdot 2 - 4 \cdot 3} = \dfrac{12 + 9\sqrt{6}}{6} =$

$\dfrac{3(4 + 3\sqrt{6})}{3 \cdot 2} = \dfrac{4 + 3\sqrt{6}}{2}$

31. $\dfrac{\sqrt{x} - \sqrt{y}}{\sqrt{x} + \sqrt{y}} = \dfrac{\sqrt{x} - \sqrt{y}}{\sqrt{x} + \sqrt{y}} \cdot \dfrac{\sqrt{x} - \sqrt{y}}{\sqrt{x} - \sqrt{y}} =$

$\dfrac{x - \sqrt{xy} - \sqrt{xy} + y}{x - y} = \dfrac{x - 2\sqrt{xy} + y}{x - y}$

33. Discussion and Writing Exercise

35. $\dfrac{1}{2} - \dfrac{1}{3} = \dfrac{5}{t}$, LCM is $6t$

$6t\left(\dfrac{1}{2} - \dfrac{1}{3}\right) = 6t\left(\dfrac{5}{t}\right)$

$\qquad 3t - 2t = 30$

$\qquad\qquad t = 30$

Check:

$$\dfrac{\dfrac{1}{2} - \dfrac{1}{3}}{\;} = \dfrac{1}{t}$$

$$\dfrac{1}{2} - \dfrac{1}{3} \;\;?\;\; \dfrac{5}{30}$$

$$\dfrac{3}{6} - \dfrac{2}{6} \;\bigg|\; \dfrac{1}{6}$$

$$\dfrac{1}{6} \;\bigg|\; \text{ TRUE}$$

The solution is 30.

37. $\dfrac{1}{x^3 - y^2} \div \dfrac{1}{(x - y)(x^2 + xy + y^2)}$

$= \dfrac{1}{(x - y)(x^2 + xy + y^2)} \cdot \dfrac{(x - y)(x^2 + xy + y^2)}{1}$

$= \dfrac{(x - y)(x^2 + xy + y^2)}{(x - y)(x^2 + xy + y^2)}$

$= 1$

39. Left to the student

41. $\sqrt{a^2 - 3} - \dfrac{a^2}{\sqrt{a^2 - 3}}$

$= \sqrt{a^2 - 3} - \dfrac{a^2}{\sqrt{a^2 - 3}} \cdot \dfrac{\sqrt{a^2 - 3}}{\sqrt{a^2 - 3}}$

$= \sqrt{a^2 - 3} - \dfrac{a^2 \sqrt{a^2 - 3}}{a^2 - 3}$

$= \sqrt{a^2 - 3} \cdot \dfrac{a^2 - 3}{a^2 - 3} - \dfrac{a^2 \sqrt{a^2 - 3}}{a^2 - 3}$

$= \dfrac{a^2 \sqrt{a^2 - 3} - 3\sqrt{a^2 - 3} - a^2 \sqrt{a^2 - 3}}{a^2 - 3}$

$= \dfrac{-3\sqrt{a^2 - 3}}{a^2 - 3}, \text{ or } -\dfrac{3\sqrt{a^2 - 3}}{a^2 - 3}$

Exercise Set 6.6

1. $\sqrt{2x-3} = 4$

$(\sqrt{2x-3})^2 = 4^2$ Principle of powers

$2x - 3 = 16$

$2x = 19$

$x = \dfrac{19}{2}$

Check: $\dfrac{\sqrt{2x-3} = 4}{}$

$\sqrt{2 \cdot \dfrac{19}{2} - 3} \; ? \; 4$

$\sqrt{19 - 3}$

$\sqrt{16}$

$4 \quad \Big| \quad$ TRUE

The solution is $\dfrac{19}{2}$.

3. $\sqrt{6x} + 1 = 8$

$\sqrt{6x} = 7$ Subtracting to isolate the radical

$(\sqrt{6x})^2 = 7^2$ Principle of powers

$6x = 49$

$x = \dfrac{49}{6}$

Check: $\dfrac{\sqrt{6x} + 1 = 8}{}$

$\sqrt{6 \cdot \dfrac{49}{6}} + 1 \; ? \; 8$

$\sqrt{49} + 1$

$7 + 1$

$8 \quad \Big| \quad$ TRUE

The solution is $\dfrac{49}{6}$.

5. $\sqrt{y+7} - 4 = 4$

$\sqrt{y+7} = 8$ Adding to isolate the radical

$(\sqrt{y+7})^2 = 8^2$ Principle of powers

$y + 7 = 64$

$y = 57$

Check: $\dfrac{\sqrt{y+7} - 4 = 4}{}$

$\sqrt{57 + 7} - 4 \; ? \; 4$

$\sqrt{64} - 4$

$8 - 4$

$4 \quad \Big| \quad$ TRUE

The solution is 57.

7. $\sqrt{5y+8} = 10$

$(\sqrt{5y+8})^2 = 10^2$ Principle of powers

$5y + 8 = 100$

$5y = 92$

$y = \dfrac{92}{5}$

Check: $\dfrac{\sqrt{5y+8} = 10}{}$

$\sqrt{5 \cdot \dfrac{92}{5} + 8} \; ? \; 10$

$\sqrt{92 + 8}$

$\sqrt{100}$

$10 \quad \Big| \quad$ TRUE

The solution is $\dfrac{92}{5}$.

9. $\sqrt[3]{x} = -1$

$(\sqrt[3]{x})^3 = (-1)^3$ Principle of powers

$x = -1$

Check: $\dfrac{\sqrt[3]{x} = -1}{}$

$\sqrt[3]{-1} \; ? \; -1$

$-1 \quad \Big| \quad$ TRUE

The solution is -1.

11. $\sqrt{x+2} = -4$

$(\sqrt{x+2})^2 = (-4)^2$

$x + 2 = 16$

$x = 14$

Check: $\dfrac{\sqrt{x+2} = -4}{}$

$\sqrt{14 + 2} \; ? \; -4$

$\sqrt{16}$

$4 \quad \Big| \quad$ FALSE

The number 14 does not check. The equation has no solution. We might have observed at the outset that this equation has no solution because the principle square root of a number is never negative.

13. $\sqrt[3]{x+5} = 2$

$(\sqrt[3]{x+5})^3 = 2^3$

$x + 5 = 8$

$x = 3$

Check: $\dfrac{\sqrt[3]{x+5} = 2}{}$

$\sqrt[3]{3 + 5} \; ? \; 2$

$\sqrt[3]{8}$

$2 \quad \Big| \quad$ TRUE

The solution is 3.

15. $\sqrt[4]{y-3} = 2$

$(\sqrt[4]{y-3})^4 = 2^4$

$y - 3 = 16$

$y = 19$

Check: $\dfrac{\sqrt[4]{y-3} = 2}{}$

$\sqrt[4]{19-3} \;?\; 2$

$\sqrt[4]{16}$

$2 \;\Big|\;$ TRUE

The solution is 19.

17. $\sqrt[3]{6x+9} + 8 = 5$

$\sqrt[3]{6x+9} = -3$

$(\sqrt[3]{6x+9})^3 = (-3)^3$

$6x + 9 = -27$

$6x = -36$

$x = -6$

Check: $\dfrac{\sqrt[3]{6x+9} + 8 = 5}{}$

$\sqrt[3]{6(-6)+9} + 8 \;?\; 5$

$\sqrt[3]{-27} + 8$

$-3 + 8$

$5 \;\Big|\;$ TRUE

The solution is -6.

19. $8 = \dfrac{1}{\sqrt{x}}$

$8 \cdot \sqrt{x} = \dfrac{1}{\sqrt{x}} \cdot \sqrt{x}$

$8\sqrt{x} = 1$

$(8\sqrt{x})^2 = 1^2$

$64x = 1$

$x = \dfrac{1}{64}$

Check: $8 = \dfrac{1}{\sqrt{x}}$

$8 \;?\; \dfrac{1}{\sqrt{\dfrac{1}{64}}}$

$\dfrac{1}{\dfrac{1}{8}}$

$8 \;\Big|\;$ TRUE

The solution is $\dfrac{1}{64}$.

21. $x - 7 = \sqrt{x-5}$

$(x-7)^2 = (\sqrt{x-5})^2$

$x^2 - 14x + 49 = x - 5$

$x^2 - 15x + 54 = 0$

$(x-6)(x-9) = 0$

$x - 6 = 0 \;\;or\;\; x - 9 = 0$

$x = 6 \;\;or\;\;\;\;\; x = 9$

Check: For 6: $\dfrac{x - 7 = \sqrt{x-5}}{}$

$6 - 7 \;?\; \sqrt{6-5}$

$-1 \;\Big|\; \sqrt{1}$

$-1 \;\Big|\; 1 \;$ FALSE

Check: For 9: $\dfrac{x - 7 = \sqrt{x-5}}{}$

$9 - 7 \;?\; \sqrt{9-5}$

$2 \;\Big|\; \sqrt{4}$

$2 \;\Big|\; 2 \;$ TRUE

The number 6 does not check, but 9 does. The solution is 9.

23. $2\sqrt{x+1} + 7 = x$

$2\sqrt{x+1} = x - 7$

$(2\sqrt{x+1})^2 = (x-7)^2$

$4(x+1) = x^2 - 14x + 49$

$4x + 4 = x^2 - 14x + 49$

$0 = x^2 - 18x + 45$

$0 = (x-3)(x-15)$

$x - 3 = 0 \;\;or\;\; x - 15 = 0$

$x = 3 \;\;or\;\;\;\;\; x = 15$

Check: For 3: $\dfrac{2\sqrt{x+1} + 7 = x}{}$

$2\sqrt{3+1} + 7 \;?\; 3$

$2\sqrt{4} + 7$

$2 \cdot 2 + 7$

$4 + 7$

$11 \;\Big|\; 3 \;$ FALSE

Check: For 15: $\dfrac{2\sqrt{x+1} + 7 = x}{}$

$2\sqrt{15+1} + 7 \;?\; 15$

$2\sqrt{16} + 7$

$2 \cdot 4 + 7$

$8 + 7$

$15 \;\Big|\; 15 \;$ TRUE

The number 3 does not check, but 15 does. The solution is 15.

25. $3\sqrt{x-1} - 1 = x$

$3\sqrt{x-1} = x + 1$

$(3\sqrt{x-1})^2 = (x+1)^2$

$9(x-1) = x^2 + 2x + 1$

$9x - 9 = x^2 + 2x + 1$

$0 = x^2 - 7x + 10$

$0 = (x-2)(x-5)$

$$x - 2 = 0 \quad or \quad x - 5 = 0$$
$$x = 2 \quad or \quad x = 5$$

Check: For 2:

$$\frac{3\sqrt{x-1}-1 = x}{3\sqrt{2-1}-1 \; ? \; 2}$$
$$3\sqrt{1}-1 \; \Big|$$
$$3 \cdot 1 - 1 \; \Big|$$
$$3 - 1 \; \Big|$$
$$2 \; \Big| \; 2 \quad \text{TRUE}$$

Check: For 5:

$$\frac{3\sqrt{x-1}-1 = x}{3\sqrt{5-1}-1 \; ? \; 5}$$
$$3\sqrt{4}-1 \; \Big|$$
$$3 \cdot 2 - 1 \; \Big|$$
$$6 - 1 \; \Big|$$
$$5 \; \Big| \; 5 \quad \text{TRUE}$$

Both numbers check. The solutions are 2 and 5.

27.
$$x - 3 = \sqrt{27 - 3x}$$
$$(x-3)^2 = (\sqrt{27-3x})^2$$
$$x^2 - 6x + 9 = 27 - 3x$$
$$x^2 - 3x - 18 = 0$$
$$(x-6)(x+3) = 0$$
$$x - 6 = 0 \quad or \quad x + 3 = 0$$
$$x = 6 \quad or \quad x = -3$$

Check: For 6:

$$\frac{x - 3 = \sqrt{27 - 3x}}{6 - 3 \; ? \; \sqrt{27 - 3 \cdot 6}}$$
$$3 \; \Big| \; \sqrt{27 - 18}$$
$$\Big| \; \sqrt{9}$$
$$3 \; \Big| \; 3 \quad \text{TRUE}$$

Check: For -3:

$$\frac{x - 3 = \sqrt{27 - 3x}}{-3 - 3 \; ? \; \sqrt{27 - 3(-3)}}$$
$$-6 \; \Big| \; \sqrt{27 + 9}$$
$$\Big| \; \sqrt{36}$$
$$-6 \; \Big| \; 6 \quad \text{FALSE}$$

The number 6 checks but -3 does not. The solution is 6.

29.
$$\sqrt{3y+1} = \sqrt{2y+6}$$
$$(\sqrt{3y+1})^2 = (\sqrt{2y+6})^2$$
$$3y + 1 = 2y + 6$$
$$y = 5$$

Check:

$$\frac{\sqrt{3y+1} = \sqrt{2y+6}}{\sqrt{3 \cdot 5 + 1} \; ? \; \sqrt{2 \cdot 5 + 6}}$$
$$\sqrt{16} \; \Big| \; \sqrt{16} \quad \text{TRUE}$$

The solution is 5.

31.
$$\sqrt{y-5} + \sqrt{y} = 5$$
$$\sqrt{y-5} = 5 - \sqrt{y} \qquad \text{Isolating one radical}$$
$$(\sqrt{y-5})^2 = (5 - \sqrt{y})^2$$
$$y - 5 = 25 - 10\sqrt{y} + y$$
$$10\sqrt{y} = 30 \qquad \text{Isolating the remaining radical}$$
$$\sqrt{y} = 3 \qquad \text{Dividing by 10}$$
$$(\sqrt{y})^2 = 3^2$$
$$y = 9$$

The number 9 checks, so it is the solution.

33.
$$3 + \sqrt{z-6} = \sqrt{z+9}$$
$$(3 + \sqrt{z-6})^2 = (\sqrt{z+9})^2$$
$$9 + 6\sqrt{z-6} + z - 6 = z + 9$$
$$6\sqrt{z-6} = 6$$
$$\sqrt{z-6} = 1 \qquad \text{Dividing by 6}$$
$$(\sqrt{z-6})^2 = 1^2$$
$$z - 6 = 1$$
$$z = 7$$

The number 7 checks, so it is the solution.

35.
$$\sqrt{20-x} + 8 = \sqrt{9-x} + 11$$
$$\sqrt{20-x} = \sqrt{9-x} + 3 \qquad \text{Isolating one radical}$$
$$(\sqrt{20-x})^2 = (\sqrt{9-x} + 3)^2$$
$$20 - x = 9 - x + 6\sqrt{9-x} + 9$$
$$2 = 6\sqrt{9-x} \qquad \text{Isolating the remaining radical}$$
$$1 = 3\sqrt{9-x} \qquad \text{Dividing by 2}$$
$$1^2 = (3\sqrt{9-x})^2$$
$$1 = 9(9-x)$$
$$1 = 81 - 9x$$
$$9x = 80$$
$$x = \frac{80}{9}$$

The number $\frac{80}{9}$ checks, so it is the solution.

37. $\sqrt{4y+1} - \sqrt{y-2} = 3$

$\sqrt{4y+1} = 3 + \sqrt{y-2}$ Isolating one radical

$(\sqrt{4y+1})^2 = (3+\sqrt{y-2})^2$

$4y+1 = 9 + 6\sqrt{y-2} + y - 2$

$3y - 6 = 6\sqrt{y-2}$ Isolating the remaining radical

$y - 2 = 2\sqrt{y-2}$ Multiplying by $\frac{1}{3}$

$(y-2)^2 = (2\sqrt{y-2})^2$

$y^2 - 4y + 4 = 4(y-2)$

$y^2 - 4y + 4 = 4y - 8$

$y^2 - 8y + 12 = 0$

$(y-6)(y-2) = 0$

$y - 6 = 0$ or $y - 2 = 0$

$y = 6$ or $y = 2$

The numbers 6 and 2 check, so they are the solutions.

39. $\sqrt{x+2} + \sqrt{3x+4} = 2$

$\sqrt{x+2} = 2 - \sqrt{3x+4}$ Isolating one radical

$(\sqrt{x+2})^2 = (2-\sqrt{3x-4})^2$

$x+2 = 4 - 4\sqrt{3x+4} + 3x + 4$

$-2x - 6 = -4\sqrt{3x+4}$ Isolating the remaining radical

$x + 3 = 2\sqrt{3x+4}$ Dividing by 2

$(x+3)^2 = (2\sqrt{3x+4})^2$

$x^2 + 6x + 9 = 4(3x+4)$

$x^2 + 6x + 9 = 12x + 16$

$x^2 - 6x - 7 = 0$

$(x-7)(x+1) = 0$

$x - 7 = 0$ or $x + 1 = 0$

$x = 7$ or $x = -1$

Check: For 7: $\dfrac{\sqrt{x+2} + \sqrt{3x+4} = 2}{\sqrt{7+2} + \sqrt{3\cdot7+4} \; ? \; 2}$

$\sqrt{9} + \sqrt{25}$

$8 \; \bigg| \;$ FALSE

Check: For -1: $\dfrac{\sqrt{x+2} + \sqrt{3x+4} = 2}{\sqrt{-1+2} + \sqrt{3(-1)+4} \; ? \; 2}$

$\sqrt{1} + \sqrt{1}$

$2 \; \bigg| \;$ TRUE

Since -1 checks but 7 does not, the solution is -1.

41. $\sqrt{3x-5} + \sqrt{2x+3} + 1 = 0$

$\sqrt{3x-5} + 1 = -\sqrt{2x+3}$

$(\sqrt{3x-5}+1)^2 = (-\sqrt{2x+3})^2$

$3x - 5 + 2\sqrt{3x-5} + 1 = 2x + 3$

$2\sqrt{3x-5} = -x + 7$

$(2\sqrt{3x-5})^2 = (-x+7)^2$

$4(3x-5) = x^2 - 14x + 49$

$12x - 20 = x^2 - 14x + 49$

$0 = x^2 - 26x + 69$

$0 = (x-23)(x-3)$

$x - 23 = 0$ or $x - 3 = 0$

$x = 23$ or $x = 3$

Neither number checks. There is no solution. (At the outset we might have observed that there is no solution since the sum on the left side of the equation must be at least 1.)

43. $2\sqrt{t-1} - \sqrt{3t-1} = 0$

$2\sqrt{t-1} = \sqrt{3t-1}$

$(2\sqrt{t-1})^2 = (\sqrt{3t-1})^2$

$4(t-1) = 3t - 1$

$4t - 4 = 3t - 1$

$t = 3$

Since 3 checks, it is the solution.

45. $D = 1.2\sqrt{h}$

$D = 1.2\sqrt{1353}$

$D \approx 44.1$

A tourist can see about 44.1 mi to the horizon.

47. $D = 1.2\sqrt{h}$

$31.3 = 1.2\sqrt{h}$

$(31.3)^2 = (1.2\sqrt{h})^2$

$979.69 = 1.44h$

$680 \approx h$

The height of Elaine's eyes is about 680 ft.

49. $D = 1.2\sqrt{h}$

$13 = 1.2\sqrt{h}$

$13^2 = (1.2\sqrt{h})^2$

$169 = 1.44h$

$117 \approx h$

The height of the steeplejack's eyes is about 117 ft.

51. $D = 1.2\sqrt{h}$

$D = 1.2\sqrt{13}$

$D \approx 4.3$

The sailor can see about 4.3 mi to the horizon.

53. At 55 mph: $r = 2\sqrt{5L}$

$$55 = 2\sqrt{5L}$$
$$27.5 = \sqrt{5L}$$
$$(27.5)^2 = (\sqrt{5L})^2$$
$$756.25 = 5L$$
$$151.25 = L$$

At 55 mph, a car will skid 151.25 ft.

At 75 mph: $r = 2\sqrt{5L}$

$$75 = 2\sqrt{5L}$$
$$37.5 = \sqrt{5L}$$
$$(37.5)^2 = (\sqrt{5L})^2$$
$$1406.25 = 5L$$
$$281.25 = L$$

At 75 mph, a car will skid 281.25 ft.

55.

$$S = 21.9\sqrt{5t + 2457}$$
$$1113 = 21.9\sqrt{5t + 2457}$$
$$\frac{1113}{21.9} = \sqrt{5t + 2457}$$
$$\left(\frac{1113}{21.9}\right)^2 = (\sqrt{5t + 2457})^2$$
$$2583 \approx 5t + 2457$$
$$126 \approx 5t$$
$$25.2 \approx t$$

The temperature was approximately 25°F.

57.

$$T = 2\pi\sqrt{\frac{L}{32}}$$
$$1 = 2(3.14)\sqrt{\frac{L}{32}}$$
$$1 = 6.28\sqrt{\frac{L}{32}}$$
$$1^2 = \left(6.28\sqrt{\frac{L}{32}}\right)^2$$
$$1 = 39.4384 \cdot \frac{L}{32}$$
$$\frac{32}{39.4384} = L$$
$$0.81 \approx L$$

The pendulum is about 0.81 ft long.

59. Discussion and Writing Exercise

61. **Familiarize**. Let t = the time it will take Julia and George to paint the room working together.

Translate. We use the work principle.

$$\frac{t}{a} + \frac{t}{b} = 1$$
$$\frac{t}{8} + \frac{t}{10} = 1$$

Solve. We first multiply by 40 to clear fractions.

$$40\left(\frac{t}{8} + \frac{t}{10}\right) = 40 \cdot 1$$
$$40 \cdot \frac{t}{8} + 40 \cdot \frac{t}{10} = 40$$
$$5t + 4t = 40$$
$$9t = 40$$
$$t = \frac{40}{9}, \text{ or } 4\frac{4}{9}$$

Check. In $\frac{40}{9}$ hr, Julia does $\frac{40}{9}\left(\frac{1}{8}\right)$, or $\frac{5}{9}$, of the job and George does $\frac{40}{9}\left(\frac{1}{10}\right)$, or $\frac{4}{9}$, of the job. Together they do $\frac{5}{9} + \frac{4}{9}$, or 1 entire job. The answer checks.

State. It will take them $4\frac{4}{9}$ hr to paint the room, working together.

63. **Familiarize**. Let d = the distance the cyclist would travel in 56 days at the same rate.

Translate. We translate to a proportion.

$$\begin{array}{l} \text{Distance} \to \\ \text{Days} \quad \to \end{array} \frac{702}{14} = \frac{d}{56} \begin{array}{l} \leftarrow \text{Distance} \\ \leftarrow \quad \text{Days} \end{array}$$

Solve. We equate cross products.

$$\frac{702}{14} = \frac{d}{56}$$
$$702 \cdot 56 = 14 \cdot d$$
$$\frac{702 \cdot 56}{14} = d$$
$$2808 = d$$

Check. We substitute in the proportion and check cross products.

$$\frac{702}{14} = \frac{2808}{56}; \; 702 \cdot 56 = 39,312; \; 14 \cdot 2808 = 39,312$$

The cross products are the same, so the answer checks.

State. The cyclist would have traveled 2808 mi in 56 days.

65.

$$x^2 + 2.8x = 0$$
$$x(x + 2.8) = 0$$
$$x = 0 \;\; or \;\; x + 2.8 = 0$$
$$x = 0 \;\; or \;\;\;\;\;\;\; x = -2.8$$

The solutions are 0 and -2.8.

67.

$$x^2 - 64 = 0$$
$$(x + 8)(x - 8) = 0$$
$$x + 8 = 0 \;\;\; or \;\; x - 8 = 0$$
$$x = -8 \;\; or \;\;\;\;\;\; x = 8$$

The solutions are -8 and 8.

69. $f(x) = x^2$

$$f(a + h) - f(a) = (a + h)^2 - a^2$$
$$= a^2 + 2ah + h^2 - a^2$$
$$= 2ah + h^2$$

71. $f(x) = 2x^2 - 3x$

$$f(a + h) - f(a)$$
$$= 2(a + h)^2 - 3(a + h) - (2a^2 - 3a)$$
$$= 2(a^2 + 2ah + h^2) - 3a - 3h - 2a^2 + 3a$$
$$= 2a^2 + 4ah + 2h^2 - 3a - 3h - 2a^2 + 3a$$
$$= 4ah + 2h^2 - 3h$$

73. Left to the student

75. $\sqrt[3]{\dfrac{z}{4}} - 10 = 2$

$$\sqrt[3]{\dfrac{z}{4}} = 12$$
$$\left(\sqrt[3]{\dfrac{z}{4}}\right)^3 = 12^3$$
$$\dfrac{z}{4} = 1728$$
$$z = 6912$$

The number 6912 checks, so it is the solution.

77. $\sqrt{\sqrt{y + 49} - \sqrt{y}} = \sqrt{7}$

$$\left(\sqrt{\sqrt{y + 49} - \sqrt{y}}\right)^2 = (\sqrt{7})^2$$
$$\sqrt{y + 49} - \sqrt{y} = 7$$
$$\sqrt{y + 49} = 7 + \sqrt{y}$$
$$(\sqrt{y + 49})^2 = (7 + \sqrt{y})^2$$
$$y + 49 = 49 + 14\sqrt{y} + y$$
$$0 = 14\sqrt{y}$$
$$0 = \sqrt{y}$$
$$0^2 = (\sqrt{y})^2$$
$$0 = y$$

The number 0 checks and is the solution.

79. $\sqrt{\sqrt{x^2 + 9x + 34}} = 2$

$$\left(\sqrt{\sqrt{x^2 + 9x + 34}}\right)^2 = 2^2$$
$$\sqrt{x^2 + 9x + 34} = 4$$
$$(\sqrt{x^2 + 9x + 34})^2 = 4^2$$
$$x^2 + 9x + 34 = 16$$
$$x^2 + 9x + 18 = 0$$
$$(x + 6)(x + 3) = 0$$
$$x + 6 = 0 \quad \text{or} \quad x + 3 = 0$$
$$x = -6 \quad \text{or} \quad x = -3$$

Both values check. The solutions are -6 and -3.

81. $\sqrt{x - 2} - \sqrt{x + 2} + 2 = 0$

$$\sqrt{x - 2} + 2 = \sqrt{x + 2}$$
$$(\sqrt{x - 2} + 2)^2 = (\sqrt{x + 2})^2$$
$$(x - 2) + 4\sqrt{x - 2} + 4 = x + 2$$
$$4\sqrt{x - 2} = 0$$
$$\sqrt{x - 2} = 0$$
$$(\sqrt{x - 2})^2 = 0^2$$
$$x - 2 = 0$$
$$x = 2$$

The number 2 checks, so it is the solution.

83. $\sqrt{a^2 + 30a} = a + \sqrt{5a}$

$$(\sqrt{a^2 + 30a})^2 = (a + \sqrt{5a})^2$$
$$a^2 + 30a = a^2 + 2a\sqrt{5a} + 5a$$
$$25a = 2a\sqrt{5a}$$
$$(25a)^2 = (2a\sqrt{5a})^2$$
$$625a^2 = 4a^2 \cdot 5a$$
$$625a^2 = 20a^3$$
$$0 = 20a^3 - 625a^2$$
$$0 = 5a^2(4a - 125)$$
$$5a^2 = 0 \quad \text{or} \quad 4a - 125 = 0$$
$$a^2 = 0 \quad \text{or} \quad 4a = 125$$
$$a = 0 \quad \text{or} \quad a = \dfrac{125}{4}$$

Both values check. The solutions are 0 and $\dfrac{125}{4}$.

85. $\dfrac{x - 1}{\sqrt{x^2 + 3x + 6}} = \dfrac{1}{4}$,

$$\text{LCM} = 4\sqrt{x^2 + 3x + 6}$$
$$4\sqrt{x^2 + 3x + 6} \cdot \dfrac{x - 1}{\sqrt{x^2 + 3x + 6}} = 4\sqrt{x^2 + 3x + 6} \cdot \dfrac{1}{4}$$
$$4x - 4 = \sqrt{x^2 + 3x + 6}$$
$$16x^2 - 32x + 16 = x^2 + 3x + 6$$
$$\text{Squaring both sides}$$
$$15x^2 - 35x + 10 = 0$$
$$3x^2 - 7x + 2 = 0 \qquad \text{Dividing by 5}$$
$$(3x - 1)(x - 2) = 0$$
$$3x - 1 = 0 \quad \text{or} \quad x - 2 = 0$$
$$3x = 1 \quad \text{or} \quad x = 2$$
$$x = \dfrac{1}{3} \quad \text{or} \quad x = 2$$

The number 2 checks but $\dfrac{1}{3}$ does not. The solution is 2.

87. $\sqrt{y^2 + 6} + y - 3 = 0$

$$\sqrt{y^2 + 6} = 3 - y$$

$$(\sqrt{y^2 + 6})^2 = (3 - y)^2$$

$$y^2 + 6 = 9 - 6y + y^2$$

$$-3 = -6y$$

$$\frac{1}{2} = y$$

The number $\frac{1}{2}$ checks and is the solution.

89.

$$\sqrt{y + 1} - \sqrt{2y - 5} = \sqrt{y - 2}$$

$$(\sqrt{y + 1} - \sqrt{2y - 5})^2 = (\sqrt{y - 2})^2$$

$$y + 1 - 2\sqrt{(y + 1)(2y - 5)} + 2y - 5 = y - 2$$

$$-2\sqrt{2y^2 - 3y - 5} = -2y + 2$$

$$\sqrt{2y^2 - 3y - 5} = y - 1$$

Dividing by -2

$$(\sqrt{2y^2 - 3y - 5})^2 = (y - 1)^2$$

$$2y^2 - 3y - 5 = y^2 - 2y + 1$$

$$y^2 - y - 6 = 0$$

$$(y - 3)(y + 2) = 0$$

$$y - 3 = 0 \quad or \quad y + 2 = 0$$

$$y = 3 \quad or \qquad y = -2$$

The number 3 checks but -2 does not. The solution is 3.

Exercise Set 6.7

1. $a = 3, \quad b = 5$

Find c.

$c^2 = a^2 + b^2$ Pythagorean equation

$c^2 = 3^2 + 5^2$ Substituting

$c^2 = 9 + 25$

$c^2 = 34$

$c = \sqrt{34}$ Exact answer

$c \approx 5.831$ Approximation

3. $a = 15, \quad b = 15$

Find c.

$c^2 = a^2 + b^2$ Pythagorean equation

$c^2 = 15^2 + 15^2$ Substituting

$c^2 = 225 + 225$

$c^2 = 450$

$c = \sqrt{450}$ Exact answer

$c \approx 21.213$ Approximation

5. $b = 12, \quad c = 13$

Find a.

$a^2 + b^2 = c^2$ Pythagorean equation

$a^2 + 12^2 = 13^2$ Substituting

$a^2 + 144 = 169$

$a^2 = 25$

$a = 5$

7. $c = 7, \quad a = \sqrt{6}$

Find b.

$c^2 = a^2 + b^2$ Pythagorean equation

$7^2 = (\sqrt{6})^2 + b^2$ Substituting

$49 = 6 + b^2$

$43 = b^2$

$\sqrt{43} = b$ Exact answer

$6.557 \approx b$ Approximation

9. $b = 1, \quad c = \sqrt{13}$

Find a.

$a^2 + b^2 = c^2$ Pythagorean equation

$a^2 + 1^2 = (\sqrt{13})^2$ Substituting

$a^2 + 1 = 13$

$a^2 = 12$

$a = \sqrt{12}$ Exact answer

$a \approx 3.464$ Approximation

11. $a = 1, \quad c = \sqrt{n}$

Find b.

$a^2 + b^2 = c^2$

$1^2 + b^2 = (\sqrt{n})^2$

$1 + b^2 = n$

$b^2 = n - 1$

$b = \sqrt{n - 1}$

13. We make a drawing and let $d =$ the length of the guy wire.

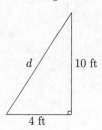

We use the Pythagorean equation to find d.

$d^2 = 4^2 + 10^2$

$d^2 = 16 + 100$

$d^2 = 116$

$d = \sqrt{116}$

$d \approx 10.770$

The wire is $\sqrt{116}$ ft, or about 10.770 ft long.

15. $L = \dfrac{0.000169 d^{2.27}}{h}$

$L = \dfrac{0.000169 (200)^{2.27}}{4}$

≈ 7.1

The length of the letters should be about 7.1 ft.

17. We add labels to the drawing in the text. We let h = the height of the bulge.

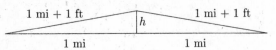

Note that 1 mi = 5280 ft, so 1 mi + 1 ft = 5280 + 1, or 5281 ft.

We use the Pythagorean equation to find h.

$$5281^2 = 5280^2 + h^2$$
$$27,888,961 = 27,878,400 + h^2$$
$$10,561 = h^2$$
$$\sqrt{10,561} = h$$
$$102.767 \approx h$$

The bulge is $\sqrt{10,561}$ ft, or about 102.767 ft high.

19. We add some labels to the drawing in the text.

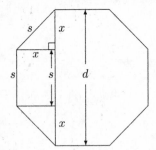

Note that $d = s + 2x$. We use the Pythagorean equation to find x.

$$x^2 + x^2 = s^2$$
$$2x^2 = s^2$$
$$x^2 = \frac{s^2}{2}$$
$$x = \sqrt{\frac{s^2}{2}}$$
$$x = \frac{s}{\sqrt{2}}$$
$$x = \frac{s\sqrt{2}}{2} \quad \text{Rationalizing the denominator}$$

Then $d = s + 2x = s + 2\left(\dfrac{s\sqrt{2}}{2}\right) = s + s\sqrt{2}$.

21. We make a drawing. Let x = the width of the rectangle. Then $x + 1$ = the length.

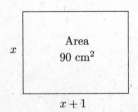

We first find the length and width of the rectangle. Recall the formula for the area of a rectangle, $A = lw$. We substitute 90 for A, $x + 1$ for l, and x for w in this formula and solve for x.

$$90 = (x + 1)x$$
$$90 = x^2 + x$$
$$0 = x^2 + x - 90$$
$$0 = (x + 10)(x - 9)$$
$$x + 10 = 0 \quad \text{or} \quad x - 9 = 0$$
$$x = -10 \quad \text{or} \quad x = 9$$

Since the width cannot be negative, we know that the width is 9 cm. Thus the length is 10 cm. (These numbers check since 9 and 10 are consecutive integers and the area of a rectangle with width 9 cm and length 10 cm is $10 \cdot 9$, or 90 cm^2.)

Now we find the length of the diagonal of the rectangle. We make another drawing, letting d = the length of the diagonal.

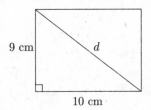

We use the Pythagorean equation to find d.

$$d^2 = 9^2 + 10^2$$
$$d^2 = 81 + 100$$
$$d^2 = 181$$
$$d = \sqrt{181}$$
$$d \approx 13.454$$

The length of the diagonal is $\sqrt{181}$ cm, or about 13.454 cm.

23. We use the drawing in the text, replacing w with 16 in.

We use the Pythagorean equation to find h.

$$h^2 + 16^2 = 20^2$$
$$h^2 + 256 = 400$$
$$h^2 = 144$$
$$h = 12$$

The height is 12 in.

25. We first make a drawing. A point on the x-axis has coordinates $(x, 0)$ and is $|x|$ units from the origin.

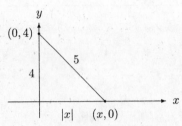

We use the Pythagorean equation to find x.

$$4^2 + |x|^2 = 5^2$$
$$16 + x^2 = 25 \quad |x|^2 = x^2$$
$$x^2 - 9 = 0 \quad \text{Subtracting 25}$$
$$(x + 3)(x - 3) = 0$$
$$x - 3 = 0 \quad \text{or} \quad x + 3 = 0$$
$$x = 3 \quad \text{or} \quad x = -3$$

The points are $(3, 0)$ and $(-3, 0)$.

27. We make a drawing, letting $d =$ the distance the wire will run diagonally, disregarding the slack.

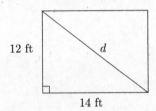

We use the Pythagorean equation to find d.

$$d^2 = 12^2 + 14^2$$
$$d^2 = 144 + 196$$
$$d^2 = 340$$
$$d = \sqrt{340}$$

Adding 4 ft of slack on each end, we find that a wire of length $\sqrt{340} + 4 + 4$, or $\sqrt{340} + 8$ ft should be purchased. This is approximately 26.439 ft of wire.

29. Referring to the drawing in the text, we let $t =$ the travel. Then we use the Pythagorean equation.

$$t^2 = (17.75)^2 + (10.25)^2$$
$$t^2 = 315.0625 + 105.0625$$
$$t^2 = 420.125$$
$$t = \sqrt{420.125}$$
$$t \approx 20.497$$

The travel is $\sqrt{420.125}$ in., or about 20.497 in.

31. Discussion and Writing Exercise

33. *Familiarize*. Let $r =$ the speed of the Carmel Crawler. Then $r + 14 =$ the speed of the Zionsville Flash. We organize the information in a table.

	Distance	Speed	Time
Crawler	230	r	t
Flash	290	$r + 14$	t

Translate. Using the formula $t = d/r$ and noting that the times are the same, we have

$$\frac{230}{r} = \frac{290}{r + 14}.$$

Solve. We first clear fractions by multiplying by the LCM of the denominators, $r(r + 14)$.

$$r(r + 14) \cdot \frac{230}{r} = r(r + 14) \cdot \frac{290}{r + 14}$$
$$230(r + 14) = 290r$$
$$230r + 3220 = 290r$$
$$3220 = 60r$$
$$\frac{161}{3} = r, \text{ or}$$
$$53\frac{2}{3} = r$$

If $r = 53\frac{2}{3}$, then $r + 14 = 67\frac{2}{3}$.

Check. At $53\frac{2}{3}$, or $\frac{161}{3}$ mph, the Crawler travels 230 mi in $\frac{230}{161/3}$, or about 4.3 hr. At $67\frac{2}{3}$, or $\frac{203}{3}$ mph, the Flash travels 290 mi in $\frac{290}{203/3}$, or about 4.3 hr. Since the times are the same, the answer checks.

State. The Carmel Crawler's speed is $53\frac{2}{3}$ mph, and the Zionsville Flash's speed is $67\frac{2}{3}$ mph.

35.
$$2x^2 + 11x - 21 = 0$$
$$(2x - 3)(x + 7) = 0$$
$$2x - 3 = 0 \quad \text{or} \quad x + 7 = 0$$
$$x = \frac{3}{2} \quad \text{or} \quad x = -7$$

The solutions are $\frac{3}{2}$ and -7.

37.
$$\frac{x + 2}{x + 3} = \frac{x - 4}{x - 5},$$
$$\text{LCM is } (x + 3)(x - 5)$$
$$(x + 3)(x - 5) \cdot \frac{x + 2}{x + 3} = (x + 3)(x - 5) \cdot \frac{x - 4}{x - 5}$$
$$(x - 5)(x + 2) = (x + 3)(x - 4)$$
$$x^2 - 3x - 10 = x^2 - x - 12$$
$$-2x = -2$$
$$x = 1$$

The number 1 checks, so it is the solution.

39.
$$\frac{x-5}{x-7} = \frac{4}{3}, \text{ LCM is } 3(x-7)$$

$$3(x-7) \cdot \frac{x-5}{x-7} = 3(x-7) \cdot \frac{4}{3}$$

$$3(x-5) = 4(x-7)$$

$$3x - 15 = 4x - 28$$

$$13 = x$$

The number 13 checks and is the solution.

41.

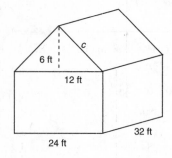

$$c^2 = 6^2 + 12^2 = 36 + 144 = 180$$

$$c = \sqrt{180} \text{ ft}$$

Area of the roof $= 2 \cdot \sqrt{180} \cdot 32 = 64\sqrt{180} \text{ ft}^2$

Number of packets $= \dfrac{64\sqrt{180}}{33\frac{1}{3}} \approx 26$

Kit should buy 26 packets of shingles.

43.

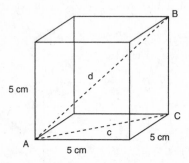

First find the length of a diagonal of the base of the cube. It is the hypotenuse of an isosceles right triangle with $a = 5$ cm. Then $c = a\sqrt{2} = 5\sqrt{2}$ cm.

Triangle ABC is a right triangle with legs of $5\sqrt{2}$ cm and 5 cm and hypotenuse d. Use the Pythagorean equation to find d, the length of the diagonal that connects two opposite corners of the cube.

$$d^2 = (5\sqrt{2})^2 + 5^2$$

$$d^2 = 25 \cdot 2 + 25$$

$$d^2 = 50 + 25$$

$$d^2 = 75$$

$$d = \sqrt{75}$$

Exact answer: $d = \sqrt{75}$ cm

Exercise Set 6.8

1. $\sqrt{-35} = \sqrt{-1 \cdot 35} = \sqrt{-1} \cdot \sqrt{35} = i\sqrt{35}$, or $\sqrt{35}i$

3. $\sqrt{-16} = \sqrt{-1 \cdot 16} = \sqrt{-1} \cdot \sqrt{16} = i \cdot 4 = 4i$

5. $-\sqrt{-12} = -\sqrt{-1 \cdot 12} = -\sqrt{-1} \cdot \sqrt{12} = -i \cdot 2\sqrt{3} = -2\sqrt{3}i$, or $-2i\sqrt{3}$

7. $\sqrt{-3} = \sqrt{-1 \cdot 3} = \sqrt{-1} \cdot \sqrt{3} = i\sqrt{3}$, or $\sqrt{3}i$

9. $\sqrt{-81} = \sqrt{-1 \cdot 81} = \sqrt{-1} \cdot \sqrt{81} = i \cdot 9 = 9i$

11. $\sqrt{-98} = \sqrt{-1 \cdot 98} = \sqrt{-1} \cdot \sqrt{98} = i \cdot 7\sqrt{2} = 7\sqrt{2}i$, or $7i\sqrt{2}$

13. $-\sqrt{-49} = -\sqrt{-1 \cdot 49} = -\sqrt{-1} \cdot \sqrt{49} = -i \cdot 7 = -7i$

15. $4 - \sqrt{-60} = 4 - \sqrt{-1 \cdot 60} = 4 - \sqrt{-1} \cdot \sqrt{60} = 4 - i \cdot 2\sqrt{15} = 4 - 2\sqrt{15}i$, or $4 - 2i\sqrt{15}$

17. $\sqrt{-4} + \sqrt{-12} = \sqrt{-1 \cdot 4} + \sqrt{-1 \cdot 12} = \sqrt{-1} \cdot \sqrt{4} + \sqrt{-1} \cdot \sqrt{12} = i \cdot 2 + i \cdot 2\sqrt{3} = (2 + 2\sqrt{3})i$

19. $\quad (7 + 2i) + (5 - 6i)$
$= (7 + 5) + (2 - 6)i \quad$ Collecting like terms
$= 12 - 4i$

21. $\quad (4 - 3i) + (5 - 2i)$
$= (4 + 5) + (-3 - 2)i \quad$ Collecting like terms
$= 9 - 5i$

23. $(9 - i) + (-2 + 5i) = (9 - 2) + (-1 + 5)i$
$= 7 + 4i$

25. $(6 - i) - (10 + 3i) = (6 - 10) + (-1 - 3)i$
$= -4 - 4i$

27. $(4 - 2i) - (5 - 3i) = (4 - 5) + [-2 - (-3)]i$
$= -1 + i$

29. $(9 + 5i) - (-2 - i) = [9 - (-2)] + [5 - (-1)]i$
$= 11 + 6i$

31. $\sqrt{-36} \cdot \sqrt{-9} = \sqrt{-1} \cdot \sqrt{36} \cdot \sqrt{-1} \cdot \sqrt{9}$
$= i \cdot 6 \cdot i \cdot 3$
$= i^2 \cdot 18$
$= -1 \cdot 18 \qquad i^2 = -1$
$= -18$

33. $\sqrt{-7} \cdot \sqrt{-2} = \sqrt{-1} \cdot \sqrt{7} \cdot \sqrt{-1} \cdot \sqrt{2}$
$= i \cdot \sqrt{7} \cdot i \cdot \sqrt{2}$
$= i^2(\sqrt{14})$
$= -1(\sqrt{14}) \qquad i^2 = -1$
$= -\sqrt{14}$

35. $-3i \cdot 7i = -21 \cdot i^2$
$$= -21(-1) \qquad i^2 = -1$$
$$= 21$$

37. $-3i(-8 - 2i) = -3i(-8) - 3i(-2i)$
$$= 24i + 6i^2$$
$$= 24i + 6(-1) \qquad i^2 = -1$$
$$= 24i - 6$$
$$= -6 + 24i$$

39. $(3 + 2i)(1 + i)$
$$= 3 + 3i + 2i + 2i^2 \quad \text{Using FOIL}$$
$$= 3 + 3i + 2i - 2 \quad i^2 = -1$$
$$= 1 + 5i$$

41. $(2 + 3i)(6 - 2i)$
$$= 12 - 4i + 18i - 6i^2 \quad \text{Using FOIL}$$
$$= 12 - 4i + 18i + 6 \quad i^2 = -1$$
$$= 18 + 14i$$

43. $(6 - 5i)(3 + 4i) = 18 + 24i - 15i - 20i^2$
$$= 18 + 24i - 15i + 20$$
$$= 38 + 9i$$

45. $(7 - 2i)(2 - 6i) = 14 - 42i - 4i + 12i^2$
$$= 14 - 42i - 4i - 12$$
$$= 2 - 46i$$

47. $(3 - 2i)^2 = 3^2 - 2 \cdot 3 \cdot 2i + (2i)^2 \quad \text{Squaring a binomial}$
$$= 9 - 12i + 4i^2$$
$$= 9 - 12i - 4 \qquad i^2 = -1$$
$$= 5 - 12i$$

49. $(1 + 5i)^2$
$$= 1^2 + 2 \cdot 1 \cdot 5i + (5i)^2 \quad \text{Squaring a binomial}$$
$$= 1 + 10i + 25i^2$$
$$= 1 + 10i - 25 \qquad i^2 = -1$$
$$= -24 + 10i$$

51. $(-2 + 3i)^2 = 4 - 12i + 9i^2 = 4 - 12i - 9 =$
$$-5 - 12i$$

53. $i^7 = i^6 \cdot i = (i^2)^3 \cdot i = (-1)^3 \cdot i = -1 \cdot i = -i$

55. $i^{24} = (i^2)^{12} = (-1)^{12} = 1$

57. $i^{42} = (i^2)^{21} = (-1)^{21} = -1$

59. $i^9 = (i^2)^4 \cdot i = (-1)^4 \cdot i = 1 \cdot i = i$

61. $i^6 = (i^2)^3 = (-1)^3 = -1$

63. $(5i)^3 = 5^3 \cdot i^3 = 125 \cdot i^2 \cdot i = 125(-1)(i) = -125i$

65. $7 + i^4 = 7 + (i^2)^2 = 7 + (-1)^2 = 7 + 1 = 8$

67. $i^{28} - 23i = (i^2)^{14} - 23i = (-1)^{14} - 23i = 1 - 23i$

69. $i^2 + i^4 = -1 + (i^2)^2 = -1 + (-1)^2 = -1 + 1 = 0$

71. $i^5 + i^7 = i^4 \cdot i + i^6 \cdot i = (i^2)^2 \cdot i + (i^2)^3 \cdot i =$
$$(-1)^2 \cdot i + (-1)^3 \cdot i = 1 \cdot i + (-1)i = i - i = 0$$

73. $1 + i + i^2 + i^3 + i^4 = 1 + i + i^2 + i^2 \cdot i + (i^2)^2$
$$= 1 + i + (-1) + (-1) \cdot i + (-1)^2$$
$$= 1 + i - 1 - i + 1$$
$$= 1$$

75. $5 - \sqrt{-64} = 5 - \sqrt{-1} \cdot \sqrt{64} = 5 - i \cdot 8 = 5 - 8i$

77. $\dfrac{8 - \sqrt{-24}}{4} = \dfrac{8 - \sqrt{-1} \cdot \sqrt{24}}{4} = \dfrac{8 - i \cdot 2\sqrt{6}}{4} =$
$$\dfrac{2(4 - i\sqrt{6})}{2 \cdot 2} = \dfrac{\cancel{2}(4 - i\sqrt{6})}{\cancel{2} \cdot 2} = \dfrac{4 - i\sqrt{6}}{2} = 2 - \dfrac{\sqrt{6}}{2}i$$

79. $\dfrac{4 + 3i}{3 - i} = \dfrac{4 + 3i}{3 - i} \cdot \dfrac{3 + i}{3 + i}$
$$= \dfrac{(4 + 3i)(3 + i)}{(3 - i)(3 + i)}$$
$$= \dfrac{12 + 4i + 9i + 3i^2}{9 - i^2}$$
$$= \dfrac{12 + 13i - 3}{9 - (-1)}$$
$$= \dfrac{9 + 13i}{10}$$
$$= \dfrac{9}{10} + \dfrac{13}{10}i$$

81. $\dfrac{3 - 2i}{2 + 3i} = \dfrac{3 - 2i}{2 + 3i} \cdot \dfrac{2 - 3i}{2 - 3i}$
$$= \dfrac{(3 - 2i)(2 - 3i)}{(2 + 3i)(2 - 3i)}$$
$$= \dfrac{6 - 9i - 4i + 6i^2}{4 - 9i^2}$$
$$= \dfrac{6 - 13i - 6}{4 - 9(-1)}$$
$$= \dfrac{-13i}{13}$$
$$= -i$$

83. $\dfrac{8 - 3i}{7i} = \dfrac{8 - 3i}{7i} \cdot \dfrac{-7i}{-7i}$
$$= \dfrac{-56i + 21i^2}{-49i^2}$$
$$= \dfrac{-21 - 56i}{49}$$
$$= -\dfrac{21}{49} - \dfrac{56}{49}i$$
$$= -\dfrac{3}{7} - \dfrac{8}{7}i$$

85. $\dfrac{4}{3+i} = \dfrac{4}{3+i} \cdot \dfrac{3-i}{3-i}$

$= \dfrac{12-4i}{9-i^2}$

$= \dfrac{12-4i}{9-(-1)}$

$= \dfrac{12-4i}{10}$

$= \dfrac{12}{10} - \dfrac{4}{10}i$

$= \dfrac{6}{5} - \dfrac{2}{5}i$

87. $\dfrac{2i}{5-4i} = \dfrac{2i}{5-4i} \cdot \dfrac{5+4i}{5+4i}$

$= \dfrac{10i+8i^2}{25-16i^2}$

$= \dfrac{10i+8(-1)}{25-16(-1)}$

$= \dfrac{-8+10i}{41}$

$= -\dfrac{8}{41} + \dfrac{10}{41}i$

89. $\dfrac{4}{3i} = \dfrac{4}{3i} \cdot \dfrac{-3i}{-3i}$

$= \dfrac{-12i}{-9i^2}$

$= \dfrac{-12i}{-9(-1)}$

$= \dfrac{-12i}{9}$

$= -\dfrac{4}{3}i$

91. $\dfrac{9-4i}{8i} = \dfrac{2-4i}{8i} \cdot \dfrac{-8i}{-8i}$

$= \dfrac{-16i+32i^2}{-64i^2}$

$= \dfrac{-16i+32(-1)}{-64(-1)}$

$= \dfrac{-32-16i}{64}$

$= -\dfrac{32}{64} - \dfrac{16}{64}i$

$= -\dfrac{1}{2} - \dfrac{1}{4}i$

93. $\dfrac{6+3i}{6-3i} = \dfrac{6+3i}{6-3i} \cdot \dfrac{6+3i}{6+3i}$

$= \dfrac{36+18i+18i+9i^2}{36-9i^2}$

$= \dfrac{36+36i-9}{36-9(-1)}$

$= \dfrac{27+36i}{45}$

$= \dfrac{27}{45} + \dfrac{36}{45}i$

$= \dfrac{3}{5} + \dfrac{4}{5}i$

95. Substitute $1-2i$ for x in the equation.

$$x^2 - 2x + 5 = 0$$

$\begin{array}{c|c} (1-2i)^2 - 2(1-2i) + 5 \ ? \ 0 & \\ 1 - 4i + 4i^2 - 2 + 4i + 5 & \\ 1 - 4i - 4 - 2 + 4i + 5 & \\ 0 & \text{TRUE} \end{array}$

$1-2i$ is a solution.

97. Substitute $2+i$ for x in the equation.

$$x^2 - 4x - 5 = 0$$

$\begin{array}{c|c} (2+i)^2 - 4(2+i) - 5 \ ? \ 0 & \\ 4 + 4i + i^2 - 8 - 4i - 5 & \\ 4 + 4i - 1 - 8 - 4i - 5 & \\ -10 & \text{FALSE} \end{array}$

$2+i$ is not a solution.

99. Discussion and Writing Exercise

101. An expression that consists of the quotient of two polynomials, where the polynomial in the denominator is nonzero, is called a <u>rational</u> expression.

103. When being graphed, the numbers in an ordered pair are called <u>coordinates</u>.

105. An equality of ratios, $A/B = C/D$, read "A is to B as C is to D" is called a <u>proportion</u>.

107. <u>Negative</u> numbers do not have real-number square roots.

109. $g(2i) = \dfrac{(2i)^4 - (2i)^2}{2i-1} = \dfrac{16i^4 - 4i^2}{-1+2i} = \dfrac{20}{-1+2i} =$

$\dfrac{20}{-1+2i} \cdot \dfrac{-1-2i}{-1-2i} = \dfrac{-20-40i}{5} = -4 - 8i;$

$g(i+1) = \dfrac{(i+1)^4 - (i+1)^2}{(i+1)-1} =$

$\dfrac{(i+1)^2[(i+1)^2 - 1]}{i} = \dfrac{2i(2i-1)}{i} = 2(2i-1) =$

$-2 + 4i;$

$g(2i-1) = \dfrac{(2i-1)^4 - (2i-1)^2}{(2i-1)-1} =$

$\dfrac{(2i-1)^2[(2i-1)^2 - 1]}{2i-2} = \dfrac{(-3-4i)(-4-4i)}{-2+2i} =$

$$\frac{(-3-4i)(-2-2i)}{-1+i} = \frac{-2+14i}{-1+i} =$$

$$\frac{-2+14i}{-1+i} \cdot \frac{-1-i}{-1-i} = \frac{16-12i}{2} = 8-6i$$

111. $\frac{1}{8}\left(-24-\sqrt{-1024}\right) = \frac{1}{8}(-24-32i) = -3-4i$

113. $7\sqrt{-64} - 9\sqrt{-256} = 7\cdot 8i - 9\cdot 16i = 56i - 144i = -88i$

115. $(1-i)^3(1+i)^3 =$

$(1-i)(1+i)\cdot(1-i)(1+i)\cdot(1-i)(1+i) =$

$(1-i^2)(1-i^2)(1-i^2) = (1+1)(1+1)(1+1) =$

$2\cdot 2\cdot 2 = 8$

117. $\frac{6}{1+\dfrac{3}{i}} = \frac{6}{\dfrac{i+3}{i}} = \frac{6i}{i+3} = \frac{6i}{i+3}\cdot\frac{-i+3}{-i+3} =$

$\frac{-6i^2+18i}{-i^2+9} = \frac{6+18i}{10} = \frac{6}{10} + \frac{18}{10}i = \frac{3}{5} + \frac{9}{5}i$

119. $\frac{i-i^{38}}{1+i} = \frac{i-(i^2)^{19}}{1+i} = \frac{i-(-1)^{19}}{1+i} = \frac{i-(-1)}{1+i} =$

$\frac{i+1}{1+i} = 1$

Chapter 6 Review Exercises

1. $\sqrt{778} \approx 27.893$

2. $\sqrt{\dfrac{963.2}{23.68}} \approx 6.378$

3. $f(x) = \sqrt{3x-16}$

$f(0) = \sqrt{3\cdot 0 - 16} = \sqrt{-16}$

> Since negative numbers do not have real-number square roots, $f(0)$ does not exist as a real number.

$f(-1) = \sqrt{3(-1)-16} = \sqrt{-3-16} = \sqrt{-19}$

> Since negative numbers do not have real-number square roots, $f(-1)$ does not exist as a real number.

$f(1) = \sqrt{3\cdot 1 - 16} = \sqrt{3-16} = \sqrt{-13}$

> Since negative numbers do not have real-number square roots, $f(1)$ does not exist as a real number.

$f\left(\dfrac{41}{3}\right) = \sqrt{3\cdot\dfrac{41}{3}-16} = \sqrt{41-16} = \sqrt{25} = 5$

4. The domain of $f(x) = \sqrt{3x-16}$ is the set of all x-values for which $3x-16 \ge 0$.

$3x - 16 \ge 0$

$3x \ge 16$

$x \ge \dfrac{16}{3}$

The domain is $\left\{x\middle|x \ge \dfrac{16}{3}\right\}$, or $\left[\dfrac{16}{3},\infty\right)$.

5. $\sqrt{81a^2} = \sqrt{(9a)^2} = |9a| = |9|\cdot|a| = 9|a|$

6. $\sqrt{(-7z)^2} = |-7z| = |-7|\cdot|z| = 7|z|$

7. $\sqrt{(c-3)^2} = |c-3|$

8. $\sqrt{x^2-6x+9} = \sqrt{(x-3)^2} = |x-3|$

9. $\sqrt[3]{-1000} = -10$ Since $(-10)^3 = -1000$

10. $\sqrt[3]{-\dfrac{1}{27}} = -\dfrac{1}{3}$ Since $\left(-\dfrac{1}{3}\right)^2 = -\dfrac{1}{27}$

11. $f(x) = \sqrt[3]{x+2}$

$f(6) = \sqrt[3]{6+2} = \sqrt[3]{8} = 2$

$f(-10) = \sqrt[3]{-10+2} = \sqrt[3]{-8} = -2$

$f(25) = \sqrt[3]{25+2} = \sqrt[3]{27} = 3$

12. $\sqrt[10]{x^{10}} = |x|$

13. $-\sqrt[13]{(-3)^{13}} = -(-3) = 3$

14. $a^{1/5} = \sqrt[5]{a}$

15. $64^{3/2} = (\sqrt{64})^3 = 8^3 = 512$

16. $\sqrt{31} = 31^{1/2}$

17. $\sqrt[5]{a^2b^3} = (a^2b^3)^{1/5}$

18. $49^{-1/2} = \dfrac{1}{49^{1/2}} = \dfrac{1}{\sqrt{49}} = \dfrac{1}{7}$

19. $(8xy)^{-2/3} = \dfrac{1}{(8xy)^{2/3}} = \dfrac{1}{8^{2/3}x^{2/3}y^{2/3}} =$

$\dfrac{1}{(2^3)^{2/3}x^{2/3}y^{2/3}} = \dfrac{1}{2^2x^{2/3}y^{2/3}} = \dfrac{1}{4x^{2/3}y^{2/3}}$

20. $5a^{-3/4}b^{1/2}c^{-2/3} = 5\cdot\dfrac{1}{a^{3/4}}\cdot b^{1/2}\cdot\dfrac{1}{c^{2/3}} =$

$\dfrac{5b^{1/2}}{a^{3/4}c^{2/3}}$

21. $\dfrac{3a}{\sqrt[4]{t}} = \dfrac{3a}{t^{1/4}}$

22. $(x^{-2/3})^{3/5} = x^{(-2/3)(3/5)} = x^{-2/5} = \dfrac{1}{x^{2/5}}$

23. $\dfrac{7^{-1/3}}{7^{-1/2}} = 7^{-1/3-(-1/2)} = 7^{-1/3+1/2} = 7^{-2/6+3/6} = 7^{1/6}$

24. $\sqrt[3]{x^{21}} = (x^{21})^{1/3} = x^7$

25. $\sqrt[3]{27x^6} = \sqrt[3]{(3x^2)^3} = [(3x^2)^3]^{1/3} = 3x^2$

26. $x^{1/3}y^{1/4} = x^{4/12}y^{3/12} = (x^4y^3)^{1/12} = \sqrt[12]{x^4y^3}$

27. $\sqrt[4]{x}\sqrt[3]{x} = x^{1/4}\cdot x^{1/3} = x^{1/4+1/3} = x^{3/12+4/12} = x^{7/12} = \sqrt[12]{x^7}$

28. $\sqrt{245} = \sqrt{49\cdot 5} = \sqrt{49}\sqrt{5} = 7\sqrt{5}$

29. $\sqrt[3]{-108} = \sqrt[3]{-27\cdot 4} = \sqrt[3]{-27}\sqrt[3]{4} = -3\sqrt[3]{4}$

30. $\sqrt[3]{250a^2b^6} = \sqrt[3]{125b^6\cdot 2a^2} = \sqrt[3]{125b^6}\sqrt[3]{2a^2}$

$5b^2\sqrt[3]{2a^2}$

31. $\sqrt{\dfrac{49}{36}} = \dfrac{\sqrt{49}}{\sqrt{36}} = \dfrac{7}{6}$

32. $\sqrt[3]{\dfrac{64x^6}{27}} = \dfrac{\sqrt[3]{64x^6}}{\sqrt[3]{27}} = \dfrac{4x^2}{3}$, or $\dfrac{4}{3}x^2$

33. $\sqrt[4]{\dfrac{16x^8}{81y^{12}}} = \dfrac{\sqrt[4]{16x^8}}{\sqrt[4]{81y^{12}}} = \dfrac{2x^2}{3y^3}$

34. $\sqrt{5x}\sqrt{3y} = \sqrt{5x \cdot 3y} = \sqrt{15xy}$

35. $\sqrt[3]{a^5b}\,\sqrt[3]{27b} = \sqrt[3]{27a^5b^2} = \sqrt[3]{27a^3 \cdot a^2b^2} =$
$3a\sqrt[3]{a^2b^2}$

36. $\sqrt[3]{a}\,\sqrt[5]{b^3} = a^{1/3}b^{3/5} = a^{5/15}b^{9/15} = (a^5b^9)^{1/15} =$
$\sqrt[15]{a^5b^9}$

37. $\dfrac{\sqrt[3]{60xy^3}}{\sqrt[3]{10x}} = \sqrt[3]{\dfrac{60xy^3}{10x}} = \sqrt[3]{6y^3} = y\sqrt[3]{6}$

38. $\dfrac{\sqrt{75x}}{2\sqrt{3}} = \dfrac{1}{2}\sqrt{\dfrac{75x}{3}} = \dfrac{1}{2}\sqrt{25x} = \dfrac{1}{2} \cdot 5\sqrt{x} = \dfrac{5}{2}\sqrt{x}$

39. $\dfrac{\sqrt[3]{x^2}}{\sqrt[4]{x}} = \dfrac{x^{2/3}}{x^{1/4}} = x^{2/3-1/4} = x^{8/12-3/12} = x^{5/12} =$
$\sqrt[12]{x^5}$

40. $5\sqrt[3]{x} + 2\sqrt[3]{x} = (5+2)\sqrt[3]{x} = 7\sqrt[3]{x}$

41. $2\sqrt{75} - 7\sqrt{3} = 2\sqrt{25 \cdot 3} - 7\sqrt{3} = 2\sqrt{25} \cdot \sqrt{3} - 7\sqrt{3} =$
$2 \cdot 5\sqrt{3} - 7\sqrt{3} = 10\sqrt{3} - 7\sqrt{3} = 3\sqrt{3}$

42. $\sqrt[3]{8x^4} + \sqrt[3]{xy^6} = \sqrt[3]{8x^3 \cdot x} + \sqrt[3]{y^6 \cdot x} =$
$\sqrt[3]{8x^3}\,\sqrt[3]{x} + \sqrt[3]{y^6}\,\sqrt[3]{x} = 2x\sqrt[3]{x} + y^2\sqrt[3]{x} =$
$(2x + y^2)\sqrt[3]{x}$

43. $\sqrt{50} + 2\sqrt{18} + \sqrt{32} = \sqrt{25 \cdot 2} + 2\sqrt{9 \cdot 2} + \sqrt{16 \cdot 2} =$
$\sqrt{25}\sqrt{2} + 2\sqrt{9}\sqrt{2} + \sqrt{16}\sqrt{2} = 5\sqrt{2} + 2 \cdot 3\sqrt{2} + 4\sqrt{2} =$
$5\sqrt{2} + 6\sqrt{2} + 4\sqrt{2} = 15\sqrt{2}$

44. $(\sqrt{5} - 3\sqrt{8})(\sqrt{5} + 2\sqrt{8})$
$= (\sqrt{5})^2 + \sqrt{5} \cdot 2\sqrt{8} - 3\sqrt{8} \cdot \sqrt{5} - 3\sqrt{8} \cdot 2\sqrt{8}$
$= 5 + 2\sqrt{40} - 3\sqrt{40} - 6 \cdot 8$
$= 5 - \sqrt{40} - 48$
$= -43 - \sqrt{4 \cdot 10}$
$= -43 - 2\sqrt{10}$

45. $(1 - \sqrt{7})^2 = 1^2 - 2 \cdot 1 \cdot \sqrt{7} + (\sqrt{7})^2$
$= 1 - 2\sqrt{7} + 7$
$= 8 - 2\sqrt{7}$

46. $(\sqrt[3]{27} - \sqrt[3]{2})(\sqrt[3]{27} + \sqrt[3]{2})$
$= (3 - \sqrt[3]{2})(3 + \sqrt[3]{2})$
$= 9 - (\sqrt[3]{2})^2$
$= 9 - \sqrt[3]{2^2}$
$= 9 - \sqrt[3]{4}$

47. $\sqrt{\dfrac{8}{3}} = \sqrt{\dfrac{8}{3} \cdot \dfrac{3}{3}} = \sqrt{\dfrac{24}{9}} = \dfrac{\sqrt{24}}{\sqrt{9}} = \dfrac{\sqrt{4 \cdot 6}}{3} = \dfrac{2\sqrt{6}}{3}$

48. $\dfrac{2}{\sqrt{a} + \sqrt{b}} = \dfrac{2}{\sqrt{a} + \sqrt{b}} \cdot \dfrac{\sqrt{a} - \sqrt{b}}{\sqrt{a} - \sqrt{b}} = \dfrac{2(\sqrt{a} - \sqrt{b})}{(\sqrt{a})^2 - (\sqrt{b})^2} =$
$\dfrac{2\sqrt{a} - 2\sqrt{b}}{a - b}$

49. $\qquad \sqrt[4]{x + 3} = 2$
$\qquad (\sqrt[4]{x + 3})^4 = 2^4$
$\qquad\qquad x + 3 = 16$
$\qquad\qquad\quad x = 13$

The number 13 checks, so it is the solution.

50. $\qquad\qquad 1 + \sqrt{x} = \sqrt{3x - 3}$
$\qquad\quad (1 + \sqrt{x})^2 = (\sqrt{3x - 3})^2$
$\qquad 1 + 2\sqrt{x} + x = 3x - 3$
$\qquad\qquad 2\sqrt{x} = 2x - 4$
$\qquad\qquad \sqrt{x} = x - 2 \qquad$ Dividing by 2
$\qquad\quad (\sqrt{x})^2 = (x - 2)^2$
$\qquad\qquad\quad x = x^2 - 4x + 4$
$\qquad\qquad\quad 0 = x^2 - 5x + 4$
$\qquad\qquad\quad 0 = (x - 1)(x - 4)$
$x - 1 = 0 \ \ or \ \ x - 4 = 0$
$\quad x = 1 \ \ or \qquad\quad x = 4$

Since 4 checks but 1 does not, the solution is 4.

51. $\qquad\qquad x - 3 = \sqrt{5 - x}$
$\qquad\quad (x - 3)^2 = (\sqrt{5 - x})^2$
$\qquad x^2 - 6x + 9 = 5 - x$
$\qquad x^2 - 5x + 4 = 0$
$\qquad (x - 1)(x - 4) = 0$
$x - 1 = 0 \ \ or \ \ x - 4 = 0$
$\quad x = 1 \ \ or \qquad\quad x = 4$

Since 4 checks but 1 does not, the solution is 4.

52. Let $s =$ the length of a side of the square, in cm. We use the Pythagorean equation.
$\qquad s^2 + s^2 = (9\sqrt{2})^2$
$\qquad\quad 2s^2 = 81 \cdot 2$
$\qquad\quad\ s^2 = 81$
$\qquad\qquad s = 9$

The length of a side of the square is 9 cm.

53. Let $w =$ the width of the bookcase, in ft. Then we can refer to the drawing in the text, replacing "?" with w. We use the Pythagorean equation.
$\qquad 5^2 + w^2 = 7^2$
$\qquad 25 + w^2 = 49$
$\qquad\qquad w^2 = 24$
$\qquad\qquad\ w = \sqrt{24}$
$\qquad\qquad\ w \approx 4.899$

The width of the bookcase is $\sqrt{24}$ ft, or approximately 4.899 ft.

54.
$$d(n) = 0.75\sqrt{2.8n}$$
$$81 = 0.75\sqrt{2.8n}$$
$$108 = \sqrt{2.8n} \quad \text{Dividing by 0.75}$$
$$(108)^2 = (\sqrt{2.8n})^2$$
$$11,664 = 2.8n$$
$$4166 \approx n$$

The engine produces peak power at about 4166 rpm's.

55.
$$d(n) = 0.75\sqrt{2.8n}$$
$$84 = 0.75\sqrt{2.8n}$$
$$112 = \sqrt{2.8n} \quad \text{Dividing by 0.75}$$
$$(112)^2 = (\sqrt{2.8n})^2$$
$$12,544 = 2.8n$$
$$4480 = n$$

The engine produces peak power at 4480 rpm's.

56. $a = 7$, $b = 24$

Find c.
$$a^2 + b^2 = c^2$$
$$7^2 + 24^2 = c^2$$
$$49 + 576 = c^2$$
$$625 = c^2$$
$$25 = c$$

57. $a = 2$, $c = 5\sqrt{2}$

Find b.
$$a^2 + b^2 = c^2$$
$$2^2 + b^2 = (5\sqrt{2})^2$$
$$4 + b^2 = 25 \cdot 2$$
$$4 + b^2 = 50$$
$$b^2 = 46$$
$$b = \sqrt{46} \quad \text{Exact answer}$$
$$b \approx 6.782 \quad \text{Approximation}$$

58. $\sqrt{-25} + \sqrt{-8} = \sqrt{-1 \cdot 25} + \sqrt{-1 \cdot 4 \cdot 2} =$
$\sqrt{-1} \cdot \sqrt{25} + \sqrt{-1} \cdot \sqrt{4} \cdot \sqrt{2} = i \cdot 5 + i \cdot 2 \cdot \sqrt{2} =$
$5i + 2\sqrt{2}i = (5 + 2\sqrt{2})i$

59. $(-4 + 3i) + (2 - 12i) = (-4 + 2) + (3 - 12)i =$
$-2 - 9i$

60. $(4 - 7i) - (3 - 8i) = (4 - 3) + [-7 - (-8)]i = 1 + i$

61. $(2 + 5i)(2 - 5i) = 2^2 - (5i)^2$
$$= 4 - 25i^2$$
$$= 4 + 25 \quad i^2 = -1$$
$$= 29$$

62. $i^{13} = i^{12} \cdot i = (i^2)^6 \cdot i = (-1)^6 \cdot i = 1 \cdot i = i$

63.
$$(6 - 3i)(2 - i)$$
$$= 12 - 6i - 6i + 3i^2 \quad \text{Using FOIL}$$
$$= 12 - 12i - 3 \quad\quad i^2 = -1$$
$$= 9 - 12i$$

64.
$$\frac{-3 + 2i}{5i} = \frac{-3 + 2i}{5i} \cdot \frac{-5i}{-5i}$$
$$= \frac{15i - 10i^2}{-25i^2}$$
$$= \frac{15i + 10}{25}$$
$$= \frac{10}{25} + \frac{15}{25}i$$
$$= \frac{2}{5} + \frac{3}{5}i$$

65.
$$\frac{6 - 3i}{2 - i} = \frac{6 - 3i}{2 - i} \cdot \frac{2 + i}{2 + i}$$
$$= \frac{12 + 6i - 6i - 3i^2}{4 - i^2}$$
$$= \frac{12 + 3}{4 + 1}$$
$$= \frac{15}{5}$$
$$= 3$$

This exercise might also be done as follows:
$$\frac{6 - 3i}{2 - i} = \frac{3(2 - i)}{2 - i} = \frac{3}{1} \cdot \frac{2 - i}{2 - i} = 3$$

66.
$$\begin{array}{c|c} \multicolumn{2}{c}{x^2 + x + 2 = 0} \\ \hline (1 + i)^2 + (1 + i) + 2 \; ? \; 0 & \\ 1 + 2i + i^2 + 1 + i + 2 & \\ 1 + 2i - 1 + 1 + i + 2 & \\ 3 + 3i & \text{FALSE} \end{array}$$

$1 + i$ is not a solution.

67. Graph: $f(x) = \sqrt{x}$.

We find some ordered pairs, plot points, and draw the curve.

x	$f(x)$	$(x, f(x))$
0	0	$(0, 0)$
1	1	$(1, 1)$
2	1.4	$(2, 1.4)$
3	1.7	$(3, 1.7)$
4	2	$(4, 2)$
5	2.2	$(5, 2.2)$

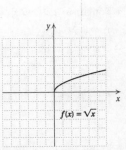

$f(x) = \sqrt{x}$

68. *Discussion and Writing Exercise.* The procedure for solving radical equations is to isolate one of the radical terms, use the principle of powers, repeat these steps if necessary until all radicals are eliminated, solve, and then check the possible solutions. A check is necessary since the principle of powers does not always yield equivalent equations.

69. $i \cdot i^2 \cdot i^3 \ldots i^{99} \cdot i^{100}$

Pairing i and i^{100}, i^2 and i^{99}, i^3 and i^{98}, and so on, we have 50 pairs of factors equivalent to i^{101}. Thus we have

$(i^{101})^{50} = i^{5050} = (i^2)^{2525} = (-1)^{2525} = -1$.

70.
$$\sqrt{11x + \sqrt{6 + x}} = 6$$
$$(\sqrt{11x + \sqrt{6 + x}})^2 = 6^2$$
$$11x + \sqrt{6 + x} = 36$$
$$\sqrt{6 + x} = 36 - 11x$$
$$(\sqrt{6 + x})^2 = (36 - 11x)^2$$
$$6 + x = 1296 - 792x + 121x^2$$
$$0 = 121x^2 - 793x + 1290$$
$$0 = (121x - 430)(x - 3)$$

$121x - 430 = 0 \quad or \quad x - 3 = 0$
$121x = 430 \quad or \qquad x = 3$
$x = \dfrac{430}{121} \quad or \qquad x = 3$

Since 3 checks but $\dfrac{430}{121}$ does not, the solution is 3.

Chapter 6 Test

1. $\sqrt{148} \approx 12.166$

2. $f(x) = \sqrt{8 - 4x}$
$f(1) = \sqrt{8 - 4 \cdot 1} = \sqrt{8 - 4} = \sqrt{4} = 2$
$f(3) = \sqrt{8 - 4 \cdot 3} = \sqrt{8 - 12} = \sqrt{-4}$

Since negative numbers do not have real-number square roots, $f(3)$ does not exist as a real number.

3. The domain of $f(x) = \sqrt{8 - 4x}$ is the set of all x-values for which $8 - 4x \geq 0$.
$8 - 4x \geq 0$
$8 \geq 4x$
$2 \geq x$

The domain is $\{x | x \leq 2\}$, or $(-\infty, 2]$.

4. $\sqrt{(-3q)^2} = |-3q| = |-3| \cdot |q| = 3|q|$

5. $\sqrt{x^2 + 10x + 25} = \sqrt{(x + 5)^2} = |x + 5|$

6. $\sqrt[3]{-\dfrac{1}{1000}} = -\dfrac{1}{10}$ \quad Since $\left(-\dfrac{1}{10}\right)^3 = -\dfrac{1}{1000}$

7. $\sqrt[5]{x^5} = x$

We do not use absolute-value notation when the index is odd.

8. $\sqrt[10]{(-4)^{10}} = |-4| = 4$

9. $a^{2/3} = \sqrt[3]{a^2}$

10. $32^{3/5} = \sqrt[5]{32^3} = \sqrt[5]{(2^5)^3} = \sqrt[5]{2^{15}} = 2^3 = 8$

11. $\sqrt{37} = 37^{1/2}$

12. $(\sqrt{5xy^2})^5 = [(5xy^2)^{1/2}]^5 = (5xy^2)^{5/2}$

13. $1000^{-1/3} = \dfrac{1}{1000^{1/3}} = \dfrac{1}{\sqrt[3]{1000}} = \dfrac{1}{10}$

14. $8a^{3/4}b^{-3/2}c^{-2/5} = 8 \cdot a^{3/4} \cdot \dfrac{1}{b^{3/2}} \cdot \dfrac{1}{c^{2/5}} = \dfrac{8a^{3/4}}{b^{3/2}c^{2/5}}$

15. $(x^{2/3}y^{-3/4})^{12/5} = (x^{2/3})^{12/5}(y^{-3/4})^{12/5} =$
$x^{24/15}y^{-36/20} = x^{8/5}y^{-9/5} = \dfrac{x^{8/5}}{y^{9/5}}$

16. $\dfrac{2.9^{-5/8}}{2.9^{2/3}} = 2.9^{-5/8-2/3} = 2.9^{-15/24-16/24} =$
$2.9^{-31/24} = \dfrac{1}{2.9^{31/24}}$

17. $\sqrt[8]{x^2} = x^{2/8} = x^{1/4} = \sqrt[4]{x}$

18. $\sqrt[4]{16x^6} = (16x^6)^{1/4} = (2^4x^6)^{1/4} = (2^4)^{1/4}(x^6)^{1/4} =$
$2x^{3/2} = 2\sqrt{x^3} = 2x\sqrt{x}$

19. $a^{2/5}b^{1/3} = a^{6/15}b^{5/15} = (a^6b^5)^{1/15} = \sqrt[15]{a^6b^5}$

20. $\sqrt[4]{2y}\sqrt[3]{y} = (2y)^{1/4}(y^{1/3}) = (2y)^{3/12}(y^{4/12}) =$
$[(2y)^3(y^4)]^{1/12} = (8y^3 \cdot y^4)^{1/12} = (8y^7)^{1/12} =$
$\sqrt[12]{8y^7}$

21. $\sqrt{148} = \sqrt{4 \cdot 37} = \sqrt{4}\sqrt{37} = 2\sqrt{37}$

22. $\sqrt[4]{80} = \sqrt[4]{16 \cdot 5} = \sqrt[4]{16}\sqrt[4]{5} = 2\sqrt[4]{5}$

23. $\sqrt[3]{24a^{11}b^{13}} = \sqrt[3]{8a^9b^{12} \cdot 3a^2b} =$
$\sqrt[3]{8a^9b^{12}}\sqrt[3]{3a^2b} = 2a^3b^4\sqrt[3]{3a^2b}$

24. $\sqrt[3]{\dfrac{16x^5}{y^7}} = \dfrac{\sqrt[3]{16x^5}}{\sqrt[3]{y^7}} = \dfrac{\sqrt[3]{8x^3 \cdot 2x^2}}{\sqrt[3]{y^6 \cdot y}} =$
$\dfrac{\sqrt[3]{8x^3}\sqrt[3]{2x^2}}{\sqrt[3]{y^6}\sqrt[3]{y}} = \dfrac{2x\sqrt[3]{2x^2}}{y^2\sqrt[3]{y}}, \text{ or } \dfrac{2x}{y^2}\sqrt[3]{\dfrac{2x^2}{y}}$

If we rationalize the denominator, we get $\dfrac{2x\sqrt[3]{2x^2y^2}}{y^3}$.

25. $\sqrt{\dfrac{25x^2}{36y^4}} = \dfrac{\sqrt{25x^2}}{\sqrt{36y^4}} = \dfrac{5x}{6y^2}$

26. $\sqrt[3]{2x}\sqrt[3]{5y^2} = \sqrt[3]{2x \cdot 5y^2} = \sqrt[3]{10xy^2}$

27. $\sqrt[4]{x^3y^2}\sqrt{xy} = (x^3y^2)^{1/4}(xy)^{1/2} = x^{3/4}y^{1/2}x^{1/2}y^{1/2} =$
$x^{3/4+1/2}y^{1/2+1/2} = x^{5/4}y = y\sqrt[4]{x^5} = xy\sqrt[4]{x}$

28. $\dfrac{\sqrt[5]{x^3y^4}}{\sqrt[5]{xy^2}} = \sqrt[5]{\dfrac{x^3y^4}{xy^2}} = \sqrt[5]{x^2y^2}$

29. $\dfrac{\sqrt{300a}}{5\sqrt{3}} = \dfrac{1}{5}\sqrt{\dfrac{300a}{3}} = \dfrac{1}{5}\sqrt{100a} = \dfrac{1}{5} \cdot 10\sqrt{a} = 2\sqrt{a}$

30. $3\sqrt{128} + 2\sqrt{18} + 2\sqrt{32}$

$= 3\sqrt{64 \cdot 2} + 2\sqrt{9 \cdot 2} + 2\sqrt{16 \cdot 2}$

$= 3\sqrt{64}\sqrt{2} + 2\sqrt{9}\sqrt{2} + 2\sqrt{16}\sqrt{2}$

$= 3 \cdot 8\sqrt{2} + 2 \cdot 3\sqrt{2} + 2 \cdot 4\sqrt{2}$

$= 24\sqrt{2} + 6\sqrt{2} + 8\sqrt{2}$

$= 38\sqrt{2}$

31. $(\sqrt{20} + 2\sqrt{5})(\sqrt{20} - 3\sqrt{5})$

$= \sqrt{20}\sqrt{20} - \sqrt{20} \cdot 3\sqrt{5} + 2\sqrt{5}\sqrt{20} - 2\sqrt{5} \cdot 3\sqrt{5}$

$\qquad\qquad\qquad\qquad$ Using FOIL

$= 20 - 3\sqrt{100} + 2\sqrt{100} - 6 \cdot 5$

$= 20 - 3 \cdot 10 + 2 \cdot 10 - 30$

$= 20 - 30 + 20 - 30$

$= -20$

32. $(3 + \sqrt{x})^2 = 3^2 + 2 \cdot 3 \cdot \sqrt{x} + (\sqrt{x})^2 = 9 + 6\sqrt{x} + x$

33. $\dfrac{1 + \sqrt{2}}{3 - 5\sqrt{2}} = \dfrac{1 + \sqrt{2}}{3 - 5\sqrt{2}} \cdot \dfrac{3 + 5\sqrt{2}}{3 + 5\sqrt{2}}$

$= \dfrac{3 + 5\sqrt{2} + 3\sqrt{2} + 5 \cdot 2}{9 - 25 \cdot 2}$

$= \dfrac{3 + 8\sqrt{2} + 10}{9 - 50}$

$= \dfrac{13 + 8\sqrt{2}}{-41}$

34. $\sqrt[5]{x - 3} = 2$

$(\sqrt[5]{x - 3})^5 = 2^5$

$x - 3 = 32$

$x = 35$

The number 35 checks, so it is the solution.

35. $\sqrt{x - 6} = \sqrt{x + 9} - 3$

$(\sqrt{x - 6})^2 = (\sqrt{x + 9} - 3)^2$

$x - 6 = x + 9 - 6\sqrt{x + 9} + 9$

$-24 = -6\sqrt{x + 9}$

$4 = \sqrt{x + 9} \qquad$ Dividing by -6

$4^2 = (\sqrt{x + 9})^2$

$16 = x + 9$

$7 = x$

The number 7 checks, so it is the solution.

36. $\sqrt{x - 1} + 3 = x$

$\sqrt{x - 1} = x - 3$

$(\sqrt{x - 1})^2 = (x - 3)^2$

$x - 1 = x^2 - 6x + 9$

$0 = x^2 - 7x + 10$

$0 = (x - 2)(x - 5)$

$x - 2 = 0 \quad or \quad x - 5 = 0$

$x = 2 \quad or \qquad x = 5$

The number 2 does not check, but 5 does. The solution is 5.

37. We make a drawing, letting $s =$ the length of a side of the square, in feet.

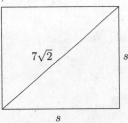

We use the Pythagorean equations.

$s^2 + s^2 = (7\sqrt{2})^2$

$2s^2 = 49 \cdot 2$

$2s^2 = 98$

$s^2 = 49$

$s = \sqrt{49} = 7$

The length of a side of the square is 7 ft.

38. $\quad V = 1.2\sqrt{h}$

$72 = 1.2\sqrt{h}$

$60 = \sqrt{h}$

$60^2 = (\sqrt{h})^2$

$3600 = h$

The airplane is 3600 ft high.

39. $a = 7$, $b = 7$; find c.

$a^2 + b^2 = c^2 \quad$ Pythagorean equation

$7^2 + 7^2 = c^2 \quad$ Substituting

$49 + 49 = c^2$

$98 = c^2$

$\sqrt{98} = c \qquad$ Exact answer

$9.899 \approx c \qquad$ Approximation

40. $a = 1$, $c = \sqrt{5}$; find b.

$a^2 + b^2 = c^2 \qquad$ Pythagorean equation

$1^2 + b^2 = (\sqrt{5})^2 \quad$ Substituting

$1 + b^2 = 5$

$b^2 = 4$

$b = 2$

41. $\sqrt{-9} + \sqrt{-64} = \sqrt{-1 \cdot 9} + \sqrt{-1 \cdot 64} = i \cdot 3 + i \cdot 8 = 3i + 8i = 11i$

42. $(5 + 8i) - (-2 + 3i) = [5 - (-2)] + (8i - 3i) = 7 + 5i$

43. $(3 - 4i)(3 + 7i) = 9 + 21i - 12i - 28i^2 \quad$ FOIL

$\qquad\qquad\qquad = 9 + 21i - 12i + 28 \quad i^2 = -1$

$\qquad\qquad\qquad = 37 + 9i$

44. $i^{95} = i^{94} \cdot i = (i^2)^{47} \cdot i = (-1)^{47} \cdot i = -i$

45. $\dfrac{-7+14i}{6-8i} = \dfrac{-7+14i}{6-8i} \cdot \dfrac{6+8i}{6+8i}$

$= \dfrac{-42-56i+84i+112i^2}{36-64i^2}$

$= \dfrac{-42-56i+84i-112}{36+64}$

$= \dfrac{-154+28i}{100}$

$= -\dfrac{154}{100} + \dfrac{28}{100}i$

$= -\dfrac{77}{50} + \dfrac{7}{25}i$

46. Substitute $1+2i$ for x in the equation.

$$x^2 + 2x + 5 = 0$$

$$\overline{(1+2i)^2 + 2(1+2i) + 5 \overset{?}{\,} 0}$$

$1 + 4i + 4i^2 + 2 + 4i + 5$

$1 + 4i - 4 + 2 + 4i + 5$ $\qquad\Big|$

$\qquad\qquad\qquad 4 + 8i \quad\Big|\quad$ FALSE

$1+2i$ is not a solution.

47. $\dfrac{1-4i}{4i(1+4i)^{-1}} = \dfrac{(1-4i)(1+4i)}{4i}$

$= \dfrac{1-16i^2}{4i}$

$= \dfrac{1+16}{4i}$

$= \dfrac{17}{4i}$

$= \dfrac{17}{4i} \cdot \dfrac{-4i}{-4i}$

$= \dfrac{-68i}{-16i^2}$

$= \dfrac{-68i}{16}$

$= -\dfrac{17i}{4}$, or $-\dfrac{17}{4}i$

48.

$\sqrt{2x-2} + \sqrt{7x+4} = \sqrt{13x+10}$

$(\sqrt{2x-2} + \sqrt{7x+4})^2 = (\sqrt{13x+10})^2$

$2x-2+2\sqrt{(2x-2)(7x+4)}+7x+4 = 13x+10$

$9x + 2 + 2\sqrt{14x^2 - 6x - 8} = 13x + 10$

$2\sqrt{14x^2 - 6x - 8} = 4x + 8$

$\sqrt{14x^2 - 6x - 8} = 2x+4 \quad$ Dividing by 2

$(\sqrt{14x^2 - 6x - 8})^2 = (2x+4)^2$

$14x^2 - 6x - 8 = 4x^2 + 16x + 16$

$10x^2 - 22x - 24 = 0$

$5x^2 - 11x - 12 = 0 \quad$ Dividing by 2

$(5x+4)(x-3) = 0$

$5x+4 = 0 \quad or \quad x-3 = 0$

$5x = -4 \quad or \qquad x = 3$

$x = -\dfrac{4}{5} \quad or \qquad x = 3$

The number $-\dfrac{4}{5}$ does not check, but 3 does. The solution is 3.

Cumulative Review Chapters R - 6

1. $N(n) = n^2 - n$

$N(6) = 6^2 - 6 = 36 - 6 = 30$ games

2. *Familiarize.* We will use the formula $N(n) = n^2 - n$.

Translate. Substitute 72 for $N(n)$.

$72 = n^2 - n$

Solve. We solve the equation.

$72 = n^2 - n$

$0 = n^2 - n - 72$

$0 = (n-9)(n+8)$

$n - 9 = 0 \quad or \quad n + 8 = 0$

$n = 9 \quad or \qquad n = -8$

Check. Since the number of teams cannot be negative, we check only 9. We have $N(9) = 9^2 - 9 = 81 - 9 = 72$, so the answer checks.

State. There are 9 teams in the league.

3. $-\dfrac{3}{5}\left(-\dfrac{7}{10}\right) = \dfrac{3 \cdot 7}{5 \cdot 10} = \dfrac{21}{50}$

4. $36.2 - 73.4 = 36.2 + (-73.4) = -37.2$

5. $\quad 7c - [4 - 3(6c - 9)]$

$= 7c - [4 - 18c + 27]$

$= 7c - [31 - 18c]$

$= 7c - 31 + 18c$

$= 25c - 31$

6. $6^2 - 3^3 \cdot 2 + 3 \cdot 8^2 = 36 - 27 \cdot 2 + 3 \cdot 64$

$= 36 - 54 + 192$

$= -18 + 192$

$= 174$

7. $\left(\dfrac{4x^4 y^{-5}}{3x^{-3}}\right)^3 = \left(\dfrac{4x^7 y^{-5}}{3}\right)^3 = \dfrac{4^3 \cdot (x^7)^3 \cdot (y^{-5})^3}{3^3} =$

$\dfrac{64x^{21} y^{-15}}{27} = \dfrac{64x^{21}}{27y^{15}}$

8. $(2x^2 - 3x + 1) + (6x - 3x^3 + 7x^2 - 4) =$

$-3x^3 + 9x^2 + 3x - 3$

9. $(2x^2 - y)^2 = (2x^2)^2 - 2 \cdot 2x^2 \cdot y + y^2$

$= 4x^4 - 4x^2 y + y^2$

10.

$\qquad\qquad 5x^2 \;- 2x + 1$

$\qquad\qquad 3x^2 \;+\; x \;- 2$

$\overline{\qquad\qquad -10x^2 + 4x - 2}$

$\qquad\qquad 5x^3 - 2x^2 + x$

$\underline{15x^4 - 6x^3 + 3x^2}$

$\overline{15x^4 - x^3 - 9x^2 + 5x - 2}$

11. $\dfrac{x^3 + 64}{x^2 - 49} \cdot \dfrac{x^2 - 14x + 49}{x^2 - 4x + 16}$

$= \dfrac{(x^3 + 64)(x^2 - 14x + 49)}{(x^2 - 49)(x^2 - 4x + 16)}$

$= \dfrac{(x + 4)(x^2 - 4x + 16)(x - 7)(x - 7)}{(x + 7)(x - 7)(x^2 - 4x + 16)}$

$= \dfrac{(x^2 - 4x + 16)(x - 7)}{(x^2 - 4x + 16)(x - 7)} \cdot \dfrac{(x + 4)(x - 7)}{x + 7}$

$= \dfrac{(x + 4)(x - 7)}{x + 7}$

12. $\dfrac{x}{x + 2} + \dfrac{1}{x - 3} - \dfrac{x^2 - 2}{x^2 - x - 6}$

$= \dfrac{x}{x + 2} + \dfrac{1}{x - 3} - \dfrac{x^2 - 2}{(x + 2)(x - 3)}$,

$\qquad\qquad$ LCM is $(x + 2)(x - 3)$

$= \dfrac{x}{x + 2} \cdot \dfrac{x - 3}{x - 3} + \dfrac{1}{x - 3} \cdot \dfrac{x + 2}{x + 2} - \dfrac{x^2 - 2}{(x + 2)(x - 3)}$

$= \dfrac{x^2 - 3x + x + 2 - (x^2 - 2)}{(x + 2)(x - 3)}$

$= \dfrac{x^2 - 3x + x + 2 - x^2 + 2}{(x + 2)(x - 3)}$

$= \dfrac{-2x + 4}{(x + 2)(x - 3)}$

13. $\dfrac{\dfrac{y^2 - 5y - 6}{y^2 - 7y - 18}}{\dfrac{y^2 + 3y + 2}{y^2 + 4y + 4}}$

$= \dfrac{y^2 - 5y - 6}{y^2 - 7y - 18} \cdot \dfrac{y^2 + 4y + 4}{y^2 + 3y + 2}$

$= \dfrac{(y^2 - 5y - 6)(y^2 + 4y + 4)}{(y^2 - 7y - 18)(y^2 + 3y + 2)}$

$= \dfrac{(y - 6)(y + 1)(y + 2)(y + 2)}{(y - 9)(y + 2)(y + 1)(y + 2)}$

$= \dfrac{(y - 6)\cancel{(y + 1)}\cancel{(y + 2)}\cancel{(y + 2)}}{(y - 9)\cancel{(y + 2)}\cancel{(y + 1)}\cancel{(y + 2)}}$

$= \dfrac{y - 6}{y - 9}$

14.

$$y + 2\ \overline{\smash{\big)}\ y^3 + 3y^2 + 0y - 5}$$

with quotient $y^2 + y - 2$:

$\ \underline{y^3 + 2y^2}$

$\ y^2$

$\ \underline{y^2 + 2y}$

$\ -2y - 5$

$\ \underline{-2y - 4}$

$\ -1$

The answer is $y^2 + y - 2 + \dfrac{-1}{y + 2}$.

15. $\sqrt[3]{-8x^3} = \sqrt[3]{(-2x)^3} = -2x$

16. $\sqrt{16x^2 - 32x + 16} = \sqrt{16(x^2 - 2x + 1)} =$
$\sqrt{16(x - 1)^2} = 4(x - 1)$

17. $9\sqrt{75} + 6\sqrt{12} = 9\sqrt{25 \cdot 3} + 6\sqrt{4 \cdot 3} =$
$9 \cdot 5\sqrt{3} + 6 \cdot 2\sqrt{3} = 45\sqrt{3} + 12\sqrt{3} = 57\sqrt{3}$

18. $\sqrt{2xy^2}\sqrt{8xy^3} = \sqrt{16x^2y^5} = \sqrt{16x^2y^4 \cdot y} =$
$4xy^2\sqrt{y}$

19. $\dfrac{3\sqrt{5}}{\sqrt{6} - \sqrt{3}} = \dfrac{3\sqrt{5}}{\sqrt{6} - \sqrt{3}} \cdot \dfrac{\sqrt{6} + \sqrt{3}}{\sqrt{6} + \sqrt{3}} = \dfrac{3\sqrt{30} + 3\sqrt{15}}{6 - 3} =$
$\dfrac{3\sqrt{30} + 3\sqrt{15}}{3} = \dfrac{3(\sqrt{30} + \sqrt{15})}{3} = \sqrt{30} + \sqrt{15}$

20. $\sqrt[6]{\dfrac{m^{12}n^{24}}{64}} = \left(\dfrac{m^{12}n^{24}}{2^6}\right)^{1/6} = \dfrac{m^2n^4}{2}$

21. $6^{2/9} \cdot 6^{2/3} = 6^{2/9} \cdot 6^{6/9} = 6^{8/9}$

22. $(6 + i) - (3 - 4i) = (6 - 3) + [1 - (-4)]i = 3 + 5i$

23. $\dfrac{2 - i}{6 + 5i} = \dfrac{2 - i}{6 + 5i} \cdot \dfrac{6 - 5i}{6 - 5i}$

$\qquad = \dfrac{12 - 10i - 6i + 5i^2}{36 - 25i^2}$

$\qquad = \dfrac{12 - 16i - 5}{36 + 25}$

$\qquad = \dfrac{7 - 6i}{61}$

$\qquad = \dfrac{7}{61} - \dfrac{16}{61}i$

24. $\dfrac{1}{5} + \dfrac{3}{10}x = \dfrac{4}{5}$

$\qquad \dfrac{3}{10}x = \dfrac{3}{5}$

$\qquad x = \dfrac{10}{3} \cdot \dfrac{3}{5}$

$\qquad x = \dfrac{30}{15} = 2$

The solution is 2.

25. $\qquad M = \dfrac{1}{8}(c - 3)$

$\qquad 8M = c - 3 \qquad$ Multiplying by 8

$\quad 8M + 3 = c$

26. $\qquad 3a - 4 < 10 + 5a$

$\qquad -2a - 4 < 10$

$\qquad\quad -2a < 14$

$\qquad\qquad a > -7 \quad$ Reversing the inequality symbol

The solution set is $\{a | a > -7\}$, or $(-7, \infty)$.

27. $\quad -8 < x + 2 < 15$

$\quad -10 < x < 13 \qquad$ Subtracting 2

The solution set is $\{x | -10 < x < 13\}$, or $(-10, 13)$.

28. $|3x - 6| = 2$

$\quad 3x - 6 = -2 \quad or \quad 3x - 6 = 2$

$\qquad\quad 3x = 4 \quad\ or \qquad\quad 3x = 8$

$\qquad\quad\ x = \dfrac{4}{3} \quad or \qquad\quad\ x = \dfrac{8}{3}$

The solutions are $\dfrac{4}{3}$ and $\dfrac{8}{3}$.

29. $3x + 5y = 30,$ (1)

$5x + 3y = 34$ (2)

We first multiply Equation (1) by 3 and multiply Equation (2) by -5 and then add.

$$9x + 15y = 90$$
$$\underline{-25x - 15y = -170}$$
$$-16x = -80$$
$$x = 5$$

Now substitute 5 for x in one of the original equations and solve for y.

$$3x + 5y = 30 \quad (1)$$
$$3 \cdot 5 + 5y = 30$$
$$15 + 5y = 30$$
$$5y = 15$$
$$y = 3$$

The solution is $(5, 3)$.

30. $3x + 2y - z = -7,$ (1)

$-x + y + 2z = 9,$ (2)

$5x + 5y + z = -1$ (3)

First we eliminate z from one pair of equations.

$$6x + 4y - 2z = -14 \quad \text{Multiplying (1) by 2}$$
$$\underline{-x + y + 2z = 9 \quad (2)}$$
$$5x + 5y = -5 \quad (4)$$

Next we add Equations (1) and (3).

$$3x + 2y - z = -7$$
$$\underline{5x + 5y + z = -1}$$
$$8x + 7y = -8 \quad (5)$$

Now we solve the system of Equations (4) and (5). We multiply Equation (4) by 7 and multiply Equation (5) by -5 and then add.

$$35x + 35y = -35$$
$$\underline{-40x - 35y = 40}$$
$$-5x = 5$$
$$x = -1$$

Next we use Equation (4) to find y.

$$5(-1) + 5y = -5 \quad \text{Substituting}$$
$$-5 + 5y = -5$$
$$5y = 0$$
$$y = 0$$

Finally we substitute -1 for x and 0 for y in Equation (3) and solve for z.

$$5(-1) + 5 \cdot 0 + z = -1$$
$$-5 + z = -1$$
$$z = 4$$

The solution is $(-1, 0, 4)$.

31. $625 = 49y^2$

$$0 = 49y^2 - 625$$
$$0 = (7y + 25)(7y - 25)$$
$$7y + 25 = 0 \quad or \quad 7y - 25 = 0$$
$$7y = -25 \quad or 7y = 25$$
$$y = -\frac{25}{7} \quad or y = \frac{25}{7}$$

The solutions are $-\dfrac{25}{7}$ and $\dfrac{25}{7}$.

32.
$$\frac{6x}{x-5} - \frac{300}{x^2 + 5x + 25} = \frac{2250}{x^3 - 125}$$
$$\frac{6x}{x-5} - \frac{300}{x^2+5x+25} = \frac{2250}{(x-5)(x^2+5x+25)}$$

LCM is $(x - 5)(x^2 + 5x + 25)$

$$(x-5)(x^2+5x+25)\left(\frac{6x}{x-5} - \frac{300}{x^2+5x+25}\right) =$$
$$(x-5)(x^2+5x+25) \cdot \frac{2250}{(x-5)(x^2+5x+25)}$$
$$6x(x^2 + 5x + 25) - 300(x - 5) = 2250$$
$$6x^3 + 30x^2 + 150x - 300x + 1500 = 2250$$
$$6x^3 + 30x^2 - 150x + 1500 = 2250$$
$$6x^3 + 30x^2 - 150x - 750 = 0$$
$$6(x^3 + 5x^2 - 25x - 125) = 0$$
$$6[x^2(x + 5) - 25(x + 5)] = 0$$
$$6(x^2 - 25)(x + 5) = 0$$
$$6(x + 5)(x - 5)(x + 5) = 0$$
$$x + 5 = 0 \quad or \quad x - 5 = 0 \quad or \quad x + 5 = 0$$
$$x = -5 \quad or x = 5 \quad or x = -5$$

The number -5 checks but 5 does not, so the solution is -5.

33.
$$\frac{3x^2}{x+2} + \frac{5x-22}{x-2} = \frac{-48}{x^2-4}$$

$$\frac{3x^2}{x+2} + \frac{5x-22}{x-2} = \frac{-48}{(x+2)(x-2)},$$
$$\text{LCM is } (x+2)(x-2)$$

$$(x+2)(x-2)\left(\frac{3x^2}{x+2} + \frac{5x-22}{x-2}\right) =$$
$$(x+2)(x-2) \cdot \frac{-48}{(x+2)(x-2)}$$

$$3x^2(x-2) + (x+2)(5x-22) = -48$$
$$3x^3 - 6x^2 + 5x^2 - 12x - 44 = -48$$
$$3x^3 - x^2 - 12x - 44 = -48$$
$$3x^3 - x^2 - 12x + 4 = 0$$
$$x^2(3x-1) - 4(3x-1) = 0$$
$$(x^2-4)(3x-1) = 0$$
$$(x+2)(x-2)(3x-1) = 0$$

$$x+2=0 \quad or \quad x-2=0 \quad or \quad 3x-1=0$$
$$x=-2 \quad or \qquad x=2 \quad or \qquad 3x=1$$
$$x=-2 \quad or \qquad x=2 \quad or \qquad x=\frac{1}{3}$$

The numbers -2 and 2 do not check but $\frac{1}{3}$ does, so the solution is $\frac{1}{3}$.

34.
$$I = \frac{nE}{R+nr}$$
$$(R+nr) \cdot I = nE \qquad \text{Multiplying by } R+nr$$
$$RI + nrI = nE$$
$$RI = nE - nrI$$
$$R = \frac{nE - nrI}{I}$$

35. $\sqrt{4x+1} - 2 = 3$
$$\sqrt{4x+1} = 5$$
$$(\sqrt{4x+1})^2 = 5^2$$
$$4x+1 = 25$$
$$4x = 24$$
$$x = 6$$

The number 6 checks, so it is the solution.

36. $2\sqrt{1-x} = \sqrt{5}$
$$(2\sqrt{1-x})^2 = (\sqrt{5})^2$$
$$4(1-x) = 5$$
$$4 - 4x = 5$$
$$-4x = 1$$
$$x = -\frac{1}{4}$$

The number $-\frac{1}{4}$ checks, so it is the solution.

37.
$$13 - x = 5 + \sqrt{x+4}$$
$$8 - x = \sqrt{x+4}$$
$$(8-x)^2 = (\sqrt{x+4})^2$$
$$64 - 16x + x^2 = x+4$$
$$x^2 - 17x + 60 = 0$$
$$(x-5)(x-12) = 0$$
$$x-5=0 \quad or \quad x-12=0$$
$$x=5 \quad or \qquad x=12$$

The number 5 checks but 12 does not, so the solution is 5.

38. Graph: $f(x) = -\frac{2}{3}x + 2$

Make a list of function values in a table.
$$f(-3) = -\frac{2}{3}(-3) + 2 = 2 + 2 = 4$$
$$f(0) = -\frac{2}{3} \cdot 0 + 2 = 2$$
$$f(3) = -\frac{2}{3} \cdot 3 + 2 = -2 + 2 = 0$$

x	$f(x)$
-3	4
0	2
3	0

Plot these points and connect them.

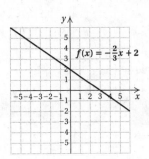

39. Graph: $4x - 2y = 8$

First we will find the intercepts. To find the x-intercept, let $y = 0$ and solve for x.
$$4x - 2 \cdot 0 = 8$$
$$4x = 8$$
$$x = 2$$

The x-intercept is $(2, 0)$.

To find the y-intercept, let $x = 0$ and solve for y.
$$4 \cdot 0 - 2y = 8$$
$$-2y = 8$$
$$y = -4$$

The y-intercept is $(0, -4)$.

We plot these points and draw the line.

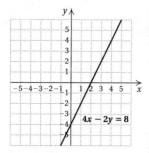

We use a third point as a check. Let $x = 4$.

$$4 \cdot 4 - 2y = 8$$
$$16 - 2y = 8$$
$$-2y = -8$$
$$y = 4$$

We plot $(4, 4)$ and note that it is on the line.

40. Graph: $4x \geq 5y + 20$

First we graph $4x = 5y + 20$. We draw the line solid since the inequality symbol is $\geq$. Test the point $(0, 0)$ to determine if it is a solution.

$$\frac{4x \geq 5y + 20}{4 \cdot 0 \ ? \ 5 \cdot 0 + 20}$$
$$0 \ \bigl| \ 20 \qquad \text{FALSE}$$

Since $0 \geq 20$ is false, we shade the half-plane that does not contain $(0, 0)$.

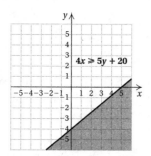

41. Graph: $y \geq -3$,
$\qquad\quad\ y \leq 2x + 3$

Graph the lines $y = -3$ and $y = 2x+3$ using solid lines. Indicate the region for each inequality by arrows, and shade the region where they overlap.

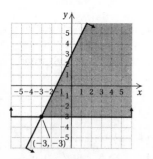

If we solve the system of related equations,

$$y = -3,$$
$$y = 2x + 3,$$

we find that the vertex is $(-3, -3)$.

42. Graph: $g(x) = x^2 - x - 2$

Make a list of function values in a table.

$$g(-2) = (-2)^2 - (-2) - 2 = 4 + 2 - 2 = 4$$
$$g(-1) = (-1)^2 - (-1) - 2 = 1 + 1 - 2 = 0$$
$$g(0) = 0^2 - 0 - 2 = -2$$
$$g(1) = 1^2 - 1 - 2 = 1 - 1 - 2 = -2$$
$$g(3) = 3^2 - 3 - 2 = 9 - 3 - 2 = 4$$

x	$g(x)$
-2	4
-1	0
0	-2
1	-2
3	4

Plot these points and connect them.

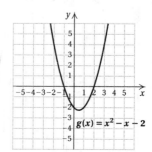

43. $f(x) = |x + 4|$

Make a list of function values in a table.

$$f(-5) = |-5 + 4| = |-1| = 1$$
$$f(-4) = |-4 + 4| = |0| = 0$$
$$f(-2) = |-2 + 4| = |2| = 2$$
$$f(0) = |0 + 4| = |4| = 4$$
$$f(1) = |1 + 4| = |5| = 5$$

x	$f(x)$
-5	1
-4	0
-2	2
0	4
1	5

Plot these points and connect them.

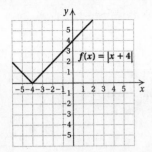

44. Graph: $g(x) = \dfrac{4}{x-3}$

Make a list of function values in a table. Note that we cannot choose $x = 3$ since it makes the denominator 0. The graph will have two branches.

$g(-5) = \dfrac{4}{-5-3} = \dfrac{4}{-8} = -\dfrac{1}{2}$

$g(-1) = \dfrac{4}{-1-3} = \dfrac{4}{-4} = -1$

$g(1) = \dfrac{4}{1-3} = \dfrac{4}{-2} = -2$

$g(4) = \dfrac{4}{4-3} = \dfrac{4}{1} = 4$

$g(5) = \dfrac{4}{5-3} = \dfrac{4}{2} = 2$

x	$g(x)$
-5	$-\dfrac{1}{2}$
-1	-1
1	-2
4	4
5	2

Plot these points and connect them.

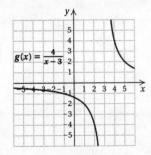

45. Graph: $f(x) = 2 - \sqrt{x}$

We find some ordered pairs, plot points, and draw the curve.

x	$f(x)$	$(x, f(x))$
0	2	$(0, 2)$
1	1	$(1, 1)$
3	0.3	$(3, 0.3)$
4	0	$(4, 0)$
5	-0.2	$(5, -0.2)$

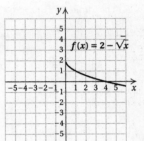

46. $12x^2y^2 - 30xy^3 = 6xy^2 \cdot 2x - 6xy^2 \cdot 5y$

$ = 6xy^2(2x - 5y)$

47. $3x^2 - 17x - 28$

We will use the *ac*-method.

1) There is no common factor (other than 1 or -1.).

2) Multiply the coefficients of the first and last terms.

$\qquad 3(-28) = -84$

3) Factor -84 so the sum of the factors is -17. The desired factorization is $4(-21)$.

4) Split the middle term as follows:

$\qquad -17x = 4x - 21x$

5) Factor by grouping:

$\qquad 3x^2 - 17x - 28 = 3x^2 + 4x - 21x - 28$
$\qquad\qquad\qquad\qquad\quad = x(3x + 4) - 7(3x + 4)$
$\qquad\qquad\qquad\qquad\quad = (x - 7)(3x + 4)$

48. $y^2 - y - 132$

We look for a pair of factors of -132 whose sum is -1. The numbers we need are 11 and -12.

$\qquad y^2 - y - 132 = (y + 11)(y - 12)$

49. $27y^3 + 8 = (3y)^3 + 2^3 = (3y + 2)(9y^2 - 6y + 4)$

50. $4x^2 - 625 = (2x)^2 - 25^2 = (2x + 25)(2x - 25)$

51. The domain is the set of all x-values in the graph. It is $\{-5, -4, -2, 0, 1, 2, 3, 4, 5\}$.

The range is the set of all y-values in the graph. It is $\{-3, 2, 3, 4\}$.

52. The set of all x-values in the graph extends from -5 to 5, so the domain is $\{x \mid -5 \le x \le 5\}$, or $[-5, 5]$.

The set of all y-values in the graph extends from -3 to 4, so the range is $\{y \mid -3 \le y \le 4\}$, or $[-3, 4]$.

53. No endpoints are indicated, so we see that the graph extends indefinitely horizontally. Thus, the domain is the set of all real numbers, or $(-\infty, \infty)$.

The smallest y-value is -5. No endpoints are indicated, so we see that the graph extends upward indefinitely from $(0, -5)$. Thus, the range is $\{y \mid y \ge -5\}$, or $[-5, \infty)$.

54. The set of all x-values in the graph extends from -5 to 5, so the domain is $\{x| -5 \le x \le 5\}$, or $[-5, 5]$.

The set of all y-values in the graph extends from 0 to 3, so the range is $\{y| 0 \le y \le 3\}$, or $[0, 3]$.

55. First we find the slope-intercept form of the equation by solving for y.

$$3x - 2y = 8$$
$$-2y = -3x + 8$$
$$y = \frac{-3x + 8}{-2}$$
$$y = \frac{3}{2}x - 4$$

The slope is $\frac{3}{2}$, and the y-intercept is $(0, -4)$.

56. First we find the slope of the given line.

$$3x - y = 5$$
$$-y = -3x + 5$$
$$y = 3x - 5$$

The slope of the given line is 3. The slope of the perpendicular line is the opposite of the reciprocal of 3, or $-\frac{1}{3}$.

We will use the point-slope equation, substituting 1 for x_1, 4 for y_1, and $-\frac{1}{3}$ for m.

$$y - y_1 = m(x - x_1)$$
$$y - 4 = -\frac{1}{3}(x - 1)$$
$$y - 4 = -\frac{1}{3}x + \frac{1}{3}$$
$$y = -\frac{1}{3}x + \frac{13}{3}$$

57. Familiarize. Let $t =$ the number of hours it will take the combines to harvest the field, working together.

Translate. We use the work principle.

$$\frac{t}{3} + \frac{t}{1.5} = 1$$

Solve. We solve the equation.

$$\frac{t}{3} + \frac{t}{1.5} = 1$$
$$3\left(\frac{t}{3} + \frac{t}{1.5}\right) = 3 \cdot 1 \quad \text{Clearing fractions}$$
$$t + 2t = 3$$
$$3t = 3$$
$$t = 1$$

Check. We verify the work principle.

$$\frac{1}{3} + \frac{1}{1.5} = \frac{1}{3} + \frac{1}{3/2} = \frac{1}{3} + 1 \cdot \frac{2}{3} = \frac{1}{3} + \frac{2}{3} = 1$$

State. It will take the combines 1 hr to harvest the field, working together.

58.
$$h = \frac{k}{b}$$
$$100 = \frac{k}{20}$$
$$2000 = k$$
$$h = \frac{2000}{b} \quad \text{Equation of variation}$$
$$h = \frac{2000}{16}$$
$$h = 125$$

When the base is 16 ft, the height is 125 ft.

We use the formula for the area of a triangle to find the fixed area.

$$A = \frac{1}{2}bh$$
$$= \frac{1}{2} \cdot 16 \cdot 125$$
$$= 1000 \text{ ft}^2$$

59. Familiarize. Let x and y represent the number of heart transplants performed at the University of Pittsburgh Medical Center and the Medical College of Virginia, respectively.

Translate. A total of 669 transplants were performed, so we have one equation:

$$x + y = 669$$

The number of transplants performed at the University of Pittsburgh was 33 more than twice the number performed at the Medical College of Virginia, so we have a second equation:

$$x = 33 + 2y$$

We have a system of equations:

$$x + y = 669, \quad (1)$$
$$x = 33 + 2y \quad (2)$$

Solve. We will use the substitution method. First we will substitute $33 + 2y$ for x in Equation (1) and solve for y.

$$(33 + 2y) + y = 669$$
$$33 + 3y = 669$$
$$3y = 636$$
$$y = 212$$

Now we substitute 212 for y in Equation (2) to find x.

$$x = 33 + 2 \cdot 212 = 33 + 424 = 457$$

Check. The total number of transplants is $457 + 212$, or 669. Also, 457 is 33 more than twice 212, so the answer checks.

State. 457 heart transplants were performed at the University of Pittsburgh Medical Center and 212 heart transplants were performed at the Medical College of Virginia.

60.
$$V = kd^2$$
$$4 = k \cdot 10^2$$
$$4 = k \cdot 100$$
$$0.04 = k$$
$$V = 0.04d^2 \qquad \text{Equation of variation}$$
$$V = 0.04(40)^2$$
$$V = 0.04(1600)$$
$$V = 64 \text{ L}$$

61. $\sqrt[5]{xy^4} = (xy^4)^{1/5}$

Answer (d) is correct.

62. Using the work principle, we translate to the equation
$$\frac{t}{3} + \frac{t}{15} = 1,$$

where t is the time required to fill the bin using both spouts together. Now we solve the equation.
$$15 \cdot \left(\frac{t}{3} + \frac{t}{15}\right) = 15 \cdot 1$$
$$5t + t = 15$$
$$6t = 15$$
$$t = \frac{5}{2}$$

This solution checks, so the answer (a) is correct.

63.
$$
\begin{array}{r|rrrr}
3 & 1 & -1 & 2 & 4 \\
  &   & 3 & 6 & 24 \\
\hline
  & 1 & 2 & 8 & 28 \\
\end{array}
$$

The answer is $x^2 + 2x + 8$, R 28, so answer (e) is correct.

64.
$$2x + 6 = 8 + \sqrt{5x + 1}$$
$$2x - 2 = \sqrt{5x + 1}$$
$$(2x - 2)^2 = (\sqrt{5x + 1})^2$$
$$4x^2 - 8x + 4 = 5x + 1$$
$$4x^2 - 13x + 3 = 0$$
$$(4x - 1)(x - 3) = 0$$
$$4x - 1 = 0 \quad or \quad x - 3 = 0$$
$$4x = 1 \quad or \quad x = 3$$
$$x = \frac{1}{4} \quad or \quad x = 3$$

The number $\frac{1}{4}$ does not check but 3 does, so answer (b) is correct.

65.
$$d = \sqrt{a^2 + b^2 + c^2}$$
$$d = \sqrt{2^2 + 4^2 + 5^2}$$
$$= \sqrt{4 + 16 + 25}$$
$$= \sqrt{45} \text{ ft}$$
$$\approx 6.708 \text{ ft}$$

66.
$$\frac{x + \sqrt{x + 1}}{x - \sqrt{x + 1}} = \frac{5}{11}$$
$$11(x - \sqrt{x+1}) \cdot \frac{x + \sqrt{x + 1}}{x - \sqrt{x + 1}} = 11(x - \sqrt{x+1}) \cdot \frac{5}{11}$$
$$11(x + \sqrt{x + 1}) = 5(x - \sqrt{x + 1})$$
$$11x + 11\sqrt{x + 1} = 5x - 5\sqrt{x + 1}$$
$$16\sqrt{x + 1} = -6x$$
$$8\sqrt{x + 1} = -3x \quad \text{Dividing by 2}$$
$$(8\sqrt{x + 1})^2 = (-3x)^2$$
$$64(x + 1) = 9x^2$$
$$64x + 64 = 9x^2$$
$$0 = 9x^2 - 64x - 64$$
$$0 = (9x + 8)(x - 8)$$

$$9x + 8 = 0 \quad or \quad x - 8 = 0$$
$$9x = -8 \quad or \quad x = 8$$
$$x = -\frac{8}{9} \quad or \quad x = 8$$

The number $-\frac{8}{9}$ checks but 8 does not, so the solution is $-\frac{8}{9}$.

Chapter 7

Quadratic Equations and Functions

Exercise Set 7.1

1. a) $6x^2 = 30$

$\quad\quad x^2 = 5$ Dividing by 6

$\quad\quad x = \sqrt{5}$ *or* $x = -\sqrt{5}$ Principle of square roots

Check: $\dfrac{6x^2 = 30}{6(\pm\sqrt{5})\ ?\ 30}$
$\quad\quad\quad\quad\quad\ \dfrac{\begin{array}{c}6\cdot 5\\30\end{array}}{}\ \bigg|$ TRUE

The solutions are $\sqrt{5}$ and $-\sqrt{5}$, or $\pm\sqrt{5}$.

b) The real-number solutions of the equation $6x^2 = 30$ are the first coordinates of the x-intercepts of the graph of $f(x) = 6x^2 - 30$. Thus, the x-intercepts are $(-\sqrt{5}, 0)$ and $(\sqrt{5}, 0)$.

3. a) $9x^2 + 25 = 0$

$\quad\quad 9x^2 = -25$ Subtracting 25

$\quad\quad x^2 = -\dfrac{25}{9}$ Dividing by 9

$\quad\quad x = \sqrt{-\dfrac{25}{9}}$ *or* $x = -\sqrt{-\dfrac{25}{9}}$ Principle of square roots

$\quad\quad x = \dfrac{5}{3}i$ *or* $x = -\dfrac{5}{3}i$ Simplifying

Check: $\dfrac{9x^2 + 25 = 0}{9\left(\pm\dfrac{5}{3}i\right) + 25\ ?\ 0}$
$\quad\quad\quad\quad\quad 9\left(-\dfrac{25}{9}\right) + 25 \quad\bigg|$
$\quad\quad\quad\quad\quad\quad\quad -25 + 25 \quad\bigg|$
$\quad\quad\quad\quad\quad\quad\quad\quad\quad\quad 0 \quad\bigg|$ TRUE

The solutions are $\dfrac{5}{3}i$ and $-\dfrac{5}{3}i$, or $\pm\dfrac{5}{3}i$.

b) Since the equation $9x^2 + 25 = 0$ has no real-number solutions, the graph of $f(x) = 9x^2 + 25$ has no x-intercepts.

5. $2x^2 - 3 = 0$

$\quad\quad 2x^2 = 3$

$\quad\quad x^2 = \dfrac{3}{2}$

$\quad\quad x = \sqrt{\dfrac{3}{2}}$ *or* $x = -\sqrt{\dfrac{3}{2}}$ Principle of square roots

$\quad\quad x = \sqrt{\dfrac{3}{2}\cdot\dfrac{2}{2}}$ *or* $x = -\sqrt{\dfrac{3}{2}\cdot\dfrac{2}{2}}$ Rationalizing denominators

$\quad\quad x = \dfrac{\sqrt{6}}{2}$ *or* $x = -\dfrac{\sqrt{6}}{2}$

Check: $\dfrac{2x^2 - 3 = 0}{2\left(\pm\dfrac{\sqrt{6}}{2}\right)^2 - 3\ ?\ 0}$
$\quad\quad\quad\quad\quad 2\cdot\dfrac{6}{4} - 3 \quad\bigg|$
$\quad\quad\quad\quad\quad\quad\quad 3 - 3 \quad\bigg|$
$\quad\quad\quad\quad\quad\quad\quad\quad 0 \quad\bigg|$ TRUE

The solutions are $\dfrac{\sqrt{6}}{2}$ and $-\dfrac{\sqrt{6}}{2}$, or $\pm\dfrac{\sqrt{6}}{2}$. Using a calculator, we find that the solutions are approximately ± 1.225.

7. $(x + 2)^2 = 49$

$\quad\quad x + 2 = 7$ *or* $x + 2 = -7$ Principle of square roots

$\quad\quad\quad x = 5$ *or* $\quad\quad x = -9$

The solutions are 5 and -9.

9. $(x - 4)^2 = 16$

$\quad\quad x - 4 = 4$ *or* $x - 4 = -4$ Principle of square roots

$\quad\quad\quad x = 8$ *or* $\quad\quad x = 0$

The solutions are 8 and 0.

11. $(x - 11)^2 = 7$

$\quad\quad x - 11 = \sqrt{7}$ *or* $x - 11 = -\sqrt{7}$

$\quad\quad\quad x = 11 + \sqrt{7}$ *or* $\quad\quad x = 11 - \sqrt{7}$

The solutions are $11 + \sqrt{7}$ and $11 - \sqrt{7}$, or $11 \pm \sqrt{7}$. Using a calculator, we find that the solutions are approximately 8.354 and 13.646.

13. $(x - 7)^2 = -4$

$\quad\quad x - 7 = \sqrt{-4}$ *or* $x - 7 = -\sqrt{-4}$

$\quad\quad x - 7 = 2i$ *or* $x - 7 = -2i$

$\quad\quad\quad x = 7 + 2i$ *or* $\quad\quad x = 7 - 2i$

The solutions are $7 + 2i$ and $7 - 2i$, or $7 \pm 2i$.

15. $(x-9)^2 = 81$

$\quad x - 9 = 9 \quad or \quad x - 9 = -9$

$\qquad x = 18 \quad or \qquad x = 0$

The solutions are 18 and 0.

17. $\left(x - \dfrac{3}{2}\right)^2 = \dfrac{7}{2}$

$x - \dfrac{3}{2} = \sqrt{\dfrac{7}{2}} \qquad or \quad x - \dfrac{3}{2} = -\sqrt{\dfrac{7}{2}}$

$x - \dfrac{3}{2} = \sqrt{\dfrac{7}{2}\cdot\dfrac{2}{2}} \quad or \quad x - \dfrac{3}{2} = -\sqrt{\dfrac{7}{2}\cdot\dfrac{2}{2}}$

$x - \dfrac{3}{2} = \dfrac{\sqrt{14}}{2} \qquad or \quad x - \dfrac{3}{2} = -\dfrac{\sqrt{14}}{2}$

$\qquad x = \dfrac{3}{2} + \dfrac{\sqrt{14}}{2} \quad or \qquad x = \dfrac{3}{2} - \dfrac{\sqrt{14}}{2}$

The solutions are $\dfrac{3}{2} + \dfrac{\sqrt{14}}{2}$ and $\dfrac{3}{2} - \dfrac{\sqrt{14}}{2}$, or $\dfrac{3}{2} \pm \dfrac{\sqrt{14}}{2}$. Using a calculator, we find that the solutions are approximately -0.371 and 3.371.

19. $x^2 + 6x + 9 = 64$

$\quad (x+3)^2 = 64$

$x + 3 = 8 \quad or \quad x + 3 = -8$

$\quad x = 5 \quad or \qquad x = -11$

The solutions are 5 and -11.

21. $y^2 - 14y + 49 = 4$

$\quad (y-7)^2 = 4$

$y - 7 = 2 \quad or \quad y - 7 = -2$

$\quad y = 9 \quad or \qquad y = 5$

The solutions are 9 and 5.

23. $x^2 + 4x \qquad = 2 \qquad$ Original equation

$x^2 + 4x + 4 = 2 + 4 \quad$ Adding 4: $\left(\dfrac{4}{2}\right)^2 = 2^2 = 4$

$\quad (x+2)^2 = 6$

$x + 2 = \sqrt{6} \qquad or \quad x + 2 = -\sqrt{6} \qquad$ Principle of square roots

$\quad x = -2 + \sqrt{6} \quad or \qquad x = -2 - \sqrt{6}$

The solutions are $-2 \pm \sqrt{6}$.

25. $x^2 - 22x \qquad = 11 \qquad$ Original equation

$x^2 - 22x + 121 = 11 + 121 \quad$ Adding 121: $\left(\dfrac{-22}{2}\right)^2 =$

$\qquad\qquad\qquad\qquad (-11)^2 = 121$

$\quad (x-11)^2 = 132$

$x - 11 = \sqrt{132} \qquad or \quad x - 11 = -\sqrt{132}$

$x - 11 = 2\sqrt{33} \qquad or \quad x - 11 = -2\sqrt{33}$

$\quad x = 11 + 2\sqrt{33} \quad or \qquad x = 11 - 2\sqrt{33}$

The solutions are $11 \pm 2\sqrt{33}$.

27. $x^2 + x \qquad = 1$

$x^2 + x + \dfrac{1}{4} = 1 + \dfrac{1}{4} \quad$ Adding $\dfrac{1}{4}$: $\left(\dfrac{1}{2}\right)^2 = \dfrac{1}{4}$

$\quad \left(x + \dfrac{1}{2}\right)^2 = \dfrac{5}{4}$

$x + \dfrac{1}{2} = \dfrac{\sqrt{5}}{2} \qquad or \quad x + \dfrac{1}{2} = -\dfrac{\sqrt{5}}{2}$

$\quad x = -\dfrac{1}{2} + \dfrac{\sqrt{5}}{2} \quad or \qquad x = -\dfrac{1}{2} - \dfrac{\sqrt{5}}{2}$

The solutions are $-\dfrac{1}{2} \pm \dfrac{\sqrt{5}}{2}$.

29. $t^2 - 5t \qquad = 7$

$t^2 - 5t + \dfrac{25}{4} = 7 + \dfrac{25}{4} \quad$ Adding $\dfrac{25}{4}$: $\left(\dfrac{-5}{2}\right)^2 = \dfrac{25}{4}$

$\quad \left(t - \dfrac{5}{2}\right)^2 = \dfrac{53}{4}$

$t - \dfrac{5}{2} = \dfrac{\sqrt{53}}{2} \qquad or \quad t - \dfrac{5}{2} = -\dfrac{\sqrt{53}}{2}$

$\quad t = \dfrac{5}{2} + \dfrac{\sqrt{53}}{2} \quad or \qquad t = \dfrac{5}{2} - \dfrac{\sqrt{53}}{2}$

The solutions are $\dfrac{5}{2} \pm \dfrac{\sqrt{53}}{2}$.

31. $x^2 + \dfrac{3}{2}x \qquad = 3$

$x^2 + \dfrac{3}{2}x + \dfrac{9}{16} = 3 + \dfrac{9}{16} \quad \left(\dfrac{1}{2}\cdot\dfrac{3}{2}\right)^2 = \left(\dfrac{3}{4}\right)^2 = \dfrac{9}{16}$

$\quad \left(x + \dfrac{3}{4}\right)^2 = \dfrac{57}{16}$

$x + \dfrac{3}{4} = \dfrac{\sqrt{57}}{4} \qquad or \quad x + \dfrac{3}{4} = -\dfrac{\sqrt{57}}{4}$

$\quad x = -\dfrac{3}{4} + \dfrac{\sqrt{57}}{4} \quad or \qquad x = -\dfrac{3}{4} - \dfrac{\sqrt{57}}{4}$

The solutions are $-\dfrac{3}{4} \pm \dfrac{\sqrt{57}}{4}$.

33. $m^2 - \dfrac{9}{2}m \qquad = \dfrac{3}{2} \qquad$ Original equation

$m^2 - \dfrac{9}{2}m + \dfrac{81}{16} = \dfrac{3}{2} + \dfrac{81}{16} \quad \left[\dfrac{1}{2}\left(-\dfrac{9}{2}\right)\right]^2 = \left(-\dfrac{9}{4}\right)^2 = \dfrac{81}{16}$

$\quad \left(m - \dfrac{9}{4}\right)^2 = \dfrac{105}{16}$

$m - \dfrac{9}{4} = \dfrac{\sqrt{105}}{4} \qquad or \quad m - \dfrac{9}{4} = -\dfrac{\sqrt{105}}{4}$

$\quad m = \dfrac{9}{4} + \dfrac{\sqrt{105}}{4} \quad or \qquad m = \dfrac{9}{4} - \dfrac{\sqrt{105}}{4}$

The solutions are $\dfrac{9}{4} \pm \dfrac{\sqrt{105}}{4}$.

35. $x^2 + 6x - 16 = 0$

$x^2 + 6x \qquad = 16 \qquad$ Adding 16

$x^2 + 6x + \ 9 = 16 + 9 \qquad \left(\dfrac{6}{2}\right)^2 = 3^2 = 9$

$(x+3)^2 = 25$

$x + 3 = 5 \ \ or \ \ x + 3 = -5$

$x = 2 \ \ or \qquad x = -8$

The solutions are 2 and -8.

37. $x^2 + 22x + 102 = 0$

$x^2 + 22x \qquad = -102 \qquad$ Subtracting 102

$x^2 + 22x + 121 = -102 + 121 \qquad \left(\dfrac{22}{2}\right)^2 = 11^2 = 121$

$(x+11)^2 = 19$

$x + 11 = \sqrt{19} \qquad or \ \ x + 11 = -\sqrt{19}$

$x = -11 + \sqrt{19} \ \ or \qquad x = -11 - \sqrt{19}$

The solutions are $-11 \pm \sqrt{19}$.

39. $x^2 - 10x - \ 4 = 0$

$x^2 - 10x \qquad = 4 \qquad$ Adding 4

$x^2 - 10x + 25 = 4 + 25 \qquad \left(\dfrac{-10}{2}\right)^2 = (-5)^2 = 25$

$(x-5)^2 = 29$

$x - 5 = \sqrt{29} \qquad or \ \ x - 5 = -\sqrt{29}$

$x = 5 + \sqrt{29} \ \ or \qquad x = 5 - \sqrt{29}$

The solutions are $5 \pm \sqrt{29}$.

41. a) $x^2 + 7x - \ 2 = 0$

$x^2 + 7x \qquad = 2 \qquad$ Adding 2

$x^2 + 7x + \dfrac{49}{4} = 2 + \dfrac{49}{4} \qquad \left(\dfrac{7}{2}\right)^2 = \dfrac{49}{4}$

$\left(x + \dfrac{7}{2}\right)^2 = \dfrac{57}{4}$

$x + \dfrac{7}{2} = \dfrac{\sqrt{57}}{2} \qquad or \ \ x + \dfrac{7}{2} = -\dfrac{\sqrt{57}}{2}$

$x = -\dfrac{7}{2} + \dfrac{\sqrt{57}}{2} \ \ or \qquad x = -\dfrac{7}{2} - \dfrac{\sqrt{57}}{2}$

The solutions are $-\dfrac{7}{2} \pm \dfrac{\sqrt{57}}{2}$.

b) The real-number solutions of the equation $x^2 + 7x - 2 = 0$ are the first coordinates of the x-intercepts of the graph of $f(x) = x^2 + 7x - 2$. Thus, the x-intercepts are $\left(-\dfrac{7}{2} - \dfrac{\sqrt{57}}{2}, 0\right)$ and $\left(-\dfrac{7}{2} + \dfrac{\sqrt{57}}{2}, 0\right)$.

43. a) $2x^2 - 5x + 8 = 0$

$\dfrac{1}{2}(2x^2 - 5x + 8) = \dfrac{1}{2} \cdot 0 \qquad$ Multiplying by $\dfrac{1}{2}$ to make the x^2-coefficient 1

$x^2 - \dfrac{5}{2}x + \ 4 = 0$

$x^2 - \dfrac{5}{2}x \qquad = -4 \qquad$ Subtracting 4

$x^2 - \dfrac{5}{2}x + \dfrac{25}{16} = -4 + \dfrac{25}{16}$

$\qquad \left[\dfrac{1}{2}\left(-\dfrac{5}{2}\right)\right]^2 = \left(-\dfrac{5}{4}\right)^2 = \dfrac{25}{16}$

$\left(x - \dfrac{5}{4}\right)^2 = -\dfrac{64}{16} + \dfrac{25}{16}$

$\left(x - \dfrac{5}{4}\right)^2 = -\dfrac{39}{16}$

$x - \dfrac{5}{4} = \sqrt{-\dfrac{39}{16}} \qquad or \ \ x - \dfrac{5}{4} = -\sqrt{-\dfrac{39}{16}}$

$x - \dfrac{5}{4} = i\sqrt{\dfrac{39}{16}} \qquad or \ \ x - \dfrac{5}{4} = -i\sqrt{\dfrac{39}{16}}$

$x = \dfrac{5}{4} + i\dfrac{\sqrt{39}}{4} \ \ or \qquad x = \dfrac{5}{4} - i\dfrac{\sqrt{39}}{4}$

The solutions are $\dfrac{5}{4} \pm i\dfrac{\sqrt{39}}{4}$.

b) Since the equation $2x^2 - 5x + 8 = 0$ has no real-number solutions, the graph of $f(x) = 2x^2 - 5x + 8$ has no x-intercepts.

45. $x^2 - \dfrac{3}{2}x - \dfrac{1}{2} = 0$

$x^2 - \dfrac{3}{2}x \qquad = \dfrac{1}{2}$

$x^2 - \dfrac{3}{2}x + \dfrac{9}{16} = \dfrac{1}{2} + \dfrac{9}{16} \qquad \left[\dfrac{1}{2}\left(-\dfrac{3}{2}\right)\right]^2 = \left(-\dfrac{3}{4}\right)^2 = \dfrac{9}{16}$

$\left(x - \dfrac{3}{4}\right)^2 = \dfrac{17}{16}$

$x - \dfrac{3}{4} = \dfrac{\sqrt{17}}{4} \qquad or \ \ x - \dfrac{3}{4} = -\dfrac{\sqrt{17}}{4}$

$x = \dfrac{3}{4} + \dfrac{\sqrt{17}}{4} \ \ or \qquad x = \dfrac{3}{4} - \dfrac{\sqrt{17}}{4}$

The solutions are $\dfrac{3}{4} \pm \dfrac{\sqrt{17}}{4}$.

47. $2x^2 - 3x - 17 = 0$

$\frac{1}{2}(2x^2 - 3x - 17) = \frac{1}{2} \cdot 0$ Multiplying by $\frac{1}{2}$ to make the x^2-coefficient 1

$x^2 - \frac{3}{2}x - \frac{17}{2} = 0$

$x^2 - \frac{3}{2}x \qquad = \frac{17}{2}$ Adding $\frac{17}{2}$

$x^2 - \frac{3}{2}x + \frac{9}{16} = \frac{17}{2} + \frac{9}{16}$ $\left[\frac{1}{2}\left(-\frac{3}{2}\right)\right]^2 =$

$\left(-\frac{3}{4}\right)^2 = \frac{9}{16}$

$\left(x - \frac{3}{4}\right)^2 = \frac{145}{16}$

$x - \frac{3}{4} = \frac{\sqrt{145}}{4}$ or $x - \frac{3}{4} = -\frac{\sqrt{145}}{4}$

$x = \frac{3}{4} + \frac{\sqrt{145}}{4}$ or $x = \frac{3}{4} - \frac{\sqrt{145}}{4}$

The solutions are $\frac{3}{4} \pm \frac{\sqrt{145}}{4}$.

49. $3x^2 - 4x - 1 = 0$

$\frac{1}{3}(3x^2 - 4x - 1) = \frac{1}{3} \cdot 0$ Multiplying to make the x^2-coefficient 1

$x^2 - \frac{4}{3}x - \frac{1}{3} = 0$

$x^2 - \frac{4}{3}x \qquad = \frac{1}{3}$ Adding $\frac{1}{3}$

$x^2 - \frac{4}{3}x + \frac{4}{9} = \frac{1}{3} + \frac{4}{9}$ $\left[\frac{1}{2}\left(-\frac{4}{3}\right)\right]^2 =$

$\left(-\frac{2}{3}\right)^2 = \frac{4}{9}$

$\left(x - \frac{2}{3}\right)^2 = \frac{7}{9}$

$x - \frac{2}{3} = \frac{\sqrt{7}}{3}$ or $x - \frac{2}{3} = -\frac{\sqrt{7}}{3}$

$x = \frac{2}{3} + \frac{\sqrt{7}}{3}$ or $x = \frac{2}{3} - \frac{\sqrt{7}}{3}$

The solutions are $\frac{2}{3} \pm \frac{\sqrt{7}}{3}$.

51. $x^2 + x + 2 = 0$

$x^2 + x \qquad = -2$ Subtracting 2

$x^2 + x + \frac{1}{4} = -2 + \frac{1}{4}$ $\left(\frac{1}{2}\right)^2 = \frac{1}{4}$

$\left(x + \frac{1}{2}\right)^2 = -\frac{7}{4}$

$x + \frac{1}{2} = \sqrt{-\frac{7}{4}}$ or $x + \frac{1}{2} = -\sqrt{-\frac{7}{4}}$

$x + \frac{1}{2} = i\sqrt{\frac{7}{4}}$ or $x + \frac{1}{2} = -i\sqrt{\frac{7}{4}}$

$x = -\frac{1}{2} + i\frac{\sqrt{7}}{2}$ or $x = -\frac{1}{2} - i\frac{\sqrt{7}}{2}$

The solutions are $-\frac{1}{2} \pm i\frac{\sqrt{7}}{2}$.

53. $x^2 - 4x + 13 = 0$

$x^2 - 4x \qquad = -13$ Subtracting 13

$x^2 - 4x + 4 = -13 + 4$ $\left(\frac{-4}{2}\right)^2 = (-2)^2 = 4$

$(x - 2)^2 = -9$

$x - 2 = \sqrt{-9}$ or $x - 2 = -\sqrt{-9}$

$x - 2 = 3i$ or $x - 2 = -3i$

$x = 2 + 3i$ or $x = 2 - 3i$

The solutions are $2 \pm 3i$.

55. $V(T) = 48T^2$

$36 = 48T^2$ Substituting 36 for $V(T)$

$\frac{36}{48} = T^2$ Solving for T^2

$0.75 = T^2$

$\sqrt{0.75} = T$

$0.866 \approx T$

The hang time is 0.866 sec.

57. $s(t) = 16t^2$

$1053 = 16t^2$ Substituting 850 for $s(t)$

$\frac{1053}{16} = t^2$ Solving for t^2

$65.8125 = t^2$

$\sqrt{65.8125} = t$

$8.1 \approx t$

It will take about 8.1 sec for an object to fall from the bridge.

59. $s(t) = 16t^2$

$745 = 16t^2$ Substituting 745 for $s(t)$

$\frac{745}{16} = t^2$ Solving for t^2

$46.5625 = t^2$

$\sqrt{46.5625} = t$

$6.8 \approx t$

It will take about 6.8 sec for an object to fall from the top of the bridge.

61. Discussion and Writing Exercise

63. a) First find the slope.

$$m = \frac{410 - 275}{5 - 1} = \frac{135}{4} = 33.75$$

Now find the function. We substitute in the point-slope equation, using the point $(1, 275)$.

$$T - T_1 = m(t - t_1)$$
$$T - 275 = 33.75(t - 1)$$
$$T - 275 = 33.75t - 33.75$$
$$T = 33.75t + 241.25$$

Using the function notation we have
$T(t) = 33.75t + 241.25$.

b) Since $2008 - 1995 = 13$, the year 2008 is 13 years since 1995. We find $T(13)$.

$$T(t) = 33.75t + 241.25$$
$$T(13) = 33.75(13) + 241.25$$
$$= 438.75 + 241.25$$
$$= 680$$

In 2008 about 680 thousand, or 680,000 people, will visit a doctor for tattoo removal.

c) Since $1,085,000 = 1085$ thousand, we substitute 1085 for $T(t)$ and solve for t.

$$T(t) = 33.75t + 241.25$$
$$1085 = 33.75t + 241.25$$
$$843.75 = 33.75t$$
$$25 = t$$

1,085,000 people will visit a doctor for tattoo removal 25 yr after 1995, or in 2020.

65. Graph $f(x) = 5 - 2x$

We find some ordered pairs $(x, f(x))$, plot them, and draw the graph.

x	$f(x)$	$(x, f(x))$
0	5	$(0, 5)$
1	3	$(1, 3)$
3	-1	$(3, -1)$
5	-5	$(5, -5)$

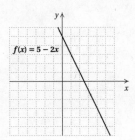

67. Graph $f(x) = |5 - 2x|$

We find some ordered pairs $(x, f(x))$, plot them, and draw the graph.

x	$f(x)$	$(x, f(x))$
0	5	$(0, 5)$
1	3	$(1, 3)$
$\frac{5}{2}$	0	$\left(\frac{5}{2}, 0\right)$
3	1	$(3, 1)$
5	5	$(5, 5)$

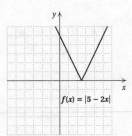

69. $\sqrt{\dfrac{2}{5}} = \sqrt{\dfrac{2}{5} \cdot \dfrac{5}{5}} = \sqrt{\dfrac{10}{25}} = \dfrac{\sqrt{10}}{\sqrt{25}} = \dfrac{\sqrt{10}}{5}$

71.
$$\sqrt{4x - 4} = \sqrt{x + 4} + 1$$
$$(\sqrt{4x - 4})^2 = (\sqrt{x + 4} + 1)^2$$
$$4x - 4 = x + 4 + 2\sqrt{x + 4} + 1$$
$$3x - 9 = 2\sqrt{x + 4}$$
$$(3x - 9)^2 = (2\sqrt{x + 4})^2$$
$$9x^2 - 54x + 81 = 4(x + 4)$$
$$9x^2 - 54x + 81 = 4x + 16$$
$$9x^2 - 58x + 65 = 0$$
$$(9x - 13)(x - 5) = 0$$
$$9x - 13 = 0 \quad or \quad x - 5 = 0$$
$$9x = 13 \quad or \qquad x = 5$$
$$x = \frac{13}{9} \quad or \qquad x = 5$$

Only 5 checks. It is the solution.

73. $-35 = \sqrt{2x + 5}$

The equation has no solution, because the principle square root of a number is always nonnegative.

75. Left to the student

77. In order for $x^2 + bx + 64$ to be a trinomial square, the following must be true:

$$\left(\frac{b}{2}\right)^2 = 64$$
$$\frac{b^2}{4} = 64$$
$$b^2 = 256$$
$$b = 16 \quad or \quad b = -16$$

79.
$$x(2x^2 + 9x - 56)(3x + 10) = 0$$
$$x(2x - 7)(x + 8)(3x + 10) = 0$$
$$x = 0 \; or \; 2x - 7 = 0 \; or \; x + 8 = 0 \; or \; 3x + 10 = 0$$
$$x = 0 \; or \qquad x = \frac{7}{2} \; or \qquad x = -8 \; or \qquad x = -\frac{10}{3}$$

The solutions are -8, $-\dfrac{10}{3}$, 0, and $\dfrac{7}{2}$.

Exercise Set 7.2

1. $x^2 + 8x + 2 = 0$

$a = 1$, $b = 8$, $c = 2$

$$x = \frac{-b \pm \sqrt{b^2 - 4ac}}{2a}$$

$$x = \frac{-8 \pm \sqrt{8^2 - 4 \cdot 1 \cdot 2}}{2 \cdot 1} = \frac{-8 \pm \sqrt{64 - 8}}{2}$$

$$x = \frac{-8 \pm \sqrt{56}}{2} = \frac{-8 \pm 2\sqrt{14}}{2}$$

$$x = \frac{2(-4 \pm \sqrt{14})}{2} = -4 \pm \sqrt{14}$$

The solutions are $-4 + \sqrt{14}$ and $-4 - \sqrt{14}$.

3. $3p^2 = -8p - 1$

$3p^2 + 8p + 1 = 0$ Finding standard form

$a = 3,\ b = 8,\ c = 1$

$p = \dfrac{-b \pm \sqrt{b^2 - 4ac}}{2a}$

$p = \dfrac{-8 \pm \sqrt{8^2 - 4 \cdot 3 \cdot 1}}{2 \cdot 3} = \dfrac{-8 \pm \sqrt{64 - 12}}{6}$

$x = \dfrac{-8 \pm \sqrt{52}}{6} = \dfrac{-8 \pm 2\sqrt{13}}{6}$

$x = \dfrac{2(-4 \pm \sqrt{13})}{2 \cdot 3} = \dfrac{-4 \pm \sqrt{13}}{3}$

The solutions are $\dfrac{-4 + \sqrt{13}}{3}$ and $\dfrac{-4 - \sqrt{13}}{3}$.

5. $x^2 - x + 1 = 0$

$a = 1,\ b = -1,\ c = 1$

$x = \dfrac{-(-1) \pm \sqrt{(-1)^2 - 4 \cdot 1 \cdot 1}}{2 \cdot 1} = \dfrac{1 \pm \sqrt{1 - 4}}{2}$

$x = \dfrac{1 \pm \sqrt{-3}}{2} = \dfrac{1 \pm i\sqrt{3}}{2} = \dfrac{1}{2} \pm i\dfrac{\sqrt{3}}{2}$

The solutions are $\dfrac{1}{2} + i\dfrac{\sqrt{3}}{2}$ and $\dfrac{1}{2} - i\dfrac{\sqrt{3}}{2}$.

7. $x^2 + 13 = 4x$

$x^2 - 4x + 13 = 0$ Finding standard form

$a = 1,\ b = -4,\ c = 13$

$x = \dfrac{-(-4) \pm \sqrt{(-4)^2 - 4 \cdot 1 \cdot 13}}{2 \cdot 1} = \dfrac{4 \pm \sqrt{16 - 52}}{2}$

$x = \dfrac{4 \pm \sqrt{-36}}{2} = \dfrac{4 \pm 6i}{2} = 2 \pm 3i$

The solutions are $2 + 3i$ and $2 - 3i$.

9. $r^2 + 3r = 8$

$r^2 + 3r - 8 = 0$ Finding standard form

$a = 1,\ b = 3,\ c = -8$

$r = \dfrac{-3 \pm \sqrt{3^2 - 4 \cdot 1 \cdot (-8)}}{2 \cdot 1} = \dfrac{-3 \pm \sqrt{9 + 32}}{2}$

$r = \dfrac{-3 \pm \sqrt{41}}{2}$

The solutions are $\dfrac{-3 + \sqrt{41}}{2}$ and $\dfrac{-3 - \sqrt{41}}{2}$.

11. $1 + \dfrac{2}{x} + \dfrac{5}{x^2} = 0$

$x^2 + 2x + 5 = 0$ Multiplying by x^2, the LCM
                              of the denominators

$a = 1,\ b = 2,\ c = 5$

$x = \dfrac{-2 \pm \sqrt{2^2 - 4 \cdot 1 \cdot 5}}{2 \cdot 1} = \dfrac{-2 \pm \sqrt{4 - 20}}{2}$

$x = \dfrac{-2 \pm \sqrt{-16}}{2} = \dfrac{-2 \pm 4i}{2} = -1 \pm 2i$

The solutions are $-1 + 2i$ and $-1 - 2i$.

13. a) $3x + x(x - 2) = 0$

$3x + x^2 - 2x = 0$

$x^2 + x = 0$

$x(x + 1) = 0$

$x = 0 \ \ or \ \ x + 1 = 0$

$x = 0 \ \ or \ \ \ \ \ \ \ x = -1$

The solutions are 0 and -1.

b) The solutions of the equation $3x + x(x - 2) = 0$ are the first coordinates of the x-intercepts of the graph of $f(x) = 3x + x(x - 2)$. Thus, the x-intercepts are $(-1, 0)$ and $(0, 0)$.

15. a) $11x^2 - 3x - 5 = 0$

$a = 11,\ b = -3,\ c = -5$

$x = \dfrac{-(-3) \pm \sqrt{(-3)^2 - 4 \cdot 11 \cdot (-5)}}{2 \cdot 11}$

$x = \dfrac{3 \pm \sqrt{9 + 220}}{22} = \dfrac{3 \pm \sqrt{229}}{22}$

The solutions are $\dfrac{3 + \sqrt{229}}{22}$ and $\dfrac{3 - \sqrt{229}}{22}$.

b) The solutions of the equation $11x^2 - 3x - 5 = 0$ are the first coordinates of the x-intercepts of the graph of $f(x) = 11x^2 - 3x - 5$. Thus, the x-intercepts are $\left(\dfrac{3 - \sqrt{229}}{22}, 0\right)$ and $\left(\dfrac{3 + \sqrt{229}}{22}, 0\right)$.

17. a) $25x^2 - 20x + 4 = 0$

$(5x - 2)(5x - 2) = 0$

$5x - 2 = 0 \ \ or \ \ 5x - 2 = 0$

$5x = 2 \ \ or \ \ \ \ \ \ 5x = 2$

$x = \dfrac{2}{5} \ \ or \ \ \ \ \ \ \ x = \dfrac{2}{5}$

The solution is $\dfrac{2}{5}$.

b) The solution of the equation $25x^2 - 20x + 4 = 0$ is the first coordinate of the x-intercept of $f(x) = 25x^2 - 20x + 4$. Thus, the x-intercept is $\left(\dfrac{2}{5}, 0\right)$.

19. $4x(x - 2) - 5x(x - 1) = 2$

$4x^2 - 8x - 5x^2 + 5x = 2$ Removing parentheses

$-x^2 - 3x = 2$

$-x^2 - 3x - 2 = 0$

$x^2 + 3x + 2 = 0$ Multiplying by -1

$(x + 2)(x + 1) = 0$

$x + 2 = 0 \ \ \ or \ \ x + 1 = 0$

$x = -2 \ or \ \ \ \ \ \ x = -1$

The solutions are -2 and -1.

21. $14(x-4)-(x+2)=(x+2)(x-4)$

$$14x-56-x-2=x^2-2x-8$$
$$13x-58=x^2-2x-8$$
$$0=x^2-15x+50$$
$$0=(x-10)(x-5)$$

$x-10=0 \quad or \quad x-5=0$

$x=10 \quad or \qquad x=5$

The solutions are 10 and 5.

23. $\qquad 5x^2=17x-2$

$5x^2-17x+2=0$

$a=5,\ b=-17,\ c=2$

$$x=\frac{-(-17)\pm\sqrt{(-17)^2-4\cdot5\cdot2}}{2\cdot5}$$

$$x=\frac{17\pm\sqrt{289-40}}{10}=\frac{17\pm\sqrt{249}}{10}$$

The solutions are $\dfrac{17+\sqrt{249}}{10}$ and $\dfrac{17-\sqrt{249}}{10}$.

25. $\qquad x^2+5=4x$

$x^2-4x+5=0$

$a=1,\ b=-4,\ c=5$

$$x=\frac{-(-4)\pm\sqrt{(-4)^2-4\cdot1\cdot5}}{2\cdot1}=\frac{4\pm\sqrt{16-20}}{2}$$

$$x=\frac{4\pm\sqrt{-4}}{2}=\frac{4\pm2i}{2}=2\pm i$$

The solutions are $2+i$ and $2-i$.

27. $\qquad x+\dfrac{1}{x}=\dfrac{13}{6}$, LCM is $6x$

$$6x\left(x+\frac{1}{x}\right)=6x\cdot\frac{13}{6}$$

$$6x^2+6=13x$$

$$6x^2-13x+6=0$$

$$(2x-3)(3x-2)=0$$

$2x-3=0 \quad or \quad 3x-2=0$

$2x=3 \quad or \qquad 3x=2$

$x=\dfrac{3}{2} \quad or \qquad x=\dfrac{2}{3}$

The solutions are $\dfrac{3}{2}$ and $\dfrac{2}{3}$.

29. $\qquad \dfrac{1}{y}+\dfrac{1}{y+2}=\dfrac{1}{3}$, LCM is $3y(y+2)$

$$3y(y+2)\left(\frac{1}{y}+\frac{1}{y+2}\right)=3y(y+2)\cdot\frac{1}{3}$$

$$3(y+2)+3y=y(y+2)$$

$$3y+6+3y=y^2+2y$$

$$6y+6=y^2+2y$$

$$0=y^2-4y-6$$

$a=1,\ b=-4,\ c=-6$

$$y=\frac{-(-4)\pm\sqrt{(-4)^2-4\cdot1\cdot(-6)}}{2\cdot1}=\frac{4\pm\sqrt{16+24}}{2}$$

$$y=\frac{4\pm\sqrt{40}}{2}=\frac{4\pm2\sqrt{10}}{2}$$

$$y=\frac{2(2\pm\sqrt{10})}{2\cdot1}=2\pm\sqrt{10}$$

The solutions are $2+\sqrt{10}$ and $2-\sqrt{10}$.

31. $\qquad (2t-3)^2+17t=15$

$$4t^2-12t+9+17t=15$$

$$4t^2+5t-6=0$$

$$(4t-3)(t+2)=0$$

$4t-3=0 \quad or \quad t+2=0$

$t=\dfrac{3}{4} \quad or \qquad t=-2$

The solutions are $\dfrac{3}{4}$ and -2.

33. $\qquad (x-2)^2+(x+1)^2=0$

$$x^2-4x+4+x^2+2x+1=0$$

$$2x^2-2x+5=0$$

$a=2,\ b=-2,\ c=5$

$$x=\frac{-(-2)\pm\sqrt{(-2)^2-4\cdot2\cdot5}}{2\cdot2}=\frac{2\pm\sqrt{4-40}}{4}$$

$$x=\frac{2\pm\sqrt{-36}}{4}=\frac{2\pm6i}{4}$$

$$x=\frac{2(1\pm3i)}{2\cdot2}=\frac{1\pm3i}{2}=\frac{1}{2}\pm\frac{3}{2}i$$

The solutions are $\dfrac{1}{2}+\dfrac{3}{2}i$ and $\dfrac{1}{2}-\dfrac{3}{2}i$.

35. $\qquad x^3-1=0$

$(x-1)(x^2+x+1)=0$

$x-1=0 \quad or \quad x^2+x+1=0$

$x=1 \quad or \qquad x=\dfrac{-1\pm\sqrt{1^2-4\cdot1\cdot1}}{2\cdot1}$

$x=1 \quad or \qquad x=\dfrac{-1\pm\sqrt{-3}}{2}$

$x=1 \quad or \qquad x=\dfrac{-1\pm i\sqrt{3}}{2}=-\dfrac{1}{2}\pm i\dfrac{\sqrt{3}}{2}$

The solutions are 1, $-\dfrac{1}{2}+i\dfrac{\sqrt{3}}{2}$, and $-\dfrac{1}{2}-i\dfrac{\sqrt{3}}{2}$.

37. $x^2+6x+4=0$

$a=1,\ b=6,\ c=4$

$$x=\frac{-6\pm\sqrt{6^2-4\cdot1\cdot4}}{2\cdot1}=\frac{-6\pm\sqrt{36-16}}{2}$$

$$x=\frac{-6\pm\sqrt{20}}{2}=\frac{-6\pm\sqrt{4\cdot5}}{2}$$

$$x=\frac{-6\pm2\sqrt{5}}{2}=\frac{2(-3\pm\sqrt{5})}{2}$$

$$x=-3\pm\sqrt{5}$$

We can use a calculator to approximate the solutions:

$-3+\sqrt{5}\approx-0.764$; $-3-\sqrt{5}\approx-5.236$

The solutions are $-3+\sqrt{5}$ and $-3-\sqrt{5}$, or approximately -0.764 and -5.236.

39. $x^2 - 6x + 4 = 0$

$a = 1, \ b = -6, \ c = 4$

$x = \dfrac{-(-6) \pm \sqrt{(-6)^2 - 4 \cdot 1 \cdot 4}}{2 \cdot 1} = \dfrac{6 \pm \sqrt{36 - 16}}{2}$

$x = \dfrac{6 \pm \sqrt{20}}{2} = \dfrac{6 \pm \sqrt{4 \cdot 5}}{2}$

$x = \dfrac{6 \pm 2\sqrt{5}}{2} = \dfrac{2(3 \pm \sqrt{5})}{2}$

$x = 3 \pm \sqrt{5}$

We can use a calculator to approximate the solutions:

$3 + \sqrt{5} \approx 5.236; \ 3 - \sqrt{5} \approx 0.764$

The solutions are $3 + \sqrt{5}$ and $3 - \sqrt{5}$, or approximately 5.236 and 0.764.

41. $2x^2 - 3x - 7 = 0$

$a = 2, \ b = -3, \ c = -7$

$x = \dfrac{-(-3) \pm \sqrt{(-3)^2 - 4 \cdot 2 \cdot (-7)}}{2 \cdot 2} = \dfrac{3 \pm \sqrt{9 + 56}}{4}$

$x = \dfrac{3 \pm \sqrt{65}}{4}$

We can use a calculator to approximate the solutions:

$\dfrac{3 + \sqrt{65}}{4} \approx 2.766; \ \dfrac{3 - \sqrt{65}}{4} \approx -1.266$

The solutions are $\dfrac{3 + \sqrt{65}}{4}$ and $\dfrac{3 - \sqrt{65}}{4}$, or approximately 2.766 and -1.266.

43. $5x^2 = 3 + 8x$

$5x^2 - 8x - 3 = 0$

$a = 5, \ b = -8, \ c = -3$

$x = \dfrac{-(-8) \pm \sqrt{(-8)^2 - 4 \cdot 5 \cdot (-3)}}{2 \cdot 5} = \dfrac{8 \pm \sqrt{64 + 60}}{10}$

$x = \dfrac{8 \pm \sqrt{124}}{10} = \dfrac{8 \pm \sqrt{4 \cdot 31}}{10}$

$x = \dfrac{8 \pm 2\sqrt{31}}{10} = \dfrac{2(4 \pm \sqrt{31})}{2 \cdot 5}$

$x = \dfrac{4 \pm \sqrt{31}}{5}$

We can use a calculator to approximate the solutions:

$\dfrac{4 + \sqrt{31}}{5} \approx 1.914; \ \dfrac{4 - \sqrt{31}}{5} \approx -0.314$

The solutions are $\dfrac{4 + \sqrt{31}}{5}$ and $\dfrac{4 - \sqrt{31}}{5}$, or approximately 1.914 and -0.314.

45. Discussion and Writing Exercise

47. $x = \sqrt{x + 2}$

$x^2 = (\sqrt{x + 2})^2$ Principle of powers

$x^2 = x + 2$

$x^2 - x - 2 = 0$

$(x - 2)(x + 1) = 0$

$x - 2 = 0 \ \ or \ \ x + 1 = 0$

$x = 2 \ \ or \ \ \ \ \ \ x = -1$

The number 2 checks but -1 does not, so the solution is 2.

49. $\sqrt{x + 2} = \sqrt{2x - 8}$

$(\sqrt{x + 2})^2 = (\sqrt{2x + 8})^2$ Principle of powers

$x + 2 = 2x - 8$

$2 = x - 8$

$10 = x$

The number 10 checks, so it is the solution.

51. $\sqrt{x + 5} = -7$

Since the square root of a number must be nonnegative, this equation has no solution.

53. $\sqrt[3]{4x - 7} = 2$

$(\sqrt[3]{4x - 7})^3 = 2^3$ Principle of powers

$4x - 7 = 8$

$4x = 15$

$x = \dfrac{15}{4}$

The number $\dfrac{15}{4}$ checks, so it is the solution.

55. The solutions of $2.2x^2 + 0.5x - 1 = 0$ are approximately -0.797 and 0.570.

57. $2x^2 - x - \sqrt{5} = 0$

$a = 2, \ b = -1, \ c = -\sqrt{5}$

$x = \dfrac{-(-1) \pm \sqrt{(-1)^2 - 4 \cdot 2 \cdot (-\sqrt{5})}}{2 \cdot 2} = \dfrac{1 \pm \sqrt{1 + 8\sqrt{5}}}{4}$

The solutions are $\dfrac{1 + \sqrt{1 + 8\sqrt{5}}}{4}$ and $\dfrac{1 - \sqrt{1 + 8\sqrt{5}}}{4}$.

59. $ix^2 - x - 1 = 0$

$a = i, \ b = -1, \ c = -1$

$x = \dfrac{-(-1) \pm \sqrt{(-1)^2 - 4 \cdot i \cdot (-1)}}{2 \cdot i} = \dfrac{1 \pm \sqrt{1 + 4i}}{2i}$

$x = \dfrac{1 \pm \sqrt{1 + 4i}}{2i} \cdot \dfrac{i}{i} = \dfrac{i \pm i\sqrt{1 + 4i}}{2i^2} = \dfrac{i \pm i\sqrt{1 + 4i}}{-2}$

$x = \dfrac{-i \pm i\sqrt{1 + 4i}}{2}$

The solutions are $\dfrac{-i + i\sqrt{1 + 4i}}{2}$ and $\dfrac{-i - i\sqrt{1 + 4i}}{2}$.

61. $\dfrac{x}{x + 1} = 4 + \dfrac{1}{3x^2 - 3}$

$\dfrac{x}{x + 1} = 4 + \dfrac{1}{3(x + 1)(x - 1)}$,

LCM is $3x(x + 1)(x - 1)$

$3(x + 1)(x - 1) \cdot \dfrac{x}{x + 1} =$

$3(x + 1)(x - 1)\left(4 + \dfrac{1}{3(x + 1)(x - 1)}\right)$

$3x(x - 1) = 12(x + 1)(x - 1) + 1$

$3x^2 - 3x = 12x^2 - 12 + 1$

$0 = 9x^2 + 3x - 11$

$a = 9, \ b = 3, \ c = -11$

$$x = \frac{-3 \pm \sqrt{3^2 - 4 \cdot 9 \cdot (-11)}}{2 \cdot 9} = \frac{-3 \pm \sqrt{9 + 396}}{18}$$

$$x = \frac{-3 \pm \sqrt{405}}{18} = \frac{-3 \pm 9\sqrt{5}}{18}$$

$$x = \frac{3(-1 \pm 3\sqrt{5})}{3 \cdot 6} = \frac{-1 \pm 3\sqrt{5}}{6}$$

The solutions are $\dfrac{-1 + 3\sqrt{5}}{6}$ and $\dfrac{-1 - 3\sqrt{5}}{6}$.

63. Replace $f(x)$ with 13.

$$13 = (x-3)^2$$
$$\pm\sqrt{13} = x - 3$$
$$3 \pm \sqrt{13} = x$$

The solutions are $3 + \sqrt{13}$ and $3 - \sqrt{13}$.

Exercise Set 7.3

1. *Familiarize*. We make a drawing and label it. We let $x =$ the length of the rectangle. Then $x - 7 =$ the width.

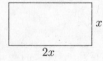

$x - 7$

x

***Translate*.** We use the formula for the area of a rectangle.

$$A = lw$$
$$18 = x(x-7) \quad \text{Substituting}$$

***Solve*.** We solve the equation.

$$18 = x^2 - 7x$$
$$0 = x^2 - 7x - 18$$
$$0 = (x-9)(x+2)$$
$$x - 9 = 0 \quad or \quad x + 2 = 0$$
$$x = 9 \quad or \qquad x = -2$$

***Check*.** We only check 9 since the length cannot be negative. If $x = 9$, then $x - 7 = 9 - 7$, or 2, and the area is $9 \cdot 2$, or 18 ft^2. The value checks.

***State*.** The length of 9 ft, and the width is 2 ft.

3. *Familiarize*. We make a drawing and label it. We let $x =$ the width of the rectangle. Then $2x =$ the length.

x

$2x$

***Translate*.**

$$A = lw$$
$$162 = 2x \cdot x \quad \text{Substituting}$$

***Solve*.** We solve the equation.

$$162 = 2x^2$$
$$81 = x^2$$
$$\pm 9 = x$$

***Check*.** We only check 9 since the width cannot be negative. If $x = 9$, then $2x = 2 \cdot 9$, or 18, and the area is $18 \cdot 9$, or 162 yd^2. The value checks.

***State*.** The length is 18 yd, and the width is 9 yd.

5. *Familiarize*. Let h represent the height of the sail. Then $h - 9$ represents the base. Recall that the formula for the area of a triangle is $A = \dfrac{1}{2} \times$ base $\times$ height.

***Translate*.** The area is 56 m^2. We substitute in the formula.

$$\frac{1}{2}(h-9)h = 56$$

***Solve*.** We solve the equation:

$$\frac{1}{2}(h-9)h = 56$$
$$(h-9)h = 112 \quad \text{Multiplying by 2}$$
$$h^2 - 9h = 112$$
$$h^2 - 9h - 112 = 0$$
$$(h-16)(h+7) = 0$$
$$h - 16 = 0 \quad or \quad h + 7 = 0$$
$$h = 16 \quad or \qquad h = -7$$

***Check*.** We check only 16, since height cannot be negative. If the height is 16 m, the base is $16 - 9$, or 7 m, and the area is $\dfrac{1}{2} \cdot 7 \cdot 16$, or 56 m^2. We have a solution.

***State*.** The height is 16 m, and the base is 7 m.

7. *Familiarize*. Let h represent the height of the sail. Then $h - 8$ represents the base. Recall that the formula for the area of a triangle is $A = \dfrac{1}{2} \times$ base $\times$ height.

***Translate*.** The area 56 ft^2. We substitute n the formula.

$$\frac{1}{2}(h-8)h = 56$$

***Solve*.** We solve the equation.

$$\frac{1}{2}(h-8)h = 56$$
$$(h-8)h = 112 \quad \text{Multiplying by 2}$$
$$h^2 - 8h = 112$$
$$h^2 - 8h - 112 = 0$$

We use the quadratic formula.

$$h = \frac{-b \pm \sqrt{b^2 - 4ac}}{2a}$$

$$= \frac{-(-8) \pm \sqrt{(-8)^2 - 4 \cdot 1 \cdot (-112)}}{2 \cdot 1}$$

$$= \frac{8 \pm \sqrt{64 + 448}}{2} = \frac{8 \pm \sqrt{512}}{2}$$

$$= \frac{8 \pm \sqrt{256 \cdot 2}}{2} = \frac{8 \pm 16\sqrt{2}}{2}$$

$$= \frac{8(1 \pm 2\sqrt{2})}{2} = 4(1 \pm 2\sqrt{2})$$

$$= 4 \pm 8\sqrt{2}$$

***Check*.** The number $4 - 8\sqrt{2}$ is negative. We do not check it since the height cannot be negative. If the height is $4 + 8\sqrt{2}$ ft, then the base is $4 + 8\sqrt{2} - 8$, or $-4 + 8\sqrt{2}$ ft, and the area is $\frac{1}{2}(-4 + 8\sqrt{2})(4 + 8\sqrt{2})$, or $\frac{1}{2}(-16 + 128)$, or $\frac{1}{2}(112)$, or 56 ft². The answer checks.

State. The base of the sail is $-4 + 8\sqrt{2}$ ft, and the height is $4 + 8\sqrt{2}$ ft.

9. Familiarize. We make a drawing and label it. We let $x =$ the width of the frame.

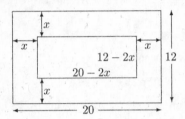

The length and width of the picture that shows are represented by $20 - 2x$ and $12 - 2x$. The area of the picture that shows is 84 cm².

Translate. Using the formula for the area of a rectangle, $A = l \cdot w$, we have

$$84 = (20 - 2x)(12 - 2x).$$

Solve. We solve the equation.

$$84 = (20 - 2x)(12 - 2x)$$
$$84 = 240 - 64x + 4x^2$$
$$0 = 156 - 64x + 4x^2$$
$$0 = 4x^2 - 64x + 156$$
$$0 = x^2 - 16x + 39 \qquad \text{Dividing by 4}$$
$$0 = (x - 13)(x - 3)$$
$$x - 13 = 0 \quad or \quad x - 3 = 0$$
$$x = 13 \quad or \qquad x = 3$$

Check. We see that 13 is not a solution, because when $x = 13$, then $20 - 2x = -6$ and $12 - 2x = -14$ and the dimensions of the picture cannot be negative. We check 3. When $x = 3$, then $20 - 2x = 14$ and $12 - 2x = 6$ and $14 \cdot 6 = 84$, the area of the picture that shows. The number 3 checks.

State. The width of the frame is 3 cm.

11. Familiarize. Using the labels on the drawing in the text, we let x and $x + 2$ represent the lengths of the legs of the right triangle.

Translate. We use the Pythagorean equation.

$$a^2 + b^2 = c^2$$
$$x^2 + (x + 2)^2 = 10^2 \qquad \text{Substituting}$$

Solve. We solve the equation.

$$x^2 + x^2 + 4x + 4 = 100$$
$$2x^2 + 4x + 4 = 100$$
$$2x^2 + 4x - 96 = 0$$
$$x^2 + 2x - 48 = 0 \qquad \text{Dividing by 2}$$
$$(x + 8)(x - 6) = 0$$

$$x + 8 = 0 \quad or \quad x - 6 = 0$$
$$x = -8 \quad or \qquad x = 6$$

Check. We only check 6 since the length of a leg cannot be negative. When $x = 6$, then $x + 2 = 8$, and $6^2 + 8^2 = 100 = 10^2$. The number 6 checks.

State. The lengths of the legs are 6 ft and 8 ft.

13. Familiarize. The page numbers on facing pages are consecutive integers. Let $x =$ the number on the left-hand page. Then $x + 1 =$ the number on the right-hand page.

Translate.

$$\underbrace{\text{The product of the page numbers}}_{x(x+1)} \quad \underset{=}{\text{is}} \quad \underset{812}{812.}$$

Solve. We solve the equation.

$$x^2 + x = 812$$
$$x^2 + x - 812 = 0$$
$$(x + 29)(x - 28) = 0$$
$$x + 29 = 0 \quad or \quad x - 28 = 0$$
$$x = -29 \quad or \qquad x = 28$$

Check. We only check 28 since a page number cannot be negative. If $x = 28$, then $x + 1 = 29$ and $28 \cdot 29 = 812$. The number 28 checks.

State. The page numbers are 28 and 29.

15. Familiarize. We make a drawing and label it. We let $x =$ the length and $x - 4 =$ the width.

$$\boxed{} \quad x - 4$$
$$x$$

Translate. We use the formula for the area of a rectangle.

$$A = lw$$
$$10 = x(x - 4) \qquad \text{Substituting}$$

Solve. We solve the equation.

$$10 = x^2 - 4x$$
$$0 = x^2 - 4x - 10$$
$$x = \frac{-b \pm \sqrt{b^2 - 4ac}}{2a} = \frac{-(-4) \pm \sqrt{(-4)^2 - 4 \cdot 1 \cdot (-10)}}{2 \cdot 1}$$
$$x = \frac{4 \pm \sqrt{16 + 40}}{2} = \frac{4 \pm \sqrt{56}}{2} = \frac{4 \pm \sqrt{4 \cdot 14}}{2}$$
$$x = \frac{4 \pm 2\sqrt{14}}{2} = 2 \pm \sqrt{14}$$

Check. We only need to check $2 + \sqrt{14}$ since $2 - \sqrt{14}$ is negative and the length cannot be negative. If $x = 2 + \sqrt{14}$, then $x - 4 = (2 + \sqrt{14}) - 4$, or $\sqrt{14} - 2$. Using a calculator we find that the length is $2 + \sqrt{14} \approx 5.742$ ft and the width is $\sqrt{14} - 2 \approx 1.742$ ft, and $(5.742)(1.742) = 10.003 \approx 10$. Our result checks.

State. The length is $2 + \sqrt{14}$ ft ≈ 5.742 ft; the width is $\sqrt{14} - 2$ ft ≈ 1.742 ft.

17. *Familiarize.* We make a drawing and label it. We let $x =$ the width of the margin.

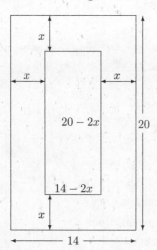

The length and width of the printed text are represented by $20 - 2x$ and $14 - 2x$. The area of the printed text is 100 in^2.

Translate. We use the formula for the area of a rectangle.
$$A = lw$$
$$100 = (20 - 2x)(14 - 2x)$$

Solve. We solve the equation.
$$100 = 280 - 68x + 4x^2$$
$$0 = 4x^2 - 68x + 180$$
$$0 = x^2 - 17x + 45 \qquad \text{Dividing by 4}$$
$$x = \frac{-b \pm \sqrt{b^2 - 4ac}}{2a} = \frac{-(-17) \pm \sqrt{(-17)^2 - 4 \cdot 1 \cdot 45}}{2 \cdot 1}$$
$$x = \frac{17 \pm \sqrt{289 - 180}}{2} = \frac{17 \pm \sqrt{109}}{2}$$
$$x \approx 13.720 \ \text{ or } \ x \approx 3.280$$

Check. If $x \approx 13.720$, then $20 - 2x \approx -7.440$ and $14 - 2x \approx -13.440$. Since the width of the margin cannot be negative, 13.720 is not a solution. If $x \approx 3.280$, then $20 - 2x \approx 13.440$ and $14 - 2x \approx 7.440$ and $(13.440)(7.440) = 99.99 \approx 100$. The number $\frac{17 - \sqrt{109}}{2} \approx 3.280$ checks.

State. The width of the margin is $\frac{17 - \sqrt{109}}{2}$ in. ≈ 3.280 in.

19. *Familiarize.* We make a drawing. We let $x =$ the length of the shorter leg and $x + 14 =$ the length of the longer leg.

Translate. We use the Pythagorean equation.
$$a^2 + b^2 = c^2$$
$$x^2 + (x + 14)^2 = 24^2 \qquad \text{Substituting}$$

Solve. We solve the equation.
$$x^2 + x^2 + 28x + 196 = 576$$
$$2x^2 + 28x - 380 = 0$$
$$x^2 + 14x - 190 = 0 \qquad \text{Dividing by 2}$$
$$x = \frac{-b \pm \sqrt{b^2 - 4ac}}{2a} = \frac{-14 \pm \sqrt{14^2 - 4 \cdot 1 \cdot (-190)}}{2 \cdot 1}$$
$$x = \frac{-14 \pm \sqrt{196 + 760}}{2} = \frac{-14 \pm \sqrt{956}}{2} = \frac{-14 \pm \sqrt{4 \cdot 239}}{2}$$
$$x = \frac{-14 \pm 2\sqrt{239}}{2} = -7 \pm \sqrt{239}$$
$$x \approx 8.460 \text{ or } x \approx -22.460$$

Check. Since the length of a leg cannot be negative, we only need to check 8.460. If $x = -7 + \sqrt{239} \approx 8.460$, then $x + 14 = -7 + \sqrt{239} + 14 = 7 + \sqrt{239} \approx 22.460$ and $(8.460)^2 + (22.460)^2 = 576.0232 \approx 576 = 24^2$. The number $-7 + \sqrt{239} \approx 8.460$ checks.

State. The lengths of the legs are $-7 + \sqrt{239}$ ft ≈ 8.460 ft and $7 + \sqrt{239}$ ft ≈ 22.460 ft.

21. *Familiarize.* We first make a drawing, labeling it with the known and unknown information. We can also organize the information in a table. We let r represent the speed and t the time for the first part of the trip.

r mph t hr $r - 10$ mph $4 - t$ hr
                    120 mi   100 mi

Trip	Distance	Speed	Time
1st part	120	r	t
2nd part	100	$r - 10$	$4 - t$

Translate. Using $r = \frac{d}{t}$, we get two equations from the table, $r = \frac{120}{t}$ and $r - 10 = \frac{100}{4 - t}$.

Solve. We substitute $\frac{120}{t}$ for r in the second equation and solve for t.
$$\frac{120}{t} - 10 = \frac{100}{4 - t}, \text{ LCD is } t(4 - t)$$
$$t(4 - t)\left(\frac{120}{t} - 10\right) = t(4 - t) \cdot \frac{100}{4 - t}$$
$$120(4 - t) - 10t(4 - t) = 100t$$
$$480 - 120t - 40t + 10t^2 = 100t$$
$$10t^2 - 260t + 480 = 0 \quad \text{Standard form}$$
$$t^2 - 26t + 48 = 0 \quad \text{Multiplying by } \frac{1}{10}$$
$$(t - 2)(t - 24) = 0$$
$$t = 2 \ \text{ or } \ t = 24$$

Check. Since the time cannot be negative (If $t = 24$, $4 - t = -20$.), we check only 2 hr. If $t = 2$, then $4 - t = 2$. The speed of the first part is $\frac{120}{2}$, or 60 mph. The speed of the second part is $\frac{100}{2}$, or 50 mph. The speed of the second part is 10 mph slower than the first part. The value checks.

State. The speed of the first part was 60 mph, and the speed of the second part was 50 mph.

23. *Familiarize*. We first make a drawing. We also organize the information in a table. We let $r =$ the speed and $t =$ the time of the slower trip.

200 mi	r mph	t hr
200 mi	$r + 10$ mph	$t - 1$ hr

Trip	Distance	Speed	Time
Slower	200	r	t
Faster	200	$r + 10$	$t - 1$

Translate. Using $t = d/r$, we get two equations from the table:

$$t = \frac{200}{r} \text{ and } t - 1 = \frac{200}{r + 10}$$

Solve. We substitute $\frac{200}{r}$ for t in the second equation and solve for r.

$$\frac{200}{r} - 1 = \frac{200}{r + 10}, \text{ LCD is}$$
$$r(r + 10)$$

$$r(r + 10)\left(\frac{200}{r} - 1\right) = r(r + 10) \cdot \frac{200}{r + 10}$$

$$200(r + 10) - r(r + 10) = 200r$$

$$200r + 2000 - r^2 - 10r = 200r$$

$$0 = r^2 + 10r - 2000$$

$$0 = (r + 50)(r - 40)$$

$$r = -50 \ \text{ or } \ r = 40$$

Check. Since negative speed has no meaning in this problem, we check only 40. If $r = 40$, then the time for the slower trip is $\frac{200}{40}$, or 5 hours. If $r = 40$, then $r + 10 = 50$ and the time for the faster trip is $\frac{200}{50}$, or 4 hours. This is 1 hour less time than the slower trip took, so we have an answer to the problem.

State. The speed is 40 mph.

25. *Familiarize*. We make a drawing and then organize the information in a table. We let $r =$ the speed and $t =$ the time of the Cessna.

600 mi	r mph	t hr
1000 mi	$r + 50$ mph	$t + 1$ hr

Plane	Distance	Speed	Time
Cessna	600	r	t
Beechcraft	1000	$r + 50$	$t + 1$

Translate. Using $t = d/r$, we get two equations from the table:

$$t = \frac{600}{r} \text{ and } t + 1 = \frac{1000}{r + 50}$$

Solve. We substitute $\frac{600}{r}$ for t in the second equation and solve for r.

$$\frac{600}{r} + 1 = \frac{1000}{r + 50},$$
$$\text{LCD is } r(r + 50)$$

$$r(r + 50)\left(\frac{600}{r} + 1\right) = r(r + 50) \cdot \frac{1000}{r + 50}$$

$$600(r + 50) + r(r + 50) = 1000r$$

$$600r + 30,000 + r^2 + 50r = 1000r$$

$$r^2 - 350r + 30,000 = 0$$

$$(r - 150)(r - 200) = 0$$

$$r = 150 \ \text{ or } \ r = 200$$

Check. If $r = 150$, then the Cessna's time is $\frac{600}{150}$, or 4 hr and the Beechcraft's time is $\frac{1000}{150 + 50}$, or $\frac{1000}{200}$, or 5 hr. If $r = 200$, then the Cessna's time is $\frac{600}{200}$, or 3 hr and the Beechcraft's time is $\frac{1000}{200 + 50}$, or $\frac{1000}{250}$, or 4 hr. Since the Beechcraft's time is 1 hr longer in each case, both values check. There are two solutions.

State. The speed of the Cessna is 150 mph and the speed of the Beechcraft is 200 mph; or the speed of the Cessna is 200 mph and the speed of the Beechcraft is 250 mph.

27. *Familiarize*. We make a drawing and then organize the information in a table. We let r represent the speed and t the time of the trip to Hillsboro.

			Hillsboro
40 mi	r mph	t hr	
40 mi	$r - 6$ mph	$14 - t$ hr	

Trip	Distance	Speed	Time
To Hillsboro	40	r	t
Return	40	$r - 6$	$14 - t$

Translate. Using $t = \frac{d}{r}$, we get two equations from the table,

$$t = \frac{40}{r} \text{ and } 14 - t = \frac{40}{r - 6}.$$

Solve. We substitute $\frac{40}{r}$ for t in the second equation and solve for r.

$$14 - \frac{40}{r} = \frac{40}{r - 6},$$

LCD is $r(r - 6)$

$$r(r - 6)\left(14 - \frac{40}{r}\right) = r(r - 6) \cdot \frac{40}{r - 6}$$

$$14r(r - 6) - 40(r - 6) = 40r$$

$$14r^2 - 84r - 40r + 240 = 40r$$

$$14r^2 - 164r + 240 = 0$$

$$7r^2 - 82r + 120 = 0$$

$$(7r - 12)(r - 10) = 0$$

$$r = \frac{12}{7} \quad or \quad r = 10$$

Check. Since negative speed has no meaning in this problem (If $r = \frac{12}{7}$, then $r - 6 = -\frac{30}{7}$.), we check only 10 mph. If $r = 10$, then the time of the trip to Hillsboro is $\frac{40}{10}$, or 4 hr. The speed of the return trip is $10 - 6$, or 4 mph, and the time is $\frac{40}{4}$, or 10 hr. The total time for the round trip is $4 \text{ hr} + 10 \text{ hr}$, or 14 hr. The value checks.

State. Naoki's speed on the trip to Hillsboro was 10 mph and it was 4 mph on the return trip.

29. Familiarize. We make a drawing and organize the information in a table. Let r represent the speed of the barge in still water, and let t represent the time of the trip upriver.

24 mi $r - 4$ mph t hr   → Upriver

Downriver ←   24 mi $r + 4$ mph $5 - t$ hr

Trip	Distance	Speed	Time
Upriver	24	$r - 4$	t
Downriver	24	$r + 4$	$5 - t$

Translate. Using $t = \frac{d}{r}$, we get two equations from the table,

$$t = \frac{24}{r - 4} \text{ and } 5 - t = \frac{24}{r + 4}.$$

Solve. We substitute $\frac{24}{r - 4}$ for t in the second equation and solve for r.

$$5 - \frac{24}{r - 4} = \frac{24}{r + 4},$$

LCD is $(r-4)(r+4)$

$$(r - 4)(r + 4)\left(5 - \frac{24}{r - 4}\right) = (r - 4)(r + 4) \cdot \frac{24}{r + 4}$$

$$5(r - 4)(r + 4) - 24(r + 4) = 24(r - 4)$$

$$5r^2 - 80 - 24r - 96 = 24r - 96$$

$$5r^2 - 48r - 80 = 0$$

We use the quadratic formula.

$$r = \frac{-(-48) \pm \sqrt{(-48)^2 - 4 \cdot 5 \cdot (-80)}}{2 \cdot 5}$$

$$r = \frac{48 \pm \sqrt{3904}}{10}$$

$$r \approx 11 \quad or \quad r \approx -1.5$$

Check. Since negative speed has no meaning in this problem, we check only 11 mph. If $r \approx 11$, then the speed upriver is about $11 - 4$, or 7 mph, and the time is about $\frac{24}{7}$, or 3.4 hr. The speed downriver is about $11 + 4$, or 15 mph, and the time is about $\frac{24}{15}$, or 1.6 hr. The total time of the round trip is $3.4 + 1.6$, or 5 hr. The value checks.

State. The barge must be able to travel about 11 mph in still water.

31.
$$A = 6s^2$$
$$\frac{A}{6} = s^2 \qquad \text{Dividing by 6}$$
$$\sqrt{\frac{A}{6}} = s \qquad \text{Taking the positive square root}$$

33.
$$F = \frac{Gm_1 m_2}{r^2}$$
$$Fr^2 = Gm_1 m_2 \qquad \text{Multiplying by } r^2$$
$$r^2 = \frac{Gm_1 m_2}{F} \qquad \text{Dividing by } F$$
$$r = \sqrt{\frac{Gm_1 m_2}{F}} \qquad \text{Taking the positive square root}$$

35.
$$E = mc^2$$
$$\frac{E}{m} = c^2 \qquad \text{Dividing by } m$$
$$\sqrt{\frac{E}{m}} = c \qquad \text{Taking the square root}$$

37.
$$a^2 + b^2 = c^2$$
$$b^2 = c^2 - a^2 \qquad \text{Subtracting } a^2$$
$$b = \sqrt{c^2 - a^2} \qquad \text{Taking the square root}$$

39.
$$N = \frac{k^2 - 3k}{2}$$
$$2N = k^2 - 3k$$
$$0 = k^2 - 3k - 2N \qquad \text{Standard form}$$
$$a = 1, \ b = -3, \ c = -2N$$
$$k = \frac{-(-3) \pm \sqrt{(-3)^3 - 4 \cdot 1 \cdot (-2N)}}{2 \cdot 1} \qquad \begin{array}{l}\text{Using the}\\ \text{quadratic formula}\end{array}$$
$$k = \frac{3 \pm \sqrt{9 + 8N}}{2}$$

Since taking the negative square root would result in a negative answer, we take the positive one.

$$k = \frac{3 + \sqrt{9 + 8N}}{2}$$

41.
$$A = 2\pi r^2 + 2\pi r h$$
$$0 = 2\pi r^2 + 2\pi r h - A \qquad \text{Standard form}$$
$$a = 2\pi, \ b = 2\pi h, \ c = -A$$
$$r = \frac{-2\pi h \pm \sqrt{(2\pi h)^2 - 4 \cdot 2\pi \cdot (-A)}}{2 \cdot 2\pi} \qquad \begin{array}{l}\text{Using the}\\ \text{quadratic formula}\end{array}$$

$$r = \frac{-2\pi h \pm \sqrt{4\pi^2 h^2 + 8\pi A}}{4\pi}$$

$$r = \frac{-2\pi h \pm 2\sqrt{\pi^2 h^2 + 2\pi A}}{4\pi}$$

$$r = \frac{-\pi h \pm \sqrt{\pi^2 h^2 + 2\pi A}}{2\pi}$$

Since taking the negative square root would result in a negative answer, we take the positive one.

$$r = \frac{-\pi h + \sqrt{\pi^2 h^2 + 2\pi A}}{2\pi}$$

43.
$$T = 2\pi \sqrt{\frac{L}{g}}$$

$$\frac{T}{2\pi} = \sqrt{\frac{L}{g}} \qquad \text{Dividing by } 2\pi$$

$$\frac{T^2}{4\pi^2} = \frac{L}{g} \qquad \text{Squaring}$$

$$gT^2 = 4\pi^2 L \qquad \text{Multiplying by } 4\pi^2 g$$

$$g = \frac{4\pi^2 L}{T^2} \qquad \text{Dividing by } T^2$$

45.
$$I = \frac{704.5W}{H^2}$$

$$H^2 I = 704.5W \qquad \text{Multiplying by } H^2$$

$$H^2 = \frac{704.5W}{I} \qquad \text{Dividing by } I$$

$$H = \sqrt{\frac{704.5W}{I}}$$

47.
$$m = \frac{m_0}{\sqrt{1 - \dfrac{v^2}{c^2}}}$$

$$m^2 = \frac{m_0^2}{1 - \dfrac{v^2}{c^2}} \qquad \text{Principle of powers}$$

$$m^2 \left(1 - \frac{v^2}{c^2}\right) = m_0^2$$

$$m^2 - \frac{m^2 v^2}{c^2} = m_0^2$$

$$m^2 - m_0^2 = \frac{m^2 v^2}{c^2}$$

$$c^2(m^2 - m_0^2) = m^2 v^2$$

$$\frac{c^2(m^2 - m_0^2)}{m^2} = v^2$$

$$\sqrt{\frac{c^2(m^2 - m_0^2)}{m^2}} = v$$

$$\frac{c\sqrt{m^2 - m_0^2}}{m} = v$$

49. Discussion and Writing Exercise

51.
$$\frac{1}{x-1} + \frac{1}{x^2 - 3x + 2}$$

$$= \frac{1}{x-1} + \frac{1}{(x-1)(x-2)}, \quad \text{LCD is } (x-1)(x-2)$$

$$= \frac{1}{x-1} \cdot \frac{x-2}{x-2} + \frac{1}{(x-1)(x-2)}$$

$$= \frac{x-2}{(x-1)(x-2)} + \frac{1}{(x-1)(x-2)}$$

$$= \frac{x-2+1}{(x-1)(x-2)}$$

$$= \frac{x-1}{(x-1)(x-2)}$$

$$= \frac{(x-1) \cdot 1}{(x-1)(x-2)}$$

$$= \frac{1}{x-2}$$

53.
$$\frac{2}{x+3} - \frac{x}{x-1} + \frac{x^2+2}{x^2+2x-3}$$

$$= \frac{2}{x+3} - \frac{x}{x-1} + \frac{x^2+2}{(x+3)(x-1)},$$
$$\text{LCD is } (x+3)(x-1)$$

$$= \frac{2}{x+3} \cdot \frac{x-1}{x-1} - \frac{x}{x-1} \cdot \frac{x+3}{x+3} + \frac{x^2+2}{(x+3)(x-1)}$$

$$= \frac{2(x-1)}{(x+3)(x-1)} - \frac{x(x+3)}{(x-1)(x+3)} + \frac{x^2+2}{(x+3)(x-1)}$$

$$= \frac{2(x-1) - x(x+3) + x^2 + 2}{(x+3)(x-1)}$$

$$= \frac{2x - 2 - x^2 - 3x + x^2 + 2}{(x+3)(x-1)}$$

$$= \frac{-x}{(x+3)(x-1)}$$

55. $\sqrt{-20} = \sqrt{-1 \cdot 4 \cdot 5} = i \cdot 2 \cdot \sqrt{5} = 2\sqrt{5}i$, or $2i\sqrt{5}$

57.
$$\frac{\dfrac{4}{a^2 b}}{\dfrac{3}{a} - \dfrac{4}{b^2}} = \frac{\dfrac{4}{a^2 b}}{\dfrac{3}{a} - \dfrac{4}{b^2}} \cdot \frac{a^2 b^2}{a^2 b^2}$$

$$= \frac{\dfrac{4}{a^2 b} \cdot a^2 b^2}{\dfrac{3}{a} \cdot a^2 b^2 - \dfrac{4}{b^2} \cdot a^2 b^2}$$

$$= \frac{4b}{3ab^2 - 4a^2}, \text{ or } \frac{4b}{a(3b^2 - 4a)}$$

59.
$$\frac{1}{a-1} = a + 1$$

$$\frac{1}{a-1} \cdot a - 1 = (a+1)(a-1)$$

$$1 = a^2 - 1$$

$$2 = a^2$$

$$\pm\sqrt{2} = a$$

61. Let s represent a length of a side of the cube, let S represent the surface area of the cube, and let A represent the surface area of the sphere. Then the diameter of the sphere is s, so the radius r is $s/2$. From Exercise 32, we know, $A = 4\pi r^2$, so when $r = s/2$ we have $A = 4\pi\left(\dfrac{s}{2}\right)^2 = 4\pi \cdot \dfrac{s^2}{4} = \pi s^2$. From the formula for the surface area of a cube (See Exercise 31.) we know that $S = 6s^2$, so $\dfrac{S}{6} = s^2$ and then $A = \pi \cdot \dfrac{S}{6}$, or $A(S) = \dfrac{\pi S}{6}$.

63.
$$\frac{w}{l} = \frac{l}{w+l}$$
$$l(w+l) \cdot \frac{w}{l} = l(w+l) \cdot \frac{l}{w+l}$$
$$w(w+l) = l^2$$
$$w^2 + lw = l^2$$
$$0 = l^2 - lw - w^2$$

Use the quadratic formula with $a = 1$, $b = -w$, and $c = -w^2$.
$$l = \frac{-(-w) \pm \sqrt{(-w)^2 - 4 \cdot 1(-w^2)}}{2 \cdot 1}$$
$$l = \frac{w \pm \sqrt{w^2 + 4w^2}}{2} = \frac{w \pm \sqrt{5w^2}}{2}$$
$$l = \frac{w \pm w\sqrt{5}}{2}$$

Since $\dfrac{w - w\sqrt{5}}{2}$ is negative we use the positive square root:
$$l = \frac{w + w\sqrt{5}}{2}$$

Exercise Set 7.4

1. $x^2 - 8x + 16 = 0$

$a = 1$, $b = -8$, $c = 16$

We compute the discriminant.
$$b^2 - 4ac = (-8)^2 - 4 \cdot 1 \cdot 16$$
$$= 64 - 64$$
$$= 0$$

Since $b^2 - 4ac = 0$, there is just one solution, and it is a real number.

3. $x^2 + 1 = 0$

$a = 1$, $b = 0$, $c = 1$

We compute the discriminant.
$$b^2 - 4ac = 0^2 - 4 \cdot 1 \cdot 1$$
$$= -4$$

Since $b^2 - 4ac < 0$, there are two nonreal solutions.

5. $x^2 - 6 = 0$

$a = 1$, $b = 0$, $c = -6$

We compute the discriminant.
$$b^2 - 4ac = 0^2 - 4 \cdot 1 \cdot (-6)$$
$$= 24$$

Since $b^2 - 4ac > 0$, there are two real solutions.

7. $4x^2 - 12x + 9 = 0$

$a = 4$, $b = -12$, $c = 9$

We compute the discriminant.
$$b^2 - 4ac = (-12)^2 - 4 \cdot 4 \cdot 9$$
$$= 144 - 144$$
$$= 0$$

Since $b^2 - 4ac = 0$, there is just one solution, and it is a real number.

9. $x^2 - 2x + 4 = 0$

$a = 1$, $b = -2$, $c = 4$

We compute the discriminant.
$$b^2 - 4ac = (-2)^2 - 4 \cdot 1 \cdot 4$$
$$= 4 - 16$$
$$= -12$$

Since $b^2 - 4ac < 0$, there are two nonreal solutions.

11. $9t^2 - 3t = 0$

$a = 9$, $b = -3$, $c = 0$

We compute the discriminant.
$$b^2 - 4ac = (-3)^2 - 4 \cdot 9 \cdot 0$$
$$= 9 - 0$$
$$= 9$$

Since $b^2 - 4ac > 0$, there are two real solutions.

13. $y^2 = \dfrac{1}{2}y + \dfrac{3}{5}$

$$y^2 - \frac{1}{2}y - \frac{3}{5} = 0 \qquad \text{Standard form}$$

$a = 1$, $b = -\dfrac{1}{2}$, $c = -\dfrac{3}{5}$

We compute the discriminant.
$$b^2 - 4ac = \left(-\frac{1}{2}\right)^2 - 4 \cdot 1 \cdot \left(-\frac{3}{5}\right)$$
$$= \frac{1}{4} + \frac{12}{5}$$
$$= \frac{53}{20}$$

Since $b^2 - 4ac > 0$, there are two real solutions.

15. $4x^2 - 4\sqrt{3}x + 3 = 0$

$a = 4$, $b = -4\sqrt{3}$, $c = 3$

We compute the discriminant.
$$b^2 - 4ac = (-4\sqrt{3})^2 - 4 \cdot 4 \cdot 3$$
$$= 48 - 48$$
$$= 0$$

Since $b^2 - 4ac = 0$, there is just one solution, and it is a real number.

17. The solutions are -4 and 4.
$$x = -4 \quad or \qquad x = 4$$
$$x + 4 = 0 \quad or \quad x - 4 = 0$$
$$(x+4)(x-4) = 0 \qquad \text{Principle of zero products}$$
$$x^2 - 16 = 0 \qquad (A+B)(A-B) = A^2 - B^2$$

19. The solutions are -2 and -7.

$$x = -2 \quad or \quad x = -7$$
$$x + 2 = 0 \quad or \quad x + 7 = 0$$
$$(x+2)(x+7) = 0 \quad \text{Principle of zero products}$$
$$x^2 + 9x + 14 = 0 \quad \text{FOIL}$$

21. The only solution is 8. It must be a double solution.

$$x = 8 \quad or \quad x = 8$$
$$x - 8 = 0 \quad or \quad x - 8 = 0$$
$$(x-8)(x-8) = 0 \quad \text{Principle of zero products}$$
$$x^2 - 16x + 64 = 0 \quad (A-B)^2 = A^2 - 2AB + B^2$$

23. The solutions are $-\frac{2}{5}$ and $\frac{6}{5}$.

$$x = -\frac{2}{5} \quad or \quad x = \frac{6}{5}$$
$$x + \frac{2}{5} = 0 \quad or \quad x - \frac{6}{5} = 0$$
$$5x + 2 = 0 \quad or \quad 5x - 6 = 0 \quad \text{Clearing fractions}$$
$$(5x+2)(5x-6) = 0 \quad \text{Principle of zero products}$$
$$25x^2 - 20x - 12 = 0 \quad \text{FOIL}$$

25. The solutions are $\frac{k}{3}$ and $\frac{m}{4}$.

$$x = \frac{k}{3} \quad or \quad x = \frac{m}{4}$$
$$x - \frac{k}{3} = 0 \quad or \quad x - \frac{m}{4} = 0$$
$$3x - k = 0 \quad or \quad 4x - m = 0 \quad \text{Clearing fractions}$$
$$(3x-k)(4x-m) = 0 \quad \text{Principle of zero products}$$
$$12x^2 - 3mx - 4kx + km = 0 \quad \text{FOIL}$$
$$12x^2 - (3m+4k)x + km = 0 \quad \text{Collecting like terms}$$

27. The solutions are $-\sqrt{3}$ and $2\sqrt{3}$.

$$x = -\sqrt{3} \quad or \quad x = 2\sqrt{3}$$
$$x + \sqrt{3} = 0 \quad or \quad x - 2\sqrt{3} = 0$$
$$(x+\sqrt{3})(x-2\sqrt{3}) = 0 \quad \text{Principle of zero products}$$
$$x^2 - 2\sqrt{3}x + \sqrt{3}x - 2(\sqrt{3})^2 = 0 \quad \text{FOIL}$$
$$x^2 - \sqrt{3}x - 6 = 0$$

29. $x^4 - 6x^2 + 9 = 0$

Let $u = x^2$ and think of x^4 as $(x^2)^2$.

$$u^2 - 6u + 9 = 0 \quad \text{Substituting } u \text{ for } x^2$$
$$(u-3)(u-3) = 0$$
$$u - 3 = 0 \quad or \quad u - 3 = 0$$
$$u = 3 \quad or \quad u = 3$$

Now we substitute x^2 for u and solve the equation:

$$x^2 = 3$$
$$x = \pm\sqrt{3}$$

Both $\sqrt{3}$ and $-\sqrt{3}$ check. They are the solutions.

31. $x - 10\sqrt{x} + 9 = 0$

Let $u = \sqrt{x}$ and think of x as $(\sqrt{x})^2$.

$$u^2 - 10u + 9 = 0 \quad \text{Substituting } u \text{ for } \sqrt{x}$$
$$(u-9)(u-1) = 0$$
$$u - 9 = 0 \quad or \quad u - 1 = 0$$
$$u = 9 \quad or \quad u = 1$$

Now we substitute $\sqrt{x}$ for u and solve these equations:

$$\sqrt{x} = 9 \quad or \quad \sqrt{x} = 1$$
$$x = 81 \quad or \quad x = 1$$

The numbers 81 and 1 both check. They are the solutions.

33. $(x^2 - 6x) - 2(x^2 - 6x) - 35 = 0$

Let $u = x^2 - 6x$.

$$u^2 - 2u - 35 = 0 \quad \text{Substituting } u \text{ for } x^2-6x$$
$$(u-7)(u+5) = 0$$
$$u - 7 = 0 \quad or \quad u + 5 = 0$$
$$u = 7 \quad or \quad u = -5$$

Now we substitute $x^2 - 6x$ for u and solve these equations:

$$x^2 - 6x = 7 \quad or \quad x^2 - 6x = -5$$
$$x^2 - 6x - 7 = 0 \quad or \quad x^2 - 6x + 5 = 0$$
$$(x-7)(x+1) = 0 \quad or \quad (x-5)(x-1) = 0$$
$$x = 7 \text{ or } x = -1 \text{ or } x = 5 \text{ or } x = 1$$

The numbers -1, 1, 5, and 7 check. They are the solutions.

35. $x^{-2} - 5^{-1} - 36 = 0$

Let $u = x^{-1}$.

$$u^2 - 5u - 36 = 0 \quad \text{Substituting } u \text{ for } x^{-1}$$
$$(u-9)(u+4) = 0$$
$$u - 9 = 0 \quad or \quad u + 4 = 0$$
$$u = 9 \quad or \quad u = -4$$

Now we substitute x^{-1} for u and solve these equations:

$$x^{-1} = 9 \quad or \quad x^{-1} = -4$$
$$\frac{1}{x} = 9 \quad or \quad \frac{1}{x} = -4$$
$$\frac{1}{9} = x \quad or \quad -\frac{1}{4} = x$$

Both $\frac{1}{9}$ and $-\frac{1}{4}$ check. They are the solutions.

37. $(1 + \sqrt{x})^2 + (1 + \sqrt{x}) - 6 = 0$

Let $u = 1 + \sqrt{x}$.

$$u^2 + u - 6 = 0 \quad \text{Substituting } u \text{ for } 1+\sqrt{x}$$
$$(u+3)(u-2) = 0$$

$$u + 3 = 0 \quad or \quad u - 2 = 0$$
$$u = -3 \quad or \quad u = 2$$
$1 + \sqrt{x} = -3 \quad or \quad 1 + \sqrt{x} = 2$ Substituting $1 + \sqrt{x}$ for u

$$\sqrt{x} = -4 \quad or \quad \sqrt{x} = 1$$
No solution $\quad\quad\quad x = 1$

The number 1 checks. It is the solution.

39. $(y^2 - 5y)^2 - 2(y^2 - 5y) - 24 = 0$

Let $u = y^2 - 5y$.

$u^2 - 2u - 24 = 0$ Substituting u for $y^2 - 5y$

$(u - 6)(u + 4) = 0$

$$u - 6 = 0 \quad or \quad u + 4 = 0$$
$$u = 6 \quad or \quad u = -4$$
$y^2 - 5y = 6 \quad or \quad y^2 - 5y = -4$

Substituting $y^2 - 5y$ for u

$y^2 - 5y - 6 = 0 \quad or \quad y^2 - 5y + 4 = 0$

$(y - 6)(y + 1) = 0 \quad or \quad (y - 4)(y - 1) = 0$

$y = 6 \quad or \quad y = -1 \quad or \quad y = 4 \quad or \quad y = 1$

The numbers -1, 1, 4, and 6 check. They are the solutions.

41. $t^4 - 10t^2 + 9 = 0$

Let $u = t^2$.

$u^2 - 10u + 9 = 0$ Substituting u for t^2

$(u - 1)(u - 9) = 0$

$$u - 1 = 0 \quad or \quad u - 9 = 0$$
$$u = 1 \quad or \quad u = 9$$
$$t^2 = 1 \quad or \quad t^2 = 9$$
$$t = \pm 1 \quad or \quad t = \pm 3$$

All four numbers check. They are the solutions.

43. $2x^{-2} + x^{-1} - 1 = 0$

Let $u = x^{-1}$.

$2u^2 + u - 1 = 0$ Substituting u for x^{-1}

$(2u - 1)(u + 1) = 0$

$$2u - 1 = 0 \quad or \quad u + 1 = 0$$
$$2u = 1 \quad or \quad u = -1$$
$$u = \frac{1}{2} \quad or \quad u = -1$$
$x^{-1} = \frac{1}{2} \quad or \quad x^{-1} = -1$ Substituting x^{-1} for u

$$\frac{1}{x} = \frac{1}{2} \quad or \quad \frac{1}{x} = -1$$
$$x = 2 \quad or \quad x = -1$$

Both 2 and -1 check. They are the solutions.

45. $6x^4 - 19x^2 + 15 = 0$

Let $u = x^2$.

$6u^2 - 19u + 15 = 0$ Substituting u for x^2

$(3u - 5)(2u - 3) = 0$

$$3u - 5 = 0 \quad or \quad 2u - 3 = 0$$
$$3u = 5 \quad or \quad 2u = 3$$
$$u = \frac{5}{3} \quad or \quad u = \frac{3}{2}$$
$x^2 = \frac{5}{3} \quad or \quad x^2 = \frac{3}{2}$ Substituting x^2 for u

$$x = \pm\sqrt{\frac{5}{3}} \quad or \quad x = \pm\sqrt{\frac{3}{2}}$$
$$x = \pm\frac{\sqrt{15}}{3} \quad or \quad x = \pm\frac{\sqrt{6}}{2}$$

Rationalizing denominators

All four numbers check. They are the solutions.

47. $x^{2/3} - 4x^{1/3} - 5 = 0$

Let $u = x^{1/3}$.

$u^2 - 4u - 5 = 0$ Substituting u for $x^{1/3}$

$(u - 5)(u + 1) = 0$

$$u - 5 = 0 \quad or \quad u + 1 = 0$$
$$u = 5 \quad or \quad u = -1$$
$x^{1/3} = 5 \quad or \quad x^{1/3} = -1$ Substituting $x^{1/3}$ for u

$(x^{1/3})^3 = 5^3 \quad or \quad (x^{1/3})^3 = (-1)^3$ Principle of powers

$x = 125 \quad or \quad x = -1$

Both 125 and -1 check. They are the solutions.

49. $\left(\dfrac{x-4}{x+1}\right)^2 - 2\left(\dfrac{x-4}{x+1}\right) - 35 = 0$

Let $u = \dfrac{x-4}{x+1}$.

$u^2 - 2u - 35 = 0$ Substituting u for $\dfrac{x-4}{x+1}$

$(u - 7)(u + 5) = 0$

$$u - 7 = 0 \quad or \quad u + 5 = 0$$
$$u = 7 \quad or \quad u = -5$$
$\dfrac{x-4}{x+1} = 7 \quad or \quad \dfrac{x-4}{x+1} = -5$ Substituting $\dfrac{x-4}{x+1}$ for u

$x - 4 = 7(x + 1) \quad or \quad x - 4 = -5(x + 1)$

$x - 4 = 7x + 7 \quad or \quad x - 4 = -5x - 5$

$$-6x = 11 \quad or \quad 6x = -1$$
$$x = -\frac{11}{6} \quad or \quad x = -\frac{1}{6}$$

Both $-\dfrac{11}{6}$ and $-\dfrac{1}{6}$ check. They are the solutions.

51. $9\left(\dfrac{x+2}{x+3}\right)^2 - 6\left(\dfrac{x+2}{x+3}\right) + 1 = 0$

Let $u = \dfrac{x+2}{x+3}$.

$9u^2 - 6u + 1 = 0$ Substituting u for $\dfrac{x+2}{x+3}$

$(3u - 1)(3u - 1) = 0$

$3u - 1 = 0 \quad or \quad 3u - 1 = 0$

$3u = 1 \quad or \quad 3u = 1$

$u = \dfrac{1}{3} \quad or \quad u = \dfrac{1}{3}$

Now we substitute $\dfrac{x+2}{x+3}$ for u and solve the equation:

$\dfrac{x+2}{x+3} = \dfrac{1}{3}$

$3(x+2) = x+3 \qquad$ Multiplying by $3(x+3)$

$3x + 6 = x + 3$

$2x = -3$

$x = -\dfrac{3}{2}$

The number $-\dfrac{3}{2}$ checks. It is the solution.

53. $\left(\dfrac{x^2 - 2}{x}\right)^2 - 7\left(\dfrac{x^2 - 2}{x}\right) - 18 = 0$

Let $u = \dfrac{x^2 - 2}{x}$.

$u^2 - 7u - 18 = 0 \qquad$ Substituting u for $\dfrac{x^2 - 2}{x}$

$(u - 9)(u + 2) = 0$

$u - 9 = 0 \quad or \quad u + 2 = 0$

$u = 9 \quad or \qquad u = -2$

$\dfrac{x^2 - 2}{x} = 9 \quad or \qquad \dfrac{x^2 - 2}{x} = -2$

$\qquad\qquad\qquad$ Substituting $\dfrac{x^2 - 2}{x}$ for u

$x^2 - 2 = 9x \quad or \qquad x^2 - 2 = -2x$

$x^2 - 9x - 2 = 0 \quad or \quad x^2 + 2x - 2 = 0$

$x = \dfrac{-(-9) \pm \sqrt{(-9)^2 - 4 \cdot 1 \cdot (-2)}}{2 \cdot 1}$

$x = \dfrac{9 \pm \sqrt{89}}{2}$

or

$x = \dfrac{-2 \pm \sqrt{2^2 - 4 \cdot 1 \cdot (-2)}}{2 \cdot 1} = \dfrac{-2 \pm \sqrt{12}}{2}$

$x = \dfrac{-2 \pm 2\sqrt{3}}{2} = -1 \pm \sqrt{3}$

All four numbers check. They are the solutions.

55. The x-intercepts occur where $f(x) = 0$. Thus, we must have $5x + 13\sqrt{x} - 6 = 0$.

Let $u = \sqrt{x}$.

$5u^2 + 13u - 6 = 0 \qquad$ Substituting

$(5u - 2)(u + 3) = 0$

$u = \dfrac{2}{5} \quad or \quad u = -3$

Now replace u with $\sqrt{x}$ and solve these equations:

$\sqrt{x} = \dfrac{2}{5} \quad or \quad \sqrt{x} = -3$

$x = \dfrac{4}{25} \qquad$ No solution

The number $\dfrac{4}{25}$ checks. Thus, the x-intercept is $\left(\dfrac{4}{25}, 0\right)$.

57. The x-intercepts occur where $f(x) = 0$. Thus, we must have $(x^2 - 3x)^2 - 10(x^2 - 3x) + 24 = 0$.

Let $u = x^2 - 3x$.

$u^2 - 10u + 24 = 0 \qquad$ Substituting

$(u - 6)(u - 4) = 0$

$u = 6 \quad or \quad u = 4$

Now replace u with $x^2 - 3x$ and solve these equations:

$x^2 - 3x = 6 \quad or \qquad x^2 - 3x = 4$

$x^2 - 3x - 6 = 0 \quad or \quad x^2 - 3x - 4 = 0$

$x = \dfrac{-(-3) \pm \sqrt{(-3)^2 - 4(1)(-6)}}{2 \cdot 1} \quad or$

$\qquad\qquad\qquad\qquad (x - 4)(x + 1) = 0$

$x = \dfrac{3 \pm \sqrt{33}}{2} \quad or \quad x = 4 \; or \; x = -1$

All four numbers check. Thus, the x-intercepts are $\left(\dfrac{3 + \sqrt{33}}{2}, 0\right)$, $\left(\dfrac{3 - \sqrt{33}}{2}, 0\right)$, $(4, 0)$, and $(-1, 0)$.

59. Discussion and Writing Exercise

61. *Familiarize.* Let $x =$ the number of pounds of Kenyan coffee and $y =$ the number of pounds of Peruvian coffee in the mixture. We organize the information in a table.

Type of Coffee	Kenyan	Peruvian	Mixture
Price per pound	\$6.75	\$11.25	\$8.55
Number of pounds	x	y	50
Total cost	\$6.75x	\$11.25y	\$8.55 × 50, or \$427.50

Translate. From the last two rows of the table we get a system of equations.

$x + y = 50,$

$6.75x + 11.25y = 427.50$

Solve. Solving the system of equations, we get $(30, 20)$.

Check. The total number of pounds in the mixture is $30 + 20$, or 50. The total cost of the mixture is \$6.75(30) + \$11.25(20) = \$427.50. The values check.

State. The mixture should consist of 30 lb of Kenyan coffee and 20 lb of Peruvian coffee.

63. $\sqrt{8x} \cdot \sqrt{2x} = \sqrt{8x \cdot 2x} = \sqrt{16x^2} = \sqrt{(4x)^2} = 4x$

65. $\sqrt[4]{9a^2} \cdot \sqrt[4]{18a^3} = \sqrt[4]{9a^2 \cdot 18a^3} =$

$\sqrt[4]{3 \cdot 3 \cdot a^2 \cdot 3 \cdot 3 \cdot 2 \cdot a^3} = \sqrt[4]{3^4 a^4 \cdot 2a} = \sqrt[4]{3^4} \sqrt[4]{a^4} \sqrt[4]{2a} =$

$3a\sqrt[4]{2a}$

67. Graph $f(x) = -\frac{3}{5}x + 4$.

Choose some values for x, find the corresponding values of $f(x)$, plot the points $(x, f(x))$, and draw the graph.

x	$f(x)$	$(x, f(x))$
-5	7	$(-5, 7)$
0	4	$(0, 4)$
5	1	$(5, 1)$

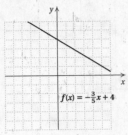

$f(x) = -\frac{3}{5}x + 4$

69. Graph $y = 4$.

The graph of $y = 4$ is a horizontal line with y-intercept $(0, 4)$.

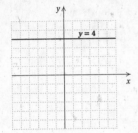

$y = 4$

71. Left to the student

73. a) $kx^2 - 2x + k = 0$; one solution is -3

We first find k by substituting -3 for x.

$$k(-3)^2 - 2(-3) + k = 0$$
$$9k + 6 + k = 0$$
$$10k = -6$$
$$k = -\frac{6}{10}$$
$$k = -\frac{3}{5}$$

b) $-\frac{3}{5}x^2 - 2x + \left(-\frac{3}{5}\right) = 0$ Substituting $-\frac{3}{5}$ for k

$$3x^2 + 10x + 3 = 0 \quad \text{Multiplying by } -5$$
$$(3x + 1)(x + 3) = 0$$

$$3x + 1 = 0 \quad or \quad x + 3 = 0$$
$$3x = -1 \quad or \quad x = -3$$
$$x = -\frac{1}{3} \quad or \quad x = -3$$

The other solution is $-\frac{1}{3}$.

75. For $ax^2 + bx + c = 0$, $-\frac{b}{a}$ is the sum of the solutions and $\frac{c}{a}$ is the product of the solutions. Thus $-\frac{b}{a} = \sqrt{3}$ and $\frac{c}{a} = 8$.

$$ax^2 + bx + c = 0$$
$$x^2 + \frac{b}{a}x + \frac{c}{a} = 0 \quad \text{Multiplying by } \frac{1}{a}$$
$$x^2 - \left(-\frac{b}{a}\right)x + \frac{c}{a} = 0$$
$$x^2 - \sqrt{3}x + 8 = 0 \quad \text{Substituting } \sqrt{3} \text{ for } -\frac{b}{a} \text{ and } 8 \text{ for } \frac{c}{a}$$

77. The graph includes the points $(-3, 0)$, $(0, -3)$, and $(1, 0)$. Substituting in $y = ax^2 + bx + c$, we have three equations.

$$0 = 9a - 3b + c,$$
$$-3 = \qquad\qquad c,$$
$$0 = a + b + c$$

The solution of this system of equations is $a = 1$, $b = 2$, $c = -3$.

79. $\dfrac{x}{x-1} - 6\sqrt{\dfrac{x}{x-1}} - 40 = 0$

Let $u = \sqrt{\dfrac{x}{x-1}}$.

$$u^2 - 6u - 40 = 0 \quad \text{Substituting for } \sqrt{\dfrac{x}{x-1}}$$
$$(u - 10)(u + 4) = 0$$

$$u = 10 \qquad or \qquad u = -4$$
$$\sqrt{\dfrac{x}{x-1}} = 10 \qquad or \qquad \sqrt{\dfrac{x}{x-1}} = -4$$

Substituting for u

$$\dfrac{x}{x-1} = 100 \qquad or \qquad \text{No solution}$$
$$x = 100x - 100 \quad \text{Multiplying by } (x - 1)$$
$$100 = 99x$$
$$\dfrac{100}{99} = x$$

This number checks. It is the solution.

81. $\sqrt{x-3} - \sqrt[4]{x-3} = 12$

$$(x-3)^{1/2} - (x-3)^{1/4} - 12 = 0$$

Let $u = (x-3)^{1/4}$.

$$u^2 - u - 12 = 0 \quad \text{Substituting for } (x-3)^{1/4}$$
$$(u - 4)(u + 3) = 0$$

$$u = 4 \qquad or \qquad u = -3$$
$$(x-3)^{1/4} = 4 \qquad or \qquad (x-3)^{1/4} = -3$$

Substituting for u

$$x - 3 = 4^4 \qquad or \qquad \text{No solution}$$
$$x - 3 = 256$$
$$x = 259$$

This number checks. It is the solution.

83. $x^6 - 28x^3 + 27 = 0$

    Let $u = x^3$.

        $u^2 - 28u + 27 = 0$ Substituting for x^3

        $(u - 27)(u - 1) = 0$

        $u = 27$ or $u = 1$

        $x^3 = 27$ or $x^3 = 1$ Substituting for u

        $x = 3$ or $x = 1$

    Both 3 and 1 check. They are the solutions.

Exercise Set 7.5

---

1. $f(x) = 4x^2$

    $f(x) = 4x^2$ is of the form $f(x) = ax^2$. Thus we know that the vertex is $(0, 0)$ and $x = 0$ is the line of symmetry.

    We know that $f(x) = 0$ when $x = 0$ since the vertex is $(0, 0)$.

        For $x = 1$, $f(x) = 4x^2 = 4 \cdot 1^2 = 4$.

        For $x = -1$, $f(x) = 4x^2 = 4 \cdot (-1)^2 = 4$.

        For $x = 2$, $f(x) = 4x^2 = 4 \cdot 2^2 = 16$.

        For $x = -2$, $f(x) = 4x^2 = 4 \cdot (-2)^2 = 16$.

    We complete the table.

x	$f(x)$	
0	0	← Vertex
1	4	
2	16	
−1	4	
−2	16	

    We plot the ordered pairs $(x, f(x))$ from the table and connect them with a smooth curve.

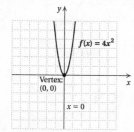

3. $f(x) = \dfrac{1}{3}x^2$ is of the form $f(x) = ax^2$. Thus we know that the vertex is $(0, 0)$ and $x = 0$ is the line of symmetry. We choose some numbers for x and find the corresponding values for $f(x)$. Then we plot the ordered pairs $(x, f(x))$ and connect them with a smooth curve.

    For $x = 3$, $f(x) = \dfrac{1}{3}x^2 = \dfrac{1}{3} \cdot 3^2 = 3$.

    For $x = -3$, $f(x) = \dfrac{1}{3}x^2 = \dfrac{1}{3} \cdot (-3)^2 = 3$.

    For $x = 6$, $f(x) = \dfrac{1}{3}x^2 = \dfrac{1}{3} \cdot 6^2 = 12$.

    For $x = -6$, $f(x) = \dfrac{1}{3}x^2 = \dfrac{1}{3} \cdot (-6)^2 = 12$.

x	$f(x)$	
0	0	← Vertex
1	$\dfrac{1}{3}$	
2	$\dfrac{4}{3}$	
−1	$\dfrac{1}{3}$	
−2	$\dfrac{4}{3}$	

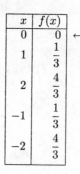

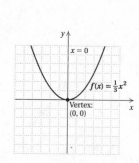

5. $f(x) = -\dfrac{1}{2}x^2$ is of the form $f(x) = ax^2$. Thus we know that the vertex is $(0, 0)$ and $x = 0$ is the line of symmetry. We choose some numbers for x and find the corresponding values for $f(x)$. Then we plot the ordered pairs $(x, f(x))$ and connect them with a smooth curve.

    For $x = 2$, $f(x) = -\dfrac{1}{2}x^2 = -\dfrac{1}{2} \cdot 2^2 = -2$.

    For $x = -2$, $f(x) = -\dfrac{1}{2}x^2 = -\dfrac{1}{2} \cdot (-2)^2 = -2$.

    For $x = 4$, $f(x) = -\dfrac{1}{2}x^2 = -\dfrac{1}{2} \cdot 4^2 = -8$.

    For $x = -4$, $f(x) = -\dfrac{1}{2}x^2 = -\dfrac{1}{2} \cdot (-4)^2 = -8$.

x	$f(x)$	
0	0	← Vertex
2	−2	
−2	−2	
4	−8	
−4	−8	

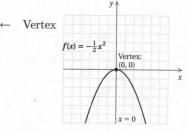

7. $f(x) = -4x^2$ is of the form $f(x) = ax^2$. Thus we know that the vertex is $(0, 0)$ and $x = 0$ is the line of symmetry. We choose some numbers for x and find the corresponding values for $f(x)$. Then we plot the ordered pairs $(x, f(x))$ and connect them with a smooth curve.

    For $x = 1$, $f(x) = -4x^2 = -4 \cdot 1^2 = -4$.

    For $x = -1$, $f(x) = -4x^2 = -4 \cdot (-1)^2 = -4$.

    For $x = 2$, $f(x) = -4x^2 = -4 \cdot 2^2 = -16$.

    For $x = -2$, $f(x) = -4x^2 = -4 \cdot (-2)^2 = -16$.

x	$f(x)$	
0	0	← Vertex
1	−4	
−1	−4	
2	−16	
−2	−16	

9. $f(x) = (x + 3)^2 = [x - (-3)]^2$ is of the form $f(x) = a(x - h)^2$.

    Thus we know that the vertex is $(-3, 0)$ and $x = -3$ is the line of symmetry. We choose some numbers for x and find the corresponding values for $f(x)$. Then we plot the ordered pairs $(x, f(x))$ and connect them with a smooth curve.

x	$f(x)$
-3	0 $\leftarrow$ Vertex
-2	1
-1	4
-4	1
-5	4

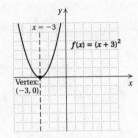

11. Graph: $f(x) = 2(x - 4)^2$

We choose some values of x and compute $f(x)$. Then we plot these ordered pairs and connect them with a smooth curve.

x	$f(x)$
4	0
5	2
3	2
6	8
2	8

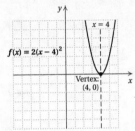

The graph of $f(x) = 2(x - 4)^2$ looks like the graph of $f(x) = 2x^2$ except that it is translated 4 units to the right. The vertex is $(4, 0)$, and the line of symmetry is $x = 4$.

13. Graph: $f(x) = -2(x + 2)^2$

We choose some values of x and compute $f(x)$. Then we plot these ordered pairs and connect them with a smooth curve.

x	$f(x)$
1	-18
0	-8
-1	-2
-2	0
-3	-2
-4	-8

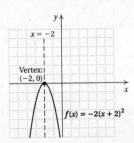

We can express the equation in the equivalent form $f(x) = -2[x - (-2)]^2$. Then we know that the graph looks like the graph of $f(x) = -2x^2$ translated 2 units to the left. The vertex is $(-2, 0)$, and the line of symmetry is $x = -2$.

15. Graph: $f(x) = 3(x - 1)^2$

We choose some values of x and compute $f(x)$. Then we plot these ordered pairs and connect them with a smooth curve.

x	$f(x)$
1	0
2	3
0	3
3	12
-1	12

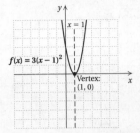

The graph of $f(x) = 3(x - 1)^2$ looks like the graph of $f(x) = 3x^2$ except that it is translated 1 unit to the right. The vertex is $(1, 0)$, and the line of symmetry is $x = 1$.

17. Graph: $f(x) = -\dfrac{3}{2}(x + 2)^2$

We choose some values of x and compute $f(x)$. Then we plot these ordered pairs and connect them with a smooth curve.

x	$f(x)$
-4	-6
-2	0
0	-6
2	-24

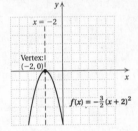

We can express the equation in the equivalent form $f(x) = -\dfrac{3}{2}[x - (-2)]^2$. Then we know that the graph looks like the graph of $f(x) = -\dfrac{3}{2}x^2$ translated 2 units to the left. The vertex is $(-2, 0)$, and the line of symmetry is $x = -2$.

19. Graph: $f(x) = (x - 3)^2 + 1$

We choose some values of x and compute $f(x)$. Then we plot these ordered pairs and connect them with a smooth curve.

x	$f(x)$
3	1
4	2
2	2
5	5
1	5

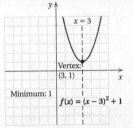

The graph of $f(x) = (x - 3)^2 + 1$ looks like the graph of $f(x) = x^2$ except that it is translated 3 units right and 1 unit up. The vertex is $(3, 1)$, and the line of symmetry is $x = 3$. The equation is of the form $f(x) = a(x - h)^2 + k$ with $a = 1$. Since $1 > 0$, we know that 1 is the minimum value.

21. Graph: $f(x) = -3(x + 4)^2 + 1$
$$f(x) = -3[x - (-4)]^2 + 1$$

We choose some values of x and compute $f(x)$. Then we plot these ordered pairs and connect them with a smooth curve.

x	$f(x)$
-4	1
$-3\frac{1}{2}$	$\frac{1}{4}$
$-4\frac{1}{2}$	$\frac{1}{4}$
-3	-2
-5	-2
-2	-11
-6	-11

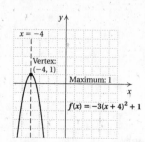

The graph of $f(x) = -3(x+4)^2 + 1$ looks like the graph of $f(x) = 3x^2$ except that it is translated 4 units left and 1 unit up and opens downward. The vertex is $(-4, 1)$, and the line of symmetry is $x = -4$. Since $-3 < 0$, we know that 1 is the maximum value.

23. Graph: $f(x) = \frac{1}{2}(x+1)^2 + 4$

$$f(x) = \frac{1}{2}[x - (-1)]^2 + 4$$

We choose some values of x and compute $f(x)$. Then we plot these ordered pairs and connect them with a smooth curve.

x	$f(x)$
1	6
2	$8\frac{1}{2}$
0	$4\frac{1}{2}$
-1	4
-2	$4\frac{1}{2}$
-3	6

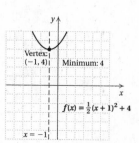

The graph of $f(x) = \frac{1}{2}(x+1)^2 + 4$ looks like the graph of $f(x) = \frac{1}{2}x^2$ except that it is translated 1 unit left and 4 units up. The vertex is $(-1, 4)$, and the line of symmetry is $x = -1$. Since $\frac{1}{2} > 0$, we know that 4 is the minimum value.

25. Graph: $f(x) = -(x+1)^2 - 2$
$$f(x) = -[x - (-1)]^2 + (-2)$$

We choose some values of x and compute $f(x)$. Then we plot these ordered pairs and connect them with a smooth curve.

x	$f(x)$
-1	-2
0	-3
-2	-3
1	-6
-3	-6

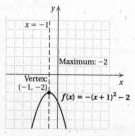

The graph of $f(x) = -(x+1)^2 - 2$ looks like the graph of $f(x) = x^2$ except that it is translated 1 unit left and 2

units down and opens downward. The vertex is $(-1, -2)$, and the line of symmetry is $x = -1$. Since $-1 < 0$, we know that -2 is the maximum value.

27. Discussion and Writing Exercise

29.
$$x - 5 = \sqrt{x+7}$$
$$(x-5)^2 = (\sqrt{x+7})^2 \quad \text{Principle of powers}$$
$$x^2 - 10x + 25 = x + 7$$
$$x^2 - 11x + 18 = 0$$
$$(x-9)(x-2) = 0$$
$$x - 9 = 0 \quad or \quad x - 2 = 0$$
$$x = 9 \quad or \qquad x = 2$$

Check: For 9:

$x - 5 = \sqrt{x+7}$	
$9 - 5$? $\sqrt{9+7}$	
4	$\sqrt{16}$
	4 TRUE

For 2:

$x - 5 = \sqrt{x+7}$	
$2 - 5$? $\sqrt{2+7}$	
-3	$\sqrt{9}$
	3 FALSE

Only 9 checks. It is the solution.

31. $\sqrt{x+4} = -11$

The equation has no solution, because the principal square root of a number is always nonnegative.

33. $\sqrt[4]{5x^3y^5} \, \sqrt[4]{125x^2y^3} = \sqrt[4]{625x^5y^8} = \sqrt[4]{625x^4y^8 \cdot x} = \sqrt[4]{625x^4y^8} \cdot \sqrt[4]{x} = 5xy^2 \sqrt[4]{x}$

35. Left to the student

Exercise Set 7.6

1. $f(x) = x^2 - 2x - 3 = (x^2 - 2x) - 3$

We complete the square inside the parentheses. We take half the x-coefficient and square it.

$$\frac{1}{2} \cdot (-2) = -1 \text{ and } (-1)^2 = 1$$

Then we add $1 - 1$ inside the parentheses.

$$f(x) = (x^2 - 2x + 1 - 1) - 3$$
$$= (x^2 - 2x + 1) - 1 - 3$$
$$= (x-1)^2 - 4$$
$$= (x-1)^2 + (-4)$$

Vertex: $(1, -4)$

Line of symmetry: $x = 1$

The coefficient of x^2 is 1, which is positive, so the graph opens up. This tells us that -4 is a minimum.

We plot a few points and draw the curve.

x	$f(x)$
1	-4
2	-3
0	-3
3	0
-1	0
4	5
-2	5

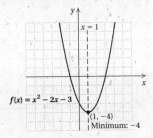

x	$f(x)$
4	2
5	5
3	5
6	14
2	14

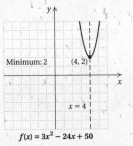

3. $f(x) = -x^2 - 4x - 2 = -(x^2 + 4x) - 2$

We complete the square inside the parentheses. We take half the x-coefficient and square it.

$\frac{1}{2} \cdot 4 = 2$ and $2^2 = 4$

Then we add $4 - 4$ inside the parentheses.

$$
\begin{aligned}
f(x) &= -(x^2 + 4x + 4 - 4) - 2 \\
&= -(x^2 + 4x + 4) + (-1)(-4) - 2 \\
&= -(x + 2)^2 + 4 - 2 \\
&= -(x + 2)^2 + 2 \\
&= -[x - (-2)]^2 + 2
\end{aligned}
$$

Vertex: $(-2, 2)$

Line of symmetry: $x = -2$

The coefficient of x^2 is -1, which is negative, so the graph opens down. This tells us that 2 is a maximum.

We plot a few points and draw the curve.

x	$f(x)$
-2	2
-4	-2
-3	1
-1	1
0	-2

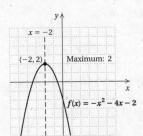

5. $f(x) = 3x^2 - 24x + 50 = 3(x^2 - 8x) + 50$

We complete the square inside the parentheses. We take half the x-coefficient and square it.

$\frac{1}{2} \cdot (-8) = -4$ and $(-4)^2 = 16$

Then we add $16 - 16$ inside the parentheses.

$$
\begin{aligned}
f(x) &= 3(x^2 - 8x + 16 - 16) + 50 \\
&= 3(x^2 - 8x + 16) - 48 + 50 \\
&= 3(x - 4)^2 + 2
\end{aligned}
$$

Vertex: $(4, 2)$

Line of symmetry: $x = 4$

The coefficient of x^2 is 3, which is positive, so the graph opens up. This tells us that 2 is a minimum.

We plot a few points and draw the curve.

7. $f(x) = -2x^2 - 2x + 3 = -2(x^2 + x) + 3$

We complete the square inside the parentheses. We take half the x-coefficient and square it.

$\frac{1}{2} \cdot 1 = \frac{1}{2}$ and $\left(\frac{1}{2}\right)^2 = \frac{1}{4}$

Then we add $\frac{1}{4} - \frac{1}{4}$ inside the parentheses.

$$
\begin{aligned}
f(x) &= -2\left(x^2 + x + \frac{1}{4} - \frac{1}{4}\right) + 3 \\
&= -2\left(x^2 + x + \frac{1}{4}\right) + (-2)\left(-\frac{1}{4}\right) + 3 \\
&= -2\left(x + \frac{1}{2}\right)^2 + \frac{1}{2} + 3 \\
&= -2\left(x + \frac{1}{2}\right)^2 + \frac{7}{2} \\
&= -2\left[x - \left(-\frac{1}{2}\right)\right]^2 + \frac{7}{2}
\end{aligned}
$$

Vertex: $\left(-\frac{1}{2}, \frac{7}{2}\right)$

Line of symmetry: $x = -\frac{1}{2}$

The coefficient of x^2 is -2, which is negative, so the graph opens down. This tells us that $\frac{7}{2}$ is a maximum.

We plot a few points and draw the curve.

x	$f(x)$
$-\frac{1}{2}$	$\frac{7}{2}$
-2	-1
-1	3
0	3
1	-1

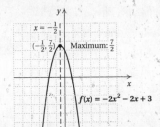

9. $f(x) = 5 - x^2 = -x^2 + 5 = -(x - 0)^2 + 5$

Vertex: $(0, 5)$

Line of symmetry: $x = 0$

The coefficient of x^2 is -1, which is negative, so the graph opens down. This tells us that 5 is a maximum.

We plot a few points and draw the curve.

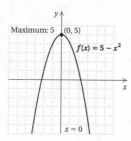

x	$f(x)$
0	5
1	4
−1	4
2	1
−2	1
3	−4
−3	−4

11. $f(x) = 2x^2 + 5x - 2 = 2\left(x^2 + \dfrac{5}{2}x\right) - 2$

We complete the square inside the parentheses. We take half the x-coefficient and square it.

$$\frac{1}{2} \cdot \frac{5}{2} = \frac{5}{4} \text{ and } \left(\frac{5}{4}\right)^2 = \frac{25}{16}$$

Then we add $\dfrac{25}{16} - \dfrac{25}{16}$ inside the parentheses.

$$f(x) = 2\left(x^2 + \frac{5}{2}x + \frac{25}{16} - \frac{25}{16}\right) - 2$$
$$= 2\left(x^2 + \frac{5}{2}x + \frac{25}{16}\right) + 2\left(-\frac{25}{16}\right) - 2$$
$$= 2\left(x + \frac{5}{4}\right)^2 - \frac{25}{8} - 2$$
$$= 2\left(x + \frac{5}{4}\right)^2 - \frac{41}{8}$$
$$= 2\left[x - \left(-\frac{5}{4}\right)\right]^2 + \left(-\frac{41}{8}\right)$$

Vertex: $\left(-\dfrac{5}{4}, -\dfrac{41}{8}\right)$

Line of symmetry: $x = -\dfrac{5}{4}$

The coefficient of x^2 is 2, which is positive, so the graph opens up. This tells us that $-\dfrac{41}{8}$ is a minimum.

We plot a few points and draw the curve.

x	$f(x)$
$-\dfrac{5}{4}$	$-\dfrac{41}{8}$
−3	1
−2	−4
−1	−5
0	−2
1	−5

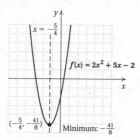

13. $f(x) = x^2 - 6x + 1$

The y-intercept is $(0, f(0))$. Since $f(0) = 0^2 - 6 \cdot 0 + 1 = 1$, the y-intercept is $(0, 1)$.

To find the x-intercepts, we solve $x^2 - 6x + 1 = 0$. Using the quadratic formula gives us $x = 3 \pm 2\sqrt{2}$.

Thus, the x-intercepts are $(3 - 2\sqrt{2}, 0)$ and $(3 + 2\sqrt{2}, 0)$, or approximately $(0.172, 0)$ and $(5.828, 0)$.

15. $f(x) = -x^2 + x + 20$

The y-intercept is $(0, f(0))$. Since $f(0) = -0^2 + 0 + 20 = 20$, the y-intercept is $(0, 20)$.

To find the x-intercepts, we solve $-x^2 + x + 20 = 0$. Factoring and using the principle of zero products gives us $x = -4$ or $x = 5$. Thus, the x-intercepts are $(-4, 0)$ and $(5, 0)$.

17. $f(x) = 4x^2 + 12x + 9$

The y-intercept is $(0, f(0))$. Since $f(0) = 4 \cdot 0^2 + 12 \cdot 0 + 9 = 9$, the y-intercept is $(0, 9)$.

To find the x-intercepts, we solve $4x^2 + 12x + 9 = 0$. Factoring and using the principle of zero products gives us $x = -\dfrac{3}{2}$. Thus, the x-intercept is $\left(-\dfrac{3}{2}, 0\right)$.

19. $f(x) = 4x^2 - x + 8$

The y-intercept is $(0, f(0))$. Since $f(0) = 4 \cdot 0^2 - 0 + 8 = 8$, the y-intercept is $(0, 8)$.

To find the x-intercepts, we solve $4x^2 - x + 8 = 0$. Using the quadratic formula gives us $x = \dfrac{1 \pm i\sqrt{127}}{8}$. Since there are no real-number solutions, there are no x-intercepts.

21. Discussion and Writing Exercise

23. a) $D = kw$
$$420 = k \cdot 28$$
$$\frac{420}{28} = k$$
$$15 = k$$

The equation of variation is $D = 15w$.

b) We substitute 42 for w and compute D.
$$D = 15w$$
$$D = 15 \cdot 42$$
$$D = 630$$

630 mg would be recommended for a child who weighs 42 kg.

25. $y = \dfrac{k}{x}$

$125 = \dfrac{k}{2}$

$250 = k$ Variation constant

$y = \dfrac{250}{x}$ Equation of variation

27. $y = kx$

$125 = k \cdot 2$

$\dfrac{125}{2} = k$ Variation constant

$y = \dfrac{125}{2}x$ Equation of variation

29. a) Minimum: −6.954

b) Maximum: 7.014

31. $f(x) = |x^2 - 1|$

We plot some points and draw the curve. Note that it will lie entirely on or above the x-axis since absolute value is never negative.

x	$f(x)$
-3	8
-2	3
-1	0
0	1
1	0
2	3
3	8

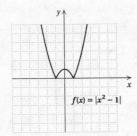

33. $f(x) = |x^2 - 3x - 4|$

We plot some points and draw the curve. Note that it will lie entirely on or above the x-axis since absolute value is never negative.

x	$f(x)$
-4	24
-3	14
-2	6
-1	0
0	4
1	6
2	6
3	4
4	0
5	6
6	14

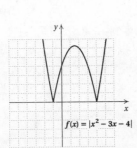

35. The horizontal distance from $(-1, 0)$ to $(3, -5)$ is $|3 - (-1)|$, or 4, so by symmetry the other x-intercept is $(3 + 4, 0)$, or $(7, 0)$. Substituting the three ordered pairs $(-1, 0)$, $(3, -5)$, and $(7, 0)$ in the equation $f(x) = ax^2 + bx + c$ yields a system of equations:

$$0 = a(-1)^2 + b(-1) + c,$$
$$-5 = a \cdot 3^2 + b \cdot 3 + c,$$
$$0 = a \cdot 7^2 + b \cdot 7 + c$$

or

$$0 = a - b + c,$$
$$-5 = 9a + 3b + c,$$
$$0 = 49a + 7b + c$$

The solution of this system of equations is $\left(\dfrac{5}{16}, -\dfrac{15}{8}, -\dfrac{35}{16}\right)$, so $f(x) = \dfrac{5}{16}x^2 - \dfrac{15}{8}x - \dfrac{35}{16}$.

37. $f(x) = \dfrac{x^2}{8} + \dfrac{x}{4} - \dfrac{3}{8}$

The x-coordinate of the vertex is $-b/2a$:

$$-\frac{b}{2a} = -\frac{\frac{1}{4}}{2 \cdot \frac{1}{8}} = -\frac{\frac{1}{4}}{\frac{1}{4}} = -1$$

The second coordinate is $f(-1)$:

$$f(-1) = \frac{(-1)^2}{8} + \frac{-1}{4} - \frac{3}{8}$$
$$= \frac{1}{8} - \frac{1}{4} - \frac{3}{8}$$
$$= -\frac{1}{2}$$

The vertex is $\left(-1, -\dfrac{1}{2}\right)$.

The line of symmetry is $x = -1$.

The coefficient of x^2 is $\dfrac{1}{8}$, which is positive, so the graph opens up. This tells us that $-\dfrac{1}{2}$ is a minimum.

We plot some points and draw the graph.

x	$f(x)$
-5	$\dfrac{3}{2}$
-3	0
-1	$-\dfrac{1}{2}$
0	$-\dfrac{3}{8}$
1	0
3	$\dfrac{3}{2}$
5	4

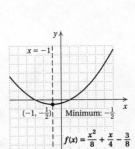

39. Graph $y_1 = x^2 - 4x + 2$ and $y_2 = 2 + x$ and use INTERSECT to find the points of intersection. They are $(0, 2)$ and $(5, 7)$.

Exercise Set 7.7

1. *Familiarize.* Referring to the drawing in the text, we let $l =$ the length of the atrium and $w =$ the width. Then the perimeter of each floor is $2l + 2w$, and the area is $l \cdot w$.

Translate. Using the formula for perimeter we have:

$$2l + 2w = 720$$
$$2l = 720 - 2w$$
$$l = \frac{720 - 2w}{2}$$
$$l = 360 - w$$

Substituting $360 - w$ for l in the formula for area, we get a quadratic function.

$$A = lw = (360 - w)w = 360w - w^2 = -w^2 + 360w$$

Carry out. We complete the square in order to find the vertex of the quadratic function.

$$A = -w^2 + 360w$$
$$= -(w^2 - 360w)$$
$$= -(w^2 - 360w + 32,400 - 32,400)$$
$$= -(w^2 - 360w + 32,400) + (-1)(-32,400)$$
$$= -(w - 180)^2 + 32,400$$

The vertex is $(180, 32,400)$. The coefficient of w^2 is negative, so the graph of the function is a parabola that opens down. This tells us that the function has a maximum value and that value occurs when $w = 180$. When $w = 180$, $l = 360 - w = 360 - 180 = 180$.

Check. We could find the value of the function for some values of w less than 180 and for some values greater than 180, determining that the maximum value we found, 32,400, is larger than these function values. We could also use the graph of the function to check the maximum value. Our answer checks.

State. Floors with dimensions 180 ft by 180 ft will allow an atrium with maximum area.

3. **Familiarize**. Let x represent the height of the file and y represent the width. We make a drawing.

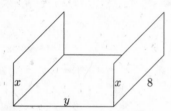

Translate. We have two equations.
$$2x + y = 14$$
$$V = 8xy$$

Solve the first equation for y.
$$y = 14 - 2x$$

Substitute for y in the second equation.
$$V = 8x(14 - 2x)$$
$$V = -16x^2 + 112x$$

Carry out. Completing the square, we get
$$V = -16\left(x - \frac{7}{2}\right)^2 + 196.$$

The maximum function value of 196 occurs when $x = \frac{7}{2}$. When $x = \frac{7}{2}$, $y = 14 - 2 \cdot \frac{7}{2} = 7$.

Check. Check a function value for x less than $\frac{7}{2}$ and for x greater than $\frac{7}{2}$.
$$V(3) = -16 \cdot 3^2 + 112 \cdot 3 = 192$$
$$V(4) = -16 \cdot 4^2 + 112 \cdot 4 = 192$$

Since 196 is greater than these numbers, it looks as though we have a maximum.

We could also use the graph of the function to check the maximum value.

State. The file should be $\frac{7}{2}$ in., or 3.5 in., tall.

5. **Familiarize and Translate**. We want to find the value of x for which $C(x) = 0.1x^2 - 0.7x + 2.425$ is a minimum.

Carry out. We complete the square.
$$C(x) = 0.1(x^2 - 7x + 12.25) + 2.425 - 1.225$$
$$C(x) = 0.1(x - 3.5)^2 + 1.2$$

The minimum function value of 1.2 occurs when $x = 3.5$.

Check. Check a function value for x less than 3.5 and for x greater than 3.5.
$$C(3) = 0.1(3)^2 - 0.7(3) + 2.425 = 1.225$$
$$C(4) = 0.1(4)^2 - 0.7(4) + 2.425 = 1.225$$

Since 1.2 is less than these numbers, it looks as though we have a minimum.

We could also use the graph of the function to check the minimum value.

State. The shop should build 3.5 hundred, or 350 bicycles.

7. **Familiarize**. We make a drawing and label it.

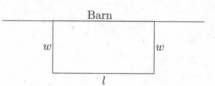

Translate. We have two equations.
$$l + 2w = 40$$
$$A = lw$$

Solve the first equation for l.
$$l = 40 - 2w$$

Substitute for l in the second equation.
$$A = (40 - 2w)w = 40w - 2w^2$$
$$= -2w^2 + 40w$$

Carry out. Completing the square, we get
$$A = -2(w - 10)^2 + 200$$

The maximum function value is 200. It occurs when $w = 10$. When $w = 10$, $l = 40 - 2 \cdot 10 = 20$.

Check. Check a function value for w less than 10 and for w greater than 10.
$$A(9) = -2 \cdot 9^2 + 40 \cdot 9 = 198$$
$$A(11) = -2 \cdot 11^2 + 40 \cdot 11 = 198$$

Since 200 is greater than these numbers, it looks as though we have a maximum. We could also use the graph of the function to check the maximum value.

State. The maximum area of 200 ft^2 will occur when the dimensions are 10 ft by 20 ft.

9. **Familiarize and Translate**. We are given the function $N(x) = -0.4x^2 + 9x + 11$.

Carry out. To find the value of x for which $N(x)$ is a maximum, we first find $-\frac{b}{2a}$:
$$-\frac{b}{2a} = -\frac{9}{2(-0.4)} = 11.25$$

Now we find the maximum value of the function $N(11.25)$:

$N(11.25) = -0.4(11.25)^2 + 9(11.25) + 11 = 61.625$

Check. We can go over the calculations again. We could also solve the problem again by completing the square. The answer checks.

State. Daily ticket sales will peak 11 days after the concert was announced. About 62 tickets will be sold that day.

11. Find the total profit:

$$P(x) = R(x) - C(x)$$
$$P(x) = (1000x - x^2) - (3000 + 20x)$$
$$P(x) = -x^2 + 980x - 3000$$

To find the maximum value of the total profit and the value of x at which it occurs we complete the square:

$$P(x) = -(x^2 - 980x) - 3000$$
$$= -(x^2 - 980x + 240,100 - 240,100) - 3000$$
$$= -(x^2 - 980x + 240,100) - (-240,100) - 3000$$
$$= -(x - 490)^2 + 237,100$$

The maximum profit of \$237,100 occurs at $x = 490$.

13. Familiarize. Let x and y represent the numbers.

Translate. The sum of the numbers is 22, so we have $x + y = 22$. Solving for y, we get $y = 22 - x$. The product of the numbers is xy. Substituting $22 - x$ for y in the product, we get a quadratic function:

$$P = xy = x(22 - x) = 22x - x^2 = -x^2 + 22x$$

Carry out. The coefficient of x^2 is negative, so the graph of the function is a parabola that opens down and a maximum exists. We complete the square in order to find the vertex of the quadratic function.

$$P = -x^2 + 22x$$
$$= -(x^2 - 22x)$$
$$= -(x^2 - 22x + 121 - 121)$$
$$= -(x^2 - 22x + 121) + (-1)(-121)$$
$$= -(x - 11)^2 + 121$$

The vertex is $(11, 121)$. This tells us that the maximum product is 121. The maximum occurs when $x = 11$. Note that when $x = 11$, $y = 22 - x = 22 - 11 = 11$, so the numbers that yield the maximum product are 11 and 11.

Check. We could find the value of the function for some values of x less than 11 and for some greater than 11, determining that the maximum value we found is larger than these function values. We could also use the graph of the function to check the maximum value. Our answer checks.

State. The maximum product is 121. The numbers 11 and 11 yield this product.

15. Familiarize. Let x and y represent the numbers.

Translate. The difference of the numbers is 4, so we have $x - y = 4$. Solve for x, we get $x = y + 4$. The product of the numbers is xy. Substituting $y + 4$ for x in the product, we get a quadratic function:

$$P = xy = (y + 4)y = y^2 + 4y$$

Carry out. The coefficient of y^2 is positive, so the graph of the function opens up and a minimum exists. We complete the square in order to find the vertex of the quadratic function.

$$P = y^2 + 4y$$
$$= y^2 + 4y + 4 - 4$$
$$= (y + 2)^2 - 4$$
$$= [y - (-2)]^2 + (-4)$$

The vertex is $(-2, 4)$. This tells us that the minimum product is -4. The minimum occurs when $y = -2$. Note that when $y = -2$, $x = y + 4 = -2 + 4 = 2$, so the numbers that yield the minimum product are 2 and -2.

Check. We could find the value of the function for some values of y less than -2 and for some greater than -2, determining that the minimum value we found is smaller than these function values. We could also use the graph of the function to check the minimum value. Our answer checks.

State. The minimum product is -4. The numbers 2 and -2 yield this product.

17. Familiarize. We let x and y represent the two numbers, and we let P represent their product.

Translate. We have two equations.

$$x + y = -12,$$
$$P = xy$$

Solve the first equation for y.

$$y = -12 - x$$

Substitute for y in the second equation.

$$P = x(-12 - x) = -12x - x^2$$
$$= -x^2 - 12x$$

Carry out. Completing the square, we get

$$P = -(x + 6)^2 + 36$$

The maximum function value is 36. It occurs when $x = -6$. When $x = -6$, $y = -12 - (-6)$, or -6.

Check. Check a function value for x less than -6 and for x greater than -6.

$$P(-7) = -(-7)^2 - 12(-7) = 35$$
$$P(-5) = -(-5)^2 - 12(-5) = 35$$

Since 36 is greater than these numbers, it looks as though we have a maximum.

We could also use the graph of the function to check the maximum value.

State. The maximum product of 36 occurs for the numbers -6 and -6.

19. The data seem to fit a linear function $f(x) = mx + b$.

21. The data fall and then rise in a curved manner fitting a quadratic function $f(x) = ax^2 + bx + c$, $a > 0$.

23. The data fall, then rise, then fall again so they do not fit a linear or a quadratic function but might fit a polynomial function that is neither quadratic nor linear.

25. The data rise and then fall in a curved manner fitting a quadratic function $f(x) = ax^2 + bx + c$, $a < 0$.

27. We look for a function of the form $f(x) = ax^2 + bx + c$. Substituting the data points, we get

$$4 = a(1)^2 + b(1) + c,$$
$$-2 = a(-1)^2 + b(-1) + c,$$
$$13 = a(2)^2 + b(2) + c,$$

or

$$4 = a + b + c,$$
$$-2 = a - b + c,$$
$$13 = 4a + 2b + c.$$

Solving this system, we get

$$a = 2,\ b = 3,\ \text{and } c = -1.$$

Therefore the function we are looking for is

$$f(x) = 2x^2 + 3x - 1.$$

29. We look for a function of the form $f(x) = ax^2 + bx + c$. Substituting the data points, we get

$$0 = a(2)^2 + b(2) + c,$$
$$3 = a(4)^2 + b(4) + c,$$
$$-5 = a(12)^2 + b(12) + c,$$

or

$$0 = 4a + 2b + c,$$
$$3 = 16a + 4b + c,$$
$$-5 = 144a + 12b + c.$$

Solving this system, we get

$$a = -\frac{1}{4},\ b = 3,\ c = -5.$$

Therefore the function we are looking for is

$$f(x) = -\frac{1}{4}x^2 + 3x - 5.$$

31. a) We look for a function of the form $A(s) = as^2 + bs + c$, where $A(s)$ represents the number of nighttime accidents (for every 200 million km) and s represents the travel speed (in km/h). We substitute the given values of s and $A(s)$.

$$400 = a(60)^2 + b(60) + c,$$
$$250 = a(80)^2 + b(80) + c,$$
$$250 = a(100)^2 + b(100) + c,$$

or

$$400 = 3600a + 60b + c,$$
$$250 = 6400a + 80b + c,$$
$$250 = 10,000a + 100b + c.$$

Solving the system of equations, we get

$$a = \frac{3}{16},\ b = -\frac{135}{4},\ c = 1750.$$

Thus, the function $A(s) = \frac{3}{16}s^2 - \frac{135}{4}s + 1750$ fits the data.

b) Find $A(50)$.

$$A(50) = \frac{3}{16}(50)^2 - \frac{135}{4}(50) + 1750 = 531.25$$

About 531 accidents per 200,000,000 km driven occur at 50 km/h.

33. *Familiarize.* Think of a coordinate system placed on the drawing in the text with the origin at the point where the arrow is released. Then three points on the arrow's parabolic path are $(0, 0)$, $(63, 27)$, and $(126, 0)$. We look for a function of the form $h(d) = ad^2 + bd + c$, where $h(d)$ represents the arrow's height and d represents the distance the arrow has traveled horizontally.

Translate. We substitute the values given above for d and $h(d)$.

$$0 = a \cdot 0^2 + b \cdot 0 + c,$$
$$27 = a \cdot 63^2 + b \cdot 63 + c,$$
$$0 = a \cdot 126^2 + b \cdot 126 + c$$

or

$$0 = c,$$
$$27 = 3969a + 63b + c,$$
$$0 = 15,876a + 126b + c$$

Solve. Solving the system of equations, we get

$$a = -\frac{1}{147} \approx -0.0068,\ b = \frac{6}{7} \approx 0.8571,\ \text{and } c = 0.$$

Check. Recheck the calculations.

State. The function $h(d) = -\frac{1}{147}d^2 + \frac{6}{7}d \approx -0.0068d^2 + 0.8571d$ expresses the arrow's height as a function of the distance it has traveled horizontally.

35. Discussion and Writing Exercise

37. In the expression $5\sqrt{2x - 9} + 3$, the symbol $\sqrt{}$ is called a <u>radical</u> and $2x - 9$ is called the <u>radicand</u>.

39. The degree of a term of a polynomial is the <u>sum</u> of the exponents of the variables.

41. The equation $y = k/x$, where k is a positive constant, is an equation of <u>inverse</u> variation.

43. If the exponents in a polynomial decrease from left to right, the polynomial is written in <u>descending</u> order.

45. We will let x represent the number of years since 1997 and y represent the gross profit, in billions of dollars. Enter the data points (x, y) in STAT lists and use the quartic regression feature to find the equation $y = -0.290x^4 + 2.699x^3 - 8.306x^2 + 9.190x + 12.235$, where x is the number of years after 1997 and y is in billions of dollars.

Exercise Set 7.8

1. $(x - 6)(x + 2) > 0$

The solutions of $(x - 6)(x + 2) = 0$ are 6 and -2. They divide the real-number line into three intervals as shown:

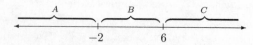

We try test numbers in each interval.

A: Test -3, $y = (-3 - 6)(-3 + 2) = 9 > 0$

B: Test 0, $y = (0 - 6)(0 + 2) = -12 < 0$

C: Test 7, $y = (7 - 6)(7 + 2) = 9 > 0$

The expression is positive for all values of x in intervals A and C. The solution set is $\{x | x < -2 \text{ or } x > 6\}$, or $(-\infty, -2) \cup (6, \infty)$.

From the graph in the text we see that the value of $(x - 6)(x + 2)$ is positive to the left of -2 and to the right of 6. This verifies the answer we found algebraically.

3. $4 - x^2 \geq 0$

$(2 + x)(2 - x) \geq 0$

The solutions of $(2 + x)(2 - x) = 0$ are -2 and 2. They divide the real-number line into three intervals as shown.

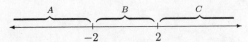

We try test numbers in each interval.

A: Test -3, $y = 4 - (-3)^2 = -5 < 0$

B: Test 0, $y = 4 - 0^2 = 4 > 0$

C: Test 3, $y = 4 - 3^2 = -5 < 0$

The expression is positive for values of x in interval B. Since the inequality symbol is $\geq$ we also include the intercepts. The solution set is $\{x | -2 \leq x \leq 2\}$, or $[-2, 2]$.

From the graph we see that $4 - x^2 \geq 0$ at the intercepts and between them. This verifies the answer we found algebraically.

5. $3(x + 1)(x - 4) \leq 0$

The solutions of $3(x + 1)(x - 4) = 0$ are -1 and 4. They divide the real-number line into three intervals as shown:

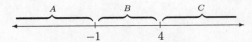

We try test numbers in each interval.

A: Test -2, $y = 3(-2 + 1)(-2 - 4) = 18 > 0$

B: Test 0, $y = 3(0 + 1)(0 - 4) = -12 < 0$

C: Test 5, $y = 3(5 + 1)(5 - 4) = 18 > 0$

The expression is negative for all numbers in interval B. The inequality symbol is $\leq$, so we need to include the intercepts. The solution set is $\{x | -1 \leq x \leq 4\}$, or $[-1, 4]$.

7. $x^2 - x - 2 < 0$

$(x + 1)(x - 2) < 0$ Factoring

The solutions of $(x + 1)(x - 2) = 0$ are -1 and 2. They divide the real-number line into three intervals as shown:

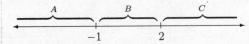

We try test numbers in each interval.

A: Test -2, $y = (-2 + 1)(-2 - 2) = 4 > 0$

B: Test 0, $y = (0 + 1)(0 - 2) = -2 < 0$

C: Test 3, $y = (3 + 1)(3 - 2) = 4 > 0$

The expression is negative for all numbers in interval B. The solution set is $\{x | -1 < x < 2\}$, or $(-1, 2)$.

9. $x^2 - 2x + 1 \geq 0$

$(x - 1)^2 \geq 0$

The solution of $(x - 1)^2 = 0$ is 1. For all real-number values of x except 1, $(x - 1)^2$ will be positive. Thus the solution set is $\{x | x \text{ is a real number}\}$, or $(-\infty, \infty)$.

11. $x^2 + 8 < 6x$

$x^2 - 6x + 8 < 0$

$(x - 4)(x - 2) < 0$

The solutions of $(x - 4)(x - 2) = 0$ are 4 and 2. They divide the real-number line into three intervals as shown:

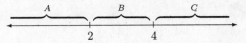

We try test numbers in each interval.

A: Test 0, $y = (0 - 4)(0 - 2) = 8 > 0$

B: Test 3, $y = (3 - 4)(3 - 2) = -1 < 0$

C: Test 5, $y = (5 - 4)(5 - 2) = 3 > 0$

The expression is negative for all numbers in interval B. The solution set is $\{x | 2 < x < 4\}$, or $(2, 4)$.

13. $3x(x + 2)(x - 2) < 0$

The solutions of $3x(x + 2)(x - 2) = 0$ are 0, -2, and 2. They divide the real-number line into four intervals as shown:

We try test numbers in each interval.

A: Test -3, $y = 3(-3)(-3 + 2)(-3 - 2) = -45 < 0$

B: Test -1, $y = 3(-1)(-1 + 2)(-1 - 2) = 9 > 0$

C: Test 1, $y = 3(1)(1 + 2)(1 - 2) = -9 < 0$

D: Test 3, $y = 3(3)(3 + 2)(3 - 2) = 45 > 0$

The expression is negative for all numbers in intervals A and C. The solution set is $\{x | x < -2 \text{ or } 0 < x < 2\}$, or $(-\infty, -2) \cup (0, 2)$.

15. $(x + 9)(x - 4)(x + 1) > 0$

The solutions of $(x + 9)(x - 4)(x + 1) = 0$ are -9, 4, and -1. They divide the real-number line into four intervals as shown:

We try test numbers in each interval.

A: Test -10, $y = (-10 + 9)(-10 - 4)(-10 + 1) = -126 < 0$

B: Test -2, $y = (-2 + 9)(-2 - 4)(-2 + 1) = 42 > 0$

C: Test 0, $y = (0 + 9)(0 - 4)(0 + 1) = -36 < 0$

D: Test 5, $y = (5 + 9)(5 - 4)(5 + 1) = 84 > 0$

The expression is positive for all values of x in intervals B and D. The solution set is $\{x|-9 < x < -1 \text{ } or \text{ } x > 4\}$, or $(-9, -1) \cup (4, \infty)$.

17. $(x+3)(x+2)(x-1) < 0$

The solutions of $(x+3)(x+2)(x-1) = 0$ are -3, -2, and 1. They divide the real-number line into four intervals as shown:

We try test numbers in each interval.

A: Test -4, $y = (-4+3)(-4+2)(-4-1) = -10 < 0$

B: Test $-\dfrac{5}{2}$, $y = \left(-\dfrac{5}{2}+3\right)\left(-\dfrac{5}{2}+2\right)\left(-\dfrac{5}{2}-1\right) = \dfrac{7}{8} > 0$

C: Test 0, $y = (0+3)(0+2)(0-1) = -6 < 0$

D: Test 2, $y = (2+3)(2+2)(2-1) = 20 > 0$

The expression is negative for all numbers in intervals A and C. The solution set is $\{x|x < -3 \text{ } or \text{ } -2 < x < 1\}$, or $(-\infty, -3) \cup (-2, 1)$.

19. $\dfrac{1}{x-6} < 0$

We write the related equation by changing the $<$ symbol to $=$:

$$\frac{1}{x-6} = 0$$

We solve the related equation.

$$(x-6) \cdot \frac{1}{x-6} = (x-6) \cdot 0$$

$$1 = 0$$

We get a false equation, so the related equation has no solution.

Next we find the numbers for which the rational expression is undefined by setting the denominator equal to 0 and solving:

$$x - 6 = 0$$

$$x = 6$$

We use 6 to divide the number line into two intervals as shown:

We try test numbers in each interval.

A: Test 0,

$$\frac{1}{x-6} < 0$$

$$\frac{1}{0-6} \text{ ? } 0$$

$$-\frac{1}{6} \text{ } \Big| \text{ } \text{TRUE}$$

The number 0 is a solution of the inequality, so the interval A is part of the solution set.

B: Test 7,

$$\frac{1}{x-6} < 0$$

$$\frac{1}{7-6} \text{ ? } 0$$

$$1 \text{ } \Big| \text{ } \text{FALSE}$$

The number 7 is not a solution of the inequality, so the interval B is not part of the solution set. The solution set is $\{x|x < 6\}$, or $(-\infty, 6)$.

21. $\dfrac{x+1}{x-3} > 0$

Solve the related equation.

$$\frac{x+1}{x-3} = 0$$

$$x + 1 = 0$$

$$x = -1$$

Find the numbers for which the rational expression is undefined.

$$x - 3 = 0$$

$$x = 3$$

Use the numbers -1 and 3 to divide the number line into intervals as shown:

Try test numbers in each interval.

A: Test -2,

$$\frac{x+1}{x-3} > 0$$

$$\frac{-2+1}{-2-3} \text{ ? } 0$$

$$\frac{-1}{-5}$$

$$\frac{1}{5} \text{ } \Big| \text{ } \text{TRUE}$$

The number -2 is a solution of the inequality, so the interval A is part of the solution set.

B: Test 0,

$$\frac{x+1}{x-3} > 0$$

$$\frac{0+1}{0-3} \text{ ? } 0$$

$$-\frac{1}{3} \text{ } \Big| \text{ } \text{FALSE}$$

The number 0 is not a solution of the inequality, so the interval B is not part of the solution set.

C: Test 4,

$$\frac{x+1}{x-3} > 0$$

$$\frac{4+1}{4-3} \text{ ? } 0$$

$$\frac{5}{1}$$

$$5 \text{ } \Big| \text{ } \text{TRUE}$$

The number 4 is a solution of the inequality, so the interval C is part of the solution set. The solution set is

$\{x | x < -1 \ or \ x > 3\}$, or $(-\infty, -1) \cup (3, \infty)$.

23. $\dfrac{3x + 2}{x - 3} \leq 0$

Solve the related equation.

$$\frac{3x + 2}{x - 3} = 0$$

$$3x + 2 = 0$$

$$3x = -2$$

$$x = -\frac{2}{3}$$

Find the numbers for which the rational expression is undefined.

$$x - 3 = 0$$

$$x = 3$$

Use the numbers $-\dfrac{2}{3}$ and 3 to divide the number line into intervals as shown:

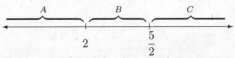

Try test numbers in each interval.

A: Test -1,

$$\frac{3x + 2}{x - 3} \leq 0$$

$$\frac{3(-1) + 2}{-1 - 3} \ ? \ 0$$

$$\frac{-1}{-4} \ \bigg|$$

$$\frac{1}{4} \ \bigg| \quad \text{FALSE}$$

The number -1 is not a solution of the inequality, so the interval A is not part of the solution set.

B: Test 0,

$$\frac{3x + 2}{x - 3} \leq 0$$

$$\frac{3 \cdot 0 + 2}{0 - 3} \ ? \ 0$$

$$\frac{2}{-3} \ \bigg|$$

$$-\frac{2}{3} \ \bigg| \quad \text{TRUE}$$

The number 0 is a solution of the inequality, so the interval B is part of the solution set.

C: Test 4,

$$\frac{3x + 2}{x - 3} \leq 0$$

$$\frac{3 \cdot 4 + 2}{4 - 3} \ ? \ 0$$

$$14 \ \bigg| \quad \text{FALSE}$$

The number 4 is not a solution of the inequality, so the interval C is not part of the solution set. The solution set includes the interval B. The number $-\dfrac{2}{3}$ is also included

since the inequality symbol is $\leq$ and $-\dfrac{2}{3}$ is the solution of the related equation. The number 3 is not included because the rational expression is undefined for 3. The solution set is $\left\{ x \bigg| -\dfrac{2}{3} \leq x < 3 \right\}$, or $\left[-\dfrac{2}{3}, 3 \right)$.

25. $\dfrac{x - 1}{x - 2} > 3$

Solve the related equation.

$$\frac{x - 1}{x - 2} = 3$$

$$x - 1 = 3(x - 2)$$

$$x - 1 = 3x - 6$$

$$5 = 2x$$

$$\frac{5}{2} = x$$

Find the numbers for which the rational expression is undefined.

$$x - 2 = 0$$

$$x = 2$$

Use the numbers $\dfrac{5}{2}$ and 2 to divide the number line into intervals as shown:

Try test numbers in each interval.

A: Test 0,

$$\frac{x - 1}{x - 2} > 3$$

$$\frac{0 - 1}{0 - 2} \ ? \ 3$$

$$\frac{1}{2} \ \bigg| \quad \text{FALSE}$$

The number 0 is not a solution of the inequality, so the interval A is not part of the solution set.

B: Test $\dfrac{9}{4}$,

$$\frac{x - 1}{x - 2} > 3$$

$$\frac{\dfrac{9}{4} - 1}{\dfrac{9}{4} - 2} \ ? \ 3$$

$$\frac{\dfrac{5}{4}}{\dfrac{1}{4}} \ \bigg|$$

$$5 \ \bigg| \quad \text{TRUE}$$

The number $\dfrac{9}{4}$ is a solution of the inequality, so the interval B is part of the solution set.

C: Test 3,

$$\frac{x - 1}{x - 2} > 3$$

$$\frac{3 - 1}{3 - 2} \ ? \ 3$$

$$2 \ \bigg| \quad \text{FALSE}$$

The number 3 is not a solution of the inequality, so the interval C is not part of the solution set. The solution set is $\left\{x \middle| 2 < x < \frac{5}{2}\right\}$, or $\left(2, \frac{5}{2}\right)$.

27. $\dfrac{(x-2)(x+1)}{x-5} < 0$

Solve the related equation.

$$\frac{(x-2)(x+1)}{x-5} = 0$$
$$(x-2)(x+1) = 0$$
$$x = 2 \text{ or } x = -1$$

Find the numbers for which the rational expression is undefined.

$$x - 5 = 0$$
$$x = 5$$

Use the numbers 2, -1, and 5 to divide the number line into intervals as shown:

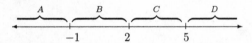

Try test numbers in each interval.

A: Test -2, $\dfrac{(x-2)(x+1)}{x-5} < 0$

$$\frac{(-2-2)(-2+1)}{-2-5} \ ? \ 0$$
$$\frac{-4(-1)}{-7} \ \Big|$$
$$-\frac{4}{7} \ \Big| \quad \text{TRUE}$$

Interval A is part of the solution set.

B: Test 0, $\dfrac{(x-2)(x+1)}{x-5} < 0$

$$\frac{(0-2)(0+1)}{0-5} \ ? \ 0$$
$$\frac{-2 \cdot 1}{-5} \ \Big|$$
$$\frac{2}{5} \ \Big| \quad \text{FALSE}$$

Interval B is not part of the solution set.

C: Test 3, $\dfrac{(x-2)(x+1)}{x-5} < 0$

$$\frac{(3-2)(3+1)}{3-5} \ ? \ 0$$
$$\frac{1 \cdot 4}{-2} \ \Big|$$
$$-2 \ \Big| \quad \text{TRUE}$$

Interval C is part of the solution set.

D: Test 6, $\dfrac{(x-2)(x+1)}{x-5} < 0$

$$\frac{(6-2)(6+1)}{6-5} \ ? \ 0$$
$$\frac{4 \cdot 7}{1} \ \Big|$$
$$28 \ \Big| \quad \text{FALSE}$$

Interval D is not part of the solution set.

The solution set is $\{x | x < -1 \ or \ 2 < x < 5\}$, or $(-\infty, -1) \cup (2, 5)$.

29. $\dfrac{x+3}{x} \le 0$

Solve the related equation.

$$\frac{x+3}{x} = 0$$
$$x + 3 = 0$$
$$x = -3$$

Find the numbers for which the rational expression is undefined.

$$x = 0$$

Use the numbers -3 and 0 to divide the number line into intervals as shown:

Try test numbers in each interval.

A: Test -4, $\dfrac{x+3}{x} \le 0$

$$\frac{-4+3}{-4} \ ? \ 0$$
$$\frac{1}{4} \ \Big| \quad \text{FALSE}$$

Interval A is not part of the solution set.

B: Test -1, $\dfrac{x+3}{x} \le 0$

$$\frac{-1+3}{-1} \ ? \ 0$$
$$-2 \ \Big| \quad \text{TRUE}$$

Interval B is part of the solution set.

C: Test 1, $\dfrac{x+3}{x} \le 0$

$$\frac{1+3}{1} \ ? \ 0$$
$$4 \ \Big| \quad \text{FALSE}$$

Interval C is not part of the solution set.

The solution set includes the interval B. The number -3 is also included since the inequality symbol is $\le$ and -3 is a solution of the related equation. The number 0 is not included because the rational expression is undefined for 0. The solution set is $\{x | -3 \le x < 0\}$, or $[-3, 0)$.

31. $\dfrac{x}{x-1} > 2$

Solve the related equation.

$$\frac{x}{x-1} = 2$$

$$x = 2x - 2$$

$$2 = x$$

Find the numbers for which the rational expression is undefined.

$$x - 1 = 0$$

$$x = 1$$

Use the numbers 1 and 2 to divide the number line into intervals as shown:

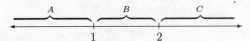

Try test numbers in each interval.

A: Test 0,

$$\frac{x}{x-1} > 2$$

$$\frac{0}{0-1} \;?\; 2$$

$$0 \;\Big|\; \text{FALSE}$$

Interval A is not part of the solution set.

B: Test $\dfrac{3}{2}$,

$$\frac{x}{x-1} > 2$$

$$\frac{\frac{3}{2}}{\frac{3}{2}-1} \;?\; 2$$

$$\frac{\frac{3}{2}}{\frac{1}{2}}$$

$$3 \;\Big|\; \text{TRUE}$$

Interval B is part of the solution set.

C: Test 3,

$$\frac{x}{x-1} > 2$$

$$\frac{3}{3-1} \;?\; 2$$

$$\frac{3}{2} \;\Big|\; \text{FALSE}$$

Interval C is not part of the solution set.

The solution set is $\{x\,|\,1 < x < 2\}$, or $(1, 2)$.

33. $\dfrac{x-1}{(x-3)(x+4)} < 0$

Solve the related equation.

$$\frac{x-1}{(x-3)(x+4)} = 0$$

$$x - 1 = 0$$

$$x = 1$$

Find the numbers for which the rational expression is undefined.

$$(x-3)(x+4) = 0$$

$$x = 3 \text{ or } x = -4$$

Use the numbers 1, 3, and −4 to divide the number line into intervals as shown:

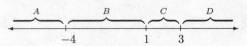

Try test numbers in each interval.

A: Test −5,

$$\frac{x-1}{(x-3)(x+4)} < 0$$

$$\frac{-5-1}{(-5-3)(-5+4)} \;?\; 0$$

$$\frac{-6}{-8(-1)}$$

$$-\frac{3}{4} \;\Big|\; \text{TRUE}$$

Interval A is part of the solution set.

B: Test 0,

$$\frac{x-1}{(x-3)(x+4)} < 0$$

$$\frac{0-1}{(0-3)(0+4)} \;?\; 0$$

$$\frac{-1}{-3\cdot 4}$$

$$\frac{1}{12} \;\Big|\; \text{FALSE}$$

Interval B is not part of the solution set.

C: Test 2,

$$\frac{x-1}{(x-3)(x+4)} < 0$$

$$\frac{2-1}{(2-3)(2+4)} \;?\; 0$$

$$\frac{1}{-1\cdot 6}$$

$$-\frac{1}{6} \;\Big|\; \text{TRUE}$$

Interval C is part of the solution set.

D: Test 4,

$$\frac{x-1}{(x-3)(x+4)} < 0$$

$$\frac{4-1}{(4-3)(4+4)} \;?\; 0$$

$$\frac{3}{1\cdot 8}$$

$$\frac{3}{8} \;\Big|\; \text{FALSE}$$

Interval D is not part of the solution set.

The solution set is $\{x\,|\,x < -4 \text{ or } 1 < x < 3\}$, or $(-\infty, -4) \cup (1, 3)$.

35. $3 < \dfrac{1}{x}$

Solve the related equation.

$$3 = \dfrac{1}{x}$$

$$x = \dfrac{1}{3}$$

Find the numbers for which the rational expression is undefined.

$$x = 0$$

Use the numbers 0 and $\dfrac{1}{3}$ to divide the number line into intervals as shown:

Try test numbers in each interval.

A: Test -1,

$$3 < \dfrac{1}{x}$$

$$3 \; ? \; \dfrac{1}{-1}$$

$$\Big| \; -1 \quad \text{FALSE}$$

Interval A is not part of the solution set.

B: Test $\dfrac{1}{6}$,

$$3 < \dfrac{1}{x}$$

$$3 \; ? \; \dfrac{1}{\frac{1}{6}}$$

$$\Big| \; 6 \quad \text{TRUE}$$

Interval B is part of the solution set.

C: Test 1,

$$3 < \dfrac{1}{x}$$

$$3 \; ? \; \dfrac{1}{1}$$

$$\Big| \; 1 \quad \text{FALSE}$$

Interval C is not part of the solution set.

The solution set is $\left\{ x \middle| 0 < x < \dfrac{1}{3} \right\}$, or $\left(0, \dfrac{1}{3} \right)$.

37. $\dfrac{(x-1)(x+2)}{(x+3)(x-4)} > 0$

Solve the related equation.

$$\dfrac{(x-1)(x+2)}{(x+3)(x-4)} = 0$$

$$(x-1)(x+2) = 0$$

$$x = 1 \text{ or } x = -2$$

Find the numbers for which the rational expression is undefined.

$$(x+3)(x-4) = 0$$

$$x = -3 \text{ or } x = 4$$

Use the numbers 1, -2, -3, and 4 to divide the number line into intervals as shown:

Try test numbers in each interval.

A: Test -4,

$$\dfrac{(x-1)(x+2)}{(x+3)(x-4)} > 0$$

$$\dfrac{(-4-1)(-4+2)}{(-4+3)(-4-4)} \; ? \; 0$$

$$\dfrac{-5(-2)}{-1(-8)} \Bigg|$$

$$\dfrac{5}{4} \Bigg| \quad \text{TRUE}$$

Interval A is part of the solution set.

B: Test $-\dfrac{5}{2}$,

$$\dfrac{(x-1)(x+2)}{(x+3)(x-4)} > 0$$

$$\dfrac{\left(-\frac{5}{2}-1\right)\left(-\frac{5}{2}+2\right)}{\left(-\frac{5}{2}+3\right)\left(-\frac{5}{2}-4\right)} \; ? \; 0$$

$$\dfrac{-\frac{7}{2}\left(-\frac{1}{2}\right)}{\frac{1}{2}\left(-\frac{13}{2}\right)} \Bigg|$$

$$-\dfrac{7}{13} \Bigg| \quad \text{FALSE}$$

Interval B is not part of the solution set.

C: Test 1,

$$\dfrac{(x-1)(x+2)}{(x+3)(x-4)} > 0$$

$$\dfrac{(0-1)(0+2)}{(0+3)(0-4)} \; ? \; 0$$

$$\dfrac{-1 \cdot 2}{3(-4)} \Bigg|$$

$$\dfrac{1}{6} \Bigg| \quad \text{TRUE}$$

Interval C is part of the solution set.

D: Test 2,

$$\dfrac{(x-1)(x+2)}{(x+3)(x-4)} > 0$$

$$\dfrac{(2-1)(2+2)}{(2+3)(2-4)} \; ? \; 0$$

$$\dfrac{1 \cdot 4}{5(-2)} \Bigg|$$

$$-\dfrac{2}{5} \Bigg| \quad \text{FALSE}$$

Interval D is not part of the solution set.

E: Test 5, $\dfrac{(x-1)(x+2)}{(x+3)(x-4)} > 0$

$$\dfrac{(5-1)(5+2)}{(5+3)(5-4)} \ ? \ 0$$

$$\left.\begin{array}{c} \dfrac{4 \cdot 7}{8 \cdot 1} \\[2mm] \dfrac{7}{2} \end{array}\right| \ \text{TRUE}$$

Interval E is part of the solution set.

The solution set is $\{x|x < -3 \ or \ -2 < x < 1 \ or \ x > 4\}$, or $(-\infty, -3) \cup (-2, 1) \cup (4, \infty)$.

39. Discussion and Writing Exercise

41. $\sqrt[3]{\dfrac{125}{27}} = \dfrac{\sqrt[3]{125}}{\sqrt[3]{27}} = \dfrac{5}{3}$

43. $\sqrt{\dfrac{16a^3}{b^4}} = \dfrac{\sqrt{16a^3}}{\sqrt{b^4}} = \dfrac{\sqrt{16a^2 \cdot a}}{\sqrt{b^4}} = \dfrac{\sqrt{16a^2}\sqrt{a}}{\sqrt{b^4}} = \dfrac{4a}{b^2}\sqrt{a}$

45. $3\sqrt{8} - 5\sqrt{2} = 3\sqrt{4 \cdot 2} - 5\sqrt{2}$

$\qquad = 3\sqrt{4}\sqrt{2} - 5\sqrt{2}$

$\qquad = 3 \cdot 2\sqrt{2} - 5\sqrt{2}$

$\qquad = 6\sqrt{2} - 5\sqrt{2}$

$\qquad = \sqrt{2}$

47. $5\sqrt[3]{16a^4} + 7\sqrt[3]{2a} = 5\sqrt[3]{8a^3 \cdot 2a} + 7\sqrt[3]{2a}$

$\qquad = 5\sqrt[3]{8a^3}\sqrt[3]{2a} + 7\sqrt[3]{2a}$

$\qquad = 5 \cdot 2a\sqrt[3]{2a} + 7\sqrt[3]{2a}$

$\qquad = 10a\sqrt[3]{2a} + 7\sqrt[3]{2a}$

$\qquad = (10a + 7)\sqrt[3]{2a}$

49. For Exercise 11, graph $y_1 = x^2 + 8$ and $y_2 = 6x$. Then determine the values of x for which the graph of y_1 lies below the graph of y_2.

For Exercise 22, graph $y_1 = \dfrac{x-2}{x+5}$ and $y_2 = 0$. Then determine the values of x for which the graph of y_1 lies below the graph of y_2. Since the graph of $y_2 = 0$ is the x-axis, this could also be done by graphing $y_1 = \dfrac{x-2}{x+5}$ and determining the values of x for which the graph of y_1 lies below the x-axis.

For Exercise 25, graph $y_1 = \dfrac{x-1}{x-2}$ and $y_2 = 3$. Then determine the values of x for which the graph of y_1 lies above the graph of y_2.

51. $x^2 - 2x \leq 2$

$x^2 - 2x - 2 \leq 0$

The solutions of $x^2 - 2x - 2 = 0$ are found using the quadratic formula. They are $1 \pm \sqrt{3}$, or about 2.7 and -0.7. These numbers divide the number line into three intervals as shown:

We try test numbers in each interval.

A: Test -1, $y = (-1)^2 - 2(-1) - 2 = 1 > 0$

B: Test 0, $y = 0^2 - 2 \cdot 0 - 2 = -2 < 0$

C: Test 3, $y = 3^2 - 2 \cdot 3 - 2 = 1 > 0$

The expression is negative for all values of x in interval B. The inequality symbol is $\leq$, so we must also include the intercepts. The solution set is $\{x|1 - \sqrt{3} \leq x \leq 1 + \sqrt{3}\}$, or $[1 - \sqrt{3}, 1 + \sqrt{3}]$.

53. $x^4 + 2x^2 > 0$

$x^2(x^2 + 2) > 0$

$x^2 > 0$ for all $x \neq 0$, and $x^2 + 2 > 0$ for all values of x. Then $x^2(x^2 + 2) > 0$ for all $x \neq 0$. The solution set is $\{x|x \neq 0\}$, or the set of all real numbers except 0, or $(-\infty, 0) \cup (0, \infty)$.

55. $\left|\dfrac{x+2}{x-1}\right| < 3$

$-3 < \dfrac{x+2}{x-1} < 3$

We rewrite the inequality using "and."

$-3 < \dfrac{x+2}{x-1} \ and \ \dfrac{x+2}{x-1} < 3$

We will solve each inequality and then find the intersection of their solution sets.

Solve: $-3 < \dfrac{x+2}{x-1}$

Solve the related equation.

$$-3 = \dfrac{x+2}{x-1}$$

$$-3x + 3 = x + 2$$

$$1 = 4x$$

$$\dfrac{1}{4} = x$$

Find the numbers for which the rational expression is undefined.

$$x - 1 = 0$$

$$x = 1$$

Use the numbers $\dfrac{1}{4}$ and 1 to divide the number line into intervals as shown:

Try test numbers in each interval.

A: Test 0,

$$-3 < \dfrac{x+2}{x-1}$$

$$-3 \ ? \ \dfrac{0+2}{0-1}$$

$$\left.\begin{array}{c} \\ -2 \end{array}\right| \ \text{TRUE}$$

Interval A is part of the solution set.

B: Test $\dfrac{1}{2}$,

$$-3 < \dfrac{x+2}{x-1}$$

$$-3 \;?\; \dfrac{\dfrac{1}{2}+2}{\dfrac{1}{2}-1}$$

$$\dfrac{\dfrac{5}{2}}{-\dfrac{1}{2}}$$

$$-5 \quad \text{FALSE}$$

Interval B is not part of the solution set.

C: Test 2,

$$-3 < \dfrac{x+2}{x-1}$$

$$-3 \;?\; \dfrac{2+2}{2-1}$$

$$\bigm|\; 4 \quad \text{TRUE}$$

Interval C is part of the solution set.

The solution set of $-3 < \dfrac{x+2}{x-1}$ is $\left\{x \middle| x < \dfrac{1}{4} \text{ or } x > 1\right\}$, or $\left(-\infty, \dfrac{1}{4}\right) \cup (1, \infty)$.

Solve: $\dfrac{x+2}{x-1} > 3$

Solve the related equation.

$$\dfrac{x+2}{x-1} = 3$$

$$x+2 = 3x-3$$

$$5 = 2x$$

$$\dfrac{5}{2} = x$$

From our work above we know that the rational expression is undefined for 1.

Use the numbers $\dfrac{5}{2}$ and 1 to divide the number line into intervals as shown:

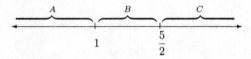

Try test numbers in each interval.

A: Test 0,

$$\dfrac{x+2}{x-1} < 3$$

$$\dfrac{0+2}{0-1} \;?\; 3$$

$$-2 \;\bigm|\; \text{TRUE}$$

Interval A is part of the solution set.

B: Test 2,

$$\dfrac{x+2}{x-1} < 3$$

$$\dfrac{2+2}{2-1} \;?\; 3$$

$$4 \;\bigm|\; \text{FALSE}$$

Interval B is not part of the solution set.

C: Test 3,

$$\dfrac{x+2}{x-1} < 3$$

$$\dfrac{3+2}{3-1} \;?\; 3$$

$$\dfrac{5}{2} \;\bigm|\; \text{TRUE}$$

Interval C is part of the solution set.

The solution set of $\dfrac{x+2}{x-1} < 3$ is $\left\{x \middle| x < 1 \text{ or } x > \dfrac{5}{2}\right\}$, or $(-\infty, 1) \cup \left(\dfrac{5}{2}, \infty\right)$.

The solution set of the original inequality is

$\left\{x \middle| x < \dfrac{1}{4} \text{ or } x > 1\right\} \cap \left\{x \middle| x < 1 \text{ or } x > \dfrac{5}{2}\right\}$, or

$\left\{x \middle| x < \dfrac{1}{4} \text{ or } x > \dfrac{5}{2}\right\}$, or $\left(-\infty, \dfrac{1}{4}\right) \cup \left(\dfrac{5}{2}, \infty\right)$.

57. a) Solve: $-16t^2 + 32t + 1920 > 1920$

$$-16t^2 + 32t > 0$$

$$t^2 - 2t < 0$$

$$t(t-2) < 0$$

The solutions of $t(t-2) = 0$ are 0 and 2. They divide the number line into three intervals as shown:

Try test numbers in each interval.

A: Test -1, $y = -1(-1-2) = 3 > 0$

B: Test 1, $y = 1(1-2) = -1 < 0$

C: Test 3, $y = 3(3-2) = 3 > 0$

The expression is negative for all values of t in interval B. The solution set is $\{t | 0 < t < 2\}$, or $(0, 2)$.

b) Solve: $-16t^2 + 32t + 1920 < 640$

$$-16t^2 + 32t + 1280 < 0$$

$$t^2 - 2t - 80 > 0$$

$$(t-10)(t+8) > 0$$

The solutions of $(t-10)(t+8) = 0$ are 10 and -8. They divide the number line into three intervals as shown:

Try test numbers in each interval.

A: Test -10, $y = (-10-10)(-10+8) = 40 > 0$

B: Test 0, $y = (0-10)(0+8) = -80 < 0$

C: Test 20, $y = (20-10)(20+8) = 80 = 280 > 0$

The expression is positive for all values of t in intervals A and C. However, since negative values of t have no meaning in this problem, we disregard interval A. Thus, the solution set is $\{t | t > 10\}$, or $(10, \infty)$.

Chapter 7 Review Exercises

1. a) $2x^2 - 7 = 0$

$2x^2 = 7$

$x^2 = \dfrac{7}{2}$

$x = \sqrt{\dfrac{7}{2}}$ or $x = -\sqrt{\dfrac{7}{2}}$

$x = \sqrt{\dfrac{7}{2} \cdot \dfrac{2}{2}}$ or $x = -\sqrt{\dfrac{7}{2} \cdot \dfrac{2}{2}}$

$x = \dfrac{\sqrt{14}}{2}$ or $x = -\dfrac{\sqrt{14}}{2}$

The solutions are $\pm \dfrac{\sqrt{14}}{2}$.

b) The real-number solutions of the equation $2x^2 - 7 = 0$ are the first coordinates of the x-intercepts of the graph of $f(x) = 2x^2 - 7$. Thus, the x-intercepts are $\left(-\dfrac{\sqrt{14}}{2}, 0 \right)$ and $\left(\dfrac{\sqrt{14}}{2}, 0 \right)$.

2. $14x^2 + 5x = 0$

$x(14x + 5) = 0$

$x = 0$ or $14x + 5 = 0$

$x = 0$ or $14x = -5$

$x = 0$ or $x = -\dfrac{5}{14}$

The solutions are 0 and $-\dfrac{5}{14}$.

3. $x^2 - 12x + 27 = 0$

$(x - 3)(x - 9) = 0$

$x - 3 = 0$ or $x - 9 = 0$

$x = 3$ or $x = 9$

The solutions are 3 and 9.

4. $4x^2 + 3x + 1 = 0$

$a = 4,\ b = 3,\ c = 1$

$x = \dfrac{-b \pm \sqrt{b^2 - 4ac}}{2a}$

$x = \dfrac{-3 \pm \sqrt{3^2 - 4 \cdot 4 \cdot 1}}{2 \cdot 4} = \dfrac{-3 \pm \sqrt{9 - 16}}{8}$

$x = \dfrac{-3 \pm \sqrt{-7}}{8} = \dfrac{-3 \pm i\sqrt{7}}{8} = -\dfrac{3}{8} \pm i\dfrac{\sqrt{7}}{8}$

The solutions are $-\dfrac{3}{8} \pm i\dfrac{\sqrt{7}}{8}$.

5. $x^2 - 7x + 13 = 0$

$a = 1,\ b = -7,\ c = 13$

$x = \dfrac{-b \pm \sqrt{b^2 - 4ac}}{2a}$

$x = \dfrac{-(-7) \pm \sqrt{(-7)^2 - 4 \cdot 1 \cdot 13}}{2 \cdot 1} = \dfrac{7 \pm \sqrt{49 - 52}}{2}$

$x = \dfrac{7 \pm \sqrt{-3}}{2} = \dfrac{7 \pm i\sqrt{3}}{2} = \dfrac{7}{2} \pm i\dfrac{\sqrt{3}}{2}$

The solutions are $\dfrac{7}{2} \pm i\dfrac{\sqrt{3}}{2}$.

6. $4x(x - 1) + 15 = x(3x + 4)$

$4x^2 - 4x + 15 = 3x^2 + 4x$

$x^2 - 8x + 15 = 0$

$(x - 3)(x - 5) = 0$

$x - 3 = 0$ or $x - 5 = 0$

$x = 3$ or $x = 5$

The solutions are 3 and 5.

7. $x^2 + 4x + 1 = 0$

$a = 1,\ b = 4,\ c = 1$

$x = \dfrac{-b \pm \sqrt{b^2 - 4ac}}{2a}$

$x = \dfrac{-4 \pm \sqrt{4^2 - 4 \cdot 1 \cdot 1}}{2 \cdot 1} = \dfrac{-4 \pm \sqrt{16 - 4}}{2}$

$x = \dfrac{-4 \pm \sqrt{12}}{2} = \dfrac{-4 \pm 2\sqrt{3}}{2}$

$x = \dfrac{2(-2 \pm \sqrt{3})}{2} = -2 \pm \sqrt{3}$

We can use a calculator to approximate the solutions:

$-2 + \sqrt{3} \approx 0.268;\ -2 - \sqrt{3} \approx -3.732$

The solutions are $-2 \pm \sqrt{3}$, or approximately 0.268 and -3.732.

8. $\dfrac{x}{x-2} + \dfrac{4}{x-6} = 0$, LCM is $(x-2)(x-6)$

$(x-2)(x-6)\left(\dfrac{x}{x-2} + \dfrac{4}{x-6} \right) = (x-2)(x-6) \cdot 0$

$x(x - 6) + 4(x - 2) = 0$

$x^2 - 6x + 4x - 8 = 0$

$x^2 - 2x - 8 = 0$

$(x - 4)(x + 2) = 0$

$x - 4 = 0$ or $x + 2 = 0$

$x = 4$ or $x = -2$

Both numbers check. The solutions are 4 and -2.

9. $\dfrac{x}{4} - \dfrac{4}{x} = 2$, LCM is $4x$

$4x\left(\dfrac{x}{4} - \dfrac{4}{x} \right) = 4x \cdot 2$

$x^2 - 16 = 8x$

$x^2 - 8x - 16 = 0$

$a = 1$, $b = -8$, $c = -16$

$$x = \frac{-(-8) \pm \sqrt{(-8)^2 - 4 \cdot 1 \cdot (-16)}}{2 \cdot 1} = \frac{8 \pm \sqrt{64 + 64}}{2}$$

$$x = \frac{8 \pm \sqrt{128}}{2} = \frac{8 \pm 8\sqrt{2}}{2}$$

$$x = \frac{2(4 \pm 4\sqrt{2})}{2} = 4 \pm 4\sqrt{2}$$

Both numbers check. The solutions are $4 \pm 4\sqrt{2}$.

10. $15 = \dfrac{8}{x+2} - \dfrac{6}{x-2}$, LCM is $(x+2)(x-2)$

$$(x+2)(x-2) \cdot 15 = (x+2)(x-2)\left(\frac{8}{x+2} - \frac{6}{x-2}\right)$$

$$15(x^2 - 4) = 8(x - 2) - 6(x + 2)$$

$$15x^2 - 60 = 8x - 16 - 6x - 12$$

$$15x^2 - 60 = 2x - 28$$

$$15x^2 - 2x - 32 = 0$$

$a = 15$, $b = -2$, $c = -32$

$$x = \frac{-(-2) \pm \sqrt{(-2)^2 - 4 \cdot 15 \cdot (-32)}}{2 \cdot 15} = \frac{2 \pm \sqrt{4 + 1920}}{30}$$

$$x = \frac{2 \pm \sqrt{1924}}{30} = \frac{2 \pm 2\sqrt{481}}{30}$$

$$x = \frac{2(1 \pm \sqrt{481})}{2 \cdot 15} = \frac{1 \pm \sqrt{481}}{15}$$

Both numbers check. The solutions are $\dfrac{1 \pm \sqrt{481}}{15}$.

11. $x^2 + 4x + 1 = 0$

$x^2 + 4x = -1$

$x^2 + 4x + 4 = -1 + 4$ $\left(\dfrac{4}{2}\right)^2 = 2^2 = 4$

$(x + 2)^2 = 3$

$x + 2 = \sqrt{3}$ or $x + 2 = -\sqrt{3}$

$x = -2 + \sqrt{3}$ or $x = -2 - \sqrt{3}$

The solutions are $-2 \pm \sqrt{3}$.

12. $V(T) = 48T^2$

$39 = 48T^2$ Substituting 39 for $V(T)$

$0.8125 = T^2$

$\sqrt{0.8125} = T$

$0.901 \approx T$

The hang time is 0.901 sec.

13. Familiarize. Let $l = $ the length of the screen, in cm. Then $l - 5 = $ the width.

Translate. We will use the formula for the area of a rectangle, $A = lw$.

$$l(l - 5) = 126$$

Solve. We solve the equation.

$$l(l - 5) = 126$$

$$l^2 - 5l = 126$$

$$l^2 - 5l - 126 = 0$$

$$(l - 14)(l + 9) = 0$$

$l - 14 = 0$ or $l + 9 = 0$

$l = 14$ or $l = -9$

Check. Since the length cannot be negative, we check only 14. If $l = 14$, then $l - 5 = 14 - 5 = 9$. If the length is 14 cm and the width is 9 cm, the width is 5 cm less than the length and the area is $14 \cdot 9$, or 126 cm². The answer checks.

State. The length is 14 cm, and the width is 9 cm.

14. Familiarize. Using the labels on the drawing in the text, we let $x = $ the width of the mat, in inches. Then the dimensions of the picture are $16 - x - x$ by $12 - x - x$, or $16 - 2x$ by $12 - 2x$.

Translate. We use the formula for the area of a rectangle, $A = lw$.

$$(16 - 2x)(12 - 2x) = 140$$

Solve. We solve the equation.

$$(16 - 2x)(12 - 2x) = 140$$

$$192 - 56x + 4x^2 = 140$$

$$4x^2 - 56x + 52 = 0$$

$$x^2 - 14x + 13 = 0 \quad \text{Dividing by 4}$$

$$(x - 1)(x - 13) = 0$$

$x - 1 = 0$ or $x - 13 = 0$

$x = 1$ or $x = 13$

Check. Since the matted picture measures 12 in. by 16 in., the width of the mat cannot be 13. If the width of the mat is 1 in., then the dimensions of the picture are $12 - 2 \cdot 1$, or 10, by $16 - 2 \cdot 1$, or 14. Thus, the area of the picture is $14 \cdot 10$, or 140 cm². The answer checks.

State. The mat is 1 in. wide.

15. Familiarize. We first make a drawing, labeling it with the known and unknown information. We can also organize the information in a table. We let r represent the speed and t the time for the first part of the trip.

$\underbrace{}$ r mph t hr $r - 10$ mph $3 - t$ hr

 50 mi 80 mi

Trip	Distance	Speed	Time
1st part	50	r	t
2nd part	80	$r - 10$	$3 - t$

Translate. Using $r = \dfrac{d}{t}$, we get two equations from the table, $r = \dfrac{50}{t}$ and $r - 10 = \dfrac{80}{3 - t}$.

Solve. We substitute $\dfrac{50}{t}$ for r in the second equation and solve for t.

$$\frac{50}{t} - 10 = \frac{80}{3-t}, \text{ LCD is } t(3-t)$$

$$t(3-t)\left(\frac{50}{t} - 10\right) = t(3-t) \cdot \frac{80}{3-t}$$

$$50(3-t) - 10t(3-t) = 80t$$

$$150 - 50t - 30t + 10t^2 = 80t$$

$$150 - 80t + 10t^2 = 80t$$

$$10t^2 - 160t + 150 = 0$$

$$t^2 - 16t + 15 = 0 \qquad \text{Dividing by 10}$$

$$(t-1)(t-15) = 0$$

$$t = 1 \quad or \quad t = 15$$

Check. Since the time cannot be negative (If $t = 15$, $3 - t = -12$.), we check only 1 hr. If $t = 1$, then $3 - t = 2$. The speed of the first part is $\dfrac{50}{1}$, or 50 mph. The speed of the second part is $\dfrac{80}{2}$, or 40 mph. The speed of the second part is 10 mph slower than the first part. The value checks.

State. The speed of the first part was 50 mph, and the speed of the second part was 40 mph.

16. $x^2 + 3x - 6 = 0$

$a = 1, b = 3, c = -6$

We compute the discriminant.

$b^2 - 4ac = 3^2 - 4 \cdot 1 \cdot (-6) = 9 + 24 = 33$

Since $b^2 - 4ac > 0$, there are two real solutions.

17. $x^2 + 2x + 5 = 0$

$a = 1, b = 2, c = 5$

We compute the discriminant.

$b^2 - 4ac = 2^2 - 4 \cdot 1 \cdot 5 = 4 - 20 = -16$

Since $b^2 - 4ac < 0$, there are two nonreal solutions.

18. $\qquad x = \dfrac{1}{5} \quad or \qquad x = -\dfrac{3}{5}$

$$x - \frac{1}{5} = 0 \quad or \quad x + \frac{3}{5} = 0$$

$$5x - 1 = 0 \quad or \quad 5x + 3 = 0 \qquad \text{Clearing fractions}$$

$$(5x - 1)(5x + 3) = 0$$

$$25x^2 + 10x - 3 = 0$$

19. Since -4 is the only solution, it must be a double solution.

$$x = -4 \quad or \qquad x = -4$$

$$x + 4 = 0 \quad or \quad x + 4 = 0$$

$$(x+4)(x+4) = 0$$

$$x^2 + 8x + 16 = 0$$

20.

$$N = 3\pi\sqrt{\frac{1}{p}}$$

$$\frac{N}{3\pi} = \sqrt{\frac{1}{p}}$$

$$\left(\frac{N}{3\pi}\right)^2 = \left(\sqrt{\frac{1}{p}}\right)^2$$

$$\frac{N^2}{9\pi^2} = \frac{1}{p}$$

$$pN^2 = 9\pi^2 \qquad \text{Multiplying by } 9\pi^2 p$$

$$p = \frac{9\pi^2}{N^2}$$

21. $\qquad 2A = \dfrac{3B}{T^2}$

$$2AT^2 = 3B \qquad \text{Multiplying by } T^2$$

$$T^2 = \frac{3B}{2A}$$

$$T = \sqrt{\frac{3B}{2A}}$$

22. $x^4 - 13x^2 + 36 = 0$

Let $u = x^2$.

$$u^2 - 13u + 36 = 0$$

$$(u-4)(u-9) = 0$$

$$u = 4 \quad or \quad u = 9$$

$$x^2 = 4 \quad or \quad x^2 = 9$$

$$x = \pm 2 \quad or \quad x = \pm 3$$

All four numbers check. The solutions are 2, -2, 3, and -3.

23. $15x^{-2} - 2x^{-1} - 1 = 0$

Let $u = x^{-1}$.

$$15u^2 - 2u - 1 = 0$$

$$(5u + 1)(3u - 1) = 0$$

$$u = -\frac{1}{5} \qquad or \qquad u = \frac{1}{3}$$

$$x^{-1} = -\frac{1}{5} \qquad or \qquad x^{-1} = \frac{1}{3}$$

$$\frac{1}{x} = -\frac{1}{5} \qquad or \qquad \frac{1}{x} = \frac{1}{3}$$

$$-5x \cdot \frac{1}{x} = -5x\left(-\frac{1}{5}\right) \quad or \quad 3x \cdot \frac{1}{x} = 3x \cdot \frac{1}{3}$$

$$-5 = x \qquad or \qquad 3 = x$$

Both numbers check. The solutions are -5 and 3.

24. $(x^2 - 4)^2 - (x^2 - 4) - 6 = 0$

Let $u = x^2 - 4$.

$$u^2 - u - 6 = 0$$

$$(u-3)(u+2) = 0$$

$$u = 3 \qquad or \qquad u = -2$$

$$x^2 - 4 = 3 \qquad or \quad x^2 - 4 = -2$$

$$x^2 = 7 \qquad or \qquad x^2 = 2$$

$$x = \pm\sqrt{7} \quad or \qquad x = \pm\sqrt{2}$$

All four numbers check. The solutions are $\pm\sqrt{7}$ and $\pm\sqrt{2}$.

25. $x - 13\sqrt{x} + 36 = 0$

Let $u = \sqrt{x}$.

$$u^2 - 13u + 36 = 0$$
$$(u - 4)(u - 9) = 0$$
$$u = 4 \quad or \quad u = 9$$
$$\sqrt{x} = 4 \quad or \quad \sqrt{x} = 9$$
$$(\sqrt{x})^2 = 4^2 \quad or \quad (\sqrt{x})^2 = 9^2$$
$$x = 16 \quad or \quad x = 81$$

Both numbers check. The solutions are 16 and 81.

26. $f(x) = -\frac{1}{2}(x - 1)^2 + 3$

a) Vertex: $(1, 3)$

b) Line of symmetry: $x = 1$

c) Since $-\frac{1}{2} < 0$, we know that 3 is a maximum value.

d) We plot a few points and draw the curve.

x	$f(x)$
-3	-5
-1	1
1	3
3	1
5	-5

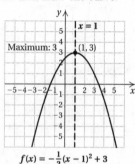

$f(x) = -\frac{1}{2}(x-1)^2 + 3$

27. $f(x) = x^2 - x + 6 = (x^2 - x) + 6$

We complete the square inside the parentheses. We take half the x-coefficient and square it.

$$\frac{1}{2}(-1) = -\frac{1}{2} \text{ and } \left(-\frac{1}{2}\right)^2 = \frac{1}{4}$$

We add $\frac{1}{4} - \frac{1}{4}$ inside the parentheses.

$$f(x) = \left(x^2 - x + \frac{1}{4} - \frac{1}{4}\right) + 6$$
$$= \left(x^2 - x + \frac{1}{4}\right) - \frac{1}{4} + 6$$
$$= \left(x - \frac{1}{2}\right)^2 + \frac{23}{4}$$

a) Vertex: $\left(\frac{1}{2}, \frac{23}{4}\right)$

b) Line of symmetry: $x = \frac{1}{2}$

c) Since the coefficient of the x^2-term, 1, is positive, we know that $\frac{23}{4}$ is a minimum.

d) We plot a few points and draw the curve.

x	$f(x)$
-2	10
-1	8
0	6
1	6
2	8

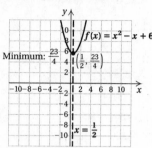

$f(x) = x^2 - x + 6$

28. $f(x) = -3x^2 - 12x - 8 = -3(x^2 + 4x) - 8$

We complete the square inside the parentheses. We take half the x-coefficient and square it.

$$\frac{1}{2} \cdot 4 = 2 \text{ and } 2^2 = 4$$

We add $4 - 4$ inside the parentheses.

$$f(x) = -3(x^2 + 4x + 4 - 4) - 8$$
$$= -3(x^2 + 4x + 4) + (-3)(-4) - 8$$
$$= -3(x + 2)^2 + 12 - 8$$
$$= -3(x + 2)^2 + 4$$
$$= -3[x - (-2)]^2 + 4$$

a) Vertex: $(-2, 4)$

b) Line of symmetry: $x = -2$

c) Since $-3 < 0$, we know that 4 is a maximum.

d) We plot a few points and draw the curve.

x	$f(x)$
-4	-8
-3	1
-2	4
-1	1
0	-8

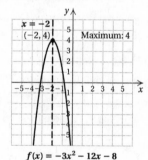

$f(x) = -3x^2 - 12x - 8$

29. $f(x) = x^2 - 9x + 14$

$f(0) = 0^2 - 9 \cdot 0 + 14 = 14$, so the y-intercept is $(0, 14)$. To find the x-intercepts we solve $x^2 - 9x + 14 = 0$. Factoring and using the principle of zero products gives us $x = 2$ or $x = 7$. Thus, the x-intercepts are $(2, 0)$ and $(7, 0)$.

30. *Familiarize*. Let x and y represent the numbers.

Translate. The difference of the numbers is 22, so we have $x - y = 22$. Solve for x, we get $x = y + 22$. The product of the numbers is xy. Substituting $y + 22$ for x in the product, we get a quadratic function:

$$P = xy = (y + 22)y = y^2 + 22y$$

Carry out. The coefficient of y^2 is positive, so the graph of the function opens up and a minimum exists. We complete the square in order to find the vertex of the quadratic function.

$$P = y^2 + 22y$$
$$= y^2 + 22y + 121 - 121$$
$$= (y + 11)^2 - 121$$
$$= [y - (-11)]^2 + (-121)$$

The vertex is $(-11, -121)$. This tells us that the minimum product is -121. The minimum occurs when $y = -11$. Note that when $y = -11$, $x = y + 22 = -11 + 22 = 11$, so the numbers that yield the minimum product are 11 and -11.

Check. We could find the value of the function for some values of y less than -11 and for some greater than -11, determining that the minimum value we found is smaller than these function values. We could also use the graph of the function to check the minimum value. Our answer checks.

State. The minimum product is -121. The numbers 11 and -11 yield this product.

31. We look for a function of the form $ax^2 + bx + c = 0$. Substituting the data points, we get

$$-2 = a \cdot 0^2 + b \cdot 0 + c,$$
$$3 = a \cdot 1^2 + b \cdot 1 + c,$$
$$7 = a \cdot 3^2 + b \cdot 3 + c,$$

or

$$-2 = c,$$
$$3 = a + b + c,$$
$$7 = 9a + 3b + c.$$

Solving this system, we get $(-1, 6, -2)$.

Thus, the desired function is $f(x) = -x^2 + 6x - 2$.

32. a) We look for a function of the form $N(x) = ax^2 + bx + c$ where $N(x)$ represents the number of live births per 1000 women and x represents the age of the woman. We substitute the data points.

$$34 = a \cdot 16^2 + b \cdot 16 + c,$$
$$113.9 = a \cdot 27^2 + b \cdot 27 + c,$$
$$35.4 = a \cdot 37^2 + b \cdot 37 + c,$$

or

$$34 = 256a + 16b + c,$$
$$113.9 = 729a + 27b + c,$$
$$35.4 = 1369a + 37b + c$$

Solving the system of equations, we get $a \approx -0.720$, $b \approx 38.211$, and $c \approx -393.127$. Thus, the desired function is $N(x) = -0.720x^2 + 38.211x - 393.127$.

b) $N(30) = -0.720(30)^2 + 38.211(30) - 393.127$
$$\approx 105$$

33. $(x + 2)(x - 1)(x - 2) > 0$

The solutions of $(x + 2)(x - 1)(x - 2) = 0$ are -2, 1, and 2. They divide the real-number line into four intervals as shown:

We try test numbers in each interval.

A: Test -3: $(-3 + 2)(-3 - 1)(-3 - 2) = -20 < 0$

B: Test 0: $(0 + 2)(0 - 1)(0 - 2) = 4 > 0$

C: Test 1.5: $(1.5 + 2)(1.5 - 1)(1.5 - 2) = -0.875 < 0$

D: Test 3: $(3 + 2)(3 - 1)(3 - 2) = 10 > 0$

The expression is positive for all values of x in intervals B and D. Thus, the solution set is $\{x \mid -2 < x < 1 \text{ or } x > 2\}$, or $(-2, 1) \cup (2, \infty)$.

34. $\dfrac{(x + 4)(x - 1)}{x + 2} < 0$

Solve the related equation.

$$\frac{(x + 4)(x - 1)}{x + 2} = 0$$
$$(x + 4)(x - 1) = 0$$
$$x = -4 \text{ or } x = 1$$

Find the numbers for which the rational expression is undefined.

$$x + 2 = 0$$
$$x = -2$$

Use the numbers -4, 1, and -2 to divide the number line into intervals as shown:

Try test numbers in each interval.

A: Test -5,
$$\frac{(x + 4)(x - 1)}{x + 2} < 0$$
$$\frac{(-5 + 4)(-5 - 1)}{-5 + 2} \; ? \; 0$$
$$\frac{-1(-6)}{-3} \;\bigg|$$
$$-2 \;\bigg|\; \text{TRUE}$$

Interval A is part of the solution set.

B: Test -3,
$$\frac{(x + 4)(x - 1)}{x + 2} < 0$$
$$\frac{(-3 + 4)(-3 - 1)}{-3 + 2} \; ? \; 0$$
$$\frac{1(-4)}{-1} \;\bigg|$$
$$4 \;\bigg|\; \text{FALSE}$$

Interval B is not part of the solution set.

C: Test 0,
$$\frac{(x + 4)(x - 1)}{x + 2} < 0$$
$$\frac{(0 + 4)(0 - 1)}{0 + 2} \; ? \; 0$$
$$\frac{4(-1)}{2} \;\bigg|$$
$$-2 \;\bigg|\; \text{TRUE}$$

Interval C is part of the solution set.

D: Test 2, $\dfrac{(x+4)(x-1)}{x+2} < 0$

$$\dfrac{(2+4)(2-1)}{2+2} \; ? \; 0$$

$$\dfrac{6 \cdot 1}{4}$$

$$\dfrac{3}{2} \quad \bigg| \quad \text{FALSE}$$

Interval D is not part of the solution set.

The solution set is $\{x | x < -4 \; or \; -2 < x < 1\}$, or $(-\infty, -4) \cup (-2, 1)$.

35. *Discussion and Writing Exercise.* The graph of $f(x) = ax^2 + bx + c$ is a parabola with vertex $\left(-\dfrac{b}{2a}, \dfrac{4ac - b^2}{4a} \right)$.

The line of symmetry is $x = -\dfrac{b}{2a}$. If $a < 0$, then the parabola opens down and $\dfrac{4ac - b^2}{4a}$ is the maximum function value. If $a > 0$, the parabola opens up and $\dfrac{4ac - b^2}{4a}$ is the minimum function value. The x-intercepts are $\left(\dfrac{-b - \sqrt{b^2 - 4ac}}{2a}, 0 \right)$ and $\left(\dfrac{-b + \sqrt{b^2 - 4ac}}{2a}, 0 \right)$, for $b^2 - 4ac > 0$. When $b^2 - 4ac = 0$, there is just one x-intercept, $\left(-\dfrac{b}{2a}, 0 \right)$. When $b^2 - 4ac < 0$, there are no x-intercepts.

36. *Discussion and Writing Exercise.* The x-coordinate of the maximum or minimum point lies halfway between the x-coordinates of the x-intercepts. The function must be evaluated for this value of x in order to determine the maximum or minimum value.

37. First we find a quadratic function $f(x) = ax^2 + bx + c$ that fits the given points. We substitute.

$$0 = a(-3)^2 + b(-3) + c,$$
$$0 = a \cdot 5^2 + b \cdot 5 + c,$$
$$-7 = a \cdot 0^2 + b \cdot 0 + c,$$

or

$$0 = 9a - 3b + c,$$
$$0 = 25a + 5b + c,$$
$$-7 = c$$

Solving the system of equations, we get $a = \dfrac{7}{15}$, $b = -\dfrac{14}{15}$, and $c = -7$. Thus, the desired function is $f(x) = \dfrac{7}{15}x^2 - \dfrac{14}{15}x - 7$. Since $\dfrac{7}{15} > 0$, we know the second coordinate of the vertex of the function is a minimum value. We first find the first coordinate of the vertex.

$$-\dfrac{b}{2a} = -\dfrac{-14/15}{2(7/15)} = -\dfrac{-14/15}{14/15} = 1$$

Now we find $f(1)$.

$$f(1) = \dfrac{7}{15} \cdot 1^2 - \dfrac{14}{15} \cdot 1 - 7 = -\dfrac{112}{15}$$

The minimum value of the function is $-\dfrac{112}{15}$.

38. For a quadratic equation $ax^2 + bx + c = 0$, the sum of the solutions is

$$\dfrac{-b + \sqrt{b^2 - 4ac}}{2a} + \dfrac{-b - \sqrt{b^2 - 4ac}}{2a} = \dfrac{-2b}{2a} = -\dfrac{b}{a}$$

and the product of the solutions is

$$\left(\dfrac{-b + \sqrt{b^2 - 4ac}}{2a} \right) \left(\dfrac{-b - \sqrt{b^2 - 4ac}}{2a} \right) = \dfrac{b^2 - (b^2 - 4ac)}{4a2} = \dfrac{4ac}{4a^2} = \dfrac{c}{a}.$$

Now consider $3x^2 - hx + 4k = 0$.

We have $-\dfrac{b}{a} = 20$, or $-\dfrac{-h}{3} = 20$, or $\dfrac{h}{3} = 20$, so $h = 60$.

We also have $\dfrac{c}{a} = 80$, or $\dfrac{4k}{3} = 80$, or $k = 60$.

39. *Familiarize.* Let x represent one of the numbers. Then $\sqrt{x}$ represents the other number.

Translate. The average of the two numbers is 171, so we have

$$\dfrac{x + \sqrt{x}}{2} = 171.$$

Solve. We solve the equation.

$$\dfrac{x + \sqrt{x}}{2} = 171$$
$$x + \sqrt{x} = 342$$
$$x + \sqrt{x} - 342 = 0$$

Let $u = \sqrt{x}$.

$$u^2 + u - 342 = 0$$
$$(u + 19)(u - 18) = 0$$
$$u = -19 \quad or \quad u = 18$$
$$\sqrt{x} = -19 \quad or \quad \sqrt{x} = 18$$

No solution *or* $x = 324$

Check. Since $\sqrt{x}$ denotes the positive square root of x, the equation $\sqrt{x} = -19$ has no solution. If $x = 324$, then $\sqrt{x} = \sqrt{324} = 18$. Since $\dfrac{324 + 18}{2} = \dfrac{342}{2} = 171$, the answer checks.

State. The numbers are 324 and 18.

Chapter 7 Test

1. a) $3x^2 - 4 = 0$

$$3x^2 = 4$$
$$x^2 = \dfrac{4}{3}$$
$$x = -\sqrt{\dfrac{4}{3}} \quad or \quad x = \sqrt{\dfrac{4}{3}}$$
$$x = -\sqrt{\dfrac{4}{3} \cdot \dfrac{3}{3}} \quad or \quad x = \sqrt{\dfrac{4}{3} \cdot \dfrac{3}{3}}$$
$$x = -\dfrac{2\sqrt{3}}{3} \quad or \quad x = \dfrac{2\sqrt{3}}{3}$$

The solutions are $-\dfrac{2\sqrt{3}}{3}$ and $\dfrac{2\sqrt{3}}{3}$, or $\pm\dfrac{2\sqrt{3}}{3}$.

b) The real-number solutions of the equation $3x^2 - 4 = 0$ are the first coordinates of the x-intercepts of the graph of $f(x) = 3x^2 - 4$. Thus, the x-intercepts are $\left(-\dfrac{2\sqrt{3}}{3}, 0\right)$ and $\left(\dfrac{2\sqrt{3}}{3}, 0\right)$.

2. $x^2 + x + 1 = 0$

$a = 1, \ b = 1, \ c = 1$

$x = \dfrac{-b \pm \sqrt{b^2 - 4ac}}{2a}$

$x = \dfrac{-1 \pm \sqrt{1^2 - 4 \cdot 1 \cdot 1}}{2 \cdot 1} = \dfrac{-1 \pm \sqrt{1 - 4}}{2}$

$x = \dfrac{-1 \pm \sqrt{-3}}{2} = \dfrac{-1 \pm i\sqrt{3}}{2} = -\dfrac{1}{2} \pm i\dfrac{\sqrt{3}}{2}$

The solutions are $-\dfrac{1}{2} \pm i\dfrac{\sqrt{3}}{2}$.

3. $x - 8\sqrt{x} + 7 = 0$

Let $u = \sqrt{x}$ and think of x as $(\sqrt{x})^2$, or u^2.

$u^2 - 8u + 7 = 0$ Substituting

$(u - 1)(u - 7) = 0$

$u - 1 = 0 \ \ or \ \ u - 7 = 0$

$u = 1 \ \ or \ \ \ \ \ \ u = 7$

Now we substitute $\sqrt{x}$ for u and solve these equations.

$\sqrt{x} = 1 \ \ or \ \ \sqrt{x} = 7$

$x = 1 \ \ or \ \ \ \ \ x = 49$

Both numbers check. The solutions are 1 and 49.

4. $4x(x - 2) - 3x(x + 1) = -18$

$4x^2 - 8x - 3x^2 - 3x = -18$

$x^2 - 11x = -18$

$x^2 - 11x + 18 = 0$

$(x - 2)(x - 9) = 0$

$x - 2 = 0 \ \ or \ \ x - 9 = 0$

$x = 2 \ \ or \ \ \ \ \ x = 9$

Both numbers check. The solutions are 2 and 9.

5. $x^4 - 5x^2 + 5 = 0$

Let $u = x^2$ and think of x^4 as $(x^2)^2$, or u^2.

$u^2 - 5u + 5 = 0$ Substituting

$u = \dfrac{-(-5) \pm \sqrt{(-5)^2 - 4 \cdot 1 \cdot 5}}{2 \cdot 1}$

$u = \dfrac{5 \pm \sqrt{25 - 20}}{2} = \dfrac{5 \pm \sqrt{5}}{2}$

Now substitute x^2 for u and solve these equations.

$x^2 = \dfrac{5 - \sqrt{5}}{2} \ \ or \ \ x^2 = \dfrac{5 + \sqrt{5}}{2}$

$x = \pm\sqrt{\dfrac{5 - \sqrt{5}}{2}} \ \ or \ \ x = \pm\sqrt{\dfrac{5 + \sqrt{5}}{2}}$

All four numbers check. They are the solutions.

6. $x^4 + 4x = 2$

$x^2 + 4x - 2 = 0$

$x = \dfrac{-4 \pm \sqrt{4^2 - 4 \cdot 1 \cdot (-2)}}{2 \cdot 1} = \dfrac{-4 \pm \sqrt{16 + 8}}{2}$

$x = \dfrac{-4 \pm \sqrt{24}}{2} = \dfrac{-4 \pm \sqrt{4 \cdot 6}}{2}$

$x = \dfrac{-4 \pm 2\sqrt{6}}{2} = \dfrac{2(-2 \pm \sqrt{6})}{2 \cdot 1} = -2 \pm \sqrt{6}$

We can use a calculator to approximate the solutions:

$-2 - \sqrt{6} \approx -4.449; \ -2 + \sqrt{6} \approx 0.449$

The solutions are $-2 \pm \sqrt{6}$, or approximately -4.449 and 0.449.

7. $\dfrac{1}{4 - x} + \dfrac{1}{2 + x} = \dfrac{3}{4}$ LCM is $(4 - x)(2 + x)$

$(4 - x)(2 + x)\left(\dfrac{1}{4 - x} + \dfrac{1}{2 + x}\right) = (4 - x)(2 + x) \cdot \dfrac{3}{4}$

$2 + x + 4 - x = \dfrac{3}{4}(8 + 2x - x^2)$

$6 = \dfrac{3}{4}(8 + 2x - x^2)$

$\dfrac{4}{3} \cdot 6 = \dfrac{4}{3} \cdot \dfrac{3}{4}(8 + 2x - x^2)$

$8 = 8 + 2x - x^2$

$x^2 - 2x = 0$

$x(x - 2) = 0$

$x = 0 \ \ or \ \ x - 2 = 0$

$x = 0 \ \ or \ \ \ \ \ \ x = 2$

Both numbers check. The solutions are 0 and 2.

8. $x^2 - 4x + 1 = 0$

$x^2 - 4x = -1$

$x^2 - 4x + 4 = -1 + 4$ Adding 4: $\left(\dfrac{-4}{2}\right)^2 = (-2)^2 = 4$

$(x - 2)^2 = 3$

$x - 2 = -\sqrt{3} \ \ \ or \ \ x - 2 = \sqrt{3}$

$x = 2 - \sqrt{3} \ or \ \ \ \ \ \ x = 2 + \sqrt{3}$

The solutions are $2 \pm \sqrt{3}$.

9. $s(t) = 16t^2$

$723 = 16t^2$

$\dfrac{723}{16} = t^2$

$45.1875 = t^2$

$\sqrt{45.1875} = t$

$6.7 \approx t$

It will take an object about 6.7 sec to fall from the top.

10. *Familiarize.* Let $r =$ the speed of the boat in still water and let $t =$ the time of the trip upriver. Then $4 - t =$ the time of the return trip downriver. We organize the information in a table.

	Distance	Speed	Time
Upriver	3	$r-2$	t
Downriver	3	$r+2$	$4-t$

Translate. Using $t = d/r$ in each row of the table, we have two equations:

$$t = \frac{3}{r-2} \text{ and } 4 - t = \frac{3}{r+2}.$$

Solve. We substitute $\frac{3}{r-2}$ for t in the second equation and solve for r.

$$4 - \frac{3}{r-2} = \frac{3}{r+2}$$
$$\text{LCM} = (r-2)(r+2)$$
$$(r-2)(r+2)\left(4 - \frac{3}{r-2}\right) = (r-2)(r+2) \cdot \frac{3}{r+2}$$
$$4(r-2)(r+2) - 3(r+2) = 3(r-2)$$
$$4(r^2 - 4) - 3r - 6 = 3r - 6$$
$$4r^2 - 16 - 3r - 6 = 3r - 6$$
$$4r^2 - 3r - 22 = 3r - 6$$
$$4r^2 - 6r - 16 = 0$$
$$2r^2 - 3r - 8 = 0 \quad \text{Dividing by 2}$$
$$r = \frac{-(-3) \pm \sqrt{(-3)^2 - 4 \cdot 2 \cdot (-8)}}{2 \cdot 2}$$
$$r = \frac{3 \pm \sqrt{9 + 64}}{4}$$
$$r = \frac{3 \pm \sqrt{73}}{4}$$
$$r \approx 2.89 \text{ or } r \approx 1.39$$

Check. Since negative speed has no meaning in this application, we check only 2.89. If $r \approx 2.89$, then the speed upriver is about $2.89 - 2$, or 0.89 mph, and the time it takes to travel 3 mi is about $3/0.89 \approx 3.37$ hr. The speed downriver is about $2.89 + 2$, or 4.89 mph, and the time it takes to travel 3 mi is about $3/4.89 \approx 0.61$ hr. The total time is about $3.37 + 0.61$, or $3.98 \approx 4$ hr, so the answer checks.

State. The speed of the boat in still water must be about 2.89 mph.

11. Familiarize. Let $l =$ the length of the board and $w =$ the width, in cm. Then the perimeter of the board is $2l + 2w$ and the area is $l \cdot w$.

Translate. Using the formula for perimeter, we have:

$$2l + 2w = 28$$
$$2l = 28 - 2w$$
$$l = \frac{28 - 2w}{2}$$
$$l = 14 - w$$

Substituting $14 - w$ for l in the formula for area, we get a quadratic function.

$$A = l \cdot w = (14 - w)w = 14w - w^2, \text{ or } -w^2 + 14w$$

Carry out. We complete the square in order to find the vertex of the quadratic function.

$$A = -w^2 + 14w$$
$$= -(w^2 - 14w)$$
$$= -(w^2 - 14w + 49 - 49)$$
$$= -(w^2 - 14w + 49) + (-1)(-49)$$
$$= -(w - 7)^2 + 49$$

The vertex is $(7, 49)$. The coefficient of w^2 is negative, so the graph of the function is a parabola that opens down. This tells us that the function has a maximum value and that value occurs when $w = 7$. When $w = 7$, $l = 14 - w = 14 - 7 = 7$.

Check. We could find the value of the function for some values of w less than 7 and for some values greater than 7, determining that the maximum value we found, 49, is larger than these function values. We could also use the graph of the function to check the maximum value. Our answer checks.

State. Dimensions of 7 cm by 7 cm will provide the maximum area.

12.
$$V(T) = 48T^2$$
$$36 = 48T^2$$
$$0.75 = T^2$$
$$\sqrt{0.75} = T$$
$$0.866 \approx T$$

Hardaway's hang time is about 0.866 sec.

13. $x^2 + 5x + 17 = 0$

$a = 1$, $b = 5$, $c = 17$

We compute the discriminant.

$$b^2 - 4ac = 5^2 - 4 \cdot 1 \cdot 17 = 25 - 68 = -43$$

Since $b^2 - 4ac < 0$, there are two nonreal solutions.

14. The solutions are $\sqrt{3}$ and $3\sqrt{3}$.

$$x = \sqrt{3} \quad or \quad x = 3\sqrt{3}$$
$$x - \sqrt{3} = 0 \quad or \quad x - 3\sqrt{3} = 0$$
$$(x - \sqrt{3})(x - 3\sqrt{3}) = 0 \quad \text{Principle of zero products}$$
$$x^2 - 3\sqrt{3}x - \sqrt{3}x + 3 \cdot 3 = 0$$
$$x^2 - 4\sqrt{3}x + 9 = 0$$

15.
$$V = 48T^2$$
$$\frac{V}{48} = T^2$$
$$\sqrt{\frac{V}{48}} = T$$

We can rationalize the denominator.

$$\sqrt{\frac{V}{48}} = \sqrt{\frac{V}{48} \cdot \frac{3}{3}} = \sqrt{\frac{3V}{144}} = \frac{\sqrt{3V}}{12}$$

Thus, we also have $T = \frac{\sqrt{3V}}{12}$.

16. $f(x) = -x^2 - 2x$

We complete the square.

$f(x) = -(x^2 + 2x)$

$$= -(x^2 + 2x + 1 - 1) \quad \left(\frac{2}{2}\right)^2 = 1^2 = 1;$$
$$\text{add } 1 - 1$$
$$= -(x^2 + 2x + 1) + (-1)(-1)$$
$$= -(x + 1)^2 + 1$$
$$= -[x - (-1)]^2 + 1$$

a) The vertex is $(-1, 1)$.

b) The line of symmetry is $x = -1$.

c) The coefficient of x^2 is negative, so the graph opens down. This tells us that 1 is a maximum value.

d) We plot some points and draw the curve.

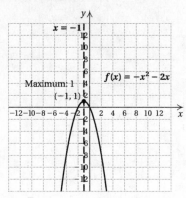

x	$f(x)$
-3	-3
-2	0
-1	1
0	0
2	-8

17. $f(x) = 4x^2 - 24x + 41$
$$= 4(x^2 - 6x) + 41$$

We complete the square inside the parentheses.

$$f(x) = 4(x^2 - 6x + 9 - 9) + 41 \quad \left(\frac{-6}{2}\right)^2 = (-3)^2 = 9;$$
$$\text{add } 9 - 9$$
$$= 4(x^2 - 6x + 9) + 4(-9) + 41$$
$$= 4(x - 3)^2 - 36 + 41$$
$$= 4(x - 3)^2 + 5$$

a) The vertex is $(3, 5)$.

b) The line of symmetry is $x = 3$.

c) The coefficient of x^2 is positive, so the graph opens up. This tells us that 5 is a minimum value.

d) We plot some points and draw the curve.

x	$f(x)$
2	9
3	5
4	9
5	21

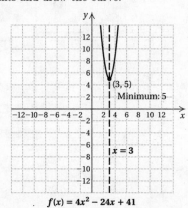

$f(x) = 4x^2 - 24x + 41$

18. $f(x) = -x^2 + 4x - 1$

The y-intercept is $(0, f(0))$. Since $f(0) = -0^2 + 4 \cdot 0 - 1 = -1$, the y-intercept is $(0, -1)$.

To find the x-intercepts, we solve $-x^2 + 4x - 1 = 0$. Using the quadratic formula we get $x = 2 \pm \sqrt{3}$. Thus, the x-intercepts are $(2 - \sqrt{3}, 0)$ and $(2 + \sqrt{3}, 0)$.

19. *Familiarize*. Let x and y represent the numbers.

Translate. The difference of the numbers is 8, so we have $x - y = 8$. Solve for x, we get $x = y + 8$. The product of the numbers is xy. Substituting $y + 8$ for x in the product, we get a quadratic function:

$$P = xy = (y + 8)y = y^2 + 8y$$

Carry out. The coefficient of y^2 is positive, so the graph of the function opens up and a minimum exists. We complete the square in order to find the vertex of the quadratic function.

$$P = y^2 + 8y$$
$$= y^2 + 8y + 16 - 16$$
$$= (y + 4)^2 - 16$$
$$= [y - (-4)]^2 + (-16)$$

The vertex is $(-4, -16)$. This tells us that the minimum product is -16. The minimum occurs when $y = -4$. Note that when $y = -4$, $x = y + 8 = -4 + 8 = 4$, so the numbers that yield the minimum product are 4 and -4.

Check. We could find the value of the function for some values of y less than -4 and for some values greater than -4, determining that the minimum value we found is smaller than these function values. We could also use the graph of the function to check the minimum value. Our answer checks.

State. The minimum product is -16. The numbers 4 and -4 yield this product.

20. We look for a function of the form $f(x) = ax^2 + bx + c$. Substituting the data points, we get

$$0 = a \cdot 0^2 + b \cdot 0 + c,$$
$$0 = a \cdot 3^2 + b \cdot 3 + c,$$
$$2 = a \cdot 5^2 + b \cdot 5 + c,$$
or
$$0 = c,$$
$$0 = 9a + 3b + c,$$
$$2 = 25a + 5b + c.$$

Solving this system, we get

$$a = \frac{1}{5}, \; b = -\frac{3}{5}, \text{ and } c = 0.$$

Then the function we are looking for is

$$f(x) = \frac{1}{5}x^2 - \frac{3}{5}x.$$

21. a) Substituting the data points, we get

$$349 = a \cdot 0^2 + b \cdot 0 + c,$$
$$780 = a \cdot 10^2 + b \cdot 10 + c,$$
$$1692 = a \cdot 20^2 + b \cdot 20 + c,$$

or

$$349 = c,$$
$$780 = 100a + 10b + c,$$
$$1692 = 400a + 20b + c.$$

Solving this system, we get

$$a = 2.405, \; b = 19.05, \; \text{and} \; c = 349.$$

Then the function we are looking for is

$$C(t) = 2.405t^2 + 19.05t + 349.$$

b) In 2007, $t = 2007 - 1980 = 27$.

$$C(27) = 2.405(27)^2 + 19.05(27) + 349$$
$$= 2616.595$$
$$\approx \$2617 \text{ billion}$$

In 2010, $t = 2010 - 1980 = 30$.

$$C(30) = 2.405(30)^2 + 19.05(30) + 349$$
$$= \$3085 \text{ billion}$$

22.
$$x^2 < 6x + 7$$
$$x^2 - 6x - 7 < 0$$
$$(x + 1)(x - 7) < 0$$

The solutions of $(x + 1)(x - 7) = 0$ are -1 and 7. They divide the real-number line into three intervals as shown:

We try test numbers in each interval.

A: Test -2, $y = (-2 + 1)(-2 - 7) = 9 > 0$

B: Test 0, $y = (0 + 1)(0 - 7) = -7 < 0$

C: Test 8, $y = (8 + 1)(8 - 7) = 9 > 0$

The expression is negative for all real numbers in interval B. The solution set is $\{x | -1 < x < 7\}$, or $(-1, 7)$.

23. $\dfrac{x - 5}{x + 3} < 0$

Solve the related equation.

$$\frac{x - 5}{x + 3} = 0$$
$$x - 5 = 0$$
$$x = 5$$

Find the numbers for which the rational expression is undefined.

$$x + 3 = 0$$
$$x = -3$$

Use the numbers 5 and -3 to divide the number line into intervals as shown:

A: Test -4, $\dfrac{x - 5}{x + 3} < 0$

$$\frac{-4 - 5}{-4 + 3} \; ? \; 0$$
$$\frac{-9}{-1}$$
$$9 \quad \bigg| \quad \text{FALSE}$$

The number -4 is not a solution of the inequality, so the interval A is not part of the solution set.

B: Test 0, $\dfrac{x - 5}{x + 3} < 0$

$$\frac{0 - 5}{0 + 3} \; ? \; 0$$
$$\frac{-5}{3}$$
$$-\frac{5}{3} \quad \bigg| \quad \text{TRUE}$$

The number 0 is a solution of the inequality, so the interval B is part of the solution set.

C: Test 6, $\dfrac{x - 5}{x + 3} < 0$

$$\frac{6 - 5}{6 + 3} \; ? \; 0$$
$$\frac{1}{9} \quad \bigg| \quad \text{FALSE}$$

The number 6 is not a solution of the inequality, so the interval C is not part of the solution set. The solution set is $\{x | -3 < x < 5\}$, or $(-3, 5)$.

24. $\dfrac{x - 2}{(x + 3)(x - 1)} \geq 0$

Solve the related equation.

$$\frac{x - 2}{(x + 3)(x - 1)} = 0$$
$$x - 2 = 0$$
$$x = 2$$

Find the numbers for which the rational expression is undefined.

$$(x + 3)(x - 1) = 0$$
$$x = -3 \; \text{or} \; x = 1$$

Use the numbers 2, -3, and 1 to divide the number line into intervals as shown:

We try test numbers in each interval.

A: Test -4, $\dfrac{x - 2}{(x + 3)(x - 1)} \geq 0$

$$\frac{-4 - 2}{(-4 + 3)(-4 - 1)} \; ? \; 0$$
$$\frac{-6}{-1(-5)}$$
$$-\frac{6}{5} \quad \bigg| \quad \text{FALSE}$$

Interval A is not part of the solution set.

B: Test 0,

$$\frac{x-2}{(x+3)(x-1)} \geq 0$$

$$\frac{0-2}{(0+3)(0-1)} \; ? \; 0$$

$$\frac{-2}{3(-1)}$$

$$\frac{2}{3} \; \bigg| \; \text{TRUE}$$

Interval B is part of the solution set.

C: Test 1.5,

$$\frac{x-2}{(x+3)(x-1)} \geq 0$$

$$\frac{1.5-2}{(1.5+3)(1.5-1)} \; ? \; 0$$

$$\frac{-0.5}{4.5(0.5)}$$

$$-0.\overline{2} \; \bigg| \; \text{FALSE}$$

Interval C is not part of the solution set.

D: Test 3,

$$\frac{x-2}{(x+3)(x-1)} \geq 0$$

$$\frac{3-2}{(3+3)(3-1)} \; ? \; 0$$

$$\frac{1}{6 \cdot 2}$$

$$\frac{1}{12} \; \bigg| \; \text{TRUE}$$

Interval D is part of the solution set.

The solution set includes intervals B and D. The number 2 is also included because the inequality symbol is $\geq$ and 2 is the solution of the related equation. The numbers -3 and 1 are not included because the rational expression is undefined for these numbers. The solution set is $\{x | -3 < x < 1 \text{ or } x \geq 2\}$, or $(-3,1) \cup [2,\infty)$.

25. We look for a function $f(x) = ax^2 + bx + c$. We substitute the known points.

$$0 = a(-2)^2 + b(-2) + c,$$
$$0 = a \cdot 7^2 + b \cdot 7 + c,$$
$$8 = a \cdot 0^2 + b \cdot 0 + c,$$

or

$$0 = 4a - 2b + c,$$
$$0 = 49a + 7b + c,$$
$$8 = c.$$

Solving this system, we get $\left(-\frac{4}{7}, \frac{20}{7}, 8\right)$. Thus, the function is $f(x) = -\frac{4}{7}x^2 + \frac{20}{7}x + 8$.

The coefficient of x^2 is negative, so the function has a maximum value. To find it we find the vertex of the function.

$$-\frac{b}{2a} = -\frac{20/7}{2(-4/7)} = -\frac{20/7}{-8/7} = \frac{20}{7} \cdot \frac{7}{8} = \frac{5}{2}$$

$$f\left(\frac{5}{2}\right) = -\frac{4}{7}\left(\frac{5}{2}\right)^2 + \frac{20}{7}\left(\frac{5}{2}\right) + 8 = \frac{81}{7}$$

The maximum value is $\frac{81}{7}$.

26. $kx^2 + 3x - k = 0$

First we substitute -2 for x and find k.

$$k(-2)^2 + 3(-2) - k = 0$$
$$4k - 6 - k = 0$$
$$3k - 6 = 0$$
$$3k = 6$$
$$k = 2$$

Now we have:

$$2x^2 + 3x - 2 = 0$$
$$(2x-1)(x+2) = 0$$
$$2x - 1 = 0 \quad \text{or} \quad x + 2 = 0$$
$$2x = 1 \quad \text{or} \quad x = -2$$
$$x = \frac{1}{2} \quad \text{or} \quad x = -2$$

The other solution is $\frac{1}{2}$.

27. $x^8 - 20x^4 + 64 = 0$

Let $u = x^4$ and think of x^8 as $(x^4)^2$.

$$u^2 - 20u + 64 = 0$$
$$(u-4)(u-16) = 0$$
$$u = 4 \quad \text{or} \quad u = 16$$
$$x^4 = 4 \quad \text{or} \quad x^4 = 16$$

Now we solve $x^4 = 4$.

$$x^4 = 4$$
$$x^4 - 4 = 0$$
$$(x^2+2)(x^2-2) = 0$$
$$x^2 + 2 = 0 \quad \text{or} \quad x^2 - 2 = 0$$
$$x^2 = -2 \quad \text{or} \quad x^2 = 2$$
$$x = \pm\sqrt{2}i \quad \text{or} \quad x = \pm\sqrt{2}$$

Next we solve $x^4 = 16$.

$$x^4 = 16$$
$$x^4 - 16 = 0$$
$$(x^2+4)(x^2-4) = 0$$
$$x^2 + 4 = 0 \quad \text{or} \quad x^2 - 4 = 0$$
$$x^2 = -4 \quad \text{or} \quad x^2 = 4$$
$$x = \pm 2i \quad \text{or} \quad x = \pm 2$$

All eight numbers check. The solutions are $\pm\sqrt{2}i$, $\pm\sqrt{2}$, $\pm 2i$, and ± 2.

Chapter 8

Exponential and Logarithmic Functions

Exercise Set 8.1

1. Graph: $f(x) = 2^x$

We compute some function values and keep the results in a table.

$f(0) = 2^0 = 1$

$f(1) = 2^1 = 2$

$f(2) = 2^2 = 4$

$f(-1) = 2^{-1} = \dfrac{1}{2^1} = \dfrac{1}{2}$

$f(-2) = 2^{-2} = \dfrac{1}{2^2} = \dfrac{1}{4}$

x	$f(x)$
0	1
1	2
2	4
3	8
-1	$\dfrac{1}{2}$
-2	$\dfrac{1}{4}$
-3	$\dfrac{1}{8}$

Next we plot these points and connect them with a smooth curve.

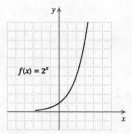

3. Graph: $f(x) = 5^x$

We compute some function values and keep the results in a table.

$f(0) = 5^0 = 1$

$f(1) = 5^1 = 5$

$f(2) = 5^2 = 25$

$f(-1) = 5^{-1} = \dfrac{1}{5^1} = \dfrac{1}{5}$

$f(-2) = 5^{-2} = \dfrac{1}{5^2} = \dfrac{1}{25}$

x	$f(x)$
0	1
1	5
2	25
-1	$\dfrac{1}{5}$
-2	$\dfrac{1}{25}$

Next we plot these points and connect them with a smooth curve.

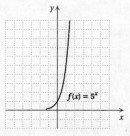

5. Graph: $f(x) = 2^{x+1}$

We compute some function values and keep the results in a table.

$f(0) = 2^{0+1} = 2^1 = 2$

$f(-1) = 2^{-1+1} = 2^0 = 1$

$f(-2) = 2^{-2+1} = 2^{-1} = \dfrac{1}{2^1} = \dfrac{1}{2}$

$f(-3) = 2^{-3+1} = 2^{-2} = \dfrac{1}{2^2} = \dfrac{1}{4}$

$f(1) = 2^{1+1} = 2^2 = 4$

$f(2) = 2^{2+1} = 2^3 = 8$

x	$f(x)$
0	2
-1	1
-2	$\dfrac{1}{2}$
-3	$\dfrac{1}{4}$
1	4
2	8

Next we plot these points and connect them with a smooth curve.

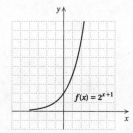

7. Graph: $f(x) = 3^{x-2}$

We compute some function values and keep the results in a table.

$f(0) = 3^{0-2} = 3^{-2} = \dfrac{1}{3^2} = \dfrac{1}{9}$

$f(1) = 3^{1-2} = 3^{-1} = \dfrac{1}{3^1} = \dfrac{1}{3}$

$f(2) = 3^{2-2} = 3^0 = 1$

$f(3) = 3^{3-2} = 3^1 = 3$

$f(4) = 3^{4-2} = 3^2 = 9$

$f(-1) = 3^{-1-2} = 3^{-3} = \dfrac{1}{3^3} = \dfrac{1}{27}$

$f(-2) = 3^{-2-2} = 3^{-4} = \dfrac{1}{3^4} = \dfrac{1}{81}$

x	$f(x)$
0	$\dfrac{1}{9}$
1	$\dfrac{1}{3}$
2	1
3	3
4	9
-1	$\dfrac{1}{27}$
-2	$\dfrac{1}{81}$

Next we plot these points and connect them with a smooth curve.

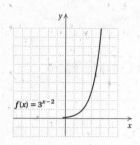

$f(x) = 3^{x-2}$

9. Graph: $f(x) = 2^x - 3$

We construct a table of values. Then we plot the points and connect them with a smooth curve.

$f(0) = 2^0 - 3 = 1 - 3 = -2$

$f(1) = 2^1 - 3 = 2 - 3 = -1$

$f(2) = 2^2 - 3 = 4 - 3 = 1$

$f(3) = 2^3 - 3 = 8 - 3 = 5$

$f(-1) = 2^{-1} - 3 = \dfrac{1}{2} - 3 = -\dfrac{5}{2}$

$f(-2) = 2^{-2} - 3 = \dfrac{1}{4} - 3 = -\dfrac{11}{4}$

x	$f(x)$
0	-2
1	-1
2	1
3	5
-1	$-\dfrac{5}{2}$
-2	$-\dfrac{11}{4}$

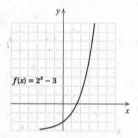

$f(x) = 2^x - 3$

11. Graph: $f(x) = 5^{x+3}$

We construct a table of values. Then we plot the points and connect them with a smooth curve.

$f(0) = 5^{0+3} = 5^3 = 125$

$f(-1) = 5^{-1+3} = 5^2 = 25$

$f(-2) = 5^{-2+3} = 5^1 = 5$

$f(-3) = 5^{-3+3} = 5^0 = 1$

$f(-4) = 5^{-4+3} = 5^{-1} = \dfrac{1}{5}$

$f(-5) = 5^{-5+3} = 5^{-2} = \dfrac{1}{25}$

x	$f(x)$
0	125
-1	25
-2	5
-3	1
-4	$\dfrac{1}{5}$
-5	$\dfrac{1}{25}$

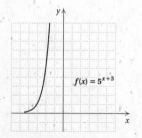

$f(x) = 5^{x+3}$

13. Graph: $f(x) = \left(\dfrac{1}{2}\right)^x$

We construct a table of values. Then we plot the points and connect them with a smooth curve.

$f(0) = \left(\dfrac{1}{2}\right)^0 = 1$

$f(1) = \left(\dfrac{1}{2}\right)^1 = \dfrac{1}{2}$

$f(2) = \left(\dfrac{1}{2}\right)^2 = \dfrac{1}{4}$

$f(3) = \left(\dfrac{1}{2}\right)^3 = \dfrac{1}{8}$

$f(-1) = \left(\dfrac{1}{2}\right)^{-1} = \dfrac{1}{\left(\dfrac{1}{2}\right)^1} = \dfrac{1}{\dfrac{1}{2}} = 2$

$f(-2) = \left(\dfrac{1}{2}\right)^{-2} = \dfrac{1}{\left(\dfrac{1}{2}\right)^2} = \dfrac{1}{\dfrac{1}{4}} = 4$

$f(-3) = \left(\dfrac{1}{2}\right)^{-3} = \dfrac{1}{\left(\dfrac{1}{2}\right)^3} = \dfrac{1}{\dfrac{1}{8}} = 8$

x	$f(x)$
0	1
1	$\dfrac{1}{2}$
2	$\dfrac{1}{4}$
3	$\dfrac{1}{8}$
-1	2
-2	4
-3	8

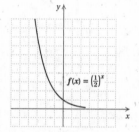

$f(x) = \left(\dfrac{1}{2}\right)^x$

15. Graph: $f(x) = \left(\dfrac{1}{5}\right)^x$

We construct a table of values. Then we plot the points and connect them with a smooth curve.

$f(0) = \left(\dfrac{1}{5}\right)^0 = 1$

$f(1) = \left(\dfrac{1}{5}\right)^1 = \dfrac{1}{5}$

$f(2) = \left(\dfrac{1}{5}\right)^2 = \dfrac{1}{25}$

$f(-1) = \left(\dfrac{1}{5}\right)^{-1} = \dfrac{1}{\dfrac{1}{5}} = 5$

$f(-2) = \left(\dfrac{1}{5}\right)^{-2} = \dfrac{1}{\dfrac{1}{25}} = 25$

x	$f(x)$
0	1
1	$\dfrac{1}{5}$
2	$\dfrac{1}{25}$
-1	5
-2	25

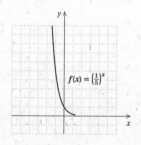

$f(x) = \left(\dfrac{1}{5}\right)^x$

17. Graph: $f(x) = 2^{2x-1}$

We construct a table of values. Then we plot the points and connect them with a smooth curve.

$f(0) = 2^{2 \cdot 0 - 1} = 2^{-1} = \dfrac{1}{2}$

$f(1) = 2^{2 \cdot 1 - 1} = 2^1 = 2$

$f(2) = 2^{2 \cdot 2 - 1} = 2^3 = 8$

$f(-1) = 2^{2(-1)-1} = 2^{-3} = \dfrac{1}{8}$

$f(-2) = 2^{2(-2)-1} = 2^{-5} = \dfrac{1}{32}$

x	$f(x)$
0	$\dfrac{1}{2}$
1	2
2	8
-1	$\dfrac{1}{8}$
-2	$\dfrac{1}{32}$

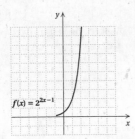

$f(x) = 2^{2x-1}$

19. Graph: $x = 2^y$

We can find ordered pairs by choosing values for y and then computing values for x.

For $y = 0$, $x = 2^0 = 1$.

For $y = 1$, $x = 2^1 = 2$.

For $y = 2$, $x = 2^2 = 4$.

For $y = 3$, $x = 2^3 = 8$.

For $y = -1$, $x = 2^{-1} = \dfrac{1}{2^1} = \dfrac{1}{2}$.

For $y = -2$, $x = 2^{-2} = \dfrac{1}{2^2} = \dfrac{1}{4}$.

For $y = -3$, $x = 2^{-3} = \dfrac{1}{2^3} = \dfrac{1}{8}$.

x	y
1	0
2	1
4	2
8	3
$\dfrac{1}{2}$	-1
$\dfrac{1}{4}$	-2
$\dfrac{1}{8}$	-3

(1) Choose values for y.

(2) Compute values for x.

We plot these points and connect them with a smooth curve.

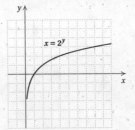

$x = 2^y$

21. Graph: $x = \left(\dfrac{1}{2}\right)^y$

We can find ordered pairs by choosing values for y and then computing values for x. Then we plot these points and connect them with a smooth curve.

For $y = 0$, $x = \left(\dfrac{1}{2}\right)^0 = 1$.

For $y = 1$, $x = \left(\dfrac{1}{2}\right)^1 = \dfrac{1}{2}$.

For $y = 2$, $x = \left(\dfrac{1}{2}\right)^2 = \dfrac{1}{4}$.

For $y = 3$, $x = \left(\dfrac{1}{2}\right)^3 = \dfrac{1}{8}$.

For $y = -1$, $x = \left(\dfrac{1}{2}\right)^{-1} = \dfrac{1}{\frac{1}{2}} = 2$.

For $y = -2$, $x = \left(\dfrac{1}{2}\right)^{-2} = \dfrac{1}{\frac{1}{4}} = 4$.

For $y = -3$, $x = \left(\dfrac{1}{2}\right)^{-3} = \dfrac{1}{\frac{1}{8}} = 8$.

x	y
1	0
$\dfrac{1}{2}$	1
$\dfrac{1}{4}$	2
$\dfrac{1}{8}$	3
2	-1
4	-2
8	-3

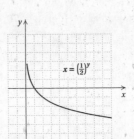

$x = \left(\dfrac{1}{2}\right)^y$

23. Graph: $x = 5^y$

We can find ordered pairs by choosing values for y and then computing values for x. Then we plot these points and connect them with a smooth curve.

For $y = 0$, $x = 5^0 = 1$.

For $y = 1$, $x = 5^1 = 5$.

For $y = 2$, $x = 5^2 = 25$.

For $y = -1$, $x = 5^{-1} = \dfrac{1}{5}$.

For $y = -2$, $x = 5^{-2} = \dfrac{1}{25}$.

x	y
1	0
5	1
25	2
$\dfrac{1}{5}$	-1
$\dfrac{1}{25}$	-2

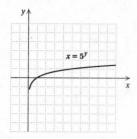

25. Graph $y = 2^x$ (see Exercise 1) and $x = 2^y$ (see Exercise 19) using the same set of axes.

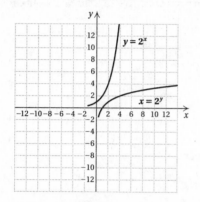

27. a) We substitute $50,000 for P and 6%, or 0.06, for r in the formula $A = P(1 + r)^t$:

$$A(t) = \$50,000(1 + 0.06)^t = \$50,000(1.06)^t$$

b)
$$A(0) = \$50,000(1.06)^0 = \$50,000$$
$$A(1) = \$50,000(1.06)^1 = \$53,000$$
$$A(2) = \$50,000(1.06)^2 = \$56,180$$
$$A(4) = \$50,000(1.06)^4 \approx \$63,123.85$$
$$A(8) = \$50,000(1.06)^8 \approx \$76,692.40$$
$$A(10) = \$50,000(1.06)^{10} \approx \$89,542.38$$
$$A(20) = \$50,000(1.06)^{20} \approx \$160,356.77$$

c)

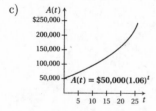

29. $N(t) = 17.7(1.018)^t$

a) As the exercise states, $t = 0$ corresponds to 2000.
$N(0) = 17.7(1.018)^0 = 17.7$ million

In 2010, $t = 2010 - 2000 = 10$.
$N(10) = 17.7(1.018)^{10} \approx 21.2$ million

In 2030, $t = 2030 - 2000 = 30$.
$N(30) = 17.7(1.018)^{30} \approx 30.2$ million

b) We use the function values computed in part (a) to draw the graph. Other values can also be computed if needed.

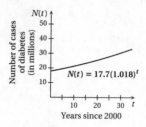

31. $A(t) = 9809(0.815)^t$

a) As the exercise states, $t = 0$ corresponds to 2000.
$A(0) = 9809(0.815)^0 = \$9809$

In 2005, $t = 2005 - 2000 = 5$.
$A(5) = 9809(0.815)^5 \approx \$3527$

In 2008, $t = 2008 - 2000 = 8$.
$A(8) = 9809(0.815)^8 \approx \$1909$

b) We use the function values computed in part (a) to draw the graph. Other values can also be computed if needed.

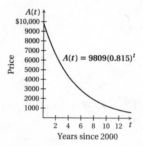

33. $N(t) = 3000(2)^{t/20}$

a) $N(10) = 3000(2)^{10/20} \approx 4243$

There will be approximately 4243 bacteria after 10 min.

$N(20) = 3000(2)^{20/20} = 6000$

There will be 6000 bacteria after 20 min.

$N(30) = 3000(2)^{30/20} \approx 8485$

There will be approximately 8485 bacteria after 30 min.

$N(40) = 3000(2)^{40/20} = 12,000$

There will be 12,000 bacteria after 40 min.

$N(60) = 3000(2)^{60/20} = 24,000$

There will be 24,000 bacteria after 60 min.

b) We use the function values computed in part (a) to draw the graph. Other values can also be computed if needed.

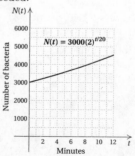

35. Discussion and Writing Exercise

37. $x^{-5} \cdot x^3 = x^{-5+3} = x^{-2} = \dfrac{1}{x^2}$

39. $9^0 = 1$ (For any nonzero number a, $a^0 = 1$.)

41. $\left(\dfrac{2}{3}\right)^1 = \dfrac{2}{3}$ (For any number a, $a^1 = a$.)

43. $\dfrac{x^{-3}}{x^4} = x^{-3-4} = x^{-7} = \dfrac{1}{x^7}$

45. $\dfrac{x}{x^0} = x^{1-0} = x^1 = x$

(This exercise could also be done as follows:

$\dfrac{x}{x^0} = \dfrac{x}{1} = x$.)

47. $(5^{\sqrt{2}})^{2\sqrt{2}} = 5^{\sqrt{2}\cdot 2\sqrt{2}} = 5^4$, or 625

49. Graph: $y = 2^x + 2^{-x}$

Construct a table of values, thinking of y as $f(x)$. Then plot these points and connect them with a curve.

$f(0) = 2^0 + 2^{-0} = 1 + 1 = 2$

$f(1) = 2^1 + 2^{-1} = 2 + \dfrac{1}{2} = 2\dfrac{1}{2}$

$f(2) = 2^2 + 2^{-2} = 4 + \dfrac{1}{4} = 4\dfrac{1}{4}$

$f(3) = 2^3 + 2^{-3} = 8 + \dfrac{1}{8} = 8\dfrac{1}{8}$

$f(-1) = 2^{-1} + 2^{-(-1)} = \dfrac{1}{2} + 2 = 2\dfrac{1}{2}$

$f(-2) = 2^{-2} + 2^{-(-2)} = \dfrac{1}{4} + 4 = 4\dfrac{1}{4}$

$f(-3) = 2^{-3} + 2^{-(-3)} = \dfrac{1}{8} + 8 = 8\dfrac{1}{8}$

x	y, or $f(x)$
0	2
1	$2\dfrac{1}{2}$
2	$4\dfrac{1}{4}$
3	$8\dfrac{1}{8}$
-1	$2\dfrac{1}{2}$
-2	$4\dfrac{1}{4}$
-3	$8\dfrac{1}{8}$

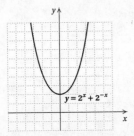

51. $y = \left|\left(\dfrac{1}{2}\right)^x - 1\right|$

Construct a table of values, thinking of y as $f(x)$. Then plot these points and connect them with a curve.

$f(-4) = \left|\left(\dfrac{1}{2}\right)^{-4} - 1\right| = |16 - 1| = |15| = 15$

$f(-2) = \left|\left(\dfrac{1}{2}\right)^{-2} - 1\right| = |4 - 1| = |3| = 3$

$f(-1) = \left|\left(\dfrac{1}{2}\right)^{-1} - 1\right| = |2 - 1| = |1| = 1$

$f(0) = \left|\left(\dfrac{1}{2}\right)^0 - 1\right| = |1 - 1| = |0| = 0$

$f(1) = \left|\left(\dfrac{1}{2}\right)^1 - 1\right| = \left|\dfrac{1}{2} - 1\right| = \left|-\dfrac{1}{2}\right| = \dfrac{1}{2}$

$f(2) = \left|\left(\dfrac{1}{2}\right)^2 - 1\right| = \left|\dfrac{1}{4} - 1\right| = \left|-\dfrac{3}{4}\right| = \dfrac{3}{4}$

$f(3) = \left|\left(\dfrac{1}{2}\right)^3 - 1\right| = \left|\dfrac{1}{8} - 1\right| = \left|-\dfrac{7}{8}\right| = \dfrac{7}{8}$

x	y, or $f(x)$
-4	15
-2	3
-1	1
0	0
1	$\dfrac{1}{2}$
2	$\dfrac{3}{4}$
3	$\dfrac{7}{8}$

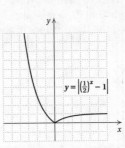

53. Construct a table of values for each equation and then draw the graphs on the same set of axes.

For $y = 3^{-(x-1)}$:

x	y
-3	81
-2	27
-1	9
0	3
1	1
2	$\dfrac{1}{3}$
3	$\dfrac{1}{9}$
4	$\dfrac{1}{27}$

For $x = 3^{-(y-1)}$:

x	y
81	-3
27	-2
9	-1
3	0
1	1
$\dfrac{1}{3}$	2
$\dfrac{1}{9}$	3
$\dfrac{1}{27}$	4

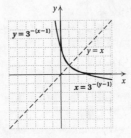

55. Left to the student

Exercise Set 8.2

1. To find the inverse of the given relation we interchange the first and second coordinates of each ordered pair. The inverse of the relation is $\{(2,1),(-3,6),(-5,-3)\}$.

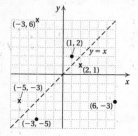

3. We interchange x and y to obtain an equation of the inverse of the relation. It is $x = 2y + 6$. The x-values in the first table become the y-values in the second table. We have

x	y
4	-1
6	0
8	1
10	2
12	3

We graph the original relation and its inverse. Since there is no horizontal line that crosses the graph more than once, the function is one-to-one.

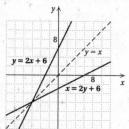

5. The graph of $f(x) = x - 5$ is shown below. Since no horizontal line crosses the graph more than once, the function is one-to-one.

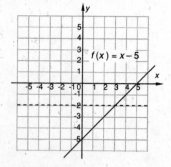

7. The graph of $f(x) = x^2 - 2$ is shown below. There are many horizontal lines that cross the graph more than once, so the function is not one-to-one.

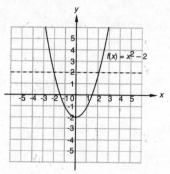

9. The graph of $g(x) = |x| - 3$ is shown below. There are many horizontal lines that cross the graph more than once, so the function is not one-to-one.

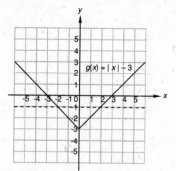

11. The graph of $g(x) = 3^x$ is shown below. Since no horizontal line crosses the graph more than once, the function is one-to-one.

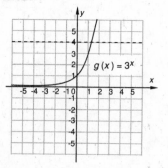

13. The graph of $f(x) = 5x - 2$ is shown below. It passes the horizontal-line test, so it is one-to-one.

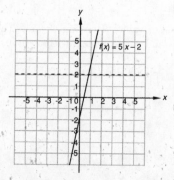

We find a formula for the inverse.

1. Replace $f(x)$ by y : $y = 5x - 2$

2. Interchange x and y : $x = 5y - 2$

3. Solve for y : $x + 2 = 5y$
$$\frac{x+2}{5} = y$$

4. Replace y by $f^{-1}(x)$: $f^{-1}(x) = \frac{x+2}{5}$

15. The graph of $f(x) = \dfrac{-2}{x}$ is shown below. It passes the horizontal-line test, so it is one-to-one.

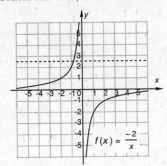

We find a formula for the inverse.

1. Replace $f(x)$ by y : $y = \dfrac{-2}{x}$

2. Interchange x and y : $x = \dfrac{-2}{y}$

3. Solve for y : $y = \dfrac{-2}{x}$

4. Replace y by $f^{-1}(x)$: $f^{-1}(x) = \dfrac{-2}{x}$

17. The graph of $f(x) = \dfrac{4}{3}x + 7$ is shown below. It passes the horizontal line test, so it is one-to-one.

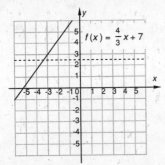

We find a formula for the inverse.

1. Replace $f(x)$ by y: $y = \dfrac{4}{3}x + 7$

2. Interchange x and y: $x = \dfrac{4}{3}y + 7$

3. Solve for y: $x - 7 = \dfrac{4}{3}y$
$$\frac{3}{4}(x - 7) = y$$

4. Replace y by $f^{(-1)}(x)$: $f^{-1}(x) = \dfrac{3}{4}(x - 7)$

19. The graph of $f(x) = \dfrac{2}{x+5}$ is shown below. It passes the horizontal line test, so it is one-to-one.

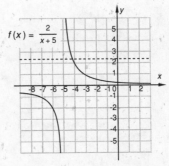

We find a formula for the inverse.

1. Replace $f(x)$ by y : $y = \dfrac{2}{x+5}$

2. Interchange x and y : $x = \dfrac{2}{y+5}$

3. Solve for y : $x(y + 5) = 2$
$$y + 5 = \frac{2}{x}$$
$$y = \frac{2}{x} - 5$$

4. Replace y by $f^{-1}(x)$: $f^{-1}(x) = \dfrac{2}{x} - 5$

21. The graph of $f(x) = 5$ is shown below. The horizontal line $y = 5$ crosses the graph more than once, so the function is not one-to-one.

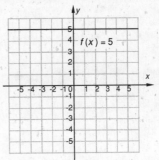

23. The graph of $f(x) = \dfrac{2x+1}{5x+3}$ is shown below. It passes the horizontal line test, so it is one-to-one.

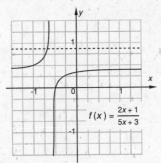

We find a formula for the inverse.

1. Replace $f(x)$ by y : $y = \dfrac{2x+1}{5x+3}$

2. Interchange x and y: $\quad x = \dfrac{2y+1}{5y+3}$

3. Solve for y: $\quad 5xy + 3x = 2y + 1$

$$5xy - 2y = 1 - 3x$$

$$y(5x - 2) = 1 - 3x$$

$$y = \dfrac{1 - 3x}{5x - 2}$$

4. Replace y by $f^{-1}(x)$: $\quad f^{-1}(x) = \dfrac{1 - 3x}{5x - 2}$

25. The graph of $f(x) = x^3 - 1$ is shown below. It passes the horizontal line test, so it is one-to-one.

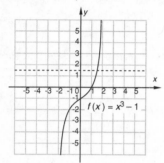

1. Replace $f(x)$ by y: $\quad y = x^3 - 1$

2. Interchange x and y: $\quad x = y^3 - 1$

3. Solve for y: $\quad x + 1 = y^3$

$$\sqrt[3]{x + 1} = y$$

4. Replace y by $f^{-1}(x)$: $\quad f^{-1}(x) = \sqrt[3]{x + 1}$

27. The graph of $f(x) = \sqrt[3]{x}$ is shown below. It passes the horizontal line test, so it is one-to-one.

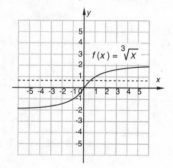

1. Replace $f(x)$ by y: $\quad y = \sqrt[3]{x}$

2. Interchange x and y: $\quad x = \sqrt[3]{y}$

3. Solve for y: $\quad x^3 = y$

4. Replace y by $f^{-1}(x)$: $\quad f^{-1}(x) = x^3$

29. We first graph $f(x) = \dfrac{1}{2}x - 3$. The graph of f^{-1} can be obtained by reflecting the graph of f across the line $y = x$.

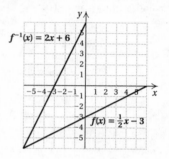

31. We first graph $f(x) = x^3$. The graph of f^{-1} can be obtained by reflecting the graph of f across the line $y = x$.

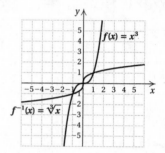

33. $f \circ g(x) = f(g(x)) = f(6 - 4x) = 2(6 - 4x) - 3 =$

$$12 - 8x - 3 = -8x + 9$$

$g \circ f(x) = g(f(x)) = g(2x - 3) = 6 - 4(2x - 3) =$

$$6 - 8x + 12 = -8x + 18$$

35. $f \circ g(x) = f(g(x)) = f(2x - 1) = 3(2x - 1)^2 + 2 =$

$$3(4x^2 - 4x + 1) + 2 = 12x^2 - 12x + 3 + 2 =$$
$$12x^2 - 12x + 5$$

$g \circ f(x) = g(f(x)) = g(3x^2 + 2) = 2(3x^2 + 2) - 1 =$

$$6x^2 + 4 - 1 = 6x^2 + 3$$

37. $f \circ g(x) = f(g(x)) = f\left(\dfrac{2}{x}\right) = 4\left(\dfrac{2}{x}\right)^2 - 1 =$

$$4\left(\dfrac{4}{x^2}\right) - 1 = \dfrac{16}{x^2} - 1$$

$g \circ f(x) = g(f(x)) = g(4x^2 - 1) = \dfrac{2}{4x^2 - 1}$

39. $f \circ g(x) = f(g(x)) = f(x^2 - 5) = (x^2 - 5)^2 + 5 =$

$$x^4 - 10x^2 + 25 + 5 = x^4 - 10x^2 + 30$$

$g \circ f(x) = g(f(x)) = g(x^2 + 5) = (x^2 + 5)^2 - 5 =$

$$x^4 + 10x^2 + 25 - 5 = x^4 + 10x^2 + 20$$

41. $h(x) = (5 - 3x)^2$

This is $5 - 3x$ raised to the second power, so the two most obvious functions are $f(x) = x^2$ and $g(x) = 5 - 3x$.

43. $h(x) = \sqrt{5x + 2}$

This is the square root of $5x + 2$, so the two most obvious functions are $f(x) = \sqrt{x}$ and $g(x) = 5x + 2$.

45. $h(x) = \dfrac{1}{x-1}$

This is the reciprocal of $x - 1$, so the two most obvious functions are $f(x) = \dfrac{1}{x}$ and $g(x) = x - 1$.

47. $h(x) = \dfrac{1}{\sqrt{7x+2}}$

This is the reciprocal of the square root of $7x + 2$. Two functions that can be used are $f(x) = \dfrac{1}{\sqrt{x}}$ and $g(x) = 7x + 2$.

49. $h(x) = (\sqrt{x} + 5)^4$

This is $\sqrt{x} + 5$ raised to the fourth power, so the two most obvious functions are $f(x) = x^4$ and $g(x) = \sqrt{x} + 5$.

51. We check to see that $f^{-1} \circ f(x) = x$ and $f \circ f^{-1}(x) = x$.

$f^{-1} \circ f(x) = f^{-1}(f(x)) = f^{-1}\left(\dfrac{4}{5}x\right) = \dfrac{5}{4} \cdot \dfrac{4}{5}x = x$

$f \circ f^{-1}(x) = f(f^{-1}(x)) = f\left(\dfrac{5}{4}x\right) = \dfrac{4}{5} \cdot \dfrac{5}{4}x = x$

53. We check to see that $f^{-1} \circ f(x) = x$ and $f \circ f^{-1}(x) = x$.

$f^{-1} \circ f(x) = f^{-1}(f(x)) = f^{-1}\left(\dfrac{1-x}{x}\right) = \dfrac{1}{\dfrac{1-x}{x}+1} = \dfrac{1}{\dfrac{1-x}{x}+1} \cdot \dfrac{x}{x} = \dfrac{x}{1-x+x} = \dfrac{x}{1} = x$

$f \circ f^{-1}(x) = f(f^{-1}(x)) = f\left(\dfrac{1}{x+1}\right) = \dfrac{1-\dfrac{1}{x+1}}{\dfrac{1}{x+1}} = \dfrac{1-\dfrac{1}{x+1}}{\dfrac{1}{x+1}} \cdot \dfrac{x+1}{x+1} = \dfrac{x+1-1}{1} = \dfrac{x}{1} = x$

55. The function $f(x) = 3x$ multiplies an input by 3, so the inverse would divide an input by 3. We have $f^{-1}(x) = \dfrac{x}{3}$.

Now we check to see that $f^{-1} \circ f(x) = x$ and $f \circ f^{-1}(x) = x$.

$f^{-1} \circ f(x) = f^{-1}(f(x)) = f^{-1}(3x) = \dfrac{3x}{3} = x$

$f \circ f^{-1}(x) = f(f^{-1}(x)) = f\left(\dfrac{x}{3}\right) = 3 \cdot \dfrac{x}{3} = x$

The inverse is correct.

57. The function $f(x) = -x$ takes the opposite of an input so the inverse would also take the opposite of an input. We have $f^{-1}(x) = -x$.

Now we check to see that $f^{-1} \circ f(x) = x$ and $f \circ f^{-1}(x) = x$.

$f^{-1} \circ f(x) = f^{-1}(f(x)) = f^{-1}(-x) = -(-x) = x$

$f \circ f^{-1}(x) = f(f^{-1}(x)) = f(-x) = -(-x) = x$

The inverse is correct.

59. The function $f(x) = \sqrt[3]{x-5}$ subtracts 5 from an input and then takes the cube root of the difference, so the inverse would cube an input and then add 5. We have $f^{-1}(x) = x^3 + 5$.

Now we check to see that $f^{-1} \circ f(x) = x$ and $f \circ f^{-1}(x) = x$.

$f^{-1} \circ f(x) = f^{-1}(f(x)) = f^{-1}(\sqrt[3]{x-5}) = (\sqrt[3]{x-5})^3 + 5 = x - 5 + 5 = x$

$f \circ f^{-1}(x) = f(f^{-1}(x)) = f(x^3 + 5) = \sqrt[3]{x^3 + 5 - 5} = \sqrt[3]{x^3} = x$

The inverse is correct.

61. a) $f(8) = 8 + 32 = 40$

Size 40 in France corresponds to size 8 in the U.S.

$f(10) = 10 + 32 = 42$

Size 42 in France corresponds to size 10 in the U.S.

$f(14) = 14 + 32 = 46$

Size 46 in France corresponds to size 14 in the U.S.

$f(18) = 18 + 32 = 50$

Size 50 in France corresponds to size 18 in the U.S.

b) The graph of $f(x) = x + 32$ is shown below. It passes the horizontal-line test, so the function is one-to-one and, hence, has an inverse that is a function.

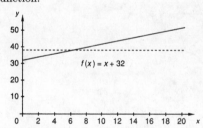

We now find a formula for the inverse.

1. Replace $f(x)$ by y: $y = x + 32$
2. Interchange x and y: $x = y + 32$
3. Solve for y: $x - 32 = y$
4. Replace y by $f^{-1}(x)$: $f^{-1}(x) = x - 32$

c) $f^{-1}(40) = 40 - 32 = 8$

Size 8 in the U.S. corresponds to size 40 in France.

$f^{-1}(42) = 42 - 32 = 10$

Size 10 in the U.S. corresponds to size 42 in France.

$f^{-1}(46) = 46 - 32 = 14$

Size 14 in the U.S. corresponds to size 46 in France.

$f^{-1}(50) = 50 - 32 = 18$

Size 18 in the U.S. corresponds to size 50 in France.

63. Discussion and Writing Exercise

65. $\sqrt[6]{a^2} = a^{2/6} = a^{1/3} = \sqrt[3]{a}$

67. $\sqrt{a^4b^6} = (a^4b^6)^{1/2} = a^2b^3$

69. $\sqrt[8]{81} = (3^4)^{1/8} = 3^{1/2} = \sqrt{3}$

71. $\sqrt[12]{64x^6y^6} = (2^6x^6y^6)^{1/12} = 2^{1/2}x^{1/2}y^{1/2} = (2xy)^{1/2} = \sqrt{2xy}$

73. $\sqrt[5]{32a^{15}b^{40}} = (2^5a^{15}b^{40})^{1/5} = 2a^3b^8$

75. $\sqrt[4]{81a^8b^8} = (3^4a^8b^8)^{1/4} = 3a^2b^2$

77. Graph the functions in a square window and determine whether one is a reflection of the other across the line $y = x$. The graphs show that these functions are not inverses of each other.

79. Graph the functions in a square window and determine whether one is a reflection of the other across the line $y = x$. The graphs show that these functions are inverses of each other.

81. (1) C; (2) A; (3) B; (4) D

83. Reflect the graph of f across the line $y = x$.

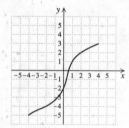

85. $f(x) = \frac{1}{2}x + 3$, $g(x) = 2x - 6$

Since $(f \circ g)(x) = x$ and $(g \circ f)(x) = x$, the functions are inverse.

Exercise Set 8.3

1. Graph: $f(x) = \log_2 x$

The equation $f(x) = y = \log_2 x$ is equivalent to $2^y = x$. We can find ordered pairs by choosing values for y and computing the corresponding x-values.

For $y = 0$, $x = 2^0 = 1$.

For $y = 1$, $x = 2^1 = 2$.

For $y = 2$, $x = 2^2 = 4$.

For $y = 3$, $x = 2^3 = 8$.

For $y = -1$, $x = 2^{-1} = \frac{1}{2}$.

For $y = -2$, $x = 2^{-2} = \frac{1}{4}$.

x, or 2^y	y
1	0
2	1
4	2
8	3
$\frac{1}{2}$	-1
$\frac{1}{4}$	-2
$\frac{1}{8}$	-3

                (1) Select y.

         (2) Compute x.

We plot the set of ordered pairs and connect the points with a smooth curve.

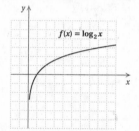

3. Graph: $f(x) = \log_{1/3} x$

The equation $f(x) = y = \log_{1/3} x$ is equivalent to $\left(\frac{1}{3}\right)^y = x$. We can find ordered pairs by choosing values for y and computing the corresponding x-values.

For $y = 0$, $x = \left(\frac{1}{3}\right)^0 = 1$.

For $y = 1$, $x = \left(\frac{1}{3}\right)^1 = \frac{1}{3}$.

For $y = 2$, $x = \left(\frac{1}{3}\right)^2 = \frac{1}{9}$.

For $y = -1$, $x = \left(\frac{1}{3}\right)^{-1} = 3$.

For $y = -2$, $x = \left(\frac{1}{3}\right)^{-2} = 9$.

x, or $\left(\frac{1}{3}\right)^y$	y
1	0
$\frac{1}{3}$	1
$\frac{1}{9}$	2
3	-1
9	-2

We plot the set of ordered pairs and connect the points with a smooth curve.

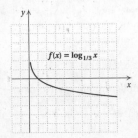

$f(x) = \log_{1/3} x$

5. Graph $f(x) = 3^x$ (see Exercise Set 10.1, Exercise 2) and $f^{-1}(x) = \log_3 x$ on the same set of axes. We can obtain the graph of f^{-1} by reflecting the graph of f across the line $y = x$.

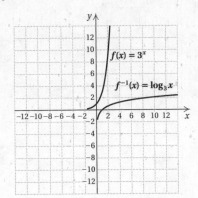

$f(x) = 3^x$

$f^{-1}(x) = \log_3 x$

7. The exponent is the logarithm.

$10^3 = 1000 \Rightarrow 3 = \log_{10} 1000$

The base remains the same.

9. The exponent is the logarithm.

$5^{-3} = \dfrac{1}{125} \Rightarrow -3 = \log_5 \dfrac{1}{125}$

The base remains the same.

11. $8^{1/3} = 2 \Rightarrow \dfrac{1}{3} = \log_8 2$

13. $10^{0.3010} = 2 \Rightarrow 0.3010 = \log_{10} 2$

15. $e^2 = t \Rightarrow 2 = \log_e t$

17. $Q^t = x \Rightarrow t = \log_Q x$

19. $e^2 = 7.3891 \Rightarrow 2 = \log_e 7.3891$

21. $e^{-2} = 0.1353 \Rightarrow -2 = \log_e 0.1353$

23. The logarithm is the exponent.

$w = \log_4 10 \Rightarrow 4^w = 10$

The base remains the same.

25. The logarithm is the exponent.

$\log_6 36 = 2 \Rightarrow 6^2 = 36$

The base remains the same.

27. $\log_{10} 0.01 = -2 \Rightarrow 10^{-2} = 0.01$

29. $\log_{10} 8 = 0.9031 \Rightarrow 10^{0.9031} = 8$

31. $\log_e 100 = 4.6052 \Rightarrow e^{4.6052} = 100$

33. $\log_t Q = k \Rightarrow t^k = Q$

35. $\log_3 x = 2$

$3^2 = x$ Converting to an exponential equation

$9 = x$ Computing 3^2

37. $\log_x 16 = 2$

$x^2 = 16$ Converting to an exponential equation

$x = 4 \text{ or } x = -4$ Principle of square roots

$\log_4 16 = 2$ because $4^2 = 16$. Thus, 4 is a solution. Since all logarithm bases must be positive, $\log_{-4} 16$ is not defined and -4 is not a solution.

39. $\log_2 16 = x$

$2^x = 16$ Converting to an exponential equation

$2^x = 2^4$

$x = 4$ The exponents are the same.

41. $\log_3 27 = x$

$3^x = 27$ Converting to an exponential equation

$3^x = 3^3$

$x = 3$ The exponents are the same.

43. $\log_x 25 = 1$

$x^1 = 25$ Converting to an exponential equation

$x = 25$

45. $\log_3 x = 0$

$3^0 = x$ Converting to an exponential equation

$1 = x$

47. $\log_2 x = -1$

$2^{-1} = x$ Converting to an exponential equation

$\dfrac{1}{2} = x$ Simplifying

49. $\log_8 x = \dfrac{1}{3}$

$8^{1/3} = x$

$2 = x$

51. Let $\log_{10} 100 = x$. Then

$10^x = 100$

$10^x = 10^2$

$x = 2$

Thus, $\log_{10} 100 = 2$.

53. Let $\log_{10} 0.1 = x$. Then

$10^x = 0.1 = \dfrac{1}{10}$

$10^x = 10^{-1}$

$x = -1$

Thus, $\log_{10} 0.1 = -1$.

55. Let $\log_{10} 1 = x$. Then

$$10^x = 1$$
$$10^x = 10^0 \qquad (10^0 = 1)$$
$$x = 0$$

Thus, $\log_{10} 1 = 0$.

57. Let $\log_5 625 = x$. Then

$$5^x = 625$$
$$5^x = 5^4$$
$$x = 4$$

Thus, $\log_5 625 = 4$.

59. Think of the meaning of $\log_7 49$. It is the exponent to which you raise 7 to get 49. That exponent is 2. Therefore, $\log_7 49 = 2$.

61. Think of the meaning of $\log_2 8$. It is the exponent to which you raise 2 to get 8. That exponent is 3. Therefore, $\log_2 8 = 3$.

63. Let $\log_9 \dfrac{1}{81} = x$. Then

$$9^x = \frac{1}{81}$$
$$9^x = 9^{-2}$$
$$x = -2$$

Thus, $\log_9 81 = -2$.

65. Let $\log_8 1 = x$. Then

$$8^x = 1$$
$$8^x = 8^0 \qquad (8^0 = 1)$$
$$x = 0$$

Thus, $\log_8 1 = 0$.

67. Let $\log_e e = x$. Then

$$e^x = e$$
$$e^x = e^1$$
$$x = 1$$

Thus, $\log_e e = 1$.

69. Let $\log_{27} 9 = x$. Then

$$27^x = 9$$
$$(3^3)^x = 3^2$$
$$3^{3x} = 3^2$$
$$3x = 2$$
$$x = \frac{2}{3}.$$

Thus, $\log_{27} 9 = \dfrac{2}{3}$.

71. 4.8970

73. -0.1739

75. Does not exist as a real number

77. 0.9464

79. $6 = 10^{0.7782}$; $84 = 10^{1.9243}$; $987,606 = 10^{5.9946}$; $0.00987606 = 10^{-2.0054}$; $98,760.6 = 10^{4.9946}$; $70,000,000 = 10^{7.8451}$; $7000 = 10^{3.8451}$

81. Discussion and Writing Exercise

83. The <u>conjugate</u> of a complex number $a + bi$ is $a - bi$.

85. In the polynomial $6x^5 - 2x^2 + 4$, $6x^5$ is called the <u>leading term.</u>

87. A system of equations that has no solution is called an <u>inconsistent</u> system.

89. For the graph of $f(x) = (x-2)^2 + 4$, the line $x = 2$ is called the <u>line of symmetry.</u>

91. Graph: $f(x) = \log_3 |x+1|$

x	$f(x)$
0	0
2	1
8	2
-2	0
-4	1
-9	2

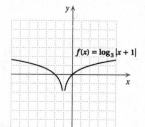

93. $\log_{125} x = \dfrac{2}{3}$

$$125^{2/3} = x$$
$$(5^3)^{2/3} = x$$
$$5^2 = x$$
$$25 = x$$

95. $\log_{128} x = \dfrac{5}{7}$

$$128^{5/7} = x$$
$$(2^7)^{5/7} = x$$
$$2^5 = x$$
$$32 = x$$

97. $\log_8 (2x + 1) = -1$

$$8^{-1} = 2x + 1$$
$$\frac{1}{8} = 2x + 1$$
$$1 = 16x + 8 \qquad \text{Multiplying by 8}$$
$$-7 = 16x$$
$$-\frac{7}{16} = x$$

99. Let $\log_{1/4} \dfrac{1}{64} = x$. Then

$$\left(\frac{1}{4}\right)^x = \frac{1}{64}$$
$$\left(\frac{1}{4}\right)^x = \left(\frac{1}{4}\right)^3$$
$$x = 3$$

Thus, $\log_{1/4} \dfrac{1}{64} = 3$.

101. $\quad \log_{10}{(\log_4{(\log_3{81}))}}$

$\quad = \log_{10}{(\log_4{4})} \qquad\qquad (\log_3{81} = 4)$

$\quad = \log_{10}{1} \qquad\qquad\qquad (\log_4{4} = 1)$

$\quad = 0$

103. Let $\log_{1/5}{25} = x$. Then

$$\left(\frac{1}{5}\right)^x = 25$$

$$(5^{-1})^x = 25$$

$$5^{-x} = 5^2$$

$$-x = 2$$

$$x = -2$$

Thus, $\log_{1/5}{25} = -2$.

Exercise Set 8.4

1. $\log_2{(32 \cdot 8)} = \log_2{32} + \log_2{8} \qquad$ Property 1

3. $\log_4{(64 \cdot 16)} = \log_4{64} + \log_4{16} \qquad$ Property 1

5. $\log_a{Qx} = \log_a{Q} + \log_a{x} \qquad$ Property 1

7. $\quad \log_b{3} + \log_b{84} = \log_b{(3 \cdot 84)} \qquad$ Property 1

$\qquad\qquad\qquad\quad = \log_b{252}$

9. $\quad \log_c{K} + \log_c{y} = \log_c{K \cdot y} \qquad$ Property 1

$\qquad\qquad\qquad = \log_c{Ky}$

11. $\log_c{y^4} = 4\ \log_c{y} \qquad$ Property 2

13. $\log_b{t^6} = 6\ \log_b{t} \qquad$ Property 2

15. $\log_b{C^{-3}} = -3\ \log_b{C} \qquad$ Property 2

17. $\log_a{\dfrac{67}{5}} = \log_a{67} - \log_a{5} \qquad$ Property 3

19. $\log_b{\dfrac{2}{5}} = \log_b{2} - \log_b{5} \qquad$ Property 3

21. $\log_c{22} - \log_c{3} = \log_c{\dfrac{22}{3}} \qquad$ Property 3

23. $\quad \log_a{x^2 y^3 z}$

$\qquad = \log_a{x^2} + \log_a{y^3} + \log_b{z} \qquad$ Property 1

$\qquad = 2\ \log_a{x} + 3\ \log_a{y} + \log_a{z} \qquad$ Property 2

25. $\quad \log_b{\dfrac{xy^2}{z^3}}$

$\qquad = \log_b{xy^2} - \log_b{z^3} \qquad\qquad$ Property 3

$\qquad = \log_b{x} + \log_b{y^2} - \log_b{z^3} \qquad$ Property 1

$\qquad = \log_b{x} + 2\ \log_b{y} - 3\ \log_b{z} \qquad$ Property 2

27. $\quad \log_c{\sqrt[3]{\dfrac{x^4}{y^3 z^2}}}$

$\qquad = \log_c{\left(\dfrac{x^4}{y^3 z^2}\right)^{1/3}}$

$\qquad = \dfrac{1}{3}\ \log_c{\dfrac{x^4}{y^3 z^2}} \qquad\qquad$ Property 2

$\qquad = \dfrac{1}{3}(\log_c{x^4} - \log_c{y^3 z^2}) \qquad$ Property 3

$\qquad = \dfrac{1}{3}[\log_c{x^4} - (\log_c{y^3} + \log_c{z^2})] \qquad$ Property 1

$\qquad = \dfrac{1}{3}(\log_c{x^4} - \log_c{y^3} - \log_c{z^2}) \qquad$ Removing parentheses

$\qquad = \dfrac{1}{3}(4\ \log_c{x} - 3\ \log_c{y} - 2\ \log_c{z}) \qquad$ Property 2

$\qquad = \dfrac{4}{3}\ \log_c{x} - \log_c{y} - \dfrac{2}{3}\log_c{z}$

29. $\quad \log_a{\sqrt[4]{\dfrac{m^8 n^{12}}{a^3 b^5}}}$

$\qquad = \log_a{\left(\dfrac{m^8 n^{12}}{a^3 b^5}\right)^{1/4}}$

$\qquad = \dfrac{1}{4}\ \log_a{\dfrac{m^8 n^{12}}{a^3 b^5}} \qquad$ Property 2

$\qquad = \dfrac{1}{4}(\log_a{m^8 n^{12}} - \log_a{a^3 b^5}) \qquad$ Property 3

$\qquad = \dfrac{1}{4}[\log_a{m^8} + \log_a{n^{12}} - (\log_a{a^3} + \log_a{b^5})]$ Property 1

$\qquad = \dfrac{1}{4}(\log_a{m^8} + \log_a{n^{12}} - \log_a{a^3} - \log_a{b^5})$

$\qquad\qquad\qquad\qquad$ Removing parentheses

$\qquad = \dfrac{1}{4}(\log_a{m^8} + \log_a{n^{12}} - 3 - \log_a{b^5}) \qquad$ Property 4

$\qquad = \dfrac{1}{4}(8\ \log_a{m} + 12\ \log_a{n} - 3 - 5\ \log_a{b}) \qquad$ Property 2

$\qquad = 2\ \log_a{m} + 3\ \log_a{n} - \dfrac{3}{4} - \dfrac{5}{4}\log_a{b}$

31. $\quad \dfrac{2}{3}\ \log_a{x} - \dfrac{1}{2}\ \log_a{y}$

$\qquad = \log_a{x^{2/3}} - \log_a{y^{1/2}} \qquad$ Property 2

$\qquad = \log_a{\dfrac{x^{2/3}}{y^{1/2}}},\ \text{or} \qquad$ Property 3

$\qquad \log_a{\dfrac{\sqrt[3]{x^2}}{\sqrt{y}}}$

33. $\quad \log_a{2x} + 3(\log_a{x} - \log_a{y})$

$\qquad = \log_a{2x} + 3\ \log_a{x} - 3\ \log_a{y}$

$\qquad = \log_a{2x} + \log_a{x^3} - \log_a{y^3} \qquad$ Property 2

$\qquad = \log_a{2x^4} - \log_a{y^3} \qquad$ Property 1

$\qquad = \log_a{\dfrac{2x^4}{y^3}} \qquad$ Property 3

35. $\log_a \dfrac{a}{\sqrt{x}} - \log_a \sqrt{ax}$

$= \log_a ax^{-1/2} - \log_a a^{1/2}x^{1/2}$

$= \log_a \dfrac{ax^{-1/2}}{a^{1/2}x^{1/2}}$　　　Property 3

$= \log_a \dfrac{a^{1/2}}{x}$, or

$\log_a \dfrac{\sqrt{a}}{x}$

37. $\log_b 15 = \log_b (3 \cdot 5)$

$= \log_b 3 + \log_b 5$　　Property 1

$= 1.099 + 1.609$

$= 2.708$

39. $\log_b \dfrac{5}{3} = \log_b 5 - \log_b 3$　　Property 3

$= 1.609 - 1.099$

$= 0.51$

41. $\log_b \dfrac{1}{5} = \log_b 1 - \log_b 5$　　Property 3

$= 0 - 1.609$　　$(\log_b 1 = 0)$

$= -1.609$

43. $\log_b \sqrt{b^3} = \log_b b^{3/2} = \dfrac{3}{2}$　　Property 4

45. $\log_b 5b = \log_b 5 + \log_b b$　　Property 1

$= 1.609 + 1$　　$(\log_b b = 1)$

$= 2.609$

47. $\log_e e^t = t$　　Property 4

49. $\log_p p^5 = 5$　　Property 4

51. $\log_2 2^7 = x$

$7 = x$　　Property 4

53. $\log_e e^x = -7$

$x = -7$　　Property 4

55. Discussion and Writing Exercise

57. $i^{29} = i^{28} \cdot i = (i^2)^{14} \cdot i = (-1)^{14} \cdot i = 1 \cdot i = i$

59. $(2+i)(2-i) = 4 - i^2 = 4 - (-1) = 4 + 1 = 5$

61. $(7 - 8i) - (-16 + 10i) = 7 - 8i + 16 - 10i = 23 - 18i$

63. $(8 + 3i)(-5 - 2i) = -40 - 16i - 15i - 6i^2 =$

$-40 - 16i - 15i + 6 = -34 - 31i$

65. Enter $y_1 = \log x^2$ and $y_2 = (\log x)(\log x)$ and show that the graphs are different and that the y-values in a table of values are not the same.

67. $\log_a (x^8 - y^8) - \log_a (x^2 + y^2)$

$= \log_a \dfrac{x^8 - y^8}{x^2 + y^2}$　　Property 3

$= \log_a \dfrac{(x^4 + y^4)(x^2 + y^2)(x+y)(x-y)}{x^2 + y^2}$　　Factoring

$= \log_a [(x^4 + y^4)(x+y)(x-y)]$　　Simplifying

$= \log_a (x^6 - x^4y^2 + x^2y^4 - y^6)$　　Multiplying

69. $\log_a \sqrt{1 - s^2}$

$= \log_a (1 - s^2)^{1/2}$

$= \dfrac{1}{2} \log_a (1 - s^2)$

$= \dfrac{1}{2} \log_a [(1-s)(1+s)]$

$= \dfrac{1}{2} \log_a (1-s) + \dfrac{1}{2} \log_a (1+s)$

71. False. For example, let $a = 10$, $P = 100$, and $Q = 10$.

$\dfrac{\log 100}{\log 10} = \dfrac{2}{1} = 2$, but

$\log \dfrac{100}{10} = \log 10 = 1$.

73. True, by Property 1

75. False. For example, let $a = 2$, $P = 1$, and $Q = 1$.

$\log_2(1+1) = \log_2 2 = 1$, but

$\log_2 1 + \log_2 1 = 0 + 0 = 0$.

Exercise Set 8.5

1. 0.6931

3. 4.1271

5. 8.3814

7. -5.0832

9. -1.6094

11. Does not exist

13. -1.7455

15. 1

17. 15.0293

19. 0.0305

21. 109.9472

23. 5

25. We will use common logarithms for the conversion. Let $a = 10$, $b = 6$, and $M = 100$ and substitute in the change-of-base formula.

$\log_b M = \dfrac{\log_a M}{\log_a b}$

$\log_6 100 = \dfrac{\log_{10} 100}{\log_{10} 6}$

$\approx \dfrac{2}{0.7782}$

≈ 2.5702

27. We will use common logarithms for the conversion. Let $a = 10$, $b = 2$, and $M = 100$ and substitute in the change-of-base formula.

$$\log_2 100 = \frac{\log_{10} 100}{\log_{10} 2}$$

$$\approx \frac{2}{0.3010}$$

$$\approx 6.6439$$

29. We will use natural logarithms for the conversion. Let $a = e$, $b = 7$, and $M = 65$ and substitute in the change-of-base formula.

$$\log_7 65 = \frac{\ln 65}{\ln 7}$$

$$\approx \frac{4.1744}{1.9459}$$

$$\approx 2.1452$$

31. We will use natural logarithms for the conversion. Let $a = e$, $b = 0.5$, and $M = 5$ and substitute in the change-of-base formula.

$$\log_{0.5} 5 = \frac{\ln 5}{\ln 0.5}$$

$$\approx \frac{1.6094}{-0.6931}$$

$$\approx -2.3219$$

33. We will use common logarithms for the conversion. Let $a = 10$, $b = 2$, and $M = 0.2$ and substitute in the change-of-base formula.

$$\log_2 0.2 = \frac{\log_{10} 0.2}{\log_{10} 2}$$

$$\approx \frac{-0.6990}{0.3010}$$

$$\approx -2.3219$$

35. We will use natural logarithms for the conversion. Let $a = e$, $b = \pi$, and $M = 200$.

$$\log_\pi 200 = \frac{\ln 200}{\ln \pi}$$

$$\approx \frac{5.2983}{1.1447}$$

$$\approx 4.6285$$

If $\ln 200$ and $\ln \pi$ are not rounded before the division is performed, the result is 4.6284.

37. Graph: $f(x) = e^x$

We find some function values with a calculator. We use these values to plot points and draw the graph.

x	e^x
0	1
1	2.7
2	7.4
3	20.1
-1	0.4
-2	0.1
-3	0.05

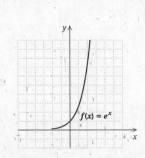

39. Graph: $f(x) = e^{-0.5x}$

We find some function values, plot points, and draw the graph.

x	$e^{-0.5x}$
0	1
1	0.61
2	0.37
-1	1.65
-2	2.72
-3	4.48
-4	7.39

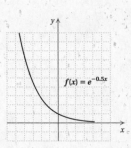

41. Graph: $f(x) = e^{x-1}$

We find some function values, plot points, and draw the graph.

x	e^{x-1}
0	0.4
1	1
2	2.7
3	7.4
4	20.1
-1	0.1
-2	0.05

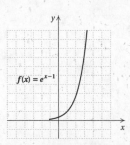

43. Graph: $f(x) = e^{x+2}$

We find some function values, plot points, and draw the graph.

x	e^{x+2}
1	20.1
0	7.4
-2	1
-3	0.4
-4	0.1

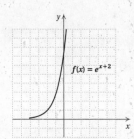

45. Graph: $f(x) = e^x - 1$

We find some function values, plot points, and draw the graph.

x	$e^x - 1$
0	0
1	1.72
2	6.39
3	19.09
-1	-0.63
-2	-0.86
-4	-0.98

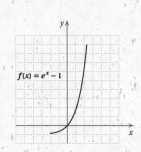

47. Graph: $f(x) = \ln(x+2)$

We find some function values, plot points, and draw the graph.

x	$\ln(x+2)$
0	0.69
1	1.10
2	1.39
3	1.61
−0.5	0.41
−1	0
−1.5	−0.69

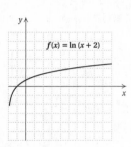

49. Graph: $f(x) = \ln(x-3)$

We find some function values, plot points, and draw the graph.

x	$\ln(x-3)$
3	Undefined
4	0
5	0.69
6	1.10
8	1.61
10	1.95

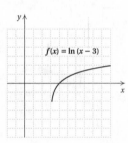

51. Graph: $f(x) = 2 \ln x$

x	$2 \ln x$
0.5	−1.4
1	0
2	1.4
3	2.2
4	2.8
5	3.2
6	3.6

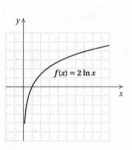

53. Graph: $f(x) = \dfrac{1}{2} \ln x + 1$

x	$\dfrac{1}{2} \ln x + 1$
1	1
2	1.35
3	1.55
4	1.69
6	1.90

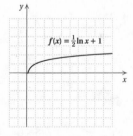

55. Graph: $f(x) = |\ln x|$

x	$\ln x$
$\dfrac{1}{4}$	1.4
$\dfrac{1}{2}$	0.7
1	0
3	1.1
5	1.6

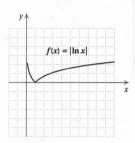

57. Discussion and Writing Exercise

59. $x^{1/2} - 6x^{1/4} + 8 = 0$

Let $u = x^{1/4}$.

$$u^2 - 6u + 8 = 0 \qquad \text{Substituting}$$
$$(u-4)(u-2) = 0$$
$$u = 4 \quad \text{or} \quad u = 2$$
$$x^{1/4} = 4 \quad \text{or} \quad x^{1/4} = 2$$
$$x = 256 \text{ or} \quad x = 16 \qquad \text{Raising both sides}$$
$$\text{to the fourth power}$$

Both numbers check. The solutions are 256 and 16.

61. $x - 18\sqrt{x} + 77 = 0$

Let $u = \sqrt{x}$.

$$u^2 - 18u + 77 = 0 \qquad \text{Substituting}$$
$$(u-7)(u-11) = 0$$
$$u = 7 \quad \text{or} \quad u = 11$$
$$\sqrt{x} = 7 \quad \text{or} \quad \sqrt{x} = 11$$
$$x = 49 \text{ or} \quad x = 121 \qquad \text{Squaring both sides}$$

Both numbers check. The solutions are 49 and 121.

63. Domain: $(-\infty, \infty)$, range: $[0, \infty)$

65. Domain: $(-\infty, \infty)$, range: $(-\infty, 100)$

67. $f(x)$ can be calculated for positive values of $2x - 5$. We have:

$$2x - 5 > 0$$
$$2x > 5$$
$$x > \frac{5}{2}$$

The domain is $\left\{ x \middle| x > \dfrac{5}{2} \right\}$, or $\left(\dfrac{5}{2}, \infty \right)$.

Exercise Set 8.6

1. $2^x = 8$

$$2^x = 2^3$$
$$x = 3 \qquad \text{The exponents are the same.}$$

3. $4^x = 256$

$$4^x = 4^4$$
$$x = 4 \qquad \text{The exponents are the same.}$$

5. $2^{2x} = 32$

$2^{2x} = 2^5$

$2x = 5$

$x = \dfrac{5}{2}$

7. $3^{5x} = 27$

$3^{5x} = 3^3$

$5x = 3$

$x = \dfrac{3}{5}$

9. $2^x = 11$

$\log 2^x = \log 11$ Taking the common logarithm on both sides

$x \, \log 2 = \log 11$ Property 2

$x = \dfrac{\log 11}{\log 2}$

$x \approx 3.4594$

11. $2^x = 43$

$\log 2^x = \log 43$ Taking the common logarithm on both sides

$x \, \log 2 = \log 43$ Property 2

$x = \dfrac{\log 43}{\log 2}$

$x \approx 5.4263$

13. $5^{4x-7} = 125$

$5^{4x-7} = 5^3$

$4x - 7 = 3$ The exponents are the same.

$4x = 10$

$x = \dfrac{10}{4}$, or $\dfrac{5}{2}$

15. $3^{x^2} \cdot 3^{4x} = \dfrac{1}{27}$

$3^{x^2+4x} = 3^{-3}$

$x^2 + 4x = -3$

$x^2 + 4x + 3 = 0$

$(x + 3)(x + 1) = 0$

$x = -3 \ or \ x = -1$

17. $4^x = 8$

$(2^2)^x = 2^3$

$2^{2x} = 2^3$

$2x = 3$ The exponents are the same.

$x = \dfrac{3}{2}$

19. $e^t = 100$

$\ln e^t = \ln 100$ Taking ln on both sides

$t = \ln 100$ Property 4

$t \approx 4.6052$ Using a calculator

21. $e^{-t} = 0.1$

$\ln e^{-t} = \ln 0.1$ Taking ln on both sides

$-t = \ln 0.1$ Property 4

$-t \approx -2.3026$

$t \approx 2.3026$

23. $e^{-0.02t} = 0.06$

$\ln e^{-0.02t} = \ln 0.06$ Taking ln on both sides

$-0.02t = \ln 0.06$ Property 4

$t = \dfrac{\ln 0.06}{-0.02}$

$t \approx \dfrac{-2.8134}{-0.02}$

$t \approx 140.6705$

25. $2^x = 3^{x-1}$

$\log 2^x = \log 3^{x-1}$

$x \, \log 2 = (x - 1) \, \log 3$

$x \, \log 2 = x \, \log 3 - \log 3$

$\log 3 = x \, \log 3 - x \, \log 2$

$\log 3 = x(\log 3 - \log 2)$

$\dfrac{\log 3}{\log 3 - \log 2} = x$

$\dfrac{0.4771}{0.4771 - 0.3010} \approx x$

$2.7095 \approx x$

27. $(3.6)^x = 62$

$\log (3.6)^x = \log 62$

$x \, \log 3.6 = \log 62$

$x = \dfrac{\log 62}{\log 3.6}$

$x \approx 3.2220$

29. $\log_4 x = 4$

$x = 4^4$ Writing an equivalent exponential equation

$x = 256$

31. $\log_2 x = -5$

$x = 2^{-5}$ Writing an equivalent exponential equation

$x = \dfrac{1}{32}$

33. $\log x = 1$ The base is 10.

$x = 10^1$

$x = 10$

35. $\log x = -2$ The base is 10.

$x = 10^{-2}$

$x = \dfrac{1}{100}$

37. $\ln x = 2$

$x = e^2 \approx 7.3891$

39. $\ln x = -1$

$x = e^{-1}$

$x = \dfrac{1}{e} \approx 0.3679$

41. $\log_3 (2x + 1) = 5$

$2x + 1 = 3^5$ Writing an equivalent exponential equation

$2x + 1 = 243$

$2x = 242$

$x = 121$

43. $\log x + \log (x - 9) = 1$ The base is 10.

$\log_{10} [x(x - 9)] = 1$ Property 1

$x(x - 9) = 10^1$

$x^2 - 9x = 10$

$x^2 - 9x - 10 = 0$

$(x - 10)(x + 1) = 0$

$x = 10 \text{ or } x = -1$

Check: For 10:

$$\begin{array}{c|c} \log x + \log (x - 9) = 1 \\ \hline \log 10 + \log (10 - 9) \ ? \ 1 \\ \log 10 + \log 1 \\ 1 + 0 \\ 1 & \text{TRUE} \end{array}$$

For -1:

$$\begin{array}{c|c} \log x + \log (x - 9) = 1 \\ \hline \log(-1) + \log (-1 - 9) \ ? \ 1 & \text{FALSE} \end{array}$$

The number -1 does not check, because negative numbers do not have logarithms. The solution is 10.

45. $\log x - \log (x + 3) = -1$ The base is 10.

$\log_{10} \dfrac{x}{x + 3} = -1$ Property 3

$\dfrac{x}{x + 3} = 10^{-1}$

$\dfrac{x}{x + 3} = \dfrac{1}{10}$

$10x = x + 3$

$9x = 3$

$x = \dfrac{1}{3}$

The answer checks. The solution is $\dfrac{1}{3}$.

47. $\log_2 (x + 1) + \log_2 (x - 1) = 3$

$\log_2[(x + 1)(x - 1)] = 3$ Property 1

$(x + 1)(x - 1) = 2^3$

$x^2 - 1 = 8$

$x^2 = 9$

$x = \pm 3$

The number 3 checks, but -3 does not. The solution is 3.

49. $\log_4 (x + 6) - \log_4 x = 2$

$\log_4 \dfrac{x + 6}{x} = 2$ Property 3

$\dfrac{x + 6}{x} = 4^2$

$\dfrac{x + 6}{x} = 16$

$x + 6 = 16x$

$6 = 15x$

$\dfrac{2}{5} = x$

The answer checks. The solution is $\dfrac{2}{5}$.

51. $\log_4 (x + 3) + \log_4 (x - 3) = 2$

$\log_4[(x + 3)(x - 3)] = 2$ Property 1

$(x + 3)(x - 3) = 4^2$

$x^2 - 9 = 16$

$x^2 = 25$

$x = \pm 5$

The number 5 checks, but -5 does not. The solution is 5.

53. $\log_3 (2x - 6) - \log_3 (x + 4) = 2$

$\log_3 \dfrac{2x - 6}{x + 4} = 2$ Property 3

$\dfrac{2x - 6}{x + 4} = 3^2$

$\dfrac{2x - 6}{x + 4} = 9$

$2x - 6 = 9x + 36$

     Multiplying by $(x + 4)$

$-42 = 7x$

$-6 = x$

Check:

$$\begin{array}{c|c} \log_3 (2x - 6) - \log_3 (x + 4) = 2 \\ \hline \log_3 [2(-6) - 6] - \log_3 (-6 + 4) \ ? \ 2 \\ \log_3 (-18) - \log_3 (-2) & \text{FALSE} \end{array}$$

The number -6 does not check, because negative numbers do not have logarithms. There is no solution.

55. Discussion and Writing Exercise

57. $x^4 + 400 = 104x^2$

$x^4 - 104x^2 + 400 = 0$

Let $u = x^2$.

$u^2 - 104u + 400 = 0$

$(u - 100)(u - 4) = 0$

$u = 100 \text{ or } u = 4$

$x^2 = 100 \text{ or } x^2 = 4$ Replacing u with x^2

$x = \pm 10 \text{ or } x = \pm 2$

The solutions are ± 10 and ± 2.

59.
$$(x^2 + 5x)^2 + 2(x^2 + 5x) = 24$$
$$(x^2 + 5x)^2 + 2(x^2 + 5x) - 24 = 0$$
Let $u = x^2 + 5x$.
$$u^2 + 2u - 24 = 0$$
$$(u + 6)(u - 4) = 0$$
$$u = -6 \quad or \quad\quad u = 4$$
$$x^2 + 5x = 6 \quad or \quad\quad x^2 + 5x = 4$$
Replacing u with $x^2 + 5x$
$$x^2 + 5x + 6 = 0 \quad or \ x^2 + 5x - 4 = 0$$
$$(x+3)(x+2) = 0 \quad or \quad\quad x =$$
$$\frac{-5 \pm \sqrt{5^2 - 4 \cdot 1(-4)}}{2 \cdot 1}$$
$$x = -3 \ or \ x = -2 \ or \quad x = \frac{-5 \pm \sqrt{41}}{2}$$
The solutions are -3, -2, and $\dfrac{-5 \pm \sqrt{41}}{2}$.

61. $(125x^3y^{-2}z^6)^{-2/3} =$
$(5^3)^{-2/3}(x^3)^{-2/3}(y^{-2})^{-2/3}(z^6)^{-2/3} =$
$5^{-2}x^{-2}y^{4/3}z^{-4} = \dfrac{1}{25}x^{-2}y^{4/3}z^{-4}$, or
$$\dfrac{y^{4/3}}{25x^2z^4}$$

63. Find the first coordinate of the point of intersection of $y_1 = \ln x$ and $y_2 = \log x$. The value of x for which the natural logarithm of x is the same as the common logarithm of x is 1.

65. a) 0.3770
   b) -1.9617
   c) 0.9036
   d) -1.5318

67.
$$2^{2x} + 128 = 24 \cdot 2^x$$
$$2^{2x} - 24 \cdot 2^x + 128 = 0$$
Let $u = 2^x$.
$$u^2 - 24u + 128 = 0$$
$$(u - 8)(u - 16) = 0$$
$$u = 8 \quad or \quad u = 16$$
$$2^x = 8 \quad or \ 2^x = 16 \quad \text{Replacing } u \text{ with } 2^x$$
$$2^x = 2^3 \quad or \ 2^x = 2^4$$
$$x = 3 \quad or \quad x = 4$$
The solutions are 3 and 4.

69.
$$8^x = 16^{3x+9}$$
$$(2^3)^x = (2^4)^{3x+9}$$
$$2^{3x} = 2^{12x+36}$$
$$3x = 12x + 36$$
$$-36 = 9x$$
$$-4 = x$$

71. $\log_6 (\log_2 x) = 0$
$$\log_2 x = 6^0$$
$$\log_2 x = 1$$
$$x = 2^1$$
$$x = 2$$

73. $\log_5 \sqrt{x^2 - 9} = 1$
$$\sqrt{x^2 - 9} = 5^1$$
$$x^2 - 9 = 25 \quad\quad \text{Squaring both sides}$$
$$x^2 = 34$$
$$x = \pm\sqrt{34}$$
Both numbers check. The solutions are $\pm\sqrt{34}$.

75. $\log (\log x) = 5 \quad\quad\quad$ The base is 10.
$$\log x = 10^5$$
$$\log x = 100,000$$
$$x = 10^{100,000}$$
The number checks. The solution is $10^{100,000}$.

77.
$$\log x^2 = (\log x)^2$$
$$2 \log x = (\log x)^2$$
$$0 = (\log x)^2 - 2 \log x$$
Let $u = \log x$.
$$0 = u^2 - 2u$$
$$0 = u(u - 2)$$
$$u = 0 \quad or \quad u = 2$$
$$\log x = 0 \quad or \ \log x = 2$$
$$x = 10^0 \quad or \quad\quad x = 10^2$$
$$x = 1 \quad or \quad\quad x = 100$$
Both numbers check. The solutions are 1 and 100.

79.
$$\log_a a^{x^2+4x} = 21$$
$$x^2 + 4x = 21 \quad\quad \text{Property 4}$$
$$x^2 + 4x - 21 = 0$$
$$(x + 7)(x - 3) = 0$$
$$x = -7 \ or \ x = 3$$
Both numbers check. The solutions are -7 and 3.

81. $3^{2x} - 8 \cdot 3^x + 15 = 0$
Let $u = 3^x$ and substitute.
$$u^2 - 8u + 15 = 0$$
$$(u - 5)(u - 3) = 0$$
$$u = 5 \quad\quad or \quad u = 3$$
$$3^x = 5 \quad\quad or \ 3^x = 3 \quad\quad \text{Substituting } 3^x \text{ for } u$$
$$\log 3^x = \log 5 \quad or \ 3^x = 3^1$$
$$x \log 3 = \log 5 \quad or \quad\quad x = 1$$
$$x = \frac{\log 5}{\log 3} \quad or \quad\quad x = 1, \text{ or}$$
$$x \approx 1.4650 \quad or \quad x = 1$$

Both numbers check. Note that we can also express $\dfrac{\log 5}{\log 3}$ as $\log_3 5$ using the change-of-base formula.

Exercise Set 8.7

1. $L = 10 \cdot \log \dfrac{I}{I_0}$

$= 10 \cdot \log \dfrac{3.2 \times 10^{-3}}{10^{-12}}$ Substituting

$= 10 \cdot \log(3.2 \times 10^9)$

$\approx 10(9.5)$

≈ 95

The sound level is about 95 dB.

3. $L = 10 \cdot \log \dfrac{I}{I_0}$

$105 = 10 \cdot \log \dfrac{I}{10^{-12}}$

$10.5 = \log \dfrac{I}{10^{-12}}$

$10.5 = \log I - \log 10^{-12}$ $(\log 10^a = a)$

$10.5 = \log I - (-12)$

$10.5 = \log I + 12$

$-1.5 = \log I$

$10^{-1.5} = I$ Converting to an exponential equation

$3.2 \times 10^{-2} \approx I$

The intensity of the sound is $10^{-1.5}$ W/m^2, or about 3.2×10^{-2} W/m^2.

5. $\text{pH} = -\log[\text{H}^+]$

$= -\log[1.6 \times 10^{-7}]$

$\approx -(-6.795880)$

≈ 6.8

The pH of milk is about 6.8.

7. $\text{pH} = -\log[\text{H}^+]$

$7.8 = -\log[\text{H}^+]$

$-7.8 = \log[\text{H}^+]$

$10^{-7.8} = [\text{H}^+]$

$1.58 \times 10^{-8} \approx [\text{H}^+]$

The hydrogen ion concentration is about 1.58×10^{-8} moles per liter.

9. $3,251,876 = 3251.876$ thousands

$w(P) = 0.37 \ln P + 0.05$

$w(3251.876) = 0.37 \ln 3251.876 + 0.05$

≈ 3.04 ft/sec

11. $311,121 = 311.121$ thousands

$w(P) = 0.37 \ln P + 0.05$

$w(311.121) = 0.37 \ln 311.121 + 0.05$

≈ 2.17 ft/sec

13. $N(t) = 3^t$

a) $N(5) = 3^5 = 243$ people

b) $6,400,000,000 = 3^t$

$\ln 6,400,000,000 = \ln 3^t$

$\ln 6,400,000,000 = t \ln 3$

$\dfrac{\ln 6,400,000,000}{\ln 3} = t$

$20.6 \approx t$

The acts of kindness will reach the entire world in about 20.6 months.

c) $2 = 3^t$

$\ln 2 = \ln 3^t$

$\ln 2 = t \ln 3$

$\dfrac{\ln 2}{\ln 3} = t$

$0.6 \approx t$

The doubling time is about 0.6 month.

15. a) In 2010, $t = 2010 - 2000$, or 10.

$C(10) = 11,054(1.06)^{10} \approx \$19,796$

b) $21,000 = 11,054(1.06)^t$

$\dfrac{21,000}{11,054} = 1.06^t$

$\log \dfrac{21,000}{11,054} = \log 1.06^t$

$\log \dfrac{21,000}{11,054} = t \log 1.06$

$\dfrac{\log \dfrac{21,000}{11,054}}{\log 1.06} = t$

$11 \approx t$

The cost will be \$21,000 11 yr after 2000, or in 2011.

c) $22,108 = 11,054(1.06)^t$

$2 = (1.06)^t$

$\log 2 = \log(1.06)^t$

$\log 2 = t \log 1.06$

$\dfrac{\log 2}{\log 1.06} = t$

$11.9 \approx t$

The doubling time is about 11.9 years.

17. a) $P(t) = 6.4e^{0.0114t}$, where $P(t)$ is in billions and t is the number of years after 2004.

b) In 2010, $t = 2010 - 2004$, or 6.

$P(6) = 6.4e^{0.0114(6)} \approx 6.9$

In 2010, the population will be about 6.9 billion.

c)
$$10 = 6.4e^{0.0114t}$$
$$\frac{10}{6.4} = e^{0.0114t}$$
$$\ln\left(\frac{10}{6.4}\right) = \ln e^{0.0114t}$$
$$\ln\left(\frac{10}{6.4}\right) = 0.0114t$$
$$\frac{\ln\left(\frac{10}{6.4}\right)}{0.0114} = t$$
$$39 \approx t$$

The population will be 10 billion 39 years after 2004, or in 2043.

d)
$$12.8 = 6.4e^{0.0114t}$$
$$2 = e^{0.0114t}$$
$$\ln 2 = \ln e^{0.0114t}$$
$$\ln 2 = 0.0114t$$
$$\frac{\ln 2}{0.0114} = t$$
$$60.8 \approx t$$

The doubling time is about 60.8 yr.

19. a) $P(t) = P_0\, e^{0.06t}$

b) To find the balance after one year, replace P_0 with 5000 and t with 1. We find $P(1)$:

$$P(1) = 5000\, e^{0.06(1)} \approx \$5309.18$$

To find the balance after 2 years, replace P_0 with 5000 and t with 2. We find $P(2)$:

$$P(2) = 5000\, e^{0.06(2)} \approx \$5637.48$$

To find the balance after 10 years, replace P_0 with 5000 and t with 10. We find $P(10)$: $P(10) = 5000e^{0.06(10)} \approx \$9110.59$

c) To find the doubling time, replace P_0 with 5000 and $P(t)$ with 10,000 and solve for t.

$$10,000 = 5000\, e^{0.06t}$$
$$2 = e^{0.06t}$$
$$\ln\ 2 = \ln\ e^{0.06t} \quad \text{Taking the natural loga-}$$
$$\text{rithm on both sides}$$
$$\ln\ 2 = 0.06t \quad \text{Finding the logarithm of}$$
$$\text{the base to a power}$$
$$\frac{\ln\ 2}{0.06} = t$$
$$11.6 \approx t$$

The investment will double in about 11.6 years.

21. a)
$$P(t) = P_0 e^{kt}$$
$$1,563,282 = 852,737 e^{k\cdot 10}$$
$$\frac{1,563,282}{852,737} = e^{10k}$$
$$\ln \frac{1,563,282}{852,737} = \ln\ e^{10k}$$
$$\ln \frac{1,563,282}{852,737} = 10k$$
$$\frac{\ln \dfrac{1,563,282}{852,737}}{10} = k$$
$$0.061 \approx k$$

The exponential growth rate is 0.061, or 6.1%.

The exponential growth function is $P(t) = 852,737e^{0.061t}$, where t is the number of years after 1990.

b) In 2010, $t = 2010 - 1990$, or 20.

$$P(20) = 852,737 e^{0.061(20)}$$
$$= 852,737 e^{1.22}$$
$$\approx 2,888,380$$

In 2010, the population of Las Vegas will be about 2,888,380.

c)
$$8,000,000 = 852,737 e^{0.061t}$$
$$\frac{8,000,000}{852,737} = e^{0.061t}$$
$$\ln \frac{8,000,000}{852,737} = \ln\ e^{0.061t}$$
$$\ln \frac{8,000,000}{852,737} = 0.061t$$
$$\frac{\ln \dfrac{8,000,000}{852,737}}{0.061} = t$$
$$37 \approx t$$

The population will reach 8,000,000 about 37 yr after 1990, or in 2027.

23. If the scrolls had lost 22.3% of their carbon-14 from an initial amount P_0, then 77.7%(P_0) is the amount present. To find the age t of the scrolls, we substitute 77.7%(P_0), or $0.777P_0$, for $P(t)$ in the carbon-14 decay function and solve for t.

$$P(t) = P_0 e^{-0.00012t}$$
$$0.777P_0 = P_0 e^{-0.00012t}$$
$$0.777 = e^{-0.00012t}$$
$$\ln 0.777 = \ln\ e^{-0.00012t}$$
$$-0.2523 \approx -0.00012t$$
$$t \approx \frac{-0.2523}{-0.00012} \approx 2103$$

The scrolls are about 2103 years old.

25. The function $P(t) = P_0\,e^{-kt}$, $k > 0$, can be used to model decay. For iodine-131, $k = 9.6\%$, or 0.096. To find the half-life we substitute 0.096 for k and $\frac{1}{2}\,P_0$ for $P(t)$, and solve for t.

$$\frac{1}{2}\,P_0 = P_0\,e^{-0.096t}, \text{ or } \frac{1}{2} = e^{-0.096t}$$

$$\ln\frac{1}{2} = \ln e^{-0.096t} = -0.096t$$

$$t = \frac{\ln 0.5}{-0.096} \approx \frac{-0.6931}{-0.096} \approx 7.2 \text{ days}$$

27. The function $P(t) = P_0 e^{-kt}$, $k > 0$, can be used to model decay. We substitute $\frac{1}{2}P_0$ for $P(t)$ and 1 for t and solve for the decay rate k.

$$\frac{1}{2}P_0 = P_0 e^{-k\cdot 1}$$

$$\frac{1}{2} = e^{-k}$$

$$\ln\frac{1}{2} = \ln e^{-k}$$

$$-0.693 \approx -k$$

$$0.693 \approx k$$

The decay rate is 0.693, or 69.3% per year.

29. a) We use the exponential decay equation $W(t) = W_0 e^{-kt}$, where t is the number of years after 1996 and $W(t)$ is in millions of tons. In 1996, at $t = 0$, 17.5 million tons of yard waste were discarded. We substitute 17.5 for W_0.

$$W(t) = 17.5e^{-kt}.$$

To find the exponential decay rate k, observe that 2 years after 1996, in 1998, 14.5 million tons of yard waste were discarded. We substitute 2 for t and 14.5 for $W(t)$.

$$14.5 = 17.5e^{-k\cdot 2}$$

$$0.8286 \approx e^{-2k}$$

$$\ln 0.8286 \approx \ln e^{-2k}$$

$$\ln 0.8286 \approx -2k$$

$$\frac{\ln 0.8286}{-2} \approx k$$

$$0.094 \approx k$$

Then we have $W(t) = 17.5e^{-0.094t}$, where t is the number of years after 1996 and $W(t)$ is in millions of tons.

b) In 2010, $t = 2010 - 1996 = 14$.

$$W(14) = 17.5e^{-0.094(14)}$$

$$= 17.5e^{-1.316}$$

$$\approx 4.7$$

In 2010, about 4.7 million tons of yard waste were discarded.

c) 1 ton is equivalent to 0.000001 million tons.

$$0.000001 = 17.5e^{-0.094t}$$

$$5.71 \times 10^{-8} \approx e^{-0.094t}$$

$$\ln(5.71 \times 10^{-8}) \approx \ln e^{-0.094t}$$

$$\ln(5.71 \times 10^{-8}) \approx -0.094t$$

$$\frac{\ln(5.71 \times 10^{-8})}{-0.094} \approx t$$

$$177 \approx t$$

Only one ton of yard waste will be discarded about 177 years after 1996, or in 2173.

31. a) We start with the exponential growth equation

$$V(t) = V_0\,e^{kt}, \text{ where } t \text{ is the number}$$
of years after 1991.

Substituting 451,000 for V_0, we have

$$V(t) = 451,000\,e^{kt}.$$

To find the exponential growth rate k, observe that the card sold for \$640,500 in 1996, or 5 years after 1991. We substitute and solve for k.

$$V(5) = 451,000\,e^{k\cdot 5}$$

$$640,500 = 451,000\,e^{5k}$$

$$1.42 = e^{5k}$$

$$\ln 1.42 = \ln e^{5k}$$

$$\ln 1.42 = 5k$$

$$\frac{\ln 1.42}{5} = k$$

$$0.07 \approx k$$

Thus the exponential growth function is $V(t) = 451,000\,e^{0.07t}$, where t is the number of years after 1991.

b) In 2015, $t = 2015 - 1991$, or 24.

$$V(24) = 451,000\,e^{0.07(24)} \approx 2,419,866$$

The card's value in 2015 will be about \$2,419,866.

c) Substitute \$902,000 for $V(t)$ and solve for t.

$$902,000 = 451,000\,e^{0.07t}$$

$$2 = e^{0.07t}$$

$$\ln 2 = \ln e^{0.07t}$$

$$\ln 2 = 0.07t$$

$$\frac{\ln 2}{0.07} = t$$

$$9.9 \approx t$$

The doubling time is about 9.9 years.

d) Substitute $1,000,000 for $V(t)$ and solve for t.

$$1,000,000 = 451,000\,e^{0.07t}$$

$$2.217 \approx e^{0.07t}$$

$$\ln 2.217 \approx \ln e^{0.07t}$$

$$\ln 2.217 \approx 0.07t$$

$$\frac{\ln 2.217}{0.07} \approx t$$

$$11.4 \approx t$$

The value of the card first exceeded $1,000,000 in $1991 + 11$, or 2002.

e) In 2001, $t = 2001 - 1991$, or 10.

$$V(10) = 451,000e^{0.07(10)}$$

$$= 451,000e^{0.7}$$

$$\approx 908,202$$

The function estimates that the card's value in 2001 would be about $908,202. According to this, the card would not have been a good buy at $1.1 million.

33. a)

$$P(t) = P_0 e^{-kt}$$

$$2,242,798 = 2,394,811e^{-k\cdot10}$$

$$\frac{2,242,798}{2,394,811} = e^{-10k}$$

$$\ln \frac{2,242,798}{2,394,811} = \ln e^{-10k}$$

$$\ln \frac{2,242,798}{2,394,811} = -10k$$

$$\frac{\ln \dfrac{2,242,798}{2,394,811}}{-10} = k$$

$$0.007 \approx k$$

The exponential decay rate is 0.007, or 0.7%. The exponential decay function is $P(t) = 2,394,811e^{-0.007t}$, where t is the number of years after 1990.

b) In 2010, $t = 2010 - 1990$, or 20.

$$P(20) = 2,394,811e^{-0.007(20)}$$

$$= 2,394,811e^{-0.14}$$

$$\approx 2,081,949$$

In 2010, the population of Pittsburgh will be about 2,081,949.

c)

$$1,000,000 = 2,394,811e^{-0.007t}$$

$$\frac{1,000,000}{2,394,811} = e^{-0.007t}$$

$$\ln \frac{1,000,000}{2,394,811} = \ln e^{-0.007t}$$

$$\ln \frac{1,000,000}{2,394,811} = -0.007t$$

$$\frac{\ln \dfrac{1,000,000}{2,394,811}}{-0.007} = t$$

$$125 \approx t$$

The population will decline to 1 million about 125 yr after 1990, or in 2115.

35. Discussion and Writing Exercise

37. $i^{46} = (i^2)^{23} = (-1)^{23} = -1$

39. $i^{53} = (i^2)^{26} \cdot i = (-1)^{26} \cdot i = i$

41. $i^{14} + i^{15} = (i^2)^7 + (i^2)^7 \cdot i = (-1)^7 + (-1)^7 \cdot i = -1 - i$

43.
$$\frac{8-i}{8+i} = \frac{8-i}{8+i} \cdot \frac{8-i}{8-i}$$

$$= \frac{64 - 16i + i^2}{64 - i^2}$$

$$= \frac{64 - 16i - 1}{64 - (-1)}$$

$$= \frac{63 - 16i}{65}$$

$$= \frac{63}{65} - \frac{16}{65}i$$

45. $(5 - 4i)(5 + 4i) = 25 - 16i^2 = 25 + 16 = 41$

47. $-0.937,\ 1.078,\ 58.770$

49. $-0.767,\ 2,\ 4$

51. We will use the exponential growth equation $V(t) = V_0 e^{kt}$, where t is the number of years after 2001 and $V(t)$ is in millions of dollars. We substitute 21 for $V(t)$, 0.05 for k, and 9 for t and solve for V_0.

$$21 = V_0 e^{0.05(9)}$$

$$21 = V_0 e^{0.45}$$

$$\frac{21}{e^{0.45}} = V_0$$

$$13.4 \approx V_0$$

George Steinbrenner needs to invest $13.4 million at 5% interest compounded continuously in order to have $21 million to pay Derek Jeter in 2010.

Chapter 8 Review Exercises

1. We interchange the coordinates of the ordered pairs. The inverse of the relation is

$$\{(2, -4), (-7, 5), (-2, -1), (11, 10)\}.$$

2. The graph of $f(x) = 4 - x^2$ fails the horizontal-line test, so it is not one-to-one.

3. The graph of $g(x) = \dfrac{2x - 3}{7}$ passes the horizontal-line test, so it is one-to-one.

We find a formula for the inverse.

1. Replace $g(x)$ by y: $y = \dfrac{2x - 3}{7}$

2. Interchange x and y: $x = \dfrac{2y - 3}{7}$

3. Solve for y: $7x = 2y - 3$

$$7x + 3 = 2y$$

$$\frac{7x + 3}{2} = y$$

4. Replace y by $g^{-1}(x)$: $g^{-1}(x) = \dfrac{7x + 3}{2}$

4. The graph of $f(x) = 8x^3$ passes the horizontal-line test, so it is one-to-one.

We find a formula for the inverse.

1. Replace $f(x)$ by y: $y = 8x^3$

2. Interchange x and y: $x = 8y^3$

3. Solve for y: $\dfrac{x}{8} = y^3$

$$\frac{\sqrt[3]{x}}{2} = y$$

4. Replace y by $f^{-1}(x)$: $f^{-1}(x) = \dfrac{\sqrt[3]{x}}{2}$, or $\dfrac{1}{2}\sqrt[3]{x}$

5. The graph of $f(x) = \dfrac{4}{3 - 2x}$ passes the horizontal-line test, so it is one-to-one.

We find a formula for the inverse.

1. Replace $f(x)$ by y: $y = \dfrac{4}{3 - 2x}$

2. Interchange x and y: $x = \dfrac{4}{3 - 2y}$

3. Solve for y: $x(3 - 2y) = 4$

$$3x - 2xy = 4$$
$$-2xy = 4 - 3x$$
$$y = \frac{4 - 3x}{-2x}, \text{ or } \frac{3x - 4}{2x}$$

4. Replace y by $f^{-1}(x)$: $f^{-1}(x) = \dfrac{3x - 4}{2x}$

6. First graph $f(x) = x^3 + 1$. Then, to graph the inverse, reflect the graph of the function across the line $y = x$.

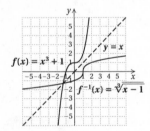

7. Graph: $f(x) = 3^{x-1}$

We make a table of values. Then we plot the points and connect them with a smooth curve.

x	$f(x)$
0	$\dfrac{1}{3}$
1	1
2	3
3	9
-1	$\dfrac{1}{9}$
-2	$\dfrac{1}{27}$
-3	$\dfrac{1}{81}$

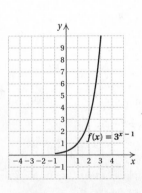

8. Graph $f(x) = \log_3 x$, or $y = \log_3 x$

The equation $f(x) = y = \log_3 x$ is equivalent to $3^y = x$. We find ordered pairs by choosing values for y and computing the corresponding x-values. Then we plot the set of ordered pairs and connect the points with a smooth curve.

x, or 3^y	y
1	0
3	1
9	2
27	3
$\dfrac{1}{3}$	-1
$\dfrac{1}{9}$	-2
$\dfrac{1}{27}$	-3

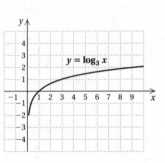

9. Graph: $f(x) = e^{x+1}$

We find some function values using a calculator. Then we use these values to plot points and draw the graph.

x	$f(x)$
0	2.7
1	7.4
2	20.1
3	54.6
-1	1
-2	0.4
-3	0.1

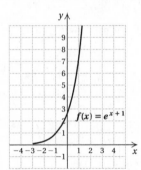

10. Graph: $f(x) = \ln(x - 1)$

We find some function values using a calculator. Then we use these values to plot points and draw the graph.

x	$f(x)$
1.5	-0.69
2	0
3	0.69
4	1.10
6	1.61
8	1.95
9	2.08

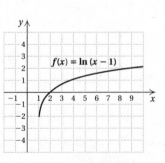

11. $f \circ g(x) = f(g(x)) = f(3x - 5) = (3x - 5)^2 = 9x^2 - 30x + 25$

$g \circ f(x) = g(f(x)) = g(x^2) = 3x^2 - 5$

12. $h(x) = \sqrt{4 - 7x}$

This is the square root of $4 - 7x$, so the two most obvious functions are $f(x) = \sqrt{x}$ and $g(x) = 4 - 7x$.

13. $10^4 = 10,000 \Rightarrow 4 = \log 10,000$

14. $25^{1/2} = 5 \Rightarrow \frac{1}{2} = \log_{25} 5$

15. $\log_4 16 = x \Rightarrow 4^x = 16$

16. $\log_{1/2} 8 = -3 \Rightarrow \left(\frac{1}{2}\right)^{-3} = 8$

17. $\log_3 9 = x$

$3^x = 9$

$3^x = 3^2$

$x = 2$

18. $\log_{10} \frac{1}{10} = x$

$10^x = \frac{1}{10}$

$10^x = 10^{-1}$

$x = -1$

19. $\log_m m = 1$ since $m^1 = m$.

20. $\log_m 1 = 0$ since $m^0 = 1$.

21. $\log\left(\frac{78}{43,112}\right) \approx -2.7425$

22. $\log(-4)$ does not exist as a real number.

23. $\log_a x^4 y^2 z^3$

$= \log_a x^4 + \log_a y^2 + \log_a z^3$

$= 4 \log_a x + 2 \log_a y + 3 \log_a z$

24. $\log \sqrt[4]{\frac{z^2}{x^3 y}}$

$= \log \left(\frac{z^2}{x^3 y}\right)^{1/4}$

$= \frac{1}{4} \log \frac{z^2}{x^3 y}$

$= \frac{1}{4}[\log z^2 - \log(x^3 y)]$

$= \frac{1}{4}[2 \log z - (\log x^3 + \log y)]$

$= \frac{1}{4}(2 \log z - \log x^3 - \log y)$

$= \frac{1}{4}(2 \log z - 3 \log x - \log y)$

$= \frac{1}{2} \log z - \frac{3}{4} \log x - \frac{1}{4} \log y$

25. $\log_a 8 + \log_a 15 = \log_a (8 \cdot 15) = \log_a 120$

26. $\frac{1}{2} \log a - \log b - 2 \log c$

$= \log a^{1/2} - \log b - \log c^2$

$= \log a^{1/2} - (\log b + \log c^2)$

$= \log a^{1/2} - \log(bc^2)$

$= \log \frac{a^{1/2}}{bc^2}$

27. $\log_m m^{17} = 17$ by Property 4

28. $\log_m m^{-7} = -7$ by Property 4

29. $\log_a 28 = \log_a (2^2 \cdot 7)$

$= \log_a 2^2 + \log_a 7$

$= 2 \log_a 2 + \log_a 7$

$= 2(1.8301) + 5.0999$

$= 8.7601$

30. $\log_a 3.5 = \log_a \left(\frac{7}{2}\right)$

$= \log_a 7 - \log_a 2$

$= 5.0999 - 1.8301$

$= 3.2698$

31. $\log_a \sqrt{7} = \log_a 7^{1/2}$

$= \frac{1}{2} \log_a 7$

$= \frac{1}{2}(5.0999)$

$= 2.54995$

32. $\log_a \frac{1}{4} = \log_a 4^{-1}$

$= \log_a (2^2)^{-1}$

$= \log_a 2^{-2}$

$= -2 \log_a 2$

$= -2(1.8301)$

$= -3.6602$

33. $\ln 0.06774 \approx -2.6921$

34. $e^{-0.98} \approx 0.3753$

35. $e^{2.91} \approx 18.3568$

36. $\ln 1 = 0$

37. $\ln 0$ does not exist.

38. $\ln e = 1$

39. $\log_5 2 = \frac{\log 2}{\log 5} \approx 0.4307$

40. $\log_{12} 70 = \frac{\ln 70}{\ln 12} \approx 1.7097$

41. $\log_3 x = -2$

$3^{-2} = x$

$\frac{1}{3^2} = x$

$\frac{1}{9} = x$

42. $\log_x 32 = 5$

$x^5 = 32$

$x^5 = 2^5$

$x = 2$

43. $\log x = -4$

$10^{-4} = x$

$\dfrac{1}{10^4} = x$

$\dfrac{1}{10,000} = x$

44. $3 \ln x = -6$

$\ln x = -2$

$e^{-2} = x$

$0.1353 \approx x$

45. $4^{2x-5} = 16$

$4^{2x-5} = 4^2$

$2x - 5 = 2$

$2x = 7$

$x = \dfrac{7}{2}$

46. $2^{x^2} \cdot 2^{4x} = 32$

$2^{x^2+4x} = 2^5$

$x^2 + 4x = 5$

$x^2 + 4x - 5 = 0$

$(x-1)(x+5) = 0$

$x = 1 \ \ or \ \ x = -5$

47. $4^x = 8.3$

$\log 4^x = \log 8.3$

$x \log 4 = \log 8.3$

$x = \dfrac{\log 8.3}{\log 4}$

$x \approx 1.5266$

48. $e^{-0.1t} = 0.03$

$\ln e^{-0.1t} = \ln 0.03$

$-0.1t = \ln 0.03$

$t = \dfrac{\ln 0.03}{-0.1}$

$t \approx 35.0656$

49. $\log_4 16 = x$

$4^x = 16$

$4^x = 4^2$

$x = 2$

50. $\log_4 x + \log_4 (x-6) = 2$

$\log_4 [x(x-6)] = 2$

$4^2 = x(x-6)$

$16 = x^2 - 6x$

$0 = x^2 - 6x - 16$

$0 = (x-8)(x+2)$

$x = 8 \ \ or \ \ x = -2$

The number 8 checks, but -2 does not. The solution is 8.

51. $\log_2 (x+3) - \log_2 (x-3) = 4$

$\log_2 \dfrac{x+3}{x-3} = 4$

$2^4 = \dfrac{x+3}{x-3}$

$16 = \dfrac{x+3}{x-3}$

$(x-3) \cdot 16 = (x-3) \cdot \dfrac{x+3}{x-3}$

$16x - 48 = x + 3$

$15x = 51$

$x = \dfrac{51}{15} = \dfrac{17}{5}$

The number $\dfrac{17}{5}$ checks. It is the solution.

52. $\log_3 (x-4) = 3 - \log_3 (x+4)$

$\log_3 (x-4) + \log_3 (x+4) = 3$

$\log_3 [(x-4)(x+4)] = 3$

$3^3 = (x-4)(x+4)$

$27 = x^2 - 16$

$43 = x^2$

$x = -\sqrt{43} \ \ or \ \ x = \sqrt{43}$

The number $-\sqrt{43}$ does not check, but $\sqrt{43}$ does. The solution is $\sqrt{43}$.

53. $L = 10 \cdot \log \dfrac{I}{I_0}$

$L = 10 \cdot \log \dfrac{10^{-3}}{10^{-12}}$

$= 10 \cdot \log 10^9$

$= 10 \cdot 9$

$= 90 \text{ dB}$

54. $S(t) = 0.15(1.4)^t$

a) In 2008, $t = 2008 - 2003 = 5$.

$S(5) = 0.15(1.4)^5 \approx 0.807$ million

In 2010, $t = 2010 - 2003 = 7$.

$S(7) = 0.15(1.4)^7 \approx 1.581$ million

In 2020, $t = 2020 - 2003 = 17$.

$S(17) = 0.15(1.4)^{17} \approx 45.737$ million

b) $20 = 0.15(1.4)^t$

$\dfrac{20}{0.15} = 1.4^t$

$\log \dfrac{20}{0.15} = \log 1.4^t$

$\log \dfrac{20}{0.15} = t \log 1.4$

$\dfrac{\log \dfrac{20}{0.15}}{\log 1.4} = t$

$14.5 \approx t$

There will be 20 million TiVo subscribers about 14.5 yr after 2003.

c) $S_0 = 0.15$, so $2\,S_0 = 2(0.15) = 0.3$.

$$0.3 = 0.15(1.4)^t$$
$$2 = 1.4^t$$
$$\log 2 = \log 1.4^t$$
$$\log 2 = t \log 1.4$$
$$\frac{\log 2}{\log 1.4} = t$$
$$2 \approx t$$

The doubling time is about 2 yr.

d) We can plot the points found in parts (a), (b), and (c) and additional points as needed to graph the function.

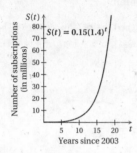

55. a) We start with the exponential growth function $V(t) = V_0\,e^{kt}$, where t is the number of years after 2005.

Substituting 40,000 for V_0, we have
$$V(t) = 40{,}000\,e^{kt}.$$

To find k observe that the value of the lot was \$53,000 in 2008, or 3 yr after 2005. We substitute and solve for k.

$$53{,}000 = 40{,}000\,e^{k \cdot 3}$$
$$1.325 = e^{3k}$$
$$\ln 1.325 = \ln e^{3k}$$
$$\ln 1.325 = 3k$$
$$\frac{\ln 1.325}{3} = k$$
$$0.094 \approx k$$

Thus the exponential growth function is $V(t) = 40{,}000\,e^{0.094t}$.

b) In 2015, $t = 2015 - 2005 = 10$.
$$V(10) = 40{,}000\,e^{0.094(10)} \approx \$102{,}399$$

c)
$$85{,}000 = 40{,}000\,e^{0.094t}$$
$$2.125 = e^{0.094t}$$
$$\ln 2.125 = \ln e^{0.094t}$$
$$\ln 2.125 = 0.094t$$
$$\frac{\ln 2.125}{0.094} = t$$
$$8 \approx t$$

The value of the lot will first reach \$85,000 about 8 yr after 2005, or in 2013.

56.
$$2\,P_0 = P_0\,e^{k \cdot 3}$$
$$2 = e^{3k}$$
$$\ln 2 = \ln e^{3k}$$
$$\ln 2 = 3k$$
$$\frac{\ln 2}{3} = k$$
$$0.231 \approx k$$

The exponential growth rate was about 0.231, or 23.1%.

57.
$$2(7600) = 7600\,e^{0.084t}$$
$$2 = e^{0.084t}$$
$$\ln 2 = \ln e^{0.084t}$$
$$\ln 2 = 0.084t$$
$$\frac{\ln 2}{0.084} = t$$
$$8.25 \text{ yr} \approx t$$

58. If the skeleton had lost 34% of its carbon-14 from an initial amount of P_0, then $66\%\,(P_0)$ is the amount present. We use the carbon-14 decay function to find the age of the skeleton.
$$P(t) = P_0\,e^{-0.00012t}$$
$$0.66\,P_0 = P_0\,e^{-0.00012t}$$
$$0.66 = e^{-0.00012t}$$
$$\ln 0.66 = \ln e^{-0.00012t}$$
$$\ln 0.66 = -0.00012t$$
$$\frac{\ln 0.66}{-0.00012} = t$$
$$3463 \text{ yr} \approx t$$

59. *Discussion and Writing Exercise.* You cannot take the logarithm of a negative number because logarithm bases are positive and there is no real-number power to which a positive number can be raised to yield a negative number.

60. *Discussion and Writing Exercise.* $\log_a 1 = 0$ because $a^0 = 1$.

61.
$$\ln(\ln x) = 3$$
$$e^3 = \ln x$$
$$x = e^{e^3}$$

62.
$$5^{x+y} = 25 \qquad\qquad 2^{2x-y} = 64$$
$$5^{x+y} = 5^2 \qquad\qquad 2^{2x-y} = 2^6$$
$$x + y = 2 \qquad\qquad 2x - y = 6$$

We have a system of equations. We solve using the elimination method.

$$\begin{array}{rl} x + y = 2, & (1) \\ \underline{2x - y = 6} & (2) \\ 3x = 8 & \text{Adding} \\ x = \dfrac{8}{3} & \end{array}$$

Now substitute $\dfrac{8}{3}$ for x in Equation (1) and solve for y.

$$\frac{8}{3} + y = 2$$

$$y = -\frac{2}{3}$$

The solution is $\left(\dfrac{8}{3}, -\dfrac{2}{3}\right)$.

Chapter 8 Test

1. Graph: $f(x) = 2^{x+1}$

We compute some function values and keep the results in a table.

$f(0) = 2^{0+1} = 2^1 = 2$

$f(-1) = 2^{-1+1} = 2^0 = 1$

$f(-2) = 2^{-2+1} = 2^{-1} = \dfrac{1}{2^1} = \dfrac{1}{2}$

$f(-3) = 2^{-3+1} = 2^{-2} = \dfrac{1}{2^2} = \dfrac{1}{4}$

$f(1) = 2^{1+1} = 2^2 = 4$

$f(2) = 2^{2+1} = 2^3 = 8$

x	$f(x)$
0	2
-1	1
-2	$\dfrac{1}{2}$
-3	$\dfrac{1}{4}$
1	4
2	8

Next we plot these points and connect them with a smooth curve.

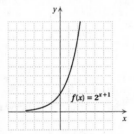

2. Graph: $y = \log_2 x$

The equation $y = \log_2 x$ is equivalent to $2^y = x$. We can find ordered pairs by choosing values for y and computing the corresponding x-values.

For $y = 0$, $x = 2^0 = 1$.

For $y = 1$, $x = 2^1 = 2$.

For $y = 2$, $x = 2^2 = 4$.

For $y = 3$, $x = 2^3 = 8$.

For $y = -1$, $x = 2^{-1} = \dfrac{1}{2}$.

For $y = -2$, $x = 2^{-2} = \dfrac{1}{4}$.

x, or 2^y	y
1	0
2	1
4	2
8	3
$\dfrac{1}{2}$	-1
$\dfrac{1}{4}$	-2
$\dfrac{1}{8}$	-3

(1) Select y.

(2) Compute x.

We plot the set of ordered pairs and connect the points with a smooth curve.

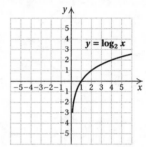

3. Graph: $f(x) = e^{x-2}$

We find some function values with a calculator. Use these values to plot points and draw the graph.

x	$f(x)$
-1	0.05
0	0.14
1	0.47
2	1
3	2.72
4	7.39

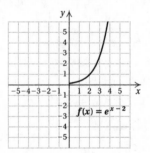

4. Graph: $f(x) = \ln(x - 4)$

We find some function values, plot points, and draw the graph.

x	$f(x)$
4.1	-2.30
4.5	-0.69
5	0
6	0.69
7	1.10

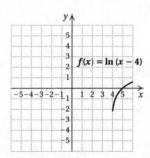

5. Interchange the first and second coordinates of each ordered pair.

$$\{(3, -4), (-8, 5), (-3, -1), (12, 10)\}$$

6. The graph of $f(x) = 4x - 3$ passes the horizontal-line test, so it is one-to-one.

We find a formula for the inverse.

1. Replace $f(x)$ by y: $\quad y = 4x - 3$

2. Interchange x and y: $\quad x = 4y - 3$

3. Solve for y: $\quad x + 3 = 4y$

$$\frac{x+3}{4} = y$$

4. Replace y by $f^{-1}(x)$: $\quad f^{-1}(x) = \dfrac{x+3}{4}$

7. The graph of $f(x) = (x + 1)^3$ passes the horizontal-line test, so it is one-to-one.

We find a formula for the inverse.

1. Replace $f(x)$ by y: $\quad y = (x+1)^3$

2. Interchange x and y: $\quad x = (y+1)^3$

3. Solve for y: $\quad \sqrt[3]{x} = y + 1$

$$\sqrt[3]{x} - 1 = y$$

4. Replace y by $f^{-1}(x)$: $\quad f^{-1}(x) = \sqrt[3]{x} - 1$

8. The graph of $f(x) = 2 - |x|$ is shown below. It fails the horizontal-line test, so it is not one-to-one.

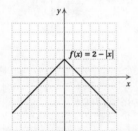

9. $f(x) = x + x^2$, $g(x) = 5x - 2$

$f(x) \circ g(x) = f(g(x)) = 5x - 2 + (5x - 2)^2 =$
$5x - 2 + 25x^2 - 20x + 4 = 25x^2 - 15x + 2$

$g(x) \circ f(x) = g(f(x)) = 5(x + x^2) - 2 = 5x + 5x^2 - 2$, or $5x^2 + 5x - 2$

10. $256^{1/2} = 16 \Rightarrow \dfrac{1}{2} = \log_{256} 16$

11. $m = \log_7 49 \Rightarrow 7^m = 49$

12. Let $\log_5 125 = x$. Then

$$5^x = 125$$
$$5^x = 5^3$$
$$x = 3$$

Thus, $\log_5 125 = 3$.

13. Since $\log_a a^k = k$, we have $\log_t t^{23} = 23$.

14. Since $\log_a 1 = 0$, we have $\log_p 1 = 0$.

15. $\log 0.0123 \approx -1.9101$

16. $\log(-5)$ does not exist as a real number.

17.
$$\log \frac{a^3 b^{1/2}}{c^2}$$
$$= \log (a^3 b^{1/2}) - \log c^2$$
$$= \log a^3 + \log b^{1/2} - \log c^2$$
$$= 3 \log a + \frac{1}{2} \log b - 2 \log c$$

18.
$$\frac{1}{3} \log_a x - 3 \log_a y + 2 \log_a z$$
$$= \log_a x^{1/3} - \log_a y^3 + \log_a z^2$$
$$= \log_a x^{1/3} + \log_a z^2 - \log_a y^3$$
$$= \log_a (x^{1/3} z^2) - \log_a y^3$$
$$= \log_a \frac{x^{1/3} z^2}{y^3}$$

19. $\log_a \dfrac{2}{7} = \log_a 2 - \log_a 7$
$$= 0.301 - 0.845$$
$$= -0.544$$

20. $\log_a 12 = \log_a (2 \cdot 6)$
$$= \log_a 2 + \log_a 6$$
$$= 0.301 + 0.778$$
$$= 1.079$$

21. $\ln 807.39 \approx 6.6938$

22. $e^{4.68} \approx 107.7701$

23. $\ln 1 = 0$

24. $\log_{18} 31 = \dfrac{\log 31}{\log 18} \approx \dfrac{1.4914}{1.2553} \approx 1.1881$

25. $\log_x 25 = 2$
$$25 = x^2$$
$$5 = x \quad \text{Taking the positive square root}$$

The solution is 5.

26. $\log_4 x = \dfrac{1}{2}$
$$x = 4^{1/2} = \sqrt{4}$$
$$x = 2$$

The solution is 2.

27. $\log x = 4$
$$x = 10^4$$
$$x = 10,000$$

The solution is 10,000.

28. $\ln x = \dfrac{1}{4}$
$$x = e^{1/4}$$
$$x \approx 1.2840$$

The solution is $e^{1/4}$, or about 1.2840.

29.
$$7^x = 1.2$$
$$\ln 7^x = \ln 1.2$$
$$x \ln 7 = \ln 1.2$$
$$x = \frac{\ln 1.2}{\ln 7}$$
$$x \approx \frac{0.1823}{1.9459}$$
$$x \approx 0.0937$$

The solution is $\dfrac{\ln 1.2}{\ln 7}$, or about 0.0937.

30.
$$\log (x^2 - 1) - \log (x - 1) = 1$$
$$\log \frac{x^2 - 1}{x - 1} = 1$$
$$\log \frac{(x+1)(x-1)}{x-1} = 1$$
$$\log (x + 1) = 1$$
$$x + 1 = 10^1$$
$$x = 9$$

The number 9 checks. It is the solution.

31.
$$\log_5 x + \log_5 (x + 4) = 1$$
$$\log_5 [x(x + 4)] = 1$$
$$x(x + 4) = 5^1$$
$$x^2 + 4x = 5$$
$$x^2 + 4x - 5 = 0$$
$$(x + 5)(x - 1) = 0$$
$$x = -5 \ or \ x = 1$$

The number -5 does not check, but 1 does. The solution is 1.

32.
$$\text{pH} = -\log [\text{H}^+]$$
$$= -\log (6.3 \times 10^{-5})$$
$$\approx -(-4.2007)$$
$$\approx 4.2$$

The pH is approximately 4.2.

33. $C(t) = 21,856(1.045)^t$

a) In 2008, $t = 2008 - 2000 = 8$.
$$C(8) = 21,856(1.045)^8 \approx \$31,081$$

b)
$$33,942 = 21,856(1.045)^t$$
$$\frac{33,942}{21,856} = 1.045^t$$
$$\log \frac{33,942}{21,856} = \log 1.045^t$$
$$\log \frac{33,942}{21,856} = t \log 1.045$$
$$\frac{\log \dfrac{33,942}{21,856}}{\log 1.045} = t$$
$$10 \approx t$$

The cost will be \$33,942 about 10 yr after 2000, or in 2010.

c) $2(21,856) = 43,712$

Then we have:
$$43,712 = 21,856(1.045)^t$$
$$2 = 1.045^t$$
$$\log 2 = \log 1.045^t$$
$$\log 2 = t \log 1.045$$
$$\frac{\log 2}{\log 1.045} = t$$
$$15.7 \approx t$$

The doubling time is about 15.7 yr.

34. a) Let t represent the number of years after 2002 and let $N(t)$ be in millions. We know that $N_0 = 31.902$, so we have
$$N(t) = 31.902e^{kt}.$$

Observe that 2 yr after 2002, in 2004, the population was 32.508 million. We find k.
$$32.508 = 31.902e^{k \cdot 2}$$
$$\frac{32.508}{31.902} = e^{2k}$$
$$\ln \frac{32.508}{31.902} = \ln e^{2k}$$
$$\ln \frac{32.508}{31.902} = 2k$$
$$\frac{\ln \dfrac{32.508}{31.902}}{2} = k$$
$$0.009 \approx k$$

Then the exponential growth function is $N(t) = 31.902e^{0.009t}$.

b) In 2008, $t = 2008 - 2002 = 6$.
$$N(6) = 31.902e^{0.009(6)}$$
$$= 31.902e^{0.054}$$
$$\approx 33.672 \text{ million}$$
In 2015, $t = 2015 - 2002 = 13$.
$$N(13) = 31.902e^{0.009(13)}$$
$$= 31.902e^{0.117}$$
$$\approx 35.862 \text{ million}$$

c)
$$50 = 31.902e^{0.009t}$$
$$\frac{50}{31.902} = e^{0.009t}$$
$$\ln \frac{50}{31.902} = \ln e^{0.009t}$$
$$\ln \frac{50}{31.902} = 0.009t$$
$$\frac{\ln \dfrac{50}{31.902}}{0.009} = t$$
$$50 \approx t$$

The population will be 50 million about 50 yr after 2002, or in 2052.

d) $2(31.902) = 63.804$

Then we have:

$63.804 = 31.902e^{0.009t}$

$2 = e^{0.009t}$

$\ln 2 = \ln e^{0.009t}$

$\ln 2 = 0.009t$

$\dfrac{\ln 2}{0.009} = t$

$77.0 \approx t$

The doubling time is about 77.0 yr.

35. Let P_0 be the amount originally invested. This grows to $2P_0$ in 15 yr. Thus we have:

$2P_0 = P_0\,e^{k \cdot 15}$

$2 = e^{15k}$

$\ln 2 = \ln e^{15k}$

$\ln 2 = 15k$

$\dfrac{\ln 2}{15} = k$

$0.046 \approx k$

The interest rate is 4.6%.

36. If the bone has lost 43% of its carbon-14 from an initial amount of P_0, then 57%(P_0), or $0.57P_0$, is the amount present. We use the carbon-14 decay function and solve for t.

$0.57P_0 = P_0\,e^{-0.00012t}$

$0.57 = e^{-0.00012t}$

$\ln 0.57 = \ln e^{-0.00012t}$

$\ln 0.57 = -0.00012t$

$\dfrac{\ln 0.57}{-0.00012} = t$

$4684 \approx t$

The bone is about 4684 years old.

37. $\log_3 |2x - 7| = 4$

$|2x - 7| = 3^4$

$|2x - 7| = 81$

$2x - 7 = -81 \quad or \quad 2x - 7 = 81$

$2x = -74 \quad or \qquad 2x = 88$

$x = -37 \quad or \qquad x = 44$

Both numbers check. The solutions are -37 and 44.

38.

$\log_a \dfrac{\sqrt[3]{x^2 z}}{\sqrt[3]{y^2 z^{-1}}}$

$= \log_a \sqrt[3]{\dfrac{x^2 z}{y^2 z^{-1}}}$

$= \log_a \left(\dfrac{x^2 z^2}{y^2} \right)^{1/3}$

$= \dfrac{1}{3} \log_a \left(\dfrac{x^2 z^2}{y^2} \right)$

$= \dfrac{1}{3} [\log_a (x^2 z^2) - \log_a y^2]$

$= \dfrac{1}{3} (\log_a x^2 + \log_a z^2 - \log_a y^2)$

$= \dfrac{1}{3} (2 \log_a x + 2 \log_a z - 2 \log_a y)$

$= \dfrac{1}{3} (2 \cdot 2 + 2 \cdot 4 - 2 \cdot 3)$

$= \dfrac{1}{3} (4 + 8 - 6)$

$= \dfrac{1}{3} (6)$

$= 2$

Cumulative Review Chapters R - 8

1. $I = 1.08 \dfrac{T}{N}$

a) 8 hr $= 8 \times 1$ hr $= 8 \times 60$ min $= 480$ min

$I = 1.08 \cdot \dfrac{480}{35} \approx 14.8$ min

b) $17.28 = 1.08 \cdot \dfrac{480}{N}$

$17.28N = 1.08(480)$

$N = \dfrac{1.08(480)}{17.28}$

$N = 30$ appointments

c) $I = 1.08 \dfrac{T}{N}$

$IN = 1.08T$

$\dfrac{IN}{1.08} = T$

d) $I = 1.08 \dfrac{T}{N}$

$IN = 1.08T$

$N = \dfrac{1.08T}{I}$, or $1.08 \dfrac{T}{I}$

e) The formula found in part (c) would be useful if we knew the interval time and the number of appointments desired and wanted to find the total number of minutes the doctor would have to spend with patients per day in order to achieve this. The formula found in part (d) would allow us to find the total number of appointments that could be scheduled given a particular interval time and the total number of minutes the doctor can spend with patients per day.

2. $\left| -\dfrac{5}{2} + \left(-\dfrac{7}{2} \right) \right| = \left| -\dfrac{12}{2} \right| = |-6| = 6$

3. $(-5x^4y^{-3}z^2)(-4x^2y^2) = (-5)(-4)x^{4+2}y^{-3+2}z^2$

$\qquad = 20x^6y^{-1}z^2$

$\qquad = \dfrac{20x^6z^2}{y}$

4. $2x - 3 - 2[5 - 3(2 - x)] = 2x - 3 - 2[5 - 6 + 3x]$

$\qquad = 2x - 3 - 2[-1 + 3x]$

$\qquad = 2x - 3 + 2 - 6x$

$\qquad = -4x - 1$

5. $\quad 3^3 + 2^2 - (32 \div 4 - 16 \div 8)$

$= 3^3 + 2^2 - (8 - 2)$

$= 3^3 + 2^2 - 6$

$= 27 + 4 - 6$

$= 31 - 6$

$= 25$

6. $8(2x - 3) = 6 - 4(2 - 3x)$

$\quad 16x - 24 = 6 - 8 + 12x$

$\quad 16x - 24 = -2 + 12x$

$\qquad\quad 4x = 22$

$\qquad\quad x = \dfrac{11}{2}$

The solution is $\dfrac{11}{2}$.

7. $\qquad x(x - 3) = 10$

$\qquad x^2 - 3x = 10$

$\quad x^2 - 3x - 10 = 0$

$\quad (x - 5)(x + 2) = 0$

$x - 5 = 0 \;\; or \;\; x + 2 = 0$

$\quad x = 5 \;\; or \qquad x = -2$

The solutions are 5 and -2.

8. $\quad x + y - 3z = -1, \quad (1)$

$\quad 2x - y + z = 4, \qquad (2)$

$\quad -x - y + z = 1 \qquad (3)$

$\quad x + y - 3z = -1 \quad (1)$

$\quad \underline{2x - y + z = 4} \quad (2)$

$\quad 3x \qquad - 2z = 3 \quad (4) \text{ Adding}$

$\quad x + y - 3z = -1 \quad (1)$

$\quad \underline{-x - y + z = 1} \quad (3)$

$\qquad\qquad - 2z = 0 \quad \text{Adding}$

$\qquad\qquad\quad z = 0$

$3x - 2 \cdot 0 = 3 \quad \text{Substituting in (4)}$

$\qquad\quad 3x = 3$

$\qquad\quad x = 1$

$1 + y - 3 \cdot 0 = -1 \quad \text{Substituting in (1)}$

$\qquad\quad 1 + y = -1$

$\qquad\qquad y = -2$

The solution is $(1, -2, 0)$.

9. $\quad 4x - 3y = 15, \quad (1)$

$\quad 3x + 5y = 4 \qquad (2)$

We multiply Equation (1) by 5 and Equation (2) by 3 and then add to eliminate y.

$\qquad 20x - 15y = 75$

$\qquad \underline{9x + 15y = 12}$

$\qquad 29x \qquad\quad = 87$

$\qquad\qquad x = 3$

Now substitute 3 for x in one of the original equations and solve for y. We use Equation (2).

$\qquad 3 \cdot 3 + 5y = 4$

$\qquad\quad 9 + 5y = 4$

$\qquad\qquad 5y = -5$

$\qquad\qquad y = -1$

The solution is $(3, -1)$.

10. $\qquad\qquad \dfrac{7}{x^2 - 5x} - \dfrac{2}{x - 5} = \dfrac{4}{x}$

$\qquad\qquad \dfrac{7}{x(x - 5)} - \dfrac{2}{x - 5} = \dfrac{4}{x}, \text{ LCD is } x(x - 5)$

$x(x - 5)\left(\dfrac{7}{x(x - 5)} - \dfrac{2}{x - 5} \right) = x(x - 5) \cdot \dfrac{4}{x}$

$\qquad\qquad\qquad 7 - 2x = 4(x - 5)$

$\qquad\qquad\qquad 7 - 2x = 4x - 20$

$\qquad\qquad\qquad -6x = -27$

$\qquad\qquad\qquad x = \dfrac{9}{2}$

The number $\dfrac{9}{2}$ checks. It is the solution.

11. $\qquad \sqrt{x - 1} = \sqrt{x + 4} - 1$

$\quad (\sqrt{x - 1})^2 = (\sqrt{x + 4} - 1)^2$

$\qquad x - 1 = x + 4 - 2\sqrt{x + 4} + 1$

$\qquad x - 1 = x + 5 - 2\sqrt{x + 4}$

$\qquad\quad -6 = -2\sqrt{x + 4}$

$\qquad\quad 3 = \sqrt{x + 4} \qquad \text{Dividing by } -2$

$\qquad\quad 3^2 = (\sqrt{x + 4})^2$

$\qquad\quad 9 = x + 4$

$\qquad\quad 5 = x$

The number 5 checks. It is the solution.

12. $x - 8\sqrt{x} + 15 = 0$

Let $u = \sqrt{x}$.

$\quad u^2 - 8u + 15 = 0$

$\quad (u - 3)(u - 5) = 0$

$u = 3 \quad or \quad u = 5$

$\sqrt{x} = 3 \quad or \quad \sqrt{x} = 5$

$x = 9 \quad or \quad x = 25$

Both numbers check. The solutions are 9 and 25.

13. $x^4 - 13x^2 + 36 = 0$

Let $u = x^2$.

$u^2 - 13u + 36 = 0$

$(u - 4)(u - 9) = 0$

$u = 4 \quad or \quad u = 9$

$x^2 = 4 \quad or \quad x^2 = 9$

$x = \pm 2 \quad or \quad x = \pm 3$

All four numbers check. The solutions are ± 2 and ± 3.

14. $\log_8 x = 1$

$8^1 = x$

$8 = x$

15. $3^{5x} = 7$

$\log 3^{5x} = \log 7$

$5x \cdot \log 3 = \log 7$

$x = \dfrac{\log 7}{5 \log 3}$

$x \approx 0.3542$

16. $\log x - \log (x - 8) = 1$

$\log \dfrac{x}{x - 8} = 1$

$10^1 = \dfrac{x}{x - 8}$

$10(x - 8) = x \qquad \text{Multiplying by } x - 8$

$10x - 80 = x$

$-80 = -9x$

$\dfrac{80}{9} = x$

The number $\dfrac{80}{9}$ checks. It is the solution.

17. $x^2 + 4x > 5$

$x^2 + 4x - 5 > 0$

$(x + 5)(x - 1) > 0$

The solutions of $(x + 5)(x - 1) = 0$ are -5 and 1. They divide the real-number line with three intervals as shown:

We try test numbers in each interval.

A: Test -6, $(-6 + 5)(-6 - 1) = 7 > 0$

B: Test 0, $(0 + 5)(0 - 1) = -5 < 0$

C: Test 2, $(2 + 5)(2 - 1) = 7 > 0$

The expression is positive for all numbers in intervals A and C. The solution set is $\{x | x < -5 \text{ or } x > 1\}$, or $(-\infty, -5) \cup (1, \infty)$.

18. $|2x - 3| \geq 9$

$2x - 3 \leq -9 \quad or \quad 2x - 3 \geq 9$

$2x \leq -6 \quad or \quad 2x \geq 12$

$x \leq -3 \quad or \quad x \geq 6$

The solution set is $\{x | x \leq -3 \text{ or } x \geq 6\}$, or $(-\infty, -3] \cup [6, \infty)$.

19. $x^2 + 6x = 11$

$x^2 + 6x - 11 = 0$

$a = 1, \ b = 6, \ c = -11$

$x = \dfrac{-b \pm \sqrt{b^2 - 4ac}}{2a}$

$x = \dfrac{-6 \pm \sqrt{6^2 - 4 \cdot 1 \cdot (-11)}}{2 \cdot 1} = \dfrac{-6 \pm \sqrt{36 + 44}}{2}$

$x = \dfrac{-6 \pm \sqrt{80}}{2} = \dfrac{-6 \pm 4\sqrt{5}}{2}$

$x = \dfrac{2(-3 \pm 2\sqrt{5})}{2} = -3 \pm 2\sqrt{5}$

The solutions are $-3 \pm 2\sqrt{5}$.

20. $D = \dfrac{ab}{b + a}$

$D(b + a) = ab \qquad \text{Multiplying by } b + a$

$Db + Da = ab$

$Db = ab - Da$

$Db = a(b - D)$

$\dfrac{Db}{b - D} = a$

21. $\dfrac{1}{p} + \dfrac{1}{q} = \dfrac{1}{f}$

$pqf\left(\dfrac{1}{p} + \dfrac{1}{q}\right) = pqf \cdot \dfrac{1}{f}$

$qf + pf = pq$

$pf = pq - qf$

$pf = q(p - f)$

$\dfrac{pf}{p - f} = q$

22. $f(x) = \dfrac{-4}{3x^2 - 5x - 2}$

The numbers excluded from the domain are those for which the denominator is 0.

$3x^2 - 5x - 2 = 0$

$(3x + 1)(x - 2) = 0$

$3x + 1 = 0 \quad or \quad x - 2 = 0$

$3x = -1 \quad or \quad x = 2$

$x = -\dfrac{1}{3} \quad or \quad x = 2$

The domain is $\left(-\infty, -\dfrac{1}{3}\right) \cup \left(-\dfrac{1}{3}, 2\right) \cup (2, \infty)$.

23. $\dfrac{x^0 + y}{-z} = \dfrac{6^0 + 9}{-(-5)} = \dfrac{1 + 9}{5} = \dfrac{10}{5} = 2$

24. Familiarize. Let $t =$ the number of minutes it will take to do the job, working together.

Translate. We use the work principle.

$$\frac{t}{10} + \frac{t}{12} = 1$$

Solve. We solve the equation.

$$60\left(\frac{t}{10} + \frac{t}{12}\right) = 60 \cdot 1$$

$$6t + 5t = 60$$

$$11t = 60$$

$$t = \frac{60}{11}, \text{ or } 5\frac{5}{11}$$

Check. We verify the work principle.

$$\frac{60/11}{10} + \frac{60/11}{12} = \frac{60}{11} \cdot \frac{1}{10} + \frac{60}{11} \cdot \frac{1}{12} = \frac{6}{11} + \frac{5}{11} = 1$$

State. It would take Anne and Clay $5\frac{5}{11}$ min to do the job, working together.

25. $S(t) = 78 - 15 \log (t + 1)$

a) $S(0) = 78 - 15 \log (0 + 1)$

$\quad = 78 - 15 \log 1$

$\quad = 78 - 15 \cdot 0$

$\quad = 78$

b) $S(4) = 78 - 15 \log (4 + 1)$

$\quad = 78 - 15 \log 5$

$\quad \approx 67.5$

26. Familiarize. Let $r =$ the speed of the stream, in km/h. Then the boat's speed downstream is $5 + r$, and the speed upstream is $5 - r$. We organize the information in a table.

	Distance	Speed	Time
Downstream	42	$5 + r$	t
Upstream	12	$5 - r$	t

Translate. Using $t = d/r$ in each row of the table, we can equate the two expressions for time.

$$\frac{42}{5 + r} = \frac{12}{5 - r}$$

Solve. We solve the equation.

$$\frac{42}{5 + r} = \frac{12}{5 - r}$$

$$(5 + r)(5 - r) \cdot \frac{42}{5 + r} = (5 + r)(5 - r) \cdot \frac{12}{5 - r}$$

$$42(5 - r) = 12(5 + r)$$

$$210 - 42r = 60 + 12r$$

$$-54r = -150$$

$$r = \frac{25}{9}, \text{ or } 2\frac{7}{9}$$

Check. If the speed of the stream is $\frac{25}{9}$ km/h, then the boat's speed downstream is $5 + \frac{25}{9}$, or $\frac{70}{9}$ km/h and it travels 42 km downstream in $\frac{42}{70/9} = 42 \cdot \frac{9}{70} = \frac{27}{5}$, or $5\frac{2}{5}$ hr. The boat's speed upstream is $5 - \frac{25}{9}$, or $\frac{20}{9}$ km/h and it travels 12 km upstream in $\frac{12}{20/9} = 12 \cdot \frac{9}{20} = \frac{27}{5}$, or $5\frac{2}{5}$ hr. Since the times are the same, the answer checks.

State. The speed of the stream is $2\frac{7}{9}$ km/h.

27. Familiarize. Let $x =$ the amount of Swim Clean and $y =$ the amount of Pure Swim that should be used, in liters. We organize the information in a table.

	Swim Clean	Pure Swim	Mixture
Amount	x	y	100 L
Percent of acid	30%	80%	50%
Amount of acid	$0.3x$	$0.8y$	$0.5(100)$, or 50 L

Translate. The first and last rows of the table yield two equations.

$$x + y = 100,$$

$$0.3x + 0.8y = 50$$

After clearing decimals we have the following system of equations.

$$x + y = 100, \quad (1)$$

$$3x + 8y = 500 \quad (2)$$

Solve. We first multiply Equation (1) by -3 and then add.

$$-3x - 3y = -300$$

$$\underline{3x + 8y = 500}$$

$$5y = 200$$

$$y = 40$$

Now we substitute in Equation (1) and solve for x.

$$x + 40 = 100$$

$$x = 60$$

Check. 60 L + 40 L = 100 L. The amount of acid in the mixture is $0.3(60) + 0.8(40) = 18 + 32 = 50$ L. The answer checks.

State. 60 L of Swim Clean and 40 L of Pure Swim should be used.

28. a) $P(t) = 19\,e^{0.012t}$, where $P(t)$ is in millions and t is the number of years after 2004.

b) In 2009, $t = 2009 - 2004 = 5$.

$P(5) = 19\,e^{0.012(5)} \approx 20.2$ million

In 2015, $t = 2015 - 2004 = 11$.

$P(11) = 19\,e^{0.012(11)} \approx 21.7$ million

c) $2\,P_0 = 2 \cdot 19 = 38$

Then we have

$$38 = 19\,e^{0.012t}$$

$$2 = e^{0.012t}$$

$$\ln 2 = \ln e^{0.012t}$$

$$\ln 2 = 0.012t$$

$$\frac{\ln 2}{0.012} = t$$

$$57.8 \text{ yr} \approx t$$

29. Familiarize. Let $x =$ the width of the sidewalk, in feet. We make a drawing.

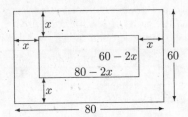

The length and width of the new lawn are represented by $80 - 2x$ and $60 - 2x$, respectively.

Translate. We use the formula for the area of a rectangle, $A = lw$.

$$2400 = (80 - 2x)(60 - 2x)$$

Solve. We solve the equation.

$$2400 = (80 - 2x)(60 - 2x)$$

$$2400 = 4800 - 280x + 4x^2$$

$$0 = 4x^2 - 280x + 2400$$

$$0 = x^2 - 70x + 600 \qquad \text{Dividing by 4}$$

$$0 = (x - 60)(x - 10)$$

$$x - 60 = 0 \quad or \quad x - 10 = 0$$

$$x = 60 \quad or \qquad x = 10$$

Check. 60 cannot be a solution because $80 - 2 \cdot 60 = -40$ and $60 - 2 \cdot 60 = -60$ and the dimensions of the new lawn cannot be negative. If $x = 10$, then the dimensions of the new lawn are $80 - 2 \cdot 10$, or 60, by $60 - 2 \cdot 10$, or 40, and the area is $60 \cdot 40$, or 2400 ft^2. The answer checks.

State. The sidewalk is 10 ft wide.

30. Graph $5x = 15 + 3y$.

We find the intercepts. To find the x-intercept we let $y = 0$ and solve for x.

$$5x = 15 + 3 \cdot 0$$

$$5x = 15$$

$$x = 3$$

The x-intercept is $(3, 0)$.

To find the y-intercept we let $x = 0$ and solve for y.

$$5 \cdot 0 = 15 + 3y$$

$$0 = 15 + 3y$$

$$-15 = 3y$$

$$-5 = y$$

The y-intercept is $(0, -5)$.

We plot these points and draw the line. A third point could be found as a check.

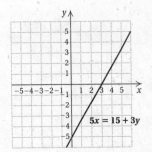

31. Graph $f(x) = 2x^2 - 4x - 1 = 2(x^2 - 2x) - 1$

We complete the square inside the parentheses. We take half the x-coefficient and square it.

$$\frac{1}{2}(-2) = -1 \text{ and } (-1)^2 = 1$$

Then we add $1 - 1$ inside the parentheses.

$$f(x) = 2(x^2 - 2x + 1 - 1) - 1$$

$$= 2(x^2 - 2x + 1) + 2(-1) - 1$$

$$= 2(x - 1)^2 - 3$$

$$= 2(x - 1)^2 + (-3)$$

The vertex is $(1, -3)$, and the line of symmetry is $x = 1$. We plot some points and draw the curve.

x	$f(x)$
1	-3
-1	5
0	-1
2	-1
3	5

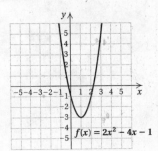

32. Graph $f(x) = \log_3 x$, or $y = \log_3 x$

The equation $f(x) = y = \log_3 x$ is equivalent to $3^y = x$. We find ordered pairs by choosing values for y and computing the corresponding x-values. Then we plot the set of ordered pairs and connect the points with a smooth curve.

x, or 3^y	y
1	0
3	1
9	2
27	3
$\frac{1}{3}$	-1
$\frac{1}{9}$	-2
$\frac{1}{27}$	-3

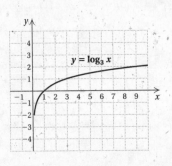

33. Graph $f(x) = 3^x$.

We find some function values, plot points, and then connect them with a smooth curve.

x	$f(x)$
0	1
1	3
2	9
3	27
-1	$\dfrac{1}{3}$
-2	$\dfrac{1}{9}$
-3	$\dfrac{1}{27}$

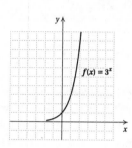

34. Graph $-2x - 3y \le 6$.

We first graph $-2x - 3y = 6$. We draw the line solid since the inequality symbol is $\le$. Then test the point $(0,0)$ to determine if it is a solution.

$$\frac{-2x - 3y \le 6}{-2 \cdot 0 - 3 \cdot 0 \ ? \ 6}$$
$$0 \ \Big| \ \text{TRUE}$$

Since $0 \le 6$ is true, we shade the half-plane that contains $(0,0)$.

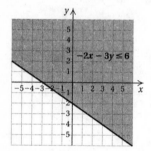

35.
$$y = \frac{kx^2}{z}$$
$$2 = \frac{k \cdot 5^2}{100}$$
$$2 = \frac{k \cdot 25}{100}$$
$$\frac{100}{25} \cdot 2 = k$$
$$8 = k \qquad \text{Constant of variation}$$
$$y = \frac{8x^2}{z} \qquad \text{Equation of variation}$$
$$y = \frac{8 \cdot 3^2}{4} = \frac{8 \cdot 9}{4} = 18$$

36. $(11x^2 - 6x - 3) - (3x^2 + 5x - 2)$
$$= 11x^2 - 6x - 3 + (-3x^2 - 5x + 2)$$
$$= 8x^2 - 11x - 1$$

37. $(3x^2 - 2y)^2 = (3x^2)^2 - 2 \cdot 3x^2 \cdot 2y + (2y)^2$
$$= 9x^4 - 12x^2y + 4y^2$$

38. $(5a + 3b)(2a - 3b)$
$$= 10a^2 - 15ab + 6ab - 9b^2 \quad \text{FOIL}$$
$$= 10a^2 - 9ab - 9b^2$$

39.
$$\frac{x^2 + 8x + 16}{2x + 6} \div \frac{x^2 + 3x - 4}{x^2 - 9}$$
$$= \frac{x^2 + 8x + 16}{2x + 6} \cdot \frac{x^2 - 9}{x^2 + 3x - 4}$$
$$= \frac{(x+4)(x+4)(x+3)(x-3)}{2(x+3)(x+4)(x-1)}$$
$$= \frac{(x+4)(x+4)(x+3)(x-3)}{2(x+3)(x+4)(x-1)}$$
$$= \frac{(x+4)(x-3)}{2(x-1)}$$

40.
$$\frac{1 + \dfrac{3}{x}}{x - 1 - \dfrac{12}{x}} = \frac{1 + \dfrac{3}{x}}{x - 1 - \dfrac{12}{x}} \cdot \frac{x}{x}$$
$$= \frac{\left(1 + \dfrac{3}{x}\right)x}{\left(x - 1 - \dfrac{12}{x}\right)x}$$
$$= \frac{x + 3}{x^2 - x - 12}$$
$$= \frac{x + 3}{(x + 3)(x - 4)}$$
$$= \frac{(x+3) \cdot 1}{(x + 3)(x - 4)}$$
$$= \frac{1}{x - 4}$$

41.
$$\frac{3}{x + 6} - \frac{2}{x^2 - 36} + \frac{4}{x - 6}$$
$$= \frac{3}{x+6} - \frac{2}{(x+6)(x-6)} + \frac{4}{x-6}, \ \text{LCM is } (x+6)(x-6)$$
$$= \frac{3}{x+6} \cdot \frac{x-6}{x-6} - \frac{2}{(x+6)(x-6)} + \frac{4}{x-6} \cdot \frac{x+6}{x+6}$$
$$= \frac{3(x - 6) - 2 + 4(x + 6)}{(x + 6)(x - 6)}$$
$$= \frac{3x - 18 - 2 + 4x + 24}{(x + 6)(x - 6)}$$
$$= \frac{7x + 4}{(x + 6)(x - 6)}$$

42. $1 - 125x^3$
$$= 1^3 - (5x)^3 \qquad \text{Difference of cubes}$$
$$= (1 - 5x)(1^2 + 1 \cdot 5x + (5x)^2)$$
$$= (1 - 5x)(1 + 5x + 25x^2)$$

43. $6x^2 + 8xy - 8y^2$
$$= 2(3x^2 + 4xy - 4y^2)$$
$$= 2(3x - 2y)(x + 2y) \quad \text{Using the FOIL method or the } ac\text{-method}$$

44.
$$x^4 - 4x^3 + 7x - 28$$
$$= x^3(x-4) + 7(x-4)$$
$$= (x^3 + 7)(x-4)$$

45.
$$2m^2 + 12mn + 18n^2$$
$$= 2(m^2 + 6mn + 9n^2)$$
$$= 2(m + 3n)^2$$

46.
$$x^4 - 16y^4$$
$$= (x^2)^2 - (4y^2)^2 \qquad \text{Difference of squares}$$
$$= (x^2 + 4y^2)(x^2 - 4y^2) \qquad \text{Difference of squares}$$
$$= (x^2 + 4y^2)(x + 2y)(x - 2y)$$

47.
$$h(x) = -3x^2 + 4x + 8$$
$$h(-2) = -3(-2)^2 + 4(-2) + 8$$
$$= -3 \cdot 4 - 8 + 8$$
$$= -12 - 8 + 8$$
$$= -12$$

48.
$$\begin{array}{r|rrrrr} 3 & 1 & -5 & 2 & 0 & -6 \\ & & 3 & -6 & -12 & -36 \\ \hline & 1 & -2 & -4 & -12 & -42 \end{array}$$

The answer is $x^3 - 2x^2 - 4x - 12 + \dfrac{-42}{x-3}$.

49.
$$(5.2 \times 10^4)(3.5 \times 10^{-6})$$
$$= (5.2 \times 3.5) \times (10^4 \times 10^{-6})$$
$$= 18.2 \times 10^{-2}$$
$$= (1.82 \times 10) \times 10^{-2}$$
$$= 1.82 \times 10^{-1}$$

50. $\dfrac{\sqrt[3]{40xy^8}}{\sqrt[3]{5xy}} = \sqrt[3]{\dfrac{40xy^8}{5xy}} = \sqrt[3]{8y^7} = \sqrt[3]{8y^6 \cdot y} =$
$\sqrt[3]{8y^6} \; \sqrt[3]{y} = 2y^2 \sqrt[3]{y}$

51. $\sqrt{7xy^3} \cdot \sqrt{28x^2y} = \sqrt{196x^3y^4} = \sqrt{196x^2y^4 \cdot x} =$
$\sqrt{196x^2y^4} \cdot \sqrt{x} = 14xy^2\sqrt{x}$

52.
$$\dfrac{3 - \sqrt{y}}{2 - \sqrt{y}} = \dfrac{3 - \sqrt{y}}{2 - \sqrt{y}} \cdot \dfrac{2 + \sqrt{y}}{2 + \sqrt{y}}$$
$$= \dfrac{6 + 3\sqrt{y} - 2\sqrt{y} - y}{4 - y}$$
$$= \dfrac{6 + \sqrt{y} - y}{4 - y}$$

53.
$$(1 + i\sqrt{3})(6 - 2i\sqrt{3})$$
$$= 6 - 2i\sqrt{3} + 6i\sqrt{3} - 2i^2 \cdot 3$$
$$= 6 - 2i\sqrt{3} + 6i\sqrt{3} - 2(-1) \cdot 3$$
$$= 6 - 2i\sqrt{3} + 6i\sqrt{3} + 6$$
$$= 12 + 4i\sqrt{3}$$

54. The function $f(x) = 7 - 2x$ passes the horizontal-line test, so it is one-to-one and thus has an inverse. We find a formula for the inverse.

1. Replace $f(x)$ by y: $y = 7 - 2x$

2. Interchange x and y: $x = 7 - 2y$

3. Solve for y: $x - 7 = -2y$
$$\dfrac{x - y}{-2} = y$$

4. Replace y by $f^{-1}(x)$: $f^{-1}(x) = \dfrac{x - 7}{-2}$

55. First solve the equation for y to determine the slope of the given line.
$$2x + y = 6$$
$$y = -2x + 6$$
The slope of the given line is -2. The slope of the perpendicular line is the opposite of the reciprocal of -2, or $\dfrac{1}{2}$. We will use the slope-intercept equation to find the desired equation. Substitute $\dfrac{1}{2}$ for m, -3 for x, and 5 for y.
$$y = mx + b$$
$$5 = \dfrac{1}{2}(-3) + b$$
$$5 = -\dfrac{3}{2} + b$$
$$\dfrac{13}{2} = b$$
Thus we have $y = \dfrac{1}{2}x + \dfrac{13}{2}$.

56.
$$3 \log x - \dfrac{1}{2} \log y - 2 \log z$$
$$= \log x^3 - \log y^{1/2} - \log z^2$$
$$= \log x^3 - (\log y^{1/2} + \log z^2)$$
$$= \log x^3 - \log (y^{1/2}z^2)$$
$$= \log \dfrac{x^3}{y^{1/2}z^2}$$

57. $\log_a 5 = x \Rightarrow a^x = 5$

58. $\log 0.05566 \approx -1.2545$

59. $10^{2.89} \approx 776.2471$

60. $\ln 12.78 \approx 2.5479$

61. $e^{-1.4} \approx 0.2466$

62. The inverse of $f(x) = a^x$ is given by $f^{-1}(x) = \log_a x$, so the inverse of $f(x) = 5^x$ is $f^{-1}(x) = \log_5 x$. Answer (d) is correct.

63.
$$f(x) = -2x^2 + 28x - 9$$
$$= -2(x^2 - 14x) - 9$$
$$= -2(x^2 - 14x + 49 - 49) - 9 \quad \dfrac{1}{2}(-14) = -7$$
$$\qquad\qquad\qquad\qquad\qquad \text{and } (-7)^2 = 49$$
$$= -2(x^2 - 14x + 49) + (-2)(-49) - 9$$
$$= -2(x - 7)^2 + 98 - 9$$
$$= -2(x - 7)^2 + 89$$
Answer (d) is correct.

64. $\log(x^2 - 9) - \log(x + 3) = 1$

$$\log \frac{x^2 - 9}{x + 3} = 1$$

$$10^1 = \frac{x^2 - 9}{x + 3}$$

$$10 = \frac{(x + 3)(x - 3)}{x + 3}$$

$$10 = x - 3$$

$$13 = x$$

The number 13 checks. Answer (e) is correct.

65. $B = 2a(b^2 - c^2)$

$$B = 2ab^2 - 2ac^2$$

$$2ac^2 = 2ab^2 - B$$

$$c^2 = \frac{2ab^2 - B}{2a}$$

$$c = \sqrt{\frac{2ab^2 - B}{2a}}$$

Answer (d) is correct.

66.
$$\frac{5}{3x - 3} + \frac{10}{3x + 6} = \frac{5x}{x^2 + x - 2}$$

$$\frac{5}{3(x - 1)} + \frac{10}{3(x + 2)} = \frac{5x}{(x + 2)(x - 1)}$$

$$\text{LCD is } 3(x - 1)(x + 2)$$

$$3(x-1)(x+2)\left(\frac{5}{3(x-1)} + \frac{10}{3(x+2)} \right) =$$

$$3(x - 1)(x + 2) \cdot \frac{5x}{(x + 2)(x - 1)}$$

$$5(x + 2) + 10(x - 1) = 3 \cdot 5x$$

$$5x + 10 + 10x - 10 = 15x$$

$$15x = 15x$$

We get an equation that is true for all values of x. Thus the solutions of the original equation are all real numbers except those for which a denominator is 0. A denominator is 0 when $x = 1$ or when $x = -2$, so all real numbers except 1 and -2 are solutions.

67.
$$\log \sqrt{3x} = \sqrt{\log 3x}$$

$$\log(3x)^{1/2} = \sqrt{\log 3x}$$

$$\frac{1}{2} \log 3x = \sqrt{\log 3x}$$

$$\left(\frac{1}{2} \log 3x \right)^2 = (\sqrt{\log 3x})^2$$

$$\frac{1}{4}(\log 3x)^2 = \log 3x$$

$$\frac{1}{4}(\log 3x)^2 - \log 3x = 0$$

Let $u = \log 3x$.

$$\frac{1}{4}u^2 - u = 0$$

$$u\left(\frac{1}{4}u - 1 \right) = 0$$

$$u = 0 \quad or \quad \frac{1}{4}u - 1 = 0$$

$$u = 0 \quad or \quad \frac{1}{4}u = 1$$

$$u = 0 \quad or \quad u = 4$$

$$\log 3x = 0 \quad or \quad \log 3x = 4$$

$$10^0 = 3x \quad or \quad 10^4 = 3x$$

$$1 = 3x \quad or \quad 10,000 = 3x$$

$$\frac{1}{3} = x \quad or \quad \frac{10,000}{3} = x$$

Both numbers check. The solutions are $\frac{1}{3}$ and $\frac{10,000}{3}$.

68. **Familiarize.** Let $r =$ the speed of the train and $t =$ the time of the trip. We organize the information in a table.

	Distance	Speed	Time
Actual trip	280	r	t
Faster trip	280	$r + 5$	$t - 1$

Translate. Using Time = Distance/Speed in each row of the table, we have two equations.

$$\frac{280}{r} = t, \quad \frac{280}{r + 5} = t - 1$$

Solve. We substitute $\frac{280}{r}$ for t in the second equation and solve for t,

$$\frac{280}{r + 5} = \frac{280}{r} - 1, \quad \text{LCD is } r(r + 5)$$

$$r(r + 5) \cdot \frac{280}{r + 5} = r(r + 5)\left(\frac{280}{r} - 1 \right)$$

$$280r = 280(r + 5) - r(r + 5)$$

$$280r = 280r + 1400 - r^2 - 5r$$

$$r^2 + 5r - 1400 = 0$$

$$(r + 40)(r - 35) = 0$$

$$r = -40 \quad or \quad r = 35$$

Check. The speed cannot be negative, so we check only 35. At 35 mph, the train would travel 280 mi in 280/35, or 8 hr. At $35 + 5$, or 40 mph, the train would travel 280 mi in 280/40, or 7 hr. Since 7 hr is 1 hr less than 8 hr, the answer checks.

State. The actual speed of the train is 35 mph.

Chapter 9

Conic Sections

---

1. Graph: $y = x^2$

The graph is a parabola. The vertex is $(0,0)$; the line of symmetry is $x = 0$. The curve opens upward. We choose some x-values on both sides of the vertex and compute the corresponding y-values. Then we plot the points and graph the parabola.

x	y
0	0
1	1
-1	1
2	4
-2	4

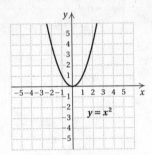

3. Graph: $x = y^2 + 4y + 1$

We complete the square.

$$x = (y^2 + 4y + 4 - 4) + 1$$
$$= (y^2 + 4y + 4) - 4 + 1$$
$$= (y + 2)^2 - 3, \text{ or}$$
$$= [y - (-2)]^2 + (-3)$$

The graph is a parabola. The vertex is $(-3, -2)$; the line of symmetry is $y = -2$. The curve opens to the right.

x	y
-3	-2
-2	-3
-2	-1
1	-4
1	0

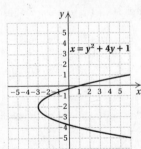

5. Graph: $y = -x^2 + 4x - 5$

We use the formula to find the first coordinate of the vertex:

$$x = -\frac{b}{2a} = -\frac{4}{2(-1)} = 2$$

Then $y = -x^2 + 4x - 5 = -(2)^2 + 4(2) - 5 = -1$.

The vertex is $(2, -1)$; the line of symmetry is $x = 2$. The curve opens downward.

x	y
2	-1
1	-2
3	-2
0	-5
4	-5

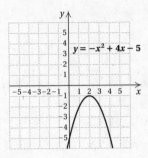

7. Graph: $x = -3y^2 - 6y - 1$

We complete the square.

$$x = -3(y^2 + 2y) - 1$$
$$= -3(y^2 + 2y + 1 - 1) - 1$$
$$= -3(y^2 + 2y + 1) + 3 - 1$$
$$= -3(y + 1)^2 + 2$$
$$= -3[y - (-1)]^2 + 2$$

The graph is a parabola. The vertex is $(2, -1)$; the line of symmetry is $y = -1$. The curve opens to the left.

x	y
2	-1
-1	-2
-1	0
-10	-3
-10	1

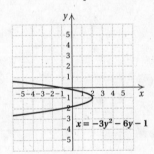

9.
$$d = \sqrt{(x_2 - x_1)^2 + (y_2 - y_1)^2}$$
$$d = \sqrt{(2 - 6)^2 + [-7 - (-4)]^2} \quad \text{Substituting}$$
$$= \sqrt{(-4)^2 + (-3)^2}$$
$$= \sqrt{25} = 5$$

11.
$$d = \sqrt{(x_2 - x_1)^2 + (y_2 - y_1)^2}$$
$$d = \sqrt{(5 - 0)^2 + [-6 - (-4)]^2}$$
$$= \sqrt{5^2 + (-2)^2}$$
$$= \sqrt{29} \approx 5.385$$

13.
$$d = \sqrt{(x_2 - x_1)^2 + (y_2 - y_1)^2}$$
$$d = \sqrt{(-9 - 9)^2 + (-9 - 9)^2}$$
$$= \sqrt{(-18)^2 + (-18)^2}$$
$$= \sqrt{648} = \sqrt{324 \cdot 2}$$
$$= 18\sqrt{2} \approx 25.456$$

15. $d = \sqrt{(x_2 - x_1)^2 + (y_2 - y_1)^2}$

$d = \sqrt{(-4.3 - 2.8)^2 + [-3.5 - (-3.5)]^2}$

$= \sqrt{(-7.1)^2 + 0^2} = \sqrt{(-7.1)^2}$

$= 7.1$

17. $d = \sqrt{(x_2 - x_1)^2 + (y_2 - y_1)^2}$

$d = \sqrt{\left(\frac{5}{7} - \frac{1}{7}\right)^2 + \left(\frac{1}{14} - \frac{11}{14}\right)^2}$

$= \sqrt{\left(\frac{4}{7}\right)^2 + \left(-\frac{5}{7}\right)^2}$

$= \sqrt{\frac{16}{49} + \frac{25}{49}}$

$= \sqrt{\frac{41}{49}}$

$= \frac{\sqrt{41}}{7} \approx 0.915$

19. $d = \sqrt{[56 - (-23)]^2 + (-17 - 10)^2}$

$= \sqrt{79^2 + (-27)^2} = \sqrt{6970} \approx 83.487$

21. $d = \sqrt{(a - 0)^2 + (b - 0)^2}$

$= \sqrt{a^2 + b^2}$

23. $d = \sqrt{(-\sqrt{7} - \sqrt{2})^2 + [\sqrt{5} - (-\sqrt{3})]^2}$

$= \sqrt{7 + 2\sqrt{14} + 2 + 5 + 2\sqrt{15} + 3}$

$= \sqrt{17 + 2\sqrt{14} + 2\sqrt{15}} \approx 5.677$

25. $d = \sqrt{[1000 - (-2000)]^2 + (-240 - 580)^2}$

$= \sqrt{3000^2 + (-820)^2}$

$= \sqrt{9,672,400} \approx 3110.048$

27. Using the midpoint formula $\left(\frac{x_1 + x_2}{2}, \frac{y_1 + y_2}{2}\right)$, we obtain

$\left(\frac{-1 + 4}{2}, \frac{9 + (-2)}{2}\right)$, or $\left(\frac{3}{2}, \frac{7}{2}\right)$.

29. Using the midpoint formula $\left(\frac{x_1 + x_2}{2}, \frac{y_1 + y_2}{2}\right)$, we obtain

$\left(\frac{3 + (-3)}{2}, \frac{5 + 6}{2}\right)$, or $\left(\frac{0}{2}, \frac{11}{2}\right)$, or $\left(0, \frac{11}{2}\right)$.

31. Using the midpoint formula $\left(\frac{x_1 + x_2}{2}, \frac{y_1 + y_2}{2}\right)$, we obtain

$\left(\frac{-10 + 8}{2}, \frac{-13 + (-4)}{2}\right)$, or $\left(\frac{-2}{2}, \frac{-17}{2}\right)$, or $\left(-1, -\frac{17}{2}\right)$.

33. Using the midpoint formula $\left(\frac{x_1 + x_2}{2}, \frac{y_1 + y_2}{2}\right)$, we obtain

$\left(\frac{-3.4 + 2.9}{2}, \frac{8.1 + (-8.7)}{2}\right)$, or $\left(\frac{-0.5}{2}, \frac{-0.6}{2}\right)$, or $(-0.25, -0.3)$.

35. Using the midpoint formula $\left(\frac{x_1 + x_2}{2}, \frac{y_1 + y_2}{2}\right)$, we obtain

$\left(\frac{\frac{1}{6} + \left(-\frac{1}{3}\right)}{2}, \frac{-\frac{3}{4} + \frac{5}{6}}{2}\right)$, or $\left(\frac{-\frac{1}{6}}{2}, \frac{\frac{1}{12}}{2}\right)$, or $\left(-\frac{1}{12}, \frac{1}{24}\right)$.

37. Using the midpoint formula $\left(\frac{x_1 + x_2}{2}, \frac{y_1 + y_2}{2}\right)$, we obtain

$\left(\frac{\sqrt{2} + \sqrt{3}}{2}, \frac{-1 + 4}{2}\right)$, or $\left(\frac{\sqrt{2} + \sqrt{3}}{2}, \frac{3}{2}\right)$.

39. $\qquad (x + 1)^2 + (y + 3)^2 = 4$

$[x - (-1)]^2 + [y - (-3)]^2 = 2^2$ Standard form

The center is $(-1, -3)$, and the radius is 2.

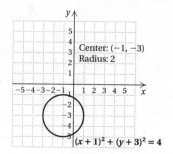

41. $\qquad (x - 3)^2 + y^2 = 2$

$(x - 3)^2 + (y - 0)^2 = (\sqrt{2})^2$ Standard form

The center is $(3, 0)$, and the radius is $\sqrt{2}$.

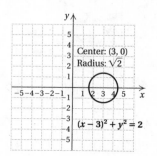

43. $\qquad x^2 + y^2 = 25$

$(x - 0)^2 + (y - 0)^2 = 5^2$ Standard form

The center is $(0, 0)$, and the radius is 5.

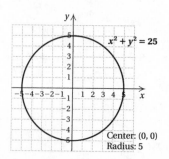

45. $(x - h)^2 + (y - k)^2 = r^2$ Standard form

$(x - 0)^2 + (y - 0)^2 = 7^2$ Substituting

$\qquad x^2 + y^2 = 49$

47. $(x - h)^2 + (y - k)^2 = r^2$ Standard form

$[x - (-5)]^2 + (y - 3)^2 = (\sqrt{7})^2$ Substituting

$\qquad (x + 5)^2 + (y - 3)^2 = 7$

49.
$$x^2 + y^2 + 8x - 6y - 15 = 0$$
$$(x^2 + 8x) + (y^2 - 6y) - 15 = 0$$
Regrouping
$$(x^2 + 8x + 16 - 16) + (y^2 - 6y + 9 - 9) - 15 = 0$$
Completing the square twice
$$(x^2 + 8x + 16) + (y^2 - 6y + 9) - 16 - 9 - 15 = 0$$
$$(x + 4)^2 + (y - 3)^2 = 40$$
$$[x - (-4)]^2 + (y - 3)^2 = (\sqrt{40})^2$$
$$[x - (-4)]^2 + (y - 3)^2 = (2\sqrt{10})^2$$
The center is $(-4, 3)$, and the radius is $2\sqrt{10}$.

51.
$$x^2 + y^2 - 8x + 2y + 13 = 0$$
$$(x^2 - 8x) + (y^2 + 2y) + 13 = 0$$
Regrouping
$$(x^2 - 8x + 16 - 16) + (y^2 + 2y + 1 - 1) + 13 = 0$$
Completing the square twice
$$(x^2 - 8x + 16) + (y^2 + 2y + 1) - 16 - 1 + 13 = 0$$
$$(x - 4)^2 + (y + 1)^2 = 4$$
$$(x - 4)^2 + [y - (-1)]^2 = 2^2$$
The center is $(4, -1)$, and the radius is 2.

53.
$$x^2 + y^2 - 4x = 0$$
$$(x^2 - 4x) + y^2 = 0$$
$$(x^2 - 4x + 4 - 4) + y^2 = 0$$
$$(x^2 - 4x + 4) + y^2 - 4 = 0$$
$$(x - 2)^2 + y^2 = 4$$
$$(x - 2)^2 + (y - 0)^2 = 2^2$$
The center is $(2, 0)$, and the radius is 2.

55. Discussion and Writing Exercise

57. We use the elimination method.
$$x - y = 7 \quad (1)$$
$$\underline{x + y = 11 \quad (2)}$$
$$2x \quad\;\; = 18 \quad \text{Adding}$$
$$x = 9$$
Substitute 9 for x in one of the original equations and solve for y.
$$x + y = 11 \quad (2)$$
$$9 + y = 11 \quad \text{Substituting}$$
$$y = 2$$
The solution is $(9, 2)$.

59.
$$y = 3x - 2, \quad (1)$$
$$2x - 4y = 50 \quad (2)$$
Substitute $3x - 2$ for y in Equation (2) and solve for y.
$$2x - 4(3x - 2) = 50$$
$$2x - 12x + 8 = 50$$
$$-10x = 42$$
$$x = -\frac{21}{5}$$

Substitute $-\dfrac{21}{5}$ for x in Equation (1) and compute y.
$$y = 3\left(-\frac{21}{5}\right) - 2 = -\frac{63}{5} - 2 = -\frac{73}{5}$$
The solution is $\left(-\dfrac{21}{5}, -\dfrac{73}{5}\right)$.

61.
$$-4x + 12y = -9, \quad (1)$$
$$x - 3y = 2 \quad (2)$$
We use the elimination method.
$$-4x + 12y = -9 \quad (1)$$
$$\underline{-4x + 12y = -8 \quad \text{Multiplying (2) by } -4}$$
$$0 = -17 \quad \text{Adding}$$
We get a false equation. The system of equations has no solution.

63. $x^2 - 16 = x^2 - 4^2 = (x + 4)(x - 4)$

65. $64p^2 - 81q^2 = (8p)^2 - (9q)^2 = (8p + 9q)(8p - 9q)$

67. We first find the length of the radius, which is the distance between $(0, 0)$ and $\left(\dfrac{1}{4}, \dfrac{\sqrt{31}}{4}\right)$.

$$r = \sqrt{\left(\frac{1}{4} - 0\right)^2 + \left(\frac{\sqrt{31}}{4} - 0\right)^2}$$
$$= \sqrt{\left(\frac{1}{4}\right)^2 + \left(\frac{\sqrt{31}}{4}\right)^2}$$
$$= \sqrt{\frac{1}{16} + \frac{31}{16}}$$
$$= \sqrt{\frac{32}{16}}$$
$$= \sqrt{2}$$
$$(x - h)^2 + (y - k)^2 = r^2 \quad \text{Standard form}$$
$$(x - 0)^2 + (y - 0)^2 = (\sqrt{2})^2$$
Substituting $(0, 0)$ for the center and $\sqrt{2}$ for the radius
$$x^2 + y^2 = 2$$

69.

The center is $(-3, -2)$ and the radius is 3.
$$(x - h)^2 + (y - k)^2 = r^2 \quad \text{Standard form}$$
$$[x - (-3)]^2 + [y - (-2)]^2 = 3^2 \quad \text{Substituting}$$
$$(x + 3)^2 + (y + 2)^2 = 9$$

71. $d = \sqrt{(x_2 - x_1)^2 + (y_2 - y_1)^2}$

$\quad d = \sqrt{(-1 - 6)^2 + (3k - 2k)^2}$ Substituting

$\quad = \sqrt{(-7)^2 + (k)^2}$

$\quad = \sqrt{49 + k^2}$

73. $d = \sqrt{(x_2 - x_1)^2 + (y_2 - y_1)^2}$

$\quad d = \sqrt{[6m - (-2m)]^2 + (-7n - n)^2}$ Substituting

$\quad = \sqrt{(8m)^2 + (-8n)^2}$

$\quad = \sqrt{64m^2 + 64n^2}$

$\quad = \sqrt{64(m^2 + n^2)}$

$\quad = 8\sqrt{m^2 + n^2}$

75. The distance between $(-8, -5)$ and $(6, 1)$ is

$\sqrt{(-8 - 6)^2 + (-5 - 1)^2} = \sqrt{196 + 36} = \sqrt{232}$.

The distance between $(6, 1)$ and $(-4, 5)$ is

$\sqrt{[6 - (-4)]^2 + (1 - 5)^2} = \sqrt{100 + 16} = \sqrt{116}$.

The distance between $(-4, 5)$ and $(-8, -5)$ is

$\sqrt{[-4 - (-8)]^2 + [5 - (-5)]^2} = \sqrt{16 + 100} = \sqrt{116}$.

Since $(\sqrt{116})^2 + (\sqrt{116})^2 = (\sqrt{232})^2$, the points are vertices of a right triangle.

77. $\left(\dfrac{(2 - \sqrt{3}) + (2 + \sqrt{3})}{2}, \dfrac{5\sqrt{2} + 3\sqrt{2}}{2} \right)$, or $\left(\dfrac{4}{2}, \dfrac{8\sqrt{2}}{2} \right)$, or

$(2, 4\sqrt{2})$

79. a) Use the fact that the center of the circle $(0, k)$ is equidistant from the points $(-575, 0)$ and $(0, 19.5)$.

$\sqrt{(-575 - 0)^2 + (0 - k)^2} = \sqrt{(0 - 0)^2 + (19.5 - k)^2}$

$\quad \sqrt{330,625 + k^2} = \sqrt{380.25 - 39k + k^2}$

$\quad 330,625 + k^2 = 380.25 - 39k + k^2$

$\qquad\qquad$ Squaring both sides

$\quad 330,244.75 = -39k$

$\quad -8467.8 \approx k$

Then the center of the circle is about $(0, -8467.8)$.

b) To find the radius we find the distance from the center, $(0, -8467.8)$ to any one of the points $(-575, 0)$, $(0, 19.5)$, or $(575, 0)$. We use $(0, 19.5)$.

$r = \sqrt{(0 - 0)^2 + [19.5 - (-8467.8)]^2} \approx$
8487.3 mm

Exercise Set 9.2

1. $\dfrac{x^2}{9} + \dfrac{y^2}{36} = 1$

$\dfrac{x^2}{3^2} + \dfrac{y^2}{6^2} = 1$

The x-intercepts are $(-3, 0)$ and $(3, 0)$, and the y-intercepts are $(0, -6)$ and $(0, 6)$. We plot these points and connect them with an oval-shaped curve.

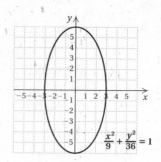

3. $\dfrac{x^2}{1} + \dfrac{y^2}{4} = 1$

$\dfrac{x^2}{1^2} + \dfrac{y^2}{2^2} = 1$

The x-intercepts are $(-1, 0)$ and $(1, 0)$, and the y-intercepts are $(0, -2)$ and $(0, 2)$. We plot these points and connect them with an oval-shaped curve.

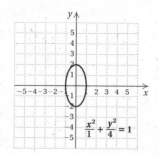

5. $4x^2 + 9y^2 = 36$

$\dfrac{x^2}{9} + \dfrac{y^2}{4} = 1$ Dividing by 36

$\dfrac{x^2}{3^2} + \dfrac{y^2}{2^2} = 1$

The x-intercepts are $(-3, 0)$ and $(3, 0)$, and the y-intercepts are $(0, -2)$ and $(0, 2)$. We plot these points and connect them with an oval-shaped curve.

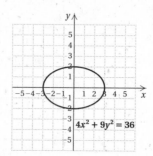

7. $x^2 + 4y^2 = 4$

$\dfrac{x^2}{4} + \dfrac{y^2}{1} = 1$ Dividing by 4

$\dfrac{x^2}{2^2} + \dfrac{y^2}{1^2} = 1$

The x-intercepts are $(-2, 0)$ and $(2, 0)$, and the y-intercepts are $(0, -1)$ and $(0, 1)$. We plot these points and connect them with an oval-shaped curve.

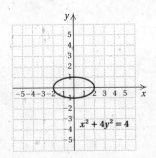

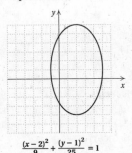

(−1, 1), (2, 6), and (2, −4). Connect these points with an oval-shaped curve.

9. $2x^2 + 3y^2 = 6$

$\dfrac{x^2}{3} + \dfrac{y^2}{2} = 1$ Multiplying by $\dfrac{1}{6}$

$\dfrac{x^2}{(\sqrt{3})^2} + \dfrac{y^2}{(\sqrt{2})^2} = 1$

The x-intercepts are $(\sqrt{3}, 0)$ and $(-\sqrt{3}, 0)$, and the y-intercepts are $(0, \sqrt{2})$ and $(0, -\sqrt{2})$. We plot these points and connect them with an oval-shaped curve.

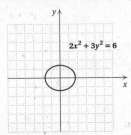

11. $12x^2 + 5y^2 - 120 = 0$

$12x^2 + 5y^2 = 120$

$\dfrac{x^2}{10} + \dfrac{y^2}{24} = 1$ Multiplying by $\dfrac{1}{120}$

$\dfrac{x^2}{(\sqrt{10})^2} + \dfrac{y^2}{(\sqrt{24})^2} = 1$

The x-intercepts are $(\sqrt{10}, 0)$ and $(-\sqrt{10}, 0)$, or about $(3.162, 0)$ and $(-3.162, 0)$. The y-intercepts are $(0, \sqrt{24})$ and $(0, -\sqrt{24})$, or about $(0, 4.899)$ and $(0, -4.899)$. We plot these points and connect them with an oval-shaped curve.

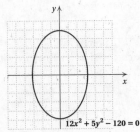

13. $\dfrac{(x-2)^2}{9} + \dfrac{(y-1)^2}{25} = 1$

$\dfrac{(x-2)^2}{3^2} + \dfrac{(y-1)^2}{5^2} = 1$

The center of the ellipse is $(2, 1)$. Note that $a = 3$ and $b = 5$. We locate the center and then plot the points $(2+3, 1)$ $(2-3, 1)$, $(2, 1+5)$, and $(2, 1-5)$, or $(5, 1)$,

$\dfrac{(x-2)^2}{9} + \dfrac{(y-1)^2}{25} = 1$

15. $\dfrac{(x+1)^2}{16} + \dfrac{(y+2)^2}{25} = 1$

$\dfrac{[x-(-1)]^2}{4^2} + \dfrac{[y-(-2)]^2}{5^2} = 1$

The center of the ellipse is $(-1, -2)$. Note that $a = 4$ and $b = 5$. We locate the center and then plot the points $(-1+4, -2)$, $(-1-4, -2)$, $(-1, -2+5)$, and $(-1, -2-5)$, or $(3, -2)$, $(-5, -2)$, $(-1, 3)$, and $(-1, -7)$. Connect these points with an oval-shaped curve.

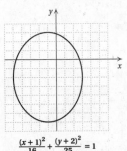

$\dfrac{(x+1)^2}{16} + \dfrac{(y+2)^2}{25} = 1$

17. $12(x-1)^2 + 3(y+2)^2 = 48$

$\dfrac{(x-1)^2}{4} + \dfrac{(y+2)^2}{16} = 1$

$\dfrac{(x-1)^2}{2^2} + \dfrac{(y-(-2))^2}{4^2} = 1$

The center of the ellipse is $(1, -2)$. Note that $a = 2$ and $b = 4$. We locate the center and then plot the points $(1+2, -2)$, $(1-2, -2)$, $(1, -2+4)$, and $(1, -2-4)$, or $(3, -2)$, $(-1, -2)$, $(1, 2)$, and $(1, -6)$. Connect these points with an oval-shaped curve.

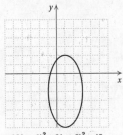

$12(x-1)^2 + 3(y+2)^2 = 48$

19.
$$(x+3)^2 + 4(y+1)^2 - 10 = 6$$
$$(x+3)^2 + 4(y+1)^2 = 16$$
$$\frac{(x+3)^2}{16} + \frac{(y+1)^2}{4} = 1$$
$$\frac{[x-(-3)]^2}{4^2} + \frac{[y-(-1)]^2}{2^2} = 1$$

The center of the ellipse is $(-3, -1)$. Note that $a = 4$ and $b = 2$. We locate the center and then plot the points $(-3+4, -1)$, $(-3-4, -1)$, $(-3, -1+2)$, and $(-3, -1-2)$, or $(1, -1)$, $(-7, -1)$, $(-3, 1)$, and $(-3, -3)$. Connect these points with an oval-shaped curve.

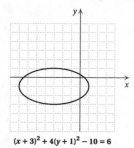

$(x+3)^2 + 4(y+1)^2 - 10 = 6$

21. Discussion and Writing Exercise

23. $3x^2 - 2x + 7 = 0$
$$a = 3,\ b = -2,\ c = 7$$
$$x = \frac{-b \pm \sqrt{b^2 - 4ac}}{2a}$$
$$x = \frac{-(-2) \pm \sqrt{(-2)^2 - 4 \cdot 3 \cdot 7}}{2 \cdot 3}$$
$$x = \frac{2 \pm \sqrt{4 - 84}}{6} = \frac{2 \pm \sqrt{-80}}{6}$$
$$x = \frac{2 \pm 4i\sqrt{5}}{6} = \frac{1 \pm 2i\sqrt{5}}{3}$$

The solutions are $\dfrac{1 + 2i\sqrt{5}}{3}$ and $\dfrac{1 - 2i\sqrt{5}}{3}$.

25. $x^2 + x + 2 = 0$
$$a = 1,\ b = 1,\ c = 2$$
$$x = \frac{-b \pm \sqrt{b^2 - 4ac}}{2a}$$
$$x = \frac{-1 \pm \sqrt{1^2 - 4 \cdot 1 \cdot 2}}{2 \cdot 1}$$
$$x = \frac{-1 \pm \sqrt{1 - 8}}{2} = \frac{-1 \pm \sqrt{-7}}{2}$$
$$x = \frac{-1 \pm i\sqrt{7}}{2}$$

The solutions are $\dfrac{-1 + i\sqrt{7}}{2}$ and $\dfrac{-1 - i\sqrt{7}}{2}$.

27. $x^2 + 2x - 17 = 0$
$$a = 1,\ b = 2,\ c = -17$$
$$x = \frac{-b \pm \sqrt{b^2 - 4ac}}{2a}$$
$$x = \frac{-2 \pm \sqrt{2^2 - 4 \cdot 1 \cdot (-17)}}{2 \cdot 1}$$
$$x = \frac{-2 \pm \sqrt{72}}{2} = \frac{-2 \pm 6\sqrt{2}}{2}$$
$$x = -1 \pm 3\sqrt{2}$$

The solutions are $-1 + 3\sqrt{2} \approx 3.2$ and $-1 - 3\sqrt{2} \approx -5.2$.

29. $3x^2 - 12x + 7 = 10 - x^2 + 5x$
$$4x^2 - 17x - 3 = 0$$
$$a = 4,\ b = -17,\ c = -3$$
$$x = \frac{-b \pm \sqrt{b^2 - 4ac}}{2a}$$
$$x = \frac{-(-17) \pm \sqrt{(-17)^2 - 4 \cdot 4 \cdot (-3)}}{2 \cdot 4}$$
$$x = \frac{17 \pm \sqrt{337}}{8}$$

The solutions are $\dfrac{17 + \sqrt{337}}{8} \approx 4.4$ and $\dfrac{17 - \sqrt{337}}{8} \approx -0.2$.

31. $a^{-t} = b \Rightarrow -t = \log_a b$

33. $\ln 24 = 3.1781 \Rightarrow e^{3.1781} = 24$

35. Plot the given points.

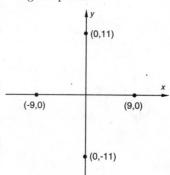

From the location of these points, we see that the ellipse that contains them is centered at the origin with $a = 9$ and $b = 11$. We write the equation of the ellipse:
$$\frac{x^2}{9^2} + \frac{y^2}{11^2} = 1$$
$$\frac{x^2}{81} + \frac{y^2}{121} = 1$$

37. Plot the given points.

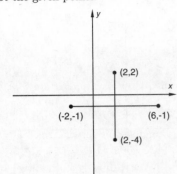

The midpoint of the segment from $(-2, -1)$ to $(6, -1)$ is $\left(\dfrac{-2 + 6}{2}, \dfrac{-1 - 1}{2}\right)$, or $(2, -1)$. The midpoint of the segment from $(2, -4)$ to $(2, 2)$ is $\left(\dfrac{2 + 2}{2}, \dfrac{-4 + 2}{2}\right)$, or $(2, -1)$. Thus, we can conclude that $(2, -1)$ is the center of the ellipse. The distance from $(-2, -1)$ to $(2, -1)$ is $\sqrt{[2 - (-2)]^2 + [-1 - (-1)]^2} = \sqrt{16} = 4$, so $a = 4$. The distance from $(2, 2)$ to $(2, -1)$ is $\sqrt{(2 - 2)^2 + (-1 - 2)^2} = \sqrt{9} = 3$, so $b = 3$. We write the equation of the ellipse.

$$\frac{(x - 2)^2}{4^2} + \frac{(y - (-1))^2}{3^2} = 1$$

$$\frac{(x - 2)^2}{16} + \frac{(y + 1)^2}{9} = 1$$

39.
$$x^2 - 4x + 4y^2 + 8y - 8 = 0$$
$$x^2 - 4x + 4y^2 + 8y = 8$$
$$x^2 - 4x + 4(y^2 + 2y) = 8$$
$$(x^2 - 4x + 4 - 4) + 4(y^2 + 2y + 1 - 1) = 8$$
$$(x^2 - 4x + 4) + 4(y^2 + 2y + 1) = 8 + 4 + 4 \cdot 1$$
$$(x - 2)^2 + 4(y + 1)^2 = 16$$
$$\frac{(x - 2)^2}{16} + \frac{(y + 1)^2}{4} = 1$$
$$\frac{(x - 2)^2}{4^2} + \frac{(y - (-1))^2}{2^2} = 1$$

41. Left to the student

Exercise Set 9.3

1. $\dfrac{y^2}{9} - \dfrac{x^2}{9} = 1$

$\dfrac{y^2}{3^2} - \dfrac{x^2}{3^2} = 1$

$a = 3$ and $b = 3$, so the asymptotes are $y = \dfrac{3}{3}x$ and $y = -\dfrac{3}{3}x$, or $y = x$ and $y = -x$. We sketch them.

Replacing x with 0 and solving for y, we get $y = \pm 3$, so the intercepts are $(0, 3)$ and $(0, -3)$.

We plot the intercepts and draw smooth curves through them that approach the asymptotes.

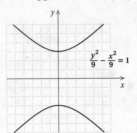

3. $\dfrac{x^2}{4} - \dfrac{y^2}{25} = 1$

$\dfrac{x^2}{2^2} - \dfrac{y^2}{5^2} = 1$

$a = 2$ and $b = 5$, so the asymptotes are $y = \dfrac{5}{2}x$ and $y = -\dfrac{5}{2}x$. We sketch them.

Replacing y with 0 and solving for x, we get $x = \pm 2$, so the intercepts are $(2, 0)$ and $(-2, 0)$.

We plot the intercepts and draw smooth curves through them that approach the asymptotes.

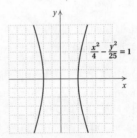

5. $\dfrac{y^2}{36} - \dfrac{x^2}{9} = 1$

$\dfrac{y^2}{6^2} - \dfrac{x^2}{3^2} = 1$

$a = 3$ and $b = 6$, so the asymptotes are $y = \dfrac{6}{3}x$ and $y = -\dfrac{6}{3}x$, or $y = 2x$ and $y = -2x$. We sketch them.

Replacing x with 0 and solving for y, we get $y = \pm 6$, so the intercepts are $(0, 6)$ and $(0, -6)$.

We plot the intercepts and draw smooth curves through them that approach the asymptotes.

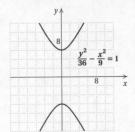

7. $y^2 - x^2 = 25$

$$\frac{y^2}{25} - \frac{x^2}{25} = 1$$

$$\frac{y^2}{5^2} - \frac{x^2}{5^2} = 1$$

$a = 5$ and $b = 5$, so the asymptotes are $y = \frac{5}{5}x$ and $y = -\frac{5}{5}x$, or $y = x$ and $y = -x$. We sketch them.

Replacing x with 0 and solving for y, we get $y = \pm 5$, so the intercepts are $(0, 5)$ and $(0, -5)$.

We plot the intercepts and draw smooth curves through them that approach the asymptotes.

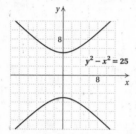

9. $x^2 = 1 + y^2$

$$x^2 - y^2 = 1$$

$$\frac{x^2}{1^2} - \frac{y^2}{1^2} = 1$$

$a = 1$ and $b = 1$, so the asymptotes are $y = \frac{1}{1}x$ and $y = -\frac{1}{1}x$, or $y = x$ and $y = -x$. We sketch them.

Replacing y with 0 and solving for x, we get $x = \pm 1$, so the intercepts are $(1, 0)$ and $(-1, 0)$.

We plot the intercepts and draw smooth curves through them that approach the asymptotes.

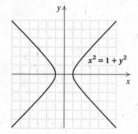

11. $25x^2 - 16y^2 = 400$

$$\frac{x^2}{16} - \frac{y^2}{25} = 1 \quad \text{Multiplying by } \frac{1}{400}$$

$$\frac{x^2}{4^2} - \frac{y^2}{5^2} = 1$$

$a = 4$ and $b = 5$, so the asymptotes are $y = \frac{5}{4}x$ and $y = -\frac{5}{4}x$. We sketch them.

Replacing y with 0 and solving for x, we get $x = \pm 4$, so the intercepts are $(4, 0)$ and $(-4, 0)$.

We plot the intercepts and draw smooth curves through them that approach the asymptotes.

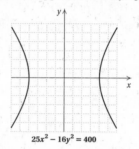

13. $xy = -4$

$$y = -\frac{4}{x} \qquad \text{Solving for } y$$

We find some solutions, keeping the results in a table.

x	y
$\frac{1}{2}$	-8
1	-4
2	-2
4	-1
8	$-\frac{1}{2}$
$-\frac{1}{2}$	8
-1	4
-2	2
-8	$\frac{1}{2}$

Note that we cannot use 0 for x. The x-axis and the y-axis are the asymptotes.

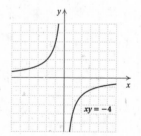

15. $xy = 3$

$y = \dfrac{3}{x}$ Solving for y

We find some solutions, keeping the results in a table.

x	y
$\dfrac{1}{3}$	9
$\dfrac{1}{2}$	6
1	3
3	1
6	$\dfrac{1}{2}$
9	$\dfrac{1}{3}$
$-\dfrac{1}{3}$	-9
$-\dfrac{1}{2}$	-6
-1	-3
-3	-1
-6	$-\dfrac{1}{2}$
-9	$-\dfrac{1}{3}$

Note that we cannot use 0 for x. The x-axis and the y-axis are the asymptotes.

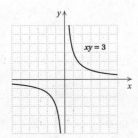

17. $xy = -2$

$y = -\dfrac{2}{x}$ Solving for y

x	y
$\dfrac{1}{2}$	-4
1	-2
2	-1
4	$-\dfrac{1}{2}$
$-\dfrac{1}{2}$	4
-1	2
-2	1
-4	$\dfrac{1}{2}$

Note that we cannot use 0 for x. The x-axis and the y-axis are the asymptotes.

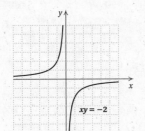

19. $xy = \dfrac{1}{2}$

$y = \dfrac{1}{2x}$ Solving for y

x	y
$\dfrac{1}{8}$	4
$\dfrac{1}{4}$	2
$\dfrac{1}{2}$	1
1	$\dfrac{1}{2}$
2	$\dfrac{1}{4}$
$-\dfrac{1}{8}$	-4
$-\dfrac{1}{4}$	-2
$-\dfrac{1}{2}$	1
1	$\dfrac{1}{2}$
2	$\dfrac{1}{4}$

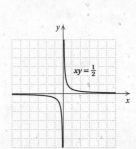

21. DIscussion and Writing Exercise

23. The expression $b^2 - 4ac$ in the quadratic formula is called the <u>discriminant</u>.

25. A graph represents a function if it is impossible to draw a <u>vertical line</u> that intersects the graph more than once.

27. The logarithm of a number is an <u>exponent</u>.

29. A <u>function</u> is a correspondence between a first set, called the domain, and a second set, called the range, such that each member of the domain corresponds to exactly one member of the range.

31. Left to the student

33.
$$x^2 + y^2 - 10x + 8y - 40 = 0$$
$$x^2 - 10x + y^2 + 8y = 40$$
$$(x^2 - 10x + 25) + (y^2 + 8y + 16) = 40 + 25 + 16$$
$$(x - 5)^2 + (y + 4)^2 = 81$$

The graph is a circle.

35.
$$1 - 3y = 2y^2 - x$$
$$x = 2y^2 + 3y - 1$$

The graph is a parabola.

37.
$$4x^2 + 25y^2 - 8x - 100y + 4 = 0$$
$$4x^2 - 8x + 25y^2 - 100y = -4$$
$$4(x^2 - 2x + 1) + 25(y^2 - 4y + 4) = -4 + 4 + 100$$
$$4(x-1)^2 + 25(y-2)^2 = 100$$
$$\frac{(x-1)^2}{25} + \frac{(y-2)^2}{4} = 1$$

The graph is an ellipse.

39. $x^2 + y^2 = 8$

The graph is a circle.

41.
$$x - \frac{3}{y} = 0$$
$$xy - 3 = 0$$
$$xy = 3$$

The graph is a hyperbola.

43.
$$3x^2 + 5y^2 + x^2 = y^2 + 49$$
$$4x^2 + 4y^2 = 49$$
$$x^2 + y^2 = \frac{49}{4}$$

The graph is a circle.

Exercise Set 9.4

1.
$$x^2 + y^2 = 100, \qquad (1)$$
$$y - x = 2 \qquad\qquad (2)$$

First solve Equation (2) for y.
$$y = x + 2 \qquad\qquad (3)$$

Then substitute $x + 2$ for y in Equation (1) and solve for x.
$$x^2 + y^2 = 100$$
$$x^2 + (x+2)^2 = 100$$
$$x^2 + x^2 + 4x + 4 = 100$$
$$2x^2 + 4x - 96 = 0$$
$$x^2 + 2x - 48 = 0 \qquad \text{Multiplying by } \frac{1}{2}$$
$$(x+8)(x-6) = 0$$
$$x + 8 = 0 \quad or \quad x - 6 = 0 \quad \text{Principle of zero}$$
$$\text{products}$$
$$x = -8 \quad or \qquad x = 6$$

Now substitute these numbers into Equation (3) and solve for y.
$$y = -8 + 2 = -6$$
$$y = 6 + 2 = 8$$

The pairs $(-8, -6)$ and $(6, 8)$ check, so they are the solutions.

3.
$$9x^2 + 4y^2 = 36, \qquad (1)$$
$$3x + 2y = 6 \qquad\qquad (2)$$

First solve Equation (2) for x.

$$3x = 6 - 2y$$
$$x = 2 - \frac{2}{3}y \qquad\qquad (3)$$

Then substitute $2 - \frac{2}{3}y$ for x in Equation (1) and solve for y.
$$9x^2 + 4y^2 = 36$$
$$9\left(2 - \frac{2}{3}y\right)^2 + 4y^2 = 36$$
$$9\left(4 - \frac{8}{3}y + \frac{4}{9}y^2\right) + 4y^2 = 36$$
$$36 - 24y + 4y^2 + 4y^2 = 36$$
$$8y^2 - 24y = 0$$
$$y^2 - 3y = 0$$
$$y(y - 3) = 0$$
$$y = 0 \quad or \quad y - 3 = 0 \quad \text{Principle of zero}$$
$$\text{products}$$
$$y = 0 \quad or \qquad y = 3$$

Now substitute these numbers into Equation (3) and solve for x.
$$x = 2 - \frac{2}{3} \cdot 0 = 2$$
$$x = 2 - \frac{2}{3} \cdot 3 = 2 - 2 = 0$$

The pairs $(2, 0)$ and $(0, 3)$ check, so they are the solutions.

5.
$$y^2 = x + 3, \qquad\qquad (1)$$
$$2y = x + 4 \qquad\qquad (2)$$

First solve Equation (2) for x.
$$2y - 4 = x \qquad\qquad (3)$$

Then substitute $y - 4$ for x in Equation (1) and solve for y.
$$y^2 = x + 3$$
$$y^2 = (2y - 4) + 3$$
$$y^2 = 2y - 1$$
$$y^2 - 2y + 1 = 0$$
$$(y-1)(y-1) = 0$$
$$y - 1 = 0 \quad or \quad y - 1 = 0$$
$$y = 1 \quad or \qquad y = 1$$

Now substitute 1 for y in Equation (3) and solve for x.
$$2 \cdot 1 - 4 = x$$
$$-2 = x$$

The pair $(-2, 1)$ checks. It is the solution.

7.
$$x^2 - xy + 3y^2 = 27, \qquad (1)$$
$$x - y = 2 \qquad\qquad (2)$$

First solve Equation (2) for y.
$$x - 2 = y \qquad\qquad (3)$$

Then substitute $x - 2$ for y in Equation (1) and solve for x.

$$x^2 - xy + 3y^2 = 27$$
$$x^2 - x(x-2) + 3(x-2)^2 = 27$$
$$x^2 - x^2 + 2x + 3x^2 - 12x + 12 = 27$$
$$3x^2 - 10x - 15 = 0$$
$$x = \frac{-(-10) \pm \sqrt{(-10)^2 - 4(3)(-15)}}{2 \cdot 3}$$
$$x = \frac{10 \pm \sqrt{100 + 180}}{6} = \frac{10 \pm \sqrt{280}}{6}$$
$$x = \frac{10 \pm 2\sqrt{70}}{6} = \frac{5 \pm \sqrt{70}}{3}$$

Now substitute these numbers in Equation (3) and solve for y.

$$y = \frac{5 + \sqrt{70}}{3} - 2 = \frac{-1 + \sqrt{70}}{3}$$
$$y = \frac{5 - \sqrt{70}}{3} - 2 = \frac{-1 - \sqrt{70}}{3}$$

The pairs $\left(\dfrac{5 + \sqrt{70}}{3}, \dfrac{-1 + \sqrt{70}}{3}\right)$ and

$\left(\dfrac{5 - \sqrt{70}}{3}, \dfrac{-1 - \sqrt{70}}{3}\right)$ check, so they are the solutions.

9. $\quad x^2 - xy + 3y^2 = 5, \qquad$ (1)

$\qquad x - y = 2 \qquad\qquad\qquad$ (2)

First solve Equation (2) for y.

$\qquad x - 2 = y \qquad\qquad\qquad$ (3)

Then substitute $x - 2$ for y in Equation (1) and solve for x.

$$x^2 - xy + 3y^2 = 5$$
$$x^2 - x(x-2) + 3(x-2)^2 = 5$$
$$x^2 - x^2 + 2x + 3x^2 - 12x + 12 = 5$$
$$3x^2 - 10x + 7 = 0$$
$$(3x - 7)(x - 1) = 0$$

$3x - 7 = 0 \quad or \quad x - 1 = 0$

$\qquad x = \dfrac{7}{3} \quad or \qquad x = 1$

Now substitute these numbers in Equation (3) and solve for y.

$$y = \frac{7}{3} - 2 = \frac{1}{3}$$
$$y = 1 - 2 = -1$$

The pairs $\left(\dfrac{7}{3}, \dfrac{1}{3}\right)$ and $(1, -1)$ check, so they are the solutions.

11. $\quad a + b = -6, \qquad$ (1)

$\qquad ab = -7 \qquad\qquad$ (2)

First solve Equation (1) for a.

$\qquad a = -b - 6 \qquad$ (3)

Then substitute $-b - 6$ for a in Equation (2) and solve for b.

$$(-b - 6)b = -7$$
$$-b^2 - 6b = -7$$
$$0 = b^2 + 6b - 7$$
$$0 = (b + 7)(b - 1)$$

$b + 7 = 0 \quad or \quad b - 1 = 0$

$\qquad b = -7 \quad or \qquad b = 1$

Now substitute these numbers in Equation (3) and solve for a.

$$a = -(-7) - 6 = 1$$
$$a = -1 - 6 = -7$$

The pairs $(-7, 1)$ and $(1, -7)$ check, so they are the solutions.

13. $\quad 2a + b = 1, \qquad$ (1)

$\qquad b = 4 - a^2 \qquad$ (2)

Equation (2) is already solved for b. Substitute $4 - a^2$ for b in Equation (1) and solve for a.

$$2a + 4 - a^2 = 1$$
$$0 = a^2 - 2a - 3$$
$$0 = (a - 3)(a + 1)$$

$a - 3 = 0 \quad or \quad a + 1 = 0$

$\qquad a = 3 \quad or \qquad a = -1$

Substitute these numbers in Equation (2) and solve for b.

$$b = 4 - 3^2 = -5$$
$$b = 4 - (-1)^2 = 3$$

The pairs $(3, -5)$ and $(-1, 3)$ check.

15. $\quad x^2 + y^2 = 5, \qquad$ (1)

$\qquad x - y = 8 \qquad\qquad$ (2)

First solve Equation (2) for x.

$\qquad x = y + 8 \qquad$ (3)

Then substitute $y + 8$ for x in Equation (1) and solve for y.

$$(y + 8)^2 + y^2 = 5$$
$$y^2 + 16y + 64 + y^2 = 5$$
$$2y^2 + 16y + 59 = 0$$
$$y = \frac{-16 \pm \sqrt{(16)^2 - 4(2)(59)}}{2 \cdot 2}$$
$$y = \frac{-16 \pm \sqrt{-216}}{4} = \frac{-16 \pm 6i\sqrt{6}}{4}$$
$$y = \frac{-8 \pm 3i\sqrt{6}}{2}, \text{ or } -4 \pm \frac{3}{2}i\sqrt{6}$$

Now substitute these numbers in Equation (3) and solve for x.

$$x = -4 + \frac{3}{2}i\sqrt{6} + 8 = 4 + \frac{3}{2}i\sqrt{6}, \text{ or } \frac{8 + 3i\sqrt{6}}{2}$$
$$x = -4 - \frac{3}{2}i\sqrt{6} + 8 = 4 - \frac{3}{2}i\sqrt{6}, \text{ or } \frac{8 - 3i\sqrt{6}}{2}$$

The pairs $\left(4 + \dfrac{3}{2}i\sqrt{6}, -4 + \dfrac{3}{2}i\sqrt{6}\right)$ and

$\left(4-\frac{3}{2}i\sqrt{6}, -4-\frac{3}{2}i\sqrt{6}\right)$, or $\left(\frac{8+3i\sqrt{6}}{2}, \frac{-8+3i\sqrt{6}}{2}\right)$ and

$\left(\frac{8-3i\sqrt{6}}{2}, \frac{-8-3i\sqrt{6}}{2}\right)$ check.

17. $x^2 + y^2 = 25,$ (1)

$y^2 = x + 5$ (2)

We substitute $x+5$ for y^2 in Equation (1) and solve for x.

$x^2 + y^2 = 25$

$x^2 + (x+5) = 25$

$x^2 + x - 20 = 0$

$(x+5)(x-4) = 0$

$x + 5 = 0 \quad or \quad x - 4 = 0$

$x = -5 \quad or \quad\quad x = 4$

We substitute these numbers for x in either Equation (1) or Equation (2) and solve for y. Here we use Equation (2).

$y^2 = -5 + 5 = 0$ and $y = 0$.

$y^2 = 4 + 5 = 9$ and $y = \pm 3$.

The pairs $(-5, 0)$, $(4, 3)$ and $(4, -3)$ check.

19. $x^2 + y^2 = 9,$ (1)

$x^2 - y^2 = 9$ (2)

Here we use the elimination method.

$x^2 + y^2 = \quad 9 \quad$ (1)

$\underline{x^2 - y^2 = \quad 9} \quad$ (2)

$2x^2 \quad\quad\quad = 18 \quad$ Adding

$x^2 = \quad 9$

$x = \pm 3$

If $x = 3$, $x^2 = 9$, and if $x = -3$, $x^2 = 9$, so substituting 3 or -3 in Equation (1) gives us

$x^2 + y^2 = 9$

$9 + y^2 = 9$

$y^2 = 0$

$y = 0.$

The pairs $(3, 0)$ and $(-3, 0)$ check.

21. $x^2 + y^2 = 20,$ (1)

$xy = 8$ (2)

First we solve Equation (2) for y.

$y = \frac{8}{x}$

Then substitute $\frac{8}{x}$ for y in Equation (1) and solve for x.

$x^2 + \left(\frac{8}{x}\right)^2 = 20$

$x^2 + \frac{64}{x^2} = 20$

$x^4 + 64 = 20x^2 \quad$ Multiplying by x^2

$x^4 - 20x^2 + 64 = 0$

$u^2 - 20u + 64 = 0 \quad$ Letting $u = x^2$

$(u-4)(u-16) = 0$

$u - 4 = 0 \quad or \quad u - 16 = 0$

$u = 4 \quad or \quad\quad u = 16$

We now substitute x^2 for u and solve for x.

$x^2 = 4 \quad or \quad x^2 = 16$

$x = \pm 2 \quad or \quad x = \pm 4$

Since $y = \frac{8}{x}$, if $x = 2$, $y = 4$; if $x = -2$, $y = -4$; if $x = 4$, $y = 2$; if $x = -4$, $y = -2$. The pairs $(2, 4)$, $(-2, -4)$, $(4, 2)$, and $(-4, -2)$ check. They are the solutions.

23. $x^2 + y^2 = 13,$ (1)

$xy = 6$ (2)

First we solve Equation (2) for y.

$y = \frac{6}{x}$

Then substitute $\frac{6}{x}$ for y in Equation (1) and solve for x.

$x^2 + \left(\frac{6}{x}\right)^2 = 13$

$x^2 + \frac{36}{x^2} = 13$

$x^4 + 36 = 13x^2 \quad$ Multiplying by x^2

$x^4 - 13x^2 + 36 = 0$

$u^2 - 13u + 36 = 0 \quad$ Letting $u = x^2$

$(u-9)(u-4) = 0$

$u - 9 = 0 \quad or \quad u - 4 = 0$

$u = 9 \quad or \quad\quad u = 4$

We now substitute x^2 for u and solve for x.

$x^2 = 9 \quad or \quad x^2 = 4$

$x = \pm 3 \quad or \quad x = \pm 2$

Since $y = \frac{6}{x}$, if $x = 3$, $y = 2$; if $x = -3$, $y = -2$; if $x = 2$, $y = 3$; if $x = -2$, $y = -3$. The pairs $(3, 2)$, $(-3, -2)$, $(2, 3)$, and $(-2, -3)$ check. They are the solutions.

25. $2xy + 3y^2 = 7,$ (1)

$3xy - 2y^2 = 4$ (2)

$6xy + 9y^2 = 21 \quad$ Multiplying (1) by 3

$\underline{-6xy + 4y^2 = -8} \quad$ Multiplying (2) by -2

$13y^2 = 13$

$y^2 = 1$

$y = \pm 1$

Substitute for y in Equation (1) and solve for x.

When $y = 1$: $2 \cdot x \cdot 1 + 3 \cdot 1^2 = 7$

$2x = 4$

$x = 2$

When $y = -1$: $2 \cdot x \cdot (-1) + 3 \cdot (-1)^2 = 7$

$-2x = 4$

$x = -2$

The pairs $(2, 1)$ and $(-2, -1)$ check. They are the solutions.

27. $4a^2 - 25b^2 = 0,$ (1)

$2a^2 - 10b^2 = 3b + 4$ (2)

$4a^2 - 25b^2 = 0$

$\underline{-4a^2 + 20b^2 = -6b - 8}$ Multiplying (2) by -2

$-5b^2 = -6b - 8$

$0 = 5b^2 - 6b - 8$

$0 = (5b + 4)(b - 2)$

$5b + 4 = 0$ or $b - 2 = 0$

$b = -\dfrac{4}{5}$ or $b = 2$

Substitute for b in Equation (1) and solve for a.

When $b = -\dfrac{4}{5}$: $4a^2 - 25\left(-\dfrac{4}{5}\right)^2 = 0$

$4a^2 = 16$

$a^2 = 4$

$a = \pm 2$

When $b = 2$: $4a^2 - 25(2)^2 = 0$

$4a^2 = 100$

$a^2 = 25$

$a = \pm 5$

The pairs $\left(2, -\dfrac{4}{5}\right)$, $\left(-2, -\dfrac{4}{5}\right)$, $(5, 2)$ and $(-5, 2)$ check. They are the solutions.

29. $ab - b^2 = -4,$ (1)

$ab - 2b^2 = -6$ (2)

$ab - b^2 = -4$

$\underline{-ab + 2b^2 = 6}$ Multiplying (2) by -1

$b^2 = 2$

$b = \pm\sqrt{2}$

Substitute for b in Equation (1) and solve for a.

When $b = \sqrt{2}$: $a(\sqrt{2}) - (\sqrt{2})^2 = -4$

$a\sqrt{2} = -2$

$a = -\dfrac{2}{\sqrt{2}} = -\sqrt{2}$

When $b = -\sqrt{2}$: $a(-\sqrt{2}) - (-\sqrt{2})^2 = -4$

$-a\sqrt{2} = -2$

$a = \dfrac{2}{\sqrt{2}} = \sqrt{2}$

The pairs $(-\sqrt{2}, \sqrt{2})$ and $(\sqrt{2}, -\sqrt{2})$ check. They are the solutions.

31. $x^2 + y^2 = 25,$ (1)

$9x^2 + 4y^2 = 36$ (2)

$-4x^2 - 4y^2 = -100$ Multiplying (1) by -4

$\underline{9x^2 + 4y^2 = 36}$

$5x^2 = -64$

$x^2 = -\dfrac{64}{5}$

$x = \pm\sqrt{-\dfrac{64}{5}} = \pm\dfrac{8i}{\sqrt{5}}$

$x = \pm\dfrac{8i\sqrt{5}}{5}$ Rationalizing the denominator

Substituting $\dfrac{8i\sqrt{5}}{5}$ or $-\dfrac{8i\sqrt{5}}{5}$ for x in Equation (1) and solving for y gives us

$-\dfrac{64}{5} + y^2 = 25$

$y^2 = \dfrac{189}{5}$

$y = \pm\sqrt{\dfrac{189}{5}} = \pm 3\sqrt{\dfrac{21}{5}}$

$y = \pm\dfrac{3\sqrt{105}}{5}.$ Rationalizing the denominator

The pairs $\left(\dfrac{8i\sqrt{5}}{5}, \dfrac{3\sqrt{105}}{5}\right)$, $\left(-\dfrac{8i\sqrt{5}}{5}, \dfrac{3\sqrt{105}}{5}\right)$,

$\left(\dfrac{8i\sqrt{5}}{5}, -\dfrac{3\sqrt{105}}{5}\right)$, and $\left(-\dfrac{8i\sqrt{5}}{5}, -\dfrac{3\sqrt{105}}{5}\right)$, check. They are the solutions.

33. *Familiarize.* Using the labels on the drawing in the text, we let $l =$ the length of the cargo area and $w =$ the width.

Translate. The area is 60 ft^2, so we have $lw = 60$. The length of a diagonal is 13 ft, so we use the Pythagorean equation to write another equation: $l^2 + w^2 = 13^2$, or $l^2 + w^2 = 169$.

Solve. We solve the system of equations.

$lw = 60,$ (1)

$l^2 + w^2 = 169$ (2)

First we solve Equation (1) for l.

$l = \dfrac{60}{w}$

Then substitute $\dfrac{60}{w}$ for l in Equation (2) and solve for w.

$l^2 + w^2 = 169$

$\left(\dfrac{60}{w}\right)^2 + w^2 = 169$

$\dfrac{3600}{w^2} + w^2 = 169$

$3600 + w^4 = 169w^2$ Multiplying by w^2

$w^4 - 169w^2 + 3600 = 0$

$u^2 - 169u + 3600 = 0$ Letting $u = w^2$

$(u - 144)(u - 25) = 0$

$u = 144$ *or* $u = 25$

Now substitute w^2 for u and solve for x.

$$w^2 = 144 \quad or \quad w^2 = 25$$
$$w = \pm 12 \quad or \quad w = \pm 5$$

Check. The dimensions cannot be negative so -12 and -5 cannot be solutions. If $w = 12$, then $l = \dfrac{60}{12}$, or 5. If $w = 5$, then $l = \dfrac{60}{5}$, or 12. Since we usually consider length to be longer than width, we check $l = 12$ and $w = 5$. The area is $12 \cdot 5$, or 60 ft^2, and $12^2 + 5^2 = 169 = 13^2$. The answer checks.

State. The dimensions are 12 ft by 5 ft.

35. Familiarize. We first make a drawing. We let $l =$ the length and $w =$ the width of the rectangle.

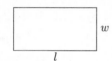

Translate.

   Area: $lw = 14$

   Perimeter: $2l + 2w = 18$, or $l + w = 9$

Solve. We solve the system.

Solve the second equation for l: $l = 9 - w$

Substitute $9 - w$ for l in the first equation and solve for w.

$$(9 - w)w = 14$$
$$9w - w^2 = 14$$
$$0 = w^2 - 9w + 14$$
$$0 = (w - 7)(w - 2)$$
$$w - 7 = 0 \quad or \quad w - 2 = 0$$
$$w = 7 \quad or \quad w = 2$$

If $w = 7$, then $l = 9 - 7$, or 2. If $w = 2$, then $l = 9 - 2$, or 7. Since length is usually considered to be longer than the width, we have the solution $l = 7$ and $w = 2$, or $(7, 2)$.

Check. If $l = 7$ and $w = 2$, the area is $7 \cdot 2$, or 14. The perimeter is $2 \cdot 7 + 2 \cdot 2$, or 18. The numbers check.

State. The length is 7 in., and the width is 2 in.

37. Familiarize. Let $l =$ the length and $w =$ the width of the rectangle. Then the length of a diagonal is $\sqrt{l^2 + w^2}$.

Translate.

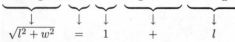

The diagonal   is   1 ft   longer than   the length.

$$\sqrt{l^2 + w^2} \quad = \quad 1 \quad + \quad l$$

The diagonal   is   3 ft   longer than   twice the width.

$$\sqrt{l^2 + w^2} \quad = \quad 3 \quad + \quad 2 \cdot \quad w$$

Solve. We solve the system of equations.

$$\sqrt{l^2 + w^2} = 1 + l, \qquad (1)$$
$$\sqrt{l^2 + w^2} = 3 + 2w \qquad (2)$$

We substitute $3 + 2w$ for $\sqrt{l^2 + w^2}$ in Equation (1) and solve for l.

$$3 + 2w = 1 + l$$
$$2 + 2w = l \qquad (3)$$

Now we substitute $2 + 2w$ for l in Equation (2).

$$\sqrt{(2 + 2w)^2 + w^2} = 3 + 2w$$
$$\sqrt{4 + 8w + 4w^2 + w^2} = 3 + 2w$$
$$\sqrt{5w^2 + 8w + 4} = 3 + 2w$$
$$(\sqrt{5w^2 + 8w + 4})^2 = (3 + 2w)^2$$
$$5w^2 + 8w + 4 = 9 + 12w + 4w^2$$
$$w^2 - 4w - 5 = 0$$
$$(w - 5)(w + 1) = 0$$
$$w - 5 = 0 \quad or \quad w + 1 = 0$$
$$w = 5 \quad or \quad w = -1$$

Since the width cannot be negative, we consider only 5. We substitute 5 for w in Equation (3) and solve for l.

$$l = 2 + 2 \cdot 5 = 12$$

Check. The length of a diagonal of a 12 ft by 5 ft rectangle is $\sqrt{12^2 + 5^2} = \sqrt{169} = 13$. This is 1 ft longer than the length, 12, and 3 ft longer than twice the width, $2 \cdot 5$, or 10. The numbers check.

State. The length is 12 ft, and the width is 5 ft.

39. Familiarize. Let $l =$ the length and $h =$ the height, in cm.

Translate. Since the ratio of the length to the height is 4 to 3, we have one equation:

$$\frac{l}{h} = \frac{4}{3}$$

The Pythagorean equation gives us a second equation:

$$l^2 + h^2 = 31^2, \text{ or } l^2 + h^2 = 961$$

We have a system of equations.

$$\frac{l}{h} = \frac{4}{3}, \qquad (1)$$
$$l^2 + h^2 = 961 \qquad (2)$$

Solve. We solve the system of equations. First we solve Eq. (1) for l:

$$\frac{l}{h} = \frac{4}{3}$$
$$l = \frac{4}{3}h \quad (3)$$

Now substitute $\frac{4}{3}h$ for w in Eq. (2) and solve for h.

$$l^2 + h^2 = 961$$
$$\left(\frac{4}{3}h\right)^2 + h^2 = 961$$
$$\frac{16}{9}h^2 + h^2 = 961$$
$$\frac{25}{9}h^2 = 961$$
$$h^2 = \frac{961 \cdot 9}{25} = \frac{8649}{25}$$

$$h = \frac{93}{5} \quad or \quad h = -\frac{93}{5}$$

Since the height cannot be negative, we consider only $\frac{93}{5}$.
Substitute $\frac{93}{5}$ for h in Eq. (3) and find l.

$$l = \frac{4}{3} \cdot \frac{93}{5} = \frac{124}{5}$$

Check. The ratio of 124/5 to 93/5 is $\frac{124/5}{93/5} =$
$\frac{124}{93} = \frac{4}{3}$. Also $\left(\frac{124}{5}\right)^2 + \left(\frac{93}{5}\right)^2 = 961 = 31^2$. The answer checks.

State. The length is $\frac{124}{5}$ cm, or 24.8 cm, and the height is $\frac{93}{5}$ cm, or 18.6 cm.

41. Familiarize. We let x = the length of a side of one peanut bed and y = the length of a side of the other peanut bed. Make a drawing.

```
┌──────────┐       ┌──────┐
│          │ x     │      │ y
│          │       │      │
└──────────┘       └──────┘
     x                 y
  Area: x²          Area: y²
```

Translate.

The sum of the areas is 832 ft².

$$x^2 + y^2 = 832$$

The difference of the areas is 320 ft².

$$x^2 - y^2 = 320$$

Solve. We solve the system of equations.

$$\begin{array}{r} x^2 + y^2 = 832 \\ x^2 - y^2 = 320 \\ \hline 2x^2 = 1152 \end{array} \quad \text{Adding}$$

$$x^2 = 576$$
$$x = \pm 24$$

Since length cannot be negative, we consider only $x = 24$. Substitute 24 for x in the first equation and solve for y.

$$24^2 + y^2 = 832$$
$$576 + y^2 = 832$$
$$y^2 = 256$$
$$y = \pm 16$$

Again, we consider only the positive value, 16. The possible solution is $(24, 16)$.

Check. The areas of the beds are 24^2, or 576, and 16^2, or 256. The sum of the areas is $576 + 256$, or 832. The difference of the areas is $576 - 256$, or 320. The values check.

State. The lengths of the beds are 24 ft and 16 ft.

43. Familiarize. We first make a drawing. We let l = the length and w = the width.

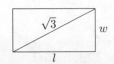

Translate.
Area: $\sqrt{2} = lw$ (1)
Using the Pythagorean equation: $l^2 + w^2 = (\sqrt{3})^2$, or
$$l^2 + w^2 = 3 \quad (2)$$

Solve. We solve the system of equations. First we solve Equation (1) for w.

$$\frac{\sqrt{2}}{l} = w$$

Then we substitute $\frac{\sqrt{2}}{l}$ for w in Equation (2) and solve for l.

$$l^2 + \left(\frac{\sqrt{2}}{l}\right)^2 = 3$$
$$l^2 + \frac{2}{l^2} = 3$$
$$l^4 + 2 = 3l^2 \quad \text{Multiplying by } l^2$$
$$l^4 - 3l^2 + 2 = 0$$
$$u^2 - 3u + 2 = 0 \quad \text{Letting } u = l^2$$
$$(u - 2)(u - 1) = 0$$

$$\begin{array}{lll} u - 2 = 0 & or & u - 1 = 0 \\ u = 2 & or & u = 1 \\ l^2 = 2 & or & l^2 = 1 \quad \text{Substituting } l^2 \text{ for } u \\ l = \pm\sqrt{2} & or & l = \pm 1 \end{array}$$

Length cannot be negative, we only need to consider $l = \sqrt{2}$ or $l = 1$. Since $w = \sqrt{2}/l$, if $l = \sqrt{2}$, $w = 1$ and if $l = 1$, $w = \sqrt{2}$. Length is usually considered to be longer than width, so we have the solution $l = \sqrt{2}$ and $w = 1$, or $(\sqrt{2}, 1)$.

Check. If $l = \sqrt{2}$ and $w = 1$, the area is $\sqrt{2} \cdot 1$, or $\sqrt{2}$. Also $(\sqrt{2})^2 + 1^2 = 2 + 1 = 3 = (\sqrt{3})^2$. The numbers check.

State. The length is $\sqrt{2}$ m, and the width is 1 m.

45. Discussion and Writing Exercise

47. $f(x) = 2x - 5$

The function passes the horizontal-line test, so it has an inverse.

1. Replace $f(x)$ by y: $y = 2x - 5$

2. Interchange x and y: $x = 2y - 5$

3. Solve for y: $x + 5 = 2y$
$$\frac{x + 5}{2} = y$$

4. Replace y by $f^{-1}(x)$: $f^{-1}(x) = \frac{x + 5}{2}$

49. $f(x) = \dfrac{x - 2}{x + 3}$

The function passes the horizontal-line test, so it has an inverse.

1. Replace $f(x)$ by y: $y = \dfrac{x - 2}{x + 3}$

2. Interchange x and y: $x = \dfrac{y - 2}{y + 3}$

3. Solve for y: $xy + 3x = y - 2$

$$3x + 2 = y - xy$$
$$3x + 2 = y(1 - x)$$
$$\frac{3x + 2}{1 - x} = y$$

4. Replace y by $f^{-1}(x)$: $f^{-1}(x) = \dfrac{3x + 2}{1 - x}$

51. $f(x) = |x|$

Since a horizontal line can cross the graph more than once, the function is not one-to-one so it does not have an inverse.

53. $f(x) = 10^x$

The function passes the horizontal-line test, so it has an inverse.

1. Replace $f(x)$ by y: $y = 10^x$

2. Interchange x and y: $x = 10^y$

3. Solve for y: $\log x = \log 10^y$

$$\log x = y \log 10$$
$$\log x = y$$

4. Replace y by $f^{-1}(x)$: $f^{-1}(x) = \log x$

55. $f(x) = x^3 - 4$

The function passes the horizontal-line test, so it has an inverse.

1. Replace $f(x)$ by y: $y = x^3 - 4$

2. Interchange x and y: $x = y^3 - 4$

3. Solve for y: $x + 4 = y^3$

$$\sqrt[3]{x + 4} = y$$

4. Replace y by $f^{-1}(x)$: $f^{-1}(x) = \sqrt[3]{x + 4}$

57. $f(x) = \ln x$

The function passes the horizontal-line test, so it has an inverse.

1. Replace $f(x)$ by y: $y = \ln x$

2. Interchange x and y: $x = \ln y$

3. Solve for y: $e^x = y$

4. Replace y by $f^{-1}(x)$: $f^{-1}(x) = e^x$

59. Left to the student

61. *Familiarize*. Let $x =$ the length of the longer piece of wire and $y =$ the length of the shorter piece. Then the lengths of the sides of the squares are $\dfrac{x}{4}$ and $\dfrac{y}{4}$.

Translate. The length of the wire is 100 cm, so we have $x + y = 100$. The area of one square is 144 cm^2 greater than the area of the other, so we have a second equation:
$$\left(\frac{x}{4}\right)^2 + \left(\frac{y}{4}\right)^2 = 144.$$

Solve.   The solution of the system of equations is $(61.52, 38.48)$.

Check. The sum of the lengths is $61.52 + 38.48$, or 100 cm. The length of a side of the larger square is $61.52/4$, or 15.38, and the area is $(15.38)^2$, or 236.5444 cm^2. The length of a side of the larger square is $38.48/4$, or 9.62, and the area is $(9.62)^2$, or 92.5444 cm^2. The larger area is 144 cm^2 greater than the smaller area. The answer checks.

State. The wire should be cut so that the length of one piece is 61.52, or $61\dfrac{13}{25}$ cm, and the other is 38.48, or $38\dfrac{12}{25}$ cm.

63.
$$R = C$$
$$100x + x^2 = 80x + 1500$$
$$x^2 + 20x - 1500 = 0$$
$$(x - 30)(x + 50) = 0$$
$$x = 30 \ or \ x = -50$$

Since the number of units cannot be negative, the solution of the problem is 30. Thus, 30 units must be sold in order to break even.

65. $a + b = \dfrac{5}{6},$ (1)

$$\frac{a}{b} + \frac{b}{a} = \frac{13}{6}   (2)$$

$b = \dfrac{5}{6} - a = \dfrac{5 - 6a}{6}$ Solving Eq. (1) for b

$\dfrac{a}{\dfrac{5 - 6a}{6}} + \dfrac{\dfrac{5 - 6a}{6}}{a} = \dfrac{13}{6}$ Substituting for b in Eq. (2)

$$\frac{6a}{5 - 6a} + \frac{5 - 6a}{6a} = \frac{13}{6}$$
$$36a^2 + 25 - 60a + 36a^2 = 65a - 78a^2$$
$$150a^2 - 125a + 25 = 0$$
$$6a^2 - 5a + 1 = 0$$
$$(3a - 1)(2a - 1) = 0$$

$a = \dfrac{1}{3} \ or \ a = \dfrac{1}{2}$

Substitute for a and solve for b.

When $a = \dfrac{1}{3}$, $b = \dfrac{5 - 6\left(\dfrac{1}{3}\right)}{6} = \dfrac{1}{2}$.

When $a = \dfrac{1}{2}$, $b = \dfrac{5 - 6\left(\dfrac{1}{2}\right)}{6} = \dfrac{1}{3}$.

The pairs $\left(\dfrac{1}{3}, \dfrac{1}{2}\right)$ and $\left(\dfrac{1}{2}, \dfrac{1}{3}\right)$ check. They are the solutions.

Chapter 9 Review Exercises

1. $d = \sqrt{(x_2 - x_1)^2 + (y_2 - y_1)^2}$
$d = \sqrt{(6-2)^2 + (6-6)^2}$
$= \sqrt{4^2 + 0^2}$
$= \sqrt{4^2} = 4$

2. $d = \sqrt{(x_2 - x_1)^2 + (y_2 - y_1)^2}$
$d = \sqrt{[-5-(-1)]^2 + (4-1)^2}$
$= \sqrt{(-4)^2 + 3^2}$
$= \sqrt{25} = 5$

3. $d = \sqrt{(x_2 - x_1)^2 + (y_2 - y_1)^2}$
$d = \sqrt{(4.7-1.4)^2 + (-5.3-3.6)^2}$
$= \sqrt{(3.3)^2 + (-8.9)^2}$
$= \sqrt{90.1} \approx 9.492$

4. $d = \sqrt{(x_2 - x_1)^2 + (y_2 - y_1)^2}$
$d = \sqrt{(-1-2)^2 + (a-3a)^2}$
$= \sqrt{(-3)^2 + (-2a)^2}$
$= \sqrt{9 + 4a^2}$

5. $\left(\dfrac{x_1 + x_2}{2}, \dfrac{y_1 + y_2}{2}\right) = \left(\dfrac{1+7}{2}, \dfrac{6+6}{2}\right) =$
$\left(\dfrac{8}{2}, \dfrac{12}{2}\right) = (4,6)$

6. $\left(\dfrac{x_1 + x_2}{2}, \dfrac{y_1 + y_2}{2}\right) = \left(\dfrac{-1+(-5)}{2}, \dfrac{1+4}{2}\right) =$
$\left(\dfrac{-6}{2}, \dfrac{5}{2}\right) = \left(-3, \dfrac{5}{2}\right)$

7. $\left(\dfrac{x_1 + x_2}{2}, \dfrac{y_1 + y_2}{2}\right) = \left(\dfrac{1 + \frac{1}{2}}{2}, \dfrac{\sqrt{3} + (-\sqrt{2})}{2}\right) =$
$\left(\dfrac{\frac{3}{2}}{2}, \dfrac{\sqrt{3} - \sqrt{2}}{2}\right) = \left(\dfrac{3}{4}, \dfrac{\sqrt{3} - \sqrt{2}}{2}\right)$

8. $\left(\dfrac{x_1 + x_2}{2}, \dfrac{y_1 + y_2}{2}\right) = \left(\dfrac{2 + (-1)}{2}, \dfrac{3a + a}{2}\right) =$
$\left(\dfrac{1}{2}, \dfrac{4a}{2}\right) = \left(\dfrac{1}{2}, 2a\right)$

9. $(x+2)^2 + (y-3)^2 = 2$
$[x-(-2)]^2 + (y-3)^2 = (\sqrt{2})^2$ Standard form
Center: $(-2, 3)$
Radius: $\sqrt{2}$

10. $(x-5)^2 + y^2 = 49$
$(x-5)^2 + (y-0)^2 = 7^2$ Standard form
Center: $(5, 0)$
Radius: 7

11. $x^2 + y^2 - 6x - 2y + 1 = 0$
$(x^2 - 6x) + (y^2 - 2y) + 1 = 0$
Regrouping
$(x^2 - 6x + 9 - 9) + (y^2 - 2y + 1 - 1) + 1 = 0$
Completing the square twice
$(x^2 - 6x + 9) + (y^2 - 2y + 1) - 9 - 1 + 1 = 0$
$(x-3)^2 + (y-1)^2 = 9$
$(x-3)^2 + (y-1)^2 = 3^2$
Center: $(3, 1)$
Radius: 3

12. $x^2 + y^2 + 8x - 6y - 10 = 0$
$(x^2 + 8x) + (y^2 - 6y) - 10 = 0$
Regrouping
$(x^2 + 8x + 16 - 16) + (y^2 - 6y + 9 - 9) - 10 = 0$
Completing the square twice
$(x^2 + 8x + 16) + (y^2 - 6y + 9) - 16 - 9 - 10 = 0$
$(x+4)^2 + (y-3)^2 = 35$
$[x-(-4)]^2 + (y-3)^2 = (\sqrt{35})^2$
Center: $(-4, 3)$
Radius: $\sqrt{35}$

13. $(x-h)^2 + (y-k)^2 = r^2$ Standard form
$[x-(-4)]^2 + (y-3)^2 = (4\sqrt{3})^2$ Substituting
$(x+4)^2 + (y-3)^2 = 48$

14. $(x-h)^2 + (y-k)^2 = r^2$
$(x-7)^2 + [y-(-2)]^2 = (2\sqrt{5})^2$
$(x-7)^2 + (y+2)^2 = 20$

15. $\dfrac{x^2}{16} + \dfrac{y^2}{4} = 1$
$\dfrac{x^2}{4^2} + \dfrac{y^2}{2^2} = 1$

The graph is an ellipse. The x-intercepts are $(-4, 0)$ and $(4, 0)$, and the y-intercepts are $(0, -2)$ and $(0, 2)$. We plot these points and connect them with an oval-shaped curve.

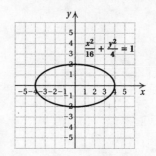

16. $\dfrac{y^2}{9} - \dfrac{x^2}{4} = 1$

$\dfrac{y^2}{3^2} - \dfrac{x^2}{2^2} = 1$

The graph is a hyperbola. $a = 2$ and $b = 3$, so the asymptotes are $y = \dfrac{3}{2}x$ and $y = -\dfrac{3}{2}x$. We sketch them. Replacing x with 0 and solving for y, we get $y = \pm 3$, so the intercepts are $(0, 3)$ and $(0, -3)$.

We plot the intercepts and draw smooth curves through them that approach the asymptotes.

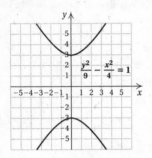

17. $x^2 + y^2 = 16$

$(x - 0)^2 + (y - 0)^2 = 4^2$

The graph is a circle. The center is $(0, 0)$, and the radius is 4.

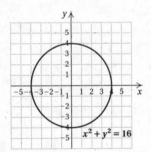

18. $x = y^2 + 2y - 2$

The graph is a parabola. We complete the square.

$$x = (y^2 + 2y + 1 - 1) - 2$$
$$= (y^2 + 2y + 1) - 1 - 2$$
$$= (y + 1)^2 - 3$$
$$= [y - (-1)]^2 + (-3)$$

The vertex is $(-3, -1)$; the line of symmetry is $y = -1$. The curve opens to the right.

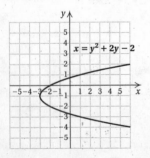

19. $y = -2x^2 - 2x + 3$

The graph is a parabola. We complete the square.

$$y = -2(x^2 + x) + 3$$
$$= -2\left(x^2 + x + \frac{1}{4} - \frac{1}{4}\right) + 3$$
$$= -2\left(x^2 + x + \frac{1}{4}\right) + (-2)\left(-\frac{1}{4}\right) + 3$$
$$= -2\left(x + \frac{1}{2}\right)^2 + \frac{7}{2}$$
$$= -2\left[x - \left(-\frac{1}{2}\right)\right]^2 + \frac{7}{2}$$

The vertex is $\left(-\dfrac{1}{2}, \dfrac{7}{2}\right)$; the line of symmetry is $x = -\dfrac{1}{2}$. The curve opens down.

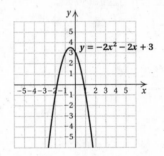

20. $x^2 + y^2 + 2x - 4y - 4 = 0$

The graph is a circle. We regroup and complete the square twice.

$$(x^2 + 2x + 1 - 1) + (y^2 - 4y + 4 - 4) - 4 = 0$$
$$(x^2 + 2x + 1) + (y^2 - 4y + 4) - 1 - 4 - 4 = 0$$
$$(x + 1)^2 + (y - 2)^2 = 9$$
$$[x - (-1)]^2 + (y - 2)^2 = 3^2$$

The center is $(-1, 2)$, and the radius is 3.

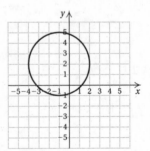

21. $\dfrac{(x - 3)^2}{9} + \dfrac{(y + 4)^2}{4} = 1$

$\dfrac{(x - 3)^2}{3^2} + \dfrac{[y - (-4)]^2}{2^2} = 1$

The graph is an ellipse with center $(3, -4)$. Note that $a = 3$ and $b = 2$. We locate the center and then plot the points $(3 - 3, 2)$, $(3 + 3, 2)$, $(3, -4 - 2)$, and $(3, -4 + 2)$, or $(0, 2)$, $(6, 2)$, $(3, -6)$, and $(3, -2)$. Connect these points with an oval-shaped curve.

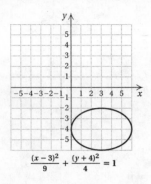

$$\frac{(x-3)^2}{9} + \frac{(y+4)^2}{4} = 1$$

22. $xy = 9$

$$y = \frac{9}{x}$$

We find some solutions. Note that x cannot be 0. The x-axis and the y-axis are the asymptotes.

x	y
-6	$-\dfrac{3}{2}$
-3	-3
-1	-9
1	9
3	3
6	$\dfrac{3}{2}$

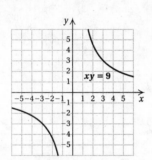

23. $x + y^2 = 2y + 1$

$$x = -y^2 + 2y + 1$$

The graph is a parabola. We complete the square.

$$x = -(y^2 - 2y + 1 - 1) + 1$$
$$= -(y^2 - 2y + 1) + (-1)(-1) + 1$$
$$= -(y - 1)^2 + 2$$

The vertex is $(2, 1)$; the line of symmetry is $y = 1$. The graph opens to the left.

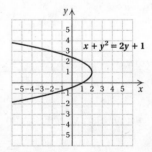

24. $\dfrac{x^2}{4} - \dfrac{y^2}{4} = 1$

$$\frac{x^2}{2^2} - \frac{y^2}{2^2} = 1$$

The graph is a hyperbola. We have $a = 2$ and $b = 2$, so the asymptotes are $y = \dfrac{2}{2}x$ and $y = -\dfrac{2}{2}x$, or $y = x$ and $y = -x$. We sketch them. Replacing y with 0 and solving for x, we get $x = \pm 2$, so the intercepts are $(2, 0)$ and $(-2, 0)$.

We plot the intercepts and draw smooth curves through them that approach the asymptotes.

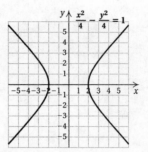

25. $x^2 - y^2 = 33,$ (1)

$x + y = 11$ (2)

First we solve Equation (2) for y.

$$y = 11 - x \quad (3)$$

Then substitute $11 - x$ for y in Equation (1) and solve for x.

$$x^2 - y^2 = 33$$
$$x^2 - (11 - x)^2 = 33$$
$$x^2 - (121 - 22x + x^2) = 33$$
$$x^2 - 121 + 22x - x^2 = 33$$
$$-121 + 22x = 33$$
$$22x = 154$$
$$x = 7$$

Now substitute 7 for x in Equation (3) and find y.

$$y = 11 - 7 = 4$$

The pair $(7, 4)$ checks, so it is the solution.

26. $x^2 - 2x + 2y^2 = 8,$ (1)

$2x + y = 6$ (2)

First we solve Equation (2) for y.

$$y = 6 - 2x \quad (3)$$

Then substitute $6 - 2x$ for y in Equation (1) and solve for x.

$$x^2 - 2x + 2(6 - 2x)^2 = 8$$
$$x^2 - 2x + 2(36 - 24x + 4x^2) = 8$$
$$x^2 - 2x + 72 - 48x + 8x^2 = 8$$
$$9x^2 - 50x + 72 = 8$$
$$9x^2 - 50x + 64 = 0$$
$$(9x - 32)(x - 2) = 0$$
$$9x - 32 = 0 \quad or \quad x - 2 = 0$$
$$x = \frac{32}{9} \quad or \quad x = 2$$

Now substitute these numbers in Equation (3) and find y.

$$y = 6 - 2 \cdot \frac{32}{9} = 6 - \frac{64}{9} = -\frac{10}{9}$$
$$y = 6 - 2 \cdot 2 = 6 - 4 = 2$$

The pairs $\left(\dfrac{32}{9}, -\dfrac{10}{9}\right)$ and $(2, 2)$ check, so they are the solutions.

27. $x^2 - y = 3$,   (1)

    $2x - y = 3$ (2)

We multiply Equation (2) by -1 and then add.

$$x^2 - y = 3$$
$$\underline{-2x + y = -3}$$
$$x^2 - 2x = 0$$

Then we have:

$$x^2 - 2x = 0$$
$$x(x - 2) = 0$$
$$x = 0 \ \ or \ \ x = 2$$

We substitute these numbers in one of the original equations and solve for y. We use Equation (2).

$$2 \cdot 0 - y = 3 \qquad\qquad 2 \cdot 2 - y = 3$$
$$-y = 3 \qquad\qquad 4 - y = 3$$
$$y = -3 \qquad\qquad -y = -1$$
$$y = -3 \qquad\qquad y = 1$$

The pairs $(0, -3)$ and $(2, 1)$ check, so they are the solutions.

28. $x^2 + y^2 = 25$ (1)

    $\underline{x^2 - y^2 = 7}$ (2)

   $2x^2 = 32$ Adding

$$x^2 = 16$$
$$x = \pm 4$$

If $x = 4$, $x^2 = 16$, and if $x = -4$, $x^2 = 16$, so substituting 4 or -4 in Equation (1) gives us

$$16 + y^2 = 25$$
$$y^2 = 9$$
$$y = \pm 3.$$

The pairs $(4, 3)$, $(4, -3)$, $(-4, 3)$, and $(-4, -3)$ check.

29. $x^2 - y^2 = 3$,   (1)

   $y = x^2 - 3$ (2)

Substitute $x^2 - 3$ for y in Equation (1) and solve for x.

$$x^2 - (x^2 - 3)^2 = 3$$
$$x^2 - (x^4 - 6x^2 + 9) = 3$$
$$x^2 - x^4 + 6x^2 - 9 = 3$$
$$0 = x^4 - 7x^2 + 12$$

Let $u = x^2$.

$$0 = u^2 - 7u + 12$$
$$0 = (u - 3)(u - 4)$$
$$u = 3 \qquad or \qquad u = 4$$
$$x^2 = 3 \qquad or \quad x^2 = 4$$
$$x = \pm\sqrt{3} \ \ or \quad x = \pm 2$$

When $x = \pm\sqrt{3}$, $x^2 = 3$ and, substituting in Equation (2), we get $y = 3 - 3 = 0$.

When $x = \pm 2$, $x^2 = 4$ and, substituting in Equation (2), we get $y = 4 - 3 = 1$.

The pairs $(\sqrt{3}, 0)$, $(-\sqrt{3}, 0)$, $(2, 1)$, and $(-2, 1)$ check.

30. $x^2 + y^2 = 18$,   (1)

   $2x + y = 3$ (2)

First we solve Equation (2) for y.

$$y = 3 - 2x \quad (3)$$

Now substitute $3 - 2x$ for y in Equation (1) and solve for x.

$$x^2 + (3 - 2x)^2 = 18$$
$$x^2 + 9 - 12x + 4x^2 = 18$$
$$5x^2 - 12x - 9 = 0$$
$$(5x + 3)(x - 3) = 0$$
$$5x + 3 = 0 \quad\ or \ \ x - 3 = 0$$
$$x = -\frac{3}{5} \ \ or \qquad x = 3$$

Substitute these numbers in Equation (3) and find y.

$$y = 3 - 2\left(-\frac{3}{5}\right) = 3 + \frac{6}{5} = \frac{21}{5}$$
$$y = 3 - 2 \cdot 3 = 3 - 6 = -3$$

The pairs $\left(-\dfrac{3}{5}, \dfrac{21}{5}\right)$ and $(3, -3)$ check.

31. $x^2 + y^2 = 100$,   (1)

   $2x^2 - 3y^2 = -120$ (2)

Multiply Equation (1) by 3 and then add.

$$3x^2 + 3y^2 = 300$$
$$\underline{2x^2 - 3y^2 = -120}$$
$$5x^2 = 180$$
$$x^2 = 36$$
$$x = \pm 6$$

When $x = 6$, $x^2 = 36$, and when $x = -6$, $x^2 = 36$, so substituting 6 or -6 in Equation (1) give us

$$36 + y^2 = 100$$
$$y^2 = 64$$
$$y = \pm 8.$$

The pairs $(6, 8)$, $(6, -8)$, $(-6, 8)$, and $(-6, -8)$ check.

32. $x^2 + 2y^2 = 12$,   (1)

   $xy = 4$ (2)

First we solve Equation (2) for y.

$$y = \frac{4}{x} \quad (3)$$

Then substitute $\dfrac{4}{x}$ for y in Equation (1) and solve for x.

$$x^2 + 2\left(\frac{4}{x}\right)^2 = 12$$
$$x^2 + 2\left(\frac{16}{x^2}\right) = 12$$
$$x^2 + \frac{32}{x^2} = 12$$
$$x^4 + 32 = 12x^2 \quad \text{Multiplying by } x^2$$
$$x^4 - 12x^2 + 32 = 0$$
$$u^2 - 12u + 32 = 0 \qquad \text{Letting } u = x^2$$
$$(u - 4)(u - 8) = 0$$

$u = 4 \quad or \quad u = 8$

$x^2 = 4 \quad or \quad x^2 = 8$

$x = \pm 2 \quad or \quad x = \pm\sqrt{8} = \pm 2\sqrt{2}$

Now we use Equation (3) to find the y-value that corresponds to each value of x.

When $x = 2$, $y = \dfrac{4}{2} = 2$.

When $x = -2$, $y = \dfrac{4}{-2} = -2$.

When $x = 2\sqrt{2}$, $y = \dfrac{4}{2\sqrt{2}} = \dfrac{2}{\sqrt{2}} = \sqrt{2}$.

When $x = -2\sqrt{2}$, $y = \dfrac{4}{-2\sqrt{2}} = -\dfrac{2}{\sqrt{2}} = -\sqrt{2}$.

The pairs $(2, 2)$, $(-2, -2)$, $(2\sqrt{2}, \sqrt{2})$, and $(-2\sqrt{2}, -\sqrt{2})$ check.

33. Familiarize. Let $l =$ the length of the garden and $w =$ the width, in meters.

Translate. The perimeter is 38 m, so we have $2l + 2w = 38$, or $l + w = 19$. The area is 84 m^2, so we have $lw = 84$.

Solve. We solve the system of equations.

$l + w = 19, \quad (1)$

$lw = 84 \qquad (2)$

First we solve Equation (1) for l.

$l = 19 - w \quad (3)$

Then substitute $19 - w$ for l in Equation (2) and solve for w.

$(19 - w)w = 84$

$19w - w^2 = 84$

$0 = w^2 - 19w + 84$

$0 = (w - 7)(w - 12)$

$w = 7 \quad or \quad w = 12$

If $w = 7$, then $l = 19 - 7 = 12$. If $w = 12$, then $l = 19 - 12 = 7$. Since length is usually considered to be longer than width, we have $l = 12$ and $w = 7$.

Check. If $l = 12$ and $w = 7$, the perimeter is $2 \cdot 12 + 2 \cdot 7$, or 38 m, and the area is $12 \cdot 7$, or 84 m^2. The answer checks.

State. The length of the garden is 12 m, and the width is 7 m.

34. Familiarize. Let x and y represent the positive integers.

Translate. The sum of the numbers is 12, so we have $x + y = 12$. The sum of the reciprocals is $\dfrac{3}{8}$, so we also have $\dfrac{1}{x} + \dfrac{1}{y} = \dfrac{3}{8}$.

Solve. We solve the system of equations.

$x + y = 12, \quad (1)$

$\dfrac{1}{x} + \dfrac{1}{y} = \dfrac{3}{8} \quad (2)$

First we solve Equation (1) for x.

$x = 12 - y$

Now substitute $12 - y$ for x in Equation (2) and solve for y.

$\dfrac{1}{12 - y} + \dfrac{1}{y} = \dfrac{3}{8}, \text{LCD is } 8y(12 - y)$

$8y(12 - y)\left(\dfrac{1}{12 - y} + \dfrac{1}{y}\right) = 8y(12 - y) \cdot \dfrac{3}{8}$

$8y + 8(12 - y) = 3y(12 - y)$

$8y + 96 - 8y = 36y - 3y^2$

$3y^2 - 36y + 96 = 0$

$y^2 - 12y + 32 = 0 \quad \text{Dividing by 3}$

$(y - 4)(y - 8) = 0$

$y = 4 \quad or \quad y = 8$

If $y = 4$, then $x = 12 - 4 = 8$; if $y = 8$, then $x = 12 - 8 = 4$. In either case, the numbers are 4 and 8.

Check. $4 + 8 = 12$ and $\dfrac{1}{4} + \dfrac{1}{8} = \dfrac{2}{8} + \dfrac{1}{8} = \dfrac{3}{8}$. The answer checks.

State. The numbers are 4 and 8.

35. Familiarize. Let l and w represent the length and width, respectively, of the carton, in inches.

Translate. The area is 108 in^2, so we have $lw = 108$. The diagonal is 15 in., so the Pythagorean theorem gives us a second equation, $l^2 + w^2 = 15^2$, or $l^2 + w^2 = 225$.

Solve. We solve the system of equations.

$lw = 108, \qquad (1)$

$l^2 + w^2 = 225 \quad (2)$

First we solve Equation (1) for l.

$l = \dfrac{108}{w}$

Then substitute $\dfrac{108}{w}$ for l in Equation (2) and solve for w.

$\left(\dfrac{108}{w}\right)^2 + w^2 = 225$

$\dfrac{11,664}{w^2} + w^2 = 225$

$11,664 + w^4 = 225w^2 \quad \text{Multiplying by } w^2$

$w^4 - 225w^2 + 11,664 = 0$

$u^2 - 225u + 11,664 = 0 \quad \text{Letting } u = w^2$

$(u - 81)(u - 144) = 0$

$u = 81 \quad or \quad u = 144$

$w^2 = 81 \quad or \quad w^2 = 144$

$w = \pm 9 \quad or \quad w = \pm 12$

Since the width cannot be negative, -9 and -12 cannot be solutions. If $w = 9$, $l = \dfrac{108}{9} = 12$; if $w = 12$, $l = \dfrac{108}{12} = 9$. Since length is usually considered to be longer than width, we let $l = 12$ and $w = 9$.

Check. If $l = 12$ and $w = 9$, the area is $12 \cdot 9$, or 108 in^2; also $12^2 + 9^2 = 225 = 15^2$, so the answer checks.

State. The length of the carton is 12 in., and the width is 9 in.

36. Familiarize. Using the labels on the drawing in the text, let $r_1 =$ the radius of the larger flower bed, in feet, and let $r_2 =$ the radius of the smaller bed.

Translate. The sum of the areas is 130π ft^2, so we have $\pi r_1{}^2 + \pi r_2{}^2 = 130\pi$, or $r_1{}^2 + r_2{}^2 = 130$. The difference of the circumferences is 16π ft, so we also have $2\pi r_1 - 2\pi r_2 = 16\pi$, or $r_1 - r_2 = 8$.

Solve. We solve the system of equations.

$$r_1{}^2 + r_2{}^2 = 130, \quad (1)$$
$$r_1 - r_2 = 8 \qquad (2)$$

First solve Equation (2) for r_1.

$$r_1 = 8 + r_2$$

Substitute $8 + r_2$ for r_1 and Equation (1) and solve for r_2.

$$(8 + r_2)^2 + r_2{}^2 = 130$$
$$64 + 16r_2 + r_2{}^2 + r_2{}^2 = 130$$
$$2r_2{}^2 + 16r_2 + 64 = 130$$
$$2r_2{}^2 + 16r_2 - 66 = 0$$
$$r_2{}^2 + 8r_2 - 33 = 0 \quad \text{Dividing by 2}$$
$$(r_2 + 11)(r_2 - 3) = 0$$
$$r_2 = -11 \ \ or \ \ r_2 = 3$$

Since the radius cannot be negative, -11 cannot be a solution. If $r_2 = 3$, then $r_1 = 8 + 3 = 11$.

Check. The sum of the area is $\pi \cdot 11^2 + \pi \cdot 3^2 = 121\pi + 9\pi = 130\pi$, and the difference of the circumferences is $2\pi \cdot 11 - 2\pi \cdot 3 = 22\pi - 6\pi = 16\pi$. The answer checks.

State. The radii are 11 ft and 3 ft.

37. *Discussion and Writing Exercise*. Earlier we studied systems of linear equation. In this chapter, we studied systems of two equations in which at least one equation is of second degree.

38. *Discussion and Writing Exercise*. The graph of a parabola has one branch whereas the graph of a hyperbola has two branches. A hyperbola has asymptotes but a parabola does not.

39. $4x^2 - x - 3y^2 = 9, \quad (1)$
$-x^2 + x + y^2 = 2 \quad (2)$

First solve Equation (2) for y^2.

$$y^2 = x^2 - x + 2 \quad (3)$$

Substitute $x^2 - x + 2$ for y^2 in Equation (1) and solve for x.

$$4x^2 - x - 3(x^2 - x + 2) = 9$$
$$4x^2 - x - 3x^2 + 3x - 6 = 9$$
$$x^2 + 2x - 6 = 9$$
$$x^2 + 2x - 15 = 0$$
$$(x + 5)(x - 3) = 0$$
$$x = -5 \ \ or \ \ x = 3$$

Now use Equation (3) to find y.

When $x = -5$:

$$y^2 = (-5)^2 - (-5) + 2$$
$$y^2 = 25 + 5 + 2$$
$$y^2 = 32$$
$$y = \pm\sqrt{32} = \pm 4\sqrt{2}$$

When $x = 3$:

$$y^2 = 3^2 - 3 + 2$$
$$y^2 = 9 - 3 + 2$$
$$y^2 = 8$$
$$y = \pm\sqrt{8} = \pm 2\sqrt{2}$$

The solutions are $(-5, 4\sqrt{2})$, $(-5, -4\sqrt{2})$, $(3, 2\sqrt{2})$, and $(3, -2\sqrt{2})$.

40. We substitute the given points in the standard form of the equation of a circle, $(x - h)^2 + (y - k)^2 = r^2$.

For $(-2, -4)$:

$$(-2 - h)^2 + (-4 - k)^2 = r^2$$
$$4 + 4h + h^2 + 16 + 8k + k^2 = r^2$$
$$h^2 + 4h + k^2 + 8k + 20 = r^2$$
$$h^2 + 4h + k^2 + 8k = r^2 - 20 \quad (1)$$

For $(5, -5)$:

$$(5 - h)^2 + (-5 - k)^2 = r^2$$
$$25 - 10h + h^2 + 25 + 10k + k^2 = r^2$$
$$h^2 - 10h + k^2 + 10k + 50 = r^2$$
$$h^2 - 10h + k^2 + 10k = r^2 - 50 \quad (2)$$

For $(6, 2)$:

$$(6 - h)^2 + (2 - k)^2 = r^2$$
$$36 - 12h + h^2 + 4 - 4k + k^2 = r^2$$
$$h^2 - 12h + k^2 - 4k + 40 = r^2$$
$$h^2 - 12h + k^2 - 4k = r^2 - 40 \quad (3)$$

We solve the system of equations (1), (2), and (3). First we will eliminate the squared terms from two pairs of equations. Multiply Equation (1) by -1 and add it to Equation (2) and to Equation (3).

$$
\begin{aligned}
-h^2 - 4h - k^2 - 8k &= -r^2 + 20 \\
\underline{h^2 - 10h + k^2 + 10k} &= \underline{r^2 - 50} \quad (2) \\
-14h \qquad\quad + 2k &= -30 \qquad (4)
\end{aligned}
$$

$$
\begin{aligned}
-h^2 - 4h - k^2 - 8k &= -r^2 + 20 \\
\underline{h^2 - 12h + k^2 - 4k} &= \underline{r^2 - 40} \\
-16h \qquad\quad - 12k &= -20 \qquad (5)
\end{aligned}
$$

Now solve the system of equations (4) and (5). Multiply Equation (4) by 6 and then add.

$$
\begin{aligned}
-84h + 12k &= -180 \\
\underline{-16h - 12k} &= \underline{-20} \\
-100h \qquad\quad &= -200 \\
h &= 2
\end{aligned}
$$

Substitute 2 for h in Equation (4) and solve for k.

$$-14 \cdot 2 + 2k = -30$$
$$-28 + 2k = -30$$
$$2k = -2$$
$$k = -1$$

Now substitute 2 for h and -1 for k in one of the original equations and solve for r^2. We will use Equation (1).

$$2^2 + 4 \cdot 2 + (-1)^2 + 8(-1) = r^2 - 20$$
$$4 + 8 + 1 - 8 = r^2 - 20$$
$$5 = r^2 - 20$$
$$25 = r^2$$

Then the equation of the circle is $(x-2)^2 + [y-(-1)]^2 = 25$, or $(x-2)^2 + (y+1)^2 = 25$.

41. The center of the ellipse is the midpoint of the segment joining the x-intercepts or the y-intercepts. We will use the x-intercepts to find the center.
$$\left(\frac{-7+7}{2}, \frac{0+0}{2}\right) = \left(\frac{0}{2}, \frac{0}{2}\right) = (0,0).$$
We also see that $a = 7$ and $b = 3$, so we have:
$$\frac{(x-0)^2}{7^2} + \frac{(y-0)^2}{3^2} = 1, \text{ or}$$
$$\frac{x^2}{49} + \frac{y^2}{9} = 1$$

42. Let $(x, 0)$ represent the coordinates of the point we want to find. We use the distance formula.
$$\sqrt{(-3-x)^2 + (4-0)^2} = \sqrt{(5-x)^2 + (6-0)^2}$$
$$\sqrt{9 + 6x + x^2 + 16} = \sqrt{25 - 10x + x^2 + 36}$$
$$\sqrt{x^2 + 6x + 25} = \sqrt{x^2 - 10x + 61}$$
$$(\sqrt{x^2 + 6x + 25})^2 = (\sqrt{x^2 - 10x + 61})^2$$
$$x^2 + 6x + 25 = x^2 - 10x + 61$$
$$16x = 36$$
$$x = \frac{9}{4}$$
The point is $\left(\frac{9}{4}, 0\right)$.

43. $-y + 4x^2 = 5 - 2x$
$$-y = -4x^2 - 2x + 5$$
$$y = 4x^2 + 2x - 5$$
The graph is a parabola.

44. $\dfrac{x^2}{23} + \dfrac{y^2}{23} = 1$
$$x^2 + y^2 = 23 \quad \text{Multiplying by 23}$$
The graph is a circle.

45. $43 - 12x^2 + y^2 = 21x^2 + 2y^2$
$$43 = 33x^2 + y^2$$
$$1 = \frac{x^2}{43/33} + \frac{y^2}{43}$$
The graph is an ellipse.

46. $3x^2 + 3y^2 = 170$
$$x^2 + y^2 = \frac{170}{3}$$
The graph is a circle.

Chapter 9 Test

1. $d = \sqrt{(x_2 - x_1)^2 + (y_2 - y_1)^2}$
$$d = \sqrt{[6 - (-6)]^2 + (8 - 2)^2}$$
$$= \sqrt{12^2 + 6^2}$$
$$= \sqrt{180} \approx 13.416$$

2. $d = \sqrt{(x_2 - x_1)^2 + (y_2 - y_1)^2}$
$$d = \sqrt{(-3 - 3)^2 + [a - (-a)]^2}$$
$$= \sqrt{(-6)^2 + (2a)^2}$$
$$= \sqrt{36 + 4a^2}$$
$$= \sqrt{4(9 + a^2)}$$
$$= 2\sqrt{9 + a^2}$$

3. We use the formula $\left(\dfrac{x_1 + x_2}{2}, \dfrac{y_1 + y_2}{2}\right)$.
$$\left(\frac{-6+6}{2}, \frac{2+8}{2}\right) = \left(\frac{0}{2}, \frac{10}{2}\right) = (0,5)$$

4. We use the formula $\left(\dfrac{x_1 + x_2}{2}, \dfrac{y_1 + y_2}{2}\right)$.
$$\left(\frac{3 + (-3)}{2}, \frac{-a+a}{2}\right) = \left(\frac{0}{2}, \frac{0}{2}\right) = (0,0)$$

5. $(x+2)^2 + (y-3)^2 = 64$
$$[x - (-2)]^2 + (y-3)^2 = 8^2 \quad \text{Standard form}$$
The center is $(-2, 3)$; the radius is 8.

6. $x^2 + y^2 + 4x - 6y + 4 = 0$
$$(x^2 + 4x) + (y^2 - 6y) + 4 = 0$$
$$(x^2 + 4x + 4 - 4) + (y^2 - 6y + 9 - 9) + 4 = 0$$
$$\text{Completing the square}$$
$$(x^2 + 4x + 4) + (y^2 - 6y + 9) - 4 - 9 + 4 = 0$$
$$(x+2)^2 + (y-3)^2 = 9$$
$$[x - (-2)]^2 + (y-3)^2 = 3^2$$
$$\text{Standard form}$$
The center is $(-2, 3)$; the radius is 3.

7. $(x-h)^2 + (y-k)^2 = r^2$
$$[x - (-2)]^2 + [y - (-5)]^2 = (3\sqrt{2})^2$$
$$(x+2)^2 + (y+5)^2 = 18$$

8. Graph: $y = x^2 - 4x - 1$

The graph is a parabola. We find the first coordinate of the vertex.
$$-\frac{b}{2a} = -\frac{-4}{2 \cdot 1} = -\frac{-4}{2} = 2$$
Then $y = 2^2 - 4 \cdot 2 - 1 = 4 - 8 - 1 = -5$.

The vertex is $(2, -5)$; the line of symmetry is $x = 2$. The curve opens up.

x	y
-1	4
0	-1
1	-4
2	-5
4	-1

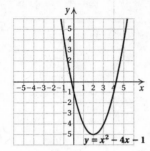

9.
$$x^2 + y^2 = 36$$
$$(x-0)^2 + (y-0)^2 = 6^2 \quad \text{Standard form}$$

The graph is a circle. The center is $(0,0)$, and the radius is 6.

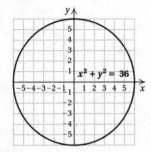

10. $\dfrac{x^2}{9} - \dfrac{y^2}{4} = 1$

$$\dfrac{x^2}{3^2} - \dfrac{y^2}{2^2} = 1$$

The graph is a hyperbola. We have $a = 3$ and $b = 2$, so the asymptotes are $y = \dfrac{2}{3}x$ and $y = -\dfrac{2}{3}x$. Replacing y with 0 and solving for x, we get $x = \pm 3$, so the intercepts are $(3,0)$ and $(-3,0)$.

We plot the intercepts and draw smooth curves through them that approach the asymptotes.

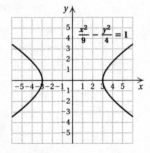

11. $\dfrac{(x+2)^2}{16} + \dfrac{(y-3)^2}{9} = 1$

$$\dfrac{[x-(-2)]^2}{4^2} + \dfrac{(y-3)^2}{3^2} = 1$$

The graph is an ellipse. The center is $(-2,3)$. Note that $a = 4$ and $b = 3$. We locate the center and then plot the points $(-2+4,3)$, $(-2-4,3)$, $(-2,3-3)$, and $(-2,3+3)$, or $(2,3)$, $(-6,3)$, $(-2,0)$, and $(-2,6)$. Connect these points with an oval-shaped curve.

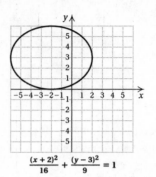

$$\dfrac{(x+2)^2}{16} + \dfrac{(y-3)^2}{9} = 1$$

12.
$$x^2 + y^2 - 4x + 6y + 4 = 0$$
$$(x^2 - 4x) + (y^2 + 6y) + 4 = 0$$
$$(x^2 - 4x + 4 - 4) + (y^2 + 6y + 9 - 9) + 4 = 0$$
$$(x^2 - 4x + 4) + (y^2 + 6y + 9) - 4 - 9 + 4 = 0$$
$$(x-2)^2 + (y+3)^2 = 9$$
$$(x-2)^2 + [y-(-3)]^2 = 3^2$$

This is the equation of a circle with center $(2,-3)$ and radius 3.

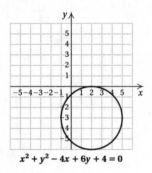

$$x^2 + y^2 - 4x + 6y + 4 = 0$$

13. $9x^2 + y^2 = 36$

$$\dfrac{x^2}{4} + \dfrac{y^2}{36} = 1 \quad \text{Dividing by 36}$$

$$\dfrac{x^2}{2^2} + \dfrac{y^2}{6^2} = 1$$

The graph is an ellipse. The x-intercepts are $(-2,0)$ and $(2,0)$ and the y-intercepts are $(0,-6)$ and $(0,6)$. We plot these points and connect them with an oval-shaped curve.

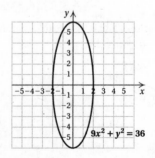

$$9x^2 + y^2 = 36$$

14. $xy = 4$

$$y = \dfrac{4}{x} \quad \text{Solving for } y$$

The graph is a hyperbola. We find some solutions, keeping the results in a table.

x	y
$\frac{1}{2}$	8
1	4
2	2
4	1
8	$\frac{1}{2}$
$-\frac{1}{2}$	-8
-1	-4
-2	-2
-8	$-\frac{1}{2}$

Note that we cannot use 0 for x. The x-axis and the y-axis are the asymptotes.

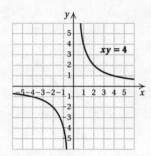

15. Graph: $x = -y^2 + 4y$

We complete the square.

$$x = -(y^2 - 4y)$$
$$= -(y^2 - 4y + 4 - 4)$$
$$= -(y^2 - 4y + 4) + (-1)(-4)$$
$$= -(y - 2)^2 + 4$$

The graph is a parabola. The vertex is $(4, 2)$; the line of symmetry is $y = 2$. The curve opens to the left.

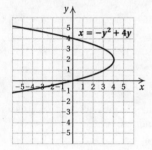

16. $\dfrac{x^2}{16} + \dfrac{y^2}{9} = 1, \quad (1)$

$3x + 4y = 12 \quad (2)$

First we multiply Equation (1) by 144 to clear the fractions.

$$144\left(\frac{x^2}{16} + \frac{y^2}{9}\right) = 144 \cdot 1$$
$$9x^2 + 16y^2 = 144 \qquad (3)$$

Now we solve Equation (2) for x.

$$3x + 4y = 12$$
$$3x = -4y + 12$$
$$x = \frac{-4y + 12}{3}$$

Substitute $\dfrac{-4y + 12}{3}$ for x in Equation (3) and solve for y.

$$9\left(\frac{-4y + 12}{3}\right)^2 + 16y^2 = 144$$
$$9\left(\frac{16y^2 - 96y + 144}{9}\right) + 16y^2 = 144$$
$$16y^2 - 96y + 144 + 16y^2 = 144$$
$$32y^2 - 96y = 0$$
$$32y(y - 3) = 0$$

$$y = 0 \; or \; y = 3$$

When $y = 0$, $x = \dfrac{-4 \cdot 0 + 12}{3} = \dfrac{12}{3} = 4$.

When $y = 3$, $x = \dfrac{-4 \cdot 3 + 12}{3} = \dfrac{-12 + 12}{3} = \dfrac{0}{3} = 0$.

Both pairs check. The solutions are $(4, 0)$ and $(0, 3)$.

17. $x^2 + y^2 = 16, \quad (1)$

$\dfrac{x^2}{16} - \dfrac{y^2}{9} = 1 \quad (2)$

First we multiply Equation (2) by 144 to clear the fractions.

$$144\left(\frac{x^2}{16} - \frac{y^2}{9}\right) = 144 \cdot 1$$
$$9x^2 - 16y^2 = 144 \qquad (3)$$

Now we multiply Equation (1) by 16 and add it to Equation (3).

$$16x^2 + 16y^2 = 256$$
$$\underline{9x^2 - 16y^2 = 144}$$
$$25x^2 \qquad\quad = 400$$
$$x^2 = 16$$
$$x = -4 \; or \; x = 4$$

When $x = -4$, we have:

$$(-4)^2 + y^2 = 16$$
$$16 + y^2 = 16$$
$$y^2 = 0$$
$$y = 0$$

When $x = 4$, we have:

$$4^2 + y^2 = 16$$
$$16 + y^2 = 16$$
$$y^2 = 0$$
$$y = 0$$

Both pairs check. The solutions are $(-4, 0)$ and $(4, 0)$.

18. ***Familiarize.*** Let $l =$ the length and $w =$ the width, in feet.

Translate. The diagonal is 20 ft, so we have $l^2 + w^2 = 20^2$ or $l^2 + w^2 = 400$. The perimeter is 56 ft, so we have $2l + 2w = 56$.

Solve. We solve the system of equations.

$$l^2 + w^2 = 400, \quad (1)$$
$$2l + 2w = 56 \qquad (2)$$

First solve Equation (2) for l.

$$2l + 2w = 56$$
$$2l = 56 - 2w$$
$$l = 28 - w$$

Now substitute $28 - w$ for l in Equation (1) and solve for w.

$$l^2 + w^2 = 400$$
$$(28 - w)^2 + w^2 = 400$$
$$784 - 56w + w^2 + w^2 = 400$$
$$2w^2 - 56w + 784 = 400$$
$$2w^2 - 56w + 384 = 0$$
$$w^2 - 28w + 192 = 0 \qquad \text{Dividing by 2}$$
$$(w - 16)(w - 12) = 0$$
$$w = 16 \; or \; w = 12$$

If $w = 16$, then $l = 28 - 16 = 12$.

If $w = 12$, then $l = 28 - 12 = 16$.

Check. Since we usually consider length to be longer than width, we check $l = 16$ and $w = 12$. Since $16^2 + 12^2 = 256 + 144 = 400 = 20^2$ and $2 \cdot 16 + 2 \cdot 12 = 32 + 24 = 56$, the answer checks.

State. The length is 16 ft, and the width is 12 ft.

19. **Familiarize**. Let $x =$ the amount Nikki invested and let $r =$ the interest rate. Then $x + 240 =$ the amount Erin invested and $\frac{5}{6}r =$ the interest rate. The amount of interest Nikki earned was $x \cdot r \cdot 1$, or xr, and Erin earned $(x + 240) \cdot \frac{5}{6}r \cdot 1$, or $(x + 240) \cdot \frac{5}{6}r$.

Translate. Each investment earned $72 interest, so we have two equations.

$$xr = 72, \qquad\qquad (1)$$
$$(x + 240) \cdot \frac{5}{6}r = 72 \quad (2)$$

Solve. First solve Equation (1) for x.

$$xr = 72$$
$$x = \frac{72}{r} \quad (3)$$

Now substitute $\frac{72}{r}$ for x in Equation (2) and solve for r.

$$\left(\frac{72}{r} + 240\right) \cdot \frac{5}{6}r = 72$$
$$60 + 200r = 72$$
$$200r = 12$$
$$r = 0.06, \text{ or } 6\%$$

Substitute 0.06 for r in Equation (3) and find x.

$$x = \frac{72}{0.06} = 1200$$

Check. If Nikki invested $1200 at 6% interest for 1 yr, the interest earned was $1200 \cdot 0.06 \cdot 1 = \$72$. The answer checks.

State. Nikki invested $1200 at 6% interest.

20. **Familiarize**. Let $l =$ the length and $w =$ the width, in yards.

Translate. The diagonal is $5\sqrt{5}$ yd, so we have $l^2 + w^2 = (5\sqrt{5})^2$, or $l^2 + w^2 = 125$. The area is 22 yd^2, so we have $lw = 22$.

Solve. We solve the system of equations.

$$l^2 + w^2 = 125, \quad (1)$$
$$lw = 22 \qquad\qquad (2)$$

First solve Equation (2) for l.

$$l = \frac{22}{w}$$

Then substitute $\frac{22}{w}$ for l in Equation (1) and solve for w.

$$\left(\frac{22}{w}\right)^2 + w^2 = 125$$
$$\frac{484}{w^2} + w^2 = 125$$
$$484 + w^4 = 125w^2 \quad \text{Multiplying by } w^2$$
$$w^4 - 125w^2 + 484 = 0$$
$$u^2 - 125u + 484 = 0 \qquad \text{Letting } u = w^2$$
$$(u - 121)(u - 4) = 0$$
$$u = 121 \quad or \quad u = 4$$
$$w^2 = 121 \quad or \quad w^2 = 4$$
$$w = \pm 11 \quad or \quad w = \pm 2$$

Check. The dimensions cannot be negative, so -11 and -2 cannot be solutions. If $w = 11$, $l = \frac{22}{11} = 2$. If $w = 2$, $l = \frac{22}{2} = 11$. Since we usually consider length to be longer than width, we check $l = 11$ and $w = 2$. We have $11^2 + 2^2 = 121 + 4 = 125 = (5\sqrt{5})^2$ and the area is $11 \cdot 2$, or 22 yd^2. The answer checks.

State. The length is 11 yd, and the width is 2 yd.

21. **Familiarize**. Let x and y represent the lengths of the sides of the squares, in meters. Then the areas are x^2 and y^2.

Translate. The sum of the areas is 8 m^2, so we have $x^2 + y^2 = 8$. The difference of the areas is 2 m^2, so we have $x^2 - y^2 = 2$.

Solve. We solve the system of equations.

$$\begin{array}{rl} x^2 + y^2 = 8, & (1) \\ \underline{x^2 - y^2 = 2} & (2) \\ 2x^2 \quad\;\; = 10 & \\ x^2 = 5 & \\ x = \pm\sqrt{5} & \end{array}$$

Since the length of a side cannot be negative we consider only $\sqrt{5}$. Substitute $\sqrt{5}$ for x in Equation (1) and solve for y.

$$(\sqrt{5})^2 + y^2 = 8$$
$$5 + y^2 = 8$$
$$y^2 = 3$$
$$y = \pm\sqrt{3}$$

Again, we consider only the positive value. The possible solution is $(\sqrt{5}, \sqrt{3})$.

Check. The areas of the squares are $(\sqrt{5})^2$, or 5 m^2, and $(\sqrt{3})^2$, or 3 m^2. The sum of the areas is $5 + 3$, or 8 m^2, and the difference is $5 - 3$, or 2 m^2. The answer checks.

State. The lengths of the sides of the squares are $\sqrt{5}$ m and $\sqrt{3}$ m.

22. The center of the ellipse is the midpoint of the segment joining the vertices. We find it.

$$\left(\frac{1+11}{2}, \frac{3+3}{2}\right) = \left(\frac{12}{2}, \frac{6}{2}\right) = (6, 3)$$

Thus, we know that $h = 6$ and $k = 3$.

Since the vertex on the left is $(1, 3)$, we know that $h - a = 6 - a = 1$, so $a = 5$. Since the points $(6, 0)$ and $(6, 6)$ lie directly below and above the center, we know that $k - b = 3 - b = 0$, or $b = 3$. We write the equation of the ellipse.

$$\frac{(x-h)^2}{a^2} + \frac{(y-k)^2}{b^2} = 1$$

$$\frac{(x-6)^2}{5^2} + \frac{(y-3)^2}{3^2} = 1, \text{ or } \frac{(x-6)^2}{25} + \frac{(y-3)^2}{9} = 1$$

23. Let (x, y) be a point whose distance from $(8, 0)$ is 10. Then we have:

$$\sqrt{(x-8)^2 + (y-0)^2} = 10$$

$$\sqrt{(x-8)^2 + y^2} = 10$$

$$(x-8)^2 + y^2 = 100 \quad \text{Squaring both sides}$$

Then the set of all points whose distance from $(8, 0)$ is 10 is $\{(x, y) | (x-8)^2 + y^2 = 100\}$.

We could also observe that the set of points whose distance from $(8, 0)$ is 10 are the points on the circle with center $(8, 0)$ and radius 10, or $(x-8)^2 + (y-0)^2 = 10^2$, or $(x-8)^2 + y^2 = 100$. This gives us the same result as above.

24. **Familiarize.** Let x and y represent the numbers.

Translate. The sum of the numbers is 36, so we have $x + y = 36$. The product of the numbers is 4, so we have $xy = 4$.

Solve. We solve the system of equations.

$$x + y = 36, \quad (1)$$
$$xy = 4 \qquad (2)$$

First we solve Equation (1) for y.

$$y = 36 - x$$

Substitute $36 - x$ for y in Equation (2) and solve for x.

$$x(36 - x) = 4$$
$$36x - x^2 = 4$$
$$0 = x^2 - 36x + 4$$

$$x = \frac{-(-36) \pm \sqrt{(-36)^2 - 4 \cdot 1 \cdot 4}}{2 \cdot 1}$$

$$= \frac{36 \pm \sqrt{1296 - 16}}{2} = \frac{36 \pm \sqrt{1280}}{2}$$

$$= \frac{36 \pm \sqrt{256 \cdot 5}}{2} = \frac{36 \pm 16\sqrt{5}}{2}$$

$$= \frac{2(18 \pm 8\sqrt{5})}{2} = 18 \pm 8\sqrt{5}$$

If $x = 18 - 8\sqrt{5}$, then $y = 36 - (18 - 8\sqrt{5}) = 36 - 18 + 8\sqrt{5} = 18 + 8\sqrt{5}$.

If $x = 18 + 8\sqrt{5}$, then $y = 36 - (18 + 8\sqrt{5}) = 36 - 18 - 8\sqrt{5} = 18 - 8\sqrt{5}$.

In either case we see that the numbers are $18 + 8\sqrt{5}$ and $18 - 8\sqrt{5}$.

Check. The sum of the numbers is $18 + 8\sqrt{5} + 18 - 8\sqrt{5} = 36$. The product is $(18 + 8\sqrt{5})(18 - 8\sqrt{5}) = 324 - 320 = 4$. The answer checks.

Now we find the sum of the reciprocals.

$$\frac{1}{18 + 8\sqrt{5}} + \frac{1}{18 - 8\sqrt{5}} = \frac{18 - 8\sqrt{5} + 18 + 8\sqrt{5}}{(18 + 8\sqrt{5})(18 - 8\sqrt{5})}$$

$$= \frac{36}{4} = 9$$

State. The sum of the reciprocals of the numbers is 9.

25. Let $(0, y)$ represent the point on the y-axis that is equidistant from $(-3, -5)$ and $(4, -7)$. Then we have:

$$\sqrt{(-3-0)^2 + (-5-y)^2} = \sqrt{(4-0)^2 + (-7-y)^2}$$

$$\sqrt{9 + 25 + 10y + y^2} = \sqrt{16 + 49 + 14y + y^2}$$

$$\sqrt{34 + 10y + y^2} = \sqrt{65 + 14y + y^2}$$

$$34 + 10y + y^2 = 65 + 14y + y^2$$

$$34 + 10y = 65 + 14y$$

$$-4y = 31$$

$$y = -\frac{31}{4}$$

The point is $\left(0, -\frac{31}{4}\right)$.

Cumulative Review Chapters R - 9

1. $\left|\frac{2}{3} - \frac{4}{5}\right| = \left|\frac{10}{15} - \frac{12}{15}\right| = \left|-\frac{2}{15}\right| = \frac{2}{15}$

2. $\frac{63x^2y^3}{-7x^{-4}y} = \frac{63}{-7} \cdot x^{2-(-4)} \cdot y^{3-1} = -9x^6y^2$

3. $1000 \div 10^2 \cdot 25 \div 4 = 1000 \div 100 \cdot 25 \div 4$

$$= 10 \cdot 25 \div 4$$

$$= 250 \div 4$$

$$= 62.5$$

4. $5x - 3[4(x-2) - 2(x+1)]$

$= 5x - 3[4x - 8 - 2x - 2]$

$= 5x - 3[2x - 10]$

$= 5x - 6x + 30$

$= -x + 30$

5. $\dfrac{1}{3}x - \dfrac{1}{5} \geq \dfrac{1}{5}x - \dfrac{1}{3}$

$15\left(\dfrac{1}{3}x - \dfrac{1}{5}\right) \geq 15\left(\dfrac{1}{5}x - \dfrac{1}{3}\right)$

$5x - 3 \geq 3x - 5$

$2x \geq -2$

$x \geq -1$

The solution set is $\{x | x \geq -1\}$, or $[-1, \infty)$.

6. $|x| > 6.4$

$x < -6.4 \ \ or \ \ x > 6.4$

The solution set is $\{x | x < -6.4 \ or \ x > 6.4\}$, or $(-\infty, -6.4) \cup (6.4, \infty)$.

7. $3 \leq 4x + 7 < 31$

$-4 \leq 4x < 24$

$-1 \leq x < 6$

The solution set is $\{x | -1 \leq x < 6\}$, or $[-1, 6)$.

8. $3x + y = \ \ 4,$ (1)

$\underline{-6x - y = -3}$ (2)

$-3x \quad\quad = \ \ 1$ Adding

$x = -\dfrac{1}{3}$

Substitute $-\dfrac{1}{3}$ for x in Equation (1) and solve for y.

$3\left(-\dfrac{1}{3}\right) + y = 4$

$-1 + y = 4$

$y = 5$

The solution is $\left(-\dfrac{1}{3}, 5\right)$.

9. $x - y + 2z = \ \ 3,$ (1)

$-x \quad\ + z = \ \ 4,$ (2)

$2x + y - \ z = -3$ (3)

First we add Equations (1) and (3).

$x - y + 2z = \ \ 3$

$\underline{2x + y - \ z = -3}$

$3x \quad\quad + z = \ \ 0$ (4)

Now we solve the system of Equations (2) and (4).

$-x + z = 4,$ (2)

$3x + z = 0$ (4)

Multiply Equation (2) by -1 and then add.

$x - z = -4$

$\underline{3x + z = \ \ 0}$

$4x \quad\quad = -4$

$x = -1$

Now substitute -1 for x in Equation (2) and solve for z.

$-(-1) + z = 4$

$1 + z = 4$

$z = 3$

Finally substitute -1 for x and 3 for z in Equation (3) and solve for y.

$2(-1) + y - 3 = -3$

$-2 + y - 3 = -3$

$y - 5 = -3$

$y = 2$

The solution is $(-1, 2, 3)$.

10. $2x^2 = x + 3$

$2x^2 - x - 3 = 0$

$(2x - 3)(x + 1) = 0$

$2x - 3 = 0 \ \ or \ \ x + 1 = 0$

$2x = 3 \ \ or \quad\quad x = -1$

$x = \dfrac{3}{2} \ \ or \quad\quad x = -1$

The solutions are $\dfrac{3}{2}$ and -1.

11. $3x - \dfrac{6}{x} = 7$

$x\left(3x - \dfrac{6}{x}\right) = x \cdot 7$

$3x^2 - 6 = 7x$

$3x^2 - 7x - 6 = 0$

$(3x + 2)(x - 3) = 0$

$3x + 2 = 0 \ \ or \ \ x - 3 = 0$

$3x = -2 \ \ or \quad\quad x = 3$

$x = -\dfrac{2}{3} \ \ or \quad\quad x = 3$

Both numbers check. The solutions are $-\dfrac{2}{3}$ and 3.

12. $\sqrt{x + 5} = x - 1$

$(\sqrt{x + 5})^2 = (x - 1)^2$

$x + 5 = x^2 - 2x + 1$

$0 = x^2 - 3x - 4$

$0 = (x - 4)(x + 1)$

$x - 4 = 0 \ \ or \ \ x + 1 = 0$

$x = 4 \ \ or \quad\quad x = -1$

The number 4 checks but -1 does not, so the solution is 4.

13. $x(x + 10) = -21$

$x^2 + 10x = -21$

$x^2 + 10x + 21 = 0$

$(x + 3)(x + 7) = 0$

$x + 3 = 0 \ \ or \ \ x + 7 = 0$

$x = -3 \ \ or \quad\quad x = -7$

The solutions are -3 and -7.

14. $2x^2 + x + 1 = 0$

$a = 2,\ b = 1,\ c = 1$

$x = \dfrac{-b \pm \sqrt{b^2 - 4ac}}{2a}$

$x = \dfrac{-1 \pm \sqrt{1^2 - 4 \cdot 2 \cdot 1}}{2 \cdot 2} = \dfrac{-1 \pm \sqrt{1 - 8}}{4}$

$\quad = \dfrac{-1 \pm \sqrt{-7}}{4} = \dfrac{-1 \pm i\sqrt{7}}{4}$

$\quad = -\dfrac{1}{4} \pm i\dfrac{\sqrt{7}}{4}$

The solutions are $-\dfrac{1}{4} \pm i\dfrac{\sqrt{7}}{4}$.

15. $x^4 - 13x^2 + 36 = 0$

Let $u = x^2$.

$u^2 - 13u + 36 = 0$

$(u - 4)(u - 9) = 0$

$u - 4 = 0 \quad or \quad u - 9 = 0$

$\quad u = 4 \quad or \quad\quad u = 9$

$\quad x^2 = 4 \quad or \quad\quad x^2 = 9 \quad$ Substituting x^2 for u

$\quad x = \pm 2 \ or \quad\quad x = \pm 3$

The solutions are 2, -2, 3, and -3.

16. $\dfrac{3}{x - 3} - \dfrac{x + 2}{x^2 + 2x - 15} = \dfrac{1}{x + 5}$

$\dfrac{3}{x - 3} - \dfrac{x + 2}{(x + 5)(x - 3)} = \dfrac{1}{x + 5}$, LCD is $(x - 3)(x + 5)$

$(x - 3)(x + 5)\left(\dfrac{3}{x - 3} - \dfrac{x + 2}{(x + 5)(x - 3)}\right) =$

$\qquad\qquad\qquad (x - 3)(x + 5) \cdot \dfrac{1}{x + 5}$

$3(x + 5) - (x + 2) = x - 3$

$3x + 15 - x - 2 = x - 3$

$2x + 13 = x - 3$

$x = -16$

The number -16 checks. It is the solution.

17. $-x^2 + 2y^2 = 7, \quad (1)$

$\underline{x^2 + y^2 = 5} \quad (2)$

$\qquad 3y^2 = 12 \quad$ Adding

$\qquad\ y^2 = 4$

$\qquad\ y = \pm 2$

When $y = -2$, $y^2 = 4$, and when $y = 2$, $y^2 = 4$, so we can substitute 4 for y^2 in Equation (2) and solve for x.

$x^2 + 4 = 5$

$x^2 = 1$

$x = \pm 1$

The solutions are $(1, 2)$, $(1, -2)$, $(-1, 2)$, and $(-1, -2)$.

18. $\log_2 x + \log_2 (x + 7) = 3$

$\log_2 x(x + 7) = 3$

$2^3 = x(x + 7)$

$8 = x^2 + 7x$

$0 = x^2 + 7x - 8$

$0 = (x + 8)(x - 1)$

$x + 8 = 0 \quad or \quad x - 1 = 0$

$x = -8 \ or \quad\quad x = 1$

The number -8 does not check but 1 does, so the solution is 1.

19. $7^x = 30$

$\log 7^x = \log 30$

$x \log 7 = \log 30$

$x = \dfrac{\log 30}{\log 7} \approx 1.748$

20. $\log_3 x = 2$

$3^2 = x$

$9 = x$

21. $x^2 - 1 \geq 0$

$(x + 1)(x - 1) \geq 0$

The solutions of $(x + 1)(x - 1) = 0$ are -1 and 1. They divide the real-number line into three intervals as shown:

We try a test number in each interval.

A: Test -2, $(-2 + 1)(-2 - 1) = 3 > 0$

B: Test 0, $(0 + 1)(0 - 1) = -1 < 0$

C: Test 2, $(2 + 1)(2 - 1) = 3 > 0$

The expression is positive for all numbers in intervals A and C. The inequality symbol is $\geq$, so we need to include the x-intercepts. The solution set is $\{x | x \leq -1 \ or \ x \geq 1\}$, or $(-\infty, -1] \cup [1, \infty)$.

22. $\dfrac{x + 1}{x - 2} > 0$

Solve the related equation.

$\dfrac{x + 1}{x - 2} = 0$

$x + 1 = 0$

$x = -1$

Find the numbers for which the rational expression is undefined.

$x - 2 = 0$

$x = 2$

Use the numbers -1 and 2 to divide the number line into intervals as shown:

Try test numbers in each interval.

A: Test -2,
$$\frac{x+1}{x-2} > 0$$

$$\frac{-2+1}{-2-2} \ ?\ 0$$
$$\frac{-1}{-4}$$
$$\frac{1}{4} \quad \bigg| \quad \text{TRUE}$$

The number -2 is a solution of the inequality, so the interval A is part of the solution set.

B: Test 0,
$$\frac{x+1}{x-2} > 0$$

$$\frac{0+1}{0-2} \ ?\ 0$$
$$-\frac{1}{2} \quad \bigg| \quad \text{FALSE}$$

The number 0 is not a solution of the inequality, so the interval B is not part of the solution set.

C: Test 4,
$$\frac{x+1}{x-2} > 0$$

$$\frac{4+1}{4-2} \ ?\ 0$$
$$\frac{5}{2} \quad \bigg| \quad \text{TRUE}$$

The number 4 is a solution of the inequality, so the interval C is part of the solution set. The solution set is $\{x | x < -1 \ or \ x > 2\}$, or $(-\infty, -1) \cup (2, \infty)$.

23.
$$P = \frac{3}{4}(M + 2N)$$
$$4P = 3(M + 2N) \quad \text{Multiplying by 4}$$
$$4P = 3M + 6N$$
$$4P - 3M = 6N$$
$$\frac{4P - 3M}{6} = N$$

24.
$$\frac{1}{p} + \frac{1}{q} = \frac{1}{f}$$
$$pqf\left(\frac{1}{p} + \frac{1}{q}\right) = pqf \cdot \frac{1}{f} \quad \text{Multiplying by } pqf$$
$$pqf \cdot \frac{1}{p} + pqf \cdot \frac{1}{q} = pq$$
$$qf + pf = pq$$
$$qf = pq - pf \quad \text{Subtracting } pf$$
$$qf = p(q - f) \quad \text{Factoring}$$
$$\frac{qf}{q - f} = p \quad \text{Dividing by } q - f$$

25. $N(t) = 65(1.018)^t$

a) In 2008, $t = 2008 - 2000 = 8$.
$N(8) = 65(1.018)^8 \approx 74.97$ billion ft^3
In 2015, $t = 2015 - 2000 = 15$.
$N(15) = 65(1.018)^{15} \approx 84.94$ billion ft^3

b) $N_0 = 65$, so $2N_0 = 2 \cdot 65 = 130$.
$$130 = 65(1.018)^t$$
$$2 = 1.018^t$$
$$\log 2 = \log 1.018^t$$
$$\log 2 = t \log 1.018$$
$$\frac{\log 2}{\log 1.108} = t$$
$$39 \approx t$$

The doubling time is about 39 yr.

c) Use the points found in parts (a) and (b) and any additional points as needed to graph the function.

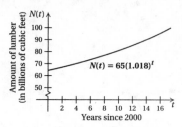

26. a) $A(t) = \$50,000(1 + 0.04)^t = \$50,000(1.04)^t$

b) $A(0) = \$50,000(1.04)^0 = \$50,000$
$A(4) = \$50,000(1.04)^4 \approx \$58,492.93$
$A(8) = \$50,000(1.04)^8 \approx \$68,428.45$
$A(10) = \$50,000(1.04)^{10} \approx \$74,012.21$

c) Use the points found in part (b) and any additional points as needed to graph the function.

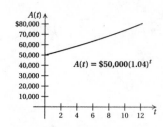

27.
$$(2x + 3)(x^2 - 2x - 1)$$
$$= (2x + 3)x^2 + (2x + 3)(-2x) + (2x + 3)(-1)$$
$$= 2x^3 + 3x^2 - 4x^2 - 6x - 2x - 3$$
$$= 2x^3 - x^2 - 8x - 3$$

28.
$$(3x^2 + x^3 - 1) - (2x^3 + x + 5)$$
$$= (3x^2 + x^3 - 1) + (-2x^3 - x - 5)$$
$$= -x^3 + 3x^2 - x - 6$$

29.
$$\frac{2m^2 + 11m - 6}{m^3 + 1} \cdot \frac{m^2 - m + 1}{m + 6}$$
$$= \frac{(2m^2 + 11m - 6)(m^2 - m + 1)}{(m^3 + 1)(m + 6)}$$
$$= \frac{(2m - 1)(m + 6)(m^2 - m + 1)}{(m + 1)(m^2 - m + 1)(m + 6)}$$
$$= \frac{(m + 6)(m^2 - m + 1)}{(m + 6)(m^2 - m + 1)} \cdot \frac{2m - 1}{m + 1}$$
$$= \frac{2m - 1}{m + 1}$$

30. $\dfrac{x}{x-1} + \dfrac{2}{x+1} - \dfrac{2x}{x^2-1}$

$= \dfrac{x}{x-1} + \dfrac{2}{x+1} - \dfrac{2x}{(x+1)(x-1)}$, LCD is $(x-1)(x+1)$

$= \dfrac{x}{x-1} \cdot \dfrac{x+1}{x+1} + \dfrac{2}{x+1} \cdot \dfrac{x-1}{x-1} - \dfrac{2x}{(x+1)(x-1)}$

$= \dfrac{x(x+1) + 2(x-1) - 2x}{(x-1)(x+1)}$

$= \dfrac{x^2 + x + 2x - 2 - 2x}{(x-1)(x+1)}$

$= \dfrac{x^2 + x - 2}{(x-1)(x+1)}$

$= \dfrac{(x+2)(x-1)}{(x-1)(x+1)}$

$= \dfrac{(x+2)\cancel{(x-1)}}{\cancel{(x-1)}(x+1)}$

$= \dfrac{x+2}{x+1}$

31. $\dfrac{1-\dfrac{5}{x}}{x-4-\dfrac{5}{x}} = \dfrac{1-\dfrac{5}{x}}{x-4-\dfrac{5}{x}} \cdot \dfrac{x}{x}$

$= \dfrac{\left(1-\dfrac{5}{x}\right)x}{\left(x-4-\dfrac{5}{x}\right)x}$

$= \dfrac{x-5}{x^2-4x-5}$

$= \dfrac{x-5}{(x+1)(x-5)}$

$= \dfrac{\cancel{(x-5)}\cdot 1}{(x+1)\cancel{(x-5)}}$

$= \dfrac{1}{x+1}$

32.

$$
\begin{array}{r}
x^3 + 2x^2 - 2x + 1 \\
x+1 \overline{)\, x^4 + 3x^3 + 0x^2 - x + 4} \\
\underline{x^4 + x^3} \\
2x^3 + 0x^2 \\
\underline{2x^3 + 2x^2} \\
-2x^2 - x \\
\underline{-2x^2 - 2x} \\
x + 4 \\
\underline{x + 1} \\
3
\end{array}
$$

The answer is $x^3 + 2x^2 - 2x + 1 + \dfrac{3}{x+1}$.

33. $\dfrac{\sqrt{75x^5y^2}}{\sqrt{3xy}} = \sqrt{\dfrac{75x^5y^2}{3xy}} = \sqrt{25x^4y} = \sqrt{25x^4} \cdot \sqrt{y} = 5x^2\sqrt{y}$

34. $4\sqrt{50} - 3\sqrt{18} = 4\sqrt{25 \cdot 2} - 3\sqrt{9 \cdot 2} = 4\sqrt{25}\sqrt{2} - 3\sqrt{9}\sqrt{2} = 4 \cdot 5\sqrt{2} - 3 \cdot 3\sqrt{2} = 20\sqrt{2} - 9\sqrt{2} = 11\sqrt{2}$

35. $(16^{3/2})^{1/2} = 16^{3/4} = (2^4)^{3/4} = 2^3 = 8$

36. $(2 - i\sqrt{2})(5 + 3i\sqrt{2}) = 10 + 6i\sqrt{2} - 5i\sqrt{2} - 3i^2 \cdot 2$

$= 10 + i\sqrt{2} + 6 \qquad (i^2 = -1)$

$= 16 + i\sqrt{2}$

37. $\dfrac{5+i}{2-4i} = \dfrac{5+i}{2-4i} \cdot \dfrac{2+4i}{2+4i}$

$= \dfrac{10 + 20i + 2i + 4i^2}{4 - 16i^2}$

$= \dfrac{10 + 22i - 4}{4 + 16}$

$= \dfrac{6 + 22i}{20}$

$= \dfrac{3}{10} + \dfrac{11}{10}i$

38. $S(t) = 18t + 344.7$

a) In 2005, $t = 2005 - 2000 = 5$.

$S(5) = 18 \cdot 5 + 344.7 = \$434.7$ billion

In 2008, $t = 2008 - 2000 = 8$.

$S(8) = 18 \cdot 8 + 344.7 = \$488.7$ billion

In 2010, $t = 2010 - 2000 = 10$.

$S(10) = 18 \cdot 10 + 344.7 = \$524.7$ billion

b) Use the points found in part (a) to graph the function.

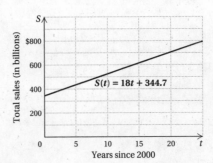

c) The equation is in slope-intercept form, so we see that the y-intercept is $(0, 344.7)$.

d) The equation is in slope-intercept form, so that the slope is 18.

e) The slope is 18, so the rate of change is \$18 billion per year.

39. $4x - 3y = 12$

We will find the intercepts. To find the x-intercept we let $y = 0$ and solve for x.

$4x - 3 \cdot 0 = 12$

$4x = 12$

$x = 3$

The x-intercept is $(3, 0)$.

To find the y-intercept, let $x = 0$ and solve for y.

$4 \cdot 0 - 3y = 12$

$-3y = 12$

$y = -4$

The y-intercept is $(0, -4)$.

We plot these points and draw the line. A third point can be plotted as a check.

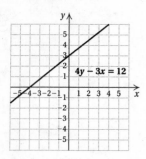

40. $y < -2$

First we graph $y = 2$. We draw the line dashed since the inequality symbol is $<$. Test the point $(0,0)$ to determine if it is a solution.

$$\frac{y < -2}{0 \ ? \ -2} \quad \text{FALSE}$$

Since $0 < -2$ is false, we shade the half-plane that does not contain $(0,0)$.

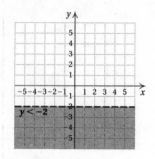

41. $x + y \leq 0$,

$\qquad x \geq -4$,

$\qquad y \geq -1$

Shade the intersection of the graphs of the three inequalities.

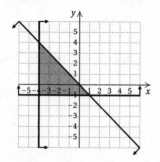

42. $f(x) = 2x^2 - 8x + 9 = 2(x^2 - 4x) + 9$

We complete the square inside the parentheses. We take half the x-coefficient and square it.

$$\frac{1}{2}(-4) = -2 \text{ and } (-2)^2 = 4$$

Then we add $4 - 4$ inside the parentheses.

$$\begin{aligned} f(x) &= 2(x^2 - 4x + 4 - 4) + 9 \\ &= 2(x^2 - 4x + 4) + 2(-4) + 9 \\ &= 2(x - 2)^2 + 1 \end{aligned}$$

Vertex: $(2, 1)$

Line of symmetry: $x = 2$

We plot a few points and draw the graph.

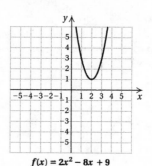

$$f(x) = 2x^2 - 8x + 9$$

43. $\qquad (x - 1)^2 + (y + 1)^2 = 9$

$\qquad (x - 1)^2 + [y - (-1)]^2 = 3^2 \quad$ Standard form

The graph is a circle with center $(1, -1)$ and radius 3.

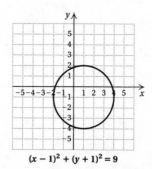

$$(x - 1)^2 + (y + 1)^2 = 9$$

44. $x = y^2 + 1 = (y - 0)^2 + 1$

The graph is a parabola with vertex $(1, 0)$ and line of symmetry $y = 0$. The curve opens to the right. We plot a few points and draw the graph.

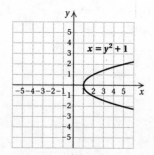

45. Graph: $f(x) = e^{-x}$

We find some function values with a calculator. We use these values to plot points and draw the graph.

x	e^{-x}
0	1
−1	2.7
−2	7.4
−3	20.1
1	0.4
2	0.1
3	0.05

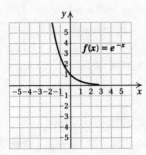

46. $f(x) = \log_2 x$

The equation $f(x) = y = \log_2 x$ is equivalent to $2^y = x$. We can find ordered pairs by choosing values for y and computing the corresponding x-values.

For $y = 0$, $x = 2^0 = 1$.

For $y = 1$, $x = 2^1 = 2$.

For $y = 2$, $x = 2^2 = 4$.

For $y = 3$, $x = 2^3 = 8$.

For $y = -1$, $x = 2^{-1} = \dfrac{1}{2}$.

For $y = -2$, $x = 2^{-2} = \dfrac{1}{4}$.

x, or 2^y	y
1	0
2	1
4	2
8	3
$\frac{1}{2}$	−1
$\frac{1}{4}$	−2

      (1) Select y.

      (2) Compute x.

We plot the set of ordered pairs and connect the points with a smooth curve.

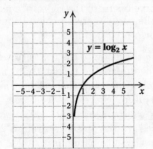

47.
$$2x^4 - 12x^3 + x - 6$$
$$= 2x^3(x - 6) + (x - 6) \quad \text{Factoring by grouping}$$
$$= (2x^3 + 1)(x - 6)$$

48.
$$3a^2 - 12ab - 135b^2$$
$$= 3(a^2 - 4ab - 45b^2)$$
$$= 3(a - 9b)(a + 5b) \quad \text{Factoring by trial and error}$$

49.
$$x^2 - 17x + 72$$
$$= (x - 8)(x - 9) \quad \text{Factoring by trial and error}$$

50.
$$81m^4 - n^4$$
$$= (9m^2)^2 - (n^2)^2 \quad \text{Difference of squares}$$
$$= (9m^2 + n^2)(9m^2 - n^2)$$
$$= (9m^2 + n^2)[(3m)^2 - n^2] \quad \text{Difference of squares}$$
$$= (9m^2 + n^2)(3m + n)(3m - n)$$

51.
$$16x^2 - 16x + 4$$
$$= 4(4x^2 - 4x + 1) \quad \text{Square of a binomial}$$
$$= 4(2x - 1)^2$$

52.
$$81a^3 - 24$$
$$= 3(27a^3 - 8)$$
$$= 3[(3a)^3 - 2^3] \quad \text{Difference of cubes}$$
$$= 3(3a - 2)[(3a)^2 + 3a \cdot 2 + 2^2]$$
$$= 3(3a - 2)(9a^2 + 6a + 4)$$

53.
$$10x^2 + 66x - 28$$
$$= 2(5x^2 + 33x - 14)$$
$$= 2(5x - 2)(x + 7) \quad \text{FOIL or } ac\text{-method}$$

54.
$$6x^3 + 27x^2 - 15x$$
$$= 3x(2x^2 + 9x - 5)$$
$$= 3x(2x - 1)(x + 5) \quad \text{FOIL or } ac\text{-method}$$

55. First we will find the slope of the line.
$$m = \frac{0 - 4}{-1 - 1} = \frac{-4}{-2} = 2$$

We will use the point-slope formula and use the point $(-1, 0)$.
$$y - y_1 = m(x - x_1)$$
$$y - 0 = 2(x - (-1))$$
$$y = 2(x + 1)$$
$$y = 2x + 2$$

56. First solve the given equation for y to determine its slope.
$$2x - y = 3$$
$$-y = -2x + 3$$
$$y = 2x - 3$$

The slope of the given line is 2. The slope of the perpendicular line is the opposite of the reciprocal of 2, or $-\dfrac{1}{2}$. We will use the slope-intercept equation.
$$y = mx + b$$
$$2 = -\frac{1}{2} \cdot 1 + b$$
$$2 = -\frac{1}{2} + b$$
$$\frac{5}{2} = b$$

Finally we substitute $-\frac{1}{2}$ for m and $\frac{5}{2}$ for b in the slope-intercept equation to find the desired equation.

$$y = -\frac{1}{2}x + \frac{5}{2}$$

57.
$$x^2 - 16x + y^2 + 6y + 68 = 0$$
$$(x^2 - 16x + 64 - 64) + (y^2 + 6y + 9 - 9) + 68 = 0$$

Completing the square twice

$$(x^2 - 16x + 64) + (y^2 + 6y + 9) - 64 - 9 + 68 = 0$$
$$(x - 8)^2 + (y + 3)^2 = 5$$
$$(x - 8)^2 + [y - (-3)]^2 = (\sqrt{5})^2$$

The center is $(8, -3)$ and the radius is $\sqrt{5}$.

58. $f(x) = 2x - 3$ passes the horizontal-line test so it is one-to-one and has an inverse that is a function. We find a formula for the inverse.

1. Replace $f(x)$ by y: $y = 2x - 3$

2. Interchange x and y: $x = 2y - 3$

3. Solve for y: $x + 3 = 2y$
$$\frac{x + 3}{2} = y$$

4. Replace y by $f^{-1}(x)$: $f^{-1}(x) = \dfrac{x + 3}{2}$, or $\dfrac{1}{2}(x + 3)$

59. $z = \dfrac{kx}{y^3}$

$$5 = \frac{k \cdot 4}{2^3}$$
$$5 = \frac{4k}{8}$$
$$10 = k \qquad \text{Variation constant}$$
$$z = \frac{10x}{y^3} \qquad \text{Equation of variation}$$
$$z = \frac{10 \cdot 10}{5^3}$$
$$z = \frac{100}{125} = \frac{4}{5}$$

60. $f(x) = x^3 - 2$
$$f(-2) = (-2)^3 - 2 = -8 - 2 = -10$$

61. $d = \sqrt{(x_2 - x_1)^2 + (y_2 - y_1)^2}$
$$d = \sqrt{(8 - 2)^2 + (9 - 1)^2}$$
$$= \sqrt{6^2 + 8^2} = \sqrt{36 + 64}$$
$$= \sqrt{100} = 10$$

62. $\left(\dfrac{x_1 + x_2}{2}, \dfrac{y_1 + y_2}{2} \right) = \left(\dfrac{-1 + 3}{2}, \dfrac{-3 + 0}{2} \right) =$
$$\left(\frac{2}{2}, \frac{-3}{2} \right) = \left(1, -\frac{3}{2} \right)$$

63. $\dfrac{5 + \sqrt{a}}{3 - \sqrt{a}} = \dfrac{5 + \sqrt{a}}{3 - \sqrt{a}} \cdot \dfrac{3 + \sqrt{a}}{3 + \sqrt{a}}$
$$= \frac{15 + 5\sqrt{a} + 3\sqrt{a} + a}{9 - a}$$
$$= \frac{15 + 8\sqrt{a} + a}{9 - a}$$

64. $f(x) = \dfrac{4x - 3}{3x^2 + x}$

To find the numbers excluded from the domain, we set the denominator equal to 0 and solve.

$$3x^2 + x = 0$$
$$x(3x + 1) = 0$$
$$x = 0 \quad or \quad 3x + 1 = 0$$
$$x = 0 \quad or \qquad 3x = -1$$
$$x = 0 \quad or \qquad x = -\frac{1}{3}$$

Then the domain is $\left(-\infty, -\dfrac{1}{3} \right) \cup \left(-\dfrac{1}{3}, 0 \right) \cup (0, \infty)$.

65.
$$3a^2 + a = 2$$
$$3a^2 + a - 2 = 0$$
$$(3a - 2)(a + 1) = 0$$
$$3a - 2 = 0 \quad or \quad a + 1 = 0$$
$$3a = 2 \quad or \qquad a = -1$$
$$a = \frac{2}{3} \quad or \qquad a = -1$$

The values of a for which $f(a) = 2$ are $\dfrac{2}{3}$ and -1.

66. *Familiarize*. Let $c =$ the number of CDs purchased annually. Then limited members would pay $10 + 10c$ and preferred members would pay $20 + 7.50c$.

***Translate*.**

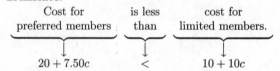

Cost for preferred members	is less than	cost for limited members.
$20 + 7.50c$	$<$	$10 + 10c$

***Solve*.** We solve the inequality.
$$20 + 7.50c < 10 + 10c$$
$$10 < 2.5c$$
$$4 < c$$

***Check*.** When 4 CDs are purchased the cost for preferred members is $20 + 7.50(4)$, or \$50, and the cost for limited members is $10 + 10 \cdot 4$, or \$50, so the costs are the same. When more than 4 CDs are purchased, say 5 CDs, preferred members pay $20 + 7.50(5)$, or \$57.50, and limited members pay $10 + 10 \cdot 5$, or \$60, so preferred members pay less. This partial check shows that the answer is probably correct.

***State*.** When more than 4 CDs are purchased annually, it is less expensive to be a preferred member.

67. *Familiarize*. Let $d =$ the distance the trains travel before the passenger train overtakes the freight train, in miles, and let $t =$ the number of hours the freight train travels before being overtaken. Since it is 9 hr from 2 A.M. to 11 A.M., $t - 9 =$ the number of hours the passenger train travels. The speed of the passenger train is $2 \cdot 34$, or 68 mph. We organize the information in a table.

Train	Distance	Speed	Time
Freight	d	34	t
Passenger	d	68	$t - 9$

Translate. Using $d = rt$ in each row of the table we have two equations.

$$d = 34t, \qquad (1)$$
$$d = 68(t - 9) \quad (2)$$

Solve. We substitute $34t$ for d in Equation (2) and solve for t.

$$34t = 68(t - 9)$$
$$34t = 68t - 612$$
$$-34t = -612$$
$$t = 18$$

When $t = 18$, $d = 34 \cdot 18 = 612$.

Check. At 34 mph, in 18 hr the freight train travels $34 \cdot 18 = 612$ mph. At 68 mph, in $18 - 9$, or 9 hr, the passenger train travels $68 \cdot 9$, or 612 mi. Since the distances are the same, the answer checks.

State. The passenger train will overtake the freight train 612 mi from the station.

68. Familiarize. Let $s =$ the length of a side of the octagon. Then $3s - 2 =$ the length of a side of the pentagon. The perimeter of the pentagon is $5(3s - 2)$ and the perimeter of the octagon is $8s$.

Translate.

Perimeter of pentagon	is the same as	perimeter of octagon.
$5(3s - 2)$	$=$	$8s$

Solve. We solve the equation.

$$5(3s - 2) = 8s$$
$$15s - 10 = 8s$$
$$-10 = -7s$$
$$\frac{10}{7} = s$$

Check. If $s = \dfrac{10}{7}$, then the perimeter of the octagon is $8 \cdot \dfrac{10}{7}$, or $\dfrac{80}{7}$, or $11\dfrac{3}{7}$. The length of a side of the pentagon is $3 \cdot \dfrac{10}{7} - 2$, or $\dfrac{30}{7} - \dfrac{14}{7}$, or $\dfrac{16}{7}$, and the perimeter is $5 \cdot \dfrac{16}{7}$, or $\dfrac{80}{7}$, or $11\dfrac{3}{7}$. Since the perimeters are the same, the answer is correct.

State. The perimeter of each figure is $11\dfrac{3}{7}$.

69. Familiarize. Let a and b represent the number of liters of solutions A and B to be used, respectively. We organize the information in a table.

	A	B	Mixture
Amount	a	b	80
Percent of ammonia	6%	2%	3.2%
Amount of ammonia	$0.06a$	$0.02b$	0.032(80), or 2.56

The first and third rows of the table give us two equations.

$$a + b = 80,$$
$$0.06a + 0.02b = 2.56$$

After clearing decimals we have the following system of equations.

$$a + b = 80, \quad (1)$$
$$6a + 2b = 256 \quad (2)$$

Solve. First multiply Equation (1) by -2 and then add.

$$-2a - 2b = -160$$
$$\underline{6a + 2b = 256}$$
$$4a = 96$$
$$a = 24$$

Now substitute 24 for a in Equation (1) and solve for b.

$$24 + b = 80$$
$$b = 56$$

Check. $24 + 56 = 80$ L. The amount of ammonia in the mixture is $0.06(24) + 0.02(56) = 1.44 + 1.12 = 2.56$ L. The answer checks.

State. 24 L of solution A and 56 L of solution B should be used.

70. Familiarize. Let $r =$ the speed of the plane in still air. Then $r + 30 =$ the speed with the wind and $r - 30 =$ the speed against the wind. Let $t =$ the time it takes to make the trip in each direction. We organize the information in a table.

	Distance	Speed	Time
With wind	190	$r + 30$	t
Against wind	160	$r - 30$	t

Translate. Using $t = d/r$ in each row of the table, we can equate the two expressions for time.

$$\frac{190}{r + 30} = \frac{160}{r - 30}$$

Solve. We solve the equation.

$$(r + 30)(r - 30) \cdot \frac{190}{r + 30} = (r + 30)(r - 30) \cdot \frac{160}{r - 30}$$
$$190(r - 30) = 160(r + 30)$$
$$190r - 5700 = 160r + 4800$$
$$30r = 10{,}500$$
$$r = 350$$

Check. If the speed of the plane in still air is 350 mph, then the speed with the wind is $350 + 30$, or 380 mph, and it takes 190/380, or 1/2 hr, to fly 190 mi. The speed against the wind is $350 - 30$, or 320 mph, and it takes 160/320, or 1/2 hr, to fly 160 mi. The times are the same, so the answer checks.

State. The speed of the plane in still air is 350 mph.

71. Familiarize. Let $t =$ the time, in minutes, it would take to do the job, working together.

Translate. We will use the work principle.

$$\frac{t}{21} + \frac{t}{14} = 1$$

Solve. We solve the equation.

$$42\left(\frac{t}{21} + \frac{t}{14}\right) = 42 \cdot 1$$
$$2t + 3t = 42$$
$$5t = 42$$
$$t = \frac{42}{5}, \text{ or } 8\frac{2}{5}$$

Check. We verify the work principle.

$$\frac{42/5}{21} + \frac{42/5}{14} = \frac{42}{5} \cdot \frac{1}{21} + \frac{42}{5} \cdot \frac{1}{14} = \frac{2}{5} + \frac{3}{5} = 1$$

State. It would take $8\frac{2}{5}$ min to do the job, working together.

72.
$$F = \frac{kv^2}{r}$$
$$8 = \frac{k \cdot 1^2}{10}$$
$$8 = \frac{k}{10}$$
$$80 = k$$
$$F = \frac{80v^2}{r} \qquad \text{Equation of variation}$$
$$F = \frac{80 \cdot 2^2}{16}$$
$$F = \frac{80 \cdot 4}{16}$$
$$F = 20$$

73. Familiarize. Let l and w represent the length and width of the rectangle, respectively, in feet.

Translate. The perimeter is 34 ft, so we have $2l + 2w = 34$, or $l + w = 17$. We use the Pythagorean theorem to get a second equation:

$$l^2 + w^2 = 13^2, \text{ or } l^2 + w^2 = 169.$$

Solve. We solve the system of equations.

$$l + w = 17, \quad (1)$$
$$l^2 + w^2 = 169 \quad (2)$$

First we solve Equation (1) for l: $l = 17 - w$. Then substitute $17 - w$ for l in Equation (2) and solve for w.

$$(17 - w)^2 + w^2 = 169$$
$$289 - 34w + w^2 + w^2 = 169$$
$$2w^2 - 34w + 120 = 0$$
$$w^2 - 17w + 60 = 0 \quad \text{Dividing by 2}$$
$$(w - 5)(w - 12) = 0$$
$$w = 5 \text{ or } w = 12$$

If $w = 5$, then $l = 17 - 5 = 12$.
If $w = 12$, then $l = 17 - 12 = 5$.

Since we usually consider length to be longer than width, we have $l = 12$ and $w = 5$.

Check. The perimeter is $2 \cdot 12 + 2 \cdot 5 = 34$ ft. Also, $12^2 + 5^2 = 144 + 25 = 169 = 13^2$, so the answer checks.

State. The dimensions are 12 ft by 5 ft.

74. Familiarize. Let l and w represent the length and width of the rug, respectively, in feet.

Translate. The diagonal is 25 ft so, using the Pythagorean theorem, we have $l^2 + w^2 = 25^2$, or $l^2 + w^2 = 625$. The area is 300 ft^2, so we have a second equation, $lw = 300$.

Solve. We solve the system of equations.

$$l^2 + w^2 = 625, \quad (1)$$
$$lw = 300 \qquad (2)$$

First we solve Equation (2) for w: $w = \frac{300}{l}$. Then substitute $\frac{300}{l}$ for w in Equation (1) and solve for l.

$$l^2 + \left(\frac{300}{l}\right)^2 = 625$$
$$l^2 + \frac{90,000}{l^2} = 625$$
$$l^4 + 90,000 = 625l^2 \quad \text{Multiplying by } l^2$$
$$l^4 - 625l^2 + 90,000 = 0$$

Let $u = l^2$.

$$u^2 - 625u + 90,000 = 0$$
$$(u - 225)(u - 400) = 0$$
$$u - 225 = 0 \quad or \quad u - 400 = 0$$
$$u = 225 \quad or \qquad\quad u = 400$$
$$l^2 = 225 \quad or \qquad\quad l^2 = 400$$
$$l = \pm 15 \quad or \qquad\quad l = \pm 20$$

Since the length cannot be negative, -15 and -20 cannot be solutions. When $l = 15$, $w = 300/15 = 20$. When $w = 20$, $l = 300/20$, or 15. Since we usually consider length to be longer than width, we have $l = 20$ and $w = 15$.

Check. We have $20^2 + 15^2 = 400 + 225 = 625 = 25^2$. Also, the area is $20 \cdot 15$, or 300 ft^2. The answer checks.

State. The length is 20 ft, and the width is 15 ft.

75. Familiarize. Using the labels on the drawing in the text, we let $w =$ the width of the rectangular region, in feet, and $100 - 2w =$ the length.

Translate. We use the formula for the area of a rectangle.

$$A = (100 - 2w)w = 100w - 2w^2 = -2w^2 + 100w$$

Carry out. The graph of the function is a parabola that opens down (since $-2 < 0$), so it has a maximum value at the vertex. We complete the square to find the vertex.

$$A = -2(w^2 - 50w)$$
$$= -2(w^2 - 50w + 625 - 625)$$
$$= -2(w^2 - 50w + 625) + (-2)(-625)$$
$$= -2(w - 25)^2 + 1250$$

The vertex is $(25, 1250)$, so the maximum value is 1250.

Check. We can do a partial check by trying some values of w and determining that each yields a value of A that is less than 1250. We could also examine the graph of the function to check the maximum value.

State. The area of the largest region that can be fenced in is 1250 ft^2.

76. If a bone has lost 25% of its carbon-14 from an initial amount P_0, then $75\%(P_0)$ is the amount present.

$$P(t) = P_0\, e^{-0.00012t}$$
$$0.75\, P_0 = P_0\, e^{-0.00012t}$$
$$0.75 = e^{-0.00012t}$$
$$\ln 0.75 = \ln e^{-0.00012t}$$
$$\ln 0.75 = -0.00012t$$
$$\frac{\ln 0.75}{-0.00012} = t$$
$$2397\text{ yr} \approx t$$

77.
$$W = \frac{k}{L}$$
$$1440 = \frac{k}{14}$$
$$20{,}160 = k$$
$$W = \frac{20{,}160}{L} \quad \text{Equation of variation}$$
$$W = \frac{20{,}160}{6}$$
$$W = 3360\text{ kg}$$

78. First we find the slope of the line.
$$m = \frac{-3 - (-4)}{2 - 5} = \frac{1}{-3} = -\frac{1}{3}$$

We will use the point-slope equation and the point $(5, -4)$.
$$y - y_1 = m(x - x_1)$$
$$y - (-4) = -\frac{1}{3}(x - 5)$$
$$y + 4 = -\frac{1}{3}x + \frac{5}{3}$$
$$y = -\frac{1}{3}x - \frac{7}{3}$$

79. We find a function of the form $f(x) = ax^2 + bx + c$. Substituting the data points we get
$$4 = a(-2)^2 + b(-2) + c,$$
$$-6 = a(-5)^2 + b(-5) + c,$$
$$-3 = a \cdot 1^2 + b \cdot 1 + c,$$

or
$$4 = 4a - 2b + c,$$
$$-6 = 25a - 5b + c,$$
$$-3 = a + b + c.$$

Solving the system of equations, we get
$$\left(-\frac{17}{18}, -\frac{59}{18}, \frac{11}{9}\right).$$

Thus, the function is $f(x) = -\frac{17}{18}x^2 - \frac{59}{18}x + \frac{11}{9}$.

80. $10^6 = r \Rightarrow \log r = 6$

81. $\log_3 Q = x \Rightarrow 3^x = Q$

82.
$$\frac{1}{5}(7 \log_b x - \log_b y - 8 \log_b z)$$
$$= \frac{1}{5}(\log_b x^7 - \log_b y - \log_b z^8)$$
$$= \frac{1}{5}[\log_b x^7 - (\log_b y + \log_b z^8)]$$
$$= \frac{1}{5}(\log_b x^7 - \log_b yz^8)$$
$$= \frac{1}{5}\log_b \frac{x^7}{yz^8}$$
$$= \log_b \left(\frac{x^7}{yz^8}\right)^{1/5}$$

83.
$$\log_b \left(\frac{xy^5}{z}\right)^{-6}$$
$$= -6 \log_b \frac{xy^5}{z}$$
$$= -6(\log_b xy^5 - \log_b z)$$
$$= -6(\log_b x + \log_b y^5 - \log_b z)$$
$$= -6(\log_b x + 5 \log_b y - \log_b z)$$
$$= -6 \log_b x - 30 \log_b y + 6 \log_b z$$

84. Familiarize. Let x and y represent the numbers.

Translate. The sum of the numbers is 26, so we have $x + y = 26$. Solving for y, we get $y = 26 - x$. The product of the numbers is xy. Substituting $26 - x$ for y in the product, we get a quadratic function:

$$P = xy = x(26 - x) = 26x - x^2 = -x^2 + 26x$$

Carry out. The coefficient of x^2 is negative, so the graph of the function is a parabola that opens down and a maximum exists. We complete the square in order to find the vertex of the quadratic function.

$$P = -x^2 + 26x$$
$$= -(x^2 - 26x)$$
$$= -(x^2 - 26x + 169 - 169)$$
$$= -(x^2 - 26x + 169) + (-1)(-169)$$
$$= -(x - 13)^2 + 169$$

The vertex is $(13, 169)$. This tells us that the maximum product is 169.

Check. We could use the graph of the function to check the maximum value.

State. The maximum product is 169.

85. The graph of $f(x) = 4 - x^2$ is shown below. We use the horizontal-line test. Since it is possible for a horizontal line to intersect the graph more than once, the function is not one-to-one.

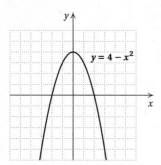

$y = 4 - x^2$

86. a) Locate 2 on the horizontal axis and the find the point on the graph for which 1 is the first coordinate. From that point, look to the vertical axis to find the corresponding y-coordinate. It is -5. Thus, $f(2) = -5$.

  b) No endpoints are indicated, so we see that the graph extends indefinitely horizontally. Thus, the domain is the set of all real numbers, or $(-\infty, \infty)$.

  c) To determine all x-values for which $f(x) = -5$, locate -5 on the vertical axis. From there look left and right to the graph to find any points for which -5 is the second coordinate. Four such points exist. Their x-coordinates are -2, -1, 1, and 2.

  d) The smallest y-value is -7. No endpoints are indicated, so we see that the graph extends upward indefinitely from $(-7, 0)$. Thus, the range is $\{y | y \geq -7\}$, or $[-7, \infty)$.

87. a)
$$P(t) = P_0 e^{kt}$$
$$251,377 = 152,099 e^{k \cdot 10}$$
$$\frac{251,377}{152,099} = e^{10k}$$
$$\ln\left(\frac{251,377}{152,099}\right) = \ln e^{10k}$$
$$\ln\left(\frac{251,377}{152,099}\right) = 10k$$
$$\frac{\ln\left(\dfrac{251,377}{152,099}\right)}{10} = k$$
$$0.05 \approx k$$

Thus, the exponential growth rate is about 0.05 and the exponential growth function is $P(t) = 152,099 e^{0.05t}$, where t is the number of years since 1990.

  b) In 2015, $t = 2015 - 1990 = 25$.
$$P(25) = 152,099 e^{0.05(25)} \approx 530,878$$

  c)
$$2,000,000 = 152,099 e^{0.05t}$$
$$\frac{2,000,000}{152,099} = e^{0.05t}$$
$$\ln\left(\frac{2,000,000}{152,099}\right) = \ln e^{0.05t}$$
$$\ln\left(\frac{2,000,000}{152,099}\right) = 0.05t$$
$$\frac{\ln\left(\dfrac{2,000,000}{152,099}\right)}{0.05} = t$$
$$52 \approx t$$

The population will reach 2 million about 52 yr after 1990, or in 2042.

88.
$$\frac{9}{x} - \frac{9}{x+12} = \frac{108}{x^2+12x}$$
$$\frac{9}{x} - \frac{9}{x+12} = \frac{108}{x(x+12)}, \text{ LCM is } x(x+12)$$
$$x(x+12)\left(\frac{9}{x} - \frac{9}{x+12}\right) = x(x+12) \cdot \frac{108}{x(x+12)}$$
$$9(x+12) - 9x = 108$$
$$9x + 108 - 9x = 108$$
$$108 = 108$$

We get an equation that is true for all values of x. Thus all real numbers except those that make a denominator 0 are solutions of the equation. A denominator is 0 when $x = 0$ or $x = -12$, so all real numbers except 0 and -12 are solutions of the equation.

89. $\log_2(\log_3 x) = 2$
$$2^2 = \log_3 x$$
$$4 = \log_3 x$$
$$x = 3^4$$
$$x = 81$$

The number 81 checks. It is the solution.

90. When $a^2 = b^2$, we have
$$\frac{x^2}{a^2} + \frac{y^2}{a^2} = 1, \text{ or } x^2 + y^2 = a^2.$$

Thus the graph is a circle with center $(0, 0)$ and radius a.

91. Let $d =$ the number of years Diaphantos lived. Then we have:
$$\frac{1}{6}d + \frac{1}{12}d + \frac{1}{7}d + 5 + 4 + \frac{1}{2}d = d$$
$$\frac{25}{28}d + 9 = d$$
$$9 = \frac{3}{28}d$$
$$\frac{28}{3} \cdot 9 = d$$
$$84 = d$$

Diaphantos lived 84 yr.

92. $x^2 + y^2 = 208,$ (1)

$xy = 96$ (2)

First solve Equation (2) for y: $y = 96/x$. Then substitute $96/x$ for y in Equation (1) and solve for x.

$$x^2 + \left(\frac{96}{x}\right)^2 = 208$$

$$x^2 + \frac{9216}{x^2} = 208$$

$$x^4 + 9216 = 208x^2 \quad \text{Multiplying by } x^2$$

$$x^4 - 208x^2 + 9216 = 0$$

Let $u = x^2$.

$$u^2 - 208u + 9216 = 0$$

$$(u - 64)(u - 144) = 0$$

$$u - 64 = 0 \quad or \quad u - 144 = 0$$

$$u = 64 \quad or \quad u = 144$$

$$x^2 = 64 \quad or \quad x^2 = 144$$

$$x = \pm 8 \quad or \quad x = \pm 12$$

If $x = 8$, $y = 96/8 = 12$; if $x = -8$, $y = 96/-8 = -12$; if $x = 12$, $y = 96/12 = 8$; and if $x = -12$, $y = 96/-12 = -8$. The solutions are $(8, 12)$, $(-8, -12)$, $(12, 8)$, and $(-12, -8)$.